Advances in Vehicle Dynamics and Road Safety: Technologies, Simulations and Applications

Advances in Vehicle Dynamics and Road Safety: Technologies, Simulations and Applications

Editors

Edgar Sokolovskij
Vidas Žuraulis

MDPI

Basel • Beijing • Wuhan • Barcelona • Belgrade • Novi Sad • Cluj • Manchester

Editors

Edgar Sokolovskij
Vilnius Gediminas Technical
University (VILNIUS TECH)
Vilnius
Lithuania

Vidas Žuraulis
Vilnius Gediminas Technical
University (VILNIUS TECH)
Vilnius
Lithuania

Editorial Office
MDPI
St. Alban-Anlage 66
4052 Basel, Switzerland

This is a reprint of articles from the Special Issue published online in the open access journal *Applied Sciences* (ISSN 2076-3417) (available at: https://www.mdpi.com/journal/applsci/special_issues/G8W2XMDZU2).

For citation purposes, cite each article independently as indicated on the article page online and as indicated below:

Lastname, A.A.; Lastname, B.B. Article Title. *Journal Name* **Year**, *Volume Number*, Page Range.

ISBN 978-3-7258-1181-6 (Hbk)
ISBN 978-3-7258-1182-3 (PDF)
doi.org/10.3390/books978-3-7258-1182-3

Contents

About the Editors

Edgar Sokolovskij

Edgar Sokolovskij is a Vice-Dean for Studies and International Relations of the Faculty of Transport Engineering and a Professor of the Department of Automobile Engineering at the Vilnius Gediminas Technical University (VILNIUS TECH). He is an author of more than 100 scientific publications and a co-author of 3 educational books and 1 methodical teaching tool. He is included into the Lithuanian Republic Judicial Experts List as a Traffic Accidents and Transport Trasology Expert. His scientific interests include vehicle dynamics; road traffic safety; and road accidents analysis, modeling, and expertise.

Vidas Žuraulis

Vidas Žuraulis is a Professor in the Department of Automobile Engineering and Chief Researcher in Transport and Logistics Competence Centre at the Vilnius Gediminas Technical University (VILNIUS TECH). He is an author of more than 60 scientific publications, a co-author of an educational book, and an editor of research issues. He is an associate editor for vehicle and traffic safety in peer-reviewed scientific journal *Transport*. His scientific interests include vehicle dynamics; road traffic safety; road accidents analysis; vehicle active and passive safety; and road user behavior.

applied sciences

MDPI

Editorial

Advances in Vehicle Dynamics and Road Safety: Technologies, Simulations, and Applications

Edgar Sokolovskij * and Vidas Žuraulis

Department of Automobile Engineering, Faculty of Transport Engineering, Vilnius Gediminas Technical University, Plytinės Str. 25, LT-10105 Vilnius, Lithuania; vidas.zuraulis@vilniustech.lt
* Correspondence: edgar.sokolovskij@vilniustech.lt

1. Introduction

Alongside various road safety aspects, vehicle dynamics play a crucial role in enhancing the quality of life in modern society within a holistic traffic safety framework [1]. According to a report by the World Health Organization, the number of road accident fatalities has been gradually declining worldwide, but the 1.19 million lives lost on the roads and the disparity in crash rates among different regions remain disproportionately high, considering technological advancements and the vast resources allocated towards road safety [2]. The risks posed by different road transport systems are particularly relevant in the context of increasing mobility and travel. The main pillars for Safe System Approach (SSA) in terms of road safety cover safe road users, vehicles, speeds, roads, and post-crash care; moreover, the potential of the SSA is further enhanced by safety systems based on modern technologies such as vehicle active and passive safety [3–5], advanced driver-assistance systems (ADAS) [6,7], computer-based accident and systems operation simulations and co-simulations [8–10]. A separate and distinct focus on achieving Vision Zero on the road [11] relates to autonomous vehicle control technologies [12,13].

While the behavior of road users is widely acknowledged as a significant factor contributing to contributory accidents [14], the interaction between vehicles, road users, and the road itself (the traffic environment) is considered to be a more rapid, engineered element affecting road safety [15–17]. It is important to recognize that the vehicle plays a key role in interactions with both the driver and the road, so the comprehensive development of safety technologies can help to further reduce road accidents. From this perspective, the technological development of various aspects of vehicle dynamics leads to both safety and sustainability aspects in road transportation [18,19].

A new approach involving modern technologies such as optimization techniques, artificial intelligence, efficient complex computational algorithms, or advanced decision-making models is needed to solve both classical automotive dynamics or road safety issues as well as the transport problems of today or the near future. In this scope, the presented Special Issue summarizes research focusing on vehicle dynamics and road safety issues, including advanced technologies, simulations, and the practical application of various technical solutions. The methods presented not only speed up the problem-solving process, but the methods developed allow the inclusion of many more variables by linking the study to modern complex problems. Depending on the specific publication of this issue, new variables used include different driver models (with a steering and braking feel approach), the effect of crosswinds on vehicle stability, specific vehicle properties (kinematics, compliance, and tire model), or the design of an artificial neural network. The proposed research articles also cover real-life issues of safe and effective cargo transportation, road accident research, vehicle technical reliability, or inspections where modern tools and intelligent process management are implemented. However, current research cannot be imagined without virtual or simulation models applicable for the development of autonomous driving, vehicle chassis design, or traffic situations.

Citation: Sokolovskij, E.; Žuraulis, V. Advances in Vehicle Dynamics and Road Safety: Technologies, Simulations, and Applications. *Appl. Sci.* **2024**, *14*, 3735. https://doi.org/10.3390/app14093735

Received: 15 April 2024
Revised: 23 April 2024
Accepted: 24 April 2024
Published: 27 April 2024

This Special Issue contains twenty-three articles, a brief overview of which is presented in the second chapter. The purpose of this review is to encourage readers to delve deeper into specific research on the topic of vehicle dynamics and road safety, sharing the latest research in the field.

2. An Overview of Published Articles

The article by Islam et al. (contribution 1) explores the correlation between annual road accidents and various indicators, such as demographics, economics, passenger and freight movement, as well as investment in road infrastructure. Their study utilizes artificial neural networks, multiple linear regression, and Poisson regression analysis techniques.

In the second article published in this Special Issue, Woo et al. (contribution 2) address the problem of the increasing complexity of vehicle development by employing a virtual prototype to set chassis design targets for high-performance cars. Their proposed methodology eliminates the process of trial and error, allowing them to achieve significant savings.

Li et al. (contribution 3) introduced a multi-object PID differential braking-control approach aimed at enhancing the yaw stability and path-tracking performance of liquid tank semi-trailers on low-adhesion roads while turning. This approach uses the tractor yaw rate, semi-trailer yaw rate, and articulation angle as control parameters. Simulation results show that the proposed approach can effectively improve the yaw stability and path-tracking performance of liquid tank semi-trailers during turns on low-adhesion roads.

El Hajjami et al. (contribution 4) developed an innovative adaptive sliding mode controller in observance of the robust law to address disturbances and uncertainties in the longitudinal dynamics of autonomous ground vehicles. The effectiveness and advantages of this approach were assessed through simulations and comparative studies.

In his article, Gorzelańczyk (contribution 5) attempts to forecast the number of road accidents by state in Poland. He conducted an analysis of the annual data of police statistics on the number of road accidents in Poland for the period from 2000 to 2021, also analyzing the number of road traffic accidents in Poland upon the prediction of the number of traffic accidents in 2022–2031. The results of this study reveal that a decrease in the number of accidents is also expected in the upcoming years.

The aim of the sixth article by Tian et al. (contribution 6) is to effectively assess the impacts of various driving conditions on the lateral stability of vehicles, producing a practical recommendation for pavement maintenance when it comes to rutting. The results show that the devised methodology demonstrates satisfactory performance in evaluating the effect of various impact factors on the lateral stability of vehicles when traversing rutted roads. The outcomes of this research may contribute to the provision of guidelines for controlling the main adverse conditions and making decisions on pavement maintenance.

The research by Batarlienė and Bazaras (contribution 7) focuses on heavy transport, which disrupts supply systems and affects timely customer service. The article aims to analyze the hurdles of the first and last mile and the role and significance of heavy transport in urban logistics. Based on the research results, the authors propose a recommendation for the development of small cargo collection and distribution services using self-service terminals located on the outskirts of cities.

Vaičiūnas et al. investigate the consistency of rail wheel damage measurements (contribution 8) in their articles. The authors analyzed the repeatability of measurements concerning the dependence of the vertical force on wheel-rolling surface damage and proposed methods to enhance consistency. Their article presents the dependence identified by the authors and discusses how the repeatability of wheel damage measurement is influenced by the speed of the rolling stock.

The article by Qin et al. (contribution 9) analyzes the lateral instability mechanism and influencing factors of the civil aircraft towing taxi-out system using computer virtual simulation technology, with Boeing 737–400 and AM210 tractors as focal points. The results reveal that the "jack-knifing" phenomenon is attributed to the lateral instability of the system when the sliding speed exceeds 15 km/h, accompanied by lateral slip conditions.

The conclusion of this study may provide a theoretical basis and guidance for the safe application of the new taxi-out departure mode.

In their article, Alomari et al. (contribution 10) investigated the potential impacts of speeding violations in the United States, encompassing the top ten states with the highest crash rates. Several variables associated with drivers, surroundings, vehicles, roads, and weather were investigated. Three different machine learning algorithms were used to predict speeding violations. The research results revealed that age, accident year, road alignment, weather conditions, the time of the accident, and speed limits are the key factors. Understanding how these factors affect speeding violations helps decision-makers devise strategies to reduce these violations and enhance road safety.

In the eleventh article, Koné et al. (contribution 11) present a novel method aimed at assessing the sensitivity of autonomous vehicles to logical situations and events before their use to generate specific scenarios for simulators. They proposed a characterization of the inputs used in sensitivity analysis and an approach to establishing a distribution function that facilitates the selection of situations and events based on their importance in terms of sensitivity.

Matijošius et al. (contribution 12) conclude that determining the priority of the forces that act on vehicles transporting wooden logs is crucial for accurately modeling the transport parameters of this cargo. The authors of the article use an analytical survey, an expert survey, and its analysis, as well as various methods to determine correlations. The research results revealed that the main forces that affect vehicles transporting logs manifest in the following order of priority: inertial forces, gravity forces, frictional forces, and centrifugal forces.

Innovative pre-crash systems are expected to contribute to mitigating unavoidable accidents. However, there are no standardized testing methods for pre-crash systems. In their work, Putter et al. (contribution 13) propose a novel validation and safety assessment strategy for a perception-based crash severity prediction function. The authors present an innovative unsupervised machine learning methodology for generating specific and logical test scenario catalogs leveraging K-Means++ and k-NN algorithms.

The article by Barbosa et al. (contribution 14) introduces a multidisciplinary approach aimed at understanding the impacts of rising temperatures on the retention of particulate pollutants, including heavy metals and microplastics, within road pavement materials. A soil with a particle size distribution similar to road dust was chosen for the analysis. The analysis results confirm that the viscoelastic properties of bituminous mixtures can contribute to retaining particles within the pavement under increased temperatures. The article paves the way for the development of innovative road pavement bituminous mixtures that can minimize the release of road particulate pollutants, enhancing resilience and sustainability under extreme weather conditions.

The work by Wang et al. (contribution 15) presents a five-degree-of-freedom nonlinear dynamic model of a vehicle with braking torque, transforming the model into an equivalent system in application of the D'Alembert principle. The equilibrium points of the equivalent system are determined by employing an improved hybrid algorithm that combines the genetic algorithm and the sequential quadratic programming method. By analyzing the bifurcation characteristics of the equilibrium points, the stability boundary at the specified initial longitudinal velocity is determined, establishing the three-dimensional stability region. The results reveal that with increasing braking torque, the number of equilibrium points increases from one to three. The proposed equilibrium bifurcation method effectively addresses the stability region of the equivalent system, with solution results aligning consistently with those of the original system stability region.

The paper by Tapak et al. (contribution 16) presents the key findings of the use of smart phone applications in the periodic technical inspection process. The authors noticed that advancements in smart technologies allow for improving the efficiency and effectiveness of vehicle inspections. They specifically focused on the Republic of Slovakia, describing the transition of technical inspections towards smart technologies in this country. The aim was to exploit new technologies and improvements without a dramatic increase in costs. The paper discusses the outcomes of the first two years of using this application at every technical inspection station in Slovakia.

The main objective of the research by Islam and Gazder (contribution 17) was to evaluate and rank the important and supporting factors influencing traffic crashes on the road. To identify the key causes of accidents, the proportion-based analytic hierarchy process was used to order the factors in terms of their relative importance. The city of Al-Ahsa (Saudi Arabia) was used in the research.

In their research, Qu et al. (contribution 18) introduce a decision model based on the lane-changing game characteristics of intelligent connected vehicles aimed at studying the lane-change interaction characteristics of intelligent connected vehicles and mitigating the risk associated with vehicle lane-changing decisions. The results show that the decision-making system based on the lane-changing characteristics of intelligent connected vehicles converges to various optimal strategy combinations under different traffic conditions. This model effectively alleviates decision-making conflicts and reduces the risk of vehicle collisions.

The research conducted by Woo et al. (contribution 19) aims to explore steering feel by examining a steering system and the logic behind electric power steering. Steering feels defined based on prior research and is discussed along with methods for its evaluation. Through a sensitivity analysis involving modeling of both our developed vehicle and a competitor renowned for its exceptional steering feel via a multi-body simulation, the authors propose a straightforward method for determining parameters associated with steering feel to achieve the desired steering characteristics. By modifying the electric power steering control system, the authors successfully achieve a steering feel in the vehicle that meets the desired steering characteristics.

The study by Li et al. (contribution 20) delved into the impact of pavement aggregate grain size on the interaction between tires and pavement during the late stages of pavement skid resistance. A three-dimensional finite element tire–pavement contact model developed using ABAQUS was used to analyze the contact interaction between each simplified pavement type and the tire under conditions of steady-state rolling and braking. The concept of occlusal depth was proposed and used to characterize pavement skid resistance. The results revealed that under steady-state rolling conditions, the maximum contact stress of the simplified pavement rose with the mean texture depth of the pavement, while the contact area decreased.

In his paper, Santarsiero (contribution 21) aims to list the technical problems and solutions in bridge refurbishment interventions for increasing traffic safety, such as, for example, installing code-conforming railings, which often require structural retrofit of bridge elements supporting the railing. The researcher describes several technical solutions, presenting their critical comparison. The economic analysis conducted highlights the impact of the slab retrofit on the total cost of the intervention.

The study by Tunay et al. (contribution 22) aimed to find out how steering inputs by drivers affect the dynamic reaction of a heavy-ground vehicle to crosswinds. They used a two-way interaction between vehicle dynamics and aerodynamic simulations in their research. The steering inputs of drivers were modeled using a driver model taken from prior research presented in the literature, allowing for the reproduction of the steering responses of a human driver. The findings of the study revealed that the steering inputs made by drivers have a significant impact on vehicle response to crosswinds.

The research conducted by Čižiūnienė et al. (contribution 23) employed theoretical methodologies to characterize the primary force exerted by vehicles in the transportation of bulk and liquid commodities. Expert assessment was used to determine the significance of indicators related to individual criteria that affect the transportation of bulk and liquid cargoes, along with the identification of their interrelationships. This involved comparing various factors in the transportation process of bulk and liquid cargoes, such as the different forces involved, criteria like centrifugal force, hazardous factors, critical tank filling levels, and factors affecting vehicle dynamics.

3. Conclusions

The topics of this Special Issue cover a wide range of modern engineering-based technologies and measures applied to tackle road safety issues or enhance ride comfort through land transport. Road accident statistics oblige us to act in all areas concerned; thus, the development of research on vehicle dynamics, together with the incorporation of new phrases such as artificial neural networks or autonomous driving, must be continuously pursued. This Special Issue allows readers to choose and explore scientific trends on the topic of their personal interest, discover a contemporary critical approach, and become familiar with the proposed research methods.

In summary, the articles collected in this Special Issue represent the latest research in the field of vehicle dynamics and road safety. The authors of the articles represent different countries in the world, and the geography of the publications is very wide. Therefore, readers have the opportunity to familiarize themselves with the progress and achievements of individual countries in this field and to assess the peculiarities, similarities, and differences of the solutions applied in different countries. Thus, this Special Issue is not limited to the context of a specific country or region but presents much broader insights into the issues of vehicle dynamics and road safety. Nevertheless, the themes with the greatest potential for scientific growth, which are also included in this publication, must be highlighted. Thus, it is important to highlight the publication's coverage of advanced solutions for autonomous driving research, climate change aspects in transport, vehicle functional safety, and driver assistance systems based on prediction or machine learning techniques.

Conflicts of Interest: The authors declare no conflicts of interest.

List of Contributions

1. Islam, M.K.; Dalhat, M.A.; Al Mamun, A. Road Infrastructure Investment Limits Based on Minimal Accidents Using Artificial Neural Network. *Appl. Sci.* **2022**, *12*, 11949. https://doi.org/10.3390/app122311949.
2. Woo, S.; Ha, Y.; Yoo, J.; Josa, E.; Shin, D. Chassis Design Target Setting for a High-Performance Car Using a Virtual Prototype. *Appl. Sci.* **2023**, *13*, 844. https://doi.org/10.3390/app13020844.
3. Li, G.; Fu, T.; Zhao, R. Research on Yaw Stability Control Method of Liquid Tank Semi-Trailer on Low-Adhesion Road under Turning Condition. *Appl. Sci.* **2023**, *13*, 39. https://doi.org/10.3390/app13010039.
4. El Hajjami, L.; Mellouli, E.M.; Žuraulis, V.; Berrada, M.; Boumhidi, I. A Robust Intelligent Controller for Autonomous Ground Vehicle Longitudinal Dynamics. *Appl. Sci.* **2023**, *13*, 501. https://doi.org/10.3390/app13010501.
5. Gorzelańczyk, P. Forecasting the Number of Road Accidents in Polish Provinces Using Trend Models. *Appl. Sci.* **2023**, *13*, 2898. https://doi.org/10.3390/app13052898.
6. Tian, G.; Jia, Y.; Chen, Z.; Gao, Y.;Wang, S.;Wei, Z.; Chen, Y.; Zhang, T. Evaluation on Lateral Stability of Vehicle: Impacts of Pavement Rutting, Road Alignment, and AdverseWeather. *Appl. Sci.* **2023**, *13*, 3250. https://doi.org/10.3390/app13053250.
7. Batarlienė, N.; Bazaras, D. Solutions to the Problem of Freight Transport Flows in Urban Logistics. *Appl. Sci.* **2023**, *13*, 4214. https://doi.org/10.3390/app13074214.
8. Vaičiūnas, G.; Bureika, G.; Steišūnas, S. Measurement Repeatability of Rail Wheel Loads Caused by Rolling Surface Damages. *Appl. Sci.* **2023**, *13*, 4474. https://doi.org/10.3390/app13074474.
9. Qin, J.; Liu, J.; Lin, Q.; Zhang, W. Research on Instability and "Jack-Knifing" of Civil Aircraft Towing Taxi-Out System. *Appl. Sci.* **2023**, *13*, 3636. https://doi.org/10.3390/app13063636.

10. Alomari, A.H.; Al-Mistarehi, B.W.; Alnaasan, T.K.; Obeidat, M.S. Utilizing Different Machine Learning Techniques to Examine Speeding Violations. *Appl. Sci.* **2023**, *13*, 5113. https://doi.org/10.3390/app13085113.
11. Koné, T.F.; Bonjour, E.; Levrat, E.; Mayer, F.; Géronimi, S. An Approach to Guide the Search for Potentially Hazardous Scenarios for Autonomous Vehicle Safety Validation. *Appl. Sci.* **2023**, *13*, 6717. https://doi.org/10.3390/app13116717.
12. Matijošius, J.; Čižiūnienė, K.; Liebuvienė, J.; Sokolovskij, E. Research Determining the Priority Order of Forces Acting on a Vehicle Transporting Logs. *Appl. Sci.* **2023**, *13*, 6174. https://doi.org/10.3390/app13106174.
13. Putter, R.; Neubohn, A.; Leschke, A.; Lachmayer, R. Predictive Vehicle Safety—Validation Strategy of a Perception-Based Crash Severity Prediction Function. *Appl. Sci.* **2023**, *13*, 6750. https://doi.org/10.3390/app13116750.
14. Barbosa, A.E.; Fontul, S.; Freire, A.C.; Simões, A.R. Road Pavement, Road Pollution, and Sustainability under Climate Change Increased Temperature. *Appl. Sci.* **2023**, *13*, 6949. https://doi.org/10.3390/app13126949.
15. Wang, X.; Li, W.; Zhang, F.; Li, Z.; Bao, W. Research on the Vehicle Steering and Braking Stability Region. *Appl. Sci.* **2023**, *13*, 7806. https://doi.org/10.3390/app13137806.
16. Tapak, P.; Kocur, M.; Rabek, M.; Matej, J. Periodical Vehicle Inspections with Smart Technology. *Appl. Sci.* **2023**, *13*, 7241. https://doi.org/10.3390/app13127241.
17. Islam, M.K.; Gazder, U. Proportion-Based Analytical Hierarchy Process for Determining Prominent Reasons Causing Severe Crashes. *Appl. Sci.* **2023**, *13*, 7814. https://doi.org/10.3390/app13137814.
18. Qu, D.; Dai, S.; Li, A.; Chen, Y.; Wei, C. Characteristic Analysis and Decision Model of Lane-Changing Game for Intelligent Connected Vehicles. *Appl. Sci.* **2023**, *13*, 8321. https://doi.org/10.3390/app13148321.
19. Woo, S.; Heo, C.; Jeong, M.-O.; Lee, J.-M. Integral Analysis of a Vehicle and Electric Power Steering Logic for Improving Steering Feel Performance. *Appl. Sci.* **2023**, *13*, 11598. https://doi.org/10.3390/app132011598.
20. Li, Z.; Chen, W.; Li, Y.;Wu, W. Finite Element Analysis of Tyre Contact Interaction Considering Simplified Pavement with Different Aggregate Sizes. *Appl. Sci.* **2023**, *13*, 12011. https://doi.org/10.3390/app132112011.
21. Santarsiero, G. Retrofitting of Bridge Slabs for Safety Railing Refurbishment in Italy: A State-of-the-Art Review. *Appl. Sci.* **2023**, *13*, 12051. https://doi.org/10.3390/app132112051.
22. Tunay, T.; Drugge, L.; O'Reilly, C.J. The Effects of Different Drivers' Steering Inputs on the Response of Heavy Ground Vehicles to Crosswind Disturbances. *Appl. Sci.* **2024**, *14*, 270. https://doi.org/10.3390/app14010270.
23. Čižiūnienė, K.; Matijošius, J.; Liebuvienė, J.; Sokolovskij, E. Comparison of the Relative Importance of Factors Affecting the Conveyance of Bulk and Liquid Cargo. *Appl. Sci.* **2024**, *14*, 1151. https://doi.org/10.3390/app14031151.

References

1. Khan, M.N.; Das, S. Advancing traffic safety through the safe system approach: A systematic review. *Accid. Anal. Prev.* **2024**, *199*, 107518. [CrossRef] [PubMed]
2. WHO. *Global Status Report on Road Safety 2023*; World Health Organisation: Geneva, Switzerland, 2023.
3. Ricciardi, V.; Travagliati, A.; Schreiber, V.; Klomp, M.; Ivanov, V.; Augsburg, K.; Faria, C. A novel semi-empirical dynamic brake model for automotive applications. *Tribol. Int.* **2020**, *146*, 106223. [CrossRef]
4. Jeong, E.; Oh, C. Evaluating the effectiveness of active vehicle safety systems. *Accid. Anal. Prev.* **2017**, *100*, 85–96. [CrossRef] [PubMed]
5. Acar, B.S. Passive Prevention Systems in Automobile Safety. In *International Encyclopedia of Transportation*; Vickerman, R., Ed.; Elsevier: Amsterdam, The Netherlands, 2021; pp. 406–414. [CrossRef]
6. Masello, L.; Castignani, G.; Sheehan, B.; Murphy, F.; McDonnell, K. On the road safety benefits of advanced driver assistance systems in different driving contexts. *Transp. Res. Interdiscip. Perspect.* **2022**, *15*, 100670. [CrossRef]
7. González-Saavedra, J.F.; Figueroa, M.; Céspedes, S.; Montejo-Sánchez, S. Survey of Cooperative Advanced Driver Assistance Systems: From a Holistic and Systemic Vision. *Sensors* **2022**, *22*, 3040. [CrossRef] [PubMed]

8. Jurecki, R.S.; Stańczyk, T.L.; Jaśkiewicz, M.J. Driver's reaction time in a simulated, complex road incident. *Transport* **2017**, *32*, 44–54. [CrossRef]
9. Dirnbach, I.; Kubjatko, T.; Kolla, E.; Ondruš, J.; Šarić, Ž. Methodology Designed to Evaluate Accidents at Intersection Crossings with Respect to Forensic Purposes and Transport Sustainability. *Sustainability* **2020**, *12*, 1972. [CrossRef]
10. Chen, P.; Ni, H.; Wang, L.; Yu, G.; Sun, J. Safety performance evaluation of freeway merging areas under autonomous vehicles environment using a co-simulation platform. *Accid. Anal. Prev.* **2024**, *199*, 107530. [CrossRef] [PubMed]
11. Björnberg, K.E.; Hansson, S.O.; Belin, M.Å.; Tingvall, C. *The Vision Zero Handbook*; Springer: Cham, Switzerland, 2020. [CrossRef]
12. Ajanović, Z.; Regolin, E.; Shyrokau, B.; Ćatić, H.; Horn, M.; Ferrara, A. Search-based task and motion planning for hybrid systems: Agile autonomous vehicles. *Eng. Appl. Artif. Intell.* **2023**, *121*, 105893. [CrossRef]
13. Skrickij, V.; Šabanovič, E.; Žuraulis, V. Autonomous road vehicles: Recent issues and expectations. *IET Intell. Transp. Syst.* **2020**, *14*, 471–479. [CrossRef]
14. Lenné, M.G. The contribution of on-road studies of road user behaviour to improving road safety. *Accid. Anal. Prev.* **2013**, *58*, 158–161. [CrossRef] [PubMed]
15. Khaliq, K.A.; Chughtai, O.; Shahwani, A.; Qayyum, A.; Pannek, J. Road Accidents Detection, Data Collection and Data Analysis Using V2X Communication and Edge/Cloud Computing. *Electronics* **2019**, *8*, 896. [CrossRef]
16. Surblys, V.; Žuraulis, V.; Sokolovskij, E. Estimation of road roughness from data of on-vehicle mounted sensors. *Eksploat. I Niezawodn. = Maint. Reliab.* **2017**, *19*, 369–374. [CrossRef]
17. Lei, Y.; Zhang, G.; Lu, S.; Qian, J. Generation Paths of Major Road Accidents Based on Fuzzy-Set Qualitative Comparative Analysis. *Int. J. Environ. Res. Public Health* **2022**, *19*, 13761. [CrossRef] [PubMed]
18. Wakabayashi, H.; Ando, M.; Kawaguchi, T.; Horie, Y.; Hotta, K. Sustainable Traffic Safety Management at Accident Black Spots Combined with Drivers' Psychology and Vehicle Engineering Using Eye Mark Recorder. *Transp. Res. Procedia* **2014**, *3*, 90–99. [CrossRef]
19. Žuraulis, V.; Sokolovskij, E. Vehicle velocity relation to slipping trajectory change: An option for traffic accident reconstruction. *Promet—Traffic Transp.* **2018**, *30*, 395–406. [CrossRef]

applied sciences

MDPI

Article

Road Infrastructure Investment Limits Based on Minimal Accidents Using Artificial Neural Network

Md Kamrul Islam [1,*], Muhammad Abubakar Dalhat [2,*] and Abdullah Al Mamun [3]

1 Department of Civil and Environmental Engineering, College of Engineering, King Faisal University, P.O. Box 380, Al-Hofuf 31982, Saudi Arabia
2 Transportation and Traffic Engineering Department, College of Engineering, Imam Abdulrahman Bin Faisal University, P.O. Box 1982, Dammam 31451, Saudi Arabia
3 Department of Civil and Environmental Engineering, University of Utah, 110 Central Campus Drive, MCE-1435, Salt Lake City, UT 84112, USA
* Correspondence: maislam@kfu.edu.sa (M.K.I.); madalhat@iau.edu.sa (M.A.D.)

Abstract: Road traffic accidents are still among the top major global causes of death, injury, and disability. Despite this cause for alarm and several preventive initiatives, global road accident statistics are not improving. This study modeled annual road accidents (ARAs) as a function of demographic, economic, passenger movement, freight movement, and road capital investment indicators. The research is based on 22 years of data from more than 36 Organization for Economic Co-operation and Development (OECD) member and partner countries. Artificial neural network (ANN), multiple linear regression (MLR), and Poisson regression (PR) analysis were employed for this purpose. The ANN model outperformed the regression models by far, thus making it possible for reliable new insights and accurate results to be obtained. The ANN's superior performance was shown to be a result of the non-linear relationship between ARA and some of the predicting variables. The average relative contribution of each variable in describing the ARA models was estimated using connection weight analysis (from the ANN model) and relative weight analysis for the regression model. The profile method was used to perform sensitivity analysis and to establish the partial variation trend of the ARA with each of the variables. The Existing Road Maintenance Investment (ERMI) and New Road Infrastructural Investment (NRII) showed a nonlinear concave-up relationship with ARA for given demography, economy, freight, and passenger movements. A combination of per capita NRII and ERMI corresponding to the minimum ARA exists. These sets of NRII and ERMI were considered safe road investment limits. The ANN-ARA model was utilized to estimate these limits with their relative proportion for diverse combinations of demography, economy, freight level, and passenger movement.

Keywords: road investments; road traffic accidents; artificial neural network; gross domestic product (GDP); demographics

Citation: Islam, M.K.; Dalhat, M.A.; Al Mamun, A. Road Infrastructure Investment Limits Based on Minimal Accidents Using Artificial Neural Network. *Appl. Sci.* **2022**, *12*, 11949. https://doi.org/10.3390/app122311949

Academic Editors: Edgar Sokolovskij and Vidas Žuraulis

Received: 25 October 2022
Accepted: 20 November 2022
Published: 23 November 2022

1. Introduction

Road traffic accidents are one of the major causes of death, injury, and disability globally [1–5]. Up to 1.3 million people lose their lives, and another 50 million sustain serious injuries due to road accidents each year, and the numbers keep increasing [6]. Annual road accidents (ARAs), which cause serious and irreparable economic and social harm to nations, are thus one of the main concerns of transportation managers around the world [7]. Successful implementation of road infrastructural safety management requires sufficient funding in addition to regulations, institutional capacity, and the availability of safety data [8]. Road management authorities are faced with the difficult task of identifying and prioritizing new projects and maintenance actions. Several optimization tools for resource allocation in road asset management have been adopted for efficient and well-informed decision-making processes [7,9–13]. Due to budget constraints, most of these optimization tools focus on the best alternative solutions within available budget constraints [14]. The question of how much is enough to tackle a single objective is usually given little attention effectively. The multi-criteria solution orientation of these decision frameworks also contributes to this lack of focus [9]. Safety, political importance, economic impact, and functional and structural performance are among the key priority functions for road asset management. However, among these factors, only the safety aspect affects human lives directly. This study focuses on safe limits for the annual maintenance of the road network and new infrastructural development investments in terms of minimum system-wide road accidents for diverse demographic, economic, and transport systems.

Numerous studies were conducted towards modeling the frequency of road accidents using artificial neural network (ANN) [15–19]. These studies focused on predicting accident frequency based on road features, environmental conditions, driver attributes, and specific types of road facilities. Traffic characteristics such as volume and vehicle mix proportions were also used as predicting variables [20]. Accident prediction models such as these are often useful in developing crash modification factors [21] and understanding and prioritizing accident risk indicators [16]. Some crash prediction models (CPMs) are incredibly helpful tools for quantitative road safety analysis and can be applied to road network screening to identify the network's most important portions and to more effectively guide in-depth studies [22]. Other road accident predictors, such as vehicle ownership, per capita gross domestic product (GDP), and the relative usage of transport modes, have also been investigated [23]. Moreover, integrating the safety of VRUs—pedestrians, cyclists, and motorcyclists, has also resulted in significant consideration when making judgments about the management of road infrastructure [24]. Previously, ANN models were applied successfully to forecast road traffic [25] and travel time [26]; to optimize signal timing at road intersections [27], road vehicle collision avoidance systems, and road sign detection and classification [28]; and to evaluate pedestrian road crossing behavior [29]. The major aim of this study is to model and critically analyze the relationship between the frequency of road accidents and investments to maintain existing roads and for new road construction using ANN and regression analysis.

Programs to improve infrastructure-related safety are typically capital-intensive and call for management and planning. Despite being constructed to engineering standards, the road network, which consists of sections and junctions, will typically need renovations and adjustments over its lifespan due to changes, most notably in traffic volumes and additional safety requirements. Different countermeasures will typically be needed as a result of these adjustments to increase safety requirements and to repair problematic areas of the road [30]. The countermeasures can be divided into capital and maintenance work categories. However, road maintenance and new construction investments (in EUR/km) were found to result in both positive and negative effects on the severity of road accidents in Spain [31]. The work suggested further study on this observed anomaly. In an earlier study on data from the same country, the maintenance investment was found to achieve consistently lower accident fatality results [32], while investments in new construction were believed to only increase the risk and fatality of road accidents. Contrarily, a study conducted in the United States showed highway capital investment to have a negative effect on road fatalities [33]. The effects of infrastructural and demographic changes on accident frequency and fatality were studied from 8-year accident data from the State of Illinois [34]. Increasing the number and width of lanes were found to result in more crashes, while changes in demography were found to have an insignificant effect on the observed results. Reports from these studies on the effect of road investment on accidents cannot be generalized due to inconsistent findings. These contradictions could be due to several different factors related to the study area, such as demography, economy, road transport system characteristics, etc. However, it can be concluded that there has been no clear general explanation as to why road investment, whether for maintenance or new construction, can result in either a decrease or increase in the crash rate or in the number of fatalities. In this study, the inconsistent effect of road infrastructure investment on traffic accidents was shown to be due to a concave nonlinear relationship between the parameters. A simple method for estimating safe limits for both maintenance and new road infrastructure investments based on minimum Annual Road Accident (ARA) using the developed ANN-ARA model was also illustrated.

2. Data and Methodology

The data utilized in this study were obtained from the Organization for Economic Co-operation and Development (OECD) database. The OECD has about 36 member countries and several partner countries. Some of these countries include the United States of America (USA), Germany, France, Canada, Spain, Australia, Japan, Italy, etc. Annual data from member and partner countries for 22 years (1994–2016) were employed in this study. Historical data that were analyzed included annual road accidents, freight and passenger transport indicators, changes in the number of new vehicles registered, investments in new road infrastructure and existing road maintenance, and demographic and economic indicators (see Table 1). Some previous studies that have employed the OECD database include the statistical modeling of road mortality [35,36], comparisons of death from road accidents with international terror [37], modeling traffic fatalities using autoregressive nonlinear time series [38], the risk of severe fatal accidents using Bayesian data analysis [39], etc. In this study, annual road accidents (ARAs) were modeled as a function of all the other listed variables using multiple linear regression (MLR), Poisson regression (PR) analysis, and artificial neural network (ANN). Because the ANN model is more accurate, it was utilized to closely examine how these variables collectively affect annual road accidents. Safe investment limits for new road infrastructure and for the maintenance of existing roads based on the minimum ARAs were estimated for a wide range of demographic, freight transport, passenger transport, and economic systems.

Table 1. Basic statistics of data.

Statistics	Mean	Standard Deviation	Median	Maximum	Minimum	25th Percentile	50th Percentile	75th Percentile
Total Accidents/yr. (1000)	202	449	41	2275	1	9	41	157
Freight (million ton-km)	269,976	776,473	37,199	4,018,805	1864	17,344	37,199	156,986
Passenger (million passenger-km)	543,779	1,162,300	124,355	5,846,312	2143	67,088	124,355	680,863
Change in Number of New Vehicles Registered (%)	2.45	18.97	1.29	152.44	−72.94	−4.82	1.29	7.47
Existing Road Maintenance Investment "ERMI" (EUR/yr.)	3,509,432,464	6,486,214,266	755,494,568	35,926,399,759	16,694,491	204,500,000	755,494,568	3,085,019,372
New Road Infrastructure Investment "NRII" (EUR/yr.)	8,118,209,606	16,109,624,922	1,893,299,904	79,312,039,934	46,000,000	653,000,000	1,893,299,904	6,034,830,090
ERMI Per Capita (EUR/yr./Capita)	84	71	73	387	0	29	73	117
NRII Per Capita (EUR/yr./Capita)	163	126	128	757	11	85	128	205
Population (Million)	44.30	68.62	10.55	318.86	1.32	5.40	10.55	58.81
Proportion of Work Group (%)	67.12	2.13	66.81	73.10	62.41	65.60	66.81	68.07
Population of Work Group (Million)	29.60	45.72	7.03	211.55	0.87	3.63	7.03	38.65
Gross Domestic Product (USD/Capita)	30,005	10,974	28,707	67,051	8608	22,784	28,707	36,519

2.1. Description and Basic Statistics of Data

The basic statistics of the variables, including the average, standard deviation, median, maximum, and minimum, are presented in Table 1. A detailed description of the dataset is given in the following paragraphs. Rows with missing data were excluded from the analysis, leaving about 435 rows.

Annual Road Accidents (ARAs) [30]: The ARAs are defined as the number of annual road accidents involving a road motor vehicle (excluding suicide using a road motor vehicle) recorded that might have resulted in death (s), injury, and or loss of property. Road motor vehicles include buses, tramways or streetcars, coaches, trolleys, and road vehicles used to transport goods and passengers.

Freight Transport [31]: This is defined as the total movement of goods using road transport on a given network expressed in million ton-km.

Passenger Transport [32] is defined as the total movement of passengers using road transport on a given network expressed in million passenger-km.

Change in Amount of New Vehicle Registrations (CNRV) [33] is defined as the annual increase or decrease in the amount of newly registered vehicles (private and commercial) expressed as a percentage.

Existing Road Maintenance Investment (ERMI) [34]: This is defined as the total annual monetary spending towards the preservation of the existing road infrastructure expressed in EUR (EUR).

New Road Infrastructure Investment (NRII) [35]: This covers the total annual monetary spending on new road construction to improve the existing road network expressed in EUR (EUR).

Population [36] is defined as all nationals present in, or temporarily absent from, a country, and aliens who have permanently settled in a country, and is expressed in millions.

The proportion of Working Group [37] is the percentage of the population between 15 and 64 years of age who are employed. The Population of Work Group was estimated from the total population and the proportion of work group and is represented in millions. The Proportion of Work Group is the percentage of the working population with respect to the total population. Different countries can have similar proportions of groups but can also have different populations.

Gross Domestic Product (GDP) Per Capita [38]: All OECD countries estimate their econometric data according to the 2008 System of National Accounts and are thus comparable for any single year. GDP per capita was measured in US dollars at current prices and purchasing power parity exchange rates.

2.2. Regression Modeling

Stepwise multiple linear regression (MLR) and Poisson regression (PR) analysis were utilized to develop simple regression models for annual road accidents as a function of other variables. Stepwise regression analysis starts with a single variable and then continuously adds and or removes the potential predicting variables to fit the model until a combination of predicting variables that yield the best model performance was achieved. A value of 0.15 for α-to-include and α-to-exclude was used, where α is the level of statistical significance. MiniTab16^{TM} statistical software was employed for this task. General model equations for the MLR and the PR are given by Equations (1) and (2), respectively. MLR is the basic statistical modeling tool, while the PR and similar discrete count models were considered more appropriate for modeling accident frequency [40].

$$Y = \sum_{i=1}^{N} \beta_i x_i \tag{1}$$

where Y represents the dependent variable ARA, x_i is the ith predicting variable, β_i is the coefficients of the ith variable, $\beta_1 = 1$, and N is the total number of terms in the final model.

$$Y = e^{(\sum_{i=1}^{N} \beta_i x_i)} \tag{2}$$

where e is natural log constant (2.718281), and all other terms are described as in Equation (1).

2.3. Artificial Neural Network (ANN) Modeling

A feed-forward 2-layer ANN with 7 neurons was developed in MatLab^{TM} (R2017a) to train the annual road accident model. Based on the balanced performance between the training and testing results, the number of neurons was optimized. This was achieved by varying the number of neurons, randomly re-splitting the training/testing data, and retraining with random initial conditions. A schematic of the final ANN structure is shown in Figure 1. The terms W and b represent the weight and bias matrices of the model, respectively. Each of the variables from the input matrix X is connected to each neuron through the weight matrix IW. In this case, a^1 is a 7-element column vector formed by "F^1" from the weighted sum of the input variables x_i and bias b_i of the neurons' outputs. The neurons' outputs serve as inputs, which are transformed to fall between [−1, 1] using the hyperbolic tangent sigmoid equivalent transfer function "F^1" given by Equation (3). This quickens the learning process. These normalized outputs from "F^1" serve as input to the output layer. "F^2" is a purelin transfer function that again normalizes the weighted sum and biases of the outputs from the hidden layer between [−1, 1] to yield a^2. Equation (4) is a post-processing function that reverses a^2 to yield a scalar y_p between $[y_{max}, y_{min}]$ that is matched and compared with the actual target (ARA) y_a. The training cycles follow this sequence while the weight and bias matrices are optimized for better model performance. Bayesian-regularized (BR) Levenberg–Marquardt optimization was selected as the training algorithm [41,42]. The BR-ANNs were found to be robust and difficult to over-train or over-fit, and the validation process is unnecessary [43]. Data partitioning of 75/25 for the training and testing data was adopted, respectively. The model performance was measured based on the mean square error (MSE) and coefficient of correlation between the actual and predicted ARA. Equation (5) represents the root mean square error. The correlation between anticipated/predicted and actual/observed ARA was estimated using Equation (6).

$$F^1(r) = 2/\left(1 + e^{-2r}\right) - 1 \tag{3}$$

where F^1 is a tangent sigmoid equivalent transfer function; r is the independent variable, while e denotes the natural log constant (2.718281).

$$y_p(u) = \frac{(y_{max} - y_{min})(u - u_{min})}{(u_{max} - u_{min})} + y_{min} \tag{4}$$

where u denotes a set of finite real numbers between [−1, 1], while the minimum and maximum values of the original target dataset is represented by y_{min} and y_{max}, respectively.

$$RMSE = \sqrt{MSE} = \sqrt{\frac{\sum_i^{n_t}(y_{ai} - y_p(u_i))^2}{n_t - n_p}} \tag{5}$$

$$R^2 = 1 - \frac{\sum_i^{n_t}(y_{ai} - y_p(u_i))^2}{\sum_i^{n_t}(y_{ai} - \overline{y_a})^2} \tag{6}$$

where $RMSE$: root mean square error, y_{ai}: actual observed ARAs, $y_p(u_i)$: modeled or predicted ARAs, n_t: total number of observed/actual ARAs, n_p: number of model parameters, $\overline{y_a}$: mean of observed/actual ARAs.

2.4. Relative Importance Analysis and Ranking

Using the connection weight matrix method, the relative significance of the predictive variables for the ANN model was calculated [44]. The concept of using the weight connection to determine the relevance of predicting variables in an ANN model was initially proposed by Garson [45]. The overall absolute weight contribution fraction of the variables with regard to each neuron was calculated using the Garson algorithm to determine the relative importance of the predicting variable, as indicated in Equation (7). Olden and colleagues later claimed that the net sum of the weight contribution for each predicting variable across the neurons yielded more accurate rankings [44], as given in Equation (8). Both approaches were utilized in this study to rank the relative relevance of the predicting variables from the ANN model.

$$Garson\ Relative\ Importance(x_i) = \frac{\sum_{j=1}^{n}\left|W_{ij}^h * W_j^o\right|}{\sum_{i=1}^{m}\left\{\sum_{j=1}^{n}\left|W_{ij}^h * W_j^o\right|\right\}} \tag{7}$$

$$Olden\ Relative\ Importance(x_i) = \frac{\left|\sum_{j=1}^{n} W_{ij}^{h} * W_{j}^{o}\right|}{\sum_{i=1}^{m}\left\{\left|\sum_{j=1}^{N} W_{ij}^{h} * W_{j}^{o}\right|\right\}} \quad (8)$$

where x_i is the ith predicting variable, W_{ij}^{h} is the weight matrix of the hidden layer of the ANN model, W_j^o is the weight matrix of the output layer, n is the maximum number of neurons in the hidden layer, and m is the number of predicting variables.

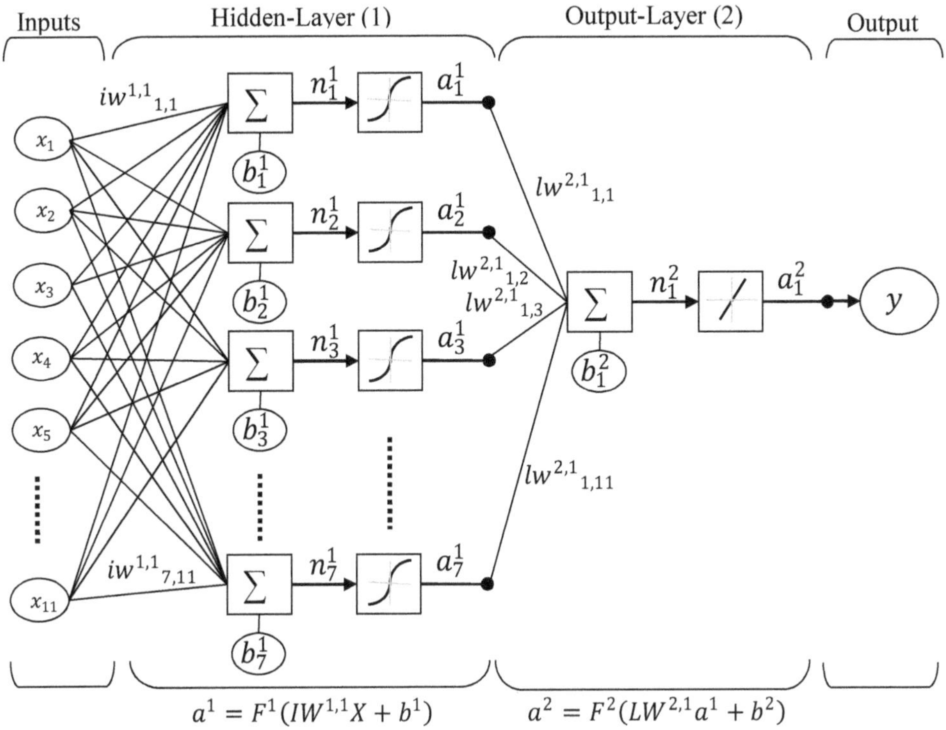

Figure 1. Structure of the ANN model.

The relative importance of the variables from the regression models was also estimated using relative weight analysis [46,47]. The traditional use of normalized coefficients was found to yield misleading results due to some level of correlation between the predicting variables [47]. The relative weight analysis eliminates the effect of variable inter-correlation by transforming the predictors into sets of orthogonal variables. The end result is a decomposed contribution of each variable to the R^2 of the model in explaining the dependent variable. The analysis was run using R code web-based software with all variables at the 5% significant level and with 10,000 bootstrapping iterations [48].

2.5. Sensitivity Analysis Using ANN Model

The profile method (Lek's profile) was employed for the sensitivity analysis of the ANN model (Gevrey et al., 2003). Lek and colleagues introduced this technique to show the general partial variation trends in the dependent variable of ANN models with each predicting variable (Lek et al., 1996). The dependent variable's profile with respect to a certain predicting variable is derived by changing the predicting variable only partially across its range while holding the minimum, mode/median, quartiles, and maximum values of all other variables constant. The various results are averaged to obtain a single curve. The disadvantage of the method is the high computational tasks, especially when dealing with numerous predicting variables. However, comparative studies using other similar ANN sensitivity analysis methods found the profile method to yield useful and reliable insight [49]. Outputs from 12 places along the range of the predicting variable were advised to reduce the amount of calculation needed (Lek et al., 1996). In this study, the other variables were held constant at their quartiles (25th, 50th, and 75th percentiles), mean, and median, while the ARAs were estimated at 50 equal intervals along the range of each predictive variable. These curves were averaged to obtain the overall profile for a predictive variable. Because some outcomes from these extreme constants produce negative ARA values that could create mistakes in the trends seen, the minimum and maximum are removed.

2.6. Annual NRII and ERMI Limits

Among all of the independent variables of the ARA model, annual NRII and EMRI are the only variables that can be easily controlled to influence the ARAs. Demographic and economic variables depend on several other factors and require a long period of time to control. The freight and passenger movement of a given country depend on the demography and economy of that country. Due to these reasons (more reasons will follow in the Results section), the annual per capita safe limits of the NRII and EMRI were considered important parameters influencing ARAs in this study. As a result, the per capita NRII and EMRI corresponding to the minimum ARAs for several sets of all other variables were estimated. This was carried out by estimating the various ARAs within the space [Minimum NRII, Maximum NRII] and [Minimum ERMI, Maximum ERMI]. The results for each run were exported to an Excel sheet, where simple algorithm searches were conducted for the minimum ARAs and the corresponding per capita NRII and ERMI. Figure 2 below shows a simplified flow chart for the safe annual per capita NRII and EMRI limits estimation process.

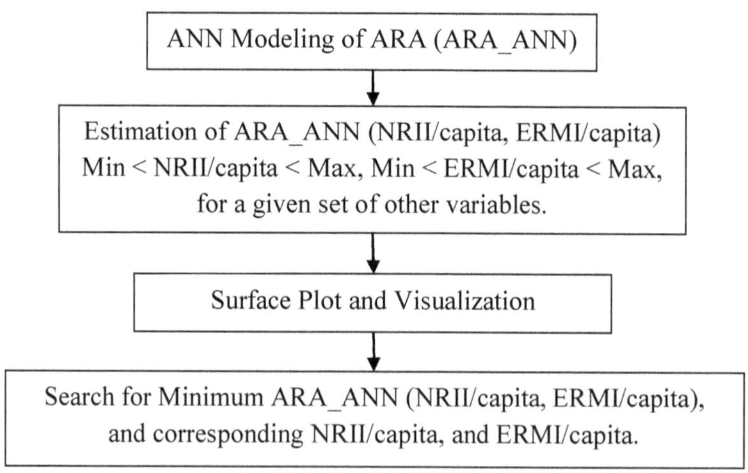

Figure 2. Flow chart for the safe NRII and EMRI limits estimation process.

3. Results and Discussion

3.1. MLR, PR and ANN ARA Models

This section discusses the fitting results of the ARA results obtained using ANN, PR, and MLR. It also highlights the need for sophisticated model fitting techniques such as ANN (for this study) instead of conventional methods such as MLR and PR.

The parameters of the ARA model from the stepwise regression analysis are summarized in Table 2. All of the included parameters showed significant influence on the ARA (p-value < 0.05). Some of the variables were excluded from the MLR models, which means that based on this model, these variables did not have a significant effect on the independent variable (ARA), with *p*-values greater than 0.15. The excluded variables include GDP/capita, CNVR, and total ERMI/yr. On the other hand, all of the variables were found to significantly affect the ARA according to the PR model. The PR model is considered more appropriate for discrete count variable statistics such as ARA [40]. It is worth mentioning that the inclusion of similar variables was obtained with either standardized or non-standardized data for both models. A comparison of the coefficients between the MLR and PR models reveals similar results, as previously reported by some studies in the literature. For example, increasing NRII/Capita will result in lower ARAs, while increasing EMRI/Capita will lead to more ARAs according to both the MLR and PR models. Based on both model structures and intercepts, any variable with a negative sign has a negative effect on the independent variable (i.e., ARA), and vice versa. This characteristic of attributing only a single effect to a variable with a single regression coefficient limits their performance when dealing with highly non-linear problems. The total NRII, ERMI/Capita, NRII/Capita, and Proportion and Population of Work Group values showed consistently similar effects on the ARAs in both the MLR and PR models. However, the freight and passenger variables alternated signs between the two models. This can be attributed to the different nature of the models. Whatever the sign or observed effect of the variables within a model, the degree of acceptability of the results depends on the overall accuracy of that model. The accuracies of the MLR and PR models are discussed in the next paragraph.

Table 2. ARA regression model parameters.

		MLR		**PR**	
Terms	Unit	Coefficient	*p*-value	Coefficient	*p*-value
Constant		−1,625,191	0.0000	1.77931	0.0000
Freight	MTkm	3.56900×10^{-01}	0.0000	-1.28461×10^{-06}	0.0000
Passenger	Million passenger km	-1.07521×10^{-01}	0.0000	1.62813×10^{-07}	0.0000
CNRV	(%)	–	–	-2.14206×10^{-03}	0.0000
Total ERMI	EUR/yr.	–	–	-1.50190×10^{-10}	0.0000
Total NRII	EUR/yr.	1.12648×10^{-05}	0.0000	4.42200×10^{-11}	0.0000
ERMI Per Capita	EUR/yr./Capital	5.40957×10^{02}	0.0000	1.08639×10^{-02}	0.0000
NRII Per Capita	EUR/yr./Capital	-2.65564×10^{02}	0.0001	-3.40675×10^{-03}	0.0000
Population	Million	4.62782×10^{04}	0.0000	1.07097×10^{-01}	0.0000
Proportion of Work Group	(%)	2.39945×10^{04}	0.0000	1.31254×10^{-01}	0.0000
Population of Work Group	Million	-6.63550×10^{04}	0.0000	-1.12287×10^{-01}	0.0000
GDP	(USD/capita)	–	–	-2.45351×10^{-05}	0.0000

Plots of the actual versus predicted ARAs for the MLR, PR, and ANN models are presented in Figures 3 and 4, respectively. Although the MLR and PR models showed good coefficients of correlation (R^2) between the actual and the predicted ARA, these models are inconsistent and low in accuracy. The inconsistency can be seen from their deviation from the Y = X line. On the other hand, the ANN model showed a higher R^2 value as well as a high degree of accuracy. The difference in the level of accuracy between these models can be clearly seen from the comparison plots of their error histograms shown in Figure 5. The MLR and PR models only have a few accurate predictions, with most of the remaining estimated ARAs being far different from the actual values. The high number of observations with deviation errors of 25 or more attest to this fact. However, the ANN model showed that most of the ARA values were within a small margin of error with the actual values. This was deduced from the higher number of observations with deviation errors closer to zero for the ANN model. Table 3 shows the summary of the various model performances in terms of root mean square error (RMSE) and R^2. Comparing the MLR, PR, and ANN ARA models in terms of RMSE, the ANN model is at least 11 times more accurate than the other models. In accordance with these observations, only the ANN model was considered reliable for the sensitivity analysis of the ARA and the model application example. Table 4 presents the weight and bias matrices of the ANN model's hidden and output layers. These results were used to estimate the relative importance of the predicting variables of the ANN model.

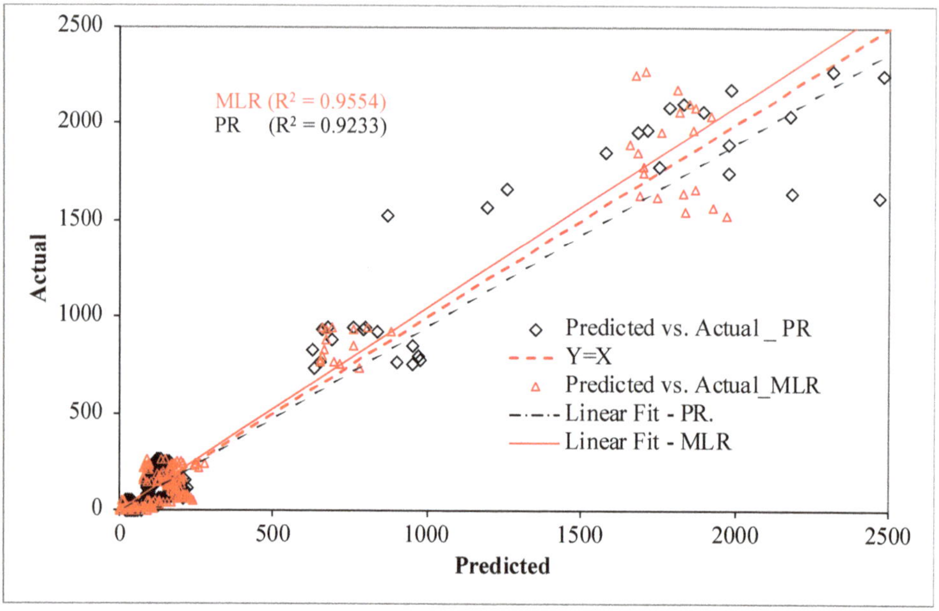

Figure 3. Predicted vs. actual annual road accident plot for MLR and PR models.

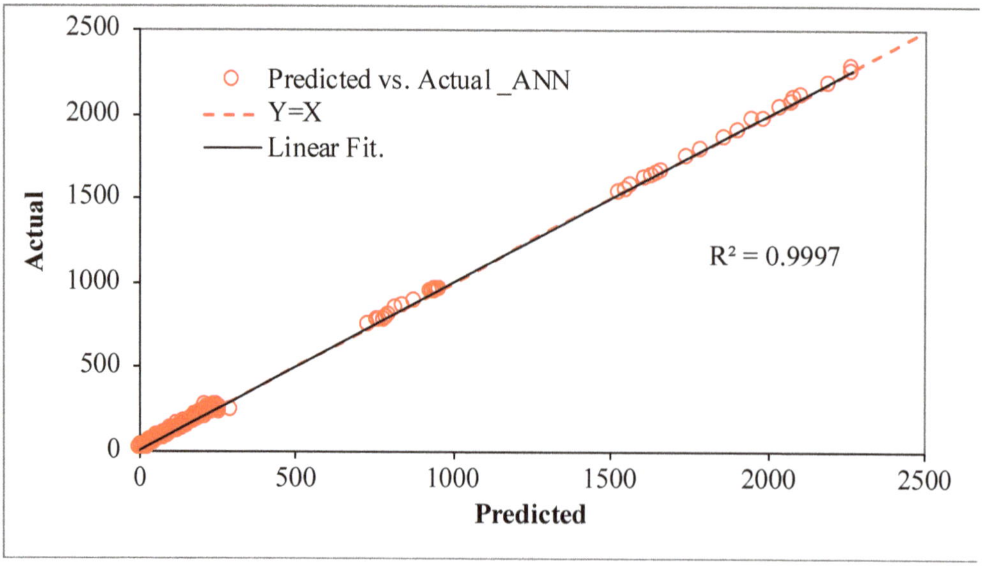

Figure 4. Predicted vs. actual annual road accident plot for ANN model.

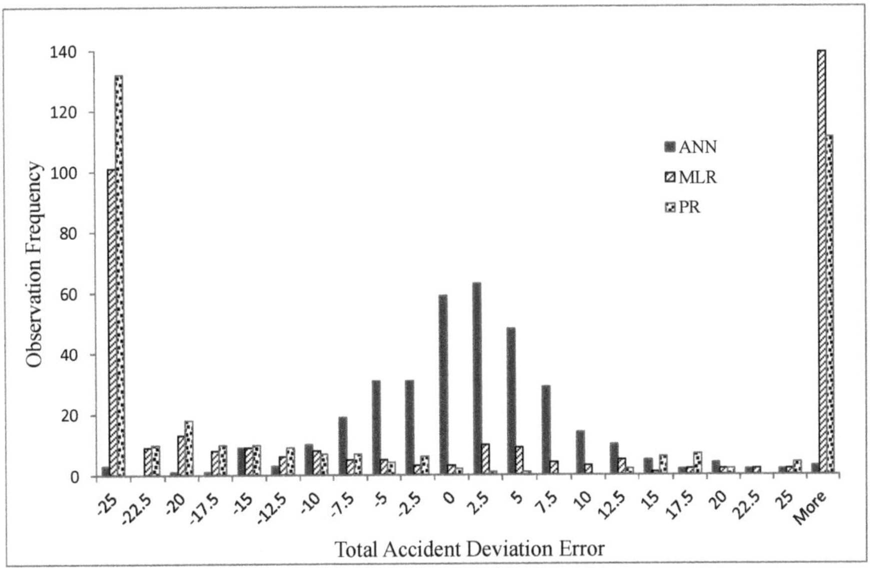

Figure 5. Error histograms of the ANN, MLR, and PR models.

Table 3. ANN model vs. MLR model performance.

	MLR	**PR**		**ANN**	
	All	All	Training	Testing	All
RMSE	96.07	127.53	7.34	9.42	8.64
R^2	0.9554	0.9233	0.9999	0.9991	0.9996

Table 4. Trained ANN model weight and bias matrices.

	Hidden Layer Weight Matrix (W_{ij})						
Inputs (x_i)/Neurons (n)	n_1^1	n_2^1	n_3^1	n_4^1	n_5^1	n_6^1	n_7^1
Freight	−0.01957	−0.03012	−0.87492	−0.39241	−0.74901	−1.96745	−0.59826
Passenger	0.50547	1.344694	0.85682	−1.09923	0.954988	0.425543	−0.31455
CNRV	1.266489	0.042526	0.137132	−0.58993	0.032595	0.035055	0.112455
Total ERMI	−1.98258	0.067577	−1.65205	2.627753	−0.38092	−0.0193	−0.99349
Total NRII	0.270451	0.579433	0.359455	1.800905	1.374441	1.101101	0.150742
ERMI Per Capita	1.404668	−1.05879	−0.25969	−2.63473	−0.83877	−1.47092	−0.68379
NRII Per Capita	−0.05638	0.63253	0.185425	−2.40709	0.636397	1.087314	0.397435
Population	−0.58079	−0.80266	0.564306	1.295473	−0.3578	0.179384	−0.0249
Proportion of Work Group	0.446989	0.523725	−0.33594	−0.08935	0.866879	1.350303	−0.04303
Population of Work Group	−1.42903	0.59404	1.841841	1.313592	1.691216	2.568981	1.323976
GDP	−2.28676	0.178019	−0.53137	2.542531	0.362622	0.493147	−0.27178
Hidden Layer Bias 'b_i'	0.763762	1.411427	0.530619	−1.21196	1.324628	0.295899	−0.81256
	Output Layer Weight Matrix						
Out Put	2.942207	1.177333	1.196882	2.90994	−1.60138	1.148312	−1.75404
Output Layer Bias 'b'	−0.889019679						

3.2. Relative Influence of Predicting Variables

The relative importance and ranking of the predicting variables of the ANN model were estimated according to Equations (7) and (8) using the data in Table 4. Similarly, the relative influence for the regression model predictors was also determined. These relative importance results for the predicting variables in the ARA model are presented in Table 5. The results obtained using the relative weight analysis for the regression showed that CNVR, ERMI per Capita, and Proportion of Work Group contributed to explaining R^2 the least in the ARA. However, these variables were found to have a certain level of relevance according to the weight contribution analysis of the ANN model. This can be associated with the different structure, accuracy, theory, method of relative importance estimation, etc., of the models. This also highlights how almost all the predicting variables included in the ANN model played some role in the model performance. Although the relative importance does not represent the level of statistical significance, it certainly can reveal high-priority variables. The relative influence results of the variables were utilized to rank them, as presented in the last column of Table 5. The #1 rank represents the variable with the highest relevance, while the rank of #11 represents the variable with the least relative influence. The overall ranking was obtained based on averages of relative importance of each variable across the different methods. This overall ranking was necessary because the various models and relative importance estimation methods did not yield similar results. Variation in the results was anticipated, but since Garson's method is still widely utilized and Olden's method was supported with compelling evidence, both results were included. The relative weight analysis of the regression has strong fundamental theoretical bases but is still affected by all the limitations of multiple regression [46]. As such, the average or the overall ranking is considered to be much closer to the fact. Overall, the variables in decreasing order of influence on ARA are total NRII (#1), population of work group (#2), total ERMI (#3), freight movement (#4), NRII/capita (#5), population (#6), ERMI/capita (#7), passengers (#8), GDP (#9), proportion of work group (#10), and CNRV (#11). The set of road infrastructure investment variables can be seen to be in the top levels with regard to the influence on ARA. Being the lowest in the ranking does imply insignificance to the dependent variable. It only means lower relative linear, nonlinear, or both influence the dependent variable with respect to the other variables.

Table 5. Relative influence/importance ranking of variables.

Input/Variable (x_i)	Regression (%)	ANN-Garson (%)	ANN-Olden (%)	Average (%)	Overall Rank/11
Freight	18.21	7.74	7.88	11.28	4
Passenger	16.48	9.67	1.40	9.18	8
CNRV	0.08	3.06	6.93	3.36	11
Total ERMI	14.64	11.84	7.72	11.40	3
Total NRII	14.63	8.29	20.45	14.46	1
ERMI Per Capita	0.42	12.68	14.57	9.22	7
NRII Per Capita	1.14	7.73	22.94	10.60	5
Population	16.87	5.86	8.99	10.57	6
Proportion of Work Group	0.08	6.04	5.18	3.77	10
Population of Work Group	16.77	18.23	1.52	12.17	2
GDP	0.68	8.86	2.43	3.99	9

3.3. Sensitivity Analysis of ANN Model

The average variation trend of ARAs with the different predicting variables is presented in Figure 6. The figure is scaled by normalizing the ranges of each input variable (X_i) between [0, 100] and their corresponding ARA outputs (Y) between [0, 100]. Each curve is an average of various curves, as previously described in the methodology. Starting from Figure 6a, it can be observed that the ARA exhibits a positive exponential growth relationship as Freight Movement increases. This trend is in line with previous findings that the number of accidents and fatalities increase as the number of heavy goods vehicles and trucks per capita increase [50,51]. It should be noted that none of these studies reported the specific nature of the ARA–Freight relationship. In addition, it can be deduced that eliminating the Freight Movement from roads (even if possible) can only enable the lowering of the ARA to a certain cap (not zero). On the other hand, the ARA increases linearly as passenger movement increases (positive linear relationship), but there is a non-zero intercept on the passenger movement axis. This observation is also in agreement with reports on the effect of passenger exposure expressed as vehicle kilometer travel [52] and gasoline sales [53] on the accident rate. A less significant logarithmic growth relationship between ARA and CNRV was also observed. These first three observations (for freight, passenger, and CNRV) are very clear and can be easily understood. The same cannot be said regarding the next four variables related to new road infrastructural investments and existing road maintenance investments (total NRII, total ERMI, NRII/capita, and EMRI/capita). Generally, the ARA decreases to a certain extent with the increase in these predicting variables; then, the ARAs begin to rise as the variables increase further. This means that monetary investment in new road infrastructure or existing road maintenance could result in either a decrease or increase in ARAs, depending on the amount invested. It also explains some of the anomalies on the effects of road investments on ARA variables, as reported by previous studies such as [31,52,53].

According to the trends seen here, an average nonlinear concave-up relationship exists, between ARA and new road infrastructure and existing road maintenance investment. The drop and rise in the ARA due to increased maintenance investment is not as pronounced as in the case of new road capital investments. The NRII curves are basic cases of constant demand (fixed freight and passenger movement) and increasing supply (additional road infrastructure). Moderate to low NRII levels will improve the system efficiency by decongesting the overall traffic. This means lesser volume, which leads to a decrease in the level of crash frequency [54]. However, high NRII will not only eliminate congestion, but will also result in a significant increase in the system-wide average speed. The accident rate was found to increase with a higher mean speed for all types of roads [55]. As for the ERMI curves: low to moderate maintenance investments are low-cost, but significant safety improvements such as signage fixing, road marking tracing, porthole patching, drainage clearing, lighting improvements, and even black spot rectification are costly. However, major road rehabilitation projects such as resurfacing and overlays require high capital expenditure. Unlike low-cost road maintenance projects such as patching, drainage clearing, etc., pavement overlays and resurfacing restore the ratings of the pavement condition and parameters such as the roughness index by 100%. Previous studies have shown that roads with better roughness indices (or semi-new roads) witness higher crash frequencies [56–58]. Thus, although major road rehabilitation results in high ride comfort and user satisfaction, it also results in the possibility of higher crash tendencies.

Both Population and Population of Work Group showed a positive S-curve relationship with the ARA. On average, the Proportion of Work Group showed a general non-linearly increasing association with ARAs. The decline in ARAs due to changes in the Proportion of Work Group prior to the increasing trend could not be fully explained. The ARA decreases linearly with an increase in the per capita GDP. The decreasing effect of GDP per capita on the accident rate was also reported in [59].

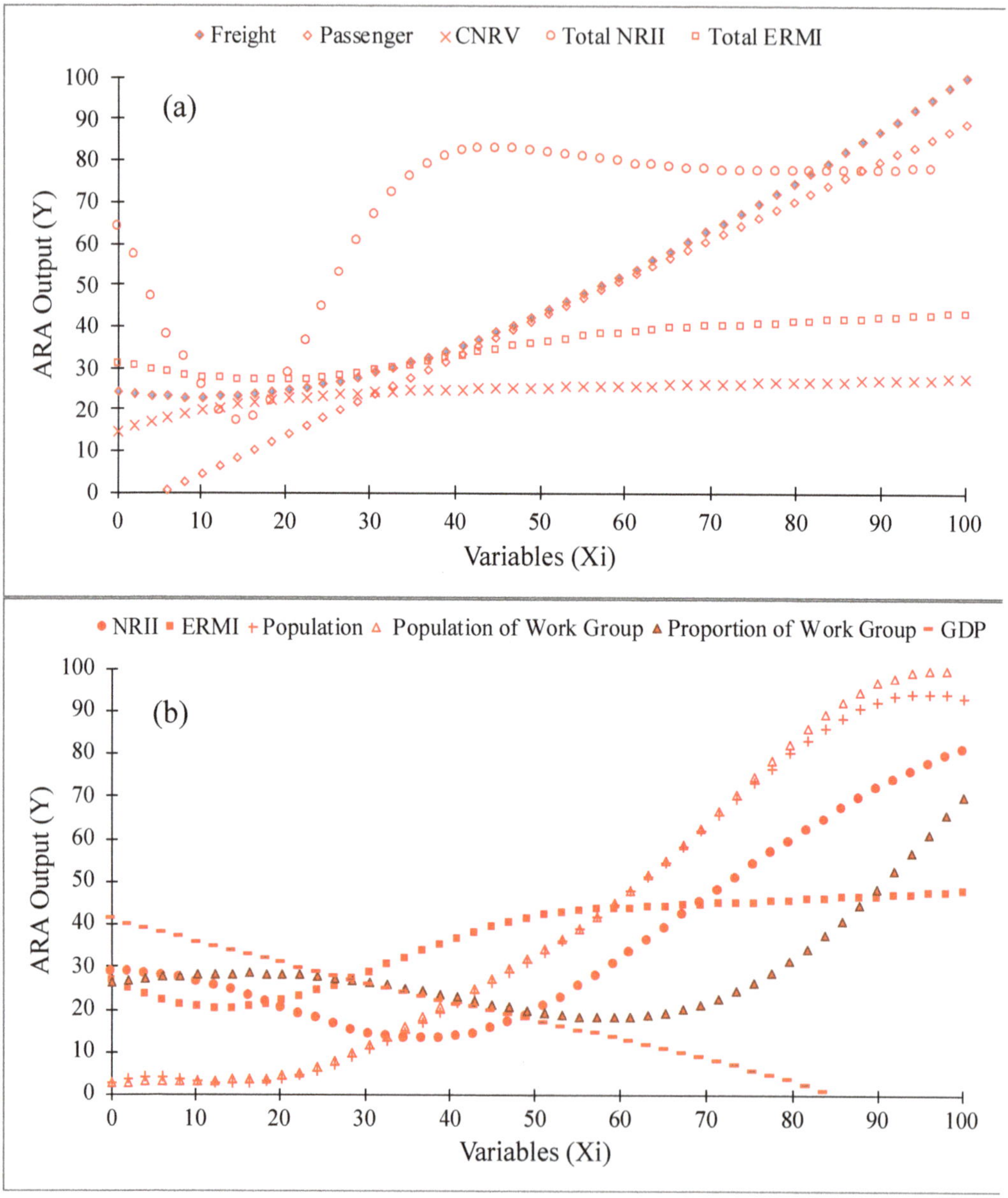

Figure 6. Sensitivity and general variation trend of ARA with predicting variables: (**a**) Part I, (**b**) Part II.

3.4. Example and Application of the ARA ANN Model

The variation in the ARA with NRII/capita along with ERMI/capita is illustrated for a medium population of 10 million. An annual freight and passenger movement of 50,000 million ton-km and 200,000 million passenger-km was assumed, respectively. A Working Population of 70 percent, no change in the number of new vehicles registered annually, and a per capita GDP of USD 40,000 was assumed. These conditions are rounded values typical of a medium population similar to that of Belgium. As of 2014, the annual number of road accidents in Belgium was around 42,000, with annual ERMI/capita of EUR 18 and NRII/capita of approximately EUR 37.

Figure 7 shows the ARA variation with NRII per capita for different per capita ERMI. Each curve represents constant ERMI allocation and varying NRII values. There is a limit to the extent by which the NRII reduces or increases the ARA for any given EMRI/capita. The ARA variation trends are similar to those previously observed from the sensitivity analysis plots for both NRII and ERMI.

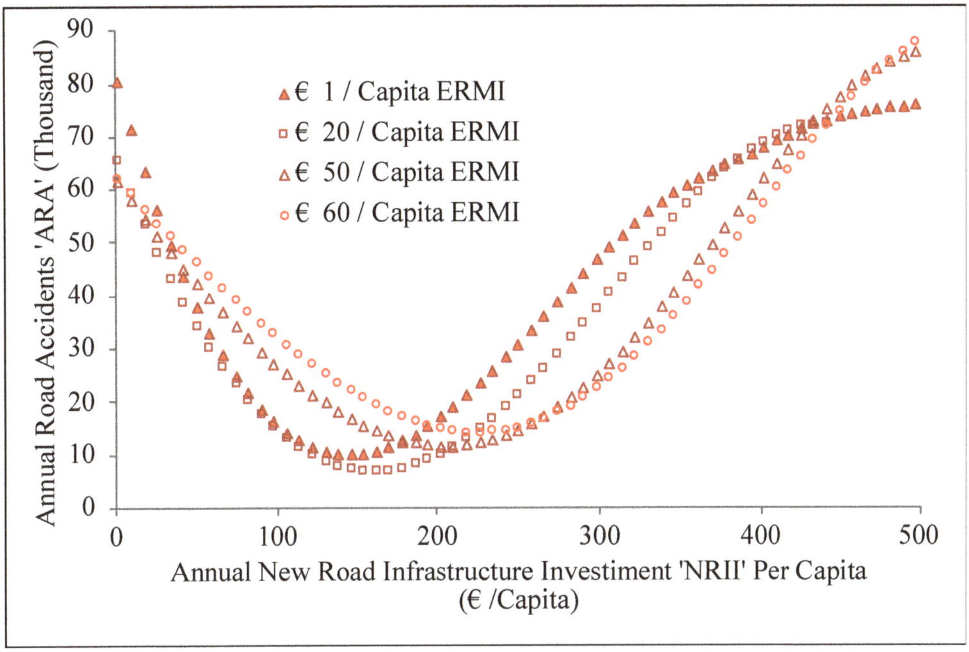

Figure 7. Annual road accident vs. ERMI and NRII per capita for a population of USD 10 million 40,000 GDP and Working Population of 70%.

3.5. NRII and ERMI per Capita Limits Estimation

The ANN model represents an accurate possible outcome of road investment interventions on ARA, for a given economy, demography, and transport system. It could serve as a useful tool for exploring effective and potential road investment limits that will result in the minimum accident rate. The best NRII/capita and ERMI/capita are the combination of NRII/capita and ERMI/capita that correspond to the lowest ARAs for a given set of demographic, economic, passenger, and freight conditions. Sample NRII/capita and ERMI/capita values corresponding to the minimum ARAs for the population of 10 million illustrated in the previous section are estimated and shown in Table 6. The per capita NRII and ERMI ratio and their relative proportions were also estimated. The NRII/ERMI ratio and their relative proportion are better when the budget is limited to below the best estimates. This is because the ratio or relative proportions are also better than random values. Safer NRII/capita and ERMI/capita limit results for populations of 5, 15, and 30 million with different demographic, economic, freight, and passenger conditions were also estimated. These results are presented in the Appendix A in Tables A1–A3, respectively. It can be seen that the safe NRII/capita limit is only greater than the safe ERMI/capita limit for cases in which there is a higher working population (approximately 70%), GDP per capita (USD 30,000 and above), or a combination of a higher CNVR and GDP. Otherwise, the ERMI/capita should be prioritized over the NRII/capita for a lower ARA. Interpolation can be used to estimate in between results with some reasonable degree of accuracy.

Table 6. Sample-minimized ARA and the corresponding NRII/capita and EMRII/capita for population of 10 million.

Freight (Million ton-km)	Passenger (Million Passenger-km)	CNVR (%)	Proportion of Working Group (%)	GDP (US $/Capita)	Minimum ARA (1000)	Opt. ERMI (EUR/Capita)	Opt. NRII (EUR/Capita)	NRII/ERMI	ERMI (%)	NRII (%)
50,000	200,000	0	60	20,000	14.922	20	2	0.1000	91%	9%
50,000	200,000	0	60	30,000	9.454	32	17	0.5313	65%	35%
50,000	200,000	0	60	40,000	4.873	47	62	1.3191	43%	57%
50,000	200,000	0	70	20,000	36.353	20	132	6.6000	13%	87%
50,000	200,000	0	70	30,000	15.937	17	142	8.3529	11%	89%
50,000	200,000	0	70	40,000	6.629	20	162	8.1000	11%	89%
50,000	200,000	20	60	20,000	15.098	20	2	0.1000	91%	9%
50,000	200,000	20	60	30,000	9.891	32	27	0.8438	54%	46%
50,000	200,000	20	60	40,000	5.164	41	72	1.7561	36%	64%
50,000	200,000	20	70	20,000	38.412	17	132	7.7647	11%	89%
50,000	200,000	20	70	30,000	18.827	17	142	8.3529	11%	89%
50,000	200,000	20	70	40,000	8.513	17	167	9.8235	9%	91%

4. Conclusions and Recommendations

The artificial neural network (ANN), MLR, and PR techniques were employed to model ARA based on 22 years of data from the OECD and its partner countries. These data include annual passenger and freight movement, demographic and economic indicators, and road network maintenance and new development investments. The ANN model was found to be more accurate and general in predicting the ARAs. The ANN model was used to illustrate the variation in ARAs with road infrastructure investments and the other variables mentioned. Safe limits per capita NRII and ERMI and their relative proportions for diverse combinations of demography, economy, freight level, and passenger movement were estimated and documented. Some other important observations are highlighted below:

- The ANN is, by far, more reliable than the MLR and PR in modeling the nonlinear relationship of ARAs with road investment variables, demographic indicators, GDP, etc.
- The results of the MLR model showed that all of the independent variables have a significant effect on the ARA, except per capita GDP and changes in the amount of newly registered vehicles.
- The PR model indicated that all predicting variables have a significant effect on the ARA.
- On average, the variables listed in decreasing order of influence on ARA are total NRII (#1), population of work group (#2), total ERMI (#3), freight movement (#4), NRII/capita (#5), population (#6), ERMI/capita (#7), passenger (#8), GDP (#9), proportion of work group (#10), and CNRV (#11).
- New road expansion and existing road maintenance investments were found to exhibit a nonlinear concave-up relationship with ARAs. This was considered to be an explanation for non-uniform conclusions on the influence of these parameters on road accident frequency, as reported by some past literature.
- A certain amount of NRII and ERMI that corresponds to minimum ARAs for a given demography, economy, freight level, and passenger movement exists.
- ARAs showed positive exponential growth with an increase in freight movement.
- ARAs exhibit a positive linear relationship as passenger movement increases but there is a non-zero intercept on the passenger movement axis.
- A positive logarithmic growth relationship between ARAs and CNRV was also observed.
- Population and population of work group showed a positive S-curve relationship with the ARAs. The proportion of the population work group showed a general non-linearly increasing association with ARAs.
- The ARAs decrease linearly as the per capita GDP increases.

Based on these observations, more and proper attention should be given to road safety as the target criteria for the capital funding of road network maintenance and infrastructural development. As a suggestion for further studies, different models for ARA prediction that will enable the estimations of safe road investment limits for specific road facilities could provide more effective alternatives for road infrastructure investments.

Author Contributions: Conceptualization, M.A.D.; methodology and software, M.A.D. and A.A.M.; validation and formal analysis, M.A.D., M.K.I. and A.A.M.; resources and data curation, M.K.I. and M.A.D.; writing—original draft preparation, review and editing, M.A.D., M.K.I. and A.A.M.; project administration, M.K.I.; funding acquisition, M.K.I. All authors have read and agreed to the published version of the manuscript.

Funding: This work was financially supported by the Deanship of Scientific Research from King Faisal University, Saudi Arabia [Grant 991].

Institutional Review Board Statement: Not applicable.

Informed Consent Statement: Not applicable.

Data Availability Statement: The data that support the findings of this study are available from the corresponding author, Muhammad Abubakar Dalhat (madalhat@iau.edu.sa), upon reasonable request.

Conflicts of Interest: The authors would like to declare that there are no conflict of interest.

Appendix A

Table A1. Optimized annual ERMI and NRII per capita for population of 5 million.

Freight (Million ton-km)	Passenger (Million Passenger-km)	CNVR (%)	Proportion of Working Group (%)	GDP (US USD/Capita)	Opt. ERMI (EUR/Capita)	Opt. NRII (EUR/Capita)	NRII/ERMI	ERMI (%)	NRII (%)
20,000	200,000	0	60	10,000	11	2	0.1818	85%	15%
20,000	200,000	0	60	20,000	17	2	0.1176	89%	11%
20,000	200,000	0	60	30,000	32	27	0.8438	54%	46%
20,000	200,000	20	60	10,000	11	2	0.1818	85%	15%
20,000	200,000	20	60	20,000	17	2	0.1176	89%	11%
20,000	200,000	20	60	30,000	29	37	1.2759	44%	56%
20,000	200,000	0	70	10,000	125	2	0.0160	98%	2%
20,000	200,000	0	70	20,000	8	112	14.0000	7%	93%
20,000	200,000	0	70	30,000	8	132	16.5000	6%	94%
20,000	200,000	20	70	10,000	131	2	0.0153	98%	2%
20,000	200,000	20	70	20,000	5	117	23.4000	4%	96%
20,000	200,000	20	70	30,000	5	137	27.4000	4%	96%
50,000	200,000	0	60	10,000	11	2	0.1818	85%	15%
50,000	200,000	0	60	20,000	17	2	0.1176	89%	11%
50,000	200,000	0	60	30,000	32	29	0.9063	52%	48%
50,000	200,000	20	60	10,000	11	2	0.1818	85%	15%
50,000	200,000	20	60	20,000	17	2	0.1176	89%	11%
50,000	200,000	20	60	30,000	29	37	1.2759	44%	56%
50,000	200,000	0	70	10,000	131	2	0.0153	98%	2%
50,000	200,000	0	70	20,000	2	127	63.5000	2%	98%
50,000	200,000	0	70	30,000	2	147	73.5000	1%	99%
50,000	200,000	20	70	10,000	137	2	0.0146	99%	1%
50,000	200,000	20	70	20,000	2	132	66.0000	1%	99%
50,000	200,000	20	70	30,000	2	152	76.0000	1%	99%
50,000	100,000	0	70	10,000	113	2	0.0177	98%	2%
50,000	100,000	0	70	20,000	5	117	23.4000	4%	96%
50,000	100,000	0	70	30,000	5	142	28.4000	3%	97%
50,000	100,000	20	70	10,000	119	2	0.0168	98%	2%
50,000	100,000	20	70	20,000	2	122	61.0000	2%	98%
50,000	100,000	20	70	30,000	2	147	73.5000	1%	99%

Table A2. Optimized annual ERMI and NRII per capita for population of 15 million.

Freight (Million ton-km)	Passenger (Million Passenger-km)	CNVR (%)	Proportion of Working Group (%)	GDP (US USD/Capita)	Opt. ERMI (EUR/Capita)	Opt. NRII (EUR/Capita)	NRII/ERMI	ERMI (%)	NRII (%)
20,000	200,000	0	60	20,000	23	2	0.0870	92%	8%
20,000	200,000	0	60	30,000	35	7	0.2000	83%	17%
20,000	200,000	0	60	40,000	50	57	1.1400	47%	53%
20,000	200,000	20	60	20,000	20	2	0.1000	91%	9%
20,000	200,000	20	60	30,000	32	17	0.5313	65%	35%
20,000	200,000	20	60	40,000	44	67	1.5227	40%	60%
20,000	200,000	0	70	20,000	44	112	2.5455	28%	72%
20,000	200,000	0	70	30,000	38	122	3.2105	24%	76%
20,000	200,000	0	70	40,000	41	142	3.4634	22%	78%
20,000	200,000	20	70	20,000	44	112	2.5455	28%	72%
20,000	200,000	20	70	30,000	35	127	3.6286	22%	78%
20,000	200,000	20	70	40,000	41	147	3.5854	22%	78%
50,000	200,000	0	60	20,000	23	2	0.0870	92%	8%
50,000	200,000	0	60	30,000	35	7	0.2000	83%	17%
50,000	200,000	0	60	40,000	50	52	1.0400	49%	51%
50,000	200,000	20	60	20,000	23	2	0.0870	92%	8%
50,000	200,000	20	60	30,000	35	17	0.4857	67%	33%
50,000	200,000	20	60	40,000	44	67	1.5227	40%	60%
50,000	200,000	0	70	20,000	41	142	3.4634	22%	78%
50,000	200,000	0	70	30,000	32	137	4.2813	19%	81%
50,000	200,000	0	70	40,000	32	152	4.7500	17%	83%
50,000	200,000	20	70	20,000	41	142	3.4634	22%	78%
50,000	200,000	20	70	30,000	32	137	4.2813	19%	81%
50,000	200,000	20	70	40,000	32	157	4.9063	17%	83%
50,000	300,000	0	60	20,000	23	2	0.0870	92%	8%
50,000	300,000	0	60	30,000	35	2	0.0571	95%	5%
50,000	300,000	0	60	40,000	53	47	0.8868	53%	47%
50,000	300,000	20	60	20,000	23	2	0.0870	92%	8%
50,000	300,000	20	60	30,000	35	12	0.3429	74%	26%
50,000	300,000	20	60	40,000	47	62	1.3191	43%	57%
50,000	300,000	0	70	20,000	41 *	162 *	3.9512 *	20% *	80% *
50,000	300,000	0	70	30,000	29 *	142 *	4.8966 *	17% *	83% *
50,000	300,000	0	70	40,000	32 *	157 *	4.9063 *	17% *	83% *
20,000	300,000	0	60	20,000	23	2	0.0870	92%	8%
20,000	300,000	0	60	30,000	35	2	0.0571	95%	5%
20,000	300,000	0	60	40,000	50	47	0.9400	52%	48%
20,000	300,000	20	60	20,000	23	2	0.0870	92%	8%
20,000	300,000	20	60	30,000	35	12	0.3429	74%	26%
20,000	300,000	20	60	40,000	47	62	1.3191	43%	57%
20,000	300,000	0	70	20,000	35	122	3.4857	22%	78%
20,000	300,000	0	70	30,000	35	127	3.6286	22%	78%
20,000	300,000	0	70	40,000	38	147	3.8684	21%	79%
20,000	300,000	20	70	20,000	38	127	3.3421	23%	77%
20,000	300,000	20	70	30,000	32	132	4.1250	20%	80%
20,000	300,000	20	70	40,000	35	152	4.3429	19%	81%

* Similar results for 20% increase in new vehicles registered.

Table A3. Optimized annual ERMI and NRII for population of 30 million.

Freight (Million ton-km)	Passenger (Million Passenger-km)	CNVR (%)	Proportion of Working Group (%)	GDP (US USD/Capita)	Opt. ERMI (EUR/Capita)	Opt. NRII (EUR/Capita)	NRII/ERMI	ERMI (%)	NRII (%)
20,000	200,000	0	60	20,000	23	2	0.0870	92%	8%
20,000	200,000	0	60	30,000	38	2	0.0526	95%	5%
20,000	200,000	0	60	40,000	59	47	0.7966	56%	44%
20,000	200,000	20	60	20,000	23	2	0.0870	92%	8%
20,000	200,000	20	60	30,000	38	2	0.0526	95%	5%
20,000	200,000	20	60	40,000	56	62	1.1071	47%	53%
20,000	200,000	0	70	20,000	65 *	2 *	0.0308 *	97% *	3% *
20,000	200,000	0	70	30,000	65 *	72 *	1.1077 *	47% *	53% *
20,000	200,000	0	70	40,000	74 *	112 *	1.5135 *	40% *	60% *
50,000	200,000	0	60	20,000	23	2	0.0870	92%	8%
50,000	200,000	0	60	30,000	38	2	0.0526	95%	5%
50,000	200,000	0	60	40,000	59	47	0.7966	56%	44%
50,000	200,000	20	60	20,000	23	2	0.0870	92%	8%
50,000	200,000	20	60	30,000	38	2	0.0526	95%	5%
50,000	200,000	20	60	40,000	56	62	1.1071	47%	53%
50,000	200,000	0	70	20,000	65	2	0.0308	97%	3%
50,000	200,000	0	70	30,000	62	97	1.5645	39%	61%
50,000	200,000	0	70	40,000	68	122	1.7941	36%	64%
50,000	200,000	20	70	20,000	56	52	0.9286	52%	48%
50,000	200,000	20	70	30,000	62	102	1.6452	38%	62%
50,000	200,000	20	70	40,000	65	127	1.9538	34%	66%
50,000	300,000	0	60	20,000	23	2	0.0870	92%	8%
50,000	300,000	0	60	30,000	41	2	0.0488	95%	5%
50,000	300,000	0	60	40,000	62	37	0.5968	63%	37%
50,000	300,000	20	60	20,000	23	2	0.0870	92%	8%
50,000	300,000	20	60	30,000	41	2	0.0488	95%	5%
50,000	300,000	20	60	40,000	59	52	0.8814	53%	47%
50,000	300,000	0	70	20,000	74	132	1.7838	36%	64%
50,000	300,000	0	70	30,000	62	107	1.7258	37%	63%
50,000	300,000	0	70	40,000	65	122	1.8769	35%	65%
50,000	300,000	20	70	20,000	65	112	1.7231	37%	63%
50,000	300,000	20	70	30,000	62	112	1.8065	36%	64%
50,000	300,000	20	70	40,000	65	132	2.0308	33%	67%
20,000	300,000	0	60	20,000	23 *	2 *	0.0870 *	92% *	8% *
20,000	300,000	0	60	30,000	41 *	2 *	0.0488 *	95% *	5% *
20,000	300,000	0	60	40,000	62	42	0.6774	60%	40%
20,000	300,000	20	60	40,000	56	57	1.0179	50%	50%
20,000	300,000	0	70	20,000	62 *	2 *	0.0323 *	97% *	3% *
20,000	300,000	0	70	30,000	68 *	92 *	1.3529 *	43% *	58% *
20,000	300,000	0	70	40,000	71 *	112 *	1.5775 *	39% *	61% *

* Similar results for a 20% increase in new vehicles registered.

References

1. Gopalakrishnan, S. A Public Health Perspective of Road Traffic Accidents. *J. Fam. Med. Prim. Care* **2012**, *1*, 144–150. [CrossRef] [PubMed]
2. Adeloye, D.; Thompson, J.Y.; Akanbi, M.A.; Azuh, D.; Samuel, V.; Omoregbe, N.; Ayo, C.K. The burden of road traffic crashes, injuries and deaths in Africa: A systematic review and meta-analysis. *Bull. World Health Organ.* **2016**, *94*, 510–521A. [CrossRef] [PubMed]
3. Islam, M.K.; Reza, I.; Gazder, U.; Akter, R.; Arifuzzaman, M.; Rahman, M.M. Predicting Road Crash Severity Using Classifier Models and Crash Hotspots. *Appl. Sci.* **2022**, *12*, 11354. [CrossRef]
4. Islam, M.K.; Gazder, U.; Akter, R.; Arifuzzaman, M. Involvement of Road Users from the Productive Age Group in Traffic Crashes in Saudi Arabia: An Investigative Study Using Statistical and Machine Learning Techniques. *Appl. Sci.* **2022**, *12*, 6368. [CrossRef]
5. Rahman, M.M.; Islam, M.K.; Al-Shayeb, A.; Arifuzzaman, M. Towards Sustainable Road Safety in Saudi Arabia: Exploring Traffic Accident Causes Associated with Driving Behavior Using a Bayesian Belief Network. *Sustainability* **2022**, *14*, 6315. [CrossRef]

6. ITF. *Road Safety Annual Report 2017*; OECD Publishing: Paris, France, 2017.
7. Abdolmanafi, S.E.; Karamad, S. A new approach for resource allocation for black spot treatment (case study: The road network of Iran). *J. Saf. Res.* **2019**, *69*, 95–100. [CrossRef]
8. ITF. *Road Infrastructure Safety Management*; OECD Publication: Paris, France, 2015.
9. Wu, Z.; Flintsch, G.; Ferreira, A.; Picado-Santos, L.D. Framework for Multiobjective Optimization of Physical Highway Assets Investments. *J. Transp. Eng.* **2012**, *138*, 1411–1421. [CrossRef]
10. Chowdhury, M.A.; Garber, N.J.; Li, D. Multiobjective Methodology for Highway Safety Resource Allocation. *J. Infrastruct. Syst.* **2000**, *6*, 138–144. [CrossRef]
11. Hsieh, T.-Y.; Liu, H.-L. Multistage Heuristic Approach for Solving Infrastructure Investment Decision Problems. *J. Infrastruct. Syst.* **1997**, *3*, 134–142. [CrossRef]
12. Loprencipe, G.; Moretti, L.; Cantisani, G.; Minati, P. Prioritization methodology for roadside and guardrail improvement: Quantitative calculation of safety level and optimization of resources allocation. *J. Traffic Transp. Eng. (Engl. Ed.)* **2018**, *5*, 348–360. [CrossRef]
13. Behnood, H.R.; Pino, S. Road safety resource allocation using the cost-effectiveness approach; Case study in Iran. In *Proceedings of the Institution of Civil Engineers-Transport*; Thomas Telford Ltd.: London, UK, 2021; pp. 1–11. [CrossRef]
14. Gharaibeh, N.G.; Chiu, Y.-C.; Gurian, P.L. Decision Methodology for Allocating Funds across Transportation Infrastructure Assets. *J. Infrastruct. Syst.* **2006**, *12*, 1–9. [CrossRef]
15. Qiang, Z.; Huang, H.; Pei, X.; Wong, S.C.; Gao, M. Rule extraction from an optimized neural network for traffic crash frequency modeling. *Accid. Anal. Prev.* **2016**, *97*, 87–95. [CrossRef]
16. Zeng, Q.; Huang, H.; Pei, X.; Wong, S.C. Modeling nonlinear relationship between crash frequency by severity and contributing factors by neural networks. *Anal. Methods Accid. Res.* **2016**, *10*, 12–25. [CrossRef]
17. Polat, K.; Durduran, S.S. Automatic determination of traffic accidents based on KMC-based attribute weighting. *Neural Comput. Appl.* **2012**, *21*, 1271–1279. [CrossRef]
18. Sowdagur, J.A.; Rozbully-Sowdagur, B.T.B.; Suddul, G. An Artificial Neural Network Approach for Road Accident Severity Prediction. In Proceedings of the 2022 IEEE Zooming Innovation in Consumer Technologies Conference (ZINC), Novi Sad, Serbia, 25–26 May 2022; pp. 267–270.
19. Shaik, M.E.; Islam, M.M.; Hossain, Q.S. A review on neural network techniques for the prediction of road traffic accident severity. *Asian Transp. Stud.* **2021**, *7*, 100040. [CrossRef]
20. Chang, L.-Y. Analysis of freeway accident frequencies: Negative binomial regression versus artificial neural network. *Saf. Sci.* **2005**, *43*, 541–557. [CrossRef]
21. Yannis, G.; Dragomanovits, A.; Laiou, A.; Richter, T.; Ruhl, S.; La Torre, F.; Domenichini, L.; Graham, D.; Karathodorou, N.; Li, H. Use of Accident Prediction Models in Road Safety Management—An International Inquiry. *Transp. Res. Procedia* **2016**, *14*, 4257–4266. [CrossRef]
22. Bonera, M.; Barabino, B.; Maternini, G. Road Network Safety Screening of County Wide Road Network. The Case of the Province of Brescia (Northern Italy). In *Advances in Road Infrastructure and Mobility. IRF 2021. Sustainable Civil Infrastructures*; Springer: Cham, Switzerland, 2022. [CrossRef]
23. Doğan, E.; Akgüngör, A.P. Forecasting highway casualties under the effect of railway development policy in Turkey using artificial neural networks. *Neural Comput. Appl.* **2013**, *22*, 869–877. [CrossRef]
24. Chang, C.M.; Vavrova, M.; Mahnaz, S.L. Integrating Vulnerable Road User Safety Criteria into Transportation Asset Management to Prioritize Budget Allocation at the Network Level. *Sustainability* **2022**, *14*, 8317. [CrossRef]
25. Yasdi, R. Prediction of Road Traffic using a Neural Network Approach. *Neural Comput. Appl.* **1999**, *8*, 135–142. [CrossRef]
26. Li, C.-S.; Chen, M.-C. Identifying important variables for predicting travel time of freeway with non-recurrent congestion with neural networks. *Neural Comput. Appl.* **2013**, *23*, 1611–1629. [CrossRef]
27. Kaedi, M.; Movahhedinia, N.; Jamshidi, K. Traffic signal timing using two-dimensional correlation, neuro-fuzzy and queuing based neural networks. *Neural Comput. Appl.* **2008**, *17*, 193–200. [CrossRef]
28. Nguwi, Y.-Y.; Kouzani, A.Z. Detection and classification of road signs in natural environments. *Neural Comput. Appl.* **2008**, *17*, 265–289. [CrossRef]
29. Kadali, B.R.; Nivedan, R.; Vedagiri, P. Evaluation of pedestrian mid-block road crossing behaviour using artificial neural network. *J. Traffic Transp. Eng. (Engl. Ed.)* **2014**, *1*, 111. [CrossRef]
30. Byaruhanga, C.B.; Evdorides, H. A budget optimisation model for road safety infrastructure countermeasures. *Cogent Eng.* **2022**, *9*, 2129363. [CrossRef]
31. Sánchez González, M.P.; Escribano Sotos, F.; Tejada Ponce, Á. Impact of provincial characteristics on the number of traffic accident victims on interurban roads in Spain. *Accid. Anal. Prev.* **2018**, *118*, 178–189. [CrossRef]
32. Albalate, D.; Fernández, L.; Yarygina, A. The road against fatalities: Infrastructure spending vs. regulation? *Accid. Anal. Prev.* **2013**, *59*, 227–239. [CrossRef] [PubMed]
33. Nguyen-Hoang, P.; Yeung, R. Dollars for lives: The effect of highway capital investments on traffic fatalities. *J. Saf. Res.* **2014**, *51*, 109–115. [CrossRef] [PubMed]
34. Noland, R.B.; Oh, L. The effect of infrastructure and demographic change on traffic-related fatalities and crashes: A case study of Illinois county-level data. *Accid. Anal. Prev.* **2004**, *36*, 525–532. [CrossRef] [PubMed]

35. Page, Y. A statistical model to compare road mortality in OECD countries. *Accid. Anal. Prev.* **2001**, *33*, 371–385. [CrossRef]
36. van Beeck, E.F.; Borsboom, G.J.J.; Mackenbach, J.P. Economic development and traffic accident mortality in the industrialized world, 1962–1990. *Int. J. Epidemiol.* **2000**, *29*, 503–509. [CrossRef] [PubMed]
37. Wilson, N.; Thomson, G. Deaths from international terrorism compared with road crash deaths in OECD countries. *Inj. Prev.* **2005**, *11*, 332–333. [CrossRef] [PubMed]
38. Yannis, G.; Antoniou, C.; Papadimitriou, E. Autoregressive nonlinear time-series modeling of traffic fatalities in Europe. *Eur. Transp. Res. Rev.* **2011**, *3*, 113–127. [CrossRef]
39. Eckle, P.; Burgherr, P. Bayesian Data Analysis of Severe Fatal Accident Risk in the Oil Chain. *Risk Anal.* **2013**, *33*, 146–160. [CrossRef] [PubMed]
40. Lord, D.; Mannering, F. The statistical analysis of crash-frequency data: A review and assessment of methodological alternatives. *Transp. Res. Part A Policy Pract.* **2010**, *44*, 291–305. [CrossRef]
41. MacKay, D.J.C. Bayesian Interpolation. *Neural Comput.* **1992**, *4*, 415–447. [CrossRef]
42. Foresee, F.D.; Hagan, M.T. Gauss-Newton Approximation to Bayesian Learning. In Proceedings of the 1997 International Joint Conference on Neural Networks, Houston, TX, USA, 12 June 1997.
43. Burden, F.; Winkler, D. Bayesian Regularization of Neural Networks. In *Artificial Neural Networks: Methods and Applications*; Livingstone, D.J., Ed.; Humana Press: Totowa, NJ, USA, 2009; pp. 23–42. [CrossRef]
44. Olden, J.D.; Joy, M.K.; Death, R.G. An accurate comparison of methods for quantifying variable importance in artificial neural networks using simulated data. *Ecol. Model.* **2004**, *178*, 389–397. [CrossRef]
45. Garson, G.D. Interpreting neural-network connection weights. *AI Expert* **1991**, *6*, 46–51.
46. Tonidandel, S.; LeBreton, J.M. Relative Importance Analysis: A Useful Supplement to Regression Analysis. *J. Bus. Psychol.* **2011**, *26*, 1–9. [CrossRef]
47. Johnson, J.W. A Heuristic Method for Estimating the Relative Weight of Predictor Variables in Multiple Regression. *Multivar. Behav. Res.* **2000**, *35*, 1–19. [CrossRef] [PubMed]
48. Tonidandel, S.; LeBreton, J.M. RWA Web: A Free, Comprehensive, Web-Based, and User-Friendly Tool for Relative Weight Analyses. *J. Bus. Psychol.* **2015**, *30*, 207–216. [CrossRef]
49. Gevrey, M.; Dimopoulos, I.; Lek, S. Review and comparison of methods to study the contribution of variables in artificial neural network models. *Ecol. Model.* **2003**, *160*, 249–264. [CrossRef]
50. Castillo-Manzano, J.I.; Castro-Nuño, M.; Fageda, X. Can cars and trucks coexist peacefully on highways? Analyzing the effectiveness of road safety policies in Europe. *Accid. Anal. Prev.* **2015**, *77*, 120–126. [CrossRef] [PubMed]
51. Ramírez, B.A.; Izquierdo, F.A.; Fernández, C.G.; Méndez, A.G. The influence of heavy goods vehicle traffic on accidents on different types of Spanish interurban roads. *Accid. Anal. Prev.* **2009**, *41*, 15–24. [CrossRef]
52. Aparicio Izquierdo, F.; Arenas Ramírez, B.; Bernardos Rodríguez, E. The interurban DRAG-Spain model: The main factors of influence on road accidents in Spain. *Res. Transp. Econ.* **2013**, *37*, 57–65. [CrossRef]
53. Fridstrom, L.; Ingebrigtsen, S. An aggregate accident model based on pooled, regional time-series data. *Accid. Anal. Prev.* **1991**, *23*, 363–378. [CrossRef]
54. Martin, J.-L. Relationship between crash rate and hourly traffic flow on interurban motorways. *Accid. Anal. Prev.* **2002**, *34*, 619–629. [CrossRef] [PubMed]
55. Taylor, M.C.; Lynam, D.C.; Baruya, A. *The Effects of Drivers' Speed on the Frequency of Road Accidents*; Transport Research Laboratory: Berkshire, UK, 2000.
56. Tehrani, S.S.; Falls, L.C.; Mesher, D. Effects of pavement condition on roadway safety in the province of Alberta. *J. Transp. Saf. Secur.* **2017**, *9*, 259–272. [CrossRef]
57. Chun Yip, C.; Baoshan, H.; Xuedong, Y.; Richards, S. Investigating effects of asphalt pavement conditions on traffic accidents in Tennessee based on the pavement management system (PMS). *J. Adv. Transp.* **2010**, *44*, 150–161. [CrossRef]
58. Chan, C.Y.; Huang, B.; Yan, X.; Richards, S. Relationship between Highway Pavement Condition, Crash Frequency, and Crash Type. *J. Transp. Saf. Secur.* **2009**, *1*, 268–281. [CrossRef]
59. He, H.; Paichadze, N.; Hyder, A.A.; Bishai, D. Economic development and road traffic fatalities in Russia: Analysis of federal regions 2004–2011. *Inj. Epidemiol.* **2015**, *2*, 19. [CrossRef] [PubMed]

applied sciences

MDPI

Article

Research on Yaw Stability Control Method of Liquid Tank Semi-Trailer on Low-Adhesion Road under Turning Condition

Gangyan Li, Teng Fu and Ran Zhao *

School of Mechanical and Electronical Engineering, Wuhan University of Technology, Wuhan 430070, China
* Correspondence: dsdwesd@163.com

Abstract: To ensure that a liquid tank semi-trailer has good yaw stability and path-following performance on low-adhesion roads under turning conditions, a multi-object PID differential braking-control method is proposed, which takes the tractor yaw rate, semi-trailer yaw rate, and articulation angle as the control parameters. According to the principle of equivalence, the trammel pendulum (TP) model is used to simulate the liquid sloshing effect in the liquid tank, and the Fluent software is used to identify the key parameters of the trammel pendulum model and verify the correctness of its simulation effect. Then, a co-simulation model is built based on TruckSim and MATLAB/Simulink. Based on the simplified six degrees of freedom model and the co-simulation model of a liquid tank semi-trailer, a multi-object PID differential braking-control method is designed, and the vehicle state responses with and without control are compared when it is turning on a low-adhesion road. The simulation results show that the proposed multi-object PID differential braking-control method can effectively improve the yaw stability and path-following performance of the liquid tank semi-trailer when turning on a low-adhesion road.

Keywords: liquid tank semi-trailer; trammel pendulum model; lateral stability; path-following; multi-object PID

Citation: Li, G.; Fu, T.; Zhao, R. Research on Yaw Stability Control Method of Liquid Tank Semi-Trailer on Low-Adhesion Road under Turning Condition. *Appl. Sci.* **2023**, *13*, 39. https://doi.org/10.3390/app13010039

Academic Editors: Edgar Sokolovskij and Vidas Žuraulis

Received: 25 November 2022
Revised: 16 December 2022
Accepted: 16 December 2022
Published: 21 December 2022

1. Introduction

With its large transportation capacity and low transportation cost, the liquid tank semi-trailer has become the main vehicle for transporting liquid chemical products. Compared with ordinary freight cars, liquid tank semi-trailers not only have the characteristics of a high center of mass position and large cargo weight but also are more prone to instability than ordinary semi-trailers, because of the coupling relationship between the lateral sloshing of the liquid in the tank and the tank body [1].

The dangerous working conditions of liquid tank semi-trailers are mainly divided into yaw instability and roll instability. When a vehicle is running on a low-adhesion road, due to the small adhesion between the tires and the road, the lateral force of the tires is easy to become saturated, which leads to the occurrence of yaw instability; when driving on a high adhesion road, a vehicle is prone to roll instability [2]. At present, most scholars' research mainly focuses on the influencing factors and improved methods of the roll stability of liquid tank semi-trailers. Li, X. [3] established a dynamic model of a liquid tank vehicle based on the trammel pendulum (TP) model and analyzed the influence of the liquid filling ratio and other influencing factors on the driving stability of the liquid tank vehicle. Peng, G. [4] established a volume of fluid (VOF) multi-phase flow model and a standard k-ε turbulence model to study the effects of transportation conditions, such as the liquid-filling ratio, the elliptical ratio of the tank section, and the number of baffles, on the vehicle's roll stability. Saeedi, M.A. [5] improved the roll stability of a liquid-carrying articulated vehicle based on the proposed two control methods to avoid rollover, shimmy, and folding accidents.

These studies analyzed the influencing factors of the roll stability of a liquid tank semi-trailer on a good road and provided the corresponding improvement methods, but there were few studies on the yaw stability of vehicles turning on low-adhesion roads. In addition, most research on the path-following problem concentrates on the field of intelligent vehicles [6], while little research focuses on the field of heavy semi-trailers or liquid tank semi-trailers. Therefore, this paper studies the yaw stability and path-following performance of liquid tank semi-trailers when turning on low-adhesion roads.

To study the yaw stability control method of liquid tank semi-trailers when driving on low-adhesion roads under turning conditions, it is necessary to study the liquid sloshing law in the tank. At present, the main research methods of liquid sloshing are hydrodynamics, the quasi-static model, the equivalent mechanical model, etc. [7]. With the ability to describe the large amplitude of liquid sloshing and the convenience of being coupled with the vehicle dynamic model, the equivalent mechanical model is widely used in practical engineering.

The most commonly used equivalent mechanical models mainly include the spring-mass model, the simple pendulum model, and the TP model [8]. However, the spring-mass model is only suitable for small amplitude shaking; for the simple pendulum model, the motion track of it is a circular arc, which is only suitable for describing the liquid sloshing effect in a cylindrical tank but not for an elliptical cylinder tank. To solve this problem, Salem [9] studied the liquid impact in the horizontal elliptical cylinder tank and proposed the TP model, which is suitable for simulating the liquid sloshing effect in the elliptical cylinder tank.

In terms of vehicle yaw stability control methods, there are mainly active steering, active suspension, differential braking, etc. [10]. Among these methods, the actuator of active steering is complex and costly, the damping adjustment range of active suspension is small, and the suspension adjustment time is too long. Differential braking generates the additional yaw moment by controlling the braking force of different wheels; since the hardware cost is low, it has been widely used [11].

In terms of control algorithms, the main methods include fuzzy PID control, optimal control, MPC control, sliding mode control, and other control algorithms [12]. Easy to use and not requiring a specific mathematical model to be obtained, the fuzzy PID control method is suitable for such highly nonlinear models as liquid tank semi-trailers.

In Section 2, according to the principle of equivalence, the TP model is taken to simulate the liquid sloshing effect in the liquid tank, and the Fluent software is used to verify the effectiveness. In Section 3, a simplified six degrees of freedom model of a liquid tank semi-trailer is built and verified based on TruckSim and MATLAB/Simulink; the control parameters and expected values are selected and determined. In Section 4, a multi-object PID differential braking-control method is designed; the calculation and distribution of the additional yaw moment are determined; and the wheel-slip-rate control method is introduced. In Section 5, the double-lane change and the step-steering-angle input working conditions are chosen to verify the effectiveness of the proposed method in improving the yaw stability and path-following performance of the liquid tank semi-trailer when running on a low-adhesion road; the robustness of the proposed method is also discussed and verified when the input disturbance pulses are included and the control parameters are changed.

2. Equivalent Model of Liquid Sloshing for the Liquid Tank Semi-Trailer and the Establishment of Its Co-Simulation Analysis Model

2.1. Equivalent Model of Liquid Sloshing for the Liquid Tank Semi-Trailer

The schematic diagram of the TP model is shown in Figure 1, and its symbols are explained in Table 1.

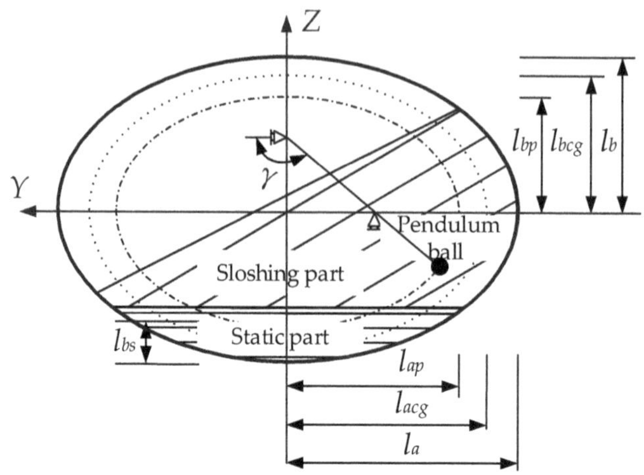

Figure 1. Schematic diagram of the TP model.

Table 1. Description and value of TP model parameters and basic parameters of the tank.

Parameter	Description	Value	Unit
l_{ap}	long semi-axes of the motion track of the pendulum ball	0.5613	m
l_{bp}	short semi-axes of the motion track of the pendulum ball	0.3742	m
l_{acg}	long semi-axes of the liquid center of mass motion track	–	m
l_{bcg}	short semi-axes of the liquid center of mass motion track	–	m
l_a	long semi-axes of the tank body section	1.0925	m
l_b	short semi-axes of the tank body section	0.7283	m
l_{bs}	distance from the centroid of the liquid at rest to the bottom of the tank	0.6939	m
m_s	mass of the liquid at rest	5631	kg
m_p	mass of the swing ball	7826	kg
m_w	total mass of liquid in the tank	13,457	kg
ζ	ellipticity of the section of the liquid tank	1.5	–
γ	liquid swing angle	–	°
Δ	liquid filling ratio in the tank	0.6	–
L	tank body length	9	m
ρ	liquid density in the tank	997	kg/m^3
a_y	lateral excitation of the tank body	1	m/s^2

For the TP model, the liquid in the tank is divided into the sloshing part and the static part. The movement of the sloshing part is equivalent to the swing of the pendulum ball. The motion track is circular or elliptical and is similar to the tank section profile, which meets the following requirements:

$$\frac{l_a}{l_b} = \frac{l_{acg}}{l_{bcg}} = \frac{l_{ap}}{l_{bp}} = \zeta \quad (1)$$

Based on Lagrange mechanics, the motion equation of the pendulum ball in the accelerated translational tank can be derived as follows [13]:

$$\ddot{\gamma}\left(l_{ap}^2 \sin^2\gamma + l_{bp}^2 \cos^2\gamma\right) + \frac{1}{2}\dot{\gamma}^2\left(l_{ap}^2 - l_{bp}^2\right)\left(\sin 2\gamma - g l_{bp}\cos\gamma - a_y l_{ap}\sin\gamma\right) = 0 \quad (2)$$

The lateral sloshing force F_W on the tank caused by liquid sloshing is

$$F_w = m_p\left(a_y - l_{ap}\ddot{\gamma}\sin\gamma - l_{ap}\dot{\gamma}^2\cos\gamma\right) + m_s a_y \quad (3)$$

The sloshing torque M_w around the lowest point of the tank caused by liquid sloshing is

$$M_w = -m_s a_y l_{bs} + m_p \begin{vmatrix} l_{ap}\cos\gamma & l_b - l_{bp}\sin\gamma \\ a_y - l_{ap}\left(\ddot{\gamma}\sin\gamma + \dot{\gamma}^2\cos\gamma\right) & l_{bp}\left(\dot{\gamma}^2\sin\gamma - \ddot{\gamma}\cos\gamma\right) + g \end{vmatrix} \tag{4}$$

The TP model parameters shall be determined using the parameter identification method. According to the equivalence principle of dynamic similarity, kinematic similarity, and geometric similarity [14], the simulation results of an elliptical cylinder tank under step excitation of unit lateral acceleration are simulated by fluid numerical simulation using the Fluent software for parameter identification, and the TP model parameters are obtained as shown in Table 1.

2.2. Verification of Equivalent Model of Liquid Sloshing for the Liquid Tank Semi-Trailer

To verify the accuracy of the TP model in simulating the liquid sloshing effect of the tank when subjected to lateral excitation, a typical liquid tank is selected, and its basic parameters are shown in Table 1. The time-varying lateral acceleration a_y applied on the tank body is

$$a_y = 0.2g\sin(0.4t) \tag{5}$$

The Fluent software and the TP model are, respectively, used to simulate the sloshing force and torque generated by the liquid sloshing in the tank. The results are shown in Figure 2a,b.

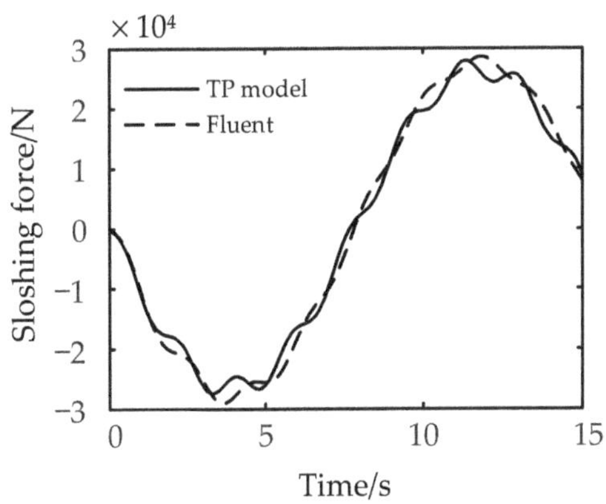

(a)

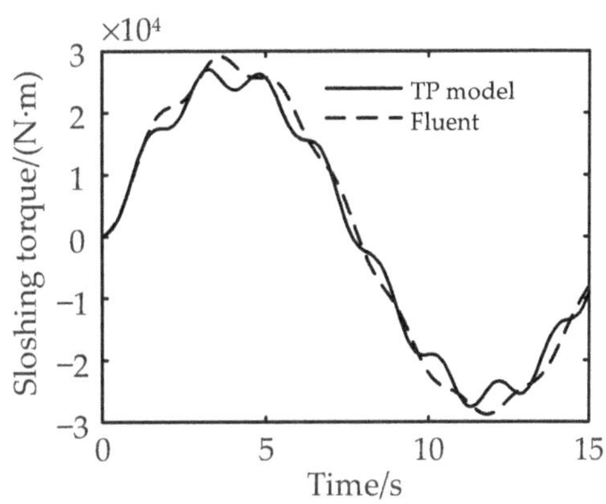

(b)

Figure 2. Comparison of liquid sloshing results: (**a**) sloshing force and (**b**) sloshing torque between Fluent software and the TP model.

It can be seen from Figure 2a,b that the magnitude and frequency of liquid sloshing force and torque obtained by the Fluent software and the TP model maintain good consistency, indicating that the TP model can simulate the liquid sloshing effect when the liquid in the tank is subjected to time-varying lateral excitation.

2.3. Co-Simulation Analysis Model of Liquid Sloshing Characteristics for the Liquid Tank Semi-Trailer

Considering the danger of the real vehicle test, a co-simulation model of the liquid tank semi-trailer is established based on TruckSim and MATLAB/Simulink, as shown in Figure 3. In this model, the lateral acceleration of the TruckSim model is input into the liquid sloshing model, and the liquid sloshing model outputs the force and torque generated by liquid sloshing to the TruckSim model [15].

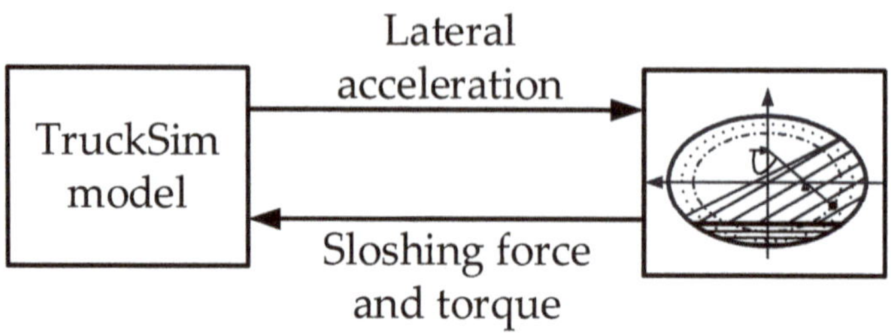

Figure 3. Schematic diagram of co-simulation model for liquid tank semi-trailer.

As a famous simulation analysis software for commercial vehicles, TruckSim provides a data communication interface with MATLAB/Simulink. When the simulation test is carried out under different working conditions, it is only necessary to set the relevant structural parameters and driving conditions of the vehicle in TruckSim, set the input parameters and output parameters, and then generate the corresponding S function for Simulink to call to achieve the data communication between the two.

The TruckSim input/output setting module is shown in Figure 4, in which the input setting module's Import Channel is the force and torque generated by liquid shaking, and the output setting module's Output Channel is the lateral acceleration of the liquid tank semi-trailer.

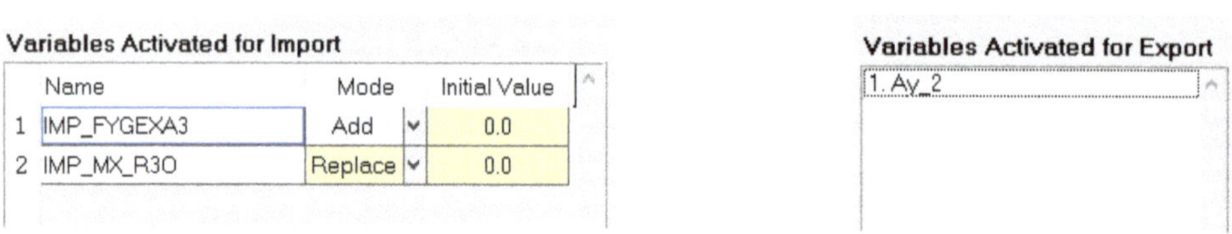

(a) (b)

Figure 4. TruckSim I/O setting module: (**a**) input module and (**b**) output module.

For the TP model, it is written in the MATLAB function module in Simulink, which uses the ode45 function to solve the simulation test in real time, so the calculation step size should be consistent with that of TruckSim.

3. Simplified Six Degrees of Freedom Model and Control Parameters of the Liquid Tank Semi-Trailer for Yaw Stability Control

3.1. Simplified Six Degrees of Freedom Model of the Liquid Tank Semi-Trailer for Yaw Stability Control

In this paper, a six-axle liquid tank semi-trailer is selected for research. The schematic diagram of the single-track model of the vehicle is shown in Figure 5. Figure 5a shows the XOY plane layout of the selected vehicle, Figure 5b,c explain the roll motion between the sprung mass and the chassis when the vehicle is turning on a road. The origin point of the tractor and semi-trailer are, respectively, the intersection of the centroid cross section and its roll axis. The X-axis points to the front of the vehicle, the Y-axis points to the left of the driver, and the Z-axis points to the top of the vehicle.

According to the definition in Figure 5, considering the lateral, yaw, and roll motions of the tractor and semi-trailer, the differential motion equations of the two are listed as follows.

For the lateral motion of the tractor:

$$m_1 u_1 \left(\dot{\beta}_1 + \dot{\psi}_1\right) - m_{1s} h_1 \ddot{\phi}_1 = F_{y1}\cos\delta + F_{y2} + F_{yH} \tag{6}$$

where m_1 is the mass of the tractor, kg; m_{1s} is the sprung mass of the tractor, kg; u_1 is the longitudinal velocity of the tractor, m/s; β_1 is the sideslip angle of the tractor's center of mass, rad; $\dot{\psi}_1$ is the yaw rate of the tractor, rad/s; $\ddot{\phi}_1$ is the tractor roll angular acceleration, rad/s^2; F_{yi} is the lateral force of wheel $i(i = 1, 2, 3)$, N; F_{yH} is the lateral force at the articulated point, N; δ is the steering wheel angle, rad; and h_1 is the distance from the tractor centroid to its roll axis, m.

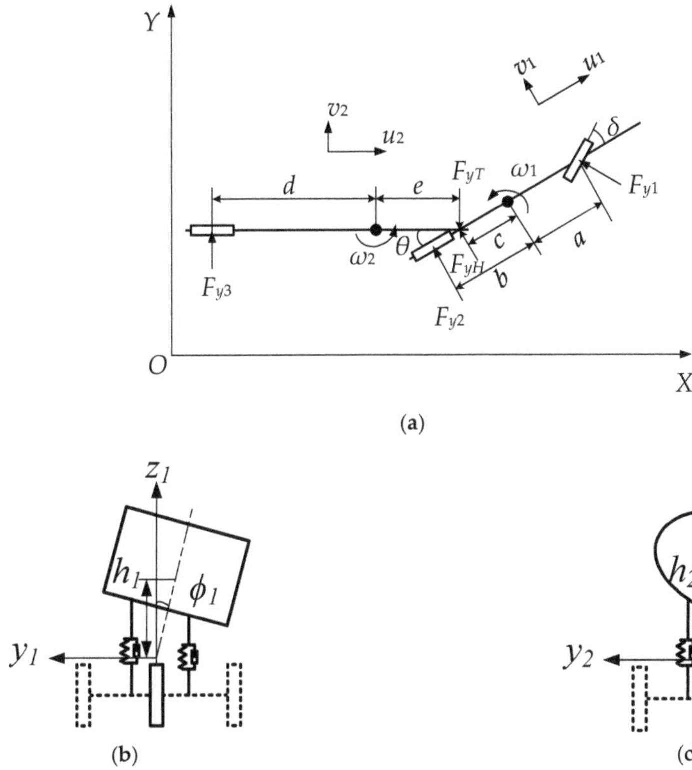

Figure 5. Schematic diagram of the liquid tank semi-trailer: (**a**) XOY plane diagram, (**b**) tractor roll model, and (**c**) semi-trailer roll model.

For the yaw motion of the tractor:

$$I_{1zz}\ddot{\psi}_1 - I_{1xz}\ddot{\phi}_1 = F_{y1}a\cos\delta - F_{y2} - F_{yH}c \tag{7}$$

where I_{1zz} is the moment of inertia of the tractor around the Z axis, kg·m^2; I_{1xz} is the centroid yaw-roll inertia product of the tractor, kg·m^2; a is the distance from the front axle of the tractor to its centroid, m; and c is the distance from the tractor centroid to the articulated point, m.

For the roll motion of the tractor:

$$\begin{gathered}(I_{1xx} + m_{1s}h_1^2)\ddot{\phi}_1 - I_{1xz}\ddot{\psi}_1 = m_{1s}gh_1\sin\phi_1 m_{1s} \\ +u_1\left(\left(\dot{\beta}_1 + \dot{\psi}_1\right) - h_1\ddot{\phi}_1\right)h_1 - k_{r1}\phi_1 - c_1\dot{\phi}_1 + k_{12}(\phi_2 - \phi_1) - F_{yH}h_{1c}\end{gathered} \tag{8}$$

where I_{1xx} is the moment of inertia of the tractor around the X-axis, kg·m^2; h_{1c} is the distance from the articulation point to the tractor roll axis, m; k_{r1} is the roll stiffness of the tractor, N·m/rad; c_1 is the roll damping of the tractor, N·m·s/rad; k_{12} is the coupling roll stiffness of the fifth wheel, N·m/rad.

For the lateral motion of the semi-trailer:

$$m_2u_2\left(\dot{\beta}_2 + \dot{\psi}_2\right) - m_{2s}h_2\ddot{\phi}_2 = -F_{yT} + F_{y3} \tag{9}$$

where: m_2 is the mass of the semi-trailer, kg; m_{2s} is the sprung mass of the semi-trailer, kg; u_2 is the longitudinal velocity of the semi-trailer, m/s; β_2 is the sideslip angle of the semi-trailer, rad; $\dot{\psi}_2$ is the yaw rate of the semi-trailer, rad/s; $\ddot{\phi}_2$ is the roll angular acceleration of semi-trailer, rad/s^2; F_{yT} is the lateral force at the articulated point, N; and h_2 is the distance from the centroid of the semi-trailer to its roll axis, m.

For the yaw motion of the semi-trailer:

$$I_{2zz}\ddot{\psi}_2 - I_{2xz}\ddot{\phi}_2 = -F_{yT}e - F_{y3}d \tag{10}$$

where I_{2zz} is the moment of inertia of the semi-trailer around the Z axis, kg·m^2; I_{2xz} is the centroid yaw-roll inertia product of the semi-trailer, kg·m^2; d is the distance from the axle of the semi-trailer to its center of mass, m; and e is the distance from the centroid of the semi-trailer to the articulated point, m.

For the semi-trailer roll movement:

$$\begin{gathered}(I_{2xx} + m_{2s}h_2^2)\ddot{\phi}_2 - I_{2xz}\ddot{\psi}_2 = m_{2s}gh_2\sin\phi_2 \\ +m_{2s}\left(u_2\left(\dot{\beta}_2 + \dot{\psi}_2\right) - h_2\ddot{\phi}_2\right)h_2 + F_{yT}h_{2c} - c_2\dot{\phi}_2 - k_{r2}\phi_2 - k_{12}(\phi_2 - \phi_1)\end{gathered} \tag{11}$$

where I_{2xx} is the moment of inertia of the semi-trailer around the X-axis, kg·m^2; h_{2c} is the distance from the articulated point to the roll axis of the semi-trailer, m; k_{r2} is the semi-trailer roll stiffness, N·m/rad; and c_2 is semi-trailer roll damping, N·m·s/rad.

The constraint equation at the fifth wheel is as follows:

$$\begin{cases} F_{yT} = F_{yH}\cos\theta \\ \dot{\theta} = \dot{\psi}_2 - \dot{\psi}_1 \\ \beta_2 = \beta_1 - \frac{h_{1c}}{u_1}\dot{\phi}_1 + \frac{h_{2c}}{u_2}\dot{\phi}_2 - \frac{c}{u_1}\dot{\psi}_1 - \frac{e}{u_2}\dot{\psi}_2 - \theta \end{cases} \tag{12}$$

The linear tire model is adopted for vehicle tires as follows:

$$\begin{cases} F_{y1} = k_1\alpha_1 = k_1\left(\beta_1 + \frac{a\dot{\psi}_1}{u} - \delta\right) \\ F_{y2} = k_2\alpha_2 = k_2\left(\beta_1 - \frac{b\dot{\psi}_1}{u}\right) \\ F_{y3} = k_3\alpha_3 = k_3\left(\beta_2 - \frac{d\dot{\psi}_2}{u}\right) \end{cases} \tag{13}$$

where θ is the articulated angle, rad; k_i is the lateral stiffness of wheel $i(i = 1,\ 2,\ 3)$, N/rad; and α_i is the side slip angle of wheel i, rad.

Combining Equations (6)–(13) and eliminating the fifth wheel force F_{yH}, a simplified dynamic model of the liquid tank semi-trailer is obtained as follows:

$$M\dot{X} = NX + KU \tag{14}$$

where X is the state variable, and $X = \left[\beta_1\ \dot{\psi}_1\ \phi_1\ \dot{\phi}_1\ \phi_2\ \dot{\phi}_2\ \dot{\theta}\ \theta\right]^T$; U is the input, and $U = \delta$;

$$
\text{Matrix } M = \begin{pmatrix}
cm_1u_1 & I_{1zz} & 0 & -I_{1xz} - ch_1m_{1s} & 0 & 0 & 0 & 0 \\
(h_{1c}m_1 - h_1m_{1s})u_1 & -I_{1xz} & 0 & m_{1s}(2h_1{}^2 - h_{1c}h_1) + I_{1xx} & 0 & 0 & 0 & 0 \\
m_1u_1 + m_2u_2 & -em_2 - \frac{cm_2u_2}{u_1} & 0 & -h_1m_{1s} - \frac{h_{1c}m_2u_2}{u_1} & 0 & h_{2c}m_2 - h_2m_{2s} & -em_2 & 0 \\
-em_2u_2 & I_{2zz} + e^2m_2 + \frac{cem_2u_2}{u_1} & 0 & \frac{eh_{1c}m_2u_2}{u_1} & 0 & e(h_2m_{2s} - h_{2c}m_2) - I_{2xz} & m_2e^2 + I_{2zz} & 0 \\
(h_{2c}m_2 - h_2m_{2s})u_2 & e(h_2m_{2s} - h_{2c}m_2) - I_{2xz} + \frac{c(h_2m_{2s} - h_{2c}m_2)u_2}{u_1} & 0 & \frac{h_{1c}(h_2m_{2s} - h_{2c}m_2)u_2}{u_1} & 0 & 2m_{2s}(h_2 - h_{2c}) + m_2h_{2c}{}^2 + I_{2xx} & e(h_2m_{2s} - h_{2c}m_2) - I_{2xz} & 0 \\
0 & 0 & 1 & 0 & 0 & 0 & 0 & 0 \\
0 & 0 & 0 & 0 & 1 & 0 & 0 & 0 \\
0 & 0 & 0 & 0 & 0 & 0 & 0 & 1
\end{pmatrix}
$$

$$
\text{Matrix } N = \begin{pmatrix}
(a+c)k_1 - (b-c)k_2 & \frac{(a^2+ac)k_1 + (b^2-bc)k_2}{u_1} - cm_1u_1 & 0 & 0 & 0 & 0 & 0 & 0 \\
h_{1c}(k_1 + k_2) & (h_1m_{1s} - h_{1c}m_1)u_1 + \frac{(ak_1 - bk_2)h_{1c}}{u_1} & gh_1m_{1s} - k_{r1} - k_{12} & -c_{r1} & k_{12} & 0 & 0 & 0 \\
k_1 + k_2 + k_3 & \frac{ak_1 - bk_2 - ck_3}{u_1} - m_2u_2 - m_1u_1 - \frac{(d+e)k_3}{u_2} & 0 & -\frac{h_{1c}k_3}{u_1} & 0 & \frac{h_{2c}k_3}{u_2} & -\frac{(d+e)k_3}{u_2} & -k_3 \\
-dk_3 - ek_3 & \frac{(d+e)^2k_3}{u_2} + em_2u_2 + \frac{c(d+e)k_3}{u_1} & 0 & \frac{(d+e)h_{1c}k_3}{u_1} & 0 & -\frac{(d+e)h_{2c}k_3}{u_2} & \frac{(d+e)^2k_3}{u_2} & (d+e)k_3 \\
h_{2c}k_3 & (h_2m_{2s} - h_{2c}m_2)u_2 - \frac{ch_{2c}k_3}{u_1} - \frac{(d+e)h_{2c}k_3}{u_2} & k_{12} & -\frac{h_{1c}h_{2c}k_3}{u_1} & gh_2m_{2s} - k_{r2} - k_{12} & \frac{h_{2c}{}^2k_3}{u_2} - c_{r2} & -\frac{(d+e)h_{2c}k_3}{u_2} & -h_{2c}k_3 \\
0 & 0 & 0 & 1 & 0 & 0 & 0 & 0 \\
0 & 0 & 0 & 0 & 0 & 1 & 0 & 0 \\
0 & 0 & 0 & 0 & 0 & 0 & 1 & 0
\end{pmatrix}
$$

$$
\text{Matrix } K = [-(a+c)k_1 - h_{1c}k_1 - k_1\ 0\ 0\ 0]^T.
$$

3.2. Verification of Simplified Six Degrees of Freedom Model of the Liquid Tank Semi-Trailer for Yaw Stability Control

To verify the accuracy of the simplified model for the liquid tank semi-trailer, the state responses of the simplified model and the TruckSim vehicle model under the double-lane shifting condition are compared. The initial vehicle speed is set to 80 km/h, the ground adhesion coefficient is 0.4, and the basic vehicle parameters are shown in Table 2.

Table 2. Basic parameters of the liquid tank semi-trailer.

Parameter	Value	Unit	Parameter	Value	Unit
m_1	5876	kg	m_{1s}	4457	kg
m_2	25,876	kg	m_{2s}	20,000	kg
a	2	m	b	2.478	m
c	2.189	m	d	4.693	m
e	5.4	m	h_1	1.175	m
h_2	2.125	m	h_{1c}	1.1	m
h_{2c}	1.1	m	I_{1xx}	2283	kg·m^2
I_{1yy}	35,402	kg·m^2	I_{1zz}	34,802	kg·m^2
I_{1xz}	1626	kg·m^2	I_{2xx}	22,330	kg·m^2
I_{2zz}	250,416	kg·m^2	I_{2xz}	0	kg·m^2

It can be seen from Figure 6a,b that the established simplified six degrees of freedom model of the liquid tank semi-trailer can reflect the vehicle state and can be used for the development of the yaw stability control method of the liquid tank semitrailer.

3.3. Selection of Control Parameters for Yaw Stability Control of the Liquid Tank Semi-Trailer

In terms of the selection of the control parameters, the yaw rate, the sideslip angle of the centroid, and the lateral acceleration are generally taken for stability control [16]. When these variables are used for the control, the vehicle's yaw stability will be greatly improved, but the driving path deviation would be produced compared with the path under steady-state driving. The magnitude of the articulated angle of the liquid tank semi-trailer directly reflects the steering characteristics and driving path of the liquid tank semi-trailer. The deviation of the driving path indicates the path-following ability of the semi-trailer to the tractor [17]. The greater the deviation is, the worse the path-following performance would be. Therefore, if the articulation angle can be adjusted to that of the steady state, while maintaining the stability of the vehicle's yaw rate, the vehicle would have both good yaw stability and good path-following performance.

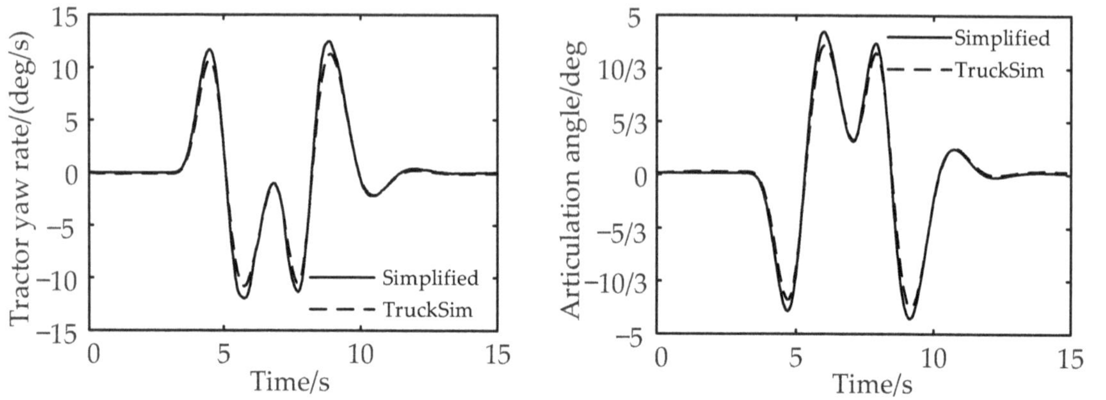

(a) (b)

Figure 6. Comparison of vehicle state responses between simplified model and TruckSim model: (**a**) comparison of yaw rate of the tractor and (**b**) comparison of articulation angle.

Therefore, the tractor yaw rate, the semi-trailer yaw rate, and the articulation angle are selected as the control parameters.

3.4. Determination of Expected Values for Yaw Stability Control Parameters of the Liquid Tank Semi-Trailer

When implementing the control method, it is necessary to obtain the vehicle state value when the vehicle is running stably, that is, the expected value of the vehicle state response. Take the derivative value of each state variable in the six degrees of freedom model of the liquid tank semi-trailer in Section 2.1 as zero, to obtain the vehicle linear reference model required by the control system, that is $\dot{\beta}_1 = \ddot{\psi}_1 = \dot{\phi}_1 = \ddot{\phi}_1 = \dot{\phi}_2 = \ddot{\phi}_2 = \dot{\theta} = 0$. Then, the expected value of each state variable under a stable state can be determined as X_d:

$$\begin{cases} X_d = \left[\beta_{1d}\ \dot{\psi}_{1d}\ \phi_{1d}\ \phi_{2d}\ \theta_d\right]^T \\ 0 = N_d X_d + K_d U_d \end{cases} \tag{15}$$

where U_d is the system input, $U_d = \delta$; K_d is a 5×1 dimensional matrix, $K_d = [-(a+c)k_1 - h_{1c}k_1 - k_1\ 0\ 0\ 0]^T$; and N_d is a 5×5 dimensional matrix.

$$N_d = \begin{bmatrix} (a+c)k_1 + (c-b)k_2 & N_{d12} & 0 & 0 & 0 \\ (k_1+k_2)h_{1c} & N_{d22} & N_{d23} & k_{12} & 0 \\ k_1+k_2+k_3 & N_{d32} & 0 & 0 & -k_3 \\ -(d+e)k_3 & N_{d42} & 0 & 0 & (d+e)k_3 \\ h_{2c}k_3 & N_{d52} & k_{12} & N_{d54} & -h_{2c}k_3 \end{bmatrix} \tag{16}$$

$$N_{d12} = \frac{(a^2+ac)k_1}{u_1} + \frac{(b^2-bc)k_2}{u_1} - m_1u_1c;\ N_{d22} = (m_{1s}h_1 - m_1h_{1c})u_1 + \frac{(ak_1-bk_2)h_{1c}}{u_1}$$

$$N_{d32} = \frac{ak_1 - bk_2 - ck_3}{u_1} - m_2u_2 - m_1u_1 - \frac{(d+e)k_3}{u_2};\ N_{d42} = \frac{(d+e)^2k_3}{u_2} + m_2u_2e + \frac{(d+e)ck_3}{u_1}$$

$$N_{d52} = (m_{2s}h_2 - m_2h_{2c})u_2 - \frac{ch_{2c}k_3}{u_1} - \frac{(d+e)h_{2c}k_3}{u_2}; N_{d23} = m_{1s}h_1g - k_{r1} - k_{12}; N_{d54} = m_{2s}gh_2 - k_{r2} - k_{12}$$

Considering the road adhesion coefficient μ, the desired yaw rate shall meet the following relations:

$$\dot{\psi}_{id} \leq \frac{\mu g}{v_i}, \ (i = 1,\ 2) \tag{17}$$

To sum up, the expected value of the yaw rate is

$$\dot{\psi}_{ir} = \min\left\{ \dot{\psi}_{id}, \left| \frac{\mu g}{v_i} \right| \right\} sign(\delta), \ (i = 1,\ 2) \tag{18}$$

where v_i is the lateral velocity of the tractor or semi-trailer, m/s.

The expected articulation angle is

$$\theta_r = \theta_d \tag{19}$$

4. Yaw Stability Control Method for the Liquid Tank Semi-Trailer and Its Implementation

4.1. General Architecture of Yaw Stability Control Method for the Liquid Tank Semi-Trailer

The general architecture of the yaw stability control method for the liquid tank semi-trailer is shown in Figure 7. The TP model calculates the force and torque generated by liquid sloshing, the TruckSim software provides the real-time state parameters of the vehicle, and the reference model provides the expected values of state parameters of the vehicle. The difference between the real-time state parameters and the stable state parameters is controlled by the tractor yaw rate fuzzy PID controller, the semi-trailer yaw rate fuzzy PID controller, and the articulation angle fuzzy PID controller, and the final additional yaw moment required by the tractor and semi-trailer is obtained after the weighted control of the above three. The additional yaw moment of the vehicle is provided by the braking torque of the wheels. Through the braking torque distribution and the wheel slip rate control, the braking torque of each wheel of the tractor and the semi-trailer is fed back to the TruckSim software to correct the state of the vehicle in real-time, to achieve the yaw stability control of the liquid tank semi-trailer.

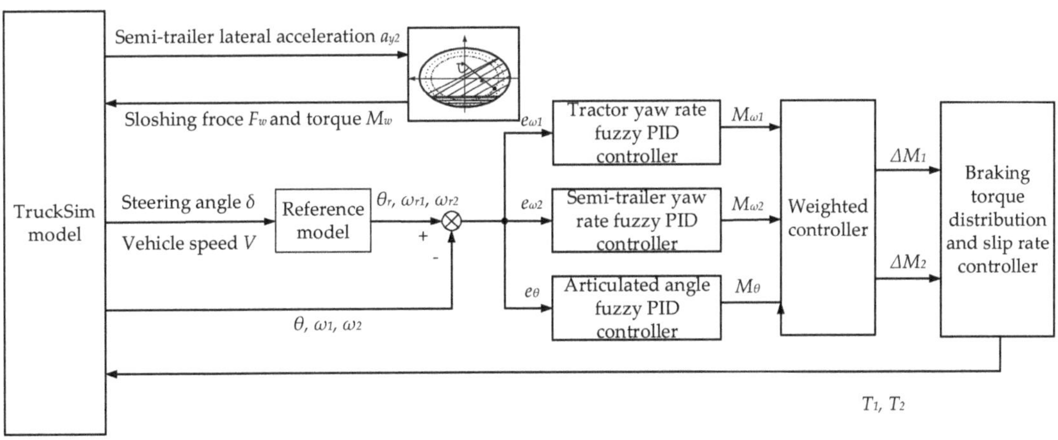

Figure 7. The general architecture of the multi-object PID differential braking-control method.

4.2. Calculation of the Additional Yaw Moment of Yaw Stability Control Method for the Liquid Tank Semi-Trailer

Considering the high nonlinearity of the liquid tank semi-trailer model, the fuzzy PID controller with a simple structure and strong robustness is taken to calculate the additional yaw moment required by the tractor and semi-trailer. Its typical structure is shown in Figure 8. The inputs of the fuzzy PID controller are deviation e and deviation change rate e_c. The outputs are the increments of the control parameters of the fuzzy PID controller, ΔK_P, ΔK_I, ΔK_D.

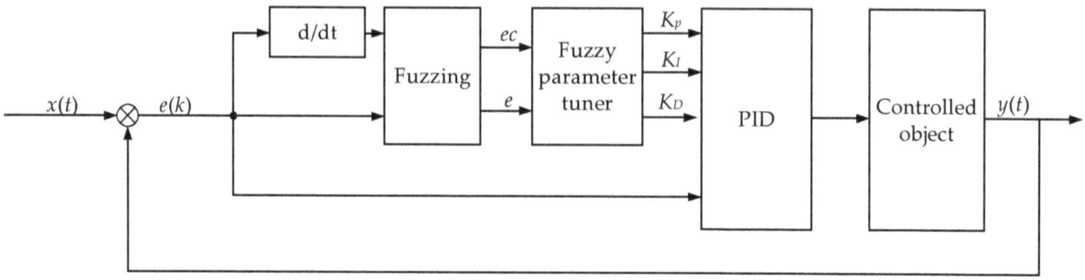

Figure 8. Structure of fuzzy PID controller.

The control parameters of the fuzzy PID controller are

$$\begin{cases} K_P = K_{P0} + \Delta K_P \cdot k_{uP} \\ K_I = K_{I0} + \Delta K_I \cdot k_{uI} \\ K_D = K_{D0} + \Delta K_D \cdot k_{uD} \end{cases} \tag{20}$$

where K_{P0}, K_{I0}, K_{D0} are the initial values of the parameters of the fuzzy PID controller; ΔK_P, ΔK_I, ΔK_D are the outputs of the fuzzy controller; k_{uP}, k_{uI}, k_{uD} are the scale factors of the fuzzy PID controller; and K_P, K_I, K_D are the actual control values of the controller.

The parameters of the fuzzy PID controller are set as follows: deviation e, deviation change rate e_c, and PID parameter variation ΔK_P, ΔK_I, ΔK_D all adopt the Gaussian membership function. After fuzzification, the fuzzy values of deviation e and deviation change rate e_c are [−6,6], while those of the PID parameters ΔK_P, ΔK_I, ΔK_D are [−1,1].

The language variables are negative big (NB), negative middle (NM), negative small (NS), zero (ZO), positive small (PS), positive middle (PM), and positive big (PB), and the corresponding fuzzy subsets are {NB, NM, NS, ZO, PS, PM, PB}.

Finally, the input and output of the fuzzy PID controller are shown in Figure 9. The surface relationships between the input value and the output value are shown in Figure 10. The fuzzy rules of parameters ΔK_P, ΔK_I, ΔK_D are shown in Tables 3–5.

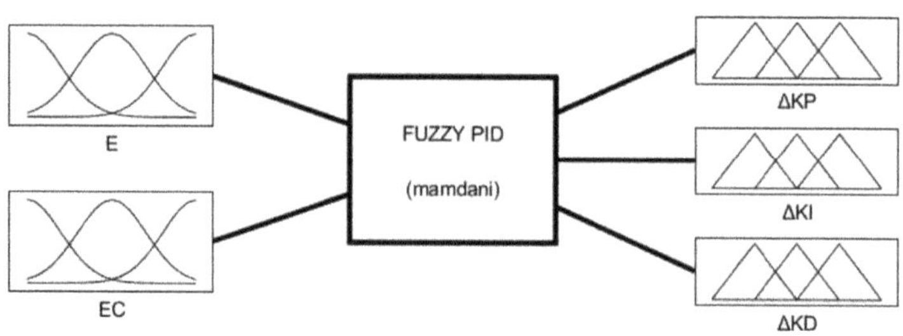

Figure 9. Input and output of the fuzzy controller.

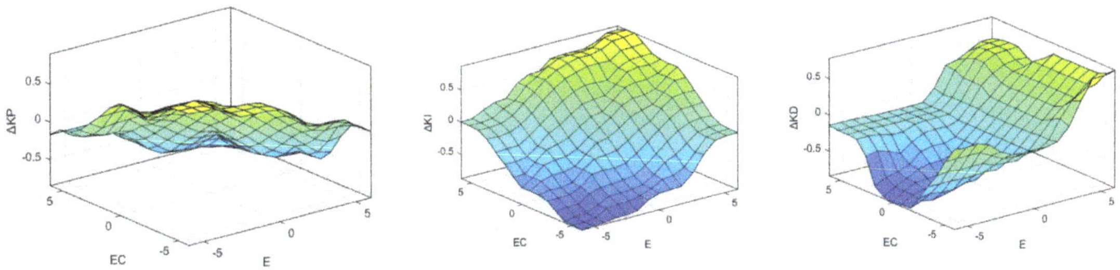

(a) (b) (c)

Figure 10. Fuzzy PID output surface relationships: (**a**) ΔK_P output surface, (**b**) ΔK_I output surface, and (**c**) ΔK_D output surface.

Table 3. Fuzzy rule table of ΔK_p.

ΔK_p		e_c						
		NB	**NM**	**NS**	**ZO**	**PS**	**PM**	**PB**
	NB	PB	PB	PM	PM	PS	ZO	ZO
	NM	PB	PB	PM	PS	PS	ZO	NS
	NS	PM	PM	PM	PS	ZO	NS	NS
e	ZO	PM	PM	PS	ZO	NS	NM	NM
	PS	PS	PS	ZO	NS	NS	NM	NM
	PM	PS	ZO	NS	NM	NM	NM	NB
	PB	ZO	ZO	NM	NM	NM	NB	NB

Table 4. Fuzzy rule table of ΔK_I.

ΔK_I		e_c						
		NB	**NM**	**NS**	**ZO**	**PS**	**PM**	**PB**
	NB	NB	NB	NM	NM	NS	ZO	ZO
	NM	NB	NB	NM	NS	NS	ZO	ZO
	NS	NB	NM	NS	NS	ZO	PS	PS
e	ZO	NM	NM	NS	ZO	PS	PM	PM
	PS	NM	NS	ZO	PS	PS	PM	PB
	PM	ZO	ZO	PS	PS	PM	PB	PB
	PB	ZO	ZO	PS	PM	PM	PB	PB

Table 5. Fuzzy rule table of ΔK_D.

ΔK_D		e_c						
		NB	**NM**	**NS**	**ZO**	**PS**	**PM**	**PB**
	NB	PS	NS	NB	NB	NB	NM	PS
	NM	PS	NS	NB	NM	NM	NS	ZO
	NS	ZO	NS	NM	NM	NS	NS	ZO
e	ZO	ZO	NS	NS	NS	NS	NS	ZO
	PS	ZO	ZO	ZO	ZO	ZO	ZO	ZO
	PM	PB	PS	PS	PS	PS	PS	PB
	PB	PB	PM	PM	PM	PS	PS	PB

Through the above analysis, the additional yaw moment provided by the tractor yaw rate fuzzy PID controller M_{ω_1}, the additional yaw moment provided by the fuzzy PID controller for the yaw rate of the semi-trailer M_{ω_2}, and the additional yaw moment provided by the articulation angle fuzzy PID controller M_θ can be obtained by Equation (21).

$$\begin{cases} M_{\omega_1} = Kpe_{\omega_1} + K_I \int_0^t e_{\omega_1} dt + K_D \frac{de_{\omega_1}}{dt} \\ M_{\omega_2} = Kpe_{\omega_2} + K_I \int_0^t e_{\omega_2} dt + K_D \frac{de_{\omega_2}}{dt} \\ M_\theta = Kpe_\theta + K_I \int_0^t e_\theta dt + K_D \frac{de_\theta}{dt} \end{cases} \quad (21)$$

Through the weighted control of the three, the additional yaw moment ΔM_1 required by the tractor and the additional yaw moment ΔM_2 required by the semi-trailer can be obtained by Equation (22).

$$\begin{cases} \Delta M_1 = M_{\omega_1} \\ \Delta M_2 = w_1 M_{\omega_2} + w_2 M_\theta \end{cases} \quad (22)$$

where w_1, w_2 are the weight coefficients of the yaw rate controller and the articulation angle controller of the semi-trailer, respectively. After several simulation comparisons, the weight coefficient w_1 is taken as 0.6, and w_2 is taken as 0.4.

4.3. Calculation and Distribution of Braking Torque of Yaw Stability Control Method for the Liquid Tank Semi-Trailer

When the additional yaw moment required for differential braking of the liquid tank semi-trailer is calculated based on the multi-object PID controller, it is necessary to calculate the braking moment generated by the wheels of the tractor and semi-trailer T_1, T_2. The additional yaw moment of the vehicle is provided by the braking force applied to the wheels. Therefore, only by determining the relationship between the additional yaw moment and the braking force of different wheels can the braking torque be accurately calculated. Take the direction in which the vehicle turns left as the positive one and the other as the negative one. The relationship between the additional yaw moment and the braking moment of the liquid tank semi-trailer under the left-turning condition is shown in Figure 11.

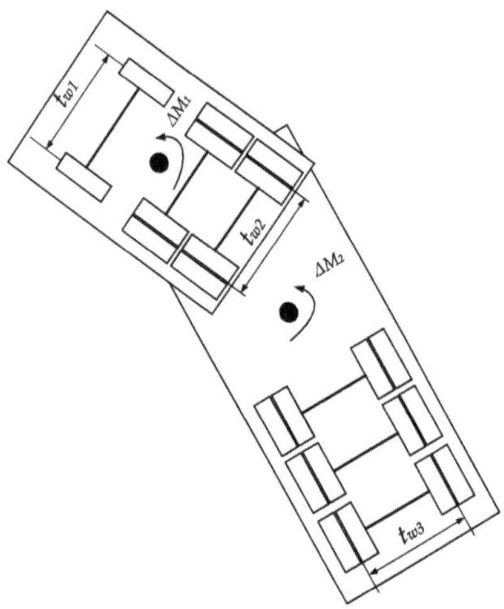

Figure 11. Relationship between additional yaw moment and braking torque of liquid tank semi-trailer.

In Figure 12, t_{w1}, t_{w2}, t_{w3} is the track width of the front axle of the tractor, the second and third axles of the tractor and the fourth, fifth, and sixth axles of the semi-trailer; $\Delta M_1, \Delta M_2$ is the additional yaw moment of the tractor and semi-trailer. When the vehicle performs differential braking, supposing that the braking force is applied to the left wheel of the first axle of the tractor and the left wheels of the fourth, fifth, and sixth axles of the semi-trailer, it can be obtained from the mechanical knowledge that

$$\begin{cases} \Delta M_1 = F_{xl1} \cdot \frac{t_{w1}}{2} \\ \Delta M_2 = (F_{xl4} + F_{xl5} + F_{xl6}) \cdot \frac{t_{w3}}{2} \end{cases} \tag{23}$$

where F_{xl1} is the braking force applied to the left wheel of the first axle of the tractor; $F_{xl4}, F_{xl5}, F_{xl6}$ are the braking forces applied to the left wheels of the fourth, fifth, and sixth axles of the semi-trailer, respectively; and t_{w1}, t_{w3} are the first axle base of the tractor and the fourth, fifth, and sixth axle bases of the semi-trailer, respectively.

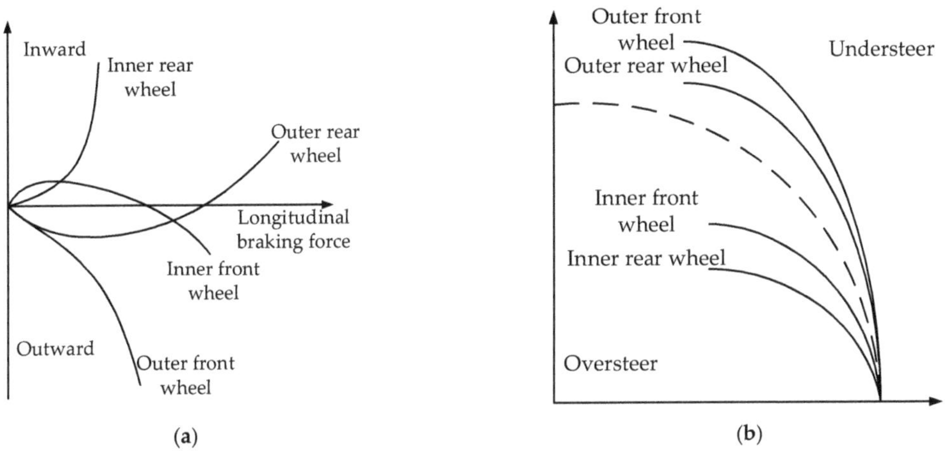

Figure 12. Braking characteristics of different brake wheels: (**a**) relationship between yaw moment generated by braking of different wheels and longitudinal braking force and (**b**) influence of single wheel braking on vehicle steering.

The cornering stiffness of the tire will change due to the change in the vertical load of the tire, and the change in the cornering stiffness of the tire will affect the yaw stability of the vehicle. Assuming that the moment of the braking wheel when the tire is not locked is approximately proportional to its vertical load, the braking force applied by the left wheel of the first axle of the tractor and the braking force distribution formula of the left wheel of the fourth, fifth, and sixth axles of the semi-trailer can be expressed, respectively, as follows:

$$F_{xfl} = \frac{2\Delta M_1}{t_{w1}} \tag{24}$$

$$\begin{cases} F_{xl4} = \frac{F_{zl4}}{F_{zl4}+F_{zl5}+F_{zl6}} \cdot \frac{2\Delta M_2}{t_{w3}} \\ F_{xl5} = \frac{F_{zl5}}{F_{zl4}+F_{zl5}+F_{zl6}} \cdot \frac{2\Delta M_2}{t_{w3}} \\ F_{xl6} = \frac{F_{zl6}}{F_{zl4}+F_{zl5}+F_{zl6}} \cdot \frac{2\Delta M_2}{t_{w3}} \end{cases} \tag{25}$$

where $F_{zl4}, F_{zl5}, F_{zl6}$ are the vertical load of the left wheel of the fourth, fifth, and sixth axles of the semi-trailer, respectively.

The braking moment of the left wheel on the first axle of the tractor and the left wheel on the fourth, fifth, and sixth axles of the semi-trailer can be expressed, respectively, as follows:

$$T_{l1} = F_{xl1} R_1 = \frac{2\Delta M_1}{t_{w1}} \cdot R_1 \tag{26}$$

$$\begin{cases} T_{l4} = F_{xl4}R_3 = \frac{F_{ll4}}{F_{zl4}+F_{zl5}+F_{zl6}} \cdot \frac{2\Delta M_2}{t_{w3}} \cdot R_3 \\ T_{l5} = F_{xl5}R_3 = \frac{F_{zl5}}{F_{zl4}+F_{ll5}+F_{zl6}} \cdot \frac{2\Delta M_2}{t_{w3}} \cdot R_3 \\ T_{l6} = F_{xl6}R_3 = \frac{F_{zl6}}{F_{zl4}+F_{zl5}+F_{zl6}} \cdot \frac{2\Delta M_2}{t_{w3}} \cdot R_3 \end{cases} \quad (27)$$

where R_1 is the wheel rolling radius of the first axle of the tractor; and R_3 is the wheel rolling radius of the fourth, fifth, and sixth axles of the semi-trailer.

From the previous analysis, the additional yaw moment required by the tractor and semi-trailer can be converted into the braking moment of the wheel. After that, the target braking wheel needs to be determined to generate the additional yaw moment efficiently.

As shown in Figure 12a, when the outer front wheels of the vehicle brake, the generated yaw moment gradually increases outwards; when the outer rear wheels of the vehicle brake, the yaw moment produced gradually increases and then decreases outwards, and finally gradually increases inwards; when the inner front wheels of the vehicle brake, the yaw moment generated gradually increases, then decreases inwards, and finally increases outwards; when the inner rear wheels of the vehicle brake, the yaw moment generated gradually increases inwards.

As shown in Figure 12b, when the liquid tank semi-trailer runs in a steady circular motion, the brake of the vehicle's outer front wheel will cause the vehicle to understeer, and the trend is large; the brake of the vehicle's outer rear wheel will cause the vehicle to understeer, and the trend is relatively small; the front wheel braking in the vehicle will cause the vehicle to oversteer, and the trend is relatively small; and the rear wheel braking in the vehicle will cause the vehicle to oversteer, and the trend is large.

Based on this, when the outer front wheel and the outer rear wheel of the vehicle brake, the brake efficiency of the outer front wheel is higher under the same conditions; when the inner front wheel and inner rear wheel of the vehicle brake, the brake efficiency of the inner rear wheel is higher under the same conditions [18].

To sum up, the wheel braking moment should be distributed by judging the deviation between the actual value and the expected value of the steering direction and yaw rate, and the deviation between the actual value and the expected value of the articulation angle. The final distribution rules are shown in Table 6.

Table 6. Rules for wheel braking moment distribution.

Front Wheel Angle δ	**Tractor/Semi-Trailer**		**Control Parameter Deviation**	**Steering Characteristics**	**Target Brake Wheel**	
	ω_{r1}, ω_{r2}	ω_1, ω_2	$e_1 = \omega_{r1} - \omega_1$ $e_2 = \omega_{r2} - \omega_2$ $e_\theta = \theta_r - \theta$		**Tractor**	**Semitrailer**
+	+	+	+	Understeer	L2, L3	L4, L5, L6
+	+	+	0	\	\	\
+	+	+	-	Oversteer	R1	R4, R5, R6
+	+	0	+	Understeer	L2, L3	L4, L5, L6
+	+	-	+	Understeer	L2, L3	L4, L5, L6
0	0	+	-	Oversteer	R1	R4, R5, R6
0	0	-	+	Oversteer	L1	L4, L5, L6
0	0	0	0	\	\	\
-	-	+	-	Understeer	R2, R3	R4, R5, R6
-	-	0	-	Understeer	R2, R3	R4, R5, R6
-	-	-	+	Oversteer	L1	L4, L5, L6
-	-	-	0	\	\	\
-	-	-	-	Understeer	R2, R3	R4, R5, R6

Note 1: "+", "-", "0", and "\" in the table represent "positive", "negative", "0", and "no control action", respectively. Note 2: "L1, R1" in the table refer to the left and right wheels of the tractor steering axle; "L2, R2, L3, R3" refer to the left and right wheels of the drive axles of the tractor; "L4, R4, L5, R5, L6, R6" refer to the left and right wheels of the semi-trailer axles.

4.4. Wheel Slip Rate Control in Yaw Stability Control Method for the Liquid Tank Semi-Trailer

The calculation formula of the wheel slip ratio is

$$\lambda = \frac{u - r_w w}{u} \tag{28}$$

where u is the longitudinal velocity of the tractor or semi-trailer; r_w is the rolling radius of the wheel; and w is the angular velocity of the wheel.

Figure 13 shows that when the slip ratio is 0.15–0.20, the longitudinal adhesion coefficient and lateral adhesion coefficient are both large, and the longitudinal force and lateral force on the tire are also sufficient, so the vehicle can maintain good dynamic performance and stability.

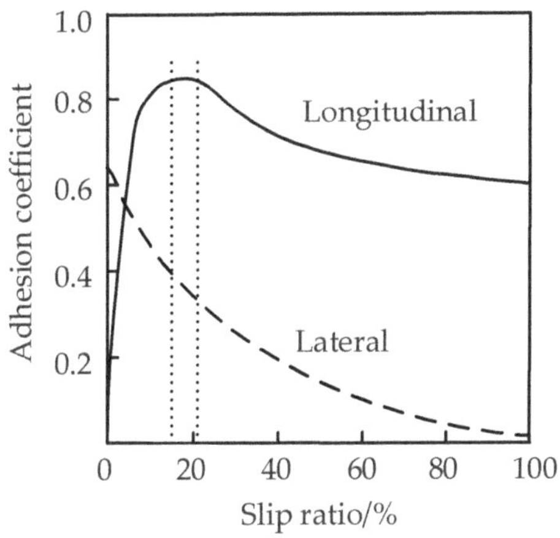

Figure 13. Relationship between longitudinal and lateral adhesion coefficient and slip ratio.

To control the vehicle tire slip rate within 0.15–0.20, when the wheel slip rate is greater than 0.20, the braking torque shall be reduced to reduce the wheel slip rate; when the wheel slip ratio is between 0.15 and 0.20, the braking torque should remain unchanged; when the wheel slip rate is lower than 0.15, the wheel braking torque shall be increased to increase the wheel slip rate. It should be noted that in the process of increasing or decreasing the braking torque, the appropriate growth rate and decline rate should be selected. When the growth rate and decline rate are too large, the slip rate will overshoot and even cause system vibration; when the values of the growth rate and the decline rate are too small, the slip rate changes slowly, and the system response is not sensitive. According to the characteristics of the wheel, the appropriate growth rate and decline rate are selected, and the corresponding braking torque can be obtained

$$\begin{cases} T_{t+1} = T_t + k_i T_s, & S < 0.15 \\ T_{t+1} = T_t, & 0.15 \le S \le 0.2 \\ T_{t+1} = T_t - k_d T_s, & S > 0.2 \end{cases} \tag{29}$$

where T_{t+1} is the output value of braking torque at time $t+1$; T_t is the output value of braking torque at time t; T_s is the calculation time interval, 0.005 s; k_i is the growth rate, which is 6000 N·m/s; and k_d is decline rate, 6000 N·m/s.

5. Verification of the Effectiveness and Robustness of the Yaw Stability Control Method for the Liquid Tank Semi-Trailer

To verify the effectiveness of the proposed control method, based on the liquid tank semi-trailer co-simulation model, the different vehicle state responses are compared while taking the multi-object PID differential braking-control (multi-object), tractor yaw rate + semi-trailer yaw rate differential braking-control ($\omega_1 + \omega_2$), tractor yaw rate + articulation angle differential braking-control ($\omega_1 + \theta$), and no control methods, respectively. For the verification of the proposed control method, the double-lane change and the step-steering angle input's working conditions are selected, the initial vehicle speed is set as 80 km/h, the road adhesion coefficient is set as 0.3, and the basic vehicle parameters can be referred to in Table 2.

To verify the robustness of the proposed method, the robustness of the proposed control method is analyzed and verified when the multi-object PID control parameters are changed.

5.1. Verification of the Effectiveness of the Yaw Stability Control Method under the Double-Lane Change Working Condition

In this section, the double-lane change working condition simulation test is performed when the liquid tank semi-trailer is running on a low-adhesion coefficient road. The comparison diagrams of the yaw rate, the vehicle track, and the articulation angle of the tractor and semi-trailer are shown in Figures 14–16, respectively.

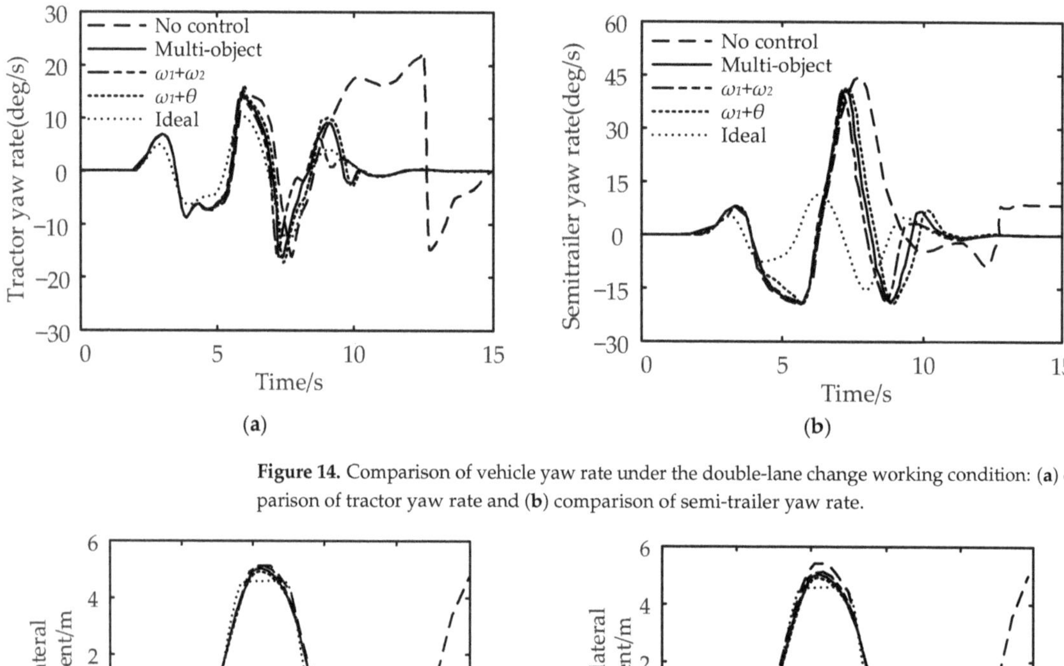

Figure 14. Comparison of vehicle yaw rate under the double-lane change working condition: (**a**) comparison of tractor yaw rate and (**b**) comparison of semi-trailer yaw rate.

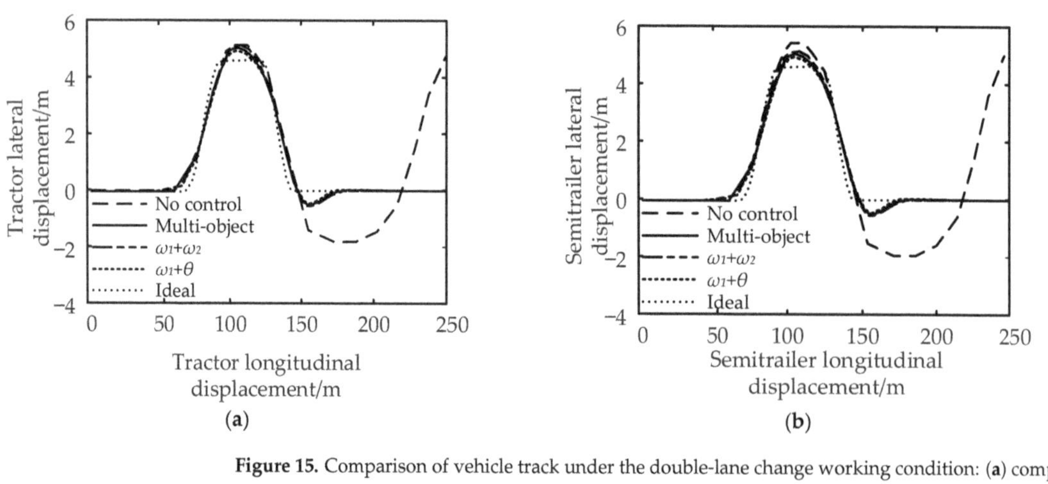

Figure 15. Comparison of vehicle track under the double-lane change working condition: (**a**) comparison of tractor track and (**b**) comparison of semi-trailer track.

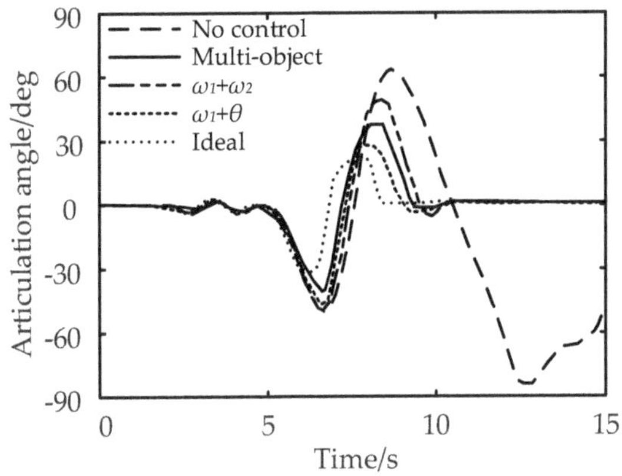

Figure 16. Comparison of articulation angle under the double-lane change working condition.

It can be seen from Figure 14a,b that compared with no control, multi-object control reduces the maximum yaw rate of the tractor by 45.45% and the maximum yaw rate of the semi-trailer by 4.55%; for $\omega_1 + \omega_2$, the maximum yaw rate of the tractor is reduced by 50.15%, and the maximum yaw rate of the semi-trailer is reduced by 6.82%; for $\omega_1 + \theta$, the control reduces the maximum yaw rate of the tractor by 40.91%, and the maximum yaw rate of the semi-trailer by 2.27%. In conclusion, compared with the no control strategy, the multi-object control and $\omega_1 + \omega_2$ control can effectively reduce the maximum yaw rate of the tractor and semi-trailer.

It can be seen from Figures 15 and 16 that compared with no control, the maximum lateral displacement deviation of the tractor under multi-object control is reduced by 3.71%, the maximum lateral displacement deviation of the semi-trailer is reduced by 5.45%, and the maximum articulation angle is reduced by 46.67%; for $\omega_1 + \omega_2$ control, the lateral displacement deviation of the tractor, semi-trailer, and the maximum articulation angle are reduced by 1.85%, 3.64%, and 33.33%, respectively; the $\omega_1 + \theta$ control reduced the lateral displacement deviation of the tractor by 5.56%, the lateral displacement deviation of the semi-trailer by 7.28%, and the maximum articulation angle by 58.33%. In conclusion, compared with the no control strategy, the multi-object control and $\omega_1 + \theta$ control can effectively reduce the maximum lateral displacement deviation and articulation angle of the tractor and semi-trailer.

5.2. Verification of the Effectiveness of the Yaw Stability Control Method under the Step-Steering-Angle Input Working Condition

In this section, the step-steering-angle input working condition simulation test is per-formed when the liquid tank semi-trailer is running on a low-adhesion coefficient road. The comparison diagrams of the yaw rate, the vehicle track, and the articulation angle of the tractor and semi-trailer are shown in Figures 17–19, respectively.

It can be seen from Figure 17a,b that compared with no control, multi-object control reduces the maximum yaw rate of the tractor by 52.5% and the maximum yaw rate of the semi-trailer by 43.3%; for $\omega_1 + \omega_2$, the maximum yaw rate of the tractor is reduced by 57.5%, and the maximum yaw rate of the semi-trailer is reduced by 46.7%; for $\omega_1 + \theta$, the control reduces the maximum yaw rate of the tractor by 37.5%, and the maximum yaw rate of the semi-trailer by 36.7%. In conclusion, compared with the no control strategy, the multi-object control and $\omega_1 + \omega_2$ control can effectively reduce the maximum yaw rate of the tractor and semi-trailer.

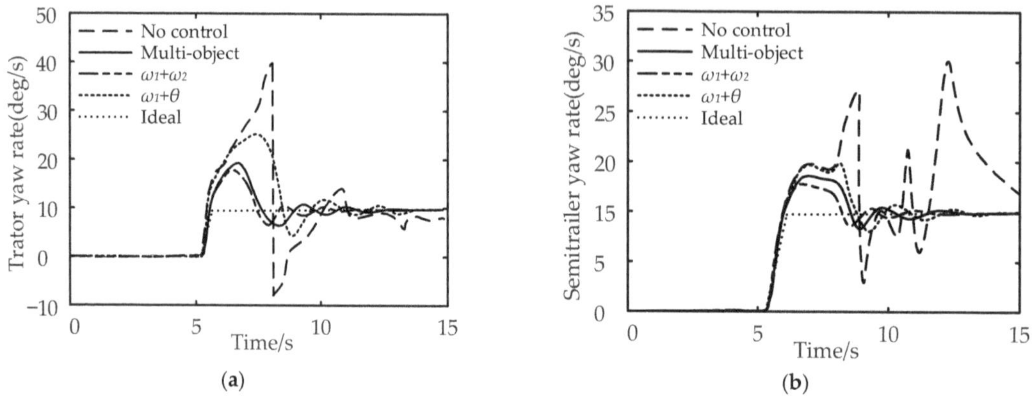

Figure 17. Comparison of vehicle yaw rate under the step-steering-angle input working condition: (**a**) comparison of tractor yaw rate and (**b**) comparison of semi-trailer yaw rate.

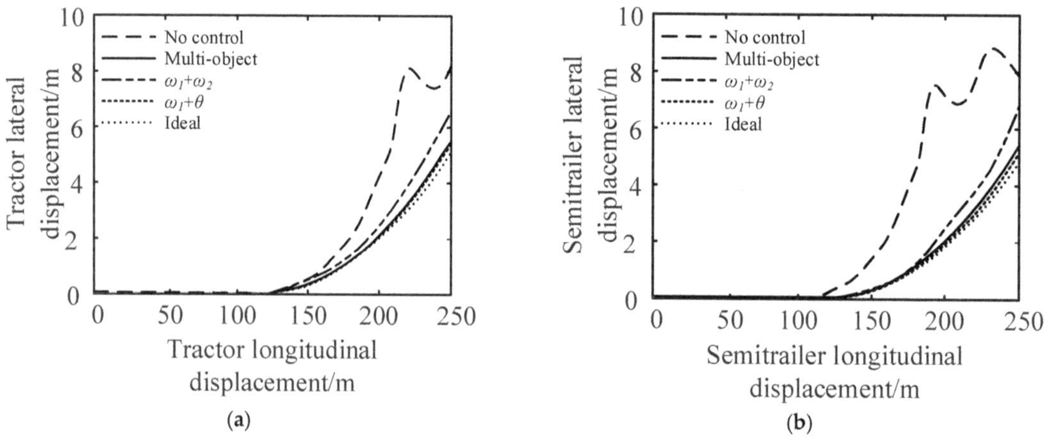

Figure 18. Comparison of vehicle track under the step-steering-angle input working condition: (**a**) comparison of tractor track and (**b**) comparison of semi-trailer track.

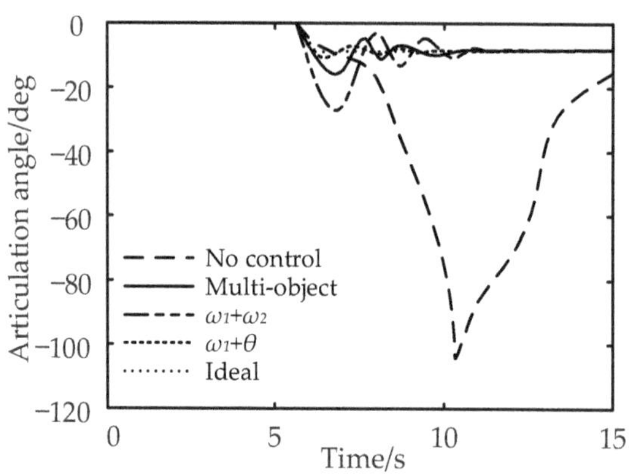

Figure 19. Comparison of articulation angle under the step-steering-angle input working condition.

It can be seen from Figures 18 and 19 that compared with no control, the maximum lateral displacement deviation of the tractor under multi-object control is reduced by 31.3%, the maximum lateral displacement deviation of the semi-trailer is reduced by 38.9%, and the maximum articulation angle is reduced by 81.5%; for $\omega_1 + \omega_2$ control, the lateral displacement deviation of the tractor, semi-trailer, and the maximum articulation angle are reduced by 25.3%, 33.3%, and 25.5% respectively; the $\omega_1 + \theta$ control reduced the lateral displacement deviation of the tractor by 37.5%, the lateral displacement deviation of the semi-trailer by 42.2%, and the maximum articulation angle by 88.2%. In conclusion, compared with the no control strategy, the multi-object control and $\omega_1 + \theta$ control can effectively reduce the maximum lateral displacement deviation and articulation angle of the tractor and semi-trailer.

To sum up, compared with the differential braking control aiming at the yaw rate or articulation angle of the tractor, the multi-object PID differential braking control can not only improve the yaw stability of the semi-trailer turning on the low-adhesion road but also improve the path-following performance.

5.3. Verification of the Robustness of the Yaw Stability Control Method

The effectiveness of the proposed control method in controlling the vehicle yaw stability is verified in the previous two sections. As another important criterion to evaluate the quality of the control system, the robustness of the proposed control method is analyzed and verified.

The robustness of control system means that the system has the ability to maintain a certain performance under the disturbance of uncertainty [19]. The proposed multi-objective PID control method is established based on the fuzzy PID control method, which has good stability and robustness [20]. To verify the robustness of the proposed control method, the tractor yaw rates under the step-steering-angle input working condition are compared when the control parameters K_P, K_I, and K_D of the multi-object PID control method are increased by 30%. The other simulation parameters are the same with those in Section 5.2, and the comparison results are shown in Figure 20.

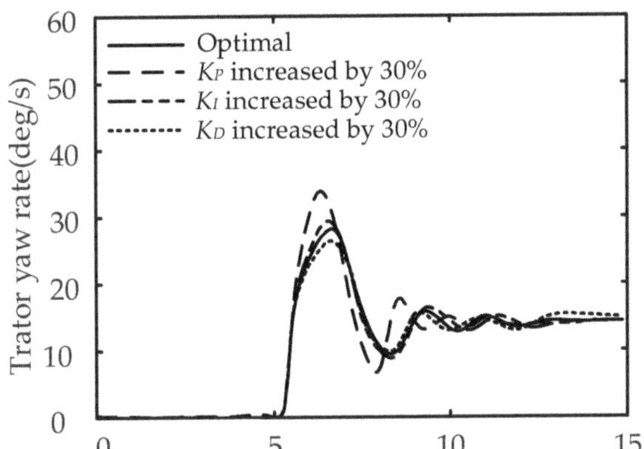

Figure 20. Comparison of the tractor yaw rates under the step-steering-angle input working condition when the control parameters K_P, K_I, and K_D of the multi-object PID control method are increased by 30%.

It can be seen from Figure 20 that compared with the control effect of the optimal parameters, the differences between the tractor yaw rate curves are not significant when these three parameters K_P, K_I, and K_D are increased by 30%, which proves the robustness of the proposed control method.

6. Conclusions

(1) The TP model is established to simulate the sloshing effect of the liquid in the elliptical cylinder tank under lateral excitation, and its simulation effect is validated using the Fluent software. Based on it, a co-simulation model is established based on TruckSim and MATLAB/Simulink.

(2) A simplified six degrees of freedom model of the liquid tank semi-trailer is established and verified using the TruckSim software. Taking the tractor yaw rate, semi-trailer yaw rate, and articulation angle as the control parameters, a multi-object PID differential braking-control method is proposed and implemented.

(3) The vehicle state responses with and without control are compared under the double lane change and the step-steering-angle input working conditions on a low-adhesion road. The simulation results show that, compared with the differential braking control which targets the yaw rate or articulation angle of the tractor, the multi-object PID differential braking control can not only improve the yaw stability of the vehicle but also improve the path-following performance of the semi-trailer.

Author Contributions: Conceptualization, G.L. and T.F.; methodology, T.F. and G.L.; software T.F. and R.Z.; validation, G.L., R.Z. and T.F.; formal analysis, T.F. and R.Z.; investigation, T.F. and R.Z.; resources, G.L.; data curation, T.F.; writing—original draft preparation, T.F.; writing—review and editing, T.F., G.L. and R.Z.; visualization, T.F. and G.L.; supervision, G.L. and R.Z.; project administration, G.L.; funding acquisition, G.L. All authors have read and agreed to the published version of the manuscript.

Funding: This research was funded by the China Postdoctoral Science Foundation, grant number 2018M642937, which is a cooperation with the SMC Corporation.

Institutional Review Board Statement: Not applicable.

Informed Consent Statement: Not applicable.

Data Availability Statement: The study did not report any data.

Acknowledgments: Thanks to Gangyan Li and Ran Zhao of Wuhan University of Technology for supporting and helping this study.

Conflicts of Interest: The authors declare no conflict of interest.

References

1. Zhao, R.; Li, G.; Yu, B.; Yang, F. The brake pressure change rate in brake chamber and its online monitoring in semitrailer transport vehicle for dangerous cargo. *Proc. Inst. Mech. Eng. Part D J. Automob. Eng.* **2022**. Available online: https://journals.sagepub.com/doi/abs/10.1177/09544070221091684 (accessed on 20 April 2022). [CrossRef]
2. Zheng, H.; Hu, J.; Ma, S. Research on Simulation and Control of Differential Braking Stability of Tractor Semi-Trailer. *SAE Tech. Pap.* 2015. [CrossRef]
3. Li, X.-S.; Zheng, X.-L.; Ren, Y.-Y.; Wang, Y.-N.; Cheng, Z.-Q. Study on Driving Stability of Tank Trucks Based on Equivalent Trammel Pendulum for Liquid Sloshing. *Discret. Dyn. Nat. Soc.* **2013**, *2013*, 659873–659887. [CrossRef]
4. Peng, G.; Zhao, Z.; Liu, T.; Hu, H.; Feng, M. Research on Dynamic Characteristics of Lateral Sloshing in Liquid Tank Semi-Trailer. In Proceedings of the 2019 3rd Conference on Vehicle Control and Intelligence (CVCI), Hefei, China, 21–22 September 2019; pp. 1–6. [CrossRef]
5. Saeedi, M.A.; Kazemi, R.; Azadi, S. Improvement in the rollover stability of a liquid-carrying articulated vehicle via a new robust controller. *Proc. Inst. Mech. Eng. Part D J. Automob. Eng.* **2017**, *231*, 322–346. [CrossRef]
6. Dewangan, D.K.; Sahu, S.P. Real Time Object Tracking for Intelligent Vehicle. In Proceedings of the 2020 First International Conference on Power, Control and Computing Technologies (ICPC2T), Raipur, India, 3–5 January 2020; pp. 134–138. [CrossRef]
7. Celebi, M.; Akyildiz, H. Nonlinear modeling of liquid sloshing in a moving rectangular tank. *Ocean Eng.* **2002**, *29*, 1527–1553. [CrossRef]
8. Salem, M.I.; Mucino, V.H.; Saunders, E.; Gautam, M.; Guzman, A.L. Lateral sloshing in partially filled elliptical tanker trucks using a trammel pendulum. *Int. J. Heavy Veh. Syst.* **2009**, *16*, 207–224. [CrossRef]
9. Salem, M.I. *Rollover Stability of Partially Filled Heavy-Duty Elliptical Tankers Using Trammel Pendulums to Simulate Fluid Sloshing*; West Virginia University: Morgantown, WV, USA, 2000.
10. Bai, Z.; Lu, Y.; Li, Y. Method of Improving Lateral Stability by Using Additional Yaw Moment of Semi-Trailer. *Energies* **2020**, *13*, 6317. [CrossRef]

11. Zhao, C.; Xiang, W.; Richardson, P. Vehicle Lateral Control and Yaw Stability Control through Differential Braking. In Proceedings of the 2006 IEEE International Symposium on Industrial Electronics, Montreal, QC, Canada, 9–13 July 2006; Volume 1, pp. 384–389. [CrossRef]
12. Zhou, H.; Liu, Z. Vehicle Yaw Stability-Control System Design Based on Sliding Mode and Backstepping Control Approach. *IEEE Trans. Veh. Technol.* **2010**, *59*, 3674–3678. [CrossRef]
13. Xue-Lian, Z.; Xian-Sheng, L.; Yuan-Yuan, R. Equivalent Mechanical Model for Lateral Liquid Sloshing in Partially Filled Tank Vehicles. *Math. Probl. Eng.* **2012**, *2012*, 162825–162846. [CrossRef]
14. Xiu-jian, Y.; Yun-xiang, X.; Xiang-ji, W.U.; Kun, Z. Multi-Mass Trammel Pendulum Model of Fluid Lateral Sloshing for Tank Vehicle. *J. Traffic Transp. Eng.* **2018**, *18*, 140–151. [CrossRef]
15. Wan, Y.; Mai, L.; Nie, Z.G. Dynamic Modeling and Analysis of Tank Vehicle under Braking Situation. *Adv. Mater. Res.* **2013**, *694*, 176–180. [CrossRef]
16. Cai, H.; Xu, X. Lateral Stability Control of a Tractor-Semitrailer at High Speed. *Machines* **2022**, *10*, 716. [CrossRef]
17. Xu, X.; Zhang, L.; Jiang, Y.; Chen, N. Active Control on Path Following and Lateral Stability for Truck–Trailer Combinations. *Arab. J. Sci. Eng.* **2019**, *44*, 1365–1377. [CrossRef]
18. Her, H.; Koh, Y.; Joa, E.; Yi, K.; Kim, K. An Integrated Control of Differential Braking, Front/Rear Traction, and Active Roll Moment for Limit Handling Performance. *IEEE Trans. Veh. Technol.* **2015**, *65*, 4288–4300. [CrossRef]
19. Sariyildiz, E.; Ohnishi, K. Stability and Robustness of Disturbance-Observer-Based Motion Control Systems. *IEEE Trans. Ind. Electron.* **2014**, *62*, 414–422. [CrossRef]
20. Duan, X.-G.; Li, H.-X.; Deng, H. Robustness of fuzzy PID controller due to its inherent saturation. *J. Process Control* **2012**, *22*, 470–476. [CrossRef]

applied sciences

MDPI

Article

A Robust Intelligent Controller for Autonomous Ground Vehicle Longitudinal Dynamics

Lhoussain El Hajjami [1], El Mehdi Mellouli [2], Vidas Žuraulis [3,*], Mohammed Berrada [1] and Ismail Boumhidi [4]

[1] Laboratory of Artificial Intelligence, Data Sciences and Emerging Systems, School of Applied Sciences, Sidi Mohamed Ben Abdellah University, BP 72, My Abdallah Avenue Km. 5 Imouzzer Road, Fez 30000, Morocco
[2] Laboratory of Engineering, Systems, and Applications, School of Applied Sciences, Sidi Mohamed Ben Abdellah University, BP 72, My Abdallah Avenue Km. 5 Imouzzer Road, Fez 30000, Morocco
[3] Department of Automobile Engineering, Vilnius Gediminas Technical University, J. Basanavičiaus St. 28, 03224 Vilnius, Lithuania
[4] LISAC Laboratory, Faculty of Science Dhar El Mahrez, Sidi Mohamed Ben Abdellah University, Fez 30000, Morocco
* Correspondence: vidas.zuraulis@vilniustech.lt; Tel.: +370-5-274-4793

Abstract: In this paper, a novel adaptive sliding mode controller (SMC) was designed based on a robust law considering disturbances and uncertainties for autonomous ground vehicle (AGV) longitudinal dynamics. The robust law was utilized in an innovative method involving the upper bounds of disturbances and uncertainties. Estimating this lumped uncertainty upper limit based on a neural network approach allowed its online knowledge. It guided the controller to withstand the disturbance and to compensate for the uncertainties. A stability analysis, according to Lyapunov, was completed to confirm the asymptotic convergence of the states to the desired state. The effectiveness and benefits of the planned approach were scrutinized by simulations and comparative studies.

Keywords: autonomous ground vehicles; robust adaptive SMC; vehicle longitudinal dynamics; neural-network-based control

Citation: El Hajjami, L.; Mellouli, E.M.; Žuraulis, V.; Berrada, M.; Boumhidi, I. A Robust Intelligent Controller for Autonomous Ground Vehicle Longitudinal Dynamics. *Appl. Sci.* **2023**, *13*, 501. https://doi.org/10.3390/app13010501

Academic Editor: Luigi Fortuna

Received: 30 November 2022
Revised: 26 December 2022
Accepted: 27 December 2022
Published: 30 December 2022

1. Introduction

The current huge jump in self-driving stems from evolutionary development since the birth of the autonomous-driving concept. This concept is based not only on the substitution of the human driver but also a reduction in fatal accidents due to human factors. Indeed, carmakers search to meet the challenge by designing sophisticated and fully automated guidance systems. The development of an autonomous vehicle involves three key phases: environment perception, path planning, and path tracking [1]. The first phase permits one to perceive the vehicle's external surroundings; the second phase exploits the data issued from the first phase, while the third phase is charged with decision-making. In other words, it tries to impose on the vehicle dynamics the path arising from the second phase. The third phase focuses on the control systems, which are in charge of tracking the target path and enforcing it on the vehicle. The fundamental mission of the control unit is to maintain the vehicle's dynamic stability. This unit has two layers; the first is dedicated to the control algorithm, and the second is dedicated to the actuators. The latter is responsible for the execution of the orders issued by the first layer. These orders are algorithms that concern lateral and longitudinal control [2].

The concepts and physics of longitudinal vehicle control strategies have been studied since the late 1960s. Various longitudinal control approaches have been elaborated [3]. Longitudinal control can be characterized as the system of control strategies that are applied to control the behavior of a vehicle on its longitudinal trajectory via the various actuators assigned to acceleration and braking [4]. These longitudinal control techniques target advanced driver assistance system (ADAS) features. Adaptive cruise control (ACC) and intelligent speed assistance (ISA) are ADAS features that are relevant to longitudinal control. As mentioned earlier, the nucleus of the control is either based on conventional control theory approaches, strategies based on automatic learning theory, or a combination of both.

Autonomous ground vehicles (AGV) longitudinal dynamics are a long-standing challenge in the automotive area. A variety of control methods are offered; the sliding mode control (SMC) is recognized as the dominant approach to controlling the dynamics of autonomous systems in general [5–8] and AGVs in particular [1,9,10]. In this light, the authors of [11] offered a second-order sliding mode controller. This technique has been used to conceive a longitudinal controller that can adjust the inter-obstacle distance. Experimental tests have been conducted to verify the effectiveness of this approach. This study addressed the improvement of the ACC and stop-and-go features of ADASs. In the same vein, the authors of [12] intended to provide the best control performances of the ACC feature based on an optimal super-twisting sliding mode control design. In a similar fashion, sliding mode control combined with a fuzzy control technique was elaborated in [13]. The SMC drove the vehicle from a random primary position to a potential parking place. At the same time, the fuzzy controller took over to park the vehicle in this desired position. This strategy was designed to improve autonomous parking algorithms. A backstepping controller merged with an extended Luenberger observer was discussed in [14]. Based on a quarter-vehicle model, the states observer provided online estimations of the brake torque, angular wheel speed, and vehicle speed. This design was intended to enhance the performance of the antilock braking system (ABS). An MPC tracking algorithm incorporating longitudinal velocity compensation to minimize tracking error and ensure vehicle stability was discussed in [15]. This method managed the lateral and longitudinal dynamics concurrently to achieve the vehicle's complete stability.

Over and above conventional proposals, combinations of deterministic and intelligent methods are omnipresent in the literature. The authors of [16] proposed beetle antenna searching (BAS) and particle swarm optimization (PSO) algorithms to determine the appropriate PID parameter settings to control a dual-wheel AGV robot. In the same vein, an intelligent approach based on an optimization technique was proposed in [17]. The butterfly optimization algorithm (BOA) was used to set the optimal parameters of a PID controller. A coupled strategy including an EMRAN architecture with an extended Kalman filter, PID cruise, and a Stanley controller was suggested in [18]. This EMRAN-based method's performance concerning RMS tracking errors has shown its superiority over the Stanley controller, fuzzy-based PID, and conventional PID. On top of providing great robustness, the EMRAN-assisted controller has demonstrated its ability to cope in extreme situations. This process focused on the improvement of ADASs' ACC feature. An intelligent robust adaptive exponential-like sliding mode was offered by [19]. A type-2 fuzzy neural network (T2FNN) was proposed to estimate the unidentified vehicle dynamics. The T2FNN-based hierarchical controller aiming at the closed-loop stability of the system has guaranteed the convergence of the heading error to zero. Intelligent high-speed trains incorporating backstepping control and deep learning were investigated in [20]. A long short-term memory network (LSTM) merged with a fully connected neural network has been utilized. The approach has presented an excellent follow-up of the high-speed profile while providing a suitable steady-state separation within the safety band. This technique aims at improving the ACC feature. To innovate within the scope of the same function, a hybrid structure combining the PID, MPC, and adaptive-network fuzzy inference system (ANFIS) was provided in [21]. This technique exhibited a good indicator in terms of acceleration comfort due to being less aggressive than the other controllers involved in this study; moreover, its tracking error reached 95% of the reference. Beyond that, the ANFIS controller showed an average error of 0.07 m/s^2 when the acceleration was at its maximum. The authors of [22] proposed a multi-objective model predictive control (MPC) design targeting improving a vehicle's active suspension control. The proposed approach has shown promising results through experimental tests related to the suspension system on a quarter car in laboratory conditions. A combination of MPC, feedforward, and PID feedback was discussed in [23]. MPC was designed to determine longitudinal acceleration under the driver's expectations. The built-in PID feedback and anticipation were adopted to ensure the desired longitudinal acceleration tracking by acting on the longitudinal actuators; this approach aimed to contribute to the ACC feature of ADASs. Along the same lines, [24] discussed a self-organizing neuro-fuzzy controller. This intelligent technique incorporated a neural network and a fuzzy system to automatically adapt the vehicle's speed by learning from the human pilot and utilizing anticipation. This process targeted the innovation at the level of the ISA feature of the ADAS. The authors of [25] sought to address the contribution to the autonomous emergency braking (AEB) feature through the proposed approach of a PID and a Kalman filter. This strategy aimed to determine adequate speed profiles to prevent collisions.

For low-friction roads, a single neural PID combined with a Kalman filter was offered to find the maximum deceleration, whereas for high-friction roads the desired deceleration was achieved according to comfort and safety. This approach was presented as a non-traditional AEB model. The authors of [26] discussed a radial basis functions neural network (RBNN)-based SMC control structure that contributed to both ACC and ISA. In this study, the upper limit of disturbance was estimated with the RBNN. On this basis, the SMC was deployed to design a longitudinal velocity controller. The combination has proven its superiority, with an average tracking error of 0.718 kph in various driving conditions. Moreover, the RBNN was deployed by [27] in order to design an intelligent braking system while contributing to the brake assist system (BA) as an ADAS feature. A self-adaptive PID using the RBNN was proposed by [28] for improving the ACC and stop-and-go features. The vehicle dynamic model utilized in this study included an engine model, a transmission system model, and a brake model. This approach has proven to be very reliable, with low-speed tracking errors of less than 0.203 (m/s), which leads to better driving comfort. In addition, a hybrid solution involving the mentioned NN algorithm and a fuzzy system was investigated in [29] by designing an ANFIS strategy to diagnose centrifugal pumps' fault types.

Famous for its ability to approximate and identify nonlinear quantities in modeling and control, the RBNN is widely involved in different aspects of AGV control. The vast majority of these studies focused on contributing to lateral stability issues [1,18,30,31]. Notwithstanding its utility for longitudinal control purposes, it remains underutilized, with a tiny portion of the research addressing the longitudinal control of the AGVs. Consequently, this study was intended to bridge this gap by suggesting a novel robust intelligent control technique for addressing AGV longitudinal dynamics.

This study proposes an adaptive SMC law taking advantage of the RBNN for the robust stability of AGV longitudinal dynamics, considering uncertainties and external disturbances. The salient distinguishing points of this research are as follows:

- The proposed design contributes to the ADASs in general and ACC with ISA in particular.
- An adaptive control design provides disturbance rejection and uncertainty compensation.
- The novel robust law is a function of the upper bounds of lumped uncertainties and does not require any prior knowledge of the latter.
- An immediate adaptation identifies the upper bounds of perturbations and uncertainties via estimating the neural architecture's parameters.

The current study is arranged as follows: The vehicle modeling is displayed in Section 2. The proposed control strategy, with the necessary derivations and stability proof, is specified in Section 3. Section 4 exemplifies the performance of the suggested method through a simulation study. A conclusion is then reached in Section 5.

2. Vehicle Modeling

As a vehicle is in motion, it is exposed to the following extraneous longitudinal forces: longitudinal tire grip forces, aerodynamic drag forces, climbing resistance forces, and rolling resistance forces (see Figure 1).

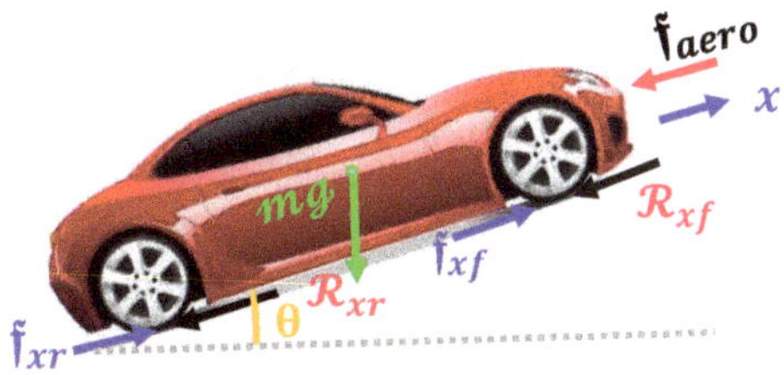

Figure 1. Vehicle longitudinal dynamics.

The aggregate effect of the forces along the vehicle's longitudinal axis is expressed as follows:

$$m\ddot{x} = \mathfrak{f}_{xf} + \mathfrak{f}_{xr} - \mathfrak{f}_{aero} - \mathcal{R}_{xf} - \mathcal{R}_{xr} - mgsin(\theta) \quad (1)$$

where $\mathfrak{f}_{xf}$ and $\mathfrak{f}_{xr}$ are the longitudinal forces arising from the encounter between the tire and the road, $\mathfrak{f}_{aero}$ is the longitudinal aerodynamic drag force, and $\mathcal{R}_{xf}$ and $\mathcal{R}_{xr}$ are the forces resulting from the rolling resistance of the front and rear tires, respectively. m is the vehicle mass, g is the acceleration caused by gravity, and θ is the road's angle of inclination.

2.1. Longitudinal Aerodynamic Drag Force

The following expression describes the aerodynamic drag force:

$$\mathfrak{f}_{aero} = \frac{1}{2}\rho c_{ad}\mathcal{A}_f(v_w \pm v_x)^2 \quad (2)$$

where ρ is the air density, c_{ad} is the coefficient of aerodynamic drag, $\mathcal{A}_f$ is the area of the vehicle's front surface, v_w is the wind velocity, and v_x denotes the vehicle's longitudinal velocity.

2.2. Rolling Resistance

Typically, rolling resistance is denominated as having an approximate proportional dependence on the vertical force exerted on each tire [32].

$$\mathcal{R}_{xf} = \mathcal{F}\left(\mathfrak{f}_{zf}, r_{sat}, v_x\right) \quad (3)$$

$$\mathcal{R}_{xr} = \mathcal{F}(\mathfrak{f}_{zr}, r_{sat}, v_x) \quad (4)$$

where $\mathfrak{f}_{zf}$ and $\mathfrak{f}_{zr}$ are the vertical forces and r_{sat} is the tire's statically loaded radius.

2.3. Longitudinal Tire Forces

The literature provides a spectrum of the longitudinal force of nonlinear tire models such as Pacejka, Dugoff, and others [32,33]. These models were developed via half-empirical or empirical studies that have given a mathematical characterization of the tire/road contact forces along the longitudinal direction and are expressed in terms of the slip rate, the coefficient of friction, and the normal load.

Pacejcka's model remains the most widely used model since it gives an in-depth description of the tire–road interaction via the magic formula described below:

$$\mathfrak{f}_{xf} = \mu\mathfrak{f}_{zf,r}\sin\left(C_{x_{f,r}}\tan^{-1}\left(B_{x_{f,r}}\mathcal{S}_{x_{f,r}} - E_{x_{fr}}\left(\begin{array}{c} B_{x_{f,r}}\mathcal{S}_{x_{f,r}} \\ -\tan^{-1}B_{x_{f,r}}\mathcal{S}_{x_{f,r}} \end{array}\right)\right)\right) \quad (5)$$

where μ is the friction coefficient, $\mathfrak{f}_z$ indicates the vertical forces, and $\mathcal{S}_{xf}$ and $\mathcal{S}_{xr}$ denote the front and rear longitudinal pure slip ratio, respectively. For a complete overview, the reader can follow [34,35].

Using Equation (5), the nonlinear behavior of the longitudinal contact forces is plotted in Figure 2 for different vertical forces and a range of longitudinal slip ratio values from −100% to +100%. This figure demonstrates that the relation between the longitudinal contact forces and the longitudinal slip is linear in the area indicated by the two dotted straight lines. The proportionality coefficients $\mathfrak{C}_{xf}$ and $\mathfrak{C}_{xr}$ are the longitudinal tire stiffness values related to the front and rear tires. These linear relationships, which will be opted for in the following section, can be expressed as:

$$\mathfrak{f}_{xf} = \mathfrak{C}_{xf}\mathcal{S}_{xf} \tag{6}$$

$$\mathfrak{f}_{xr} = \mathfrak{C}_{xr}\mathcal{S}_{xr} \tag{7}$$

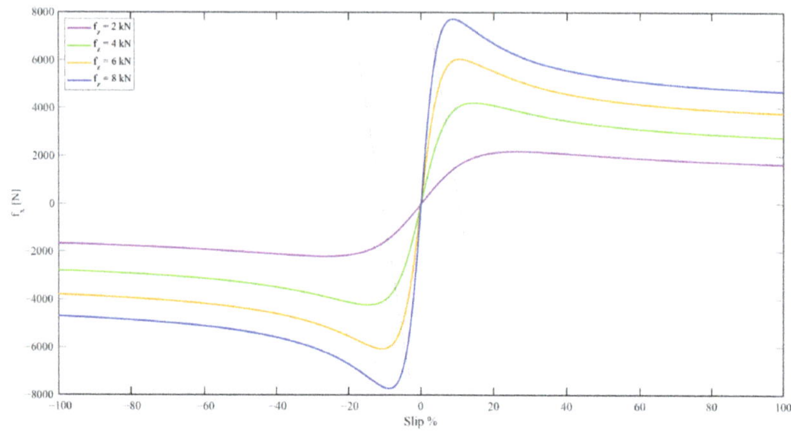

Figure 2. The vehicle's longitudinal tire–road contact forces.

2.4. Simplified Vehicle Longitudinal Model

To synthesize the controllers, the bicycle model, which presents a simplified version of the four-wheel model, is often used (see Figure 3).

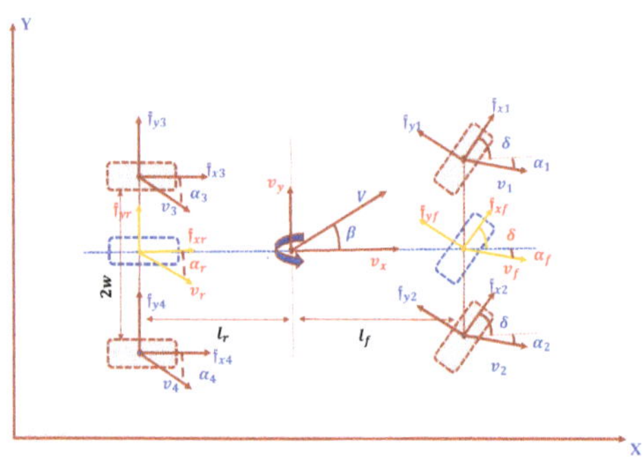

Figure 3. Four-wheel vehicle model and bicycle model.

By considering only of the vehicle's front wheel, using Equation (1) the bicycle model front-wheel dynamics are expressed as [34]:

$$\mathfrak{I}_{\mathcal{W}}\dot{\mathcal{W}}_f = -\mathfrak{f}_{xf}\,\mathfrak{r}_w + \mathcal{T}_p - \mathcal{T}_b \tag{8}$$

where $\mathcal{W}_f$ is front-wheel rotation speed, $\mathfrak{I}_{\mathcal{W}}$ is the front-wheel moment of inertia, $\mathfrak{r}_w$ is the front-wheel radius, and $\mathcal{T}_p$ and $\mathcal{T}_b$ are the propulsion and braking torque, respectively.

From Equation (9), we can write $\mathfrak{f}_{xf}$ as:

$$\mathfrak{f}_{xf} = \frac{1}{\mathfrak{r}_w}\left(\mathcal{T}_p - \mathcal{T}_b - \mathfrak{I}_{\mathcal{W}}\dot{\mathcal{W}}_f\right) \tag{9}$$

Based on the hypothesis defined in [34], Equations (10) and (11) are given:

$$v_x = \mathfrak{r}_w \mathcal{W}_f$$

$$\mathcal{W}_f = \mathfrak{r}_g \mathcal{W}_e \quad (10)$$

$$\mathcal{T}_e = \mathfrak{r}_g \mathcal{T}_p \quad (11)$$

where $\mathfrak{r}_g$ denotes the transmission gear ratio, $\mathcal{W}_e$ is the engine speed, and $\mathcal{T}_e$ is the engine torque.

Based on Equations (9)–(11), Equation (1), which represents the longitudinal vehicle dynamics, is written as:

$$\ddot{x} = -\frac{\mathfrak{r}_w}{m\mathfrak{r}_w^2 + \mathfrak{l}_W}\left(\mathcal{T}_p + \mathfrak{r}_w\left(\mathfrak{f}_{aero} + \mathcal{R}_{xf} + mg\sin\theta\right)\right) + \frac{\mathfrak{r}_w}{\mathfrak{r}_g(m\mathfrak{r}_w^2 + \mathfrak{l}_W)}\mathcal{T}_e \quad (12)$$

3. Controller Development

3.1. Motivation

In the nonlinear systems control area, neural network controllers (NNCs) are generally made by a combination of classical or modern control techniques and neural network algorithms. This hybrid structure was initially proposed by Kawato et al. and Psaltis et al. in 1988 [36–38]. Since then, a torrent of NNCs have been introduced in the literature, especially in robotics [39–41]. Based on the ANN algorithms, the NNCs are often used to identify or approximate unmodeled dynamics, nonlinearities, unmeasurable noise, and multi-loops [42]. However, for autonomous vehicle control, the challenges of designing an intelligent control structure are attracting considerable attention from researchers. In this context, a design of longitudinal control approaches based on neural networks is elaborated in the following subsections.

3.2. Problem Formulation

In view of preparing the vehicle model for the controller development as well as the expression arrangement, the following re-expression was chosen:

$$\begin{aligned} \dot{x}_1 &= \dot{x} \\ \dot{x}_2 &= \mathcal{F}(x,t) + \mathfrak{G}(x,t)\mathfrak{U} + \mathfrak{D}(x,t) \end{aligned} \quad (13)$$

with

$$\begin{aligned} \mathcal{F}(x,t) &= -\frac{\mathfrak{r}_w}{m\mathfrak{r}_w^2+\mathfrak{l}_W}\left(\mathcal{T}_p + \mathfrak{r}_w\left(\tfrac{1}{2}\rho c_{ad}\dot{x}_1{}^2 + \mathcal{R}_{xf} + mgsin\theta\right)\right) \\ \mathfrak{G}(x,t) &= \frac{\mathfrak{r}_w}{\mathfrak{r}_g(m\mathfrak{r}_w^2+\mathfrak{l}_W)} \end{aligned}$$

Remark 1. *The $\mathfrak{G}$ function depends on the vehicle's speed via the change in the transmission ratio ($\mathfrak{r}_g$), as described in Equations (11) and (12). Therefore, it is set, $\mathfrak{G}(x,t)$, as a function of longitudinal position and time.*

x is the longitudinal position, $\mathfrak{U}$ is the control input, and $\mathfrak{D}(x,t)$ is the vehicle uncertainty and the outdoor disturbances. This lumped uncertainty of the vehicle, $\mathfrak{D}(x,t)$, is considered to be constrained by a positive upper limit value ($\mathfrak{M}$), as follows:

$$\mathfrak{D}(x,t) \le \mathfrak{M} \quad (14)$$

3.3. Robust Adaptive RBNN-SMC

3.3.1. Sliding Surfaces

Based on the tracking error (e), the sliding surface ($\mathcal{Z}$) is given as:

$$\begin{aligned} e(t) &= x_d - x \\ \mathcal{Z}(t) &= \mathfrak{p}e + \dot{e} \end{aligned} \quad (15)$$

where $\mathfrak{p}$ is a positive constant.

3.3.2. Equivalent Control Law

After forcing the sliding surface's time derivative to zero, the equivalent law is given as

$$\mathfrak{U}_{eq} = -\frac{\mathfrak{p}\dot{e} + \mathcal{F}(x,t)}{\mathfrak{S}(x,t)} \tag{16}$$

3.3.3. Robust Law

To accommodate disturbances and parametric uncertainties, a robust law is given in terms of the upper limit of lumped uncertainty, as follows:

$$\mathfrak{U}_r = \frac{1}{\mathfrak{S}(x,t)}(-q\mathcal{Z} - \mathcal{K}sign(\mathcal{Z})) \tag{17}$$

where q and $\mathcal{K}$ indicate the switching gains and $\mathcal{K}$ is chosen as [43,44] $\mathcal{K} = \mathfrak{M} + o$, where $\mathfrak{M}$ is the upper limit of the lumped uncertainty, given in [43], and o is a positive constant.

The idea of designing a control law considering the upper limits of disturbances and uncertainties is insufficient to guarantee high performance against real non-static disturbances. To solve this problem, a neuronal architecture was adopted to approximate the upper limit ($\mathfrak{M}$) depending on the variation in these critical factors. Hence, the subdivision of the robust law ($\mathfrak{U}_r$) into two terms, $\mathfrak{U}_{sw}$ and $\mathfrak{U}_{rbnn}$, was as follows:

$$\begin{cases} \mathfrak{U}_r = \mathfrak{U}_{sw} + \mathfrak{U}_{rbnn} \\ \mathfrak{U}_{sw} = \frac{1}{\mathfrak{S}(x,t)}(-q\mathcal{Z} - osign(\mathcal{Z})) \\ \mathfrak{U}_{rbnn} = \frac{-\mathfrak{M}sign(\mathcal{Z})}{\mathfrak{S}(x,t)} \end{cases} \tag{18}$$

As it is expressed in the Equation (18), the term $\mathfrak{U}_{rbnn}$ contains the approximation of the lumped uncertainty upper bound by the RBNN, as described in Figure 4.

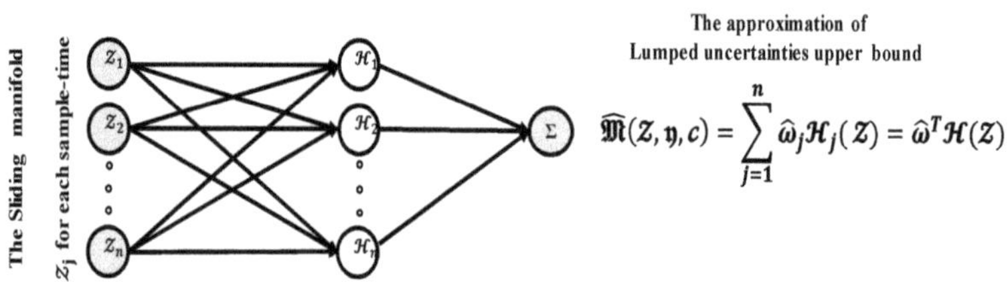

Figure 4. The RBNN structure.

$\hat{\mathfrak{M}}$ is the output of RBNN and presents the estimation of $\mathfrak{M}$ according to the expression

$$\hat{\mathfrak{M}}(\mathcal{Z},\mathfrak{y},c) = \sum_{j=1}^{n} \hat{\omega}_j\mathcal{H}_j(\mathcal{Z}) = \hat{\omega}^T\mathcal{H}(\mathcal{Z}) \tag{19}$$

where $\mathcal{H}$ is the Gaussian activation function constituted by the radial basis function defined as

$$\mathcal{H}_j(\mathcal{Z}) = e^{\left(-\frac{(\mathcal{Z}-\mathfrak{y}_j)^2}{c_j^2}\right)},\ j = 1,2\ldots n \tag{20}$$

The input of the RBNN structure is the sliding variable $\mathcal{Z}$. Each hidden node has a center parameter vector and an input vector of the same dimension.

On the other hand, $\mathfrak{M}$ can be expressed using RBNN's ideal parameters ω^*, $\mathfrak{y}^*$, and c^*, given as [42,45]:

$$\mathfrak{M} = \mathfrak{M}^* + \mathfrak{I}_1 = \omega^{*T}\mathcal{H}^*(\mathcal{Z}) + \mathfrak{I}_2 \tag{21}$$

$\beth_1$ denotes the approximation error of the RBNN, which is assumed to be limited by the non-zero positive constant $\beth_n$ as $|\beth_1| < \beth_n$ [42,45].

The estimated weight vector ($\hat{\omega}$), the estimated center vector ($\hat{\mathfrak{h}}$), and the estimated width vector ($\hat{c}$) are defined as

$$\hat{\omega} = \begin{bmatrix} \hat{\omega}_1 \\ \hat{\omega}_2 \\ .. \\ \hat{\omega}_n \end{bmatrix}; \mathcal{H} = \begin{bmatrix} \mathcal{H}_1 \\ \mathcal{H}_2 \\ .. \\ \mathcal{H}_n \end{bmatrix}; \hat{c} = \begin{bmatrix} \hat{c}_1 \\ \hat{c}_2 \\ .. \\ \hat{c}_n \end{bmatrix} \tag{22}$$

where n is the hidden-layer node number.

Obtaining the expression allows the adaptation of the parameters $\hat{\omega}$, $\hat{\mathfrak{h}}$, and $\hat{c}$, which provides the adjustment of $\mathfrak{M}$ according to external factors and uncertainties. To achieve this, the Gaussian function is linearized as in [46]. This leads to the differentiation of Equation (20) as:

$$\Delta\hat{\mathfrak{M}}_j(\mathcal{Z}) = \Delta\hat{\omega}_j \mathcal{H}_j(\mathcal{Z}) + \hat{\omega}_j \frac{\partial \mathcal{H}_j}{\partial \hat{\mathfrak{h}}_j} \Delta\hat{\mathfrak{h}}_j + \hat{\omega}_j \frac{\partial \mathcal{H}_j}{\partial \hat{c}_j} \Delta\hat{c}_j + \beth_{2j} \tag{23}$$

The linearization neglects the high-order terms of the Taylor expansion of $\hat{\mathfrak{M}}(\mathcal{Z})$, which causes a $\beth_2$ error quantity of $\beth_2 = \sum_{j=1}^{n} \beth_{2j}$, which is assumed to be constrained by the positive constant $\beth_n$ as $|\beth_2| < \beth_{2n}$ [46].

The vectors of the partial derivative of $\mathcal{H}$ with respect to $\hat{\mathfrak{h}}$ and $\hat{c}$, $\frac{\partial \mathcal{H}_j}{\partial \hat{\mathfrak{h}}_j}$ and $\frac{\partial \mathcal{H}_j}{\partial \hat{c}_j}$, are expressed as

$$\mathcal{H}'_{\hat{c}} = \begin{bmatrix} \frac{\partial \mathcal{H}_1}{\partial \hat{c}_1} & 0 & \cdots & 0 \\ 0 & \frac{\partial \mathcal{H}_2}{\partial \hat{c}_2} & \cdots & 0 \\ \vdots & 0 & \ddots & \vdots \\ 0 & 0 & \cdots & \frac{\partial \mathcal{H}_n}{\partial \hat{c}_n} \end{bmatrix} \in \mathbb{R}^{n \times n}$$

$$\mathcal{H}'_{\hat{\mathfrak{h}}} = \begin{bmatrix} \frac{\partial \mathcal{H}_1}{\partial \hat{\mathfrak{h}}_1} & 0 & \cdots & 0 \\ 0 & \frac{\partial \mathcal{H}_2}{\partial \hat{\mathfrak{h}}_2} & \cdots & 0 \\ \vdots & 0 & \ddots & \vdots \\ 0 & 0 & \cdots & \frac{\partial \mathcal{H}_n}{\partial \hat{\mathfrak{h}}_n} \end{bmatrix} \in \mathbb{R}^{n \times n}$$

$\widetilde{\omega}$, $\widetilde{\mathfrak{h}}$, and $\widetilde{c}$ denote the estimation error as

$$\begin{cases} \widetilde{\mathfrak{M}} = \mathfrak{M}^* - \hat{\mathfrak{M}} \\ \widetilde{\omega} = \omega^* - \hat{\omega} \\ \widetilde{\mathfrak{h}} = \mathfrak{h}^* - \hat{\mathfrak{h}} \\ \widetilde{c} = c^* - \hat{c} \end{cases} \tag{24}$$

Subsequently, $\Delta\widetilde{\mathfrak{M}}$ can be written as

$$\Delta\widetilde{\mathfrak{M}} = \mathfrak{M}^* - \hat{\mathfrak{M}} + \beth_1 = \Delta\hat{\mathfrak{M}} + \beth_1 \tag{25}$$

where $\Delta\hat{\mathfrak{M}} = \sum_{j=1}^{n} \Delta\hat{\mathfrak{M}}(\mathcal{Z})$.

Then, based on Equations (23) and (25), we can obtain the following:

$$\widetilde{\mathfrak{M}}(\mathcal{Z}) = \widetilde{\omega}^T \mathcal{H}(\mathcal{Z}) + \omega^T \mathcal{H}'_{\hat{\mathfrak{h}}}(\widetilde{\mathfrak{h}})\widetilde{\mathfrak{h}} + \omega^T \mathcal{H}'_{\hat{c}}(\widetilde{c})\widetilde{c} + \beth_1 + \beth_2 \tag{26}$$

Under the condition $|\beth_1 + \beth_2| \leq o$, the parameters $\hat{\omega}$, $\hat{\mathfrak{h}}$, and $\hat{\mathfrak{c}}$ can be adjusted by following adaptive laws:

$$\begin{cases} \dot{\hat{\omega}} &= -a_1|\mathcal{Z}|H(\mathcal{Z}) \\ \dot{\hat{\mathfrak{h}}} &= -a_2|\mathcal{Z}|\mathcal{H}'_{\hat{\mathfrak{h}}}\omega \\ \dot{\hat{c}} &= -a_3|\mathcal{Z}|\mathcal{H}'_{\hat{c}}\omega \end{cases} \begin{cases} \dot{\hat{\omega}}_j = -a_1|\mathcal{Z}|\mathcal{H}_j(\mathcal{Z}) \\ \dot{\hat{\mathfrak{h}}}_j = -a_2|\mathcal{Z}|\frac{\partial \mathcal{H}_j}{\partial \hat{\mathfrak{h}}_j}\omega_j \\ \dot{\hat{c}}_j = -a_3|\mathcal{Z}|\frac{\partial \mathcal{H}_j}{\partial \hat{c}_j}\omega_j \end{cases} \tag{27}$$

where $a_k, k = \{1, 2, 3\}$ are non-zero-positive constants.

Proposition 1. *Considering the system expressed by Equation (13), the lumped uncertainties are coherent with the inequality* $\mathfrak{D}(x,t) \leq \mathfrak{M}$.

If the sliding surface is set as

$$\mathcal{Z}(t) = \mathfrak{p}e + \dot{e} \tag{28}$$

the control law is defined as

$$\mathfrak{U} = \mathfrak{U}_{eq} + \mathfrak{U}_{sw} + \mathfrak{U}_{rbnn} \tag{29}$$

with

$$\begin{aligned} \mathfrak{U}_{eq} &= -\frac{\mathfrak{p}\dot{e} + \mathcal{F}(x,t)}{\mathfrak{G}(x,t)} \\ \mathfrak{U}_{sw} &= \frac{1}{\mathfrak{G}(x,t)}(-q\mathcal{Z} - o\,sign(\mathcal{Z})) \\ \mathfrak{U}_{rbnn} &= \frac{-\hat{\mathfrak{M}}sign(\mathcal{Z})}{\mathfrak{G}(x,t)} \end{aligned}$$

Accordingly, by choosing this sliding surface, given this control law, the tracking error converges to zero.

Proof. Consider the candidate Lyapunov function as

$$\mathcal{L} = \frac{1}{2}\left(\mathcal{Z}^2 + \frac{1}{a_1}\tilde{\omega}^T\tilde{\omega} + \frac{1}{a_2}\tilde{\mathfrak{h}}^T\tilde{\mathfrak{h}} + \frac{1}{a_3}\tilde{c}^T\tilde{c}\right) \tag{30}$$

The time derivative of $\mathcal{L}$ is

$$\dot{\mathcal{L}} = \mathcal{Z}\dot{\mathcal{Z}} + \frac{1}{a_1}\tilde{\omega}^T\dot{\tilde{\omega}} + \frac{1}{a_2}\tilde{\mathfrak{h}}^T\dot{\tilde{\mathfrak{h}}} + \frac{1}{a_3}\tilde{c}^T\dot{\tilde{c}}$$

$$\dot{\mathcal{L}} = \mathcal{Z}(\mathfrak{p}\dot{e} + \mathcal{F}(x,t) + \mathfrak{G}(x,t)\mathfrak{U} + \mathfrak{D}(x,t)) + \frac{1}{a_1}\tilde{\omega}^T\dot{\tilde{\omega}} + \frac{1}{a_2}\tilde{\mathfrak{h}}^T\dot{\tilde{\mathfrak{h}}} + \frac{1}{a_3}\tilde{c}^T\dot{\tilde{c}} \tag{31}$$

By substituting (29) into (31), $\dot{\mathcal{L}}$ becomes:

$$\dot{\mathcal{L}} = \mathcal{Z}(\mathfrak{D}(x,t) - q\mathcal{Z} - (\hat{\mathfrak{M}} + o)sign(\mathcal{Z})) + \frac{1}{a_1}\tilde{\omega}^T\dot{\tilde{\omega}} + \frac{1}{a_2}\tilde{\mathfrak{h}}^T\dot{\tilde{\mathfrak{h}}} + \frac{1}{a_3}\tilde{c}^T\dot{\tilde{c}} \tag{32}$$

□

Remark 2. *It is important to mention this mathematical property:* $\forall\ (\alpha, \beta) \in \mathbb{R}^2, (\alpha + \beta) \leq |\alpha + \beta| \leq |\alpha| + |\beta|$.

Using Remark 2, we can obtain the following:

$$\dot{\mathcal{L}} \leq (|\mathfrak{D}(\mathcal{X},t)| - |\hat{\mathfrak{M}}|)|\mathcal{Z}| - q\mathcal{Z}^2 - o|\mathcal{Z}| + \left|\frac{1}{a_1}\tilde{\omega}^T\dot{\tilde{\omega}} + \frac{1}{a_2}\tilde{\mathfrak{h}}^T\dot{\tilde{\mathfrak{h}}} + \frac{1}{a_3}\tilde{c}^T\dot{\tilde{c}}\right| \tag{33}$$

From Equations (25) and (26), we can obtain the following:

$$\begin{aligned}\dot{\mathcal{L}} &\leq -q\mathcal{Z}^2 - o|\mathcal{Z}| + \left|\widetilde{\mathfrak{M}}\right||\mathcal{Z}| + \left|\tfrac{1}{q_1}\widetilde{\omega}^T\dot{\widetilde{\omega}} + \tfrac{1}{q_2}\widetilde{\mathfrak{y}}^T\dot{\widetilde{\mathfrak{y}}} + \tfrac{1}{q_3}\widetilde{c}^T\dot{\widetilde{c}}\right| \\ &= -q\mathcal{Z}^2 - o|\mathcal{Z}| + \left|\widetilde{\omega}^T\mathcal{H}(\mathcal{Z}) + \omega^T\mathcal{H}'_{\hat{\mathfrak{y}}}(\widetilde{\mathfrak{y}})\widetilde{\mathfrak{y}} + \omega^T\mathcal{H}'_{\hat{c}}(\widetilde{c})\widetilde{c} + \daleth_1 + \daleth_2\right||\mathcal{Z}| \\ &+ \left|\tfrac{1}{a_1}\widetilde{\omega}^T\dot{\widetilde{\omega}} + \tfrac{1}{a_2}\widetilde{\mathfrak{y}}^T\dot{\widetilde{\mathfrak{y}}} + \tfrac{1}{a_3}\widetilde{c}^T\dot{\widetilde{c}}\right|\end{aligned} \tag{34}$$

According to Equation (24), Equation (34) becomes:

$$\begin{aligned}\dot{\mathcal{L}} &\leq -\mathrm{q}\mathcal{Z}^2 - o|\mathcal{Z}| + |\daleth_1 + \daleth_2||\mathcal{Z}| + |\widetilde{\omega}^T|\left(|\mathcal{H}(\mathcal{Z})||\mathcal{Z}| - \tfrac{1}{\mathrm{a}_1}\left|\dot{\hat{\omega}}\right|\right) \\ &+ |\widetilde{\mathfrak{y}}^T|\left(\omega^T\left|\mathcal{H}'_{\hat{\mathfrak{y}}}(\widetilde{\mathfrak{y}})\right||\mathcal{Z}| - \tfrac{1}{\mathrm{a}_2}\left|\dot{\hat{\mathfrak{y}}}\right|\right) + \widetilde{\mathrm{c}}^T\left(\omega^T\mathcal{H}'_{\hat{\mathrm{c}}}(\widetilde{\mathrm{c}})|\mathcal{Z}| - \tfrac{1}{\mathrm{a}_3}\left|\dot{\widetilde{c}}\right|\right)\end{aligned} \tag{35}$$

From Equation (27), we can achieve:

$$\begin{aligned}\dot{\mathcal{L}} &\leq -\mathrm{q}\mathcal{Z}^2 - o|\mathcal{Z}| + |\daleth_1 + \daleth_2||\mathcal{Z}| \\ &= -q\mathcal{Z}^2 - (o - |\daleth_1 + \daleth_2|)|\mathcal{Z}|\end{aligned} \tag{36}$$

Using the condition $|\daleth_1 + \daleth_2| \leq o$ results in $\dot{\mathcal{L}} \leq 0$. Asymptotic stability is guaranteed under Lyapunov's theory.

The proof is thus complete.

To conclude this conceptual section, the new proposed process is outlined visually in Figure 5.

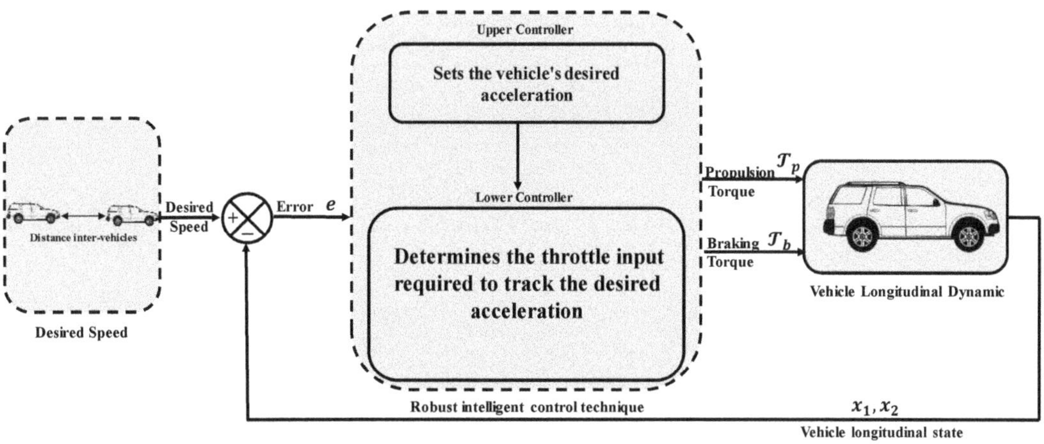

Figure 5. The overall strategy that is suggested.

4. Results and Discussion

The performance of the proposed strategy was evaluated by tracking a speed scenario while accounting for unknown disturbances and uncertainties. A comparison with the super-twisting SMC [12] was made in this context.

Since the tire is the primary ground contact, the parametric uncertainty considered in this simulation was 50% and acted on the front-wheel moment of inertia ($\mathrm{I}_{\mathcal{W}}$) [10,43]. Moreover, the disturbance's nature was random and frequent. The vehicle parameters used were similar to [12]. The proposed control scheme settings were given as follows: $\mathfrak{p} = 0.001$, $o = 0.0001$, $q = 100$, $a_1 = 15$, $a_{1,2} = 0,5$, and $n = 6$.

All simulations were performed using MATLAB R2021a software, with a fixed sample time of 10 ms and an automatic solver on a Lenovo machine characterized by a 2.7 GHz processor, 16 GB of RAM, and a 512 GB SSD.

The simulation outcomes of the proposed method and that of [12] are presented in Figures 6 and 7.

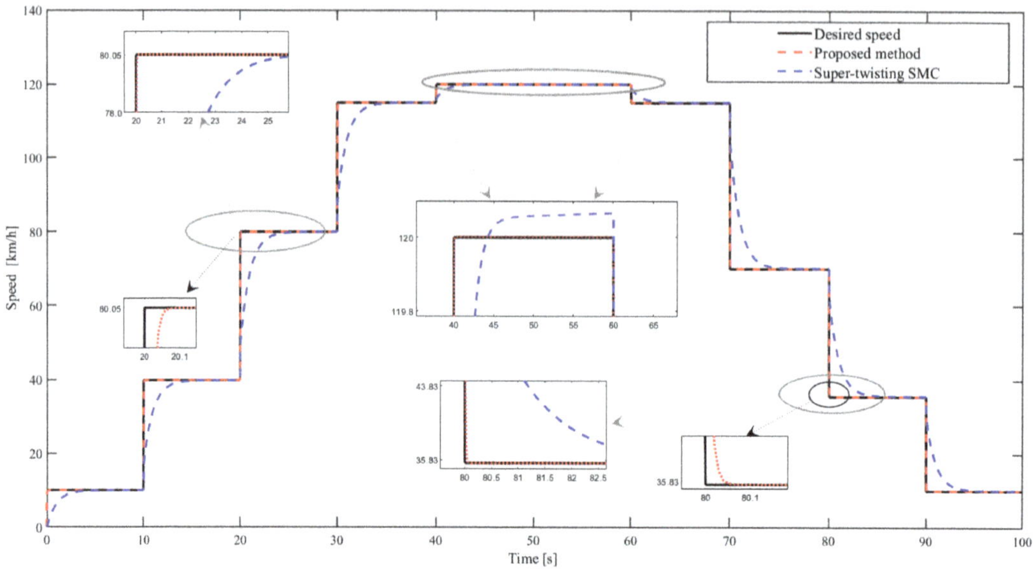

Figure 6. Speed profile tracking responses.

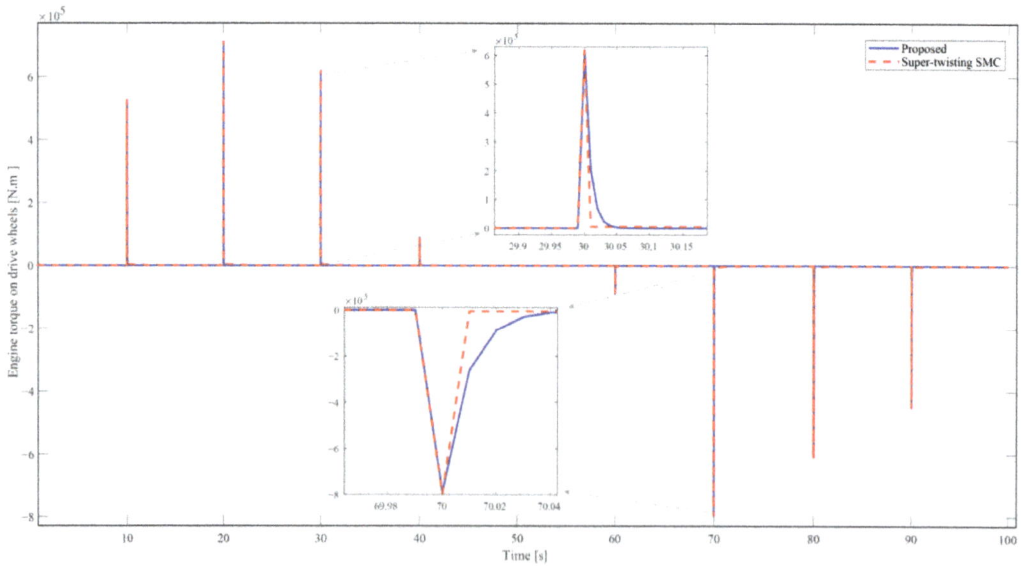

Figure 7. Controller output.

In Figure 6, it can be observed that the suggested longitudinal controller directed the vehicle speed to perfectly reach the desired speed. However, at first sight, the simulation results of the super-twisting SMC approach were not as satisfactory as the proposed approach. To further analyze the performance of both approaches, it was important to study the results in transient and steady states. Indeed, in the transient regime the proposed approach showed zero rise time, whereas the response time was extremely insignificant. However, for the super-twisting SMC the rise time and the response time were highly significant and the overshoot was considerable, even in the steady state, particularly in the speed range, indicating the onset of slow motion. Moreover, a quantitative study was undertaken to underline the offered method's success utilizing the integral of the absolute error ($\mathfrak{iae}$), as written in Equation (37), as a performance index of the absolute error of the longitudinal displacement. These numerical values are presented in Table 1.

$$\mathfrak{iae} = \int_0^{t_f} |e|dt \quad (37)$$

where t_f indicates the final simulation time.

Table 1. Quantitative performance assessment.

Approaches	$\mathfrak{iae}$ Performance Index
Proposed method	0.825
Super-twisting SMC	64.419

Figure 7 shows the engine torque acting on the wheel; this parameter exactly indicates the controller output. In fact, it is obvious that, at each speed step, the discussed control structure reacted simultaneously by imposing a torque that caused the vehicle's wheels to turn.

In the acceleration steps, the torque resulting from the proposed method and that resulting from the super-twisting SMC did not occur in an increasing sense. As can be seen, the torque reached its maximum in the second speed step, and then decreased, returning to a lower value, even when the requested speed in the next step was higher than in the previous step. It can be stated that the controllers reacted intelligently by adding or subtracting the requested quantities to achieve the desired speed. Nevertheless, when the speed decreased, the controllers reacted in a decreasing rhythm, which is logical since the vehicle started to reduce its speed.

The execution of the orders issued by the control algorithm was performed by the actuators, requiring sufficient energy for their operation. To this end, a comparative study gave numerical data regarding the energy consumption of the control strategies involved in this study. The integral of the square value (ISV) of the absolute error of the longitudinal acceleration provided this information related to the energy consumed by the acceleration/braking actuators [43]. The ISV is expressed in Equation (38). These results are illustrated in Table 2.

$$ISV = \int_0^{t_f} |e_{acc}|^2 dt \quad (38)$$

Table 2. Quantitative performance analysis.

Approaches	ISV Performance Index
Proposed method	27.719
Super-twisting SMC	196.8197

Table 2 also shows that the proposed approach is optimal in terms of the power usage by the actuators moving the vehicle forward and backward and braking.

Remark 3. *This note mentions and describes the equipment for the experimental validation of the proposed control approach for a potential implementation on the lateral level. The following are the principal elements, shown in Figure 8, involved in the process of carrying out the field experiment:*

1. *A Toyota Prius first-level autonomous vehicle*
2. *An automatic steering wheel robot.*
3. *A Lenovo computer.*

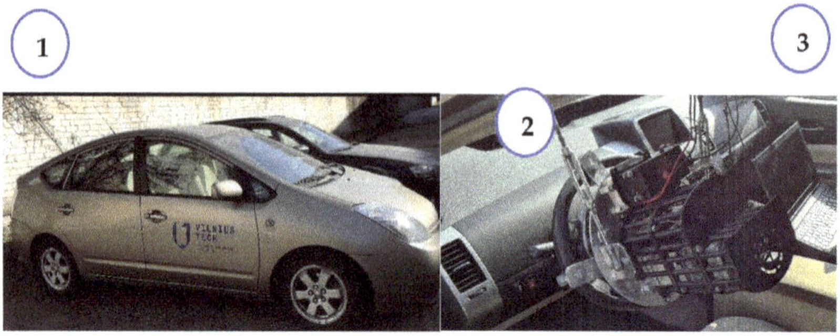

(a) (b)

Figure 8. (**a**) The prototype of the first-level autonomous ground vehicle. (**b**) The automated steering device with the control computer.

5. Conclusions

ADAS are vehicle control systems that improve driving comfort and road safety by using the vehicle's sensing capabilities and control algorithms. In this sense, this work has contributed to longitudinal autonomous control. Throughout this research, a novel approach based on a conventional sliding mode and RBNN has been developed, verified theoretically in the sense of Lyapunov, and validated in a simulation compared to the super-twisting SMC. Accordingly, it is necessary to summarize the following conclusions:

- Offering a robust law considering the disturbance and uncertainty upper limit as well as adapting it to each sampling time has been proven to provide outstanding results over and above the super-twisting SMC that often dominated the conventional SMC in terms of performance.
- Providing an algorithm that was able to estimate the upper limit of uncertainties and disturbances allowed the proposed design scheme to be utilized with no need for information regarding upper-limit external disturbances.
- Orienting the proposed design toward a real nonlinear system, such as longitudinal vehicle dynamics, can not only confirm the proposed method's sovereignty but can also be embedded to contribute to the ADASs in general and ACC and ISA in particular.

In short, since the proposed technique has revealed its superiority, this will lead to further examination and testing using other driving conditions and scenarios while performing this technique for lateral control objectives, followed by implementation and testing using the above-mentioned steering robot, all of which constitute the perspectives of this manuscript.

Author Contributions: Conceptualization, M.B. and I.B.; methodology, L.E.H. and E.M.M.; formal analysis, E.M.M. and V.Ž.; investigation, L.E.H. and E.M.M.; writing—original draft, L.E.H.; writing—review & editing, V.Ž. and E.M.M.; supervision, M.B. and I.B. All authors have read and agreed to the published version of the manuscript.

Funding: This research received no external funding.

Institutional Review Board Statement: Not applicable.

Informed Consent Statement: Not applicable.

Data Availability Statement: Data available by request from vidas.zuraulis@vilniustech.lt.

Conflicts of Interest: The authors declare no conflict of interest.

References

1. El Hajjami, L.; Mellouli, E.M.; Berrada, M. Neural network based sliding mode lateral control for autonomous vehicle. In Proceedings of the 2020 1st International Conference on Innovative Research in Applied Science, Engineering and Technology (IRASET), Meknes, Morocco, 16–19 April 2020; pp. 1–6.
2. Cao, H.; Song, X.; Huang, Z.; Pan, L. Simulation research on emergency path planning of an active collision avoidance system combined with longitudinal control for an autonomous vehicle. *Proc. Inst. Mech. Eng. Part J. Automob. Eng.* **2016**, *230*, 1624–1653. [CrossRef]
3. Shladover, S.E.; Desoer, C.A.; Hedrick, J.K.; Tomizuka, M.; Walrand, J.; Zhang, W.-B.; Mcmahon, D.H.; Peng, H.; Sheikholeslam, S.; Mckeown, N. Automated vehicle control developments in the path program. *IEEE Trans. Veh. Technol.* **1991**, *40*, 114–130. [CrossRef]
4. Skrickij, V.; Šabanovič, E.; Žuraulis, V. Autonomous road vehicles: Recent issues and expectations. *IET Intell. Transp. Syst.* **2020**, *14*, 471–479. [CrossRef]
5. Hassani, H.; Mansouri, A.; Ahaitouf, A. Robust finite-time tracking control based on disturbance observer for an uncertain quadrotor under external disturbances. *J. Robot.* **2022**, *2022*, 1–20. [CrossRef]
6. Mellouli, E.M.; Boumhidi, I. Direct adaptive fuzzy sliding mode controller without reaching phase for an uncertain three-tank-system. *Int. J. Model. Identif. Control* **2016**, *25*, 335–342.
7. Li, A.; Niu, C.; Li, S.; Huang, X.; Xu, C.; Liu, G. Research on intelligent vehicle trajectory planning and control based on an improved terminal sliding mode. *Appl. Sci.* **2022**, *12*, 2446. [CrossRef]
8. Du, Y.; Jiang, B.; Ma, Y.; Cheng, Y. Robust adp-based sliding-mode fault-tolerant control for nonlinear systems with application to spacecraft. *Appl. Sci.* **2022**, *12*, 1673. [CrossRef]
9. El Hajjami, L.; Mellouli, E.M.; Berrada, M. Autonomous vehicle lateral control for the lane-change maneuver. In *WITS 2020*; Springer: Berlin/Heidelberg, Germany, 2022; pp. 285–295.
10. El Hajjami, L.; Mellouli, E.M.; Berrada, M. Robust adaptive non-singular fast terminal sliding-mode lateral control for an uncertain ego vehicle at the lane-change maneuver subjected to abrupt change. *Int. J. Dyn. Control* **2021**, *9*, 1765–1782. [CrossRef]
11. Nouveliere, L. Experimental vehicle longitudinal control using a second order sliding mode technique. *Control Eng. Pract.* **2007**, *15*, 943–954. [CrossRef]
12. El Hajjami, L.; Mellouli, E.M.; Žuraulis, V.; Berrada, M. Vehicle adaptive cruise controller based on an optimal super-twisting sliding mode control. In Proceedings of the 2022 2nd International Conference of Smart Systems and Emerging Technologies (SMARTTECH), Riyadh, Saudi Arabia, 9–11 May 2022; pp. 160–165.
13. Xu, Y.; Lu, Z.; Shan, X.; Jia, W.; Wei, B.; Wang, Y. Study on an automatic parking method based on the sliding mode variable structure and fuzzy logical control. *Symmetry* **2018**, *10*, 523. [CrossRef]
14. El-Bakkouri, J.; Ouadi, H.; Saad, A. Output feedback control of antilock braking system. *Int. Rev. Autom. Control (IREACO)* **2021**, *14*, 214–223. [CrossRef]
15. Yao, Q.; Tian, Y. A model predictive controller with longitudinal speed compensation for autonomous vehicle path tracking. *Appl. Sci.* **2019**, *9*, 4739. [CrossRef]
16. Moshayedi, A.J.; Li, J.; Sina, N.; Chen, X.; Liao, L.; Gheisari, M.; Xie, X. Simulation and validation of optimized pid controller in agv (automated guided vehicles) model using pso and bas algorithms. *Comput. Intell. Neurosci.* **2022**, *2022*, 1–22. [CrossRef]
17. El Hajjami, L.; Mellouli, E.M.; Berrada, M. Optimal PID control of an autonomous vehicle using butterfly optimization algorithm boa. In Proceedings of the 4th International Conference on Big Data and Internet of Things, Rabat, Morocco, 23–24 October 2019; pp. 1–5.
18. Debarshi, S.; Sundaram, S.; Sundararajan, N. Robust EMRAN-aided coupled controller for autonomous vehicles. *Eng. Appl. Artif. Intell.* **2022**, *110*, 104717. [CrossRef]
19. Taghavifar, H.; Rakheja, S. Path-tracking of autonomous vehicles using a novel adaptive robust exponential-like-sliding-mode fuzzy type-2 neural network controller. *Mech. Syst. Signal Process.* **2019**, *130*, 41–55. [CrossRef]
20. Wang, X.; Li, S.; Cao, Y.; Xin, T.; Yang, L. Dynamic speed trajectory generation and tracking control for autonomous driving of intelligent high-speed trains combining with deep learning and backstepping control methods. *Eng. Appl. Artif. Intell.* **2022**, *115*, 105230. [CrossRef]
21. Marcano, M.; Matute, J.A.; Lattarulo, R.; Martí, E.; Pérez, J. Low Speed longitudinal control algorithms for automated vehicles in simulation and real platforms. *Complexity* **2018**, *2018*, 1–12. [CrossRef]
22. Reddipogu, J.; Elumalai, V. Multi-objective model predictive control for vehicle active suspension system. *Int. Rev. Autom. Control (IREACO)* **2020**, *13*, 255. Available online: https://Www.Praiseworthyprize.Org/Jsm/Index.Php?Journal=Ireaco&Page=Article&Op=View&Path%5B%5D=24828 (accessed on 2 December 2022). [CrossRef]
23. Gao, B.; Cai, K.; Qu, T.; Hu, Y.; Chen, H. Personalized adaptive cruise control based on online driving style recognition technology and model predictive control. *IEEE Trans. Veh. Technol.* **2020**, *69*, 12482–12496. [CrossRef]
24. Partouche, D.; Pasquier, M.; Spalanzani, A. Intelligent speed adaptation using a self-organizing neuro-fuzzy controller. In Proceedings of the 2007 IEEE Intelligent Vehicles Symposium, Istanbul, Turkey, 13–15 June 2007; pp. 846–851.
25. Wang, X.; Wang, J.; Sun, W.; Wang, Y.; Xie, F.; Guo, D. Development of aeb control strategy for autonomous vehicles on snow-asphalt joint pavement. *Int. J. Crashworthiness* **2021**, *27*, 1601–1621. [CrossRef]

26. Jo, A.; Lee, H.; Seo, D.; Yi, K. Model-Reference Adaptive sliding mode control of longitudinal speed tracking for autonomous vehicles. *Proc. Inst. Mech. Eng. Part J. Automob. Eng.* 2022, *Online First*. [CrossRef]
27. He, H.; Wang, C.; Jia, H.; Cui, X. An intelligent braking system composed single-pedal and multi-objective optimization neural network braking control strategies for electric vehicle. *Appl. Energy* **2020**, *259*, 114172. [CrossRef]
28. Nie, L.; Guan, J.; Lu, C.; Zheng, H.; Yin, Z. Longitudinal speed control of autonomous vehicle based on a self-adaptive pid of radial basis function neural network. *IET Intell. Transp. Syst.* **2018**, *12*, 485–494. [CrossRef]
29. Maniyan, H.; Eftekhari, S.A. Automatic Defect Analysis of Pumps Using Adaptive Neuro- Fuzzy Inference System and Vibrational Features. In Proceedings of the 9th National Mechanical Engineering Conference, Khomeyni Shahr, Iran, 1395. Available online: https://civilica.com/doc/661301 (accessed on 29 November 2022).
30. Fan, B.; Zhang, Y.; Chen, Y.; Meng, L. Intelligent vehicle lateral control based on radial basis function neural network sliding mode controller. *CAAI Trans. Intell. Technol.* **2022**, *7*, 455–468. [CrossRef]
31. Chen, L.; Tang, L. Yaw stability control for steer-by-wire vehicle based on radial basis network and terminal sliding mode theory. *Proc. Inst. Mech. Eng. Part J. Automob. Eng.* 2022, *Online First*. [CrossRef]
32. Rajamani, R. *Vehicle Dynamics and Control*; Springer Science & Business Media: Berlin/Heidelberg, Germany, 2011; ISBN 1-4614-1432-6.
33. Jazar, R.N. *Vehicle Dynamics*; Springer: Berlin/Heidelberg, Germany, 2008; Volume 1.
34. Attia, R.; Orjuela, R.; Basset, M. Longitudinal control for automated vehicle guidance. *IFAC Proc. Vol.* **2012**, *45*, 65–71. [CrossRef]
35. Eskandarian, A. *Handbook of Intelligent Vehicles*; Springer: Berlin/Heidelberg, Germany, 2012; Volume 2.
36. Xia, J.S.; Khabaz, M.K.; Patra, I.; Khalid, I.; Alvarez, J.R.; Rahmanian, A.; Eftekhari, S.A.; Toghraie, D. Using feedforward perceptron artificial neural network (ann) model to determine the rolling force, power and slip of the tandem cold rolling. *ISA Trans.* 2022, *in press*. [CrossRef]
37. Kawato, M.; Uno, Y.; Isobe, M.; Suzuki, R. Hierarchical neural network model for voluntary movement with application to robotics. *IEEE Control Syst. Mag.* **1988**, *8*, 8–15. [CrossRef]
38. Psaltis, D.; Sideris, A.; Yamamura, A.A. A multilayered neural network controller. *IEEE Control Syst. Mag.* **1988**, *8*, 17–21. [CrossRef]
39. Chen, Y.; Cheng, C.; Zhang, Y.; Li, X.; Sun, L. A neural network-based navigation approach for autonomous mobile robot systems. *Appl. Sci.* **2022**, *12*, 7796. [CrossRef]
40. Agand, P.; Shoorehdeli, M.A.; Khaki-Sedigh, A. Adaptive recurrent neural network with lyapunov stability learning rules for robot dynamic terms identification. *Eng. Appl. Artif. Intell.* **2017**, *65*, 1–11. [CrossRef]
41. Truong, H.-V.-A.; Tran, D.-T.; Ahn, K.K. A neural network based sliding mode control for tracking performance with parameters variation of a 3-dof manipulator. *Appl. Sci.* **2019**, *9*, 2023. [CrossRef]
42. Liu, J. *Radial Basis Function (RBF) Neural Network Control for Mechanical Systems: Design, Analysis and Matlab Simulation*; Springer Science & Business Media: Berlin/Heidelberg, Germany, 2013; ISBN 3-642-34816-5.
43. Boukattaya, M.; Mezghani, N.; Damak, T. Adaptive nonsingular fast terminal sliding-mode control for the tracking problem of uncertain dynamical systems. *ISA Trans.* **2018**, *77*, 1–19. [CrossRef] [PubMed]
44. Asl, S.B.F.; Moosapour, S.S. Adaptive backstepping fast terminal sliding mode controller design for ducted fan engine of thrust-vectored aircraft. *Aerosp. Sci. Technol.* **2017**, *71*, 521–529.
45. Ji, X.; He, X.; Lv, C.; Liu, Y.; Wu, J. Adaptive-neural-network-based robust lateral motion control for autonomous vehicle at driving limits. *Control Eng. Pract.* **2018**, *76*, 41–53. [CrossRef]
46. Wai, R.-J. Tracking control based on neural network strategy for robot manipulator. *Neurocomputing* **2003**, *51*, 425–445. [CrossRef]

applied sciences

MDPI

Communication

Chassis Design Target Setting for a High-Performance Car Using a Virtual Prototype

Seunghoon Woo [1], Yunchul Ha [2], Jinwoo Yoo [3], Esteve Josa [4] and Donghoon Shin [5,*]

1 Department of Automotive Engineering, Kookmin University, Seoul 02707, Republic of Korea
2 Graduate School of Automotive Engineering, Kookmin University, Seoul 02707, Republic of Korea
3 Department of Automobile and IT Convergence, Kookmin University, Seoul 02707, Republic of Korea
4 Technical Center GmbH, Hyundai Motor Europe, 65428 Russelheim, Germany
5 Future Mobility Technology Center, Seoul National University, Siheung 15011, Republic of Korea
* Correspondence: fusioni@snu.ac.kr; Tel.: +82-10-8507-1707

Abstract: In this study, the chassis design target setting for a high-performance car was performed using a virtual prototype to solve the problem of increasing complexity of vehicle development. To achieve efficient handling performance of Hyundai Avante N, a high-performance vehicle, the kinematic and compliance (K&C) characteristics of the chassis corresponding to the design target were set prior to the design process using virtual simulation, thereby facilitating the efficient and systematic development of the actual vehicle. In order to overcome the limitations of existing research and apply it to the actual development of mass-production vehicles, the following major tasks were performed. The first is setting quantitative factors that match the sensibility evaluation. The second is building a virtual model to ensure consistency in performance predictions. The third is optimizing the chassis characteristics to achieve the vehicle performance goal. When all optimization results were applied, the average of the performance items increased by 0.5 points and the standard deviation improved by 0.4 points compared to the existing Civic Type-R, which was the best. In the case of the final specification considering design constraints, the average of performance items increased by 0.1 point and the standard deviation improved by 0.5 point compared to the existing Civic Type-R. Therefore, the design target of the chassis systems that could achieve the vehicle handling performance goal could be established prior to the design. Using this virtual development, it is possible to eliminate the trial and error process that the first and second test cars needed. This could save more than 500,000 USD (per unit trim) of the first and second test vehicles.

Keywords: virtual prototype; function-based model; chassis design; vehicle dynamics; handling performance

Citation: Woo, S.; Ha, Y.; Yoo, J.; Josa, E.; Shin, D. Chassis Design Target Setting for a High-Performance Car Using a Virtual Prototype. *Appl. Sci.* **2023**, *13*, 844. https://doi.org/10.3390/app13020844

Academic Editors: Edgar Sokolovskij, Vidas Žuraulis and Adel Razek

Received: 4 November 2022
Revised: 16 December 2022
Accepted: 3 January 2023
Published: 7 January 2023

1. Introduction

The complexity of vehicles and their development process has significantly increased owing to the increase in electronic control systems installed in vehicles, shortening of the product life cycle, diversification of markets, and the application of various power sources. In addition, the pressure to reduce costs has increased because of the need for resources for the development of eco-friendly and autonomous driving technologies and the increase in material costs for the application of related parts.

Chassis specifications have a considerable influence on the vehicle's platform; therefore, it is necessary to consider the changes in chassis specifications in the early stages of development. This requires the chassis specification to be set in an appropriate direction before commencing the design process. In particular, if a goal for the design direction can be set in advance before the design, inefficiency owing to future design changes can be minimized. In the past, the chassis characteristics of vehicles with good chassis system-related vehicle performance were measured and set as a target, but this method is not suitable because of the following reasons.

- To satisfy future development goals such as fuel efficiency, crashes, and cost, different power sources, specifications, and tires from competitors are considered.
- When transitioning from a 'Fast Follower' to a 'First Mover', there cannot be a competitor to follow blindly.
- The performance goals to be pursued are different because of the differences in the situations faced by different companies.

Ultimately, it is necessary to set the chassis system characteristics to achieve the desired vehicle handling performance based on the correlation between the vehicle's handling performance and the chassis system characteristics. A simulation model can be used to check in advance whether the vehicle handling performance target can be achieved by changing the chassis-related characteristic values. However, as the multi-body-based model (ADAMS, etc.), which has been widely used in chassis system simulation, requires design data such as hard point and bush characteristics, these characteristics should be verified after the design is completed. However, after the design is completed, the process for manufacturing and evaluating the actual vehicle would be well established, making it difficult to effectively validate the proposal by design verification simulation. Therefore, in this study, the design goals are set in advance by using a function-based model (1 Dimensional model) that can use the kinematic and compliance (K&C) characteristics corresponding to the design goals as model input factors before designing (Figure 1).

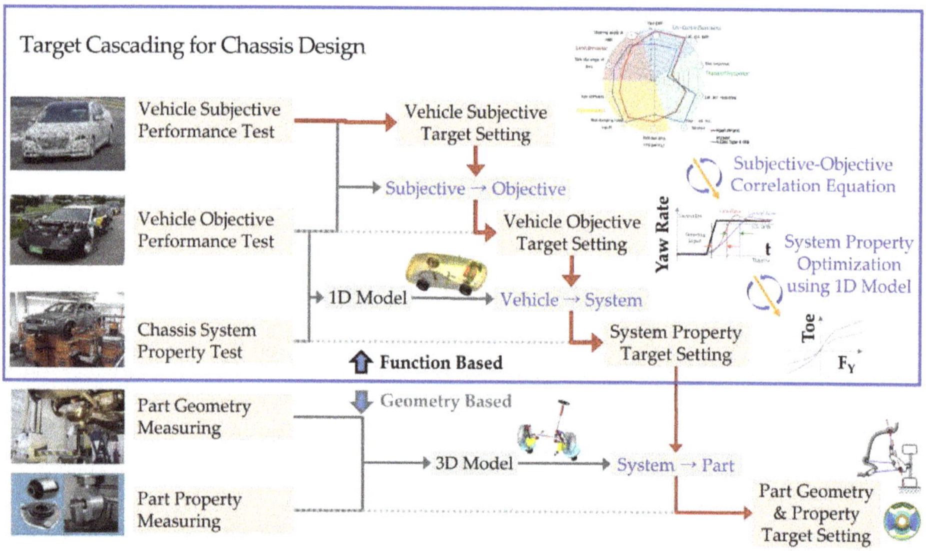

Figure 1. Optimal setting of model-based design factors before design.

Some previous studies have attempted to perform sensitivity analysis to determine the key design factors in the chassis design target establishment stage and to perform optimization based on this from various perspectives. A. Afkar et al. (2012) optimized geometric parameters of suspension systems using genetic algorithms (GAs) to improve the ride comfort, handling, and stability of vehicles. By optimizing the geometric parameters of the suspension system, it is observed that the vehicle can follow the target path with minimum deviation, along with maintaining stability and improving ride comfort conditions [1]. That is, since a model consisting of hard points, which are component-related characteristics that can be given after chassis platform design, is used, it can be reviewed only for the purpose of improving the existing design. Therefore, in order to set the design goal before chassis platform design, a model composed of K&C characteristics, which are characteristics related to the suspension system, is required. Sh. Lajqi et al. (2012) replaced the suspension model with a simplified mathematical model with adequate accuracy to effectively optimize the stiffness and damping coefficient of the passenger car's suspension [2]. However, parts-related characteristics such as spring stiffness and shock absorber damping are tuning factors and can be tuned even after the test vehicle is released. J. Tey et al. (2014) proposed an efficient approach to identify the unknown parameters through optimization based on experimental results, where the covariance matrix adaptation–evolutionary strategy (CMA-es) is utilized to improve the simulation and experimental results against the kinematic and compliance tests. Once accurate representation of the vehicle suspension model is achieved, further analysis, such as ride and handling performances, can be implemented for further optimization [3]. However, since it used a model that can be made only by testing a real vehicle, it is difficult to use it for developing a new platform where a test vehicle does not exist. Park et al. (2015) analyzed the sensitivity of chassis design factors to cross winds using simulation and driving simulator equipment [4]. That is, since a model consisting of hard points and joint stiffness, which are component-related characteristics that can be given after chassis platform design, is used, it can be reviewed only for the purpose of improving the existing design. Therefore, in order to set the design goal before chassis platform design, a model composed of K&C characteristics, which are characteristics related to the suspension system, is required. J. Tey et al. (2016) proposed a new method to simplify and reduce the conventional approach in vehicle suspension optimization for ride comfort, and the handling performance design objectives are formulated into a multi-objective optimization problem together with the suspension packaging dimensions as the design constraints. These objectives are minimized using a multi-objective optimization workflow, which involves a sampling technique and a regularity-model-based multi-objective estimation of the distribution algorithm to solve greater than 100-dimensional spaces of the design parameters by the optimization process [5]. However, optimization to achieve the goals required for vehicle development is difficult because it is not a model capable of predicting an emotionally developed ride and handling performance. Karimi Eskandary et al. (2016) optimized the air suspension system to independently control the stiffness and ride height [6]. The stiffness and ride height of the air spring are tuning parameters of the controller, and it is a general vehicle development process to tune mainly after the test vehicle is released. On the other hand, in order to set the design goal before designing the chassis platform, a model consisting of system-level parameters that reflect the geometry characteristics of the suspension system, such as K&C characteristics, is required. Y. Liu et al. (2018) established a geometric design and a virtual prototype model of the suspension system. Through Adams simulation analysis, the hardness of the suspension frame was used as a variable, and changes in a series of positioning parameters such as the toe value, camber angle, and roll rigidity of the suspension were compared and refined. This provided certain references for actual vehicle manufacturing [7]. Because a model consisting of characteristics that can be given after designing the platform, it can be reviewed only for the purpose of improving the existing design, and it is not suitable for the stage of setting goals for the chassis design. Choi et al. (2018) proposed an optimal design to determine the optimal spring setting position to minimize the side load acting on the damper of the McPherson strut suspension to improve ride comfort. Flexible multi-body dynamics (FMBD) was used for sophisticated simulation, and a sequential approximate optimal design was performed by repeating sequential sampling and meta-modeling to increase the optimization efficiency [8]. As a result, it is limited to improving the existing design because it uses a model consisting of hard points and joint stiffness, which are part-level factors that can be given after designing the platform. Therefore, in order to set design goals before platform design, a model composed of K&C characteristics, which are suspension system level factors, is required. Qi Gao et al. (2019) constructed a parameterized vehicle dynamic model using the vehicle's measurement data and performed sensitivity analysis and optimization for McPherson strut suspension to improve seven normal/abnormal handling and stability performance indicators. A generalized multi-dimension adaptive learning particle swarm optimization technique was applied to increase the convergence speed of optimization [9].

Since it also used a model that can be made only by testing a real vehicle, it is difficult to use it for developing a new platform where a test vehicle does not exist. On the other hand, G. Papaioannou et al. (2021) conducted a study on the optimization of parameters for passenger comfort and improved vehicle handling while minimizing tire wear. In addition to finding the optimal parameters, optimization is also used as a tool to identify and emphasize the potential balance between the objectives and the various design parameters [10]. However, it is difficult to optimize to achieve the objectives required in vehicle development because it is not a model that can predict ride and handling performance that is actually being developed emotionally. In order to improve the overall performance of the chassis system, T. Cui et al. (2020) proposed a hierarchical dynamic constraint analytical target cascading (DCATC) optimization method for the integrated system. Based on the dynamic model of the steering and suspension-integrated system, the evaluation indexes of the integrated system were proposed and deduced, which involved steering energy consumption, steering feel, steering sensitivity, ride comfort, and steering stability. Then, the optimization mathematical model of the integrated system was established. The simulation results showed that the proposed DCATC can improve the overall performance of the integrated system [11]. It is also difficult to optimize to achieve a goal required during vehicle development, as it is not a model capable of predicting ride comfort and steering performance, which are actually developed on an emotional basis. Yoon et al. (2021) proposed an optimal design to improve the ride comfort of a commercial vehicle. As the efficiency of the cabin suspension model needs to be increased to handle numerous design variables, a 3-Degrees of freedom (DOF) tabular-based cabin equivalent model with bound, roll, and pitch modes was developed [12]. However, component-level factors such as the stiffness of the spring mounted in the cabin and the damping of the shock absorber are tuning factors and can be tuned even after the test vehicle is released. On the other hand, in order to set design goals prior to platform design, a model composed of system-level factors that reflect suspension geometry characteristics, such as K&C characteristics, is required. Chenyu Wang et al. (2021) proposed an integrated multi-field optimal design with a collaborative optimization and approximation technique to improve ride comfort, reducing the fluctuating range of the front wheel and improving the steering experience, using a mathematical model of a 7-degrees of freedom electric power steering (EPS) system and a simulation model of the front wheel suspension [13]. However, since it is not a model capable of predicting ride comfort and steering performance, which are actually developed on an emotional basis, it is difficult to optimize to achieve the goal required during vehicle development.

As can be seen from the results of previous research, vehicle manufacturers do not rely solely on simulation to establish a chassis platform design goal to satisfy ride and handling performances. For this reason, most vehicle manufacturers perform ride and handling performance tests by manufacturing test vehicles with chassis parts designed using various methods. In order to improve the performance lacking in the test results, a test vehicle is manufactured with redesigned chassis parts, and a ride and handling performance test is performed. Vehicle manufacturers repeat these processes (design, manufacture, and test) at least three times in order to achieve the performance target of a vehicle to which a new chassis is applied. As the number of repetitions of this process increases, the cost of manufacturing a test vehicle and the number of work hours required for development increase, but a test vehicle is a prototype made in small quantities. As for the manufacturing cost, the first test vehicle costs more than about $300,000, the second test vehicle costs more than about $200,000, and the third test vehicle costs more than about $100,000. However, since it is necessary to manufacture a test car for each trim and sales region, it generally costs several million dollars. Therefore, many manufacturers, including Hyundai Motor Company, are trying to introduce a virtual development process to minimize the number of iterations of the development process dependent on the real object.

The tasks to be solved to increase the accuracy of setting design goals using virtual models are summarized as follows.

- With a simulation model that predicts quantitative values, it is difficult to predict ride and handling performances, which are emotional evaluation items.
- Insufficient analysis of the predictive consistency of the simulation model for quantitative measurement items related to ride and handling performance.
- Lack of consideration for direct application to actual design.

In this study, to efficiently achieve the handling performance goal of Hyundai Avante N, a high-performance vehicle, the following steps were followed for setting the K&C characteristics of the chassis corresponding to the design target before the design process.

- Establish a quantitative measurement item that matches the handling rating, which is an emotional evaluation item, and develop a relationship between them.
- Build a virtual model with predictive consistency for quantitative measurement items related to handling performance.
- Optimize the chassis characteristics to achieve the vehicle performance goal: Optimize step-by-step using a factor sequence with a smaller impact on other fields.
- The cause of performance improvement is analyzed when applying the design factors obtained by optimization to clarify the direction of improvement.
- Provide priority information that needs to be applied through analysis of the performance impact by a factor that changes as a result of optimization.

2. Quantitative Measurement Item That Matches Well with the Handling Rating

The target of developing the handling performance of the actual vehicle is determined by emotional evaluation, whereas the values predicted by the simulation model are quantitative physical values; therefore, it is difficult to accurately express the handling performance of the actual vehicle using the simulation results. Therefore, in this study, a relational expression between the related quantitative measurement items and emotional evaluation rating was established so that the handling development performance, which is dependent on emotional evaluation, could be expressed quantitatively. This could quantitatively set the handling performance development goal for the actual vehicle.

To establish a target for developing quantitative handling performances suitable for high-performance vehicles, a relational expression was derived through linear regression by statistically analyzing subjective evaluation results and measurement result data for high-performance reference vehicles. In general, the scores used by manufacturers for vehicle development are given by trained professional drivers by performing emotional evaluation in a prescribed manner, and they have a 10-point system. In order to convert this emotional evaluation into a measurement evaluation, after conducting an emotional evaluation and measurement test on dozens of vehicles, the correlation between the emotional evaluation score of each item and the measurement result data of the corresponding item was established as a mathematical model. To secure consistent data for statistical analysis, a relational expression was derived based on data accumulated in the same place by the same driver with the same procedure (Figure 2). In the table, the Min Rate and Max Rate columns are the minimum and maximum values of the emotional evaluation rate for each item. The column on the left of the Min Rate and Max Rate columns is the minimum and maximum values of measurement data for each item.

In the case of high-performance cars, the performance of braking-in-turn is important when entering a corner to improve the racing track-limit driving performance. Entering the corner as fast as possible to shorten the lap time causes understeer to be pushed out of the corner by the inertia of the vehicle; therefore, the vehicle speed must be reduced in advance only to the extent that it does not leave the course. Therefore, in high-performance cars, when braking and turning simultaneously occur in the corner entry section (trail braking), it is advantageous to improve the course traceability and lap time when developing in the direction of increasing the yaw motion gain by reducing the understeer. That is, the driver performs additional steering using braking, and it is advantageous for the driver to exercise complete control when the increase in yaw motion gain by braking occurs gradually without being too urgent. This is because, in race-track driving, abrupt steering input is avoided to maintain the tire grip characteristics. Moreover, when the increase in braking force occurs, the increase in yaw movement must also occur in proportion. Therefore, it is advantageous to suppress any additional over-shoot or vibration as much as possible.

Handling Parameters (Steering Input Only)					
Parameter	Value (x)	Min Rate	Value (x)	Max Rate	Function (Rate)
Yaw-rate Gain	0.2	0	0.4	10	45.45x − 9.1
Ay Gain	0.8	0	1.6	10	11.76x-9.41
Yaw-rate Delay	90	0	30	10	-0.1429x+17.143
Ay Delay	160	0	60	10	-0.1x+16
Ay Delay - Yaw Delay	100	0	0	10	-0.1x+10
Roll Damping	20	0	8	10	-0.83x+16.66
Roll Gain	7	0	1.5	10	-1.81x+12.72
Side-slip at Limit	4	0	0	10	-2.5x+10
Str.-angle at Limit	120	0	40	10	-0.125x+15

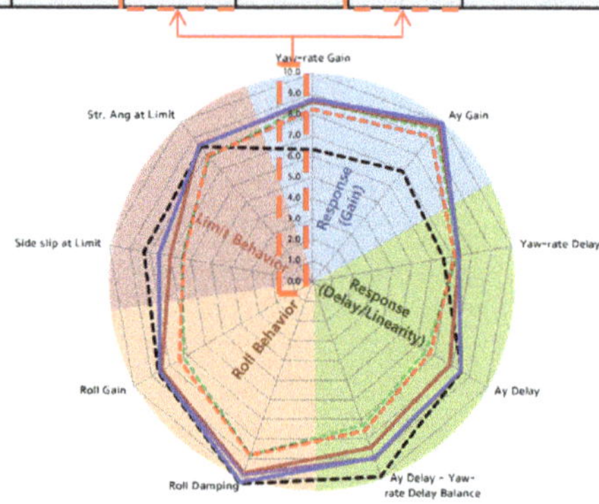

Figure 2. Relational expression between the handling subjective test items and objective test factors.

Therefore, a relational expression was additionally set up to express the development factors for the race-track-driving of high-performance cars as quantitative measurement factors in this study (Figure 3).

Handling Parameters (Steering+Braking Input)					
Parameter	Value (x)	Min Rate	Value (x)	Max Rate	Function (Rate)
Braking Yaw-rate Gradient	8	0	5	10	-3.33 x + 26.67
Braking Yaw-rate Gain (Max)	0.3	0	1	10	14.29 x − 4.29
Braking Yaw-rate Overshoot	0.23	0	0	10	-43.48 x + 10.00

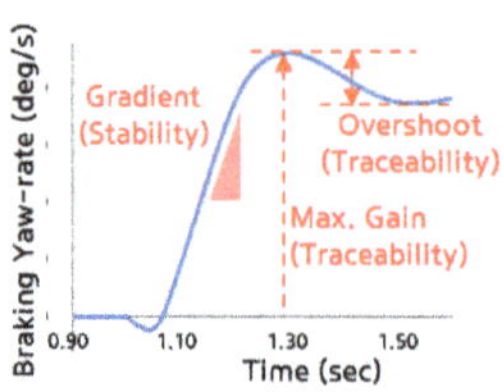

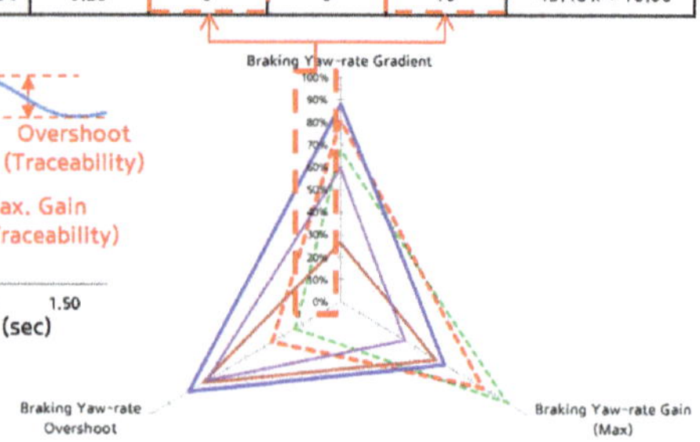

Figure 3. Additional settings for race-track-driving performance indicators.

3. A Framework for the Optimization Tool and Simulation Tool

This chapter describes how to change variables in a virtual vehicle model using an optimization tool. In this paper, we use this method for two purposes:

- Virtual vehicle modeling with predictive consistency;
- Optimizing chassis characteristics for the development goal.

Adjustment of model parameters using the optimization tool was performed using three simulation tools: PiAnO, Matlab/Simulink, and IPG CarMaker (Figure 4)

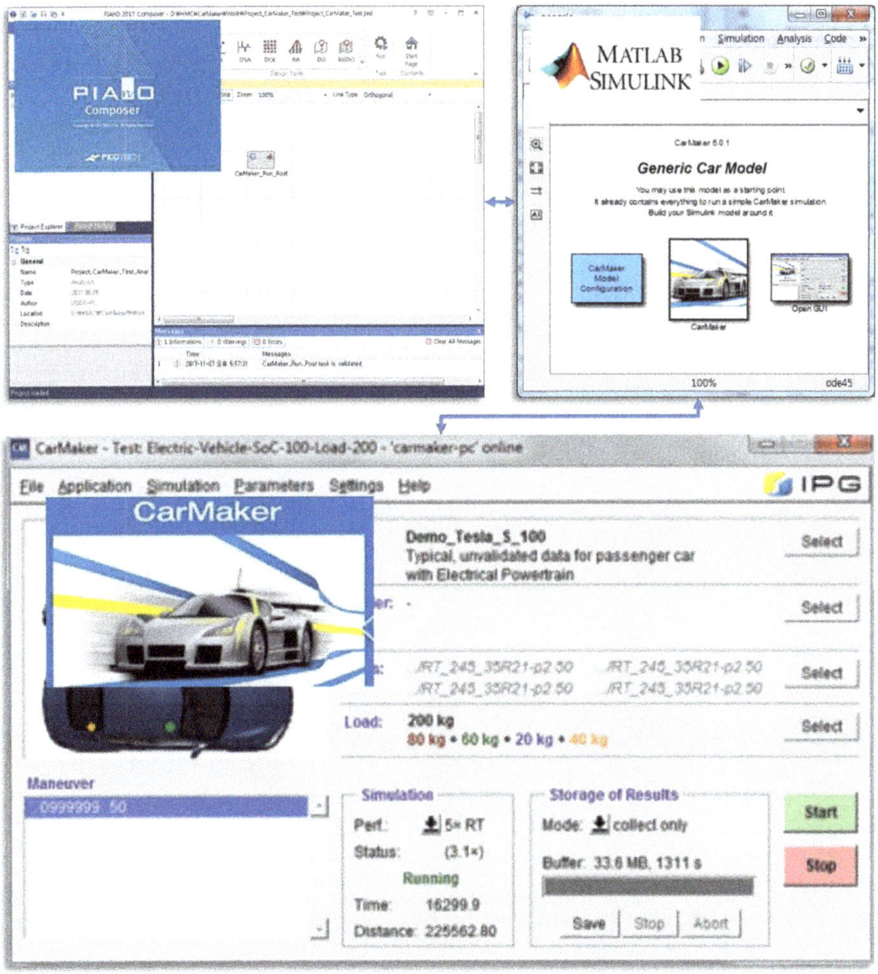

Figure 4. Adjustment of model parameters using optimization tools.

PIAnO (Process integration, automation, and optimization) is software for process integration and design optimization, which can derive optimized designs based on engineering data obtained through simulation [14]. IPG Carmaker is simulation software for virtual test driving of vehicles [15]. In this paper, IPG Carmaker was used to construct the vehicle model. This software can test vehicles modeled with user-set parameters and easily test driving simulations for test conditions to understand the dynamic characteristics of the vehicle.

Adjustment of model parameters using optimization tools is performed through a process defined in PIAnO. The process is divided into a pretreatment step, a CarMaker simulation step, and a post-treatment step. First, in the preprocessing step of PIAnO, the carmaker model file is updated to change the K&C curve of the vehicle model or to change the vehicle specifications by inputting the design variable. Through the preprocessing step, the parameters of the CarMaker vehicle model are applied according to the design variables, as shown in Figure 5. Thereafter, in the CarMaker simulation step with PIAnO the test items determined sequentially are simulated. CarMaker simulation is performed by executing Matlab's simulation code running the Simulink model linked to CarMaker in PIAnO. The driving simulation of the vehicle model is performed as shown in Figure 6. Finally, in the post-processing step of PIAnO, the performance index is calculated from the Ascii-type CarMaker driving simulation result data file. The performance index is then used for optimization, and the design variables of the first step can be obtained through optimization. This process runs repeatedly to finally set the model parameters.

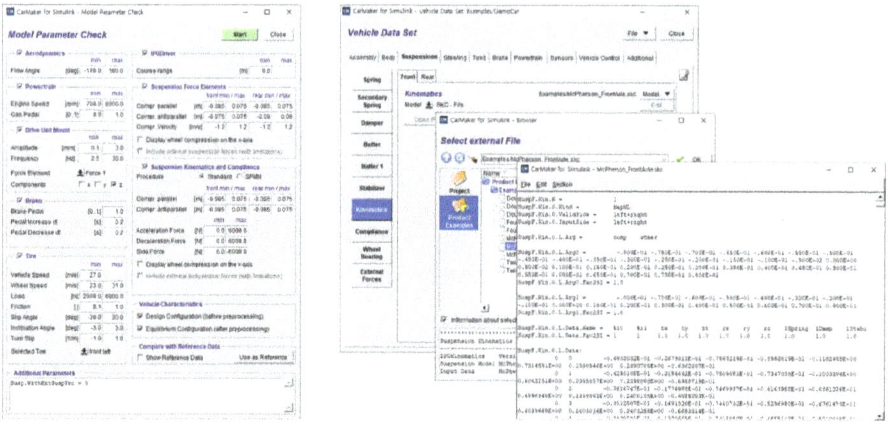

Figure 5. IPG CarMaker Software Vehicle Model Parameter setting.

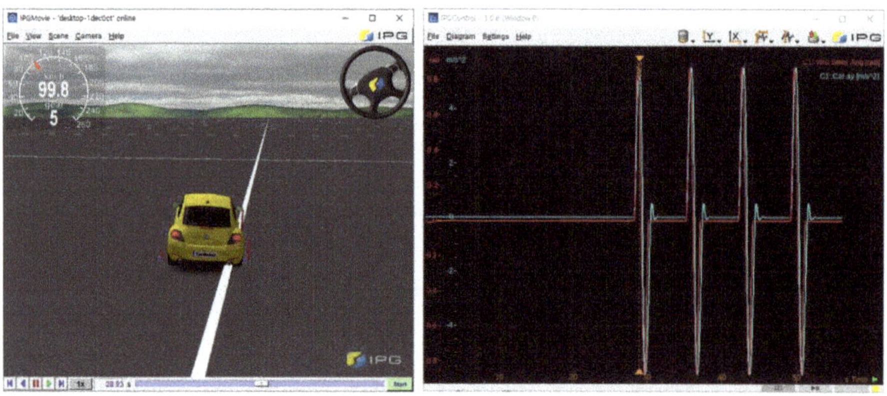

Figure 6. IPG CarMaker Software Simulation.

In this paper, progressive quadratic response surface approximation (RQRSA) was used as an optimization method for the objective function [10]. This optimization method was proposed by Hong et al. (2021). This approximation uses only 2n + 1 design points for one approximation optimization and uniquely determines main effect terms such as linear and quadratic regression coefficients. We then gradually updated the two interaction terms using the main effect terms between successive iterations. Thus, PQRSA is an efficient approximation method for constructing a quadratic reaction surface model.

4. Virtual Model with Predictive Consistency

In virtual development, the construction of a vehicle model with consistency in performance prediction should be considered. In this study, to develop a coherent model, the mother car model and the measurement data of the mother car are utilized. That is, the calibration factor of the model is set using the measurement data. Then, by reflecting the specifications of the vehicle to be developed, it is used as a base model to change the chassis characteristic factor to satisfy the performance target.

First, moment of inertia information, which is important for realizing the dynamic characteristics of the vehicle, was estimated. Moment of inertia information is important in realizing the dynamic characteristics of a vehicle. If there is no measured value, the regression model result is used. For an existing vehicle, measurement data were available and were used. However, as the moment of inertia of the existing vehicle was measured to be smaller than that of the regression model, the moment of inertia of the vehicle to be developed in the same class was also corrected and used by the same ratio. However, while the moment of inertia data of the existing and developed vehicles were managed based on 1 UP, only the data measured in the 0 UP state existed in the case of the competitive car. Therefore, in the trend model, the difference between the 0 UP and 1 UP states was corrected and compared with the vehicle to be developed. Therefore, it can be confirmed that the moment of inertia of the existing model is smaller than that of the vehicle to be developed, which is advantageous for handling performance (Figure 7). In this study, the chassis characteristics are optimized to secure the same or better handling performance even under conditions such as an unfavorable moment of inertia compared to competitors owing to the high development targets for collision safety and space utilization of the vehicle.

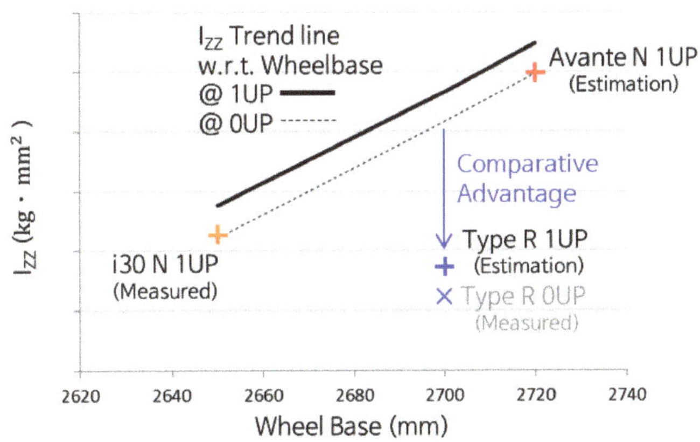

Figure 7. Izz (moment of inertia about the z-axis) Estimation: Trend Model + Measured Data.

To check the consistency of the model for each steering input speed, the lateral acceleration response according to the increase in the steering input frequency was compared with the measurement results. For reference, the frequency corresponding to the turn-in input during normal turns is approximately 0.5 Hz, and when changing lanes very quickly, it can reach the level of 2 Hz; therefore, 0.5–2 Hz was considered as the necessary guaranteed range. It can be observed that the response gain of the initial model is similar to the measurement result near 2 Hz, but it becomes larger than the measurement result as the frequency decreases. The main reason is that the actual car has a larger understeer value than the model. Analyzing the related phenomenon, the model is used for concept setting in the early stage of development, and therefore, the EPS logic that is tuned later for each vehicle was not reflected. However, the toe compliance characteristic of the lateral force of the front wheel that affects the understeer of the vehicle is proportional to the deformation of the torque sensor's torsion bar in the EPS. Here, the torsion bar torsion is inversely proportional to the motor assist torque, and the motor assist torque varies according to the vehicle or vehicle speed. Therefore, the level of understeer by EPS is different for each vehicle or each vehicle speed. Therefore, this paper proposes a method of estimating the toe compliance characteristics due to the lateral force of the front wheel when the motor assist torque is applied using the handling measurement results of the preceding vehicle (Figure 8). For estimating the toe compliance characteristics to meet the understeer level of real car, we utilized the method introduced in Section 2.

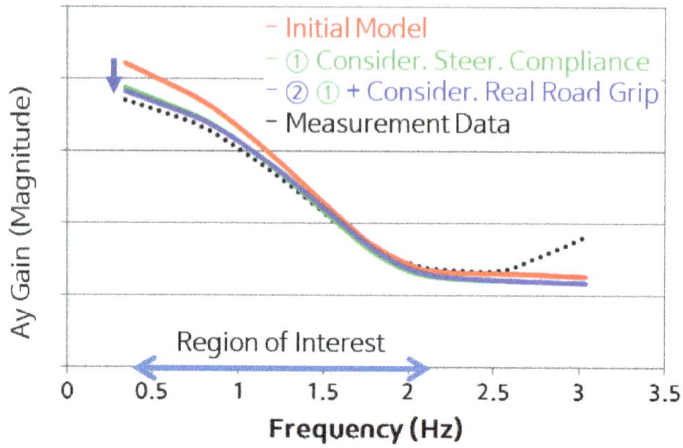

Figure 8. Estimation results of the toe compliance characteristics.

Next, the lateral acceleration response according to the increase in the steering angle was compared with the measurement results to check the consistency of the model for each steering input size. In the case of the initial model, it can be observed that the lateral acceleration at the same steering angle is larger than the measurement result. However, it can be confirmed that the value of the model becomes similar to the measurement result when the correction of the steering system deformation characteristics is considered according to the driving conditions performed earlier. However, it can still be observed that the maximum lateral acceleration is larger than the measurement result in the limit region. The tire model applied to this vehicle model is constructed by measuring the characteristics of tire from the tire flat track. In general, the friction coefficient of the external asphalt road surface including the general test proving ground tends to be smaller than the flat track belt road surface. Therefore, in this paper, we propose a method to estimate the tire lateral friction force characteristics using the method introduced in Chapter 2, so that the maximum lateral acceleration characteristics of the vehicle can be reproduced using the handling measurement results of the preceding vehicle (Figure 9).

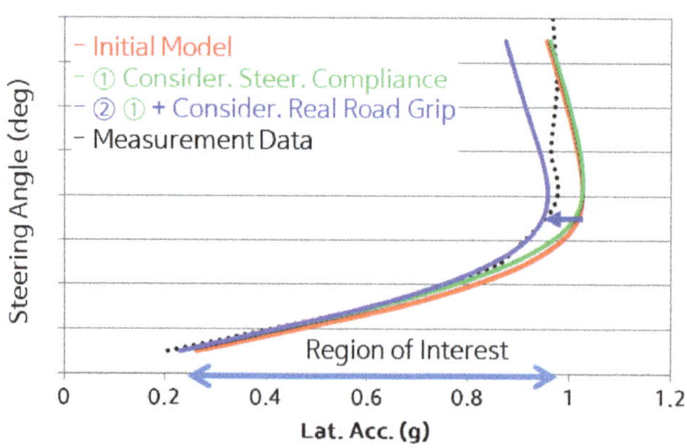

Figure 9. Estimation results of the tire lateral frictional force characteristics.

When there are two or more factors to be adjusted using the optimization tool, there is a tendency to preferentially adjust the factor of high sensitivity to achieve the purpose. In other words, the tire lateral force characteristics may have a significant influence on the model characteristics similar to the result of the frequency response measurement, and in this case, the tire lateral force characteristics could be changed more than necessary to ensure consistency in the frequency response characteristics. In that case, the toe compliance characteristics due to the lateral force of the front wheel are changed less significantly, and the probability that the model implemented in this manner, configured in a different combination from the actual vehicle, increases. As the lateral acceleration of the vehicle in the frequency response characteristic test is less than 0.4 g and the tire lateral force characteristic is a linear range, it can be considered that the tire model is reliable to some extent. However, the toe compliance characteristic of the lateral force of the front wheel is affected by the EPS motor torque, and therefore, this factor needs to be adjusted anyway. Therefore, to reproduce the frequency response characteristic of the linear section, the optimization process of adjusting only the toe compliance characteristic by the lateral force of the front wheel was performed first. Subsequently, the second optimization was performed so that the maximum lateral acceleration characteristics realized in the nonlinear region of the tire characteristics were reproduced by adjusting the tire lateral force characteristics. However, if the tire lateral force characteristics are adjusted, the frequency response characteristics, which are in a linear range, may be partially changed. Therefore, it may be necessary to partially adjust the toe compliance value by the lateral force to compensate for this. Therefore, for the second optimization, not only the maximum lateral acceleration but also the frequency response characteristics were used to define multiple objective functions, and factors requiring adjustment included not only the tire lateral force characteristics but also the toe compliance characteristics by the front wheel. However, in the second optimization process, even though the final optimization is performed by including all factors and objective functions, the first optimization is performed to adjust the toe compliance characteristic that needs adjustment in the linear response section close to an appropriate value. When performing the second integrated optimization, the toe compliance characteristic already starts from an appropriate value, and therefore, it is possible to fundamentally avoid the problem that the tire lateral force characteristic may considerably change to reproduce the linear-response region-response characteristic. In the end, fine tuning of the remaining factors is performed only to the extent of compensating for the changed tire lateral force characteristics to reproduce the maximum lateral acceleration characteristics.

5. Optimizing Chassis Characteristics

Avante N's handling performance development goal was set as Best-In-Class. Therefore, we conducted a measurement related to the handling performance of Civic Type-R and i30 N, which were well evaluated in terms of the handling performance at that time. Civic Type-R was excellent in most items because of its smaller weight and excellent tire friction characteristics. However, i30 N was excellent in items of response gain for cornering in the racing track (Figure 8). Setting the best goal for all items at the same time can be a challenging task, but to verify the potential of virtual model-based optimization technology, the goal function is set so that the best handling performance can be achieved for all items. The scope of the factors to be changed was set to the scope of the K&C design guide (Figure 10). The changeable range of high-performance vehicles is limited considering the investment costs. Therefore it has to be selectively applied among the optimization results. However the fundamental solution was found to help not only other development vehicles but also current development vehicles. Instead, a vehicle performance impact analysis was also performed to help selectively apply the changes according to this optimization result.

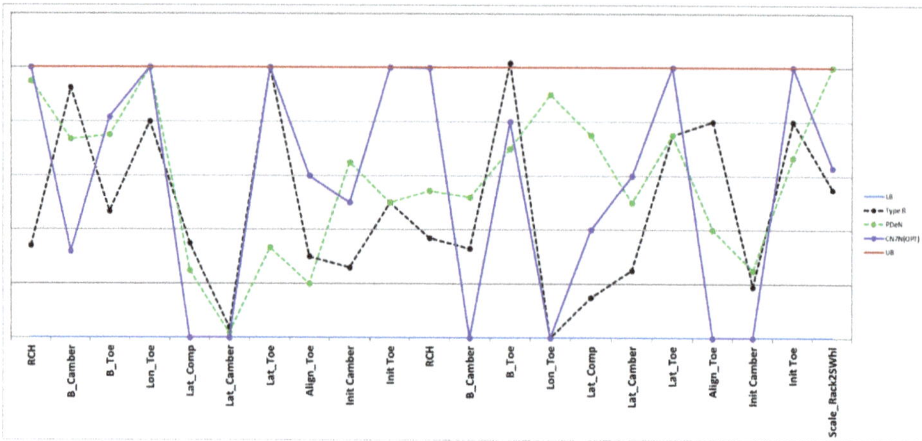

Figure 10. Optimizing within the K&C design guide.

There are several chassis system factors that need to be set to achieve the vehicle handling performance goal, and as there are interactions between them, it takes many work hours to find a set that meets the vehicle handling performance goal while changing those factors. Multiple optimization theories may be used to automatically find numerous factors to satisfy several items of the vehicle handling performance simultaneously. However, when performing multiple optimizations, the vehicle's handling performance target is usually determined by changing the sensitive chassis factors first. Mainly, there was a tendency to passively change the remaining factors accordingly with the initial toe/camber and bump toe set to the maximum value. However, when the initial toe/camber has a large value, performance in terms of tire wear and fuel efficiency becomes disadvantageous, and the toe change due to the bump stroke considerably affects the turning braking performance. Therefore, to reduce the dependence on initial toe/camber and toe changes due to bump stroke that significantly affect its performance in other parameters, the first optimization is performed without the corresponding factor, and the remaining factors are actively changed to achieve the goal. After that, the initial toe/camber is added to perform secondary optimization so that it is fine-tuned. The third optimization was performed by adding the purpose function of the turning braking performance, the toe compliance factor due to the bump stroke, and the toe compliance factor due to the braking force (Figure 11).

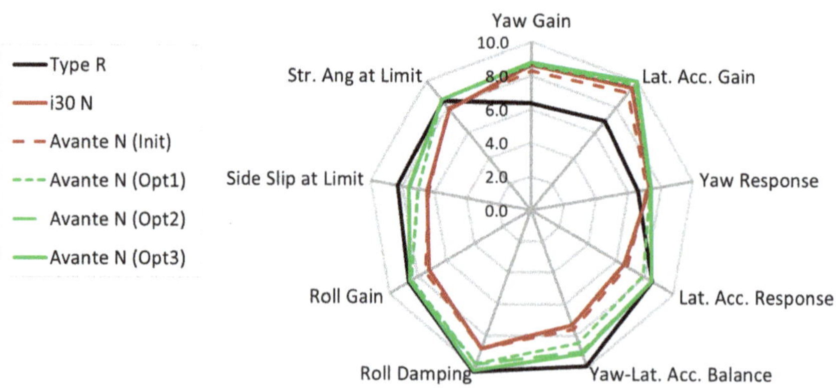

Figure 11. Optimizing handling performance (steering input).

6. Analyzing the Cause of Performance Improvement

No matter how good a tool makes the optimal factor change to improve performance, there is a high risk to apply it to the actual design unconditionally without knowing the principle. Therefore, in this study, the actual application was proposed only after analyzing the change direction for each factor found in the optimization tool in principle.

First, Avante N has the larger weight of the two vehicles considered, and in particular, Civic Type-R is equipped with a tire with excellent lateral force characteristics. For Avante N to achieve the best handling performance, it is important to secure a centripetal force (tire lateral force) corresponding to the centrifugal force (weight) (Figure 12).

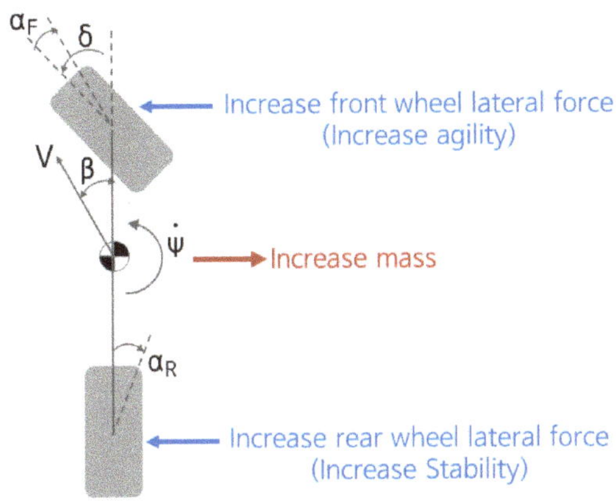

Figure 12. Centripetal force increase setting to cope with increased centrifugal force.

Therefore, it was confirmed that the chassis characteristic was set in the following manner to actively generate the lateral force of the tire.

- Increase the negative camber properties;
- Increase the roll center height;
- Increase the positive toe properties.

In particular, Figure 13 shows the principle by which the handling performance is advantageous according to the increase in the height of the roll center.

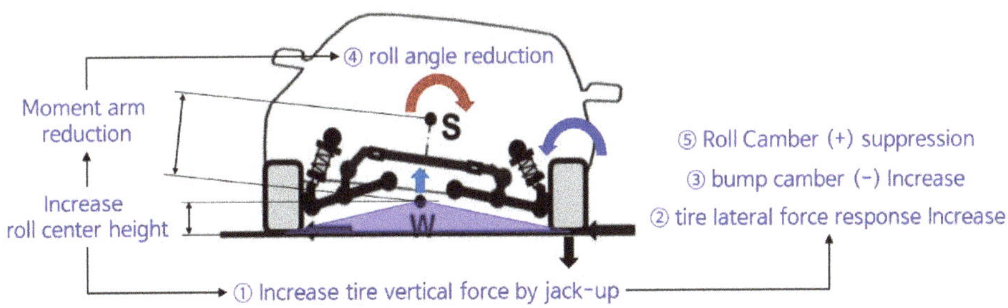

Figure 13. Principle of the lateral force increase when the roll center is raised.

7. Priority Information Required for Application

As high-performance cars are basically derivatives of basic cars, the range that can be changed is limited considering the sharing and investment costs. Moreover, during development, modifications may occur compared to the initial development goals for various reasons. Therefore, there are many reasons why the optimization results cannot be unconditionally applied in the design. Therefore, we analyze the contribution of vehicle handling performance for each factor that changes as a result of optimization and provide a solution to help developers make judgments regarding which factors and how to change them. Accordingly, the results of the final specifications developed by the development officials are shown in Figure 14.

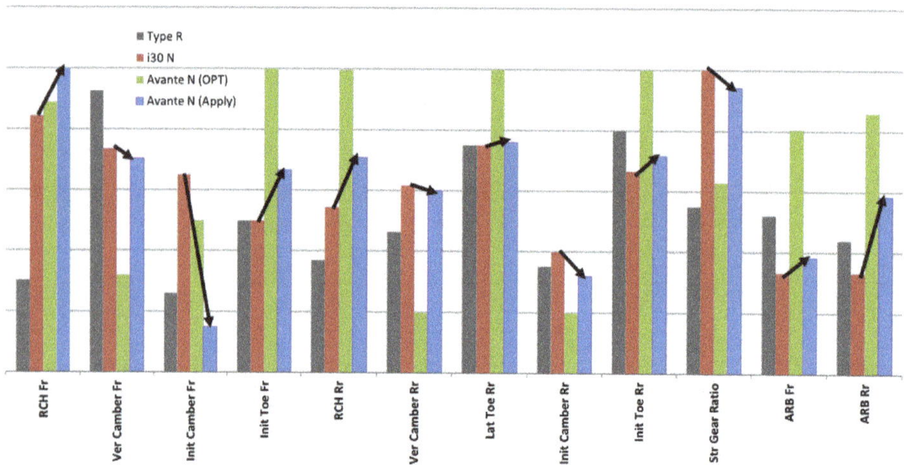

Figure 14. Performance impact factor recommendation and application results.

Among the factors changed as a result of optimization, 12 factors with high overall contribution were applied in the recommended direction. Among them, 10 factors had to be changed in smaller amounts than the optimization result due to common use with existing vehicles. Instead, the two factors with a high degree of freedom of change were applied in a more active way to make up for the insufficient amount of change in the above 10 factors.

Despite unfavorable conditions such as the wheel base increase, weight increase, investment cost minimization, and common use with existing vehicles, Avante N achieved high scores evenly in all handling performance items (Figure 15).

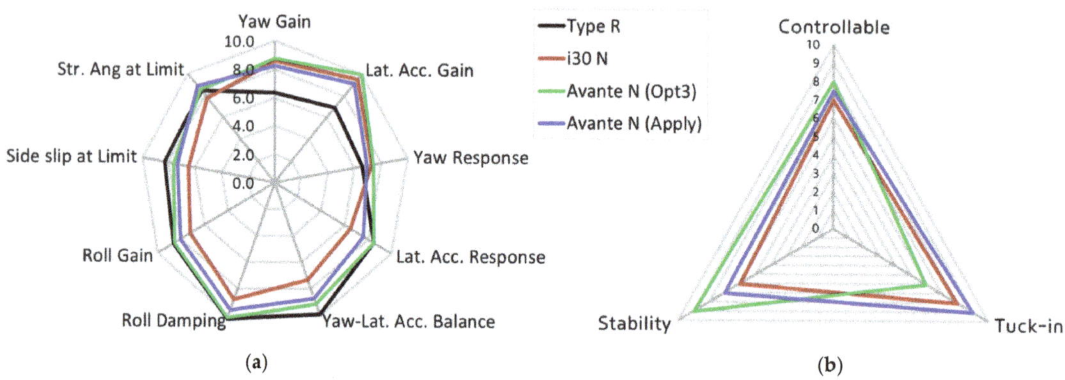

Figure 15. Avante N handling performance: (**a**) Handling performance (steering input); (**b**) Composite handling performance (steering/breaking).

In order to confirm the overall handling performance, the scores for each item were statistically analyzed (Table 1). As a result, the average score was excellent in the order of Avante N (Opt3), Avante N (Apply), Civic Type-R, and i30 N, and the standard deviation was excellent in the order of Avante N (Apply), Avante N (Opt3), i30 N, and Civic Type-R. The optimization result, Avante N (Opt3), improved by 0.5 points in the average and

0.4 points in the standard deviation compared to Civic Type-R, which was previously the best. Avante N (Apply), which can be applied to the design, also improved by 0.1 point in the average and by 0.5 points in the standard deviation compared to Civic Type-R, which was the best. Therefore, the design target of the chassis systems that could achieve the vehicle handling performance goal could be established prior to the design.

Table 1. Average and standard variation of scores by handling performance item.

Vehicle	Average	Std. Deviation
Honda Civic Type-R	8.2	1.3
Hyundai i30 N	7.7	1.0
Hyundai Avante N (Opt3)	8.7	0.9
Hyundai Avante N (Apply)	8.3	0.8

8. Conclusions and Discussion

In this study, to solve the recently increased complexity of vehicle development, we developed a high-performance car chassis concept before the design process using the virtual prototype.

To efficiently achieve the handling performance goal of Avante N, a high-performance vehicle, the following tasks were performed in setting the K&C characteristics of the chassis corresponding to the design goal before the design process.

- Set quantitative factors that match the emotional evaluation.
- Build a virtual model to ensure consistency in performance predictions.
- Optimize the chassis characteristics to achieve the vehicle performance goal: Optimize step-by-step by factor sequence with a smaller impact on other performance goals.
- The cause of performance improvement is analyzed when applying the design factors obtained by optimization to clarify the direction of improvement.
- Provide priority information that needs to be applied through analysis of the performance impact by a factor that changes as a result of optimization.

When all optimization results were applied, the average of the performance items increased by 0.5 points and the standard deviation improved by 0.4 points compared to the existing Civic Type-R, which was the best. In the case of the final specification considering design constraints, the average of performance items increased by 0.1 points and the standard deviation improved by 0.5 points compared to the existing Civic Type-R. As a result, despite unfavorable conditions such as the wheel base increase, weight increase, investment cost minimization, and common use with existing vehicles, Avante N achieved high scores evenly in all handling performance items. Therefore, the design target of the chassis systems that could achieve the vehicle handling performance goal could be established prior to the design. Using this virtual development, it is possible to eliminate the trial and error process that the first and second test cars needed. This could save more than 500,000 USD (per unit trim) of the first and second test vehicles. Considering the combination of three types by region and two types by trim, 3,000,000 USD can be saved.

The future plan is to ensure that these model-based pre-concept setup tasks continue to be carried out in an efficient and consistent manner so that most vehicle development tasks can be carried out more systematically and efficiently. In other words, to establish a framework that provides consistent and reliable data and simulation results that are needed to effectively make decisions based on data or models from the beginning of the development process. The following steps should be followed in this regard.

- Analysis of required data and outputs by development task.
- Life-cycle management for modeling and simulation result data.
- Standardization and automation of data, modeling, and simulation-related tasks.

Author Contributions: Conceptualization, S.W. and D.S.; Methodology, S.W.; Software, S.W.; Validation, S.W. and D.S.; Formal analysis, Y.H.; Investigation, S.W.; Resources, J.Y.; Data curation, E.J.; Writing—original draft preparation, S.W.; Writing—review and editing, Y.H.; Visualization, S.W.; Supervision, S.W.; Project administration, S.W.; Funding acquisition, J.Y. All authors have read and agreed to the published version of the manuscript.

Funding: This work is supported by the Korea Agency for Infrastructure Technology Advancement (KAIA) grant funded by the Ministry of Land, Infrastructure and Transport (Grant 22AMDP-C162182-02)

Institutional Review Board Statement: Not applicable.

Informed Consent Statement: Not applicable.

Data Availability Statement: Not applicable.

Conflicts of Interest: The authors declare no conflict of interest.

References

1. Afkar, A.; Mahmoodi-Kaleibar, M.; Paykani, A. Geometry optimization of double wishbone suspension system via genetic algorithm for handling improvement. *J. Vibroengineering* **2012**, *14*, 827–837.
2. Lajqi, S.; Gugler, J.; Lajqi, N.; Shala, A.; Likaj, R. Possible experimental method to determine the suspension parameters in a simplified model of a passenger car. *Int. J. Automot. Technol.* **2012**, *13*, 615–621. [CrossRef]
3. Tey, J.; Ramli, R.; Kheng, C.; Chong, S.; Abidin, M. Identification of vehicle suspension parameters by design optimization. *Eng. Optim.* **2014**, *46*, 669–686. [CrossRef]
4. Park, J.; Park, J.; Lee, J.; Yi, Y.S.; Park, T. Analysis of sensitivity and optimization for chassis design parameters on the X-wind stability. In Proceedings of the 18th Asia Pacific Automotive Engineering Conference, Melbourne, Australia, 10–12 March 2015.
5. Tey, J.; Ramli, R.; Abdullah, A. A new multi-objective optimization method for full-vehicle suspension systems. *Proc. Inst. Mech. Eng. Part D J. Automob. Eng.* **2016**, *230*, 1443–1458. [CrossRef]
6. Karimi Eskandary, P.; Khajepour, A.; Wong, A.; Ansari, M. Analysis and optimization of air suspension system with independent height and stiffness tuning. *Int. J. Automot. Technol.* **2016**, *17*, 807–816. [CrossRef]
7. Liu, Y.; Chen, B.; Shi, W. Suspension Optimization Design and Virtual Prototype Simulation Analysis of FSAE Racing Car. In Proceedings of the 2018 IEEE 4th Information Technology and Mechatronics Engineering Conference (ITOEC), Chongqing, China, 14–16 December 2018.
8. Choi, B.C.; Cho, S.; Kim, C.W. Sequential approximate optimization of Macpherson strut suspension for minimizing side load by using progressive meta-model method. *Int. J. Automot. Technol.* **2018**, *19*, 455–461. [CrossRef]
9. Gao, Q.; Feng, J.; Zheng, S. Optimization design of the key parameters of McPherson suspension systems using generalized multi-dimension adaptive learning particle swarm optimization. *Proc. Inst. Mech. Eng. Part D J. Automob. Eng.* **2019**, *233*, 3403–3423. [CrossRef]
10. Papaioannou, G.; Jerrelind, J.; Drugge, L. Multi-Objective Optimisation of Tyre and Suspension Parameters during Cornering for Different Road Roughness Profiles. *Appl. Sci.* **2021**, *11*, 5934. [CrossRef]
11. Cui, T.; Zhao, W.; Wang, C.; Guo, Y.; Zheng, H. Design optimization of a steering and suspension integrated system based on dynamic constraint analytical target cascading method. *Struct. Multidiscip. Optim.* **2020**, *62*, 419–437. [CrossRef]
12. Yoon, J.; Hyun, M.; Heo, S.; Park, J.; Kang, D. Development of a Commercial Vehicle Ride Comfort Performance Optimization Design Process Using the 3-Dof Equivalent Cabin Suspension Model. *Trans. Publ. Korean Soc. Automot. Eng.* **2021**, *29*, 515–522. [CrossRef]
13. Wang, C.; Chen, X.; Shi, G.; Yang, Z. Collaborative optimization of vehicle chassis integration system. In Proceedings of the IEEE 12th International Conference on Mechanical and Intelligent Manufacturing Technologies (ICMIMT), Cape Town, South Africa, 13–15 May 2021.
14. PIDOTECH Inc. Process Intergration, Automation, and Optimization. In *PIAnO User's Manual*; PIDOTECH Inc.: Seoul, Republic of Korea, 2021.
15. IPG Automotive GmbH. *IPG CarMaker. Reference Manual Version 9.0.1*; IPG Automotive GmbH: Karlsruhe, Germany, 2021.

applied sciences

MDPI

Article

Forecasting the Number of Road Accidents in Polish Provinces Using Trend Models

Piotr Gorzelańczyk

Transport Department Podchorazych 10 Street, Stanislaw Staszic State University of Applied Sciences in Pila, 64-920 Pila, Poland; piotr.gorzelanczyk@puss.pila.pl

Abstract: Many people die on the streets every year. The value is declining year by year, but there are still plenty of them. Although the COVID-19 pandemic reduced the number of traffic accidents, it is still very high. For this reason, in order to do everything possible to minimize the number of road accidents, it is important to know the federal states with the most road accidents and what the accident forecast is for the next few years. The purpose of this article is to predict the number of road accidents by state in Poland. The survey was divided into two parts. The first is an analysis of the annual data of police statistics on the number of road accidents in Poland for the period 2000–2021, upon the prediction of the number of traffic accidents from 2022 to 2031 was decided. The second part of the study looked at monthly data from 2000 to 2021. Again, the forecasts analyzed were determined for the period from January 2022 to December 2023. The results of this study indicate that a decrease in the number of accidents is also expected in the coming years, which becomes especially clear when analyzing the annual data. It is worth noting that the prevailing COVID-19 pandemic has distorted the results obtained. The study was performed in MS Excel using the selected propensity model.

Keywords: traffic accident; forecasting; trend models; province

Citation: Gorzelańczyk, P. Forecasting the Number of Road Accidents in Polish Provinces Using Trend Models. *Appl. Sci.* **2023**, *13*, 2898. https://doi.org/10.3390/app13052898

Academic Editors: Edgar Sokolovskij, Vidas Žuraulis and Luís Picado Santos

Received: 16 January 2023
Revised: 17 February 2023
Accepted: 18 February 2023
Published: 23 February 2023

1. Introduction

A traffic accident is an event that causes damage to property, as well as injury or death to traffic participants. According to the World Health Organization (WHO), about 1.3 million people die each year in road accidents. Those accidents account for about 3% of GDP in most countries around the world. Automobile accidents are the leading cause of death for minors and young people between the ages of 5 and 29 [1]. The United Nations General Assembly has set an ambitious goal of halving the number of traffic accident victims by 2030 [2,3].

Road accident severity is an attribute used to determine the severity of traffic collisions. Predicting accident severity is important for relevant authorities to develop road safety policies to prevent accidents and reduce injuries, fatalities and property damage [4–8]. Identifying key factors affecting accident severity is a prerequisite for taking measures to eliminate and reduce accident severity [9]. Yang et al. [10] proposed a DNN (deep neutral network) multicarbon structure for predicting various degrees of injury, death and severity of property loss. This allows for a comprehensive and accurate analysis of the severity of traffic accidents.

There are several sources of accident data. They are usually collected and evaluated by government agencies through responsible leadership agencies. Data are collected from police reports, insurance databases or hospital records. In some cases, traffic accident information is processed on a large scale for the transportation sector [11].

Intelligent transportation systems are currently the most important source of data for analyzing and predicting traffic accidents. These data can be processed using the vehicle's GPS device [12]. In addition, vehicle information can be detected using a roadside vehicle detection system, which can continuously record vehicle information (speed, traffic volume, vehicle type, etc.) [13]. License plate recognition systems can also collect large amounts of traffic data during surveillance [14]. Another source of traffic and accident information is social media, but its accuracy can be insufficient due to the incompetence of reporters [15].

In this article, we chose a trend model to predict the number of traffic accidents by state. Exponential smoothing and neural networks for predicting the number of traffic accidents have been used by authors in other studies [16,17].

2. Literature Review

For accident data to be relevant, multiple data sources must be handled properly. Combining data from many sources, by merging various road accident data, increases the accuracy of analytical results [18].

A statistical study to estimate the severity of road accidents and clarify the relationship between the accidents and road users was conducted by Vilaca et al. [19]. The result of this research was suggestions for improving traffic safety standards and adopting other traffic safety measures.

Buck et al. [20] conducted a statistical study of traffic safety in selected regions of Poland based on the speed at which the number of road accidents and their causes were determined. The study used multivariate statistical analysis to examine the safety aspects of accident perpetrators.

The choice of the source of accident data for analysis depends on the type of traffic problem under consideration. Combining statistical models with other natural driving data or other data acquired by intelligent transportation systems can improve the accuracy of accident prediction and contribute to the elimination of accidents [21].

Various methods for predicting the number of accidents can be found in the literature. Time series methods are most commonly used to predict the number of traffic accidents [22,23]. The disadvantages are the inability to assess the quality of the prediction based on outdated forecasts and the frequent autocorrelation of the residual components of the accident rates [24]. Prochazka et al. [25] used multiple seasonality models for forecasting, and Sunny et al. [26] used Holt–Winters exponential smoothing. Its limitations include the inability to introduce exogenous variables into the model [27,28].

The vector autoregression model has also been used to predict the number of traffic accidents. The disadvantage is that it requires a large number of variable observations to properly estimate the parameters [29]. To analyze the number of deaths [30] and the Al-Madani curve-fitting regression model [31], they require only a simple linear relationship [32] and an order of autoregression (assuming the series is already stationary) [33].

Biswas et al. [34] used random forest regression to predict the number of traffic accidents. In this case, the data contain clusters of correlated features with similar validity to the original data, with small clusters favored over large clusters [35] and method instability and spike prediction [36]. Chudy-Laskowska and Pisula [37] applied autoregressive quadratic trend models, one-dimensional cyclical trend models and exponentially adjusted models to the forecasting problem at hand. Moving average models can also be used to predict the problem at hand. Its drawbacks are poor prediction accuracy, the resulting loss of data, and failure to account for trends and seasonal effects [38]. Prochozka and Camej [39] used the GARMA method. This method imposes some restrictions on the parameter space to ensure the stationarity of the process. ARMA models for steady-state processes and ARIMA or SARIMA models for transient processes are often used for forecasting [25,40,41]. Although these models lead to great flexibility of the models in question, they are also a disadvantage, as identifying good models requires more experience from the researcher than, for example, regression analysis [42]. Another disadvantage is the linearity of ARIMA models [43].

Chudy-Laskowska and Pisula used the ANOVA method in a study [44] to predict the number of traffic accidents. The disadvantage of this method is additional assumptions, especially the assumption of sphericity, the violation of which can lead to erroneous conclusions [45]. Neural network models are also used to predict the number of traffic accidents.

The disadvantages of SSNs are the need for experience in this area [44,46] and the dependence of the final solution on the initial conditions of the network and the lack of interpretability in the traditional way. As a black box, the model provides results as they come in, without any analytical knowledge [47].

A new method of prediction is the use of the Hadoop model by Kumar et al. [48]. The disadvantage of this method is that it does not support small data files [49]. Karlaftis and Vlahogianni [41] used the GARCH model for prediction. The disadvantage of this method is its complex formalism and complex model [50,51]. On the other hand, McIlroy and his team used the ADF test [52], which suffers from low power due to autocorrelation of random components [53]. Authors in other publications [54,55] have also used data mining techniques for prediction, but those usually suffer from a large number of general explanations [56]. Sevego et al. also found combinations of different models [57]. Parametric models have also been proposed in Bloomfield's work [58].

Given the above literature review, a quick and simple method for determining the forecast of the number of traffic accidents offered by MS Excel, such as trend models, was used for the study. Despite many studies using trend models, they have not been used to forecast the number of traffic accidents in Poland. For this reason, the author addressed the subject under discussion. Despite some limitations—it does not take into account the influence of seasonality in road accidents—it can be used to forecast the number of road accidents.

3. Materials and Methods

More than 38 million people live in Poland (Figure 1). It occupies an area of 312,705 km^2 and is divided into 16 provinces (Table 1, Figure 2). In the states studied, the average reduction in traffic accidents between 2001 and 2021 was more than 56%. Strongest in Kuyavsko–Pomorskie (70%) and Podraskiye (69%), lowest in Rubskiye (32%). The number of automobile accidents depends on the number of residents living in a particular state [59,60]. The number of accidents in Poland is still very high compared to the rest of the European Union [61]. Therefore, every effort should be made to reduce this value and identify the states with the most traffic accidents (Figures 3–5). Based on Figures 1 and 3, it can be concluded that the highest number of traffic accidents is in the Mazowieckie Province and the highest number of accidents per 10,000 inhabitants is in the Lodz Province.

In addition, since this study only used historical data on the number of traffic accidents and did not consider other factors, there is a relationship between the data used, and for this reason the method discussed can be used.

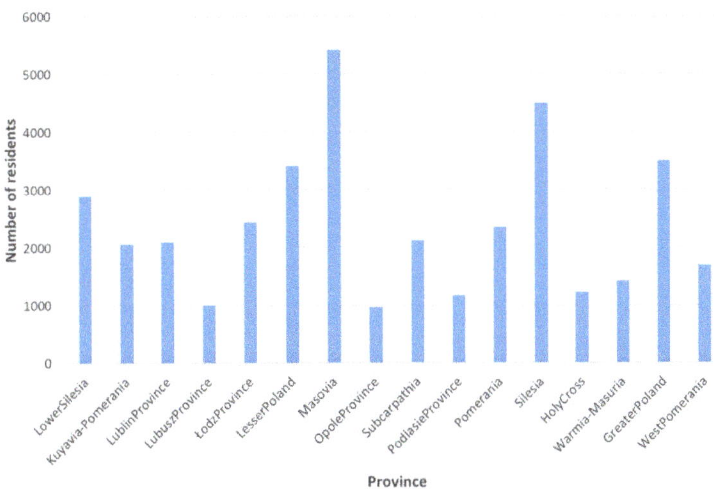

Figure 1. Population of Poland from 2001 to 2020 (thousand people) [62].

Table 1. Area, population by province in Poland in 2020 [62].

Province	Area	Population	
	in km^2	Total	Persons/km^2
POLAND	312,705	38,265,013	122
Lower Silesia	19,947	2,891,321	145
Kuyavia–Pomerania	17,971	2,061,942	115
Lublin Province	25,123	2,095,258	83
Lubusz Province	13,988	1,007,145	72
Łodz Province	18,219	2,437,970	134
Lesser Poland	15,183	3,410,441	225
Masovia	35,559	5,425,028	153
Opole Province	9412	976,774	104
Subcarpathia	17,846	2,121,229	119
Podlasie Province	20,187	1,173,286	58
Pomerania	18,323	2,346,671	128
Silesia	12,333	4,492,330	364
Holy Cross	11,710	1,224,626	105
Warmia–Masuria	24,173	1,416,495	59
Greater Poland	29,826	3,496,450	117
West Pomerania	22,905	1,688,047	74

Figure 2. Location of provinces in Poland [62]. Scale: 1:1,000,000.

Considering Figures 3 and 4, it can be seen that the number of accidents is decreasing, and accidents are seasonal. The highest number of accidents occurs in the summer months and the lowest in the winter months. In addition, we can see that the beginning of the COVID-19 pandemic, 2020, disrupted the number of traffic accidents, when there was a huge decrease. Between 2001 and 2021, the largest decrease in the number of traffic accidents, more than 300%, is seen in the following provinces: kujawsko–pomorskie, lubelskie, podlaskie. On the other hand, the smallest decrease, during the analyzed period, was seen in Lubuskie province and amounted to less than 50%.

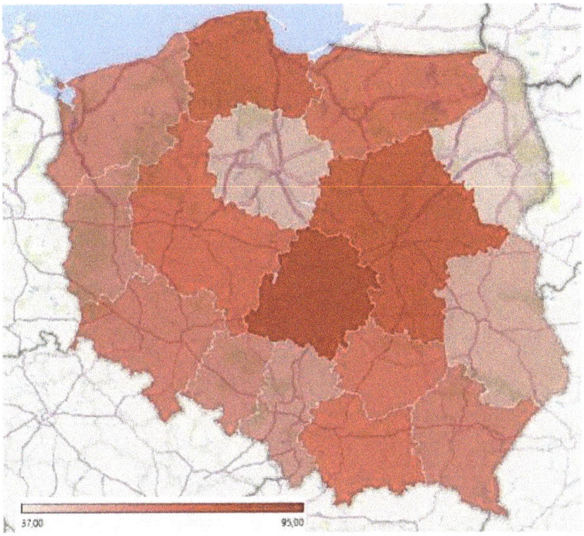

Figure 3. Accident number per 100,000 population in 2020 [62].

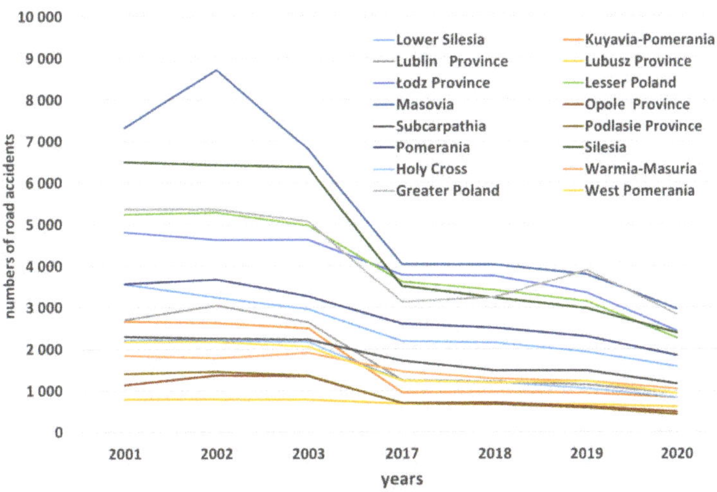

Figure 4. Number of road accidents in Poland by province from 2001 to 2021 [60].

The purpose of this article is to forecast the number of road accidents in Poland in each province. Statistical data of the Police from 2007 to 2021 were used as input data. The following trend models available in Excel software were used to forecast the number of road accidents in each province:

- Exponential;
- Linear;
- Logarithmic;
- 2nd degree polynomial;
- 3rd degree polynomial;
- Polynomial of the 4th degree;
- Polynomial of the 5th degree;
- Polynomial of the 6th degree;
- Potentiometric.

Then, for the obtained forecasts, the errors of expired forecasts were determined based on Equations (1)–(5):

- *ME*—mean error

$$ME = \frac{1}{n}\sum_{i=1}^{n}(Y_i - Y_p) \quad (1)$$

- *MAE*—mean average error

$$MAE = \frac{1}{n}\sum_{i=1}^{n}|Y_i - Y_p| \quad (2)$$

- *MPE*—mean percentage error

$$MPE = \frac{1}{n}\sum_{i=1}^{n}\frac{Y_i - Y_p}{Y_i} \quad (3)$$

- *MAPE*—mean absolute percentage error

$$MAPE = \frac{1}{n}\sum_{i=1}^{n}\frac{|Y_i - Y_p|}{Y_i} \quad (4)$$

- *MSE*—mean square error

$$MSE = \frac{1}{n}\sum_{i=1}^{n}(Y_i - Y_p)^2 \quad (5)$$

where

n—the length of the forecast horizon,
Y—observed value of road accidents,
Y_p—forecasted value of road accidents.

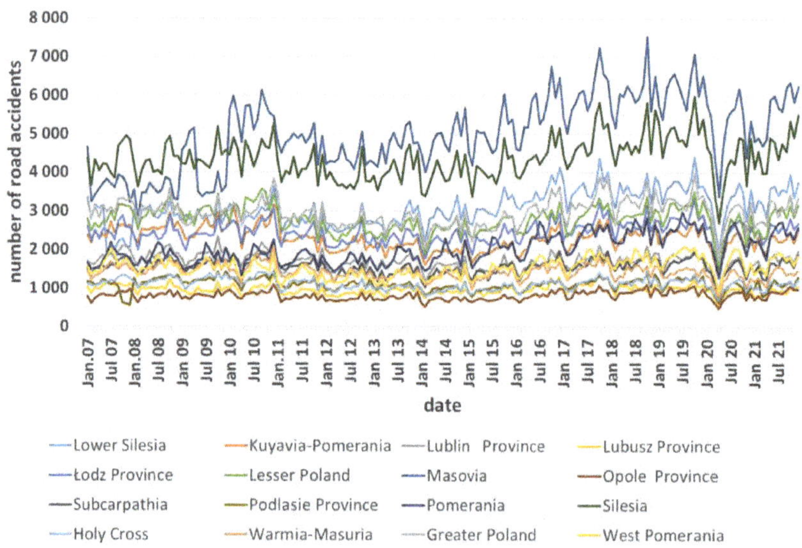

Figure 5. Number of road accidents in Poland by province from 2007 to 2021 [60].

4. Results

For the analyzed trend models, in the first step, the formulas for the statistical data analyzed on an annual and monthly basis for each federal state were determined using Excel software. For annual (2001 to 2021) and monthly (January 2007 to December 2021) data, serial graphs were drawn, and trend lines and R-squared values were determined. As can be seen, the R-squared coefficient, a measure of the quality of the model fit, is mostly good or fair for annual data, and mostly poor or satisfactory for monthly data, regardless of the model used. This is mainly due to the seasonality of traffic accident numbers in each state, which is not well accounted for by the methods used. Tables 2 and 3 and Appendix A shows trend model formulas with annual and monthly data for each state analyzed.

Table 2. Best trend models for annual data.

Province	Model	Annual Data
Lower Silesia	Exponential	$y = 3660.6e^{-0.032x}$
		$R^2 = 0.7993$
Kuyavia–Pomerania	Exponential	$y = 2968.8e^{-0.066x}$
		$R^2 = 0.9776$
Lublin Province	Polynomial of 2nd degree	$y = 1.1254x^2 - 128.19x + 3070.1$
		$R^2 = 0.9649$
Lubusz Province	Exponential	$y = 937.95e^{-0.019x}$
		$R^2 = 0.6083$
Łodz Province	Exponential	$y = 5462.8e^{-0.027x}$
		$R^2 = 0.6749$
Lesser Poland	Exponential	$y = 5751.5e^{-0.034x}$
		$R^2 = 0.7969$
Masovia	Logarithmic	$y = -1744\ln(x) + 8908.6$
		$R^2 = 0.8619$
Opole Province	Exponential	$y = 1384.8e^{-0.044x}$
		$R^2 = 0.9257$
Subcarpathia	Exponential	$y = 2661.8e^{-0.032x}$
		$R^2 = 0.849$
Podlasie Province	Exponential	$y = 1605.6e^{-0.057x}$
		$R^2 = 0.9468$
Pomerania	Exponential	$y = 3663.6e^{-0.026x}$
		$R^2 = 0.7854$
Silesian	Polynomial of 2nd degree	$y = -6.7177x^2 - 86.325x + 6865.4$
		$R^2 = 0.9765$
Warmia–Masuria	Exponential	$y = 2545.8e^{-0.047x}$
		$R^2 = 0.9269$
Holy Cross	Polynomial of 2nd degree	$y = -0.2957x^2 - 63.544x + 2323$
		$R^2 = 0.9656$
Greater Poland	Potentiometric	$y = 6695.8x^{-0.301}$
		$R^2 = 0.6259$
West Pomerania	Exponential	$y = 2381.2e^{-0.04x}$
		$R^2 = 0.939$

Table 3. Best trend models for monthly data.

Province	Model	Monthly Data
Lower Silesia	Polynomial of 2nd degree	$y = 8E\text{-}06x^2 - 0.4824x + 9334.2$
		$R^2 = 0.3703$
Kuyavia–Pomerania	Linear	$y = -0.0491x + 4343.8$
		$R^2 = 0.0651$
Lublin Province	Linear	$y = -0.0316x + 3035.6$
		$R^2 = 0.0557$
Lubusz Province	Exponential	$y = 900.21e^{2E-06x}$
		$R^2 = 0.0005$
Łodz Province	Exponential	$y = 1286.6e^{2E-05x}$
		$R^2 = 0.0459$
Lesser Poland	Linear	$y = -0.0187x + 3570.8$
		$R^2 = 0.0077$
Masovia	Exponential	$y = 121.05e^{9E-05x}$
		$R^2 = 0.4903$
Opole Province	Exponential	$y = 313.34e^{2E-05x}$
		$R^2 = 0.0597$
Subcarpathia	Linear	$y = -0.0013x + 1628.5$
		$R^2 = 0.0001$
Podlasie Province	Exponential	$y = 1308.3e^{-4E-06x}$
		$R^2 = 0.0019$
Pomerania	Exponential	$y = 84.959e^{8E-05x}$
		$R^2 = 0.417$
Silesian	Exponential	$y = 1795.9e^{2E-05x}$
		$R^2 = 0.0739$
Holy Cross	Linear	$y = -0.0142x + 1713.4$
		$R^2 = 0.0243$
Warmia-Masuria	Linear	$y = -0.0281x + 2606.2$
		$R^2 = 0.0597$
Greater Poland	Exponential	$y = 2269.7e^{6E-06x}$
		$R^2 = 0.0067$
West Pomerania	Exponential	$y = 389.87e^{3E-05x}$
		$R^2 = 0.0946$

We then used the data in Tables 2 and 3 and Appendix A to determine the number of traffic accident probabilities. For yearly data, this is the period from 2022 to 2031, and for monthly data, it is from January 2022 to December 2023. In this case, the prediction is based on data from police statistics. Prediction results using this method depend on the choice of model and its tuning.

A trend model with the lowest mean percentage error and mean absolute percentage error was chosen to predict the number of traffic accidents in the analyzed federal states. Based on this, we found that for annual statistical data, the best model is often the exponential model, which has the smallest analyzed error. In addition, polynomial, logarithmic and power models were also used for annual data. In this case, the average MAPE error was 0.52%. On the other hand, for monthly data, linear and exponential models produce the smallest MAPE errors depending on the states studied. However, the average for this error was 65%. This is a very large value and suggests that the propensity model should not be used to predict the number of traffic accidents. Tables 4 and 5 summarize the errors between the lowest modeled annual and monthly data. Using these models, the predicted number of accidents for the next year was determined on a monthly and yearly basis (Figures 6 and 7). Based on Figures 4 and 5, the number of traffic accidents is expected to decline further in the next few years. Please note that the COVID-19 pandemic has changed our forecast significantly. As can be seen from Figure 6, the trend model does not account for the seasonality that occurs in traffic accidents and should not be used in the case under consideration.

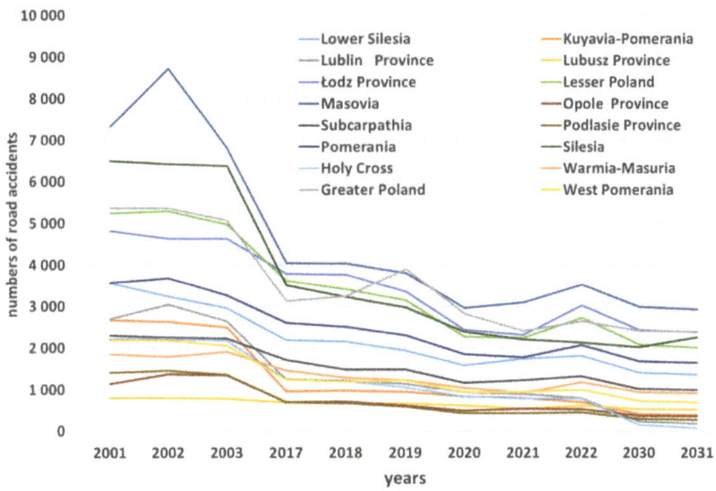

Figure 6. Forecasting annual number of road accidents for 2022–2031.

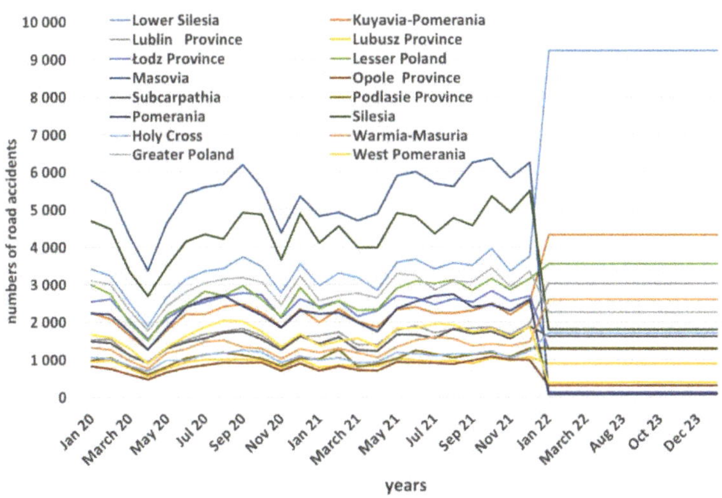

Figure 7. Forecasting monthly number of road accidents for 2022–2023.

Table 4. Summary of errors for annual data.

Province/Model	Model	ME	MPE	The Sum of the Squares	MSE	MAPE (%)	MAE (%)
Lower Silesia	Exponential	14.26	184.47	1,305,179.18	62,151.39	0.26	7.22
Kuyavia–Pomerania	Exponential	10.95	70.32	149,986.76	7142.23	0.35	5.00
Lublin Province	Polynomial of 2nd degree	0.07	93.93	301,100.51	14,338.12	0.32	5.09
Lubusz Province	Exponential	2.76	55.03	119,113.40	5672.07	0.47	6.93
Łodz Province	Exponential	6.86	317.71	3,302,784.34	157,275.44	1.01	8.72
Lesser Poland	Exponential	26.74	290.75	2,392,974.36	11,3951.16	0.14	8.17
Masovia	Logarithmic	176.78	497.81	10,533,523.06	501,596.34	1.77	8.75
Opole Province	Exponential	4.25	54.32	107,325.17	5110.72	0.02	5.98
Subcarpathia	Exponential	11.57	120.66	495,205.41	23,581.21	0.11	6.41
Podlasie Province	Exponential	2.39	52.82	81,724.62	3891.65	0.20	6.66
Pomerania	Exponential	12.34	168.89	853,023.82	40,620.18	0.14	6.61
Silesia	Polynomial of 3rd degree	0.02	144.78	755,411.76	35,971.99	0.17	3.24
Holy Cross	Polynomial of 2nd degree	0.01	72.91	134,675.61	6413.12	0.41	5.26
Warmia–Masuria	Exponential	4.82	150.77	765,403.02	36,447.76	1.00	9.60
Greater Poland	Exponential	48.45	563.32	9,570,955.89	455,759.80	1.81	16.12
West Pomerania	Exponential	1.99	75.89	166,539.04	7930.43	0.16	5.12

Table 5. Summary of errors for monthly data.

Province/Model	Model	ME	MPE	The Sum of the Squares	MSE	MAPE (%)	MAE (%)
Lower Silesia	Polynomial of 2nd degree	6201.42	6201.42	6,965,812,277.61	38,698,957.10	208.40	208.40
Kuyavia–Pomerania	Linear	2046.27	2046.27	770,270,675.15	4,279,281.53	92.75	92.75
Lublin Province	Linear	1316.42	1316.42	319,951,089.84	1,777,506.05	79.68	79.68
Lubusz Province	Exponential	75.75	111.95	3,708,472.78	20,602.63	6.29	10.97
Łodz Province	Exponential	1175.17	1175.17	262,350,703.45	1,457,503.91	47.01	47.01
Lesser Poland	Linear	778.52	778.59	129,389,638.72	718,831.33	29.95	29.95
Masovia	Exponential	4894.72	4894.72	4,479,240,358.37	24,884,668.66	97.47	97.47
Opole Province	Exponential	502.58	502.58	47,967,460.50	266,485.89	60.72	60.72
Subcarpathia	Linear	56.11	156.75	7,374,357.20	40,968.65	5.31	10.83
Podlasie Province	Exponential	198.30	206.68	11,259,059.55	62,550.33	20.66	21.26
Pomerania	Exponential	1937.33	1937.33	701,204,745.03	3,895,581.92	95.63	95.63
Silesia	Exponential	2575.68	2575.68	1,245,839,964.63	6,921,333.14	58.25	58.25
Holy Cross	Linear	591.69	591.69	66,739,299.40	370,773.89	55.47	55.47
Warmia–Masuria	Linear	1171.74	1171.74	253,066,000.67	1,405,922.23	85.00	85.00
Greater Poland	Exponential	699.20	707.79	110,227,747.58	612,376.38	22.44	22.90
West Pomerania	Exponential	1157.77	1157.77	252,984,847.29	1,405,471.37	74.04	74.04

5. Conclusions

Forecasts of the number of accidents in Poland for individual provinces were determined by selected trend models using Excel. The results show that we can still expect a decrease in the number of traffic accidents in the coming years. It should be noted that the COVID-19 pandemic has distorted the results obtained, and if it continues and traffic restrictions are introduced, the proposed model may not be adequate. The value of the average error of 0.52% for annual data can testify to the choice of an effective forecasting method. As we can see, trend models fail for forecasting the monthly number of traffic accidents in which there is seasonality. On the other hand, for annual data, the results are at a high level. The advantage of trend models is their speed of determining the forecast.

The forecast number of traffic accidents obtained in the article can be used, in the future, to formulate further measures to minimize the number of accidents in the analyzed country. These measures may include, for example, the introduction of higher fines for traffic offenses on Polish roads from 1 January 2022.

In his further research, the author plans to take into account more factors affecting the accident rate in Poland and use other methods to forecast the number of road accidents. These may include, but are not limited to traffic volume, day of the week or the age of the perpetrator of the accident and also conduct research using other methods for forecasting the number of road accidents in Poland, such as neural networks and exponential smoothing.

Funding: The article was financed by the university's own funds.

Institutional Review Board Statement: Not applicable.

Informed Consent Statement: Not applicable.

Data Availability Statement: The article was written on the basis of public data available on the pages of the Police Department.

Conflicts of Interest: The author declares no conflict of interest.

Appendix A

Table A1. Trend models for the Lower Silesian province.

Data/Model	Annual Data	Monthly Data
Exponential	$y = 3660.6e^{-0.032x}$	$y = 1E\text{-}08x^{2.4702}$
	$R^2 = 0.7993$	$R^2 = 0.3518$
Linear	$y = -79.014x + 3506.4$	$y = 0.1826x - 4546.7$
	$R^2 = 0.8145$	$R^2 = 0.3689$
Logarithmic	$y = -539.9\ln(x) + 3804.1$	$y = 7618.2\ln(x) - 77969$
	$R^2 = 0.6604$	$R^2 = 0.3679$
Polynomial of 2nd degree	$y = -2.662x^2 - 20.451x + 3282$	$y = 8E\text{-}06x^2 - 0.4824x + 9334.2$
	$R^2 = 0.8415$	$R^2 = 0.3703$
Polynomial of 3rd degree	$y = -0.3402x^3 + 8.5652x^2 - 121.56x + 3488.5$	$y = -3E\text{-}08x^3 + 0.0041x^2 - 171.09x + 2E{+}06$
	$R^2 = 0.8537$	$R^2 = 0.4156$
Polynomial of 4th degree	$y = 0.0557x^4 - 2.789x^3 + 43.763x^2 - 303.3x + 3729.9$	$y = -2E\text{-}11x^4 + 3E\text{-}06x^3 - 0.1807x^2 + 4971.6x - 5E{+}07$
	$R^2 = 0.8625$	$R^2 = 0.4408$
Polynomial of 5th degree	$y = -0.0019x^5 + 0.1617x^4 - 4.8892x^3 + 61.752x^2 - 365.45x + 3790.3$	$y = 2E\text{-}14x^5 - 3E\text{-}09x^4 + 0.0003x^3 - 12.199x^2 + 255831x - 2E{+}09$
	$R^2 = 0.8628$	$R^2 = 0.4826$
Polynomial of 6th degree	$y = 0.002x^6 - 0.1329x^5 + 3.4719x^4 - 44.893x^3 + 296.45x^2 - 961.09x + 4248$	$y = 2E\text{-}17x^6 - 4E\text{-}12x^5 + 4E\text{-}07x^4 - 0.0233x^3 + 727.3x^2 - 1E{+}07x + 8E{+}10$
	$R^2 = 0.8703$	$R^2 = 0.5586$
Potentiometric	$y = 4062.9x^{-0.21}$	$y = 1E\text{-}08x^{2.4702}$
	$R^2 = 0.6058$	$R^2 = 0.3518$

Table A2. Trend models for the Kuyavian–Pomeranian province.

Data/Model	Annual Data	Monthly Data
Exponential	$y = 2968.8e^{-0.066x}$	$y = 5563.1e^{-2E-05x}$
	$R^2 = 0.9776$	$R^2 = 0.0624$
Linear	$y = -100.67x + 2672.2$	$y = -0.0491x + 4343.8$
	$R^2 = 0.9481$	$R^2 = 0.0651$
Logarithmic	$y = -749.4\ln(x) + 3184.3$	$y = -2093\ln(x) + 24565$
	$R^2 = 0.9122$	$R^2 = 0.068$
Polynomial of 2nd degree	$y = 3.7309x^2 - 182.75x + 2986.9$	$y = 4E\text{-}05x^2 - 3.7996x + 82631$
	$R^2 = 0.9861$	$R^2 = 0.174$
Polynomial of 3rd degree	$y = 0.2451x^3 - 4.357x^2 - 109.91x + 2838.1$	$y = -2E\text{-}09x^3 + 0.0003x^2 - 14.473x + 231109$
	$R^2 = 0.9906$	$R^2 = 0.1745$
Polynomial of 4th degree	$y = -0.0404x^4 + 2.0229x^3 - 29.91x^2 + 22.025x + 2662.8$	$y = -3E\text{-}11x^4 + 5E\text{-}06x^3 - 0.313x^2 + 8706.8x - 9E+07$
	$R^2 = 0.994$	$R^2 = 0.3524$
Polynomial of 5th degree	$y = -0.0044x^5 + 0.1989x^4 - 2.7169x^3 + 10.689x^2 - 118.23x + 2799.1$	$y = 1E\text{-}14x^5 - 2E\text{-}09x^4 + 0.0002x^3 - 7.3524x^2 + 155643x - 1E+09$
	$R^2 = 0.995$	$R^2 = 0.3874$
Polynomial of 6th degree	$y = -0.0001x^6 + 0.0026x^5 + 0.0235x^4 - 0.5973x^3 - 1.7467x^2 - 86.666x + 2774.9$	$y = 1E\text{-}17x^6 - 3E\text{-}12x^5 + 3E\text{-}07x^4 - 0.0159x^3 + 494.62x^2 - 8E+06x + 6E+10$
	$R^2 = 0.995$	$R^2 = 0.4732$
Potentiometric	$y = 3938.4x^{-0.464}$	$y = 4E+07x^{-0.914}$
	$R^2 = 0.8524$	$R^2 = 0.0652$

Table A3. Trend models for Lublin province.

Data/Model	Annual Data	Monthly Data
Exponential	$y = 3305.9e^{-0.059x}$	$y = 3789.7e^{-2E-05x}$
	$R^2 = 0.9677$	$R^2 = 0.0549$
Linear	$y = -103.43x + 2975.2$	$y = -0.0316x + 3035.6$
	$R^2 = 0.9615$	$R^2 = 0.0557$
Logarithmic	$y = -733.2\ln(x) + 3421.7$	$y = -1340\ln(x) + 15972$
	$R^2 = 0.8388$	$R^2 = 0.0576$
Polynomial of 2nd degree	$y = 1.1254x^2 - 128.19x + 3070.1$	$y = 2E\text{-}05x^2 - 1.8755x + 41525$
	$R^2 = 0.9649$	$R^2 = 0.1102$
Polynomial of 3rd degree	$y = 0.2624x^3 - 7.5334x^2 - 50.211x + 2910.7$	$y = 7E\text{-}10x^3 - 7E\text{-}05x^2 + 1.9618x - 11858$
	$R^2 = 0.9699$	$R^2 = 0.1103$
Polynomial of 4th degree	$y = -0.0316x^4 + 1.6547x^3 - 27.546x^2 + 53.118x + 2773.5$	$y = -1E\text{-}11x^4 + 2E\text{-}06x^3 - 0.1304x^2 + 3628.8x - 4E+07$
	$R^2 = 0.9718$	$R^2 = 0.174$
Polynomial of 5th degree	$y = -0.0058x^5 + 0.2879x^4 - 4.6744x^3 + 26.667x^2 - 134.16x + 2955.5$	$y = 7E\text{-}15x^5 - 1E\text{-}09x^4 + 0.0001x^3 - 5.1059x^2 + 107484x - 9E+08$
	$R^2 = 0.9736$	$R^2 = 0.2102$
Polynomial of 6th degree	$y = -0.0012x^6 + 0.0765x^5 - 1.7937x^4 + 20.481x^3 - 120.92x^2 + 240.4x + 2667.7$	$y = 6E\text{-}18x^6 - 1E\text{-}12x^5 + 2E\text{-}07x^4 - 0.0086x^3 + 267.23x^2 - 4E+06x + 3E+10$
	$R^2 = 0.9756$	$R^2 = 0.2624$
Potentiometric	$y = 4104.7x^{-0.402}$	$y = 1E+07x^{-0.812}$
	$R^2 = 0.7726$	$R^2 = 0.0567$

Table A4. Trend models for Lubuskie province.

Data/Model	Annual Data	Monthly Data
Exponential	$y = 937.95e^{-0.019x}$	$y = 900.21e^{2E-06x}$
	$R^2 = 0.6083$	$R^2 = 0.0005$
Linear	$y = -13.829x + 920.97$	$y = 0.0027x + 862.02$
	$R^2 = 0.5766$	$R^2 = 0.0013$
Logarithmic	$y = -73.48\ln(x) + 927.65$	$y = 106.88\ln(x) - 161.11$
	$R^2 = 0.2827$	$R^2 = 0.0011$
Polynomial of 2nd degree	$y = -1.6013x^2 + 21.401x + 785.93$	$y = 7E\text{-}06x^2 - 0.5564x + 12532$
	$R^2 = 0.8019$	$R^2 = 0.0163$
Polynomial of 3rd degree	$y = 0.1435x^3 - 6.3358x^2 + 64.039x + 698.81$	$y = -1E\text{-}08x^3 + 0.0017x^2 - 69.977x + 978278$
	$R^2 = 0.8521$	$R^2 = 0.1304$
Polynomial of 4th degree	$y = 0.0061x^4 - 0.1246x^3 - 2.4829x^2 + 44.145x + 725.24$	$y = -5E\text{-}12x^4 + 8E\text{-}07x^3 - 0.0477x^2 + 1303.7x - 1E{+}07$
	$R^2 = 0.8546$	$R^2 = 0.1579$
Polynomial of 5th degree	$y = -0.0067x^5 + 0.3758x^4 - 7.4473x^3 + 60.24x^2 - 172.54x + 935.78$	$y = 4E\text{-}15x^5 - 9E\text{-}10x^4 + 8E\text{-}05x^3 - 3.3252x^2 + 69716x - 6E{+}08$
	$R^2 = 0.9327$	$R^2 = 0.2052$
Polynomial of 6th degree	$y = -0.0003x^6 + 0.0147x^5 - 0.1649x^4 - 0.9136x^3 + 21.908x^2 - 75.253x + 861.02$	$y = 4E\text{-}18x^6 - 1E\text{-}12x^5 + 1E\text{-}07x^4 - 0.0056x^3 + 174.52x^2 - 3E{+}06x + 2E{+}10$
	$R^2 = 0.9373$	$R^2 = 0.2721$
Potentiometric	$y = 949.27x^{-0.102}$	$y = 481.83x^{0.0656}$
	$R^2 = 0.3059$	$R^2 = 0.0004$

Table A5. Trend models for Łódź Province.

Data/Model	Annual Data	Monthly Data
Exponential	$y = 5462.8e^{-0.027x}$	$y = 1286.6e^{2E-05x}$
	$R^2 = 0.6749$	$R^2 = 0.0459$
Linear	$y = -101x + 5231.5$	$y = 0.0403x + 778.52$
	$R^2 = 0.7447$	$R^2 = 0.053$
Logarithmic	$y = -624.7\ln(x) + 5470.4$	$y = 1658.5\ln(x) - 15182$
	$R^2 = 0.4947$	$R^2 = 0.0515$
Polynomial of 2nd degree	$y = -7.4751x^2 + 63.449x + 4601.1$	$y = 2E\text{-}05x^2 - 2.0299x + 43991$
	$R^2 = 0.8636$	$R^2 = 0.0931$
Polynomial of 3rd degree	$y = -0.5521x^3 + 10.745x^2 - 100.64x + 4936.3$	$y = -1E\text{-}08x^3 + 0.0017x^2 - 73.911x + 1E{+}06$
	$R^2 = 0.8816$	$R^2 = 0.1168$
Polynomial of 4th degree	$y = -0.1513x^4 + 6.1071x^3 - 84.97x^2 + 393.56x + 4279.9$	$y = -2E\text{-}11x^4 + 3E\text{-}06x^3 - 0.1873x^2 + 5189.5x - 5E{+}07$
	$R^2 = 0.9181$	$R^2 = 0.195$
Polynomial of 5th degree	$y = -0.0197x^5 + 0.9343x^4 - 15.398x^3 + 99.229x^2 - 242.77x + 4898.2$	$y = 4E\text{-}15x^5 - 9E\text{-}10x^4 + 7E\text{-}05x^3 - 3.1414x^2 + 66850x - 6E{+}08$
	$R^2 = 0.9344$	$R^2 = 0.2024$
Polynomial of 6th degree	$y = 0.0044x^6 - 0.3069x^5 + 8.1925x^4 - 103.11x^3 + 613.84x^2 - 1548.8x + 5901.8$	$y = 1E\text{-}17x^6 - 3E\text{-}12x^5 + 3E\text{-}07x^4 - 0.0154x^3 + 479.91x^2 - 8E{+}06x + 6E{+}10$
	$R^2 = 0.9546$	$R^2 = 0.2982$
Potentiometric	$y = 5778.5x^{-0.165}$	$y = 2.9338x^{0.6322}$
	$R^2 = 0.428$	$R^2 = 0.0444$

Table A6. Trend models for the Lesser Poland province.

Data/Model	Annual Data	Monthly Data
Exponential	$y = 5751.5e^{-0.034x}$	$y = 3818.4e^{-8E-06x}$
	$R^2 = 0.7969$	$R^2 = 0.0093$
Linear	$y = -125.74x + 5453.3$	$y = -0.0187x + 3570.8$
	$R^2 = 0.8736$	$R^2 = 0.0077$
Logarithmic	$y = -856.4\ln(x) + 5920.7$	$y = -816\ln(x) + 11472$
	$R^2 = 0.7035$	$R^2 = 0.0085$
Polynomial of 2nd degree	$y = -4.2552x^2 - 32.128x + 5094.5$	$y = 4E\text{-}05x^2 - 3.0715x + 67295$
	$R^2 = 0.9027$	$R^2 = 0.0669$
Polynomial of 3rd degree	$y = -0.9622x^3 + 27.499x^2 - 318.11x + 5678.7$	$y = -4E\text{-}09x^3 + 0.0005x^2 - 24.164x + 360727$
	$R^2 = 0.9441$	$R^2 = 0.0683$
Polynomial of 4th degree	$y = -0.0701x^4 + 2.1228x^3 - 16.844x^2 - 89.152x + 5374.6$	$y = -3E\text{-}11x^4 + 5E\text{-}06x^3 - 0.3065x^2 + 8523.8x - 9E{+}07$
	$R^2 = 0.9501$	$R^2 = 0.2084$
Polynomial of 5th degree	$y = -0.0057x^5 + 0.2409x^4 - 4.0384x^3 + 35.93x^2 - 271.46x + 5551.8$	$y = 2E\text{-}14x^5 - 4E\text{-}09x^4 + 0.0003x^3 - 12.543x^2 + 263934x - 2E{+}09$
	$R^2 = 0.9511$	$R^2 = 0.2952$
Polynomial of 6th degree	$y = 0.0014x^6 - 0.1005x^5 + 2.6393x^4 - 33.023x^3 + 205.98x^2 - 703.03x + 5883.4$	$y = 1E\text{-}17x^6 - 3E\text{-}12x^5 + 3E\text{-}07x^4 - 0.0176x^3 + 547.22x^2 - 9E{+}06x + 6E{+}10$
	$R^2 = 0.9527$	$R^2 = 0.3826$
Potentiometric	$y = 6386.4x^{-0.219}$	$y = 96700x^{-0.334}$
	$R^2 = 0.5906$	$R^2 = 0.0101$

Table A7. Trend models for the Masovia province.

Data/Model	Annual Data	Monthly Data
Exponential	$y = 8271.5e^{-0.048x}$	$y = 121.05e^{9E-05x}$
	$R^2 = 0.9283$	$R^2 = 0.4903$
Linear	$y = -244.75x + 7786.2$	$y = 0.4231x - 12672$
	$R^2 = 0.8992$	$R^2 = 0.483$
Logarithmic	$y = -1744\ln(x) + 8908.6$	$y = 17743\ln(x) - 183769$
	$R^2 = 0.8619$	$R^2 = 0.4869$
Polynomial of 2nd degree	$y = 4.749x^2 - 344.48x + 8151.9$	$y = -8E\text{-}05x^2 + 6.7035x - 143764$
	$R^2 = 0.9082$	$R^2 = 0.5134$
Polynomial of 3rd degree	$y = -0.1364x^3 + 9.0453x^2 - 381.46x + 8224.4$	$y = -2E\text{-}08x^3 + 0.0026x^2 - 104.17x + 1E{+}06$
	$R^2 = 0.9084$	$R^2 = 0.5181$
Polynomial of 4th degree	$y = 0.0347x^4 - 1.593x^3 + 29.046x^2 - 480.3x + 8350.7$	$y = -4E\text{-}11x^4 + 6E\text{-}06x^3 - 0.3799x^2 + 10544x - 1E{+}08$
	$R^2 = 0.9086$	$R^2 = 0.5446$
Polynomial of 5th degree	$y = -0.0841x^5 + 4.4507x^4 - 85.146x^3 + 713.52x^2 - 2747.9x + 10479$	$y = 8E\text{-}15x^5 - 2E\text{-}09x^4 + 0.0001x^3 - 6.0186x^2 + 128241x - 1E{+}09$
	$R^2 = 0.9493$	$R^2 = 0.5468$
Polynomial of 6th degree	$y = 0.0162x^6 - 1.1063x^5 + 29.128x^4 - 370.28x^3 + 2315.8x^2 - 6656.5x + 13390$	$y = 4E\text{-}17x^6 - 1E\text{-}11x^5 + 1E\text{-}06x^4 - 0.0603x^3 + 1885.7x^2 - 3E{+}07x + 2E{+}11$
	$R^2 = 0.984$	$R^2 = 0.6682$
Potentiometric	$y = 9945.1x^{-0.324}$	$y = 3E\text{-}14x^{3.7203}$
	$R^2 = 0.8059$	$R^2 = 0.4954$

Table A8. Trend models for Opole province.

Data/Model	Annual Data	Monthly Data
Exponential	$y = 1384.8e^{-0.044x}$	$y = 313.34e^{2E-05x}$
	$R^2 = 0.9257$	$R^2 = 0.0597$
Linear	$y = -37.996x + 1306.3$	$y = 0.0201x - 25.195$
	$R^2 = 0.909$	$R^2 = 0.0728$
Logarithmic	$y = -270\ln(x) + 1471.8$	$y = 826.22\ln(x) - 7974.5$
	$R^2 = 0.797$	$R^2 = 0.0703$
Polynomial of 2nd degree	$y = 0.5472x^2 - 50.034x + 1352.4$	$y = 1E\text{-}05x^2 - 1.1854x + 25138$
	$R^2 = 0.9145$	$R^2 = 0.1476$
Polynomial of 3rd degree	$y = 0.0029x^3 + 0.451x^2 - 49.168x + 1350.7$	$y = -3E\text{-}09x^3 + 0.0004x^2 - 15.737x + 227570$
	$R^2 = 0.9145$	$R^2 = 0.153$
Polynomial of 4th degree	$y = -0.0389x^4 + 1.7163x^3 - 24.176x^2 + 77.988x + 1181.8$	$y = -8E\text{-}12x^4 + 1E\text{-}06x^3 - 0.0787x^2 + 2184.9x - 2E{+}07$
	$R^2 = 0.9353$	$R^2 = 0.2282$
Polynomial of 5th degree	$y = 0.0044x^5 - 0.281x^4 + 6.5105x^3 - 65.24x^2 + 219.85x + 1043.9$	$y = 4E\text{-}15x^5 - 8E\text{-}10x^4 + 7E\text{-}05x^3 - 2.8588x^2 + 60215x - 5E{+}08$
	$R^2 = 0.9423$	$R^2 = 0.2645$
Polynomial of 6th degree	$y = -0.0006x^6 + 0.0425x^5 - 1.2443x^4 + 18.152x^3 - 133.54x^2 + 393.19x + 910.74$	$y = 4E\text{-}18x^6 - 1E\text{-}12x^5 + 1E\text{-}07x^4 - 0.0064x^3 + 199.59x^2 - 3E{+}06x + 2E{+}10$
	$R^2 = 0.9454$	$R^2 = 0.3572$
Potentiometric	$y = 1636x^{-0.3}$	$y = 0.0415x^{0.9283}$
	$R^2 = 0.7549$	$R^2 = 0.0575$

Table A9. Trend models for the Subcarpathia province.

Data/Model	Annual Data	Monthly Data
Exponential	$y = 2661.8e^{-0.032x}$	$y = 1690.1e^{-2E-06x}$
	$R^2 = 0.849$	$R^2 = 0.0006$
Linear	$y = -56.701x + 2542.6$	$y = -0.0013x + 1628.5$
	$R^2 = 0.8799$	$R^2 = 0.0001$
Logarithmic	$y = -359.2\ln(x) + 2695$	$y = -74.05\ln(x) + 2360.1$
	$R^2 = 0.613$	$R^2 = 0.0002$
Polynomial of 2nd degree	$y = -2.7746x^2 + 4.3389x + 2308.6$	$y = 2E\text{-}05x^2 - 1.4932x + 32768$
	$R^2 = 0.9413$	$R^2 = 0.0423$
Polynomial of 3rd degree	$y = 0.1297x^3 - 7.0544x^2 + 42.883x + 2229.9$	$y = -2E\text{-}09x^3 + 0.0002x^2 - 9.9627x + 150593$
	$R^2 = 0.945$	$R^2 = 0.0429$
Polynomial of 4th degree	$y = -0.022x^4 + 1.0973x^3 - 20.961x^2 + 114.69x + 2134.5$	$y = -2E\text{-}11x^4 + 3E\text{-}06x^3 - 0.1615x^2 + 4490.4x - 5E{+}07$
	$R^2 = 0.9479$	$R^2 = 0.1587$
Polynomial of 5th degree	$y = -0.0062x^5 + 0.32x^4 - 5.676x^3 + 37.055x^2 - 85.735x + 2329.3$	$y = 8E\text{-}15x^5 - 2E\text{-}09x^4 + 0.0001x^3 - 5.7557x^2 + 121260x - 1E{+}09$
	$R^2 = 0.954$	$R^2 = 0.2129$
Polynomial of 6th degree	$y = 0.0012x^6 - 0.0856x^5 + 2.3262x^4 - 29.922x^3 + 179.3x^2 - 446.75x + 2606.7$	$y = 8E\text{-}18x^6 - 2E\text{-}12x^5 + 2E\text{-}07x^4 - 0.0116x^3 + 360.56x^2 - 6E{+}06x + 4E{+}10$
	$R^2 = 0.9598$	$R^2 = 0.3245$
Potentiometric	$y = 2875.5x^{-0.197}$	$y = 4146.2x^{-0.092}$
	$R^2 = 0.5705$	$R^2 = 0.0007$

Table A10. Trend models for Podlaskie province.

Data/Model		Annual Data	Monthly Data
Exponential		$y = 1605.6e^{-0.057x}$	$y = 1308.3e^{-4E-06x}$
		$R^2 = 0.9468$	$R^2 = 0.0019$
Linear		$y = -49.516x + 1456.8$	$y = -0.0061x + 1364.2$
		$R^2 = 0.9544$	$R^2 = 0.004$
Logarithmic		$y = -359.1\ln(x) + 1688.1$	$y = -260.2\ln(x) + 3878.5$
		$R^2 = 0.8715$	$R^2 = 0.0042$
Polynomial of 2nd degree		$y = 0.8747x^2 - 68.759x + 1530.5$	$y = 6E\text{-}06x^2 - 0.5033x + 11742$
		$R^2 = 0.963$	$R^2 = 0.0116$
Polynomial of 3rd degree		$y = -0.0076x^3 + 1.1243x^2 - 71.007x + 1535.1$	$y = 3E\text{-}09x^3 - 0.0003x^2 + 12.994x - 176022$
		$R^2 = 0.9631$	$R^2 = 0.0144$
Polynomial of 4th degree		$y = -0.0326x^4 + 1.4274x^3 - 19.501x^2 + 35.49x + 1393.7$	$y = -1E\text{-}11x^4 + 2E\text{-}06x^3 - 0.1254x^2 + 3495.8x - 4E+07$
		$R^2 = 0.9721$	$R^2 = 0.1273$
Polynomial of 5th degree		$y = -0.0041x^5 + 0.1948x^4 - 3.0764x^3 + 19.076x^2 - 97.781x + 1523.2$	$y = 3E\text{-}15x^5 - 6E\text{-}10x^4 + 5E\text{-}05x^3 - 2.1244x^2 + 45221x - 4E+08$
		$R^2 = 0.9759$	$R^2 = 0.1385$
Polynomial of 6th degree		$y = 0.0003x^6 - 0.0258x^5 + 0.7421x^4 - 9.6906x^3 + 57.881x^2 - 196.26x + 1598.8$	$y = 6E\text{-}18x^6 - 2E\text{-}12x^5 + 2E\text{-}07x^4 - 0.0091x^3 + 282.88x^2 - 5E+06x + 3E+10$
		$R^2 = 0.9765$	$R^2 = 0.2485$
Potentiometric		$y = 2004.2x^{-0.392}$	$y = 7319.3x^{-0.178}$
		$R^2 = 0.7814$	$R^2 = 0.002$

Table A11. Trend models for Pomeranian province.

Data/Model		Annual Data	Monthly Data
Exponential		$y = 3663.6e^{-0.026x}$	$y = 84.959e^{8E-05x}$
		$R^2 = 0.7854$	$R^2 = 0.417$
Linear		$y = -69.164x + 3559.7$	$y = 0.1573x - 4551.4$
		$R^2 = 0.8237$	$R^2 = 0.4341$
Logarithmic		$y = -496.7\ln(x) + 3872.3$	$y = 6541.5\ln(x) - 67579$
		$R^2 = 0.7377$	$R^2 = 0.4306$
Polynomial of 2nd degree		$y = -1.0377x^2 - 46.334x + 3472.2$	$y = 2E\text{-}05x^2 - 1.8844x + 38064$
		$R^2 = 0.8291$	$R^2 = 0.455$
Polynomial of 3rd degree		$y = -0.8089x^3 + 25.655x^2 - 286.73x + 3963.3$	$y = -3E\text{-}08x^3 + 0.0033x^2 - 139.08x + 2E+06$
		$R^2 = 0.9203$	$R^2 = 0.5015$
Polynomial of 4th degree		$y = -0.0349x^4 + 0.7259x^3 + 3.5961x^2 - 172.83x + 3812.1$	$y = -2E\text{-}11x^4 + 3E\text{-}06x^3 - 0.1737x^2 + 4788.7x - 5E+07$
		$R^2 = 0.9249$	$R^2 = 0.5384$
Polynomial of 5th degree		$y = -0.002x^5 + 0.0738x^4 - 1.4261x^3 + 22.029x^2 - 236.51x + 3873.9$	$y = 1E\text{-}14x^5 - 2E\text{-}09x^4 + 0.0002x^3 - 7.3691x^2 + 154981x - 1E+09$
		$R^2 = 0.9253$	$R^2 = 0.5622$
Polynomial of 6th degree		$y = 1E\text{-}04x^6 - 0.0085x^5 + 0.2378x^4 - 3.4082x^3 + 33.657x^2 - 266.02x + 3896.6$	$y = 8E\text{-}18x^6 - 2E\text{-}12x^5 + 2E\text{-}07x^4 - 0.0108x^3 + 336.03x^2 - 6E+06x + 4E+10$
		$R^2 = 0.9253$	$R^2 = 0.5882$
Potentiometric		$y = 4053x^{-0.178}$	$y = 6E\text{-}12x^{3.1379}$
		$R^2 = 0.6494$	$R^2 = 0.4138$

Table A12. Trend models for the Silesian province.

Data/Model		Annual Data	Monthly Data
Exponential		$y = 8278.3e^{-0.053x}$	$y = 1795.9e^{2E-05x}$
		$R^2 = 0.9134$	$R^2 = 0.0739$
Linear		$y = -234.11x + 7432$	$y = 0.1003x + 182.21$
		$R^2 = 0.9536$	$R^2 = 0.0874$
Logarithmic		$y = -1535\ln(x) + 8173.7$	$y = 4135.4\ln(x) - 39626$
		$R^2 = 0.7117$	$R^2 = 0.0852$
Polynomial of 2nd degree		$y = -6.7177x^2 - 86.325x + 6865.4$	$y = 5E\text{-}05x^2 - 4.2277x + 90523$
		$R^2 = 0.9765$	$R^2 = 0.134$
Polynomial of 3rd degree		$y = 0.6756x^3 - 29.012x^2 + 114.46x + 6455.2$	$y = -2E\text{-}08x^3 + 0.0029x^2 - 124.35x + 2E{+}06$
		$R^2 = 0.9829$	$R^2 = 0.1517$
Polynomial of 4th degree		$y = -0.1059x^4 + 5.337x^3 - 96.012x^2 + 460.4x + 5995.8$	$y = -4E\text{-}11x^4 + 6E\text{-}06x^3 - 0.3704x^2 + 10268x - 1E{+}08$
		$R^2 = 0.9872$	$R^2 = 0.2328$
Polynomial of 5th degree		$y = -0.0031x^5 + 0.0633x^4 + 1.9849x^3 - 67.299x^2 + 361.21x + 6092.1$	$y = 2E\text{-}14x^5 - 4E\text{-}09x^4 + 0.0003x^3 - 13.992x^2 + 294603x - 2E{+}09$
		$R^2 = 0.9873$	$R^2 = 0.275$
Polynomial of 6th degree		$y = 0.0019x^6 - 0.1298x^5 + 3.2675x^4 - 36.739x^3 + 159.89x^2 - 215.37x + 6535.2$	$y = 2E\text{-}17x^6 - 5E\text{-}12x^5 + 5E\text{-}07x^4 - 0.0288x^3 + 897.12x^2 - 1E{+}07x + 1E{+}11$
		$R^2 = 0.9882$	$R^2 = 0.3657$
Potentiometric		$y = 9554.3x^{-0.337}$	$y = 0.4145x^{0.87}$
		$R^2 = 0.6368$	$R^2 = 0.0719$

Table A13. Trend models for the Holy Cross province.

Data/Model		Annual Data	Monthly Data
Exponential		$y = 2545.8e^{-0.047x}$	$y = 1906e^{-1E-05x}$
		$R^2 = 0.9269$	$R^2 = 0.0241$
Linear		$y = -70.049x + 2347.9$	$y = -0.0142x + 1713.4$
		$R^2 = 0.9651$	$R^2 = 0.0243$
Logarithmic		$y = -489.1\ln(x) + 2634.3$	$y = -610.5\ln(x) + 7615.9$
		$R^2 = 0.8169$	$R^2 = 0.0258$
Polynomial of 2nd degree		$y = -0.2957x^2 - 63.544x + 2323$	$y = 2E\text{-}05x^2 - 1.5293x + 33338$
		$R^2 = 0.9656$	$R^2 = 0.1037$
Polynomial of 3rd degree		$y = -0.0844x^3 + 2.4883x^2 - 88.617x + 2374.2$	$y = -4E\text{-}10x^3 + 7E\text{-}05x^2 - 3.8804x + 66046$
		$R^2 = 0.9667$	$R^2 = 0.1037$
Polynomial of 4th degree		$y = -0.0538x^4 + 2.2813x^3 - 31.514x^2 + 86.947x + 2141$	$y = -1E\text{-}11x^4 + 2E\text{-}06x^3 - 0.1485x^2 + 4133.1x - 4E{+}07$
		$R^2 = 0.9791$	$R^2 = 0.2824$
Polynomial of 5th degree		$y = -0.0025x^5 + 0.083x^4 - 0.4272x^3 - 8.3145x^2 + 6.8028x + 2218.9$	$y = 4E\text{-}15x^5 - 8E\text{-}10x^4 + 7E\text{-}05x^3 - 2.8272x^2 + 60046x - 5E{+}08$
		$R^2 = 0.9798$	$R^2 = 0.3051$
Polynomial of 6th degree		$y = 0.0006x^6 - 0.0449x^5 + 1.1552x^4 - 13.385x^3 + 67.706x^2 - 186.13x + 2367.2$	$y = 5E\text{-}18x^6 - 1E\text{-}12x^5 + 1E\text{-}07x^4 - 0.0079x^3 + 245.49x^2 - 4E{+}06x + 3E{+}10$
		$R^2 = 0.981$	$R^2 = 0.3988$
Potentiometric		$y = 2996.8x^{-0.316}$	$y = 412018x^{-0.556}$
		$R^2 = 0.7194$	$R^2 = 0.0257$

Table A14. Trend models for the Warmian–Masurian province.

Data/Model	Annual Data	Monthly Data
Exponential	$y = 2216.7e^{-0.029x}$	$y = 3353.1e^{-2E-05x}$
	$R^2 = 0.6897$	$R^2 = 0.0613$
Linear	$y = -43.531x + 2119.9$	$y = -0.0281x + 2606.2$
	$R^2 = 0.699$	$R^2 = 0.0597$
Logarithmic	$y = -256.3\ln(x) + 2194.8$	$y = -1178\ln(x) + 13969$
	$R^2 = 0.4206$	$R^2 = 0.0602$
Polynomial of 2nd degree	$y = -3.7986x^2 + 40.037x + 1799.5$	$y = 5E\text{-}06x^2 - 0.4703x + 11837$
	$R^2 = 0.8541$	$R^2 = 0.0639$
Polynomial of 3rd degree	$y = -0.0283x^3 - 2.8652x^2 + 31.631x + 1816.7$	$y = -1E\text{-}09x^3 + 0.0002x^2 - 7.8062x + 113891$
	$R^2 = 0.8543$	$R^2 = 0.0645$
Polynomial of 4th degree	$y = -0.033x^4 + 1.4255x^3 - 23.761x^2 + 139.52x + 1673.4$	$y = -1E\text{-}11x^4 + 2E\text{-}06x^3 - 0.1233x^2 + 3430.8x - 4E{+}07$
	$R^2 = 0.8631$	$R^2 = 0.1418$
Polynomial of 5th degree	$y = -0.0037x^5 + 0.1709x^4 - 2.6132x^3 + 10.833x^2 + 20.015x + 1789.5$	$y = 7E\text{-}15x^5 - 1E\text{-}09x^4 + 0.0001x^3 - 5.2879x^2 + 111231x - 9E{+}08$
	$R^2 = 0.866$	$R^2 = 0.1945$
Polynomial of 6th degree	$y = 0.0013x^6 - 0.0879x^5 + 2.2988x^4 - 28.33x^3 + 161.71x^2 - 362.89x + 2083.8$	$y = 7E\text{-}18x^6 - 2E\text{-}12x^5 + 2E\text{-}07x^4 - 0.0095x^3 + 297.52x^2 - 5E{+}06x + 3E{+}10$
	$R^2 = 0.8748$	$R^2 = 0.2818$
Potentiometric	$y = 2322.2x^{-0.17}$	$y = 1E{+}07x^{-0.862}$
	$R^2 = 0.4065$	$R^2 = 0.0618$

Table A15. Trend models for Greater Poland province.

Data/Model	Annual Data	Monthly Data
Exponential	$y = 5383.4e^{-0.039x}$	$y = 2269.7e^{6E-06x}$
	$R^2 = 0.6153$	$R^2 = 0.0067$
Linear	$y = -146.33x + 5263.9$	$y = 0.0232x + 1999.8$
	$R^2 = 0.6566$	$R^2 = 0.0109$
Logarithmic	$y = -1137\ln(x) + 6111$	$y = 924.51\ln(x) - 6869.2$
	$R^2 = 0.6881$	$R^2 = 0.0099$
Polynomial of 2nd degree	$y = 10.982x^2 - 387.94x + 6190$	$y = 4E\text{-}05x^2 - 3.4456x + 74403$
	$R^2 = 0.7643$	$R^2 = 0.0807$
Polynomial of 3rd degree	$y = 1.0563x^3 - 23.874x^2 - 74.018x + 5548.7$	$y = -2E\text{-}08x^3 + 0.0031x^2 - 129.52x + 2E{+}06$
	$R^2 = 0.792$	$R^2 = 0.126$
Polynomial of 4th degree	$y = -0.247x^4 + 11.924x^3 - 180.07x^2 + 732.49x + 4477.5$	$y = -2E\text{-}11x^4 + 4E\text{-}06x^3 - 0.2567x^2 + 7101.8x - 7E{+}07$
	$R^2 = 0.8328$	$R^2 = 0.2176$
Polynomial of 5th degree	$y = -0.0718x^5 + 3.7046x^4 - 66.349x^3 + 490.37x^2 - 1583.6x + 6728$	$y = 8E\text{-}15x^5 - 2E\text{-}09x^4 + 0.0001x^3 - 6.2467x^2 + 132133x - 1E{+}09$
	$R^2 = 0.9236$	$R^2 = 0.2366$
Polynomial of 6th degree	$y = -0.0021x^6 + 0.0687x^5 + 0.1521x^4 - 23.418x^3 + 238.5x^2 - 944.4x + 6236.8$	$y = 2E\text{-}17x^6 - 4E\text{-}12x^5 + 4E\text{-}07x^4 - 0.0231x^3 + 723.06x^2 - 1E{+}07x + 8E{+}10$
	$R^2 = 0.9256$	$R^2 = 0.3722$
Potentiometric	$y = 6695.8x^{-0.301}$	$y = 215.04x^{0.246}$
	$R^2 = 0.6259$	$R^2 = 0.006$

Table A16. Trend models for the West Pomeranian province.

Data/Model	Annual Data	Monthly Data
Exponential	$y = 2381.2e^{-0.04x}$	$y = 389.87e^{3E-05x}$
	$R^2 = 0.939$	$R^2 = 0.0946$
Linear	$y = -60.835x + 2250.1$	$y = 0.0534x - 685.14$
	$R^2 = 0.9558$	$R^2 = 0.1096$
Logarithmic	$y = -430\ln(x) + 2510.2$	$y = 2186.6\ln(x) - 21717$
	$R^2 = 0.8291$	$R^2 = 0.1053$
Polynomial of 2nd degree	$y = 0.022x^2 - 61.319x + 2252$	$y = 4E\text{-}05x^2 - 3.6238x + 76071$
	$R^2 = 0.9558$	$R^2 = 0.2583$
Polynomial of 3rd degree	$y = -0.0448x^3 + 1.4998x^2 - 74.628x + 2279.2$	$y = -2E\text{-}08x^3 + 0.0029x^2 - 121.55x + 2E{+}06$
	$R^2 = 0.9562$	$R^2 = 0.3334$
Polynomial of 4th degree	$y = -0.0181x^4 + 0.7514x^3 - 9.9445x^2 - 15.538x + 2200.7$	$y = -1E\text{-}11x^4 + 2E\text{-}06x^3 - 0.1535x^2 + 4230.3x - 4E{+}07$
	$R^2 = 0.9581$	$R^2 = 0.3963$
Polynomial of 5th degree	$y = -0.0034x^5 + 0.1701x^4 - 2.9764x^3 + 21.986x^2 - 125.85x + 2307.9$	$y = 9E\text{-}15x^5 - 2E\text{-}09x^4 + 0.0002x^3 - 6.3676x^2 + 133940x - 1E{+}09$
	$R^2 = 0.9598$	$R^2 = 0.4351$
Polynomial of 6th degree	$y = 0.0002x^6 - 0.0185x^5 + 0.5508x^4 - 7.5768x^3 + 48.976x^2 - 194.34x + 2360.5$	$y = 5E\text{-}18x^6 - 1E\text{-}12x^5 + 1E\text{-}07x^4 - 0.0066x^3 + 205.46x^2 - 3E{+}06x + 2E{+}10$
	$R^2 = 0.96$	$R^2 = 0.4567$
Potentiometric	$y = 2763.2x^{-0.272}$	$y = 0.001x^{1.3345}$
	$R^2 = 0.7568$	$R^2 = 0.0905$

References

1. WHO. *The Global Status on Road Safety*; 2018; p. 403 p. Available online: https://www.who.int/publications/i/item/9789241565684 (accessed on 17 April 2022)ISBN 9789241565684.
2. Raza, A.; Safdar, M.; Zhong, M.; Hunt, J.D. Analyzing Spatial Location Preference of Urban Activities with Mode-Dependent Accessibility Using Integrated Land Use–Transport Models. *Land* **2022**, *11*, 1139. [CrossRef]
3. Raza, A.; Zhong, M.; Safdar, M. Evaluating Locational Preference of Urban Activities with the Time-Dependent Accessibility Using Integrated Spatial Economic Models. *Int. J. Environ. Res. Public Health* **2022**, *19*, 8317. [CrossRef]
4. Tambouratzis, T.; Souliou, D.; Chalikias, M.; Gregoriades, A. Maximising accuracy and efficiency of traffic accident prediction combining information mining with computational intelligence approaches and decision trees. *J. Artif. Intell. Soft Comput. Res.* **2014**, *4*, 31–42. [CrossRef]
5. Zhu, L.; Lu, L.; Zhang, W.; Zhao, Y.; Song, M. Analysis of accident severity for curved roadways based on bayesian networks. *Sustainability* **2019**, *11*, 2223. [CrossRef]
6. Wang, Z.; Safdar, M.; Zhong, S.; Liu, J.; Xiao, F. Public Preferences of Shared Autonomous Vehicles in Developing Countries: A Cross-National Study of Pakistan and China. *J. Adv. Transp.* **2021**, *2021*, 5141798. [CrossRef]
7. Safdar, M.; Jamal, A.; Al-Ahmadi, H.M.; Rahman, M.T.; Almoshaogeh, M. Analysis of the Influential Factors towards Adoption of Car-Sharing: A Case Study of a Megacity in a Developing Country. *Sustainability* **2022**, *14*, 2778. [CrossRef]
8. Waiss, A.A.; Jianfeng, A.; Irfan, U.; Muhammad, A. Public Preferences Towards Car Sharing Service: The Case of Djibouti. *Front. Environ. Sci.* **2022**, *10*. [CrossRef]
9. Arteaga, C.; Paz, A.; Park, J. Injury severity on traffic crashes: A text mining with an interpretable machine-learning approach. *Saf. Sci.* **2020**, *132*, 104988. [CrossRef]
10. Yang, Z.; Zhang, W.; Feng, J. Predicting multiple types of traffic accident severity with explanations: A multi-task deep learning framework. *Saf. Sci.* **2022**, *146*, 105522. [CrossRef]
11. Gorzelanczyk, P.; Pyszewska, D.; Kalina, T.; Jurkovic, M. Analysis of road traffic safety in the Pila poviat. *Sci. J. Sil. Univ. Technol. Ser. Transp.* **2020**, *107*, 33–52. [CrossRef]
12. Chen, C. Analysis and forecast of traffic accident big data. In Proceedings of the 4th Annual International Conference on Information Technology and Applications (ITA 2017), online, 5 September 2017; Volume 12, p. 04029. [CrossRef]
13. Khaliq, K.A.; Chughtai, O.; Shahwani, A.; Qayyum, A.; Pannek, J. Road accidents detection, data collection and data analysis using V2X communication and edge/cloud computing. *Electronics* **2019**, *8*, 896. [CrossRef]
14. Rajput, H.; Som, T.; Kar. S. An automated vehicle license plate recognition system. *Computer* **2015**, *48*, 56–61. [CrossRef]

15. Zheng, Z.; Wang, C.; Wang, P.; Xiong, Y.; Zhang, F.; Lv, Y. Framework for fusing traffic information from social and physical transportation data. *PLoS ONE* **2018**, *13*, e0201531. [CrossRef]
16. Jurkovic, M.; Gorzelanczyk, P.; Kalina, T.; Jaros, J.; Mohanty, M. Impact of the COVID-19 pandemic on road traffic accident forecasting in Poland and Slovakia. *Open Eng.* **2022**, *12*, 578–589. [CrossRef]
17. Gorzelańczyk, P.; Jurkovič, M.; Kalina, T.; Mohanty, M. Forecasting the road accident rate and the impact of the COVID 19 on its frequency in the polish provinces. *Commun.—Sci. Lett. Univ. Zilina* **2022**, *24*, A216–A231. [CrossRef]
18. Abdullah, E.; Emam, A. Traffic accidents analyzer using big data. In Proceedings of the 2015 International Conference on Computational Science and Computational Intelligence (CSCI), Las Vegas, NV, USA, 7–9 December 2015; Institute of Electrical and Electronics Engineers Inc., 2016; pp. 392–397. [CrossRef]
19. Vilaça, M.; Silva, N.; Coelho, M.C. Statistical analysis of the occurrence and severity of crashes involving vulnerable road users. *Transp. Res. Procedia* **2017**, *27*, 1113–1120. [CrossRef]
20. Bąk, I.; Cheba, K.; Szczecińska, B. The statistical analysis of road traffic in cities of Poland. *Transp. Res. Procedia* **2019**, *39*, 14–23. [CrossRef]
21. Chand, A.; Jayesh, S.; Bhasi, A.B. Road traffic accidents: An overview of data sources, analysis techniques and contributing factors. *Mater. Today: Proc.* **2021**, *47*, 5135–5141. [CrossRef]
22. Helgason, A. Fractional integration methods and short Time series: Evidence from asimulation study. *Polit. Anal.* **2016**, *24*, 59–68. Available online: http://www.jstor.org/stable/24573204 (accessed on 17 April 2022). [CrossRef]
23. Lavrenz, S.; Vlahogianni, E.; Gkritza, K.; Ke, Y. Time series modeling in traffic safetyresearch. *Accid. Anal. Prev.* **2018**, *117*, 368–380. [CrossRef]
24. Forecasting Based on Time Series. Available online: http://pis.rezolwenta.eu.org/Materialy/PiS-W-5.pdf (accessed on 17 April 2022).
25. Procházka, J.; Flimmel, S.; Čamaj, M.; Bašta, M. *Modelling the Number of Road Accidents*; Publishing house of the University of Economics in Wrocław: Wrocław, Poland, 2017. [CrossRef]
26. 26. Sunny, C.M.; Nithya, S.; Sinshi, K.S.; Vinodini, V.M.D.; Lakshmi, A.K.G.; Anjana, S.; Manojkumar, T.K. Forecasting of Road Accident in Kerala: A Case Study. In Proceedings of the 2018 International Conference on Data Science and Engineering (ICDSE), Kochi, India, 7–9 August 2018. [CrossRef]
27. Dudek, G. Forecasting Time Series with Multiple Seasonal Cycles Using Neural Networks with Local Learning. In *Artificial Intelligence and Soft Computing. ICAISC 2013. Lecture Notes in Computer Science*; Rutkowski, L., Korytkowski, M., Scherer, R., Tadeusiewicz, R., Zadeh, L.A., Zurada, J.M., Eds.; Springer: Berlin/Heidelberg, Germany, 2013; Volume 7894. [CrossRef]
28. Szmuksta-Zawadzka, M.; Zawadzki, J. *Forecasting on the Basis of Holt-Winters Models for Complete and Incomplete Data*; Research Papers of the Wrocław University of Economics: Poland, 2009; Volume 38.
29. Wójcik, A. *Autoregressive Vector Models as a Response to the Critique of Multi-Equation Structural Econometric Models*; Publishing house of the University of Economics in Katowice: Katowice, Poland, 2014; Volume 193.
30. Monederoa, B.D.; Gil-Alanaa, L.A.; Martínezaa, M.C.V. Road accidents in Spain: Are they persistent? *IATSS Res.* **2021**, *45*, 317–325. [CrossRef]
31. Al-Madani, H. Global road fatality trends'estimations based on country-wise microlevel data. *Accid. Anal. Prev.* **2018**, *111*, 297–310. [CrossRef]
32. Mamczur, M. Machine Learning How Does Linear Regression Work? And is it WORTH Using? Available online: https://miroslawmamczur.pl/jak-dziala-regresja-liniowa-i-czy-warto-ja-stosowac/ (accessed on 17 April 2022).
33. Piłatowska, M. The choice of the order of autoregression depending on the parameters of the generating model. *Econometrics* **2012**, *4*, 16–35.
34. Biswas, A.A.; Mia, J.; Majumder, A. Forecasting the Number of Road Accidents and Casualties using Random Forest Regression in the Context of Bangladesh. In Proceedings of the 2019 10th International Conference on Computing, Communication and Networking Technologies (ICCCNT), Kanpur, India, 6–8 July 2019.
35. Random Forest. Available online: https://pl.wikipedia.org/wiki/Las_losowy (accessed on 17 April 2022).
36. Fijorek, K.; Mróz, K.; Niedziela, K.; Fijorek, D. Forecasting electricity prices on the day-ahead market using data mining methods. *Energy Mark.* **2010**.
37. Chudy-Laskowska, K.; Pisula, T. Forecast of the Number of Road Accidents in Poland. *Logistics* **2014**, *6*, 2710–2721.
38. Kashpruk, N. *Comparative Research of Statistical Models and Soft Computing for Identification of Time Series and Forecasting*; Opole University of Technology: Opole, Poland, 2010.
39. Prochazka, J.; Camaj, M. Modelling the number of road accidents of uninsured drivers and their severity. In Proceedings of the International Academic Conferences 5408040, International Institute of Social and Economic Sciences, Geneva, Switzerland, 27 June 2017.
40. Dutta, B.; Barman, M.P.; Patowary, A.N. Application of Arima model for forecasting road accident deaths in India. *Int. J. Agric. Stat. Sci.* **2020**, *16*, 607–615.
41. Karlaftis, M.; Vlahogianni, E. Memory properties and fractional integration in trans-portation time-series. *Transp. Res. C* **2009**, *17*, 444–453. [CrossRef]
42. Łobejko, S. *Time Series Analysis and Forecasting with SAS*; Main business school in Warsaw: Warsaw, Poland, 2015.
43. Dudek, G. Exponential smoothing models for short-term power system load forecasting. *Energy Mark.* **2013**, *3*, 14—19.

44. Chudy-Laskowska, K.; Pisula, T. Prognozowanie Liczby Wypadków Drogowych na Podkarpaciu; Logistics: 2015. *Logistics* **2015**, *4*, 2782–2796.
45. Road Safety Assessment Handbook. Available online: https://www.iung.pl/PJA/wydane/11/PJA11_3.pdf (accessed on 17 April 2022).
46. Wrobel, M.S. Application of Neural Fuzzy Systems in Chemistry. Ph.D. Thesis, University of Silesia, Katowice, Poland, 2017.
47. Data Mining Techniques StatSoft. Available online: https://www.statsoft.pl/textbook/stathome_stat.html?https%3A%2F%2Fwww.statsoft.pl%2Ftextbook%2Fstdatmin.html (accessed on 17 April 2022).
48. Kumar, S.; Viswanadham, V.; Bharathi, B. Analysis of road accident. IOP Conference Series Materials Science and Engineering. In Proceedings of the International Conference on Frontiers in Materials and Smart System Technologies, Tamil Nadu, India, 10 April 2019; Volume 590, p. 012029. [CrossRef]
49. Top Advantages and Disadvantages of Hadoop 3 DataFlair. Available online: https://data-flair.training/blogs/advantages-and-disadvantages-of-hadoop/ (accessed on 17 April 2022).
50. Perczak, G.; Fiszeder, P. GARCH model—Using additional information on minimum and maximum prices. *Bank Credit.* **2014**, *2*, 105–132.
51. Fiszeder, P. *GARCH Class Models in Empirical Financial Research*; Scientific Publishers of the Nicolaus Copernicus University: Torun, Poland, 2009.
52. McIlroy, R.C.; Plant, K.A.; Hoque, M.S.; Wu, J.; Kokwaro, G.O.; Nam, V.H.; Stanton, N.A. Who is responsible for global road safety? A cross-cultural comparison ofactor maps. *Accid. Anal. Prev.* **2019**, *122*, 8–18. [CrossRef]
53. Muck, J. Econometrics. Modeling of Time Series. Stationary. Unit Root Tests. ARDL models. Co-integration. Available online: http://web.sgh.waw.pl/~jmuck/Ekonometria/EkonometriaPrezentacja5.pdf (accessed on 17 April 2022).
54. Shetty, P.; Sachin, P.C.; Kashyap, V.K.; Madi, V. Analysis of road accidents using data mining techniques. *Int. Res. J. Eng. Technol.* **2017**, *4*, 1494–1496.
55. Li, L.; Shrestha, S.; Hu, G. Analysis of road traffic fatal accidents using data mining techniques. In Proceedings of the 2017 IEEE 15th International Conference on Software Engineering Research, Management and Applications (SERA), London, UK, 7–9 June 2017; pp. 363–370. [CrossRef]
56. Marcinkowska, J. *Statistical Methods and Data Mining in Assessing the Occurrence of Syncope in the Group of Narrow-QRS Tachycardia (AVNRT and AVRT)*; Medical University of Karol Marcinkowski in Poznań: Poznań, Poland, 2015; Available online: http://www.wbc.poznan.pl/Content/373785/index.pdf (accessed on 17 April 2022).
57. Sebego, M.; Naumann, R.B.; Rudd, R.A.; Voetsch, K. Dellinger, A.M., Ndlovu, C., The impact of alcohol and road traffic policies on crash rates in Botswana, 2004–2011: Atime-series analysis. *Accid. Anal. Prev.* **2008**, *70*, 33–39. [CrossRef]
58. Bloomfield, P. An exponential model in the spectrum of a scalar time series. *Biometrika* **1973**, *60*, 217–226. Available online: https://www.jstor.org/stable/2334533 (accessed on 17 April 2022). [CrossRef]
59. Central Statistical Office. Available online: www.gus.pl (accessed on 17 April 2022).
60. Statistic Road Accident. Available online: https://statystyka.policja.pl/ (accessed on 17 April 2022).
61. Eurostat Statistics. Available online: https://ec.europa.eu/eurostat/data/database (accessed on 17 April 2022).
62. Province Names in English. Available online: https://www.facebook.com/Pogoda-Na-Przygod%C4%99-687347091288220/photos/a.1077421475614111.1073741838.687347091288220/1077421502280775 (accessed on 17 April 2022).

applied sciences

MDPI

Article

Evaluation on Lateral Stability of Vehicle: Impacts of Pavement Rutting, Road Alignment, and Adverse Weather

Gang Tian [1], Yanshun Jia [2], Zeqi Chen [1], Ying Gao [1,*], Shaoquan Wang [1], Ziyao Wei [1], Yufei Chen [3] and Tianshuo Zhang [1]

[1] School of Transportation, Southeast University, Nanjing 211189, China
[2] School of Traffic and Transportation, Shijiazhuang Tiedao University, Shijiazhuang 050043, China
[3] College of Transportation Engineering, Tongji University, Shanghai 201804, China
* Correspondence: gy@seu.edu.cn

Abstract: The combination of pavement rutting, poor road alignment, and extreme adverse weather will seriously threaten the driving safety of vehicles, whereas only a few of these factors are commonly concerned. This study aims to efficiently evaluate the impacts of various driving conditions on the lateral stability of the vehicle and produce a practical recommendation for pavement maintenance in what concerns rutting. A systematic framework was, thus, developed to conduct a comprehensive evaluation of the lateral stability of the vehicle, which incorporates a single-factor test and multi-factor test based on the stability indicators obtained from Carsim simulations. The vehicle road weather model was established in the Carsim software by considering seven factors, including driving speed, width–height ratio (WHR) of rutting sidewall, radius of circular curve, superelevation, crosswind angle, crosswind speed, and friction coefficient, respectively. The results show that the established framework behaves with satisfactory performance, regarding evaluating the effect of various impact factors on the lateral stability of the vehicle while driving across rutting. Stability indicators suddenly fluctuate in a short time, due to the instantaneous wandering behavior of crossing rutting. Additionally, the sudden fluctuation phenomenon is greatly enlarged, and the vehicle is inclined to occur with lateral instability when WHR equals 5, particularly in roll-over instability. It is recommended to concurrently confine the WHR greater than 10 and friction coefficient greater than 0.4, in order to ensuring driving stability. The multi-factor test revealed that the vehicle speed and WHR of the rutting are leading factors that affect driving stability, followed by the radius of circular curve, superelevation, crosswind angle, crosswind speed and friction coefficient, respectively, which are both essential factors for driving stability. The outcomes of this study may contribute to supplying guidelines for controlling key adverse conditions and making decisions on pavement maintenance.

Keywords: asphalt pavement; lateral stability; evaluation framework; rutting; curve segment; adverse weather

Citation: Tian, G.; Jia, Y.; Chen, Z.; Gao, Y.; Wang, S.; Wei, Z.; Chen, Y.; Zhang, T. Evaluation on Lateral Stability of Vehicle: Impacts of Pavement Rutting, Road Alignment, and Adverse Weather. *Appl. Sci.* **2023**, *13*, 3250. https://doi.org/10.3390/app13053250

Academic Editors: Edgar Sokolovskij and Vidas Žuraulis

Received: 8 February 2023
Revised: 26 February 2023
Accepted: 28 February 2023
Published: 3 March 2023

1. Introduction

Rutting is a common distress of asphalt pavements, which adversely impacts the serviceability of the pavement and driving stability of the vehicle [1,2]. When a vehicle drives through a rut, it will produce a phenomenon known as wandering, and the vehicle may show potentially unstable, oscillatory lateral, or yaw motions with a low frequency. Drivers may feel discomfort, due to bumps, while negotiating the rutting or even cause serious traffic accidents due to operational faults. The number of accidents per 100 million vehicle-miles of highways in the states of Arizona, North Carolina, and Maryland significantly increases when the rut depth approaches 10 mm [3]. It has been identified that the presence of rutting would cause lateral instability in vehicles and impact driving safety when the rutting depth exceeds 20 mm [4]. So, driving safety is remarkably influenced by pavement rutting.

Exactly as the findings about the threats of rutting to driving safety, rutting distress is more crucial for influencing driving safety than roughness or cracking, due to the large height differences and features of easy ponding and frozen in rutting. In order to improve the rutting maintenance from the perspective of driving safety, quite a few investigators have studied the effects of different rutting shape indicators on driving comfort or the safety of vehicles and obtained their allowable values. Zheng et al. [5] possessed the view that the rut depth and rut side angle obviously affected the maximum vertical acceleration of the vehicle, while the rut average width had small effects. A mathematical model regarding the interaction between automobiles, wheels, and road surfaces with ruts was developed by Vansauskas et al. [6]. They found that the vehicle became unstable if the rut is over 5 cm and the width over 50 cm, when driving at a speed of 60 km/h. Guo et al. [7] suggested lateral offset and lateral acceleration as rutting indicators for evaluating the influence of rutting length. Jia et al. [8] found that driving across rutting had a greater influence than driving on the rutting sidewall and stated that rutting depth should be less than 20 mm to ensure driving quality when across a rut on dry pavement. Kuznetsov et al. [9] determined the maximum allowable rutting depth via a computer program and accordingly defined 14 mm as the allowable value of rutting depth to ensure safety when driving over 160 km/h.

In recent years, in addition to the road safety affected by pavement rutting, unfavorable conditions with adverse weather have been frequent events and constantly cause serious impacts on traffic safety. Approximately 300 traffic accidents happened due to adverse weather in Germany in January 2016 [10]. More than 60% of traffic accidents caused by fierce winds and other adverse weather were accounted for in the south-eastern coast of China [11]. Adverse weather mainly includes strong winds, heavy rain, ice, and snow [12], and even their simultaneous occurrence. On the one hand, the influence of adverse weather can be considered as a kind of action subject to the vehicle and, thus, affecting its driving stability. On the other hand, the driving risk possibly intensifies, due to the coupling effect of adverse weather and rutting distress. The ponding formed in the rutting under adverse weather may decrease the friction coefficient of road surface and change the mechanical characteristics of the vehicle, resulting in losing lateral stability [13], while driving in heavy rains, snows, and strong winds. Moreover, a water film on the pavement surface will generate a hydroplaning phenomenon on the rolling tire, as a result of the joint impact of rutting and ponding. Due to the hydroplaning phenomenon, the water pressure in the front of the tire pushes the water under the tire, and then the water film is formed and separates the tire from the pavement, thus resulting in a loss of road friction coefficient [14,15]. Confronting such adverse circumstances, smart tires, a promising technology, are highlighted in detecting distressed rutting sections and road friction, then interacting with follow-up vehicles to avoid unfavorable positions, in order to ensure driving stability [14,16].

In addition to adverse weather, traffic accidents more frequently happen in horizontal curve segments than tangent segments of the road [17,18]. Fatal accidents commonly suddenly occur due to loss of lateral sideslip stability or roll-over stability as a result of unfavorable external factors. Exceeding the limit value of side friction is a key reason for lateral instability in curve segment [19,20], but it connects closer relationship with weather and pavement material. The instability of the vehicle is generally induced by inappropriate geometry parameters of curve segments. The relevant studies have demonstrated that the accident frequency dramatically increases as the radius of the curve decreases [21]. In addition, the vehicle on the curve segment is prone to occur kidding and roll-over with the increase of road slope and the decrease of superelevation [22]. However, the sideslip risk is not remarkably affected by superelevation of the road [23]. As a result, driving stability analysis is a complex and comprehensive topic, which shall follow the integrated consideration of adverse weather, road, alignment, and rutting in need.

As a result, recent studies have highlighted the significance of researching driving stability under adverse weather, as well as focusing on adverse effects caused by alignment of curve segment. It is generally accepted that strong winds and heavy rains lead to low lateral stability [24–27], which is susceptible to resulting in sideslip, yawing, and roll-over instability. The evaluations of these instabilities under unfavorable situations were performed by means of analyzing the dynamic indicators of vehicles [28], including lateral displacement, lateral acceleration, load transfer rate, yaw angle rate, and roll-over angle [29–34]. Chen et al. [35,36] found that the car is more likely to sideslip and the cargo truck has higher roll-over risks under crosswind. Wang et al. [37] simulated driving experiments under snowy weather in a bridge and tunnel connection segment, and finally found that the large lateral offset can be prevented by increasing the friction coefficient of pavement and radius of circular curve. Yin et al. [38] investigated the skidding, roll-over, and lateral slip of the vehicle by accounting radius of circular curve and superelevation and found that the safety margin of the vehicle's skidding, roll-over, and lateral slip increased when the radius of the circular curve and superelevation increases. Alrejjal et al. [39] focused on the roll-over propensity influenced by horizontal and vertical alignments under different weather conditions, thus revealing that the lateral acceleration was amplified due to a tight degree of curvature and steep downgrades.

The previous findings mainly investigated the effect of single adverse weather or road alignment on driving stability, but neglected rutting conditions accordingly, or solely analyzed the unfavorable conditions of vehicle stability due to rutting distress, with only a few research on the composite effects of rutting and wet condition. However, the dynamic response process of the vehicle driving across rutting coupled in curve segments with adverse weather is a crucial issue that determines driving safety. So, relevant evaluation remains to be proposed, so as to implement the management and control of adverse situations, as well as make pavement maintenance decisions.

Given the research shortcomings and gaps above, this study intends to figure out the dynamic responses of the vehicle driving across rutting on the curve segment under adverse weather and further evaluate the effects of various driving conditions on the lateral stability of the vehicle. So, a framework for evaluating lateral stability of the vehicle was developed to efficiently describe the impacts of different factors. The Carsim software was utilized to simulate the process of driving across rutting, considering various road alignments and adverse weather. The vehicle's dynamic responses that represent lateral stability were analyzed via indicator lateral acceleration (LA), slip angle (SA), load transfer ratio (LTR), and roll-over angle (RA). Considering the likelihood of poor correlation between LA, SA, LTR, and RA, the integrated lateral stability caused by various factors was compared with entropy-weighted grey relational analysis (GRA), based on orthogonal tests. The outcomes of this study will contribute to establishing criteria for controlling key adverse conditions and making decisions on pavement maintenance.

2. A Systematic Framework for Evaluating Lateral Stability of the Vehicle under Various Driving Conditions

Pavement rutting, road alignment, adverse weather, and vehicle speed are significant factors that potentially influence the driving stability of the vehicle. However, the combined effects may be neglected, due to incomplete consideration [40]. Therefore, the framework for evaluating the lateral stability of the vehicle was developed to efficiently describe the impacts of different factors. Figure 1 is the technical framework for evaluating the lateral stability of the vehicle under various driving conditions. Four indicators that separately represent sideslip stability and roll-over stability were output to preliminary evaluate the lateral stability of the vehicle. Single-factor analysis was first conducted by controlling the constant single factor. After that, the correlations of LA, SA, LTR, and RA were examined through Pearson's correlation coefficient to validate their consistency in evaluating lateral stability of the vehicle. GRA and the entropy weight method were applied to derive comprehensive relation grade (CRG) based on an orthogonal test, in which multi-factor analysis was performed, thus evaluating the lateral stability of the vehicle. The details about the developed framework are described in the following parts.

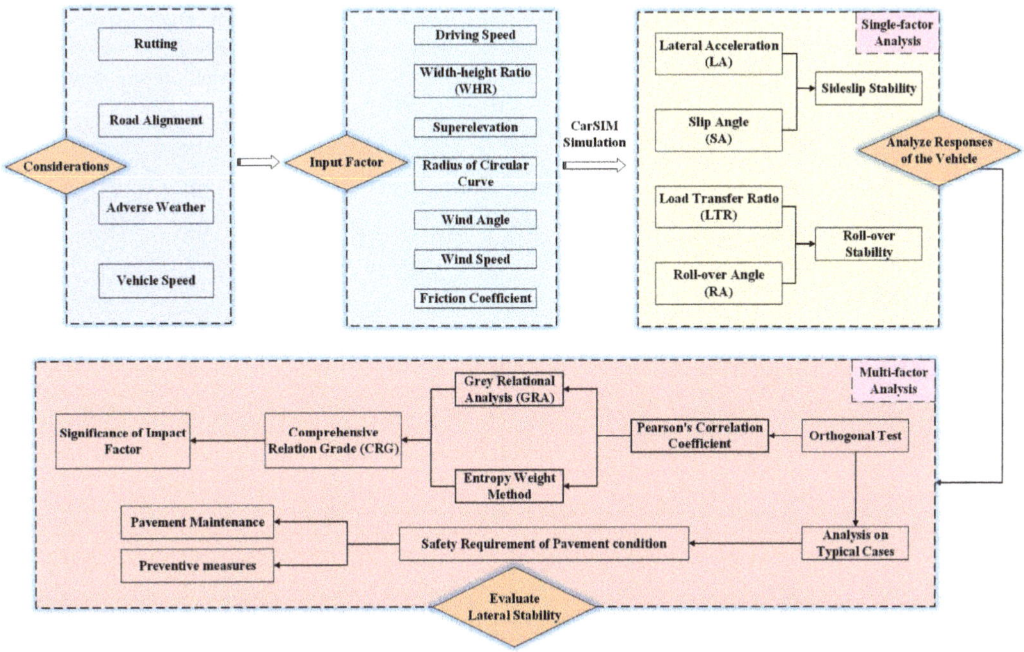

Figure 1. Technical framework of this study.

2.1. Factors in Consideration

As stated before, the combined factors of pavement rutting, road alignment, and adverse weather denote various driving conditions, which may cause unacceptable vehicle instability. In order to figure out the effects of various driving conditions on the lateral stability of the vehicle, totally, seven factors are taken into consideration, including the driving speed (V_c), width–height ratio (WHR) of rutting sidewall, radius of circular curve (R), superelevation (e), lateral wind angle (δ_{lw}), wind speed (V_{lw}), and friction coefficient (f). V_c is on behalf of the driving condition. Factor f mainly relates to rainy and snowy weather. δ_{lw} and V_{lw} indicate lateral wind action. R and e are parameters of road alignment. WHR characterizes the effect of rutting distress.

2.2. Evaluation Indicators of Lateral Stability

2.2.1. Sideslip Stability

Vehicles will be subjected to lateral force when driving on curve segments, especially in adverse weather with wind and rain. If the vehicle drives through the curve at a pretty high speed, the grip of the wheels may dramatically decrease, thus resulting in the drift of the vehicle and large lateral sideslip [23,41]. Figure 2 shows the vertical view of a dynamic model, while the vehicle drives on curve segments. F_{syl} and F_{syr} denote lateral forces exerted on the left and right tires of s th axle (s = 1, 2). V_{w1}, V_{w2}, V_{w3}, and V_{w4} are the velocities of four wheels, respectively. V_{vb} is the forward velocity of the vehicle body. α_1, α_2, α_3, and α_4 denote the slip angles of front and rear wheels, which are defined as angles between V_{wi} (i =1, 2, 3, 4) and wheel direction, respectively. β denotes the slip angle of vehicle body, which is defined as the angle between V_{vb} and longitudinal axis. δ_s is the steering angle of the front wheel.

In addition, the superposition of lateral force caused by lateral wind and the centrifugal effect will significantly increase the lateral acceleration (LA), resulting in the vehicle laterally deviating from the original trajectory and increasing sideslip risks. It was specified that the LA for ordinary cars shall be less than 0.4 g.

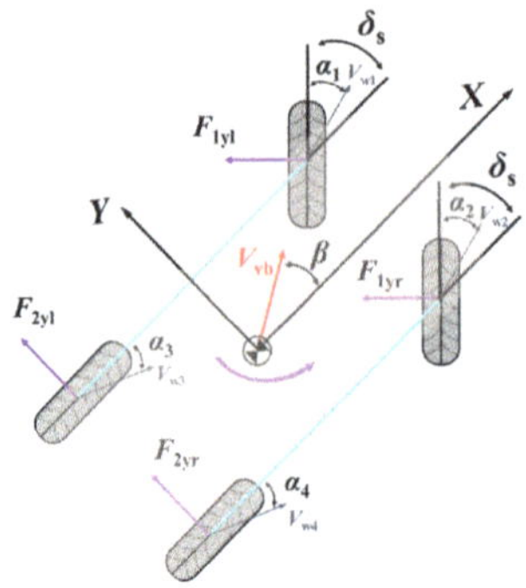

Figure 2. Vertical view of vehicle model on curve segments.

Slip angle (SA) of the vehicle body is the angle its velocity vector at the center of gravity makes with the longitudinal axis (*X* axis) of the vehicle, which is also used to characterize the risk of side slip, as described by [42]:

$$\beta = \arctan(v_y, v_x) \tag{1}$$

where v_y and v_x denote the lateral and longitudinal velocity of vehicle body, respectively. The safety threshold for slip angle is 5° [43].

2.2.2. Roll-Over Stability

Vehicle roll-over has been acknowledged as a serious vehicle accident with the highest fatality [44]. The vehicle perhaps rolls over as a result of large rotating degrees, due to an unstable mechanical state caused by cross slope of road or other external factors. Figure 3 presents the front view of vehicle model on curve segments with cross slope θ. The perpendicular forces and tangent forces exerted on the tires are F_{ZL}, F_{ZR}, F_{YL}, and F_{YR}. m is sprung mass, and g and a_y denote gravitational acceleration and lateral acceleration. The ϕ is the roll-over angle of the car. It is clearly illustrated that the forces of the car on the curve segments differ from that of the tangent segments.

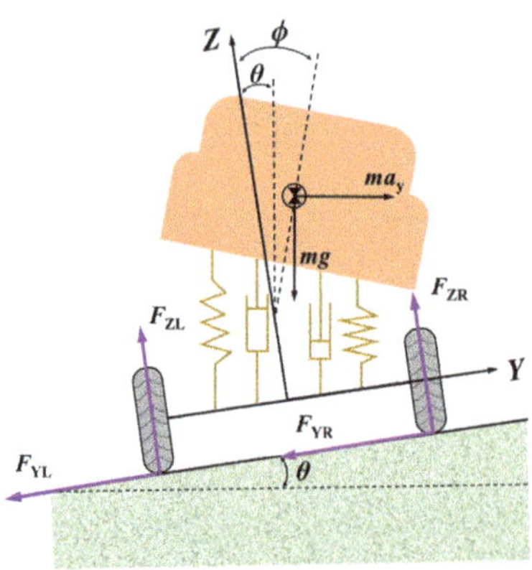

Figure 3. Front view of vehicle model on curve segments.

Load transfer ratio (LTR) was selected to characterize the roll-over stability of a car, which is defined as the ratio of the load transferred from the inside wheels to the outside wheels to the total load of the vehicle during travel [45], as given by:

$$LTR(t) = \left| \frac{\sum_{s=1}^{2} [F_{ZR}(t)_s - F_{ZL}(t)_s]}{\sum_{s=1}^{2} [F_{ZR}(t)_s + F_{ZL}(t)_s]} \right| \tag{2}$$

where *LTR(t)* denotes load transfer ratio at time t, and $F_{ZR}(t)_s$ and $F_{ZL}(t)_s$ are the vertical forces acting on the outer and inner tires of the s-th axle at time t, respectively. When the vehicle is driving in a stable state, the forces on the left and right tires are equal, that is, LTR = 0. When the inner wheel is off the ground, all the loads are transferred to the outer wheels, that is, LTR = 1. In this case, the vehicle is traveling in a limit state of roll-over instability. In general, the vehicle has a certain risk of roll-over when $0.6 \leq \text{LTR} < 0.8$ and is vulnerable to roll-over when $0.8 \leq \text{LTR} < 1$ [46,47].

As shown in Figure 3, ϕ is the roll-over angle. The vehicle will lean laterally, while turning or encountering impact, so the vehicle may roll over if the roll-over angle (RA) is too large. It is generally accepted that the assessment of RA should be carried out in conjunction with LA and concurrently not exceed the maximum value of 6°. Combining the characteristics of LA in this study, the safety threshold for RA is considered to be 6°.

2.3. Single-Factor Test Design

In order to figure out how vehicle stability varies with a single factor, a single-factor test was conducted by controlling for the single factor being constant and changing other factors. Simulation schemes are shown in Table 1. The vehicle was set to run at a high speed ranging from 90 km/h to 120 km/h, according to the common operating speeds of passenger cars in China expressways and design speed specified in Chinese Standard JTG D20-2017 [48]. This would contribute to an understanding, regarding safe operating speeds and whether the existing speed limit has to be updated in order to avoid dangerous circumstances. The wind speeds were selected according to class 7 to class 9 wind and values from previous studies, because these may affect the lateral stability of the vehicle [11,49]. The friction coefficient was determined based on the wet conditions of the road surface, as introduced in Section 3.2. By conducting designed simulations, the indicator-time curve was determined to analyze the vehicle's responses under various driving conditions.

Table 1. Simulation schemes design of single-factor analysis.

Experiment Number	V_c/(km/h)	WHR	R/m	e/%	δ_{lw} /	V_{lw}/(km/h)	f
1	90, 100, 110, 120	10	1000	4	60	80	0.4
2	110	20, 15, 10, 5	1000	4	60	80	0.4
3	110	10	1200, 1000, 800, 600	4	60	80	0.4
4	110	10	1000	8, 6, 4, 2	60	80	0.4
5	110	10	1000	4	0, 30, 60, 90	80	0.4
6	110	10	1000	4	60	60, 70, 80, 90	0.4
7	110	10	1000	4	60	80	0.8, 0.6, 0.4, 0.18

2.4. Multi-Factor Test Design

2.4.1. Orthogonal Test

The previous section is about the design of simulation schemes with the purpose of studying every single factor. However, driving safety is significantly influenced by extreme conditions where multiple factors are applied. So, an orthogonal test was conducted in this section. Seven factors were still investigated, consistent with the above section, including V_c, WHR, R, e, δ_{lw}, V_{lw}, and f, representing vehicle velocity, WHR of rutting, the radius of circular curve, superelevation, lateral wind angle, lateral wind speed, and friction coefficient, respectively. Four levels of each factor were designed to represent the degree that is unfavorable to driving safety. Interactions between various factors were ignored, and the simulation schemes and results are shown in Table 2.

Table 2. Orthogonal test design.

Case	V_c	WHR	R	e	δ_{lw}	V_{lw}	f
	km/h	-	m	%	°	km/h	-
1	90	20	1200	8	0	60	0.8
2	90	15	1000	6	30	70	0.6
3	90	10	800	4	60	80	0.4
4	90	5	600	2	90	90	0.18
5	100	20	1200	6	30	80	0.4
6	100	15	1000	8	0	90	0.18
7	100	10	800	2	90	60	0.8
8	100	5	600	4	60	70	0.6
9	110	20	1000	4	90	60	0.6
10	110	15	1200	2	60	70	0.8
11	110	10	600	8	30	80	0.18
12	110	5	800	6	0	90	0.4
13	120	20	1000	2	60	80	0.18
14	120	15	1200	4	90	90	0.4
15	120	10	600	6	0	60	0.6
16	120	5	800	8	30	70	0.8
17	90	20	600	8	90	70	0.4
18	90	15	800	6	60	60	0.18
19	90	10	1000	4	30	90	0.8
20	90	5	1200	2	0	80	0.6
21	100	20	600	6	60	90	0.8
22	100	15	800	8	90	80	0.6
23	100	10	1000	2	0	70	0.4
24	100	5	1200	4	30	60	0.18
25	110	20	800	4	0	70	0.18
26	110	15	600	2	30	60	0.4
27	110	10	1200	8	60	90	0.6
28	110	5	1000	6	90	80	0.8
29	120	20	800	2	30	90	0.6
30	120	15	600	4	0	80	0.8
31	120	10	1200	6	90	70	0.18
32	120	5	1000	8	60	60	0.4

2.4.2. Evaluation Method

- Pearson's Correlation Coefficient

As mentioned above, LA, SA, LTR, and RA were generally utilized to evaluate the lateral stability of the vehicle. However, the results determined from these four indicators may show diverse states of stability. So, the Pearson's correlation analysis was conducted to examine the consistency in evaluating the sideslip stability and roll-over stability of the vehicle. The calculation of Pearson's correlation r_{xy} was expressed as:

$$r_{xy} = \frac{\text{cov}(x,y)}{\sqrt{\text{var}(x)} \cdot \sqrt{\text{var}(y)}} \tag{3}$$

where x and y denote any two of LA, SA, LTR, and RA. cov(x, y) is the covariance of x and y. var(x) and var(y) are the variances of x and y, respectively.

- Grey Relational Analysis (GRA)

Grey relational analysis (GRA) is generally used to evaluate the correlation between factors in multi-factor systems. In this section, grey relational coefficient (GRC) and grey relational grade (GRG) are adopted to characterize the correlation between the testing factors and targeted indicators LA, SA, LTR, and RA. The basic process for grey relational analysis mainly includes the following four steps [50].

(1) Confirmation of reference sequence and comparison sequence

As expressed in the Equations below, four indicators, i.e., LA, SA, LTR, and RA, were taken as reference sequences, respectively. The comparison sequence was composed of above seven factors.

$$\Phi_0(i) = \{ \Phi_0(1), \Phi_0(2), \Phi_0(3), \ldots, \Phi_0(i)\} \quad (4)$$

$$X_k(i) = \{X_1(1), X_2(2), X_3(3), \ldots, X_k(i)\} \quad (5)$$

where $\Phi_0(i)$ denotes targeted output (LA, SA, LTR, and RA) at case i in Table 2 (i = 1, 2, 3, . . . , 32), and $X_k(i)$ denotes inputting factors (V_c, WHR, R, e, δ_{lw}, V_{lw}, and f), k = 1, 2, 3, 4, 5, 6, 7.

(2) Normalization of the original test data

The units and orders of magnitude for seven factors are different, so they need to be compared in the same range. Therefore, the original test data were processed by normalization. The calculation processes were expressed as Equations (6) and (7).

$$Y_0(i) = \frac{\Phi_0(i) - \Phi_0(\min)}{\Phi_0(\max) - \Phi_0(\min)} \quad (6)$$

$$X'_k(i) = \frac{X_k(i) - X_k(\min)}{X_k(\max) - X_k(\min)} \quad (7)$$

where $Y_0(i)$ denotes normalized value of LA, SA, LTR, and RA at case i, and $X'_k(i)$ denotes normalized value at case i of experimental factor k. $\Phi_0(\min)$ and $\Phi_0(\max)$ denote the maximum and minimum value of targeted output. $X_k(\min)$ and $X_k(\max)$ denote the maximum and minimum vale of inputting factor k.

(3) Calculation of the proximity

The proximity matrix can be obtained through calculation absolute value of the difference of $Y_0(i)$ and $X'_k(i)$, as expressed in Equation (8).

$$\delta_{0,k}(i) = |Y_0(i) - X'_k(i)| \quad (8)$$

where $\delta_{0,k}(i)$ denotes the absolute value of the difference of $Y_0(i)$ and $X'_k(i)$.

(4) Calculation of GRC and GRG

GRC is calculated as Equation (9).

$$\xi_i(k) = \frac{\delta_{0,k}(\min) + \rho\delta_{0,k}(\max)}{\delta_{0,k}(i) + \rho\delta_{0,k}(\max)} \quad (9)$$

And GRG is express as:

$$\zeta_k = \frac{\sum_{i=1}^{m} \xi_i(k)}{m} \quad (10)$$

where $\xi_i(k)$ denotes grey relational coefficient (GRC), $\delta_{0,k}(\min)$ and $\delta_{0,k}(\max)$ denote minimum and maximum of the absolute value of $\delta_{0,k}(i)$, respectively. ρ represents distinguishing coefficient, which usually takes 0.5. ζ_k denotes grey relational grade (GRG), and m is the number of experiment case (equal to 32).

- Comprehensive Relation Grade (CRG)

LA, SA, LTR, and RA were regarded as targeted indicators to characterize the lateral stability of the vehicle. However, these indicators varied in unit concurrently with a difference in evaluating instability mode (sideslip and roll-over). So, a comprehensive indicator that indicates the effect of seven factors on driving stability should be proposed. In this section, a comprehensive relation grade (CRG) is proposed based on the entropy weight method and grey relational analysis.

The entropy weight method was used to calculate the weight of each indicator (LA, SA, LTR, and RA). An initial matrix A formed by four indicators from case 1 to case 32 was constructed as follows. The corner mark j = 1, 2, 3, 4, which represents LA, SA, LTR, and RA, respectively.

$$A = \left(a_{ij}\right)_{m\times n} = \begin{bmatrix} a_{11} & a_{12} & a_{13} & \cdots & a_{1n} \\ a_{21} & a_{22} & a_{23} & \cdots & a_{1n} \\ a_{31} & a_{32} & a_{33} & \cdots & a_{1n} \\ \vdots & \vdots & \vdots & \ddots & \vdots \\ a_{m1} & a_{m2} & a_{m3} & \cdots & a_{mn} \end{bmatrix} \tag{11}$$

Then, the removal of the indicator dimension was conducted by normalization, as given by:

$$p_{ij} = \frac{a_{ij}}{\sum\limits_{i=1}^{m} a_{ij}} \tag{12}$$

Entropy value of j th indicator as calculated as:

$$e_j = -\frac{1}{\ln m}\sum_{i=1}^{m} p_{ij}\ln p_{ij} \tag{13}$$

Entropy weight of j th indicator was calculated as [51]:

$$w_j = \frac{1-e_j}{\sum\limits_{j=1}^{n} 1-e_j} \tag{14}$$

Finally, the targeted comprehensive relation grade (CRG) was calculated as:

$$CRG_k = \sum_{j=1}^{n} w_j\zeta_{jk} \tag{15}$$

3. Simulation Model

3.1. Modelling Vehicle

In this section, a simulation model was established with the Carsim software. It was generally agreed that the types of vehicles involved in accidents were mainly small vehicles (cars, minibuses) [52]. Since small-size passenger cars generally have a faster-running speeds, compared with SUVs or large trucks, the driving safety affected by rutting and adverse conditions is worth considering indeed. So, a small-size E-class vehicle was selected in the datasets provided by Carsim and modified manually based on actual vehicle parameters [37], as shown in Figure 4. The main parameters are shown in Table 3. The aerodynamics coefficients were the functions of aerodynamic slip angle, which were selected as default values in Carsim.

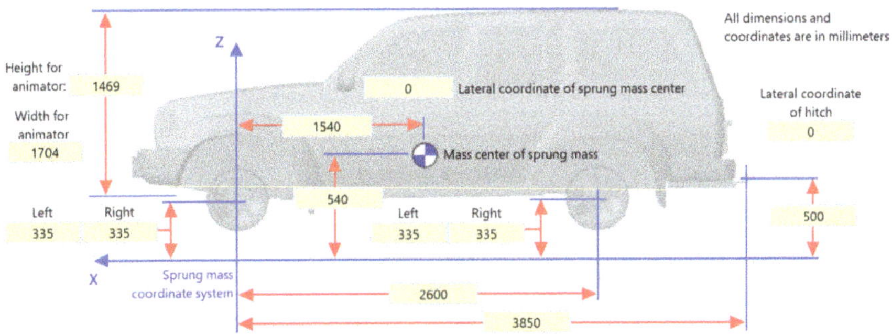

Figure 4. Vehicle parameters setting.

Table 3. Main parameters of vehicle model.

Parameters	Input Value
Sprung mass	1350 kg
Spread of axles	2600 mm
Wheel tread	1550 mm
Front overhang	1540 mm
Height of mass center	540 mm
Tire type	225/60 R18

3.2. Modelling Adverse Weather

The safety of automobile driving is significantly influenced by adverse weather. The ponding caused by strong rainfall on the pavement may induce the hydroplaning potential, thus resulting in an obvious reduction of the friction coefficient of pavements [15,53]. The variation of real road friction due to hydroplaning is not only related to the depth of water accumulation and driving speed [15,54], but also concerned with factors such as the tire properties and pavement surface properties [55], which is a comprehensive research issue. Additionally, it will become worse if the impact of side wind on the lateral stability of automobiles is considered, which may form the combined effect of rainfall, ice, and crosswind, thereby greatly increasing the complexity of the loading characteristics of automobiles and the possibility of safety accidents. Additionally, the present study focuses on the analysis of different factors affecting vehicle stability, so the impact of hydroplaning phenomenon under a rain scenario is simplified by reducing the friction coefficient. As a result, regarding the effects of rain, ice, and snow on road surface conditions, they generally increase the driving risks by decreasing the friction coefficient between the pavement surface and tire. So, different friction coefficients were input to indicate adverse weather, according to findings in [13], as given in Table 4.

In terms of modeling wind velocity and wind direction in CarSim aerodynamics, the linear interpolation and extrapolation function was used to model wind velocity and wind angle. Wind velocity is defined as zero in 0–2 s, increases linearly to the target speed in 2–4 s, maintains a constant wind speed, starts to decrease linearly at 23 s, and decreases to 0 at 25 s. Additionally, the wind angle is set as consistent with wind speed.

Table 4. Moisture conditions of pavement under different weather conditions.

Weather Condition	Normal Condition	Light Rain	Heavy Rain	Ice and Snow
Pavement surface condition	Rather dry	Wet with less ponding	Wet with massive ponding	Frozen
Friction coefficient	0.8	0.6	0.4	0.18

3.3. Modelling Road

Road conditions highly determine vehicle stability, which connects to its interaction with the wheel [56]. In terms of modeling the road, it mainly comprises a radius of circular curve, superelevation, and rutting. A set of horizontal curves with different values of radius were established, in order to test vehicle stability on curve segments, with a total length of 1200 m. Superelevation was set in the form of a lateral pavement gradient.

Concerning modeling pavement rutting, the selection of rutting parameters is of particular importance. When a vehicle drives parallel to the rutting direction, minor effects on driving safety may occur. However, in the case of changing lanes or directions, a vehicle directly drives across a rut, thereby wandering behavior occurs [57], in which the vehicle may show unstable, lateral, or yaw motions, as if the vehicle went through a hole.

As presented in Figure 5, h denotes the rutting depth, and w is the rutting width. The rutting shape is commonly described with trigonometric functions [58], as given by Equation (16). The inclination angle of the tire is expressed as Equation (17).

$$z = \frac{h}{2}\left(1 - \cos\left(\frac{2\pi y}{w}\right)\right) \tag{16}$$

$$\psi = \frac{dz}{dy} = \frac{\pi h}{w}\sin\left(\frac{2\pi y}{w}\right) \tag{17}$$

Though conjoint force analysis with a rut, the equation of vehicle motion is given by:

$$m\ddot{y} + \frac{C_{F\alpha}}{V_x}\dot{y} + \frac{2\pi^2 h}{w^2}(F_{zi} - C_{F\psi})y = 0 \tag{18}$$

where $C_{F\alpha}$ denotes the cornering stiffness of the tire, $C_{F\psi}$ denotes the camber stiffness of the tire, F_z is the vertical force exerted at tire, m equals tire mass, and V_x is the driving speed along x axis. Equation (18) presents free vibration equation with damping when the tire negotiates a rut. It can be found that the damping coefficient is proportional to $C_{F\alpha}$, but inversely proportional to V_x. Additionally, stiffness $\frac{2\pi^2 h}{w^2}(F_z - C_{F\psi})$ is obviously decided by rutting depth and rutting width. So, a higher driving speed will lead to a large vibration frequency while crossing a rut. Meanwhile, the rutting dimension has an impact on the vibration properties of the vehicle.

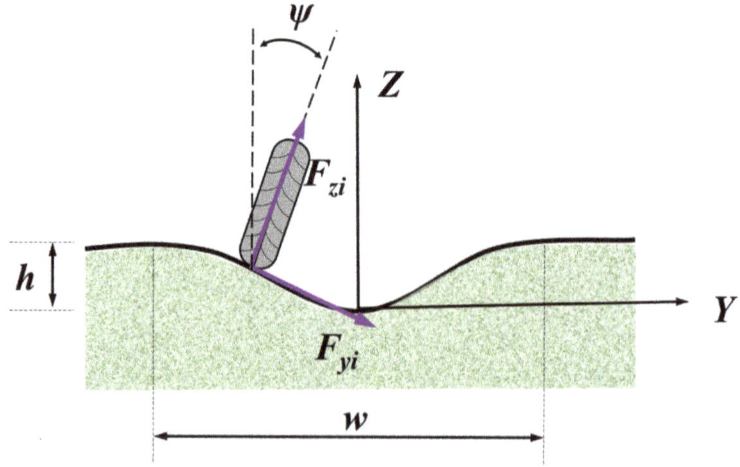

Figure 5. Diagram of the wheel driving across a rut.

The lateral vehicle dynamics when instantaneously crossing a rut are rather complex, not only for a single tire or a single axle, but also for the whole vehicle. So, evaluating the dynamic condition of the vehicle considering rutting is more crucial than roughness, cracking, and other pavement distresses. Further, from the equation of vehicle motion (18), it implies that the rutting depth and rutting width show strong connection with the dynamic vibration properties of the vehicle.

The three main parameters of rutting are generally used to characterize driving safety. Normally, the maximum rutting depth (RD) is used to evaluate the rutting conditions, which affects the driving safety by the pattern of height difference and fluctuation. The rutting width (RW) is also an important parameter to characterize the rutting shapes. When the ruts are narrow, the wheels experience large changes in elevation in a very short period of time, and the vehicle's recovery adaptations are unable to quickly follow the changes, making it inclined to lose stability. A smaller RW will present a greater impact on the driving stability. The width–height ratio (WHR) of the rutting sidewall denotes the dip angle of rutting sidewall, which refers to the ratio of the width-to-height of the rut sidewall. A higher WHR indicates a greater resistance to be overcome when changing the trajectory of the vehicle, and the greater the resistance to steering that the vehicle will experience. Therefore, in order to excellently characterize the effects of height fluctuation and lateral crossing when driving across rutting, WHR is selected as a factor that influences driving stability in different simulation cases.

In terms of modeling pavement rutting, the cross-sectional rutting shape was regularly simplified as a trapezoid because actual irregular shapes of the rutting were hard to directly describe [2,59], as shown in Figure 6a. It has been demonstrated that the WHR of the rutting sidewall had a greater impact on the lateral stability of the vehicle, compared with the rutting depth and rutting width, when it comes to discussing the case of vehicles negotiating the rutting sidewall [5,8]. Therefore, different WHRs (20, 15, 10, 5) were imported in Carsim software, which correspond to different heights of the rut sidewall (10 mm, 13.33 mm, 20 mm, 40 mm), respectively, while maintaining the width of 0.2 m unchanged. Figure 6b shows the case of WHR = 10.

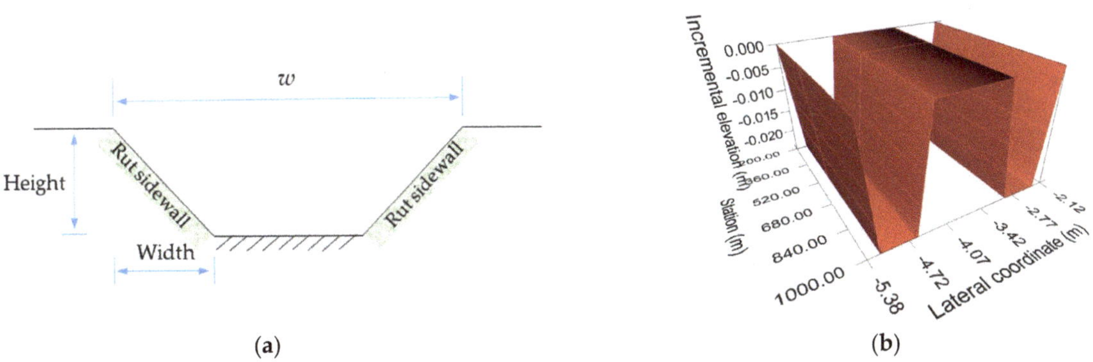

Figure 6. Schematic view of Rutting model: (**a**) Simplified trapezoid rutting shape; (**b**) rutting model in Carsim software (in case of WHR = 10).

3.4. Modelling Lane Change

To accurately simulate the process of changing lanes as much as possible, a lane changing model was established based on hyperbolic tangent function [60], which is expressed as:

$$\Delta(t) = \zeta \tanh\left(\frac{\tau}{L_d/V_d} \cdot \left(t - \frac{L_d}{2V_d}\right)\right) + \eta \tag{19}$$

$$\zeta = \begin{cases} \frac{|\Delta_0|+|\Delta_T|}{2}, & \text{LLC} \\ -\frac{|\Delta_0|+|\Delta_T|}{2}, & \text{RLC} \end{cases} \tag{20}$$

$$\eta = \frac{\Delta_0 + \Delta_T}{2} \tag{21}$$

where t denotes the travel time, $\Delta(t)$ denotes the lateral offset of the vehicle at time t, and Δ_0 and Δ_T denote lateral offset of the vehicle at the start and end point relative to the initial centerline of lane, respectively. L_d denotes the longitudinal length during lane change. V_d denotes the average longitudinal velocity during lane change. ζ and η denote the coefficients related to Δ_0 and Δ_T, respectively. τ is the parameter used to characterize the urgency of a lane change in this model. In addition, LLC and RLC mean left lane change and right lane change, respectively. The vehicle changes lanes in RLC mode, as shown in Figure 7.

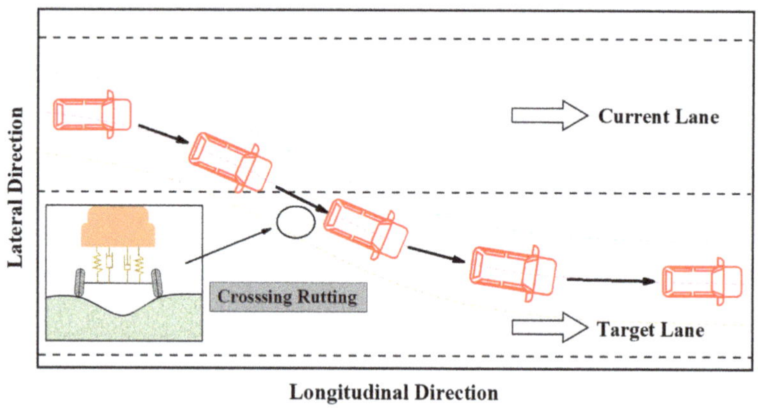

Figure 7. Diagram of lane-change trajectory.

4. Results and Discussion

4.1. Driving Stability Analysis on Single-Factor

4.1.1. Effects of Various Driving Conditions on LA

Figure 8 shows the effects of different factors on the LA of the vehicle. During the driving process, it mainly contains four stages: smooth driving, initiating lane change, crossing rutting, and ending lane change with smooth driving. It was found that the LA maintains at around 0.1 g in the smooth driving state, with only a minor variation before and after. For a short period of time, after starting a lane change, the lateral tractive force generated during lane change demonstrates a remarkable decrease of LA. When the vehicle reaches around 500 m distance travelled (approximate 16 s), the vehicle starts to cross the rutting, and the LA increases in the opposite direction. When it came to the termination of a lane change, LA developed back to a steady trend. The LA fluctuated up and down, due to the compound effects of lateral wind and rutting.

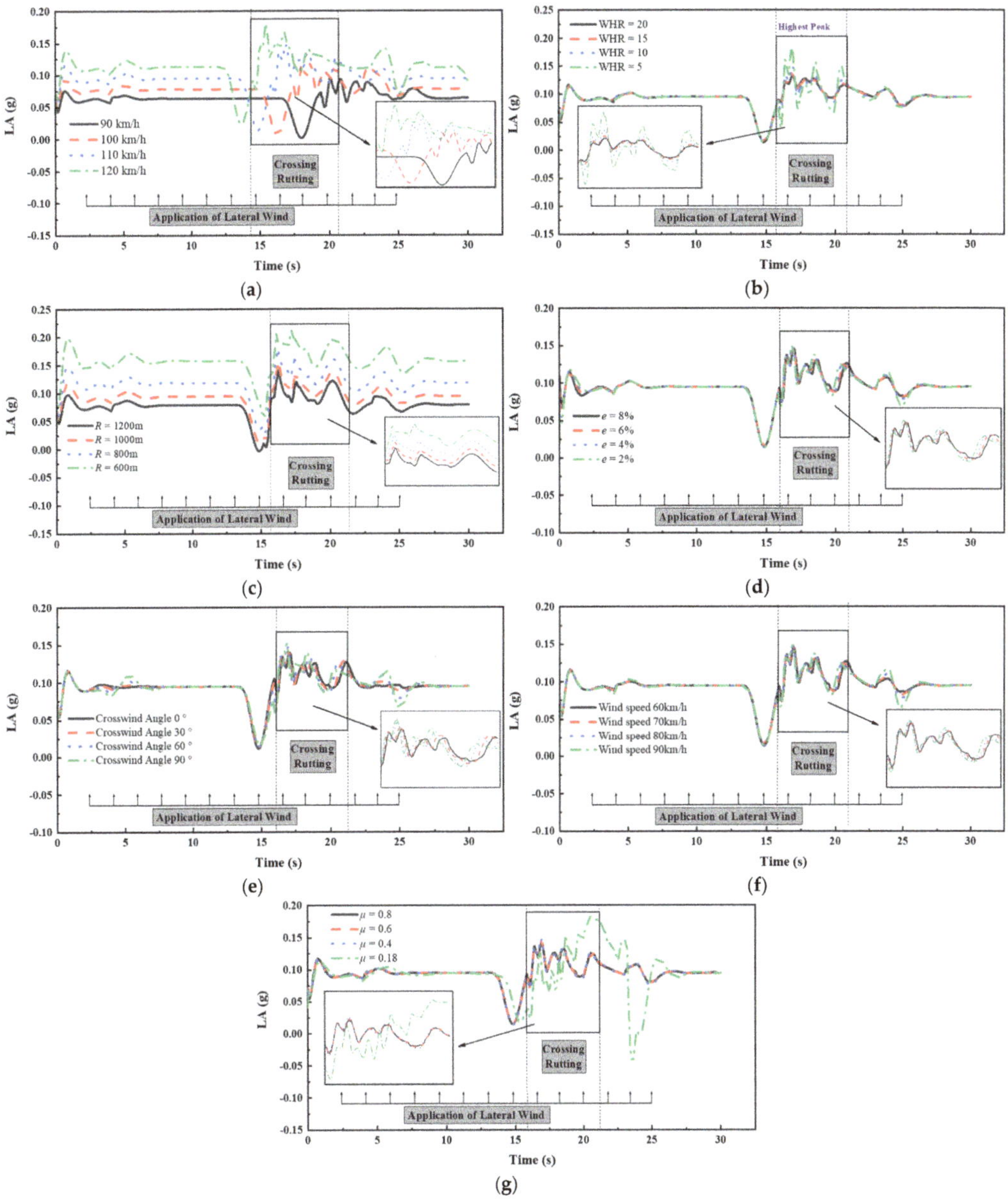

Figure 8. Change of LA over time under various driving conditions: (**a**) Vehicle speed; (**b**) WHR of the rutting; (**c**) Radius of circular curve; (**d**) Superelevation; (**e**) Crosswind angle; (**f**) Wind speed; (**g**) Friction coefficient between road & tire.

To compare the effects of different factors from the perspective of spacing between the curves in Figure 8, it is indicated that the vehicle speed, WHR of rutting, radius of the circular curve, and friction coefficient of the road surface have a greater impact on LA, while the superelevation, lateral wind angle, and speed have a relativly weak influence on LA. As shown in Figure 8a,c, LA increases with the increase of driving speed, concurrently with the decrease of the radius of circular curve. The LA reaches a high level, with a maximum value approximately 0.2 g, when the driving speed and radius of the circular curve are

120 km/h and 600 m, respectively. From Figure 8b, it is found that the LA dramatically and suddenly rises to a peak when the WHR reaches 5. In addition, the maximum LA is far greater than that of other width-to-height ratios, comprising a larger amplitude of variation. As a result, the vehicle shows a higher risk of lateral offset when driving across rutting in adverse weather.

4.1.2. Effects of Various Driving Conditions on SA

Figure 9 shows the influence of different factors on SA. The SA is always smaller than the threshold value during the whole process. It is concluded that the variation of SA generally consists of four stages: rapid increase, slow increase, fluctuating increase, and the stable stage. An initial rapid increase in SA happens from 0 to 2 s, mainly due to centrifugal force during curve driving, while the slow increase in SA before the first wave crest of curve is caused by continuous application of lateral wind. The fluctuation of the SA from 15 s to 25 s is influenced by a compound effect of the lane change, across rutting, and lateral wind action. A more significant fluctuation effect on SA is shown in Figure 9b, in which the rutting is extremely severe (WHR = 5), with the largest simultaneous wave crest.

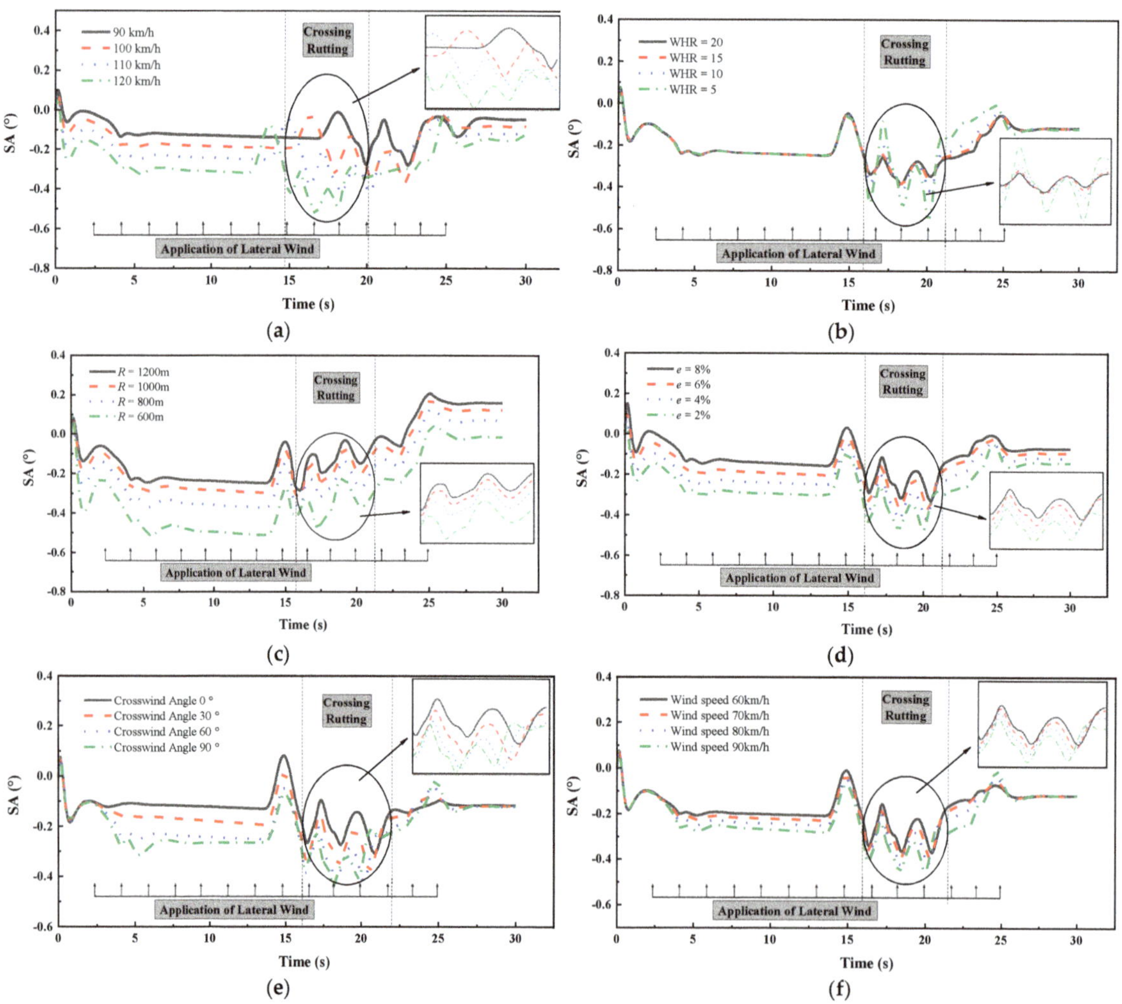

Figure 9. *Cont.*

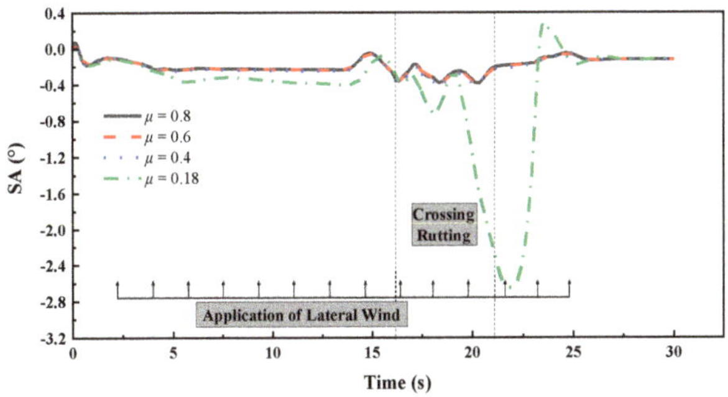

(g)

Figure 9. Change of SA over time under various driving conditions: (**a**) Vehicle speed; (**b**) WHR of the rutting; (**c**) Radius of circular curve; (**d**) Superelevation; (**e**) Crosswind angle; (**f**) Wind speed; (**g**) Friction coefficient between road & tire.

As presented in Figure 9a–g, with the increase of vehicle speed, crosswind angle, and crosswind speed, a sharp decrease of the radius of the circular curve, superelevation, and pavement friction coefficient, the SA gradually increases in the negative direction. Particularly, when driving on the pavement, when the friction coefficient decreases to 0.18 (in snowy weather), the absolute value of SA increases to 2.6, and shortly afterward, it increases in the opposite direction. The vehicle appears to swing back and forth with a pretty high degree of risk, even though the SA is still smaller than the threshold value.

4.1.3. Effects of Various Driving Conditions on LTR

The influence of different factors on LTR is shown in Figure 10. The load will be transferred from the inside wheels to the outside wheels, while the vehicle is traveling in a curve segment, so the LTR firstly increases over time and sharply fluctuates up and down while driving across rutting. It is concluded that the LTRs are generally less than 0.4 in this section, which implies that the risk of roll-over is low. During initial 0–2 s, the vehicle velocity increases from 0 to constant speed, so the LTR changes in a transitional phase. In the range of 2–15 s, the LTR gradually increases, due to the effect of lateral wind, which is consistent with the previous findings of Yu et al. [61]. In the range of 15–20 s, the LTR significantly increases and then rapidly decreases, due to crossing rutting along with the effect of lateral wind. Afterward, the lateral wind gradually dissipates, so the LTR decreases and remains unchanged.

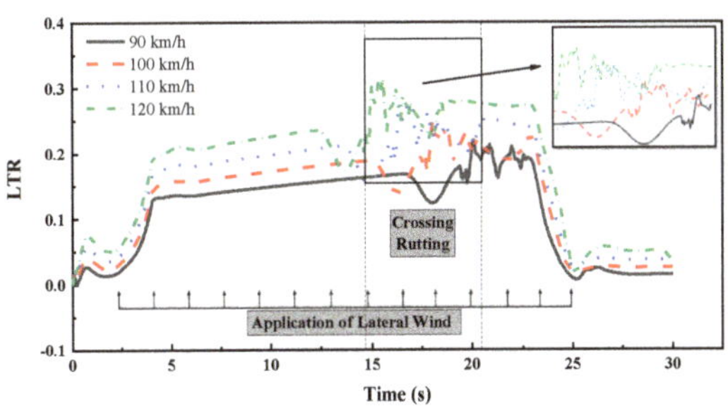

(a)

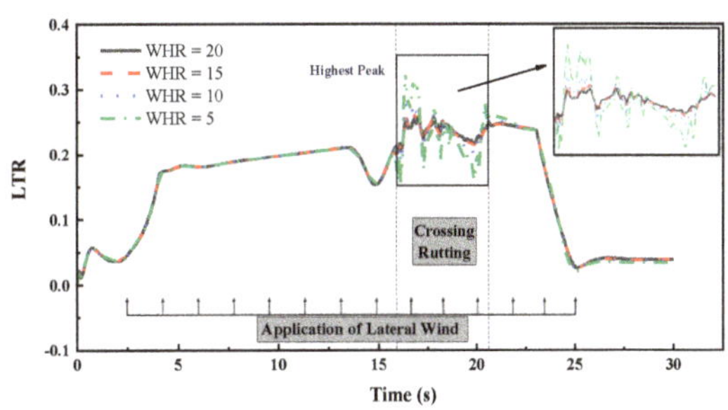

(b)

Figure 10. *Cont.*

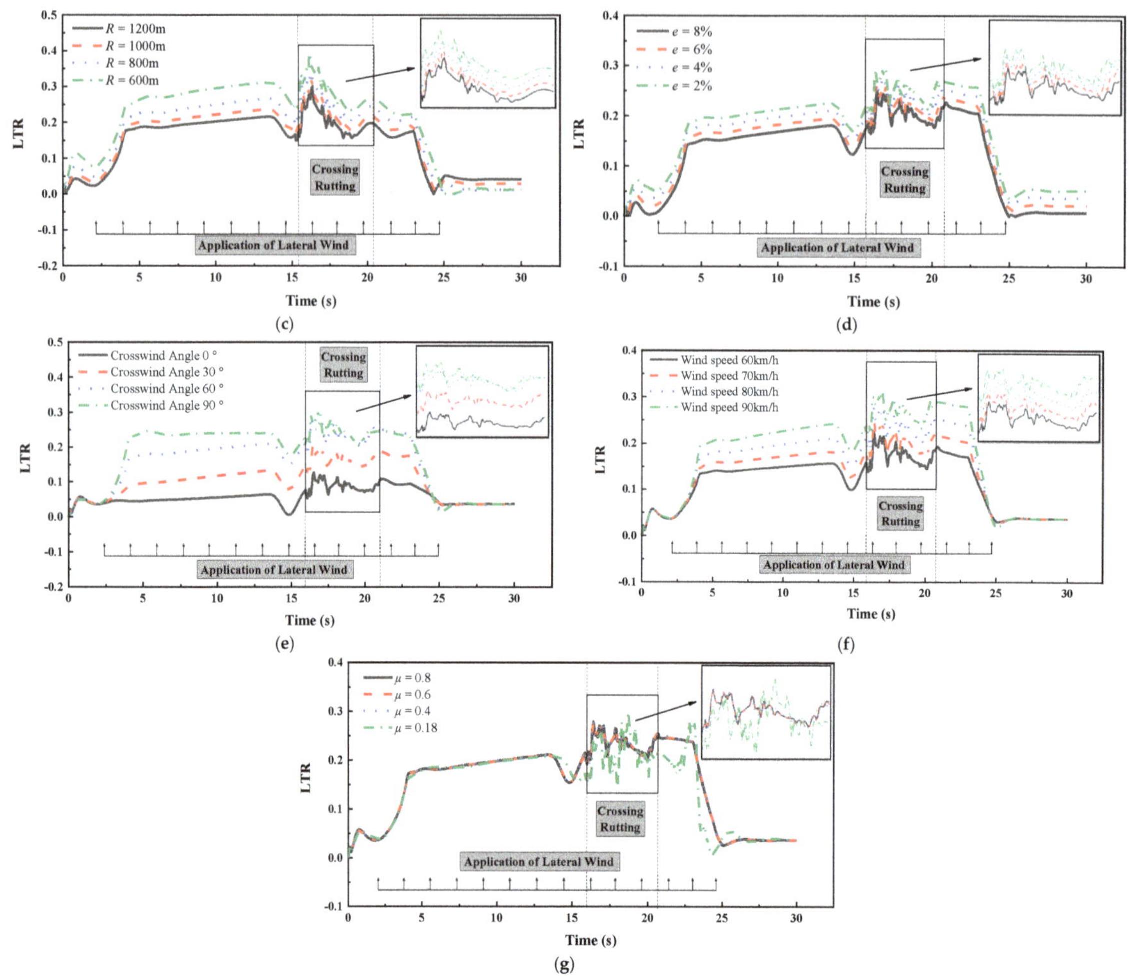

Figure 10. Change of LTR over time under various driving conditions: (**a**) Vehicle speed; (**b**) WHR of the rutting; (**c**) Radius of circular curve; (**d**) Superelevation; (**e**) Crosswind angle; (**f**) Wind speed; (**g**) Friction coefficient between road and tire.

The effects of vehicle speed, radius of circular curve, superelevation, crosswind angle, and crosswind speed on LTR can be recognized as "quantitative changes but consistent with dynamic mode", that is, vehicles are driven with similar modes in the above cases, but only varies in exact values. Note that the sudden change in LTR occurs when the WHR of rutting is 5 in Figure 10b, which is consistent with the sudden change in LA and SA, sharply fluctuating up and down.

4.1.4. Effects of Various Driving Conditions on RA

The variations of RA over time under the influence of different factors are shown in Figure 11. The simulation results for the period of time without outside adverse disturbance indicate that the RA of the vehicle is generally less than 2° and varies slightly, even when driving in a curve segment with a lateral wind. However, when it crosses a rut, the RA changes rapidly, and the instantaneous changes in RA happen to the wheels crossing the rutting, one after another, respectively. In the case of WHR equal to 5, the maximum RA of the vehicle is about 14° when driving across the rutting, which exceeds the threshold value of RA, while the maximum RA under the rest of WHR conditions does not exceed 6°, which is less than the limit value of RA. Thus, it is concluded that severe rutting damage will significantly reduce lateral stability, thus heightening the possibility of rolling over.

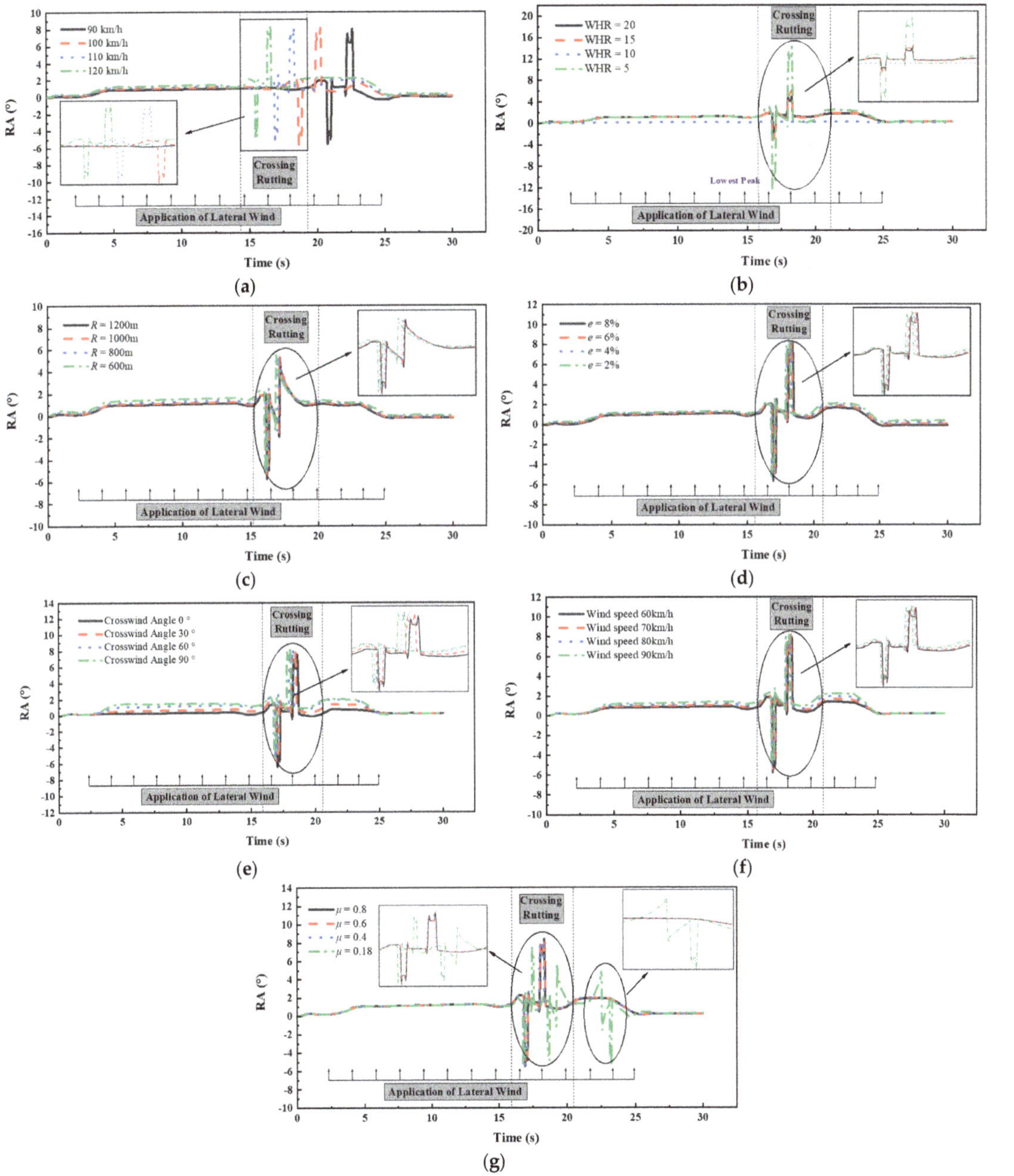

Figure 11. Change of RA over time under various driving conditions: (**a**) Vehicle speed; (**b**) WHR of the rutting; (**c**) Radius of circular curve; (**d**) Superelevation; (**e**) Crosswind angle; (**f**) Wind speed; (**g**) Friction coefficient between road & tire.

Based on the results and discussions in Sections 4.1.1–4.1.4, it is finally found that all LAs and SAs are within the threshold, but the maximum LA and maximum SA dramatically fluctuate up and down in certain cases. Regarding evaluating the roll-over stability of the vehicle, all LTRs are less than 1, showing satisfactory roll-over stability. However, the maximum RA far exceeded the threshold of 6°, which is not compliant with LTRs. So, it seems that these two indicators do not show consistency in evaluating the roll-over stability of the vehicle. Obviously, it is not thorough and accurate enough to evaluate the lateral stability with only the aid of single and independent indicators.

To overcome the limitations of a single indicator, a comprehensive indicator accounting for multiple indicators is essential for evaluating the lateral stability of the vehicle. So, in the subsequent section, Pearson's coefficient was applied to verify the correlation between LA, SA, LTR, and RA. Then, the entropy weight method was utilized to weigh these four indicators. Finally, the comprehensive relation grade (CRG) was proposed to evaluate the effects of different factors on the lateral stability of the vehicle.

4.2. *Driving Stability Analysis on Multi-Factor*

4.2.1. Analysis on Safety Requirement Based on Various Driving Conditions

Figure 12 is the bar chart of the simulation results from case 1 to case 32, which presents the maximum LA, SA, LTR, and RA determined from 32 cases. From Figure 12a, it was found that the LAs were less than 0.4 g for all 32 cases, which did not exceed the safety threshold. As shown in Figure 12b, the SAs of cases 4, 11, 13, and 31 far exceeded the threshold value. These results caused the drift of the vehicle, thus leading to steering problems and large sideslips, whereas the SAs of other cases were within the threshold. From Figure 12c, it was found that all the LTRs were smaller than 1.0, presenting a high roll-over stability. However, the RAs in a large number of cases dramatically exceeded the threshold, showing high roll-over risks. Through contrastive analysis, it is known that the results of RAs were not consistent with that of LTRs, and the results of LAs were not in compliance with that of SAs, though they equally characterized the sideslip stability and roll-over stability.

As mentioned in the results of cases 4, 11, 13, and 31, the SAs were 59.22°, 23.98°, 58.81°, and 15.80°, which far exceeded the threshold, resulting in a vehicle with a great tendency of swaying and slip. The common characteristic of the four cases was the extremely low friction coefficient of the road surface, which was as low as 0.18, regardless of other favorable factors. So, the considerable decrease in the friction coefficient, due to snowy weather, is the leading factor that gives rise to sideslip instability [62]. In terms of RAs, their values in cases 4, 8, 12, 16, 20, 24, 28, and 32 were far beyond the threshold, even reaching over 13°. It was found that the WHRs of these cases were all equal to 5, with a rutting opening width of 1 m and a rutting depth of 4 cm. The loss of instantaneous stability, when driving across the ruts, made the vehicle susceptible to rolling over. So, the rutting condition can be accounted as a dominant factor that leads to the roll-over instability of the vehicle, which verifies the relevant vehicle dynamics simulations concerning pavement rutting [4,5,63].

The worst dynamic response occurred in case 4, with super high SA and RA. In this case, the vehicle was considered to be in a completely unstable state, which showed high susceptibility to traffic accidents. The vehicle crossed a rut with a WHR equal to 5 under the poorest road alignment and weather conditions, even with relative low driving speed. Those conditions made the vehicle have insufficient resistance to lateral slip, so it had difficulties in quickly recovering back to a stable state after encountering a rut impact, resulting in a severe loss of control and continuous slip. So, it reveals that the composite effect of adverse weather and rutting conditions shows a high probability of causing the lateral instability of the vehicle. Limiting driving speed alone is still not an effective way to ensure driving safety. Instead, preventive measures on pavement maintenance are expected to be implemented.

To determine the safety requirement of various driving conditions, it was demonstrated that dangerous results always occurred when the WHR = 5 or f = 0.18. However, several instability phenomena still happened, even solely increasing the WHR to 10 or increasing the f to 0.4. Thus, it can be deduced that multiple factors need to be controlled simultaneously. By comparing 32 cases in this study, the lateral stability of the vehicle is able to be guaranteed when the WHR is greater than 10 and the f is greater than 0.4, concurrently.

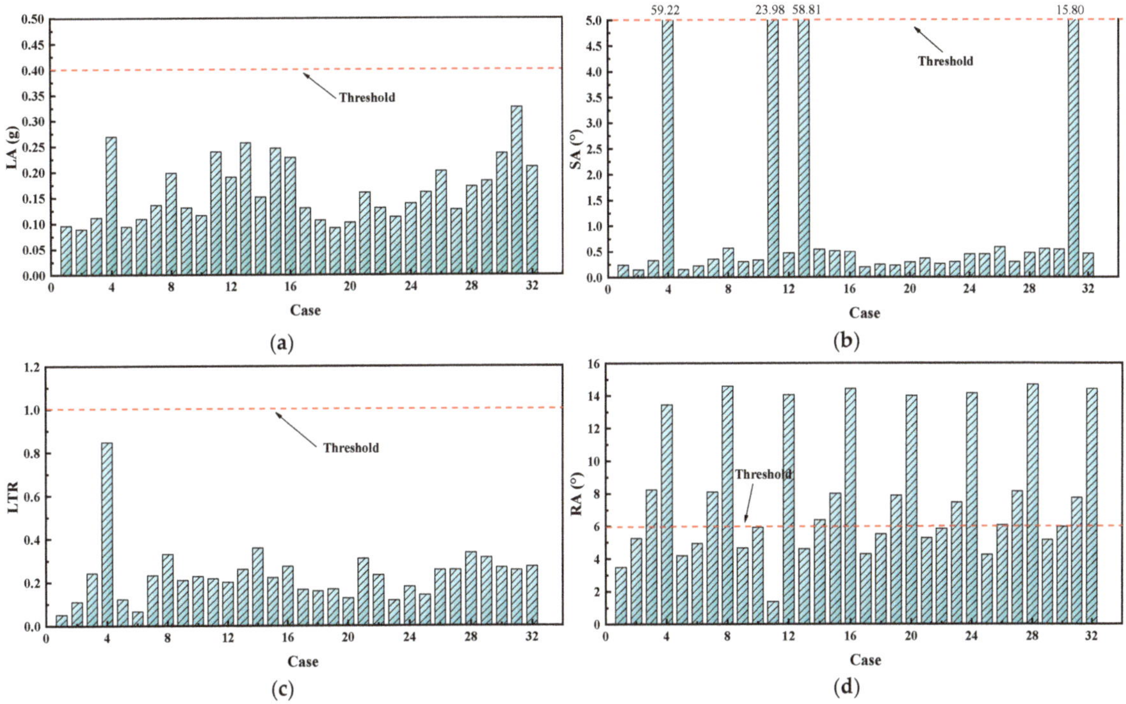

Figure 12. Driving stability indicator based on orthogonal simulation results: (**a**) LA; (**b**) SA; (**c**) LTR; (**d**) RA.

4.2.2. Correlation Analysis of Stability Indicators

Figure 13 presents the coefficient matrix of Pearson's Correlation between indicators LA, SA, LTR, and RA. It demonstrated that the LA and SA had a strong correlation to evaluate the sideslip stability of the vehicle, with a Pearson's coefficient of 0.8573. The LA and LTR, as well as the SA and LTR, both had acceptable correlations with each other, and their Pearson's coefficients were greater than 0.6. Nevertheless, LTR was generally used to evaluate the roll-over stability of the vehicle, which is not incompatible with LA and SA, as given in previous studies. Likewise, the LTR and RA were normally used to characterize the roll-over stability of the vehicle. However, the Pearson's correlation between the LTR and RA was only 0.3427, which implied the single indicator (LTR or RA) was unable to efficiently evaluate the roll-over stability of the vehicle. So, a comprehensive indicator CRG was proposed with the grey relational method to enhance this defect.

Figure 13. Coefficient matrix of Pearson's Correlation.

4.2.3. Evaluation of Lateral Stability Based on GRG

The grey relational grade (GRG) of each factor was obtained in Table 5 with Equations (9) and (10). In terms of LA, higher GRG values for vehicle speed, WHR, circle curve radius and friction coefficient are obtained, and all of which exceed 0.6, indicating that the vehicle speed had the most significant effect on LA, followed by the radius of circular curve, friction coefficient, WHR, superelevation, crosswind angle, and crosswind speed, respectively. For SA, the sequence of relational grades, ranked in descending order, is vehicle speed, radius of circular curve, superelevation, WHR, friction coefficient, crosswind speed, and crosswind angle, respectively. Regarding LTR, the degree of influence of the factors, in descending order, are crosswind angle, vehicle speed, radius of circular curve, WHR, wind speed, superelevation, and friction coefficient, respectively. For RA, WHR has the most obvious effect on RA, followed by vehicle speed, friction coefficient, crosswind angle, radius of circular curve, superelevation, and crosswind speed, respectively. It is notable that the CRG of WHR for RA is over 0.8, demonstrating the highest relation grade among the various factors. It implies that the pavement rutting is inclined to lead to roll-over instability.

Table 5. Grey relational grade (GRG) of each factor.

Indicators of Lateral Stability	Driving Speed	Rutting	Road Alignment		Adverse Weather		
	V_c	WHR	R	e	δ_{lw}	V_{lw}	f
LA	0.7696	0.6189	0.6940	0.5950	0.5911	0.5762	0.6263
SA	0.7417	0.6444	0.6963	0.6467	0.5758	0.5913	0.6072
LTR	0.7052	0.6618	0.6793	0.6499	0.7131	0.6568	0.5998
RA	0.6374	0.8171	0.6020	0.5984	0.6064	0.5880	0.6111

4.2.4. Evaluation of Lateral Stability Based on CRG

The entropy weight of each indicator was obtained in Table 6. It was found that the entropy weight of LA was 0.1806, and it has the lowest weight among the four indicators, which means LA has the poorest efficiency to characterize driving safety. The entropy weight of RA is 0.3112, so it indicates the greatest significance to driving safety. Then, it is followed by SA and LTR, and their entropy weights are 0.2904 and 0.2179, respectively. So, it is deduced that roll-over instability is more likely to occur due to the variation of seven factors, relative to sideslip instability. However, Chen et al. [35] identified that passenger cars tended to experience sideslips, which is not identical to this study. The distinction may be attributed to the extended consideration of pavement rutting in the present study.

Table 6. Entropy weight of each indicator.

Parameters of Entropy Weight Method	Sideslip Stability		Roll-Over Stability	
	LA	SA	LTR	RA
Entropy (e)	0.9569	0.9307	0.9480	0.9257
Entropy weight (w_j)	0.1806	0.2904	0.2179	0.3112

Based on the GRG and entropy weight that have been obtained, a comprehensive relation grade (CRG) was calculated from Equation (15), and the results are shown in Figure 14. It indicates that vehicle speed and WHR of rutting have the leading effects on driving stability, followed by the radius of circular curve, superelevation, crosswind angle, crosswind speed, and friction coefficient, respectively. The CRGs of driving speed and WHR are both close to 0.7, which shows high significance in affecting the lateral stability of the vehicle. The radius of the circular curve and superelevation are also important factors that influence the lateral stability of the vehicle, which confirms the findings of Peng et al. [64]. The differences in CRGs between crosswind angle, crosswind speed, and friction coefficient are rather small, three factors which have nearly identical effects on the lateral stability of the vehicle.

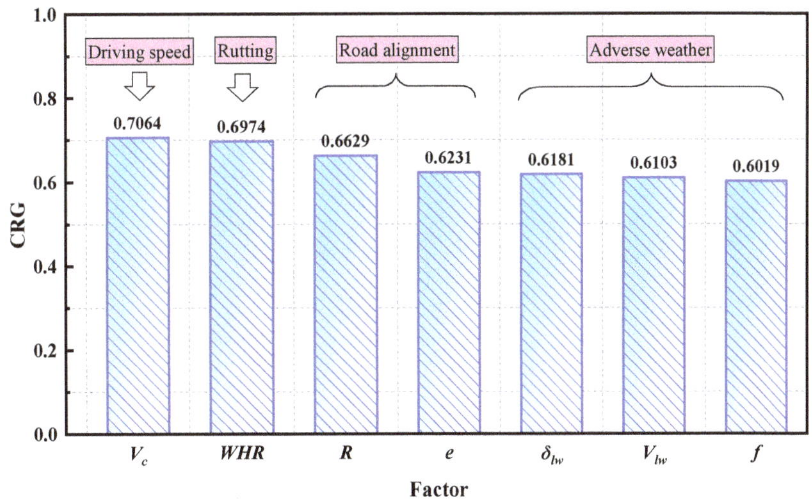

Figure 14. The impact of driving speed, rutting, road alignment, and adverse weather on CRG.

Seven factors obviously influence driving safety. Firstly, the driving speed of the vehicle is generally the first, and most significant, factor when crossing rutting in adverse weather. Secondly, pavement rutting is the secondary factor that influences driving safety. In particular, the roll-over of the vehicle may take place while crossing the rutting immediately under poor road surface conditions with serious rutting distresses. Thirdly, road alignment is still the key consideration for driving safety, including the radius of the circular curve and superelevation in this study. If such an occasion happens with a fierce wind, the degree of risks may increase correspondingly. Moreover, the change in the road friction coefficient, due to the influence of adverse weather increases the possibility of occurring driving accidents. These unfavorable factors should be taken into account.

5. Conclusions and Prospects

In this study, a novel framework was established to implement a systematic evaluation of the lateral stability of the vehicle by considering various impact factors. The software Carsim was utilized to perform simulations. Targeted indicators were adopted to evaluate the effects of seven factors on the lateral stability of the vehicle, including LA, SA, LTR, and

RA, as well as the comprehensive relation grade (CRG). The most important findings of the present study are concluded below:

(1) Four targeted indicators generally experienced several stages, due to the composite effect of lateral wind, rutting, and curve driving, including a stable increasing stage, sharp increase, sharp decrease, and fluctuation. The sharp fluctuation, up and down, of four indicators was due to the process of the wheel driving into a rut and then out of a rut during the lane change. Lateral wind created a relatively weak influence on the indicators' tendencies during the stable increasing stage.
(2) LTRs were generally within the threshold, while the maximum RA far exceeded the threshold in the results of single-factor analysis, despite the fact that they both characterized the roll-over stability of the vehicle. Consequently, the application of a single indicator showed limitations in evaluating the lateral stability of the vehicle, so a comprehensive indicator was expected to enhance this defect.
(3) The targeted indicators suddenly changed in a short time, due to the wandering behavior caused by rutting. Especially, this sudden change phenomenon was greatly enlarged in the case of WHR = 5. The indicator RA exceeded the threshold after driving across the rutting immediately, i.e., behaving with high driving risks. So, the instantaneous state of crossing severe rutting could be defined as a dangerous moment, which shall be cautiously prevented from the perspectives of both pavement and vehicle.
(4) The vehicle was out of control, due to the joint adverse conditions with severe rutting, poor road alignment, strong wind, and low skid resistance, obviously presenting high roll-over risks. Controlling a single factor was unable to ensure driving safety, due to the notable significance of the other factors. The combined requirement of WHR > 10 and $f > 0.4$ was recommended to control the pavement condition for ensuring driving safety in this study.
(5) LA and SA had a strong correlation to jointly evaluate the sideslip stability of the vehicle. However, the Pearson's correlation between LTR and RA was small, so a single indicator is weak in performing good efficiency on the roll-over stability of the vehicle. A comprehensive indicator remains to be proposed to enhance evaluation efficiency.
(6) A comprehensive relational grade (CRG) was, thus, proposed in the developed framework to improve evaluation performance, and it revealed that vehicle speed and the WHR of rutting had leading effects on driving safety, followed by radius of circular curve, superelevation, crosswind angle, crosswind speed, and friction coefficient, respectively.

The findings of this study revealed that the established framework performs well on evaluating the effect of different impact factors on the lateral stability of the vehicle, while driving across rutting, and particularly provides guidance to control pavement conditions from the perspective of driving safety. Despite the aforementioned contributions, a few shortcomings of this study remain to be optimized and enhanced. The major limitations lie in the quality and accuracy of the simulation, due to lack of real-world data. The consideration of factors is relative comprehensive, but in modeling the road, due to the uncertainty and variability of the hydroplaning potential, influenced by complex multi-factors, there is a case of simplified modeling. Further, only small-size passenger cars are discussed, on account of the limitation of the length of the article. Therefore, extensive in-depth investigations on the optimization of the developed framework, based on practical measured data, are envisaged in the follow-up studies. Moreover, the comparative study on the lateral stability of different vehicle types will be systematically performed, in conjunction with road maintenance recommendations for future consideration.

Author Contributions: Conceptualization, G.T., Y.J. and Y.G.; methodology, G.T. and Y.J.; software, Z.C. and Y.C.; validation, Y.G., S.W. and T.Z.; formal analysis, Z.W.; investigation, Y.C.; resources, G.T.; data curation, S.W.; writing—original draft preparation, G.T.; writing—review and editing, G.T., Z.C. and Y.J.; visualization, Y.J.; supervision, Y.G.; project administration, Y.G.; funding acquisition, Y.G. All authors have read and agreed to the published version of the manuscript.

Funding: This research was funded by National Natural Science Foundation of China (51878168).

Data Availability Statement: The data presented in this study are available on request from the corresponding author.

Acknowledgments: Thanks to funding for supporting this study. The useful support given by the National Demonstration Center for Experimental Road and Traffic Engineering Education (Southeast University) is acknowledged. The authors are sincerely grateful for their support.

Conflicts of Interest: The authors declare no conflict of interest.

References

1. Xue, B.; Xu, J.; Pei, J.; Zhang, J.; Li, R. Investigation on the Micromechanical Response of Asphalt Mixture during Permanent Deformation Based on 3D Virtual Wheel Tracking Test. *Constr. Build. Mater.* **2021**, *267*, 121031. [CrossRef]
2. Chen, L.; Liu, G.; Qian, Z.; Zhang, X. Determination of Allowable Rutting Depth Based on Driving Safety Analysis. *J. Transp. Eng. Part B Pavements* **2020**, *146*, 2. [CrossRef]
3. Mamlouk, M.; Vinayakamurthy, M.; Underwood, B.S.; Kaloush, K.E. Effects of the International Roughness Index and Rut Depth on Crash Rates. *Transp. Res. Rec.* **2018**, *2672*, 418–429. [CrossRef]
4. Cong, L.; Yang, J. Control Standard for Rut Depth Based on Simulation of Vehicle-Pavement System Dynamics. *Eng. Mech.* **2010**, *27*, 191–195. Available online: https://www.webofscience.com/wos/alldb/full-record/CSCD:4056960 (accessed on 25 February 2023).
5. Zheng, B.; Huang, X.; Zhao, R.; Hong, Z.; Chen, J.; Zhu, S. Study on the Rut Control Threshold of Asphalt Pavement Considering Steering Stability of Autonomous Vehicles Based on Fuzzy Control Theory. *Adv. Civ. Eng.* **2021**, 1–13. [CrossRef]
6. Vansauskas, V.; Bogdevičius, M. Investigation into the Stability of Driving An Automobile on the Road Pavement with Ruts. *Transport* **2009**, *24*, 170–179. [CrossRef]
7. Guo, X.; Zhou, B.; Zhang, C. Analysis of Rutting Index for Pavement Maintenance Based on Driving Safety on Surface Gathered Water Consideration. In *CICTP 2014: Safe, Smart, and Sustainable Multimodal Transportation Systems*; American Society of Civil Engineers: Reston, VI, USA, 2014; pp. 909–918. [CrossRef]
8. Jia, Y.; Wang, S.; Peng, J.; Gao, Y.; Hu, D.; Zhao, X. Evaluation of Pavement Rutting Based on Driving Safety of Vehicles. *Int. J. Pavement Res. Technol.* **2021**, *15*, 457–469. [CrossRef]
9. Kuznetsov, Y.V.; Moiseenko, D.A.; Plotnikov, P.V. Method for Determining the Maximum Allowable Pavement Rutting from the Condition of Ensuring the Safety of Traffic in Rainy Weather. In Proceedings of the 2021 Intelligent Technologies and Electronic Devices in Vehicle and Road Transport Complex (TIRVED), Moscow, Russia, 11–12 November 2021; pp. 1–6. [CrossRef]
10. Zhao, L.; Liu, H.; Wang, L.; Wei, H. Relationship Between Driver's Physiological Reaction and Driving Safety in Bad Weather. *China J. Highw. Transp.* **2016**, *29*, 147–152. [CrossRef]
11. Jiang, K.; Yin, H.; Feng, Z. Simulation Analysis of the Influence of Crosswind on Traffic Safety of Long-Span Bridges. *J. Hefei Univ. Technol. (Nat. Sci.)* **2019**, *42*, 145–150.
12. Leonaviciene, T.; Pukalskas, S.; Pumputis, V.; Kulesiene, E.; Zuraulis, V. Investigation of Factors that Have Affected the Outcomes of Road Traffic Accidents on Lithuanian Roads. *Balt. J. Road Bridge Eng.* **2020**, *15*, 1–20. [CrossRef]
13. Huang, X.; Cao, Q.; Liu, X.; Chen, J.; Zhou, X. Simulation of Vehicle Braking Performance on Rainy Days Based on Pavement Surface Fractal Friction Theory. *J. Jilin Univ. (Eng. Technol. Ed.)* **2019**, *49*, 757–765. [CrossRef]
14. Pomoni, M. Exploring Smart Tires as a Tool to Assist Safe Driving and Monitor Tire–Road Friction. *Vehicles* **2022**, *4*, 744–765. [CrossRef]
15. Zhang, Y.; Yuan, B.; Chou, Y. Analyzing Driving Safety Using Vehicle-Water-Filled Rutting Dynamics Model and Simulation. *Adv. Mater. Sci. Eng.* **2022**, *2022*, 1–16. [CrossRef]
16. Lee, H.; Taheri, S. Intelligent Tires? A Review of Tire Characterization Literature. *IEEE Intell. Transp. Syst. Mag.* **2017**, *9*, 114–135. [CrossRef]
17. Jeong, H.; Liu, Y. Horizontal Curve Driving Performance and Safety Affected by Road Geometry and Lead Vehicle. *Proc. Hum. Factors Ergon. Soc. Annu. Meet.* **2017**, *61*, 1629–1633. [CrossRef]
18. Žuraulis, V.; Surblys, V. Assessment of Risky Cornering on A Horizontal Road Curve by Improving Vehicle Suspension Performance. *Balt. J. Road Bridge Eng.* **2021**, *16*, 1–27. [CrossRef]
19. Cvitanić, D.; Maljković, B. Determination of Applicable Adjacent Horizontal Curve Radii Using Operating Speed. *Promet* **2019**, *31*, 443–452. [CrossRef]
20. Maljković, B.; Cvitanić, D. Improved Horizontal Curve Design Consistency Approach Using Steady-State Bicycle Model Combined with Realistic Speeds and Path Radii. *J. Transp. Eng. Part A Syst.* **2022**, *148*, 04022069. [CrossRef]
21. Hauer, E. Safety and the Choice of Degree of Curve. *Transp. Res. Rec.* **1999**, *1665*, 22–27. [CrossRef]

22. You, K.; Sun, L. Reliability Analysis of Vehicle Stability on Combined Horizontal and Vertical Alignments: Driving Safety Perspective. *J. Transp. Eng.* **2013**, *139*, 804–813. [CrossRef]
23. Zhang, C.; Liang, M.; Wang, S.; Chen, J.; Shao, D. Sideslip Risk Simulation Analysis of Passenger Car Braking Behavior on Expressway Curved Sections. *China J. Highw. Transp.* **2015**, *28*, 134–142. [CrossRef]
24. Jackson, T.L.; Sharif, H.O. Rainfall Impacts on Traffic Safety: Rain-related Fatal Crashes in Texas. *Geomat. Nat. Hazards Risk* **2016**, *7*, 843–860. [CrossRef]
25. Peng, T.; Guan, Z.; Zhang, R.; Dong, J.; Li, K.; Xu, H. Bifurcation of Lane Change and Control on Highway for Tractor-Semitrailer under Rainy Weather. *J. Adv. Transp.* **2017**, *2017*, 1–9. [CrossRef]
26. Wang, C.; Huang, H.; Xu, S. The Study on the Influence of Gust Wind on Vehicle Stability. In *Applied Mechanics and Materials*; Trans Tech Publications Ltd.: Bäch, Switzerland, 2014; Volume 598, pp. 198–201. [CrossRef]
27. Žuraulis, V.; Sivilevičius, H.; Šabanovič, E.; Ivanov, V.; Skrickij, V. Variability of Gravel Pavement Roughness: An Analysis of the Impact on Vehicle Dynamic Response and Driving Comfort. *Appl. Sci.* **2021**, *11*, 7582. [CrossRef]
28. Sekulic, D.; Vdovin, A.; Jacobson, B.; Simone, S.; Moe, J.S. Effects of Wind Loads and Floating Bridge Motion on Intercity Bus Lateral Stability. *J. Wind. Eng. Ind. Aerodyn.* **2021**, *212*, 104589. [CrossRef]
29. Wang, B.; Xu, Y. Safety Analysis of a Road Vehicle Passing by a Bridge Tower under Crosswinds. *J. Wind. Eng. Ind. Aerodyn.* **2015**, *137*, 25–36. [CrossRef]
30. Chen, F.; Peng, H.; Ma, X.; Liang, J.; Hao, W.; Pan, X. Examining the Safety of Trucks under Crosswind at Bridge-Tunnel Section: A Driving Simulator Study. *Tunn. Undergr. Space Technol. Inc. Trenchless Technol. Res.* **2019**, *92*, 103034. [CrossRef]
31. Xu, D.; Wang, G.; Qu, L.; Ge, C. Robust Control with Uncertain Disturbances for Vehicle Drift Motions. *Appl. Sci.* **2021**, *11*, 4917. [CrossRef]
32. Mansor, S.; Passmore, M.A. Effect of Rear Slant Angle on Vehicle Crosswind Stability Simulation on A Simplified Car Model. *Int. J. Automot. Technol.* **2013**, *14*, 701–706. [CrossRef]
33. Hassan, M.A.; Abdelkareem, M.A.A.; Moheyeldein, M.M.; Elagouz, A.; Tan, G. Advanced study of tire characteristics and their influence on vehicle lateral stability and untripped rollover threshold. *Alex. Eng. J.* **2020**, *59*, 1613–1628. [CrossRef]
34. Xiao, F.; Hu, J.; Jia, M.; Zhu, P.; Deng, C. A novel integrated control framework of AFS, ASS, and DYC based on ideal roll angle to improve vehicle stability. *Adv. Eng. Inform.* **2022**, *54*, 101764. [CrossRef]
35. Chen, F.; Peng, H.; Shao, X.; Pan, X. Comparative analysis of driving stability between truck and car under crosswind. *J. Harbin Inst. Technol.* **2021**, *53*, 10–16. [CrossRef]
36. Wang, Y.; Hu, X. Simulation of roll stability of a van in cross-wind. *Appl. Mech. Mater.* **2013**, *397–400*, 401–404. [CrossRef]
37. Wang, L.; Chen, X.; Chen, H. Influencing factors on vehicles lateral stability on tunnel section in mountainous expressway under strong wind: A case of Xi-Han highway. *Adv. Civ. Eng.* **2020**, *2020*, 1–11. [CrossRef]
38. Yin, Y.; Wen, H.; Sun, L.; Hou, W. Study on the influence of road geometry on vehicle lateral instability. *J. Adv. Transp.* **2020**, *2020*, 1–15. [CrossRef]
39. Alrejjal, A.; Ksaibati, K. Impact of Combined Alignments and Different Weather Conditions on Vehicle Rollovers. *KSCE J. Civ. Eng.* **2022**, *26*, 893–906. [CrossRef]
40. Tian, L.; Li, Y.; Li, J.; Lv, W. A simulation based large bus side slip and rollover threshold study in slope-curve section under adverse weathers. *PLoS ONE* **2021**, *16*, e0256354. [CrossRef] [PubMed]
41. Zheng, B.; Huang, X.; Tang, J.; Chen, J.; Zhao, R.; Hong, Z.; Tang, T.; Han, M. Evaluation on Braking Stability of Autonomous Vehicles Running along Curved Sections Based on Asphalt Pavement Adhesion Properties. *J. Adv. Transp.* **2022**, *2022*, 1–20. [CrossRef]
42. Siringoringo, D.M.; Fujino, Y.; Yabe, M. Investigation on vehicle lateral instability when crossing a curved highway bridge during an earthquake. *Struct. Infrastruct. Eng.* **2021**, *17*, 720–740. [CrossRef]
43. Zhang, C.; Xia, Q.; He, L. A Study on the Influence of Sideslip Angle at Mass Center on Vehicle Stability. *Automot. Eng.* **2011**, *4*, 5–10. [CrossRef]
44. Wang, Z.; Dong, M.; Gu, L.; Rath, J.; Qin, Y.; Bai, B. Influence of Road Excitation and Steering Wheel Input on Vehicle System Dynamic Responses. *Appl. Sci.* **2017**, *7*, 570. [CrossRef]
45. You, K.; Sun, L.; Gu, W. Risk Analysis-based Identification of Road Hazard Locations Using Vehicle Dynamic Simulation. *J. Southeast Univ. Nat. Sci. Ed.* **2012**, *42*, 150–155. [CrossRef]
46. Wu, X.; Fu, S.; Guo, L. Study on Highway Alignment Optimization Considering Rollover Stability Based on Two-Dimensional Point Collision Dynamics. *Appl. Sci.* **2022**, *13*, 509. [CrossRef]
47. Zhang, X.; Yang, Y.; Guo, K.; Lv, J.; Peng, T. Contour line of load transfer ratio for vehicle rollover prediction. *Veh. Syst. Dyn.* **2017**, *55*, 1748–1763. [CrossRef]
48. Ministry of Transport of the People's Republic of China. *Design Specification for Highway Alignment (JTG D20-2017)*; China Communications Press: Beijing, China, 2018. (In Chinese)
49. Kwonhee, S.; Hiseak, Y. Numerical Investigations of the Crosswind Stability of the Korean Light Tactical Vehicle During Airlift. *J. Mech. Sci. Technol.* **2017**, *31*, 1067–1072. [CrossRef]
50. Wang, F.; Yu, Y.; Zhao, X.; Xu, J.; Xie, T.; Deresa, S. Performance Evaluation of Reinforced Recycled Aggregate Concrete Columns under Cyclic Loadings. *Appl. Sci.* **2019**, *9*, 1460. [CrossRef]

51. Jiang, R.; Sun, T.; Liu, D.; Pan, Z.; Wang, D. Multi-Objective Reliability-Based Optimization of Control Arm Using MCS and NSGA-II Coupled with Entropy Weighted GRA. *Appl. Sci.* **2021**, *11*, 5825. [CrossRef]
52. Wang, L.; Li, R.; Wang, C.; Liu, Z. Driver injury severity analysis of crashes in a western China's rural mountainous county: Taking crash compatibility difference into consideration. *J. Traffic Transp. Eng.* **2021**, *8*, 703–714. [CrossRef]
53. Li, P.; Sun, C.; Huang, M.; Jiang, S.; Khan, M.D. Water Accumulation and Anti-Sliding Decay Characteristics of Freeway Pavement at Superelevation Transitions. *Road Mater. Pavement Des.* **2022**, 1–16. [CrossRef]
54. Fwa, T.F.; Pasindu, H.R.; Ong, G.P. Critical Rut Depth for Pavement Maintenance Based on Vehicle Skidding and Hydroplaning Consideration. *J. Transp. Eng.* **2012**, *138*, 423–429. [CrossRef]
55. Rana, M.M.; Hossain, K. Impact of Autonomous Truck Implementation: Rutting and Highway Safety Perspectives. *Road Mater. Pavement Des.* **2022**, *23*, 2205–2226. [CrossRef]
56. Žuraulis, V.; Surblys, V.; Šabanovič, E. Technological Measures of Forefront Road Identification for Vehicle Comfort and Safety Improvement. *Transport* **2019**, *34*, 363–372. [CrossRef]
57. Hui, B.; Tsai, Y.J.; Guo, M.; Liu, X. Critical assessment of the impact of vehicle wandering on rut depth measurement accuracy using 13-point based lasers. *Measurement* **2018**, *123*, 246–253. [CrossRef]
58. Besselink, I.J.M.; Achrifi, S.; Nijmeijer, H. Lateral vehicle dynamics on rutted roads. In *The IAVSD International Symposium on Dynamics of Vehicles on Roads and Tracks*; Springer: Cham, Switzerland, 2019; pp. 1242–1251. [CrossRef]
59. Yan, J.; Zhang, H.; Hui, B. Driving Safety Analysis Using Grid-Based Water-Filled Rut Depth Distribution. *Adv. Mater. Sci. Eng.* **2021**, *2021*, 1–13. [CrossRef]
60. Zhou, B.; Wang, Y.; Yu, G.; Wu, X. A lane-change trajectory model from drivers' vision view. *Transp. Res. Part C Emerg. Technol.* **2017**, *85*, 609–627. [CrossRef]
61. Yu, H.; Wang, B.; Li, Y.; Zhang, M. Driving Risk of Road Vehicle Shielded by Bridge Tower under Strong Crosswind. *Nat. Hazards* **2019**, *96*, 497–519. [CrossRef]
62. Chen, S.; Chen, F. Simulation-Based Assessment of Vehicle Safety Behavior under Hazardous Driving Conditions. *J. Transp. Eng.* **2010**, *136*, 304–315. [CrossRef]
63. Chen, D.; Chen, L.; Qian, Z. Impact of Pavement Rutting on Vehicle Safety: A Closed-Loop Assessment Method. *Road Mater. Pavement Des.* **2022**, 1–16. [CrossRef]
64. Peng, J.; Chu, L.; Wang, T.; Fwa, T.F. Analysis of Vehicle Skidding Potential on Horizontal Curves. *Accid. Anal. Prev.* **2021**, *152*, 105960. [CrossRef]

applied sciences

Article

Research on Instability and "Jack-Knifing" of Civil Aircraft Towing Taxi-Out System

Jiahao Qin [1,2,3], Jinwang Liu [1,2,3], Qiwei Lin [1,2,3] and Wei Zhang [1,2,3,*]

1 College of Aeronautical Engineering, Civil Aviation University of China, Tianjin 300300, China
2 Aviation Special Ground Equipment Research Base, College of Aeronautical Engineering, Civil Aviation University of China, Tianjin 300300, China
3 Key Laboratory of Smart Airport Theory and System, College of Aeronautical Engineering, Civil Aviation University of China, Tianjin 300300, China
* Correspondence: weizhang@cauc.edu.cn

Abstract: A new towing taxi-out departure mode has become the preferred choice to replace the traditional departure mode of civil aircraft due to its obvious advantages of low energy consumption, low emissions, and high efficiency. However, the inertia may lead to potential lateral instability and even "Jack-knifing", as the mass of the system is concentrated in the rear. In this paper, the lateral instability mechanism and influencing factors of the civil aircraft towing taxi-out system were studied using computer virtual simulation technology and taking Boeing 737–400 and AM210 tractors as research objects. The results show that the "Jack-knifing" phenomenon was caused by the lateral instability of the system when the sliding speed of the system was more than 15 km/h, and the system was accompanied by the lateral slip condition. Furthermore, the safety zone, side-show zone and jack-knifing zone of the system were defined in terms of towing speed and steering angle. Additionally, a towing safety zone was created under different airport track conditions by analyzing the lateral stability of the system. The conclusion of this study may provide a theoretical basis and reference for the safe application of the new taxi-out departure mode.

Keywords: towing taxi-out system; multibody dynamics; simulation modeling; lateral stability

Citation: Qin, J.; Liu, J.; Lin, Q.; Zhang, W. Research on Instability and "Jack-Knifing" of Civil Aircraft Towing Taxi-Out System. *Appl. Sci.* **2023**, *13*, 3636. https://doi.org/10.3390/app13063636

Academic Editors: Edgar Sokolovskij and Vidas Žuraulis

Received: 28 February 2023
Revised: 9 March 2023
Accepted: 10 March 2023
Published: 13 March 2023

1. Introduction

During the embarkation of a flight from the airport, subsequent to a civil aircraft being towed out of the berth by the tractor (indicated by the red dashed line in Figure 1), it ignites its engines and proceeds to the takeoff runway under its own power, awaiting takeoff (represented by the red solid line in Figure 1). The civil aircraft engine remains idle and operates ineffectively in this departure mode, leading to a significant amount of fuel consumption [1], waste discharge [2], and engine maintenance cost [3]. Additionally, the low efficiency of flight departures due to the large number of participants [4] no longer satisfies the demands for high-quality and sustainable development in the global civil aviation industry.

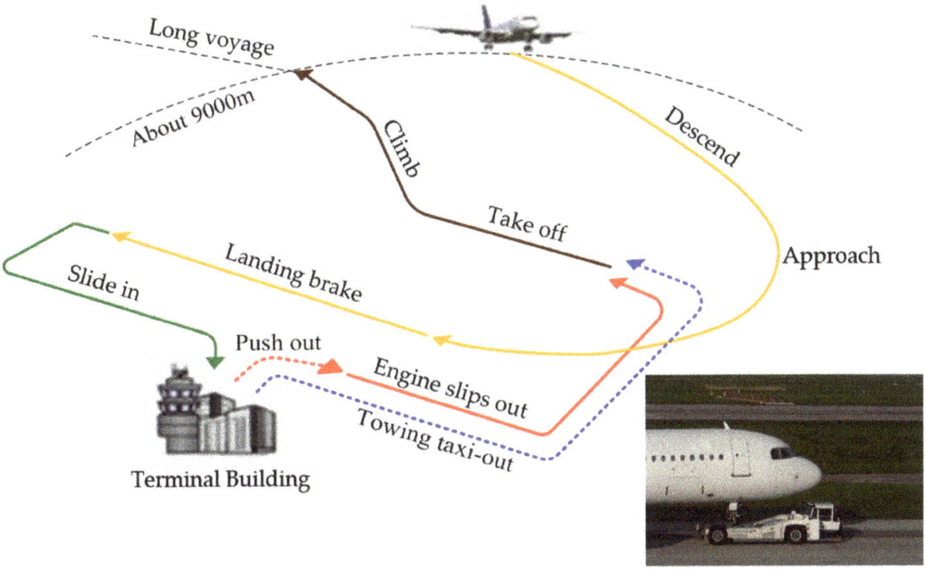

Figure 1. Schematic diagram of civil aircraft departure and taxiing process [5,6].

In this respect, the novel towing departure mode (indicated by the blue dotted line in Figure 1), whereby the tractor clamps and holds the nose landing gear of the civil aircraft through the holding wheel mechanism as an integrated civil aircraft towing system (represented by the narrow figure at the bottom right of Figure 1), has become a focal point of research both domestically and internationally. However, the towing system of civil aircraft under the new departure mode possesses typical characteristics of high speed and heavy load. High speed implies that the system speed increases from a low speed (about 2–5 km/h) of the traditional mode to a high speed (about 30–44 km/h) [7]. Heavy load denotes that the mass ratio between the civil aircraft and the tractor is significant, with the weight concentrated in the rear of the system. The aforementioned characteristics give rise to the lateral instability phenomenon, including shimmy, sideshow, and even "jack-knifing" (where the longitudinal angle α of the front longitudinal of the tractor and the front longitudinal angle of the civil aircraft surpass 90°) when the system traverses turns, emergency obstacle avoidance conditions, or extreme airport roads (slippery and icy) caused by inclement weather.

The jack-knifing phenomenon is a result of the relative rotation between the hinge point and semi-trailer when a multi-body system with a similar structure to a semi-trailer undergoes corner braking, emergency obstacle avoidance, or high-speed steering due to the synergistic effect of single or multiple factors. This rotation forms an angle similar to a folding knife, causing serious lateral instability of the tractor and making it difficult to control [8], as shown in Figure 2a. The occurrence of this phenomenon is often accompanied by adverse effects, as shown in Figure 2b, which depicts a jack-knifing accident that resulted in severe fire damage to the system's oil and electrical circuits [9], causing significant human and economic losses.

(**a**) (**b**)

Figure 2. Accident diagram of lateral instability of civil aircraft towing system. (Note: (**a**) An accident scene of the "Jack-knifing"; (**b**) A scene of a fire accident in airport).

Thus, the examination of lateral stability within the civil aircraft towing system, avoidance of jack-knifing during extreme lateral instability events, and enhancement of driving safety are vital prerequisites to ensure successful application of the new departure mode. Nonetheless, given that this model remains in its exploratory phase, there exist scarce reports regarding relevant studies. Refs. [10–13] comprise over 15 years of tracking research on aircraft tractors, encompassing the kinematic and dynamic characteristics of both rod-tractors and rod-less tractors, optimization of body structure and ride comfort, design and development of multifunctional roller mechanisms, and more. Ref. [14] analyzed the ride comfort of rod-less tractors on B-class road surfaces, concluding that low-speed towing work can aid in enhancing ride comfort. Taking the Boeing 777 as an example, [15] scrutinized the braking performance of rod-less tractors and concluded that the use of tractor braking can effectively reduce braking risks. Ref. [16] established a six-degrees-of-freedom aircraft dynamics model for the fuselage, obtaining the ultimate sideslip speed during turns through Adams/Simulink co-simulation. Ref. [17] simplified the aircraft towing system into a three-degrees-of-freedom dynamics model, concentrating on the impact of the hinged angle between the wheel embrace mechanism and the nose landing gear on system handling stability, and proposed improving the system stability through active steering control of the front wheel. Ref. [18] constructed a model of the aircraft towing system and deduced that the braking effect is optimal when the rear wheel braking force of the tractor accounts for 0.6 of the total braking force. Refs. [19,20] established a nonlinear dynamic model of the towing system of ships on the plane, with the model taking into full account the irregular movement of the system, the effect of the deck on the movement of the towing system of ship motion, and the coupling relationship accurately described, based on which the straight-line driving stability and turning driving stability of the system were examined. Refs. [7,21] utilized the nonlinear calculation model of A320 developed by Airbus to study the steering nonlinear dynamics of the aircraft under various airport pavement conditions and used the generalized continuation method to solve and depict the instability process of the system, providing the system bifurcation diagram and the stability/instability zone.

Furthermore, it is noteworthy that the key connection component, known as the "fifth wheel," in the trailer–semitrailer system, bears resemblance to the pickaback longitudinal three-point clamping connection mechanism present in the civil aircraft towing system.

Both systems also exhibit structural features that concentrate mass towards the rear. Hence, research into the trailer–semitrailer system is of relevance to this project. Ref. [22] explored the stability parameters of the heavy-duty towing system and determined that vehicles with shorter wheelbases were more susceptible to lateral instability than those with longer wheelbases. Ref. [23] constructed a three-axle semitrailer model and verified that independent control of each wheel could enhance the driving stability of the semitrailer. Ref. [24] established a model of the trailer–semitrailer system and achieved real-time monitoring of jack-knifing stability by observing the angle between the trailer and semitrailer, as well as its rate of change. Ref. [25] developed a mathematical model of the trailer–semitrailer system through Lagrange and Newton methods and compared it with the simulation modelto verify its accuracy. Addressing the driving safety issues, such as roll, yaw, and jack-knifing, that are prone to occur when the trailer–semitrailer system is operating at high speeds, [26] conducted a meticulous study on the rollover–yaw model of the system, determined the dynamic instability threshold value, and tested the early warning controller on the self-designed trailer–semitrailer rollover–yaw warning vehicle test platform. The experimental outcomes indicate that the controller can efficiently detect the onset of instability accidents. Ref. [27] researched the driving conditions of the trailer–semitrailer system, such as lane changes, snake-shaped trajectories, and steady-state rotations, that are inevitable during the driving process. The primary factors influencing the lateral stability of the system were assessed and evaluated from a multi-body dynamics perspective. Based on the outcomes of the single lane change test, recommendations and cautionary notes for lateral stability evaluations were proposed. Table 1 shows the faults in the previous research. The analysis of the aforementioned literature illustrates that, despite the fact that research has been conducted on aircraft towing systems and semi-trailers, it is not applicable to novel towing and taxiing circumstances characterized by high speed and heavy loads. As such, targeted research is imperative to effectively promote its implementation.

Table 1. The faults in the previous research.

Reference No.	Authors	Work and Contribution	Deficiencies of Work
[8–11]	Wang, L.W; Xie, B.M.	Load simulation of landing gear, analysis of clamping lifting mechanism of aircraft tractor.	Have not studied the overall dynamics of the traction system.
[12]	Zhang, J.H.	Low-speed traction operation helps improve ride comfort.	Analyzed only from the perspective of comfort.
[13]	Gordon, F.	Using tractor braking to reduce accident risk.	Does not take the high-speed traction condition into account.
[14]	Huang, M	The ultimate sideslip speed of the aircraft during turning is obtained.	Does not take the side slip of the tractor at high speed into account.
[15]	Wang, N	The wheel holding mechanism and the front landing gear have an impact on the stability of the traction system is concluded.	Does not consider the role of multi-body dynamics composed of tractor and aircraft.
[16]	Shi, H.T.	The braking effect is best when the rear wheel braking force of the tractor accounts for 0.6 of the total braking force.	Does not take the impact of the braking of the tractor on the traction system at high speed into account.
[17,18]	Zhou, L.J; Xiao, H	Define the influence of deck on the motion of aircraft towing system.	The research is not applicable to the traction conditions of large aircraft.
[19]	Coetzee, E	The stable and unstable regions of the aircraft under various airport pavement conditions are obtained.	The research object is the aircraft individual, and the multi-body dynamics effect of the traction system is not considered.

In summary, in the towing taxi-out departure system, which consists of the tractor and the aircraft, the mass is concentrated at the rear of the system because the takeoff weight of the aircraft is much larger than that of the rodless tractor, and the inertia may lead to potential lateral instability and even "Jack-knifing" which is a potential major accident. Thus, there is great value for the safe application of towing taxi-out mode with an in-depth analysis of the mechanization of lateral instability and the definition of the safe working area. This paper investigates the civil aircraft towing system, comprising a Boeing 737–400 aircraft (The Boeing Company, Arlington, USA)and AM210 tractor (Weihai Guangtai Airport Equipment Co., Ltd, Weihai, China), as the research object. The multi-body dynamics model and mathematical model of the towing taxiing system are established through Adams dynamics simulation software, which is verified for its effectiveness and reliability. Based on these models, the preliminary investigation into the impact of towing speed, system steering angle, and airport ground adhesion coefficient on the lateral stability of the system is conducted, providing a theoretical foundation and reference for subsequent active stabilization control and intelligent taxiing of the system.

2. Model Construction and Experimentation

2.1. Adams Dynamics Model of Towing Taxiing System

For this research, we have chosen to focus on the Boeing 737–400 and AM210 rod-less tractor [28] as the primary subject matter. The essential parameters for these models are presented in Tables 2 and 3.

Table 2. Model parameters of Boeing 737–400 aircraft.

Parameter	Value
Aircraft length/mm	36,400
Wingspan/mm	28,880
Aircraft height/mm	10,567
Horizontal distance between nose and main landing gear/mm	11,070
Horizontal distance between main landing gears/mm	5230
Spacing between two wheels of nose landing gear/mm	674
Spacing between two wheels of main landing gear/mm	338.16
Weight of equipment/kg	60,555
Moment of inertia of mass center around X-axis/(kg·m^2)	8.12×10^{11}
Moment of inertia of mass center around Y-axis/(kg·m^2)	2.854×10^{12}
Moment of inertia of mass center around Z-axis/(kg·m^2)	2.704×10^{12}

Table 3. Model parameters of rod-less aircraft tractor.

Parameter	Value
Length/mm	8000
Width/mm	3750
Height/mm	1975
Wheelbase/mm	4500
Front wheel base/mm	2100
Rear wheel base/mm	2700
Weight of equipment/kg	12,900
Moment of inertia of mass center around X-axis/(kg·m^2)	6114
Moment of inertia of mass center around Y-axis/(kg·m^2)	52,545
Moment of inertia of mass center around Z-axis/(kg·m^2)	53,675

During the modeling process, the rod-less tractor body was simplified and replaced by a frame with steering to ensure authenticity in simulation. The location of the center of mass was adjusted accordingly. The connection between the civil aircraft and tire is a rigid connection, disregarding the landing gear strut, while the connection between the civil aircraft and the tractor is a rigid connection of two rotating pairs to ensure accurate lateral stability during simulation, replacing the holding wheel mechanism. The model road surface was smoothed using 2D software and different influencing factors were transformed by adjusting the parameters of the tire and road surface. The Adams model created is depicted in Figure 3.

Figure 3. Model drawing of aircraft and aircraft tractor.

Furthermore, Tables 4 and 5 present the essential parameters of the tractor and aircraft tire model, respectively [28].

Table 4. Tire parameters of rod-less aircraft tractor.

Parameter	Value
Front tire specification	285/70 R19.5
Rear tire specification	385/65 R22.5
Tire weight/kg	190
Coefficient of normal stiffness	2000
Normal damping coefficient	28
Rolling resistance coefficient	0.0165
Longitudinal stiffness (N/mm)	5000
Lateral stiffness of tire	80,000
Static friction factor of tire	0.8
Dynamic friction factor of tire	1

Table 5. Tire parameters of Boeing 737–400.

Parameter	Value
Front tire weight/kg	50
Rear tire weight/kg	150
Front tire radius of freedom/mm	340
Rear tire radius of freedom/mm	551
Front tread width mm	196.8
Rear tread width/mm	411.44
Front wheel normal stiffness coefficient	2400
Rear wheel normal stiffness coefficient	2400
Normal damping coefficient	28
Rolling resistance coefficient	0.0165
Longitudinal stiffness of front wheels (N/mm)	2000
Longitudinal stiffness of rear wheels (N/mm)	6000
Lateral stiffness of tire	80,000
Static friction factor of tire	0.8
Dynamic friction factor of tire	1

With the incorporation of the aforementioned tire parameters into Adams, the multibody dynamic model of the civil aircraft and rod-less tractor towing and taxiing system can be obtained.

2.2. Mathematical Model Construction and Verification

To ensure the reliability and validity of the dynamic model of the aircraft towing and taxiing system established in this study, a four-degrees-of-freedom dynamic model was constructed. This included both the lateral and yaw motions of the tractor, as well as the lateral and yaw motions of the aircraft. A schematic diagram of the two-centroid mathematical model is shown in Figure 4, and the parameters of the model is shown in table of nomenclature.

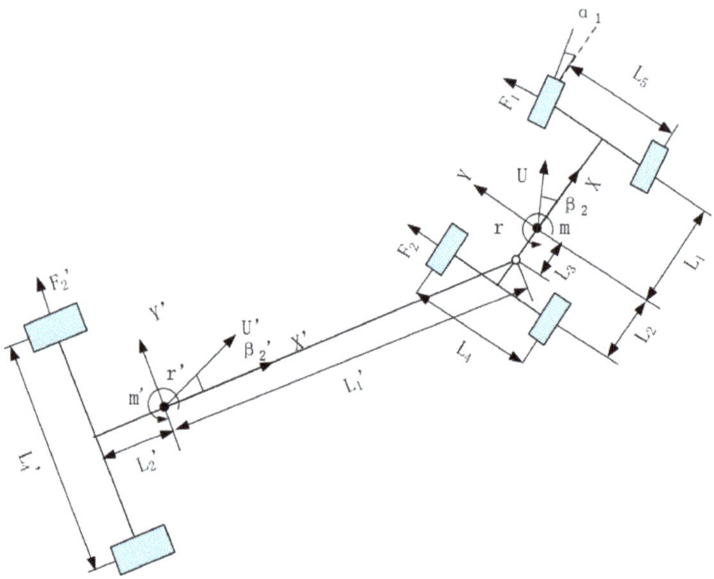

Figure 4. Double centroid model of aircraft and tractor.

One of the dynamic models that can be constructed from the system is that of the lateral and yaw motion of the tractor, which can be expressed as follows:

$$\begin{cases} m(\dot{U}\sin\beta_2 + rU\cos\beta_2) = c_f[\alpha_1 - \frac{(U\sin\beta_2 + L_1 r)}{U\cos\beta_2}] + c_r[\alpha_2 - \frac{(U\sin\beta_2 - L_2 r)}{U\cos\beta_2}] - F_Y \\ I_Z\dot{r} = L_1 c_f[\alpha_1 - \frac{(U\sin\beta_2 + L_1 r)}{U\cos\beta_2}] - L_2 c_r[\alpha_2 - \frac{(U\sin\beta_2 - L_2 r)}{U\cos\beta_2}] + L_3 F_Y \end{cases} \tag{1}$$

The model describing the aircraft's two degrees of freedom can be expressed as follows:

$$\begin{cases} m'(\dot{U}'\sin\beta_2' + U'\cos\beta_2' r') = c_r'[\alpha_2' - \frac{(U'\sin\beta_2' - L_2' r')}{U'\cos\beta_2'}] + F_Y' \\ I_Z'\dot{r}' = L_1' F_Y' - L_2' c_r'[\alpha_2' - \frac{(U'\sin\beta_2' - L_2' r')}{U'\cos\beta_2'}] \end{cases} \tag{2}$$

During the process of constructing the model, the following plane coordinate transformation is utilized to convert the model into a unified coordinate system [29,30]:

$$\begin{bmatrix} X' \\ Y' \end{bmatrix} = \begin{bmatrix} \cos\theta & -\sin\theta \\ \sin\theta & \cos\theta \end{bmatrix} \begin{bmatrix} X \\ Y \end{bmatrix} \tag{3}$$

$$\begin{bmatrix} U' \sin \beta_2{}' \\ U' \cos \beta_2{}' + L_1{}'r' \end{bmatrix} = \begin{bmatrix} \cos\theta & -\sin\theta \\ \sin\theta & \cos\theta \end{bmatrix} \begin{bmatrix} U \sin \beta_2 \\ U \cos \beta_2 - L_3 r \end{bmatrix} \tag{4}$$

Based on the above model, the Matlab-Simulink mathematical model was constructed. Taking the towing speed of 30 km/h and the ground adhesion coefficient of 0.8 as the experimental conditions, the change trends of the angle between the system tractor and the aircraft in the Adams dynamic model and the mathematical model when the system steering angle was 4° and 15° were compared, respectively. The comparison test results are shown in Figure 5a,b.

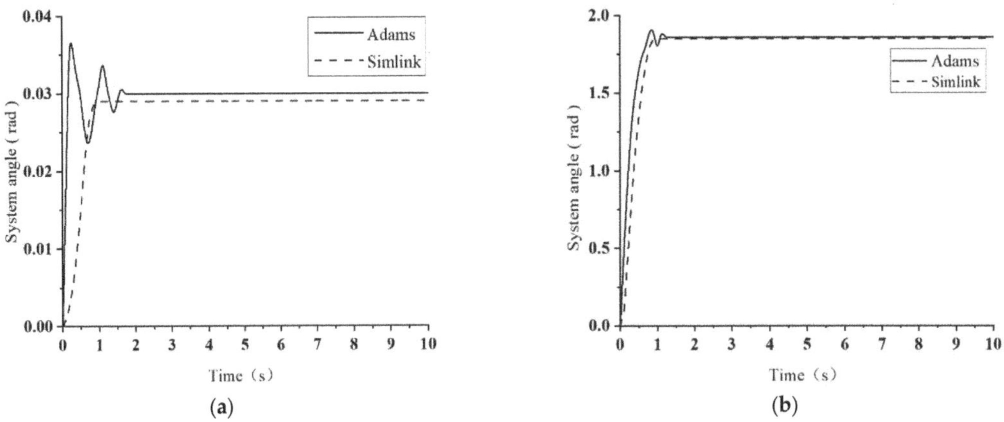

Figure 5. Comparison test between Adams model and Simulink model of 30 km/h towing system. (Note: (**a**) the comparison of system stability; (**b**) the comparison of "Jack-knifing").

Figure 5 demonstrate that the dynamic model and mathematical model accurately depict the changes in the angle between the aircraft and the tractor under two steering conditions. Specifically, as time progresses, the relative motion between the two bodies remains relatively stable, resulting in a stable angle. Hence, the verification process of the dynamic model and mathematical model supports the conclusion that the multi-body dynamic system of civil aircraft towing and taxiing is accurately described, indicating that the model is both reliable and valid.

3. Results and Discussion

Building upon the aforementioned model, this paper investigates the influence of various factors on the lateral stability of the system under idle running conditions, such as system speed, tractor steering angle, and friction factor, as well as the instability of jack-knifing. Furthermore, the study examines the impact of such differences on the safety domain of the towing system and defines the system's warning data boundary. These findings provide a theoretical basis and support threshold for the lateral stability control of civil aircraft towing systems operating under different working conditions.

3.1. System Jack-Knifing Mechanism

To investigate the impact of the system's velocity on the lateral stability of the civil aircraft towing and sliding system in an idle running state, the road adhesion coefficient and rolling resistance coefficient were established as 0.8 and 0.018, respectively. Then, the initial velocity of the system was gradually reduced from 45 km/h to 15 km/h. At every velocity, the steering angle of the tractor's front wheel was elevated in increments of 0.1° up to the point of the jack-knifing phenomenon. The findings of the experiment are presented in Figure 6.

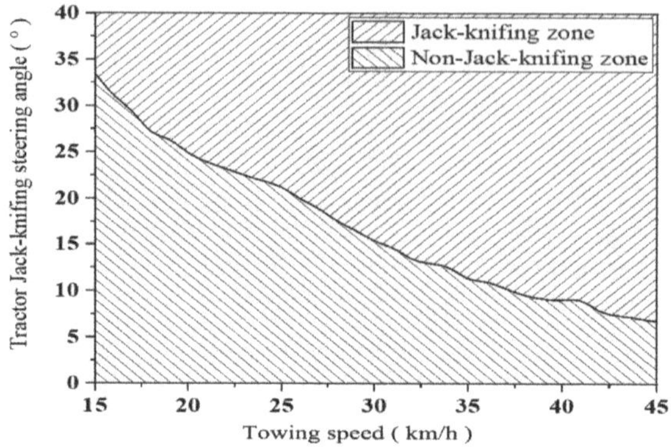

Figure 6. Jack-knifing zone diagram of 15–45 km/h towing speed and tractor steering angle.

The present study employs the safety calibration curve to determine the curve wherein the angle between the civil aircraft and the tractor is less than 90°. Figure 6 depicts the jack-knifing zone diagram of the towing speed and tractor steering angle, with the initial velocity of the towing system ranging from 15–45 km/h. The diagram is divided into two segments; namely, the non-jack-knifing zone on the lower side of the curve and the jack-knifing zone on the upper side. Based on the image data obtained from Figure 6, it can be inferred that the jack-knifing phenomenon is more likely to occur with an increase in the initial speed of the system, thereby necessitating a smaller steering angle for the tractor.

In low-speed scenarios, excessive steering angle may result in the jack-knifing phenomenon. Despite the fact that the longitudinal angle between the civil aircraft and the tractor exceeds 90°, the system does not manifest instability. Thus, this cannot be the criterion to assess the loss of lateral stability of the system. Consequently, this paper proposes a methodology to evaluate whether the system has lateral instability by combining two factors: whether the tractor produces lateral slip and the angle of the system.

Figure 7 depicts the temporal variation of the system angle as the tractor's steering angle increases from 9.0° to 9.1° during the taxiing of the towing system at an initial speed of 40 km/h. The sudden rise in the angle signifies the occurrence of the jack-knifing phenomenon due to the lateral instability of the system. Consequently, it can be inferred that the lateral instability of the system arises during the jack-knifing phenomenon at 40 km/h, and this continues until 25 km/h. As demonstrated in Figure 8, while the entire system is sliding at an initial speed of 25 km/h, there is no significant change in the system's included angle as the tractor's steering angle increases from 21.2° to 21.3°, making it impossible to determine the lateral instability of the system from this representation. Therefore, the most straightforward approach to detect the lateral instability of the system is to detect whether the lateral force of the tire surpasses the maximum static friction force of the lateral slip.

The crucial threshold for lateral slippage is determined by multiplying the maximum adhesion coefficient of the tractor tire by the pressure on the road surface. If the resulting value is less than zero, it is deemed that the tire is slipping laterally and the towing system is unstable.

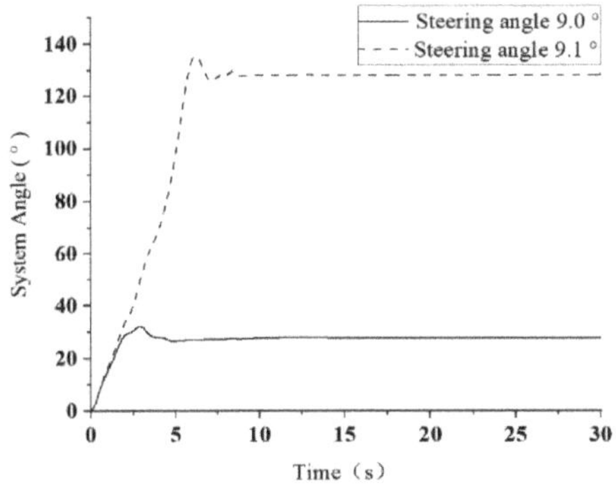

Figure 7. Variation diagram of system angle under 40 km/h.

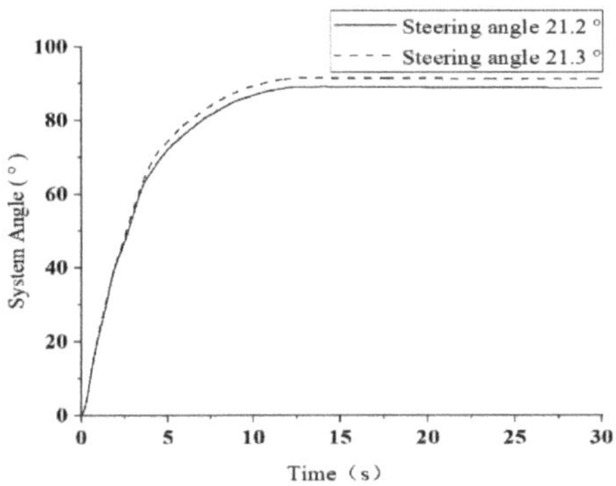

Figure 8. Variation diagram of system angle under 25 km/h.

As depicted in Figure 9, the lateral force diagram of the front and rear wheels of the tractor is illustrated when the towing system slides in at an initial speed of 25 km/h with a tractor steering angle of 21.2°. Due to the significant influence of the quality of the civil aircraft on the rear wheel load of the tractor, the front wheel is more susceptible to lateral slip. The portion of the front wheel force curve that is less than zero in Figure 6 indicates the presence of lateral instability in the towing system. Thus, in the 25 km/h state, the occurrence of the jack-knifing phenomenon is accompanied by lateral instability in the system.

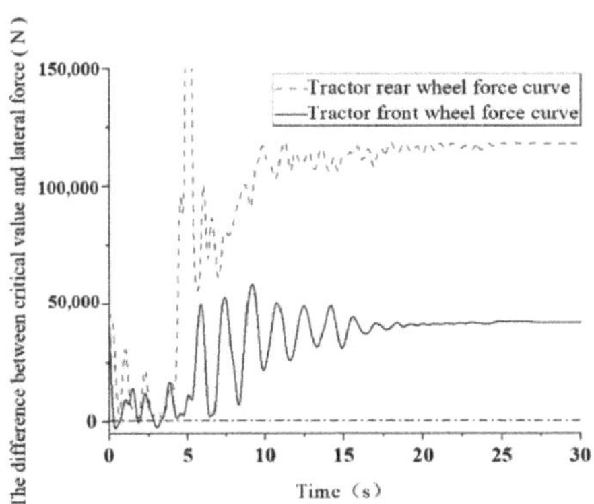

Figure 9. Stress diagram of front and rear wheels of tractor at steering angle of 21.2° at a speed of 25 km/h.

3.2. Lateral Slip Instability

In practical towing operations, jack-knifing of the towing system is a rare occurrence. However, the system often struggles to maintain stable tire operation due to tire side deviation, leading to more frequent tire lateral slipping. Additionally, it should be noted that the boundary of the jack-knifing zone discussed earlier is based on the lateral instability resulting from tire lateral slip. Consequently, there is a need to investigate the non-jack-knifing zone further.

The lateral force diagram presented in Figure 10 depicts the behavior of the front and rear wheels of the tractor at a speed of 40 km/h when the tractor steering angle is 9.0°. It is apparent that as long as the tractor steering angle does not exceed 9°, the tire does not experience lateral slip, thus ensuring the lateral stability of the towing system.

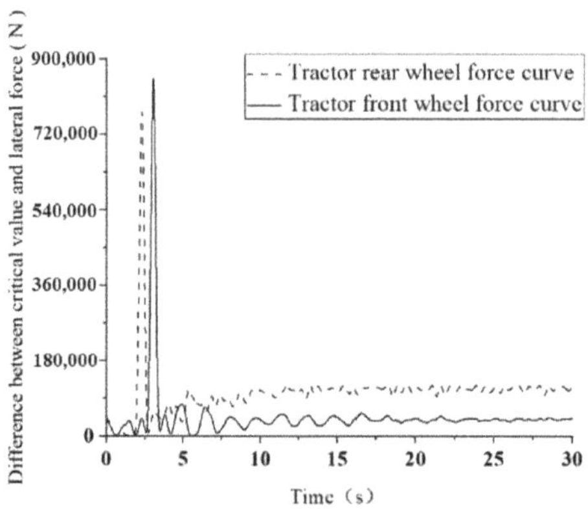

Figure 10. Stress diagram of front and rear wheels of tractor at a steering angle of 9.0° and speed of 40 km/h.

As depicted in Figure 11, the lateral force diagram of the front and rear wheels of the tractor is illustrated when the steering angle of the tractor reaches 9.1° under the velocity of 40 km/h. It was discerned that the emergence of lateral slip on the tire and consequent lateral instability of the towing system occurs only when the steering angle of the tractor equals or exceeds 9.1°.

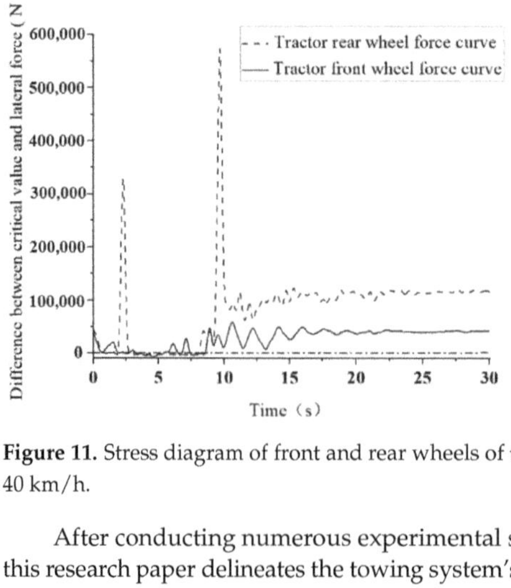

Figure 11. Stress diagram of front and rear wheels of tractor at a steering angle of 9.1° and speed of 40 km/h.

After conducting numerous experimental simulations and analyzing the outcomes, this research paper delineates the towing system's range of stability and instability between 15 and 45 km/h, based on the presence or absence of lateral slip in the tractor tire, as depicted in Figure 12. The findings suggest that as the towing speed increases, the steering angle of the tractor causing the instability of the towing system decreases, thereby rendering the towing system more susceptible to lateral instability.

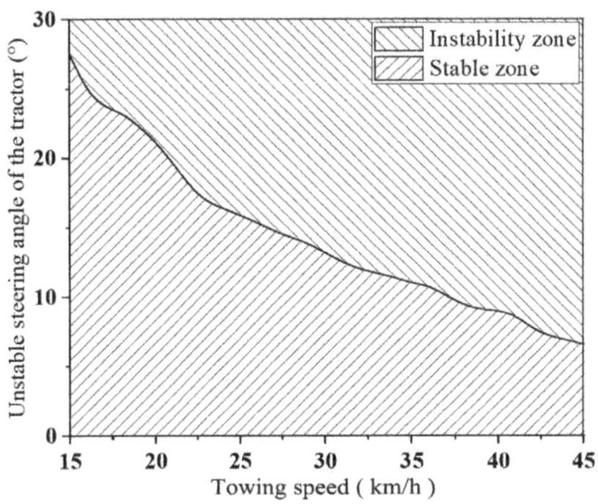

Figure 12. Stability zone diagram of 15–45 km/h towing speed and tractor steering angle.

After combining Figure 12 with Figure 6, we can obtain the safety zone diagram for the towing system within the speed range of 15–45 km/h, as illustrated in Figure 13.

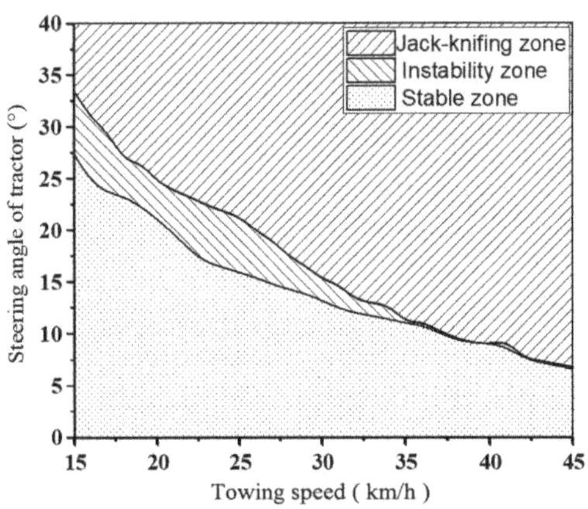

Figure 13. Safety zone diagram of 15–45 km/h towing speed and tractor steering angle.

3.3. Definition and Analysis of Allowable Steering Range in Different Pavement Environments

To further investigate the impact of different factors on the lateral stability of the towing system within the instability jack-knifing zone, the system's initial speed was fixed at 40 km/h, as this is when lateral instability is most severe. We examined the effect of changing the steering angle of the tractor and the road conditions of the towing system separately to evaluate the influence of various factors on the lateral stability of the towing system.

The pavements listed in Table 6 have been categorized into two classifications: ordinary pavement and extreme pavement. Using the same methodology as before, we depicted the safety zone diagram of the towing system moving at 40 km/h, with friction coefficients ranging from 0.45 to 0.8 and a tire rolling resistance coefficient of 0.018. The diagram, presented in Figure 14, showcases the effect of pavement wetness on the lateral stability of the towing system under regular pavement conditions that are predominantly comprised of concrete or asphalt.

Table 6. Adhesion coefficient and rolling resistance of tires on different roads.

Road Conditions	Adhesion Coefficient	Rolling Resistance
Concrete or asphalt (dry)	0.55–0.80	0.010–0.018
Concrete or asphalt (wet)	0.45–0.70	0.018–0.020
Flat ice road	0.10–0.25	0.015–0.030
Snow-compacted pavement	0.30–0.40	0.030–0.050

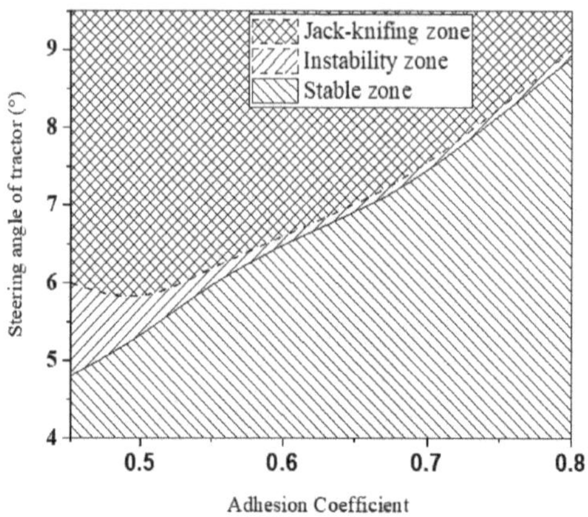

Figure 14. Safety zone map of 40 km/h asphalt or concrete pavement.

On asphalt or concrete pavement, the lateral stability of the towing system is divided into three zones; namely, the jack-knifing zone, instability zone, and stability zone. It can be inferred that a higher adhesion coefficient of the ground leads to better lateral stability of the towing system and reduces the likelihood of instability.

Due to the snow and ice road surface being an extreme road surface, the ground adhesion coefficient is small, making it less likely for the jack-knifing phenomenon to occur and more prone to tire lateral slip. Therefore, the analysis of the jack-knifing phenomenon was not conducted, and the definition of the tire lateral slip instability zone was introduced, as depicted in Figure 15.

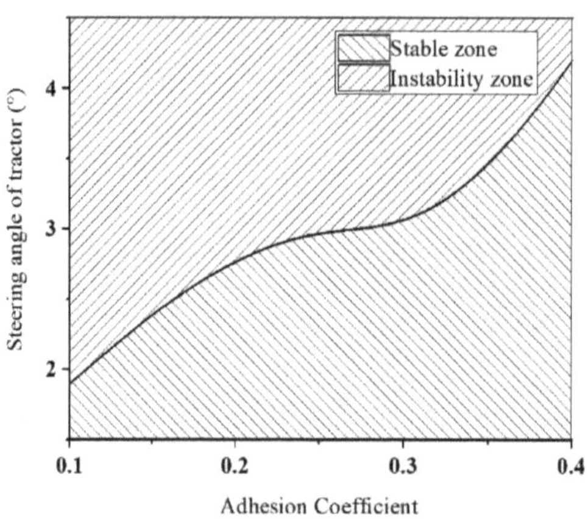

Figure 15. Stability zone map of 40 km/h ice or snow pavement.

The lateral stability of the towing system under snow and ice road conditions can be categorized into two zones: the stability zone and the instability zone. It can be inferred that when operating on snow- and ice-covered roads at high speeds, the tractor should steer clear of causing any lateral instability of the towing system. Furthermore, it can be deduced that the stability of the towing system will increase as the adhesion coefficient of the snow- and ice-covered road increases.

4. Conclusions

This paper utilizes the Adams simulation platform to establish a multi-body dynamics model that examines the interaction between a Boeing 737–400 civil aircraft and the Weihai Guangtai AM210 rod-less tractor. The study conducts a simulation analysis of the system's idle running condition. The experimental results reveal that as the system speed increases, the steering angle of the tractor required to produce the jack-knifing phenomenon is smaller. Moreover, the jack-knifing phenomenon in the towing speed range of 15–45 km/h is the lateral slip instability of the tire. The paper provides a safety zone diagram of the towing system within this speed range. The study further explores the lateral stability of the towing system under different road conditions, dividing road conditions under 40 km/h towing speed into ordinary and extreme roads, and determining the influence of the steering angle of the tractor and the adhesion coefficient of different roads on the lateral stability of the towing system. This paper presents the safety zone diagram of the lateral stability of the towing system on asphalt or concrete pavement and the stability zone diagram of the towing system on snow- and ice-covered pavement. An active control to recover the instability of system from a potential accident will be carried out as the aim of the research is ensuring the safety of towing taxi-out, and the results in this research will provide the control threshold in future research. The research findings can offer theoretical and practical support for the lateral stability control of civil aircraft towing and taxiing systems, and promote the secure landing application of the new towing and taxiing mode.

Author Contributions: Conceptualization, J.Q. and W.Z.; methodology, J.Q.; software, Jinwang Liu; validation, J.Q., Q.L. and W.Z.; formal analysis, Jinwang Liu; investigation, Q.L.; resources, W.Z.; data curation, Jinwang Liu; writing—original draft preparation, J.Q. and Jinwang Liu; writing—review and editing, J.Q. and W.Z.; visualization, Jinwang Liu and Q.L.; supervision, W.Z.; project administration, J.Q.; funding acquisition, J.Q. All authors have read and agreed to the published version of the manuscript.

Funding: This research was funded by National Natural Science Foundation of China, grant number 52102446 and Fundamental Research Funds for The Central Universities, grant number 3122021050.

Data Availability Statement: All research data are supported for sharing. Please do not hesitate to email jhqin@cauc.edu.cn if necessary.

Conflicts of Interest: The authors declare no conflict of interest.

Nomenclature

m	Mass of tractor (kg)
L_1	Distance from front axle to barycenter of tractor (mm)
L_2	Distance from rear axle to barycenter of tractor (mm)
L_3	Distance from towing point to barycenter of tractor (mm)
L_4	Rear wheel pitch of tractor (mm)
L_5	Front wheel pitch of tractor (mm)
r	Rotational angular velocity of tractor around barycenter (rad/s)
U	Center of barycenter velocity of tractor (m/s)
F_1	Side-deviation force of tractor front wheel (N)
α_1	Side-deviation angle of tractor front wheel (deg)
α_2	Side-deviation angle of tractor rear wheel (deg)
F_2	Side-deviation force of the rear wheel of the tractor (deg)

β_2	Lateral-deviation angle of tractor barycenter (deg)
C_f	Lateral-deviation stiffness of the front wheel of the tractor (N/rad)
C_r	Lateral-deviation stiffness of the rear wheel of the tractor (N/rad)
F_X	Force in X-direction of tractor towing point (N)
F_Y	Force in Y-direction of tractor towing point (N)
m'	Mass of aircraft (kg)
$L_1{}'$	Distance from front axle to barycenter of aircraft (mm)
$L_2{}'$	Distance from rear axle to barycenter of aircraft (mm)
$L_4{}'$	rear wheel Pitch of aircraft (mm)
r'	Angular velocity of aircraft rotating around the center of mass (rad/s)
U'	Velocity of aircraft barycenter (m/s)
$F_2{}'$	Side-deviation force of aircraft rear wheel (N)
$\beta_2{}'$	Lateral-deviation angle of aircraft barycenter (deg)
$C_r{}'$	Lateral-deviation stiffness of aircraft rear wheel (N/rad)
$\alpha_2{}'$	Lateral-deviation angle of aircraft rear wheel (deg)
$F_X{}'$	Force in X-direction of aircraft towing point (N)
$F_Y{}'$	Force in Y-direction of aircraft towing point (N)
θ	Angle between aircraft and tractor (deg)
φ	Angle between tractor and ground coordinate system (deg)
φ'	Angle between aircraft and ground coordinate system (deg)
I_Z	Moment of inertia of tractor around barycenter (kg·m^2)
$I_Z{}'$	Moment of inertia of the aircraft around barycenter (kg·m^2)

References

1. Ministry of Transport of the People's Republic of China. *Statistical Bulletin on the Development of Civil Aviation Industry in 2019*; Development Planning Division Beijing: Beijing, China, 2019.
2. Airbus MoU with IAI to Explore Eco-Efficient "Engines-Off" Taxiing[EB/OL]. Available online: https://www.airbus.com/newsroom/press-releases/en/2009/06/airbus-mou-with-iai-to-explore-eco-efficient-apos-engines-off-apos-taxiing.html (accessed on 11 January 2022).
3. Postorino, M.N.; Mantecchini, L.; Paganelli, F. Improving taxi-out operations at city airports to reduce CO_2 emissions. *Transp. Policy* **2019**, *80*, 167–176. [CrossRef]
4. Guo, R.; Zhang, Y.; Wang, Q. Comparison of emerging ground propulsion systems for electrified aircraft taxi operations. *Transp. Res. Part C* **2014**, *44*, 98–109. [CrossRef]
5. Zhao, G.; Xing, W. Research on Airport Operation Efficiency Based on Big Data Analysis Technology. *J. Civ. Aviat.* **2022**, *6*, 33–37+97.
6. Deng, S.W.; Bin, Y.P.; Ren, Y. Research on Lean Operation Management Methods for Large Airports. *J. Civ. Aviat.* **2022**, *6*, 18–21+46.
7. Coetzee, E.; Krauskopf, B.; Lowenberg, M. Analysis of Medium-Speed Runway Exit Maneuvers. *J. Aircr.* **2011**, *48*, 1553–1564. [CrossRef]
8. Gu, Y.M. *Development and Research of Train Jack-Knifing Instability Simulation System for Semitrailer*; Chang'an University: Xi'An, China, 2019.
9. The Most Expensive Spontaneous Combustion in History? The Aircraft Tractor Caught Fire and Burned the Nose of the Airliner. Available online: https://www.sohu.com/a/235270086_240515 (accessed on 11 January 2022).
10. Zhang, W.; Ning, Z.H.; Wang, W.; Wang, L.W. Study on the influence of airport pavement depression on the dynamic performance of aircraft and tractor wheel holding system. *Mech. Des. Manuf. Eng.* **2019**, *48*, 15–19.
11. Wang, L.W.; Liu, B.; Sun, Y.K.; Zhang, W. Load simulation of landing gear under uneven road condition of rod-less traction system. *J. Civ. Aviat. Univ. China* **2019**, *37*, 33–37+43.
12. Wang, L.W.; Wu, Z.H.; Zhang, W. Kinematic analysis of clamping lifting mechanism of bar aircraft tractor. *Mach. Tools Hydraul.* **2015**, *43*, 54–57+15.
13. Xie, B.M.; Zhu, J.W.; Wang, W.; Zhang, W. Ride comfort simulation and optimization of aircraft tractor under multi working conditions. *Mech. Des. Manuf.* **2020**, *10*, 15. [CrossRef]
14. Zhang, J.H.; Li, G.Y. Simulation analysis of ride comfort of aircraft tractor based on ADAMS. *Mach. Des.* **2005**, *3*, 68–70.
15. Hayhoe, G.F. Braking Performance of Towbarless Towing Vehicles During Maintenance Tows. In Proceedings of the First Congress of Transportation and Development Institute (TDI), Chicago, IL, USA, 13–16 March 2011.
16. Huang, M.; Nie, H.; Zhang, M. Analysis of ground handling characteristic of aircraft with electric taxi system. *Proc. Inst. Mech. Eng. Part D J. Automob. Engineering.* **2019**, *233*, 1546–1561. [CrossRef]
17. Wang, N.; Liu, H. Research on trajectory tracking behavior in traction task for tractor-aircraft system. In *IEEE International Conference on Mechatronics & Automation*; IEEE: Piscataway, NJ, USA, 2013; pp. 1251–1255.
18. Shi, H.T. *Braking Analysis of Aircraft Tractor*; Harbin Engineering University: Harbin, China, 2011.

19. Zhou, L.J. *Study on Stability of Tractor Aircraft System on Ship*; Harbin Engineering University: Harbin, China, 2012.
20. Xiao, H. *Research on Traction Operation Simulation Technology of Traction Aircraft System*; Huazhong University of Science and Technology: Wuhan, China, 2008.
21. Rankin, J.; Coetzee, E.; Krauskopf, B.; Lowenberg, M. Bifurcation and Stability Analysis of Aircraft Turning on the Ground. *J. Guid. Control. Dyn.* **2009**, *32*, 500–511. [CrossRef]
22. Winkle, C.B. *Rollover of Heavy Commercial Vehicles*; SAE Paper: Warrendale, PA, USA, 2001; ISBN 0-7680-0626-0.
23. Chieh, C. Lateral Control of Commercial Heavy Vehicles. *Veh. Syst. Dyn.* **2000**, *33*, 91–420.
24. Ashley, L.D. *Jackknife Stability of Articulated Tractor Semitrailer Vehicles With High-Output Brakes and Jackknife Detection on Low Coefficient Surfaces*; The Ohio State University: Columbus, OH, USA, 2003.
25. Hyun, D. *Predictive Modeling and Active Control of Rollover in Heavy Vehicles*; Texas A&M University: Texas, TX, USA, 2001.
26. Zhu, T.J. *Research on Rollover Warning and Multi-Objective Stability Control Algorithm of Heavy Vehicle Based on Improved TTR*; Jilin University: Changchun, China, 2010.
27. Zhuo, K.M.; Wei, D.B. Lateral stability test and simulation analysis of semitrailer train. *Spec. Purp. Veh.* **2020**, *6*, 80–85.
28. Zhu, H.; Li, J.F. Braking stability analysis of aircraft tractor under turning traction condition. *Mach. Des.* **2020**, *37*, 72–76. [CrossRef]
29. Liu, H.F. *Simulation and Control Strategy Research on Lateral Dynamics of Semi-Trailer Train*; Jilin University: Changchun, China, 2005.
30. Gu, Y.M. *Development of a Simulation System for Jack-Knifing Instability of Semi-Trailer Train*; Chang'an University: Xi'an, China, 2019.

applied sciences

MDPI

Article

Solutions to the Problem of Freight Transport Flows in Urban Logistics

Nijolė Batarlienė * and Darius Bazaras

Department of Logistics and Transport Management, Vilnius Gediminas Technical University (VILNIUS TECH), Plytinės 27, LT-10105 Vilnius, Lithuania
* Correspondence: nijole.batarliene@vilniustech.lt; Tel.: +370-5-2370-634

Abstract: The operation of a city's logistics system is associated with many problematic issues, for example, intense pollution and negative impacts on the environment, uneven intensity of traffic flows, and other problems. This article focuses on heavy transport because it causes disruptions in the supply system and affects timely customer service. Optimization processes are associated with route selection, deployment of logistics centers, and the need for cargo consolidation–de-consolidation, which leads to the possibility of using the PPP (public–private partnership) method in practice. A unique aspect of this research is related to the problems of the first and last mile and the use of elements of the "sharing economy". Therefore, this article aims to analyze the problematic challenges of the first and last mile and the role and significance of heavy transport in urban logistics. For that purpose, in addition to an analysis of scientific sources, an expert survey was conducted and responses related to the questions on optimizing heavy traffic flows in city logistics were analyzed. The study data were processed using Kendall's correlation coefficient, the concordance method, and the consistency coefficient. During analysis, using the average rank transformation to weights (ARTIW) method, the subjective normalized weights of the main factors that have the greatest influence on a city's logistics system were determined. Groups of interested parties are also named separately, for whom the results of the study and the formulated decision may be relevant. Based on the results of the research, a recommendation was formulated for the development of small cargo collection and distribution services through self-service terminals located on the outskirts of cities. Proposals are presented to reorganize the system, provide details of new infrastructure elements and suburban terminals, and encourage for the use of environmentally friendly vehicles with a lower carrying capacity.

Keywords: urban logistics; road freight transport; traffic flows; first and last mile

Citation: Batarlienė, N.; Bazaras, D. Solutions to the Problem of Freight Transport Flows in Urban Logistics. *Appl. Sci.* **2023**, *13*, 4214. https://doi.org/10.3390/app13074214

Academic Editor: Luís Picado Santos

Received: 14 February 2023
Revised: 22 March 2023
Accepted: 23 March 2023
Published: 26 March 2023

1. Introduction

As cities grow, new challenges arise for mobility. It is estimated that 10–15% of the total mileage of a city comes from freight transport. When freight flows increase in cities, many different problems arise, such as loading and delivery delays, increasing environmental pollution and noise. Other problems also arise, such as lack of heavy transport drivers, wear and tear of roads, etc.

As the population grows, urban logistics has become the subject of scientific research and has gained great importance. In recent decades, more and more people live in cities. From 1950 to 2018, the proportion of the world's population in cities increased to 55 percent [1]. Rapid urbanization has also brought unprecedented challenges. The future sustainability of cities depends on the compatibility of development and the environment and the applied innovations [2]. Increasing traffic flows cause a number of problems [3–5]. Therefore, many researchers are looking for various solutions that can be applied in urban logistics. Dutch scientists estimate that the amount of greenhouse gas emissions across the country was as high as 0.9 million cubic meters [6].

As early as 2001, researchers [7] defined urban logistics as a process that allows complete optimization of logistics and transport activities of companies based on advanced information systems in urban areas, including traffic environment, traffic congestion, traffic safety, and energy saving, in the conditions of a market economy [8].

With the expansion of cities, access of commercial transport to the central part of a city is an increasing problem, which affects overall mobility in a city, especially during rush hours.

The main problems for the mobility of cargo traffic flow are due to outdated infrastructure, slow implementation of information technologies to create a common informational field of urban logistics, and underutilization of car capacity in order to reduce the flow of partially loaded cars in cities.

The intensity of heavy and freight transport flows in a city is uneven. The negative impact of this transport on city logistics is manifested when all traffic flows are intensive in city logistics. Freight transport negatively affects the speed of traffic flows and can be the cause of both technical accidents and traffic jams [5]. In this case, logistics companies suffer losses due to late or non-delivered cargo. The quality of service drops due to congestion.

The main function of freight transport is the physical transport of goods, including the delivery of goods to receivers within the limits of a city's logistics infrastructure [8]. Optimization of the distribution of freight transport flows can be performed by modelling the consequences of emergency situations.

The purpose of this study is to determine the subjective normalized weights of the main factors that have the greatest influence on a city's logistics system, using Kendall's correlation coefficient method and the average rank transformation of weights (ARTIW) method [9]. At the same time, the main problems of the topic under analysis are identified, along with the dependencies of the phenomena and possible trends and insights. After systematizing the obtained results, a possible solution for the optimization of freight transport flows in a city is formulated, which would provide a real basis for the optimal distribution of freight transport flows in cities.

It can be said that the study and its results are relevant for potentially interested parties, which are city municipalities, transport and logistics companies, service providers, and users located in the city boundaries. The study is universal enough and its process can be easily replicated in different cities, and the results of the existing study are applicable in different cities as well. In any case, it should be taken into account that the current study was carried out based on a city with a population of 700,000, a developed transport and logistics infrastructure and the city covers an area of 401 km^2. Each stakeholder group, in the context of this research, has individual objectives. These goals could be noted as follows:

- Municipalities: to reduce the burden on a city's logistics system, improve the quality of life, and optimize heavy traffic flows;
- Logistics companies: use available resources more efficiently, reduce the cost of services, reduce hazardous substances, and CO_2 emissions;
- Service recipients and users: to receive the necessary loads promptly and plan their activities;
- City dwellers: to live in a safe environment where traffic flows are optimized and a decent standard of living is ensured.

2. Materials and Methods

2.1. Theoretical Aspects of the Flows of Freight Road Traffic in the Urban Logistics System

These days, urban logistics planning is based on high-quality and fast cargo transportation, with various ecological solutions. Service providers must pay attention to shorter delivery times, flexibility, and service reliability [10].

Analyzing the sources of the scientific literature, it can be stated that city logistics can be seen as a tool that allows efficient management of the movement of material flow in the city territory and forms innovative solutions [11]. In most cases, the content of logistics solutions consists of analysis, planning, integrated information management, and decision making. The uniqueness of a city's logistics system is the consolidation of cargo from various senders and receivers and the management of the cargo transportation process in a city's infrastructure [12].

Analyzing the existing terminology, it is stated that the objective of city logistics is to support and develop the harmonious development of urban areas, increase the efficiency of transport and logistics processes, and significantly reduce the negative impact on the environment [13].

The fast pace of life and the pursuit of prosperity led to the need to find new solutions for the development of cargo distribution and transportation efficiency [14]. These goals can be achieved by organizing distribution activities in a more targeted and qualitative way and by properly forming an infrastructure network, and cooperation between carrier companies is also important in this area [15]. Scientists and researchers emphasize the most important areas of urban logistics development as planning, information technology, communication, public–private partnership, support measures, and regulations [16].

Heavy transport causes many problems in cities. These include traffic congestion, air and noise pollution, and traffic accidents [17]. However, freight transport is an inevitable part of city life, as trade takes place in cities and various services are provided.

In their publications, scientists and researchers have widely analyzed the problems of the first and last mile and their practical expression forms, which depend on the complexity of the city infrastructure, the location of the terminals, and their distance to the city border [18]. The last mile is defined as the part of the logistics supply chain from the cargo terminal to the final destination of the shipment (cargo) [19].

Scientific works emphasize that to optimize the last mile, it is necessary to harmonize different solutions related to the operation of cargo terminals and warehouses, as well as transport solutions in the logistic environment of a city. However, new transport solutions are associated with the transformation of a city's logistics infrastructure, such as a network of charging stations for the use of electric cars [20]. The social aspect of the problem should also be noted, because cities have a significant impact on social and economic development, as well as an impact on the environment in general. Therefore, when analyzing city problems, it is necessary to emphasize topics related to the appropriate use of resources, optimization of the transport system, information and data management, social change, the impact on the environment, and conservation of nature [21,22].

According to researchers, great attention is paid to innovations in the field of information and communication technologies in modern city logistics:

- "Green technologies" allow for a more responsible use of a city's main resources (gas, water, CO_2 emissions, and introducing electric transport) and the development of renewable energy sources;
- Implementation of broadband internet 5G, which ensures the development of modern information technologies and the development of closed and secure networks;
- Artificial intelligence that makes certain decisions;
- The spread of eSim technology to control the movement of goods from the production process to sales [20,21].

Another researcher [22] studied the impact of freight transport on CO_2 emissions. In their research, they relied on data from the French Environment and Energy Agency (ADEME). Table 1 shows a table of ADEME data showing the distribution of CO_2 emissions by weight of freight cars.

Table 1. CO_2 emissions of different types of freight vehicles [22].

Cars by Weight	CO_2 Emission (g/tkm)
Small commercial vehicle (1.5–3.5 t)	1103
Cars (6.1–10.9 t)	435
Cars (11–21 t)	221
Cars (21–32.6 t)	196

A total of 82% of the cargo was also found to reach the city by low-capacity transport; the rest is transported by heavy duty freight transport [22].

The examples presented show that freight transport is unevenly distributed in a city logistics system. The use of freight transport in urban logistics is due more to the capacity and legal regulations in the urban area. In a city's logistics system, there are possible restrictions on the movement of goods and heavy vehicles and restrictions on driving on weekends and during appropriate hours. The load on the axle may also be limited, taking into account the existing state of a city's infrastructure, such as street surfaces, bridges, historical heritage, etc. City logistics is dominated by freight transport with a carrying capacity of up to 3.5 t, but the use of heavy-duty transport is also unavoidable, causing the most problems, especially during peak hours, of possible traffic jams, noise, and increased emissions due to braking, increasing traffic jams or the number of controlled traffic light intersections.

Researchers [21,23] have proposed the application of a two-tier hyperconnected urban logistics system, which, according to the authors, would help reduce heavy traffic in cities. This proposed optimization model is based on the idea that cargo from various customers could be consolidated in terminals near cities and transported by heavy transport to small distribution centers in urban areas, from where it would be distributed to recipients by urban commercial transport.

According to the researchers' proposal, large terminals in the countryside could be connected with smaller terminals in urban areas, to which transport transporting consolidated cargo would go, and distribution would be carried out by city-friendly, ecological, and small-capacity urban freight transport. According to the researchers, this would reduce heavy traffic in cities [23].

Another way to optimize transport flows in a city's logistics system is the "sharing economy". In this context, it is a very broadly interpreted concept, as it can cover both the sharing of vehicles and serviced routes. Such an "exchange" would enable transportation and logistics companies to transport more cargo faster in a more environmentally friendly way at a lower cost. Favorable conditions would be created for better utilization of resources, optimization of routes, simplification of schedules, and reduction in carbon dioxide emissions. It should be noted that, in addition to greater operational efficiency, the sharing of available and underutilized resources would allow solving problems related to the overloading of the transport system and the lack of drivers. In the area of warehousing, the sharing economy aims to increase the utilization of existing shared customer warehouses. Finally, the sharing economy presents new and creative ways to start a business and increase efficiency through on-demand staffing models and shared logistics services [24].

The distribution of freight transport in a city depends largely on the characteristics of the traffic network, for example, the road category, number of movement lanes, organization and regulation of movement, organization of stopping places, parking capacity, etc. These parameters affect the structure of vehicles in the realization of goods flows, the intensity of traffic congestion by commercial vehicles, route planning, duration of loading–unloading operations, operational costs of carriers and service providers, etc. [25].

Thus, summarizing the opinions of different researchers, it can be stated that the distribution of freight transport flows in cities is mainly influenced by the geographical location of the city, the location of industrial and commercial places, and the locations of distribution centers in the territory and periphery. Furthermore, the mobility of freight transport in a city is influenced by road infrastructure, traffic regulations and restrictions, and parking capacity. The solution to the problem of improving the distribution of freight transport flows in urban logistics could be the optimization of these transport flows in the first and last mile stages, eliminating heavy transport, and using self-service terminals in rural logistics centers and ecological transport for the collection or distribution of small loads.

2.2. Research Methodology

After analyzing the examined studies and the proposed solutions for the optimization of cargo transport flows, it is clear that this is relevant today and will be relevant in the future. The problems of city logistics, which are related to transport, life in cities, and environmental protection, are solved at various levels, from the strategic level to the cooperation of municipal administrations and business entities.

A survey of experts was conducted in the feasibility study to optimize cargo flows by road transport. An in-depth interview or other method can be an alternative to a survey but a survey is more acceptable to both respondents and interviewers. This study aims to gain a deeper understanding of the studied phenomena and to determine the possibilities of optimizing freight transport and how these possibilities can be practically applied in real life. The survey was conducted according to a rated scale of questions [26]. Questions are present as statements or criteria and then ranked.

In the survey, ten respondents were interviewed, who were logistics experts who have a university degree—at least a bachelor's degree—and at least 5 years of experience in the field of international cargo transportation and cargo distribution city logistics. Experts with an average logistics experience of almost 14 years were selected.

The survey was conducted in May–June 2022. The respondents of the survey were experts in the Lithuanian logistics sector and heads of companies or company divisions The survey was sent to respondents electronically via ref [27]. The apklausa.lt system was designed to create and conduct online surveys with unlimited quantities of questionnaires and questions and distribute it to respondents [28]. Answers to questionnaires are provided in a simple, understandable form. The results can be saved in a file that can be opened with popular office applications (OO Calc, MS Excel, or SPSS).

After evaluating the provisions and recommendations of multi-criteria decision making (MCDM) methods and concepts [29], the number of experts was determined. This number must be greater than the number of multiple-choice questions. In the case of our study, there were ten experts and the number of optional answers is eight, so it can be said that the number of experts was appropriate.

During the research, selected experts were asked questions aimed at obtaining information on city logistics, environmental protection and environmental impact, freight traffic flows, traffic throughput, and infrastructure impact. At the same time, an attempt was made to obtain the prognostic information required for the study on possible optimization measures and solutions, possible financial aspects of the use of electricity or other ecological transport, and the use of information technologies.

The data from the expert survey, expressed in numerical values, were analyzed and calculated for the reliability of the data. The processing of the data was aimed at obtaining generalized data to study the possibilities of optimizing the flow of freight transport in a city logistics system.

Research data were processed using Kendall's concordance coefficient method of agreement [9] and the average rank transformation to weights (ARTIW) method [30,31]. All expert assessments were ranked.

Kendall's coefficient of concordance was applied to calculate the survey results and determine the distribution and concordance of opinions. The values of the concordance coefficient (W) are in the range from 0 to 1. The higher the obtained value (W), the less the opinions differ and are aligned.

To find Kendall's concordance coefficient, the sum of the ranks, R_j, assigned to each j-th criterion by n experts is calculated according to the formula:

$$R_j = \sum_{i=1}^{n} R_{ij} (j = 1, 2, \ldots m). \tag{1}$$

More precisely, it is based on the sum, S, of the squared deviations, R_j (the variance analogue), from the mean rank, $\overline{R}$:

$$S = \sum_{j=1}^{m} (R_j - \overline{R})^2. \quad (2)$$

The following calculates the average criterion rank, $\overline{R}$. It is obtained by dividing the sum of the ranks, assigned to the criterion by the experts, by the number of criteria, m:

$$\overline{R} = \frac{\sum_{j=1}^{m} R_{ij}}{m} = \frac{\sum_{i=1}^{n}\sum_{j=1}^{m} R_{ij}}{m}, \quad (3)$$

where R_{ij} is the rank assigned by the i-th expert to the j-th criterion; n is the number of experts ($i = 1, 2, \ldots, n$); and m is the number of criteria ($j = 1, 2, \ldots, m$).

When S (true sum of squared values) is calculated according to the formula (2), then the correlation coefficient, W, can be calculated according to the formula:

$$W = \frac{12 \cdot S}{n^2 \cdot (m^3 - m)} \quad (4)$$

When the opinions of the experts are similar, the value of the concordance coefficient, W is about one, and if the opinions differ greatly, the value of W is about zero.

The value of the random number S is calculated by summing the squared values of all criteria enclosed in square brackets (Formula (5)).

$$S = \sum_{j=1}^{m} \left[\sum_{i=1}^{n} R_{ij} - \frac{1}{2} n \cdot (m+1) \right]^2, \quad (5)$$

where m is the number of criteria ($j = 1, 2, \ldots, m$) and n is the number of experts ($i = 1, 2, \ldots, n$).

The following rule has been proven [30]: when the number of criteria is $m > 7$, the significance of the correlation coefficient, W, can be determined using Pearson's criteria (chi-square test). The random value is distributed according to χ^2:

$$\chi^2 = n \cdot (m-1) \cdot W = \frac{12 \cdot S}{n \cdot m \cdot (m+1)}, \quad (6)$$

with the degree of freedom $\nu = m - 1$. Based on the selected confidence level α (which is assumed to be 0.05 or 0.01), the critical value $\chi^2_{\nu,\alpha}$ is found from the table of χ^2 distribution with the degree of freedom $\nu = m - 1$. If the value of χ^2 calculated by formula (6) is larger than $\chi^2_{\nu,\alpha}$, it shows that the experts' estimates are consistent.

The smallest value of the concordance coefficient, $W_{\min}$, can be estimated by applying the following formula (7):

$$W_{\min} = \frac{\chi^2_{v,a}}{n \cdot (m-1)} \quad (7)$$

where n is the expert opinions; m is the number of comparative criteria that indicate the quality of an object under analysis with the selected levels of significance α, and the degree of freedom $\nu = m - 1$. Having calculated this value, if it is not possible to assert that the experts' opinions are in agreement, $\chi^2_{\nu,\alpha}$ (the critical Pearson's statistic) at the degree of freedom and significant level are taken [30].

Next, the criteria of the ARTIW method [31], which describe the significances (weights), are calculated. For that, the following Formula (8) is applied:

$$\omega_j = \frac{(m+1) - \overline{R}_j}{\sum\limits_{j=1}^{m} \overline{R}_j} \quad (8)$$

where m is the number of criteria (j = 1, 2,..., m) and $\overline{R}_j$ is the average rank of the j-th criterion calculated according to Formula (9):

$$\overline{R}_j = \frac{\sum\limits_{i=1}^{n} R_{ij}}{n} (j = 1, 2, \ldots m), \quad (9)$$

where R_{ij} is the rank of the criteria granted by the experts and n is the number of experts.

3. Results

During the research, 10 respondents, who were experts in the field of logistics with at least 5 years of experience in international cargo transportation and cargo consolidation, as well as distribution in urban logistics, were interviewed. All experts who participated in the survey have a higher education degree, and their total average work experience in the field of logistics is almost 14 years.

Experts assessed the influence of each factor on a scale of eight points (1—the least influence; 8—the greatest influence). Answers could not be repeated.

The following factors were presented for the survey:

A1—Urban road infrastructure;
A2—Traffic regulations;
A3—Legal regulations;
A4—Geographical location of the city;
A5—Geographical distribution of economic operators;
A6—Innovation and information technology;
A7—Cooperation between economic operators;
A8—Cooperation between the city administration and economic operators.

The ratings provided by each expert for each factor were summed. The responses' evaluations were converted into ranks according to the formula $R_{ij} = (n + 1) - B_{ij}$ and listed in Table 2.

Table 2. Calculations of the factors that have the greatest impact on a city's logistics system.

Codes of Experts	Factors That Have the Greatest Impact on a City's Logistics System (j = 1, 2, . . . , 8)							
	A1	A2	A3	A4	A5	A6	A7	A8
E1	3	1	7	2	5	6	8	4
E2	1	2	3	4	5	6	7	8
E3	2	1	8	3	6	7	5	4
E4	1	2	3	8	7	6	5	4
E5	1	3	8	2	7	6	5	4
E6	1	2	6	7	8	4	5	3
E7	1	2	7	4	5	8	6	3
E8	2	3	4	5	6	8	6	7
E9	4	5	7	1	2	8	6	3

Table 2. Cont.

Codes of Experts	Factors That Have the Greatest Impact on a City's Logistics System (j = 1, 2, . . . , 8)							
	A1	A2	A3	A4	A5	A6	A7	A8
E10	2	4	7	1	8	3	6	5
Sum of the Ranks $R_j = \sum_{i=1}^{n} R_{ij}(j = 1,2,\ldots m)$.	18	25	60	37	59	62	59	45
Average Rank $\bar{R} = \frac{\sum_{j=1}^{m} R_{ij}}{m} = \frac{\sum_{i=1}^{n}\sum_{j=1}^{m} R_{ij}}{m}$,	1.8	2.5	6	3.7	5.9	6.2	5.9	4.5
Subtraction of the sum of ranks and a constant $\sum_{i=1}^{n} R_{ij} - \frac{1}{2}n(m+1)$	−27	−20	15	−8	14	17	14	0
$\left[\sum_{i=1}^{n} R_{ij} - \frac{1}{2}n(m+1)\right]^2$	729	400	225	64	196	289	196	0

Kendall's concordance coefficient was calculated according to Equation (4).

W = 12 × 2099/100 × (512 − 8) = 0.500, since the obtained number is equal to the value of 0.5, it can be said that the opinions of the respondents are aligned.

The weight of the concordance coefficient was calculated according to Equation (6):

χ^2 = 12 × 2099/10 × 8(8 + 1) = 34.98. Since the calculated value of χ^2 is greater than the value of χ^2_{kr} (18.4753), the opinion of the respondents is considered to be consistent and the average ranks show a common opinion.

According to Equation (7), the lowest value of the concordance coefficient, W_{min}, was calculated:

W_{min} = 18.4753/10 (8 − 1) = 0.264 << 0.500.

The calculations showed that the opinions of ten respondents agree on eight factors that influence the optimization of a city's logistics system, and the opinions of all respondents are summarized.

Next, the indicators of the importance of factors influencing the mobility of urban logistics and ecology, Q_j, were calculated according to the ARTIW method.

The correlation of the criteria is calculated according to the equation:

$$\bar{q} = \frac{\bar{R_j}}{\sum_{j=1}^{m} R_J}. \quad (10)$$

The inverse is calculated using the equation $d_j = 1 - \bar{q}_j = 1 - \frac{\bar{R_j}}{\sum_{j=1}^{m} R_j}$.

A weight indicator was obtained using this equation $Q_j = \frac{d_j}{\sum_{j=1}^{m} d_j} = \frac{d_j}{m-1}$.

Indicators according to all criteria were calculated in a similar sequence and are given in Table 3.

According to the equation $\tilde{Q}_j = \frac{\sum_{i=1}^{n} B_{ij}}{\sum_{i=1}^{n}\sum_{j=1}^{m} B_{ij}}$, the indicators of the importance of each factor were calculated and are also summarized in Table 3.

The distribution of factors that influence the mobility and ecology according to importance is shown graphically in Figure 1.

Table 3. Distribution of factors according to importance.

Indicators	A1	A2	A3	A4	A5	A6	A7	A8
$\bar{q}_j$	0.049	0.068	0.164	0.101	0.162	0.170	0.162	0.123
d_j	0.951	0.932	0.836	0.899	0.838	0.830	0.838	0.877
Q_j	0.136	0.133	0.119	0.128	0.120	0.119	0.120	0.125
$\tilde{Q}_j$	0.200	0.181	0.083	0.147	0.086	0.078	0.086	0.125
Importance	1	2	7	3	5	8	6	4

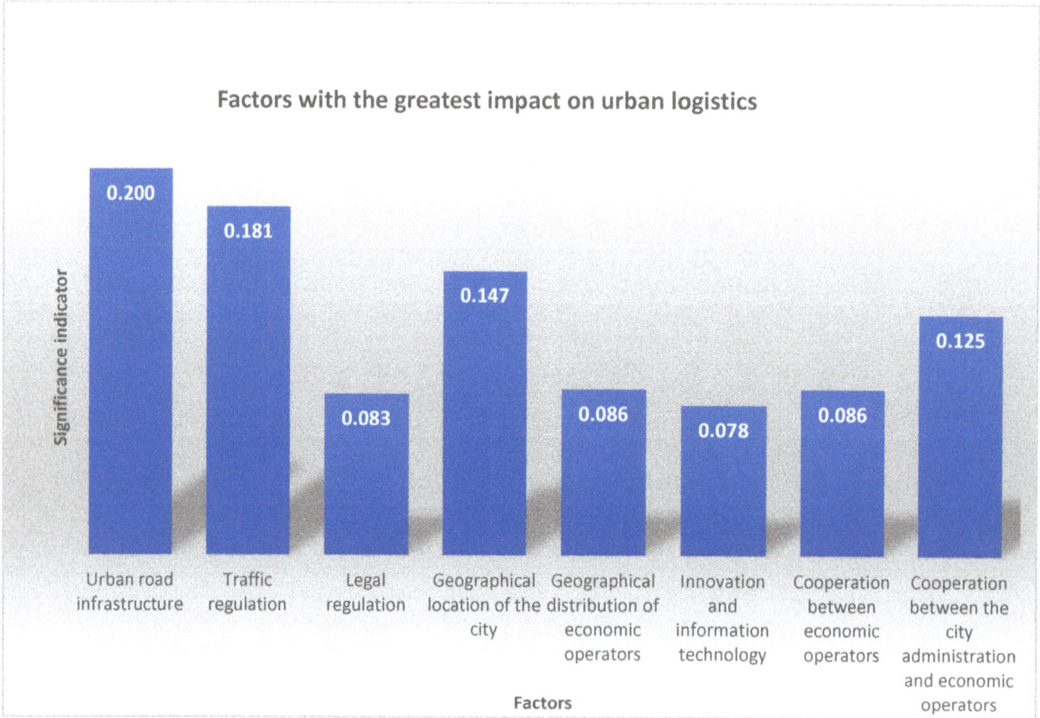

Figure 1. Factors that have the greatest impact on a city's logistics system.

After calculating the importance indicators, it can be concluded that legal regulations, innovation and information technology, geographic location of economic entities, and cooperation of economic entities currently have a smaller influence on city logistics (importance indicators of 0.083, 0.078, 0.086, and 0.086, respectively). On the contrary, according to experts, traffic regulation and city road infrastructure have the greatest influence on city logistics (indicators of 0.181 and 0.2, respectively).

The same methodology was used for the calculations.

The ratings provided by each expert for each factor were summed. Respondents' evaluations were converted into ranks and listed in Table 4.

Table 4. Calculations of factors that have the greatest influence on the optimization of a city's logistics system.

Codes of Experts	Factors That Have the Greatest Influence on the Optimization of a City Logistics System ($j = 1, 2, \ldots, 8$)							
	B1	**B2**	**B3**	**B4**	**B5**	**B6**	**B7**	**B8**
E1	3	1	5	4	8	7	2	6
E2	1	2	4	6	8	5	7	3
E3	2	3	1	8	5	7	6	4
E4	1	3	2	8	6	5	7	4
E5	1	2	8	3	6	4	7	5
E6	2	3	1	4	6	8	7	5
E7	1	2	3	5	7	8	6	4
E8	1	4	7	6	8	5	3	2
E9	5	6	3	7	8	2	4	1
E10	1	2	4	5	7	8	6	3
Sum of the Ranks $R_j = \sum_{i=1}^{n} R_{ij} (j = 1, 2, \ldots m)$.	18	28	38	56	69	59	55	37
Average Rank $\overline{R} = \frac{\sum_{j=1}^{m} R_{ij}}{m} = \frac{\sum_{i=1}^{n}\sum_{j=1}^{m} R_{ij}}{m}$,	1.8	2.8	3.8	5.6	6.9	5.9	5.5	3.7
Subtraction of the sum of ranks and a constant $\sum_{i=1}^{n} R_{ij} - \frac{1}{2}n(m+1)$	−27	−17	−7	11	24	14	10	−8
$\left[\sum_{i=1}^{n} R_{ij} - \frac{1}{2}n(m+1)\right]^2$	729	289	49	121	576	196	100	64

The following factors were presented for the survey:

B1—Prohibition of heavy transport entering the central part of a city;

B2—Ecological fee for freight transport in the city;

B3—State promotion for the purchase of environmentally friendly vehicles;

B4—Relocation of large economic entities to the periphery of the city;

B5—Promotion of the state by implementing information technology and connecting to the smart city system;

B6—Tax incentives for business entities cooperating with each other for the transportation of goods in the city territory;

B7 Connecting logistics companies to a unified information system for city cargo transportation;

B8—Creation of a network of small self-service terminals on the outskirts of the city.

Kendall's concordance coefficient was calculated according to Formula (4).

W = 12 × 2124/100 × (512 − 8) = 0.506, and since the number obtained is greater than the value of 0.5, it can be said that the opinions of the respondents are aligned.

The weight of the concordance coefficient was calculated according to Equation (6):

χ^2 = 12 × 2124/10 × 8(8 + 1) = 35.40. Since the calculated value of χ^2 is greater than the value of χ^2_{kr} (18.4753), the opinion of the respondents is considered to be consistent and the average ranks show a common opinion.

According to Formula (7), the lowest value of the concordance coefficient, W_{min}, was calculated:

W_{min} = 18.4753/10(8 − 1) = 0.264 << 0.506

The results of these calculations are presented in Table 4.

The calculations showed that the opinions of the ten respondents agree on eight factors that influence the optimization of a city's logistics system, and the opinions of all respondents are summarized.

According to the same methodology, indicators of the importance, Q_j, of factors influencing the optimization of a city's logistics system were calculated.

The normalized subjective weights of the main factors with the greatest influence on the optimization of a city logistics system were calculated in the same way as the previous calculations and are summarized in Table 5.

Table 5. Distribution of indicators according to importance.

Indicators	B1	B2	B3	B4	B5	B6	B7	B8
$\bar{q}_j$	0.050	0.078	0.106	0.156	0.192	0.164	0.153	0.103
d_j	0.950	0.922	0.894	0.844	0.808	0.836	0.847	0.897
Q_j	0.136	0.132	0.128	0.121	0.115	0.119	0.121	0.128
$\tilde{Q}_j$	0.200	0.172	0.144	0.094	0.058	0.086	0.097	0.147
Importance	1	2	4	6	8	7	6	3

The obtained results show that when optimizing s city's logistics system, it is necessary to take into account the ban on entering the city center (indicator 0.2) and the ecological tax (indicator 0.172), because these indicators have the greatest importance. The second most important indicators are the creation of a network of self-service terminals on the outskirts of the city (index 0.147) and support for the purchase of environmentally friendly vehicles (index 0.144). The third is the integration of logistics companies into a unified one, followed by an urban freight transport information system (index 0.097) (Figure 2). The results obtained from the research show that the role of the state is strategically important in optimizing freight transport flows in a city logistics system.

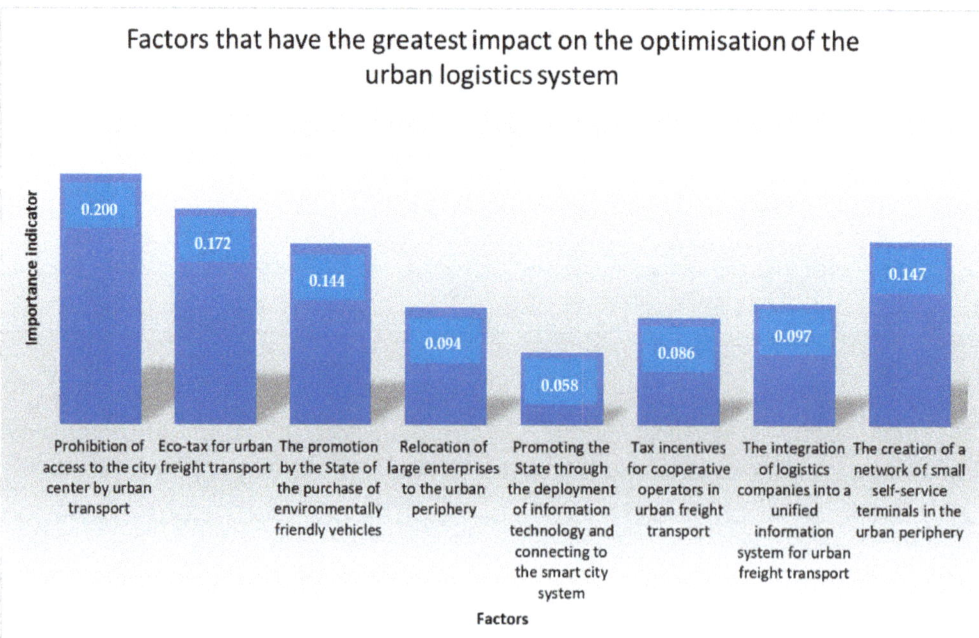

Figure 2. Factors that have the greatest impact on the optimization of an urban logistics system.

The study also aimed to assess which cargo vehicles in urban logistics pose the greatest problem of mobility and urban ecology, taking into account the carrying capacity of the vehicles. Respondents rated the impact of each vehicle on a scale of eight points (1—the least impact; 8—the greatest impact).

The indicators according to all criteria were calculated in a similar sequence and are shown in Table 6.

Table 6. Distribution of factors according to importance.

Indicators	Up to 999	1000–1499	1500–2999	3000–4999	5000–6999	7000–9999	10,000–14,999	15,000 and Above
$\bar{q}_j$	0.189	0.175	0.153	0.125	0.122	0.117	0.069	0.050
d_j	0.811	0.825	0.847	0.875	0.878	0.883	0.931	0.950
Q_j	0.116	0.118	0.121	0.125	0.125	0.126	0.133	0.136
$\tilde{Q}_j$	0.061	0.075	0.097	0.125	0.128	0.133	0.181	0.200
Importance	8	7	6	5	4	3	2	1

The distribution of factors that influence the infrastructure and the entire city's logistics system by importance is shown graphically in Figure 3.

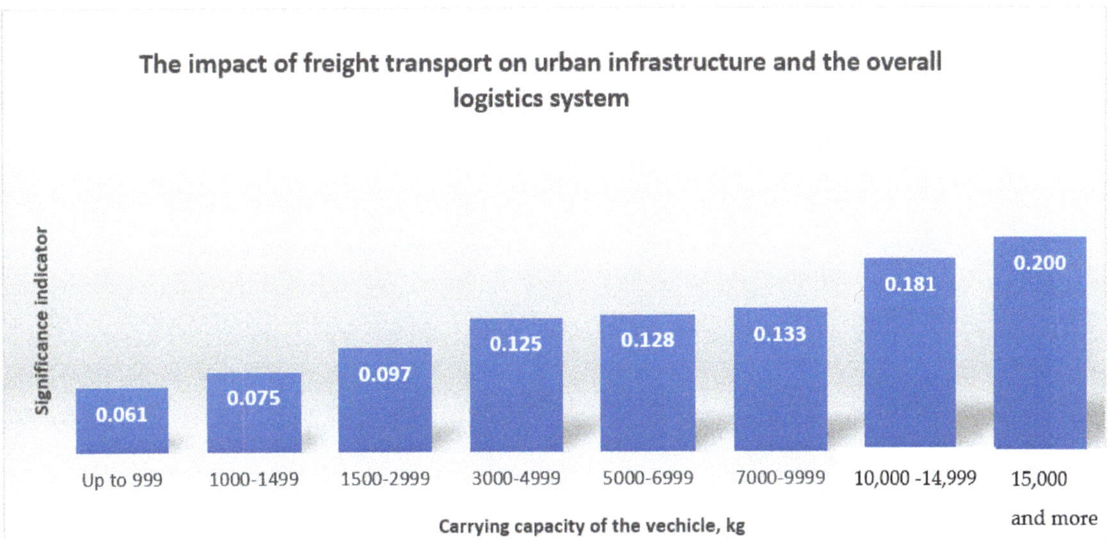

Figure 3. The impact of road vehicles on the infrastructure and the whole logistics system of a city.

The results of the study show that vehicles with a carrying capacity of up to 999 kg and 1000–1499 kg have the least impact on mobility and ecology in the city (significance indices of 0.061 and 0.075). However, vehicles with a carrying capacity of 15,000 kg or more are the most problematic and have the greatest impact on mobility and ecology in cities (importance index 0.200). At the same time, it can be assumed that vehicles up to 999 kg and 1000–1499 kg can be more easily replaced by electric vehicles.

Experts were asked how many companies would be willing to use small cargo terminals on the outskirts of a city. The vast majority of experts were in favor of such an initiative (seventy percent), ten percent were against it, and twenty percent of experts were likely to use it.

4. Discussion

From the analysis of scientific literature sources, several solutions for the optimization of cargo transport flows can be distinguished:

- The use and development of information technologies to manage, coordinate, and control information;
- Management efficiency through innovation in both urban regulations and environmental solutions;
- Placement of large terminals and small distribution centers in the countryside, taking into account the location of shopping centers and industrial districts;
- Infrastructure development using innovative solutions and information technologies;
- Collaboration, cooperation, and information and resource sharing among urban logistics participants.

The function of freight transport and the physical movement of goods from one place to another depends on the carrying capacity and regulation of the urban area. The problem is that these parameters are often not used if there is no cooperation between the public and private sectors and between business entities operating in one economic space in a certain territory in a city's logistics system.

Evaluating the purpose of this research, the chosen methodology, and the circumstances of the applicability of the obtained data, it can be said that the research covered all the research questions and tasks required by researchers. The study was focused on identifying possible optimization opportunities for city logistics by evaluating freight traffic and it was implemented.

In the studies conducted by researchers, it was not possible to find how the use of self-service terminals in peripheral logistics centers for small cargoes in city logistics, as is done when transporting small parcels, could affect freight traffic flows. Research has focused on the cooperation of logistics companies in the first or last mile and the use of information technology, ecological transport, and mobility, but has not examined the contribution of the shippers or receivers themselves to the planning and sustainable transportation process in urban logistics.

In summarizing the results of the previous research and the conducted research, it can be said that ideally, heavy transport should only be used for the direct transportation of bulky and heavy loads. In an urban logistics system, it should only be applied in the long or split first or last mile. However, due to the restriction of heavy freight transport in the central part of a city, transshipment points are needed for transshipment to commercial transport of smaller dimensions and carrying capacity.

Freight transport is least desirable in the central part of a city, because that is where it causes the most problems due to its dimensions, pollution, acceleration, and braking. Traffic flows are determined by the developed urban infrastructure, road capacity, traffic regulations, and the location of shopping centers, industrial centers, and logistics centers. In order for freight transport to move smoothly through the urban area, a suitable infrastructure is needed to optimize freight transport flows. This study revealed that the creation of a network of small self-service terminals on the outskirts of a city would benefit from such a function. Terminals of this type should be located on the outskirts of the central part of the city, taking into account the concentration and geographical location of business entities. Terminal functioning and servicing should be delegated to logistics companies with logistics warehouses outside of a city's territory. Business entities could transport small loads to the self-service terminals in the central part of the city through established transport channels.

Transport companies are looking for different ways to attract professional drivers, both from the local labor market and through migrants. In this case, drivers with category B driver's licenses would be sufficient for the first or last mile transportation in urban logistics, and heavy-duty vehicles driven by professional drivers would be better used to transport goods between the logistics center and self-service terminals. This should reduce the need for such transport and at the same time help to control the lack of human resources in transport companies.

Freight transport, especially heavy freight transport, increases urban pollution and creates noise. In this paper, it is assumed that the use of self-service terminals for small loads in the periphery of cities would help reduce heavy traffic flows in cities. There would be a clear opportunity for the adoption of light-duty green vehicles, such as CO_2-free electric cars that would reduce pollution in cities. In addition, non-professional drivers can drive cars with a low carrying capacity up to 3.5 t. Such a vehicle could easily be driven not only by the driver of the transport company but also by the sender or recipient of the cargo. This would help solve another current problem: the lack of professional drivers.

The limitations of the study are given as follows. It can be said that the current study covers most of the important components of the urban logistics system, which are related to environmental impact and freight transport. The research aims to determine the general possibilities and trends in freight transport optimization by evaluating the individual factors that are related to this process; therefore, it can be said that there is an open possibility to additionally analyze each factor separately, analyze its significance, and determine the individual characteristics of the factor. Due to the limitations of the study, the potential vehicle exchange systems, information technologies, and other aspects are not detailed, but they are evaluated in the general context of the study.

5. Conclusions

The authors of scientific sources emphasize that city logistics has unique features, namely sensitivity to the impact on the environment, intensive delivery of cargo and parcels, and the need for continuous optimization of processes. Optimization processes are associated with routing, deployment of logistics centers, and the need for cargo consolidation–deconsolidation, with the possibility of using the PPP (public–private partnership) method in practice.

When analyzing the problems of urban logistics and freight transport flows, the authors of the scientific literature emphasize the challenges of the last mile related to the operation of the distribution system and make assumptions for developing the problem of the first mile, which is not widely analyzed. In summary, it can be stated that the distribution of freight transport flows in cities is mainly influenced by the geographical location of the city, the location of industrial and commercial places, and the locations of distribution centers in the territory and periphery. Furthermore, the mobility of freight transport in a city is influenced by road infrastructure, traffic regulations and restrictions, and parking capacity.

The expert survey method was used during this investigation; experts with sufficient education and professional practical experience were interviewed, taking into account the MCDM methodology, a sufficient number of experts were selected, and the data processing methodology used allowed us to state that the research results were reliable.

The results of the research showed that the greatest influences on the optimization of a city logistics system are the ban on entering the city center (indicator 0.2) and the ecological tax (0.172). The strategic role of the state and municipalities in the formation of rules for the use of a city logistics system and the promotion of insurance measures has been observed and evaluated. The study also showed that in city logistics, to optimize freight transport flows and improve the ecological climate, it is necessary to use cars weighing up to 999 kg, 1000–1499 kg, and 1500–2999 kg.

Taking into account the results of the study and modelling the possible actions, it is recommended to organize a network of self-service terminals to consolidate small amounts of cargo, which would solve the optimization problems of city logistics cargo flows raised during the analysis. This would allow shippers or consignees in the central part of the city to independently pick up or deliver goods to self-service terminals. This would make the transportation process easier to plan and prevent delays. It would also create a niche for the targeted use of commercial electric vehicles in cities, helping to reduce pollution and noise.

A city's logistics system is mostly dependent on a city's infrastructure and traffic regulations. To optimize the flow of freight transport, a city's infrastructure must be constantly improved. It can be said that companies are ready to pick up or deliver small loads themselves to terminals on the outskirts of the city. Cooperation between institutions in consolidating or distributing cargo from terminals located on the outskirts of the city is important for customers. Additionally, companies take care of the optimization of cargo transportation flows by implementing information technologies to organize transport in the first or last mile.

The idea of self-service terminals on the outskirts of the city is attractive, but its novelty is questionable. Doubts are related to the fact that state support is not clear, its size is not clear, and it is not clear whether it will attract enough users.

A separate, additional topic of discussion, which this study reveals, is the competitive environment and its management possibilities. During commercial activities, representing clients may raise additional questions about cargo delivery times, order processing, consolidation, and waiting issues. This aspect needs to be further analyzed.

A separate area of additional research could be the details of the impact on the environment, assessing the elements of a city's logistics infrastructure, and the impact on them through emissions, noise, and accidents. However, it can be said that the current study is fully sufficient to create an initial reasonable picture of the situation confirmed by objective research methods, as well as the possibility of possible optimization methods.

An additional topic of research and to continue existing research can be the impact of state or municipal authorities based on the principle of a PPP (public–private partnership), their place and role in the development of possible cooperation systems, the establishment of new terminals, and issues of infrastructure development and use. This research topic is related to the current study, but has independent goals and requires a deeper analysis of specific issues.

Author Contributions: Conceptualization, N.B. and D.B.; methodology, D.B.; validation, N.B. and D.B.; formal analysis, N.B.; investigation, D.B.; resources, N.B.; data curation, N.B.; writing—original draft preparation, N.B.; writing—review and editing, N.B. and D.B.; visualization, N.B.; supervision, N.B.; project administration, N.B. and D.B.; funding acquisition, N.B. All authors have read and agreed to the published version of the manuscript.

Funding: This research was funded by Vilnius Gediminas Technical University.

Institutional Review Board Statement: Not applicable.

Informed Consent Statement: Not applicable.

Data Availability Statement: Not applicable.

Acknowledgments: The authors thank their colleague Šlajus Š. for their cooperation.

Conflicts of Interest: The authors declare no conflict of interest.

References

1. Population Facts. United Nations. Population Division. World Urbanization Prospects 2018. United Nations, New York: United Nations Publications. Available online: https://population.un.org/wup/Publications/Files/WUP2018-PopFacts_2018-1.pdf (accessed on 14 September 2021).
2. United Nations. Department of Economic and Social Affairs. Meeting Challenges for the Urban Future. Available online: https://sdgs.un.org/publications/meeting-challenges-urban-future-17398 (accessed on 15 September 2021).
3. European Commission. Mobility and Transport, Sustainable Urban Mobility Plans. Available online: https://transport.ec.europa.eu/transport-themes/clean-transport-urban-transport/urban-mobility/urban-mobility-actions/sustainable-urban-mobility-plans_en (accessed on 25 October 2021).
4. Commission of the European Communities. Roadmap to a Single European Transport Area. Brussels, 28.3.2011 COM (2011) 144 Final. Available online: http://eur-lex.europa.eu/LexUriServ/LexUriServ.do?uri=COM:2011:0144:FIN:EN:PDF (accessed on 25 October 2021).

5. Absi, N.; Cattaruzza, D.; Gonzalez-Feliu, J. Vehicle Routing Problems for City Logistics. *EURO J. Transp. Logist.* **2017**, *6*, 51–79. [CrossRef]
6. Anand, N.; Van den Band Van Duin, J.H.R.; Quak, H.J. Designing sharing logistics as a disruptive innovation in city logistics. In Proceedings of the 4th International Conference GREEN CITIES, Szczecin, Poland, 3–5 June 2020.
7. Taniguchi, E.; Thompson, R. Modeling City Logistics. *Transp. Res. Rec.* **2002**, *1790*, 45–51. [CrossRef]
8. Taniguchi, E. Concepts of City Logistics for Sustainable and Liveable Cities. *Procedia-Soc. Behav. Sci. Procedia.* **2014**, *151*, 310–317. [CrossRef]
9. Kendall, M.; Gibbons, J.D. *Rank Correlation Methods*, 5th ed.; Edward Arnold: London, UK, 1990; 260p.
10. Hülsmann, M.; Windt, K. *Understanding Autonomous Cooperation & Control—The Impact of Autonomy on Management, Information, Communication, and Material Flow*; Springer: Berlin/Heidelberg, Germany, 2007. [CrossRef]
11. Dablanc, L. *City Logistics*; International Encyclopedia of Geography: People, the Earth, Environment and Technology; Wiley Online Library: Hoboken, NJ, USA, 2019. [CrossRef]
12. Crainic, T.G. *City Logistics*; School of Management, Université du Québec à : Montrea, QC, Canada, 2008. [CrossRef]
13. Tadić, S. City Logistics Performance. Conference: LOGIC. In Proceedings of the 2nd Logistics International Conference, Belgrade, Serbia, 21–23 May 2015.
14. Mancini, S. *Multi-Echelon Freight Distribution Systems: A Smart and Innovative Tool for Increasing Logistic Operations Efficiency*; Alpen-Adria-Universität Klagenfurt: Klagenfurt, Austria, 2013.
15. Anand, N.; Van Duin Quak, H. *Development of City Logistics Maturity Model for Municipality Performance Measurement*; Breda University of Applied Sciences: Rotterdam, Belgium, 2021.
16. Carrillo, J.; Cortes, P.; Grosso, R.; Muñuzuri, J. City Logistics: Are Sustainability Policies Really Sustainable? *Dir. Y Organ.* **2014**, *53*, 45–50. [CrossRef]
17. Ploos van Amstel, W. *City Logistics*; Amsterdam University of Applied Sciences (HvA), Faculty of Technology Urban Technology Research Program: Amsterdam, The Netherlands, 2015.
18. Ploos van Amstel, W.; Quak, H. *Annual Outlook City Logistics 2050*; Amsterdam University of Applied Sciences, Centre of Applied Research Technology, TU Delft: Amsterdam, the Netherlands, 2017.
19. Özbekler, T.M.; Akgül, A.K. Last Mile Logistics in the Framework of Smart Cities: A Typology of City Logistics Schemes. *Int. Arch. Photogramm. Remote Sens. Spat. Inf. Sci.* **2020**, 335–337. [CrossRef]
20. Savin, G. *Smart City Logistics*; Ural State University of Economics: Ekaterinburg, Russia, 2020.
21. Crainic, T.G.; Gendreaua, M.; Jemai, L. Planning hyperconnected, urban logistics systems. In Proceedings of the 22nd EURO Working Group on Transportation Meeting, Barcelona, Spain, 18–20 September 2019; Available online: http://www.sciencedirect.com/science/article/pii/S2352146520302490 (accessed on 22 February 2021).
22. Andriankaja, D. The impacts of logistics sprawl: How does the location of parcel transport terminals affect the energy efficiency of goods' movements in Paris and what can we do about it? *Procedia-Soc. Behav. Sci.* **2010**, *2*, 6087–6096. [CrossRef]
23. Crainic, T.G.; Gonzalez-Feliu, J.; Mancini, S. Planning and Optimization Methods for Advanced Urban Logistics Systems at Tactical Level. In *Sustainable Urban Logistics: Concepts, Methods and Information Systems*; Gonzalez-Feliu, J., Semet, F., Routhier, J.L., Eds.; Springer: Berlin, Germany, 2014; pp. 145–164. [CrossRef]
24. Strulak-Wójcikiewicz, R.; Wagner, N. Exploring opportunities of using the sharing economy in sustainable urban freight transport. *Sustain. Cities Soc.* **2021**, *68*, 102778. [CrossRef]
25. Tadić, R.S. Differences between Typical Logistics Systems and Other City Logistics Generators. *Tehnika* **2021**, *76*, 485–489. [CrossRef]
26. Kardelis, K. *Mokslinių Tyrimų Metodologija ir Metodai*; Technologija: 2-asis pataisytas ir papildytas leidimas: Kaunas, Lithuania, 2002.
27. Apklausa.lt: Surveys and Questionnaires, Online. Available online: https://apklausa.lt (accessed on 12 February 2022).
28. Šlajus, Š.; Batarlienė, N. Investigation of the Possibilities of Optimization of Freight Road Traffic Flows in the Urban Logistics System. In Proceedings of the 13th International Conference TRANSBALTICA, Vilnius, Lithuania, 15–16 September 2022.
29. Taherdoost, H.; Madanchian, M. Multi-Criteria Decision Making (MCDM) Methods and Concepts. *Encyclopedia* **2023**, *3*, 77–87. [CrossRef]
30. Montgomery, D.C. *Statistical Quality Control: A Modern Introduction*, 7th Edition International Student Version ed; John Willey and Sons, Inc.: New York, NY, USA, 2012; 768p, ISSN 978-1-118-53137-2.
31. Sivilevičius, H.; Maskeliūnaitė, L. Multiple Criteria Evaluation and the Inverse Hierarchy Model for Justifying the Choice of Rail Transport Mode. *Promet-Traffic Transp.* **2018**, *30*, 57–69. [CrossRef]

applied sciences

MDPI

Article

Measurement Repeatability of Rail Wheel Loads Caused by Rolling Surface Damages

Gediminas Vaičiūnas *, Gintautas Bureika and Stasys Steišūnas

Faculty of Transport Engineering, Vilnius Gediminas Technical University (VILNIUS TECH), Plytinės Str. 25, 10105 Vilnius, Lithuania
* Correspondence: gediminas.vaiciunas@vilniustech.lt; Tel.: +37-0610-15376

Abstract: The repeatability of rail wheel damage measurements is considered in this study. The authors investigated the measurement repeatability of vertical force dependence on wheel-rolling surface damage nature and suggested ways to reach higher repeatability. To investigate wheel-rolling surface damage impact on vertical force measurement repeatability, the results of measuring forces with different measurement systems were compared first. Another critical issue was estimating the deviation field for the measured force values. The box and whisker principles were used. Different types (shapes) of rail wheel damages and rolling stock operating conditions were examined by field tests. The article presents the dependence determined by the authors, and how the repeatability of the wheel damage measurement depends on the speed of the rolling stock.

Keywords: rail transport; wheel-rolling surface; wheel vertical loads; wheel flat; measurement repeatability; box and whisker methods

Citation: Vaičiūnas, G.; Bureika, G.; Steišūnas, S. Measurement Repeatability of Rail Wheel Loads Caused by Rolling Surface Damages. *Appl. Sci.* **2023**, *13*, 4474. https://doi.org/10.3390/app13074474

Academic Editor: Diogo Ribeiro

Received: 16 February 2023
Revised: 9 March 2023
Accepted: 15 March 2023
Published: 31 March 2023

1. Introduction

Track failures or rail vehicle derailments represent approximately 50% of overall accidents in railway transport [1]. The majority of the track integrity problems are related to the track geometry or gauge distortion. Other factors that harmfully impact on properness of wheel–rail contact are badly repaired running gear, discrepancy of rails, and critical dynamical impacts (such as high tractive and braking forces of trains). Furthermore, some railway traffic accidents happen due to the diminishing of reliable wheel/rail contact and due to the unsatisfactory stability of rolling-stock running [2].

The parameters and shape of the rolling surface of the rail wheel mainly influence the technical condition of rolling stock running gear. The "wheel-rail" contact transfers high vertical static and dynamic loads to the vehicle and the track [3]. As the wheel rolls, high-impact loads are created by track irregularities and rail junctions [4]. During rolling-stock exploitation, the interaction between the wheel and the rail may cause damage to the rolling surface of the rail wheel [5]. They can also be caused by manufacturing inaccuracies [6] or poor machining [7]. It should be noted that the occurrence of damage is most affected by braking and track surface irregularities [8]. The damage of the wheel rolling surface is very destructive to the upper track structure. It can also cause wheel cracks and derailment of the wagons [9]. Polish scientists Burdzik et al. defined the dynamical properties of the driving train and evaluated the technical state of the rail track by using proprietary measurement software and a mobile device. Authors proposed useful metrics to describe the dynamical properties of the driving train [10]. Strikes of damaged wheels produce high noise and vibrations that impair passengers' comfort [11]. Vertical interactions of the wheel with the rail acquire an impact character when the wheel is damaged. These studies require unique methods since the processes under investigation have a higher frequency [12]. Polish scientists Celinski et al. defined the dynamical properties of the driving train and the evaluated the technical state of the rail track by using proprietary measurement software and a mobile device [13]. Furthermore, authors proposed useful metrics to describe the dynamical properties of the driving train. R. Burdzik et al. considered the impact of rail damage on vibration waveforms and vibroacoustic spectrum images of vibrations. The major source of railway vibration is the structural vibration of the wheel and rail, which is generated by the combination of small-scale undulations on the wheel and rail contact surfaces. The inevitable vibration transmitted by running rail vehicles through the rails and wheels causes and propagates various cracks in the metal of running gear elements. The authors revealed the amplification of the signal from the Ai3 sensor located near the damaged place where the frequency oscillated around 4000 Hz, while the spectral analysis of the signal for the Ai2 sensor located 5.4 m away to the damage indicated oscillating values in the range from 50 to 500 Hz [14]. Depending on the train's speed, the maximum duration of the force is in the order of several milliseconds. The ATLAS-LG subsystem of wayside equipment is used by Lithuanian Railways (LTG) to detect and prevent damage due to their impact. ATLAS-LG type systems are used by Lithuanian and Swiss Railways and are designed to avoid traffic accidents. These systems detect passing wagon running gear that causes vertical impact loads on the rail track, which can be caused by damage to the wheel rolling surface (e.g., flat, crack, uneven wear, and "out-of-roundness").

Worldwide rigorous research activities are targeting not only security. A high level of safety and reliability of infrastructure systems also aims to diminish the problems associated with the degradation of performance in terms of train-ride quality and stability, passenger comfort, etc. To ensure the railway transport safety, decision-makers of rail companies need, inevitably, to monitor the geometrical parameters of the track and the condition of the vehicle wheel rolling surfaces. Wheel damage (flats, crumbling, cavities, cracks, wear, out-of-roundness, etc.) endangers traffic safety, i. e., the risk of derailment. The intensive damaging of the rails and track as a whole track geometry also increases.

Mathematical models of the impact of wheel flat damage on the rail usually developed using the reverse Laplace transformation methodology. These models were later developed by researchers from Ukraine and other countries [15]. It is noted in the literature that it is not entirely clear how short-term loads affect the systemic wear of the wheel surface. It is preliminarily known that the wear processes are influenced by such factors as surface irregularities and roughness, material hardness, and elasticity [11]. Due to the large number of factors, processes are modeled by simplifying them [16]. As an example, one can give the fact that elastic and plastic deformations and the formation of metal cracks as a result of them are not always taken into account when modelling processes [17]. In order to create a more reliable wheel wear prediction tree, it is necessary to evaluate not only one or two factors, but many more, perhaps a dozen or a few tens [18].

Both when examining wheel damage and when examining the causes of traffic accidents, it can be seen that wheel damage frequently causes traffic accidents (on the railway, traffic accidents are considered not only accidents with drastic consequences, but also minor incidents when train traffic is stopped for a short time) [19]. Flats and cracks are the most common damage of the rolling surface of the wheel to rail [20]. As rolling stock speeds increase, increasing dynamic forces affect the development of damages [21]. According to the impact on traffic safety, it is accepted to preliminarily divide wheel-rolling surface damage into two types [22]:

1. Damages that cause a short-term loss of wheel–rail adhesion (cracks, flats, bends);
2. Damages that do not cause loss of wheel–rail adhesion (e.g., uneven wear).

Some of the wheel damage may go unnoticed during the external inspection due to the human factor, so it is necessary to improve the automatic damage detection systems [23]. Different diagnostic systems associate the magnitude of the wheel surface damage differently with the impact force (the data processing algorithms used for this purpose may differ).

This study aimed to determine whether the repeatability of force measurement is more determined by the nature of the wheel damage or the driving speed, and to determine whether it is possible to detect the laws of the dependence of the repeatability of this measurement on the cause.

2. Analysis of Rolling-Stock Wheel-Rolling Surface Damage

During rolling stock inspections at railway stations, wheel damage is detected by templates and other devices (Figure 1). An external inspection identifies visible damages, such as wheel flats or cracks, and templates are used to inspect the wheel profile. Ultrasonic and magnetic defectoscopes are used to detect cracks in the wheel.

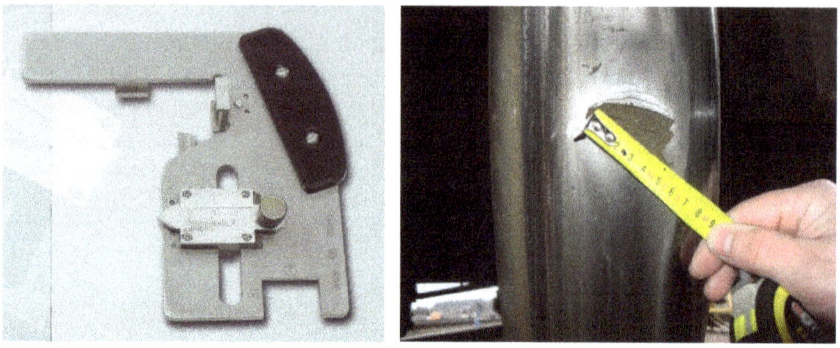

Figure 1. Wheel measurement template and measurement with a ruler.

The main damages of the rail wheels are the wear of the rolling surface, thinning of the flange, flats, cracks, and metal folds of the surface. The main types (shapes) of wheel-rolling surface damages are shown in Figure 2.

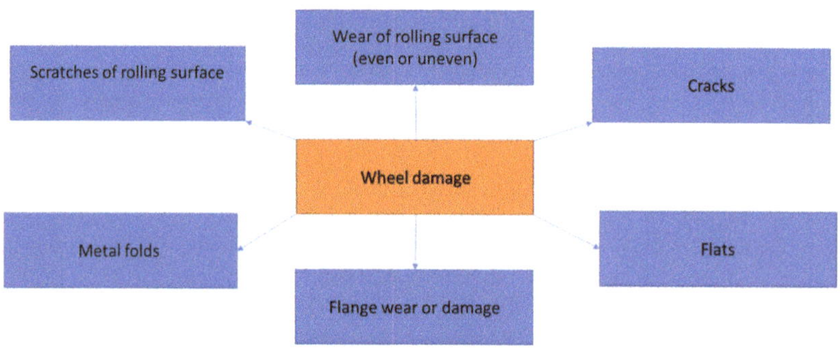

Figure 2. Main nature of damage of wheel rolling surface.

Flange damage. The wheel flange prevents the wheel from derailing. A wheel is considered damaged when its flange is critically thinned (up to 25 mm).

Metal folds. Metal folds occur in the presence of thermomechanical damage [24]. Intense plastic deformation of the metal occurs due to sudden braking, short-term wheel slip, wheel jumping, or sudden heating of the wheel metal, and then sudden cooling [25]. Several metal folds can be formed on the rolling surface of the wheel, as well as being in one or several layers. When the wheel of a passenger car has a metal fold up to 0.5 mm, or a freight wagon up to 1 mm, operation of the wagon is prohibited.

Rolling surface wear. Most publications on the long-term interaction between rolling stock and track deal with the wear of rail and wheel [26–28]. The phenomenon of wheel wear has been extensively studied [29]. With the wear of the rolling surface of the wheel, train resistance to movement and the wheel/rail adhesion increase [30]. The wear of the rolling surface of the wheel is divided into even and uneven.

Uneven wear. This wear is special in that it does not have a dimensional extremity (such as the deepest point), so sometimes this type of wear is difficult to detect by the values of the dynamic diagnostic indicators [31]. The consequence of such wear is isolated wheel irregularities [32].

Flats. These are the most common wheel damage caused by wheel slip or a brake pad stuck [33], as shown in Figure 3.

Figure 3. Sample of wagon wheel damage—flat.

Wheel flats also occur from wheel slippage, wheel jamming, or braking equipment failures. Flats occur much more often in winter than in summer. The main reasons for the formation of the wheel flat are shown in Figure 4.

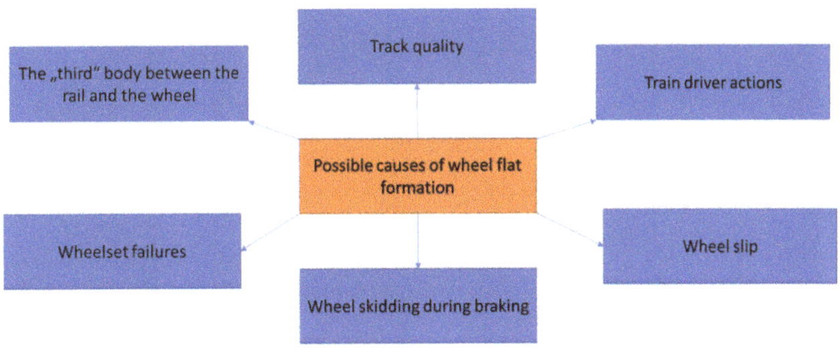

Figure 4. Main reasons of wheel flat formation.

Dynamic measurement methods include the vibration acceleration method, machine vision method, and YOLO deep learning network methods for wheel-rolling surface damage discovery. The operation of automatic crack detection system implemented based on YOLOv2 target detection, which improves the detection accuracy, is analyzed in the study [34]. Chinese researchers developed [35] a YOLOv3-based railway track line detection method that improved detection accuracy and speed. Furthermore, it preferred to incorporate lasers with images and, therefore, resulted in an structured light vision sensor (SLVS) sensor for assessing localized faults on wheel, such as a crack or wheel flat.

Various diagnostic systems can determine the magnitude of the damage based on the impact force. There are a lot of methods in the scientific literature to determine the correlation between these values [12] and the principles by which these systems operate [36]. On the other hand, rolling-stock wheelsets oscillate when rolling on a rail track [37] (see Figure 5), so the wheel–rail contact points are not always on the same circle of the wheel rolling surface [38].

Different detection methods have their advantages and disadvantages. Some methods have low detection accuracy or high actual operating costs. Therefore, the search for a more precise method has become a relevant issue.

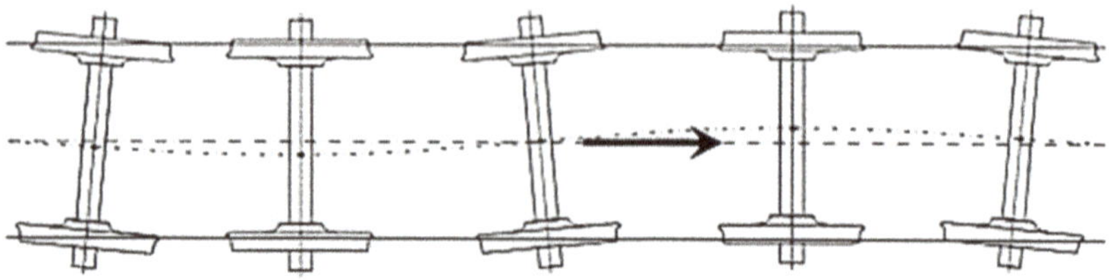

Figure 5. Wheelset oscillation of on the track.

Due to oscillation processes, the damaged wheel impacts on the rail at different surface points at each time, and the impact force varies. In operation, it is impossible to determine which of the measured values of the impact force is the most accurate (actual). It needs to be clarified on which basis the diagnostic equipment should estimate the magnitude of the damage.

3. Research Methodology

At first, the study compares the results of measuring vertical forces by different systems. In the same graph, the force values obtained from the ATLAS LG system are plotted on one axis and the special designed measurement system IC VEIP on the other. The view of both used measurement systems is presented in Figure 6.

Figure 6. Used measurement systems: (**a**) regular system ATLAS-LG; (**b**) specially designed system IC VEIP.

The measuring range consists of 14 sleeper reaction measuring points (R-points) and 12 axle load measuring points (T-points). The last T-point is used for identifying wheelsets. Vertical impact on rail and static axle load of wheels is measured by strain gauges mounted on the rail neck. The main particularities are: ATLAS-LG system measures directly the vertical force, while the IC-VEIP system measures the rail acceleration and then converts it into a force value.

The best result is when the readings of both systems coincide. This result would indicate that both systems are equally suitable for studies on the influence of wagon wheelset damage on the nature of vertical forces. If all values are in a straight line and the line is formed at a different angle with the axes, it would mean that there is a systematic error in at least one of the systems. If the values are not in a straight line (scattered in the plane), then the measurement results are not related to different systems, and at least one system is unsuitable for the test. Another important issue is estimating the scattering field for the measured force values. The box and whisker principle is used in the study [39].

The box and whisker graph is convenient in that it shows where the median of the values is (see Figure 7). It also shows where the 50% of values closest to the median (upper quartile and lower quartile) are distributed, the range of the remaining values (upper extreme and lower extreme). In addition, this type of graph can show individual values that stand out from the whole (outlier). A median value is used, taking into account the specific algorithm of the box and whisker method. This type of graph is very suitable for evaluating the repeatability of experimental results. Understandably, the better the repeatability of the results, the smaller the scatter fields of the values. An example of the application of the box and whisker principle is shown in Figure 7.

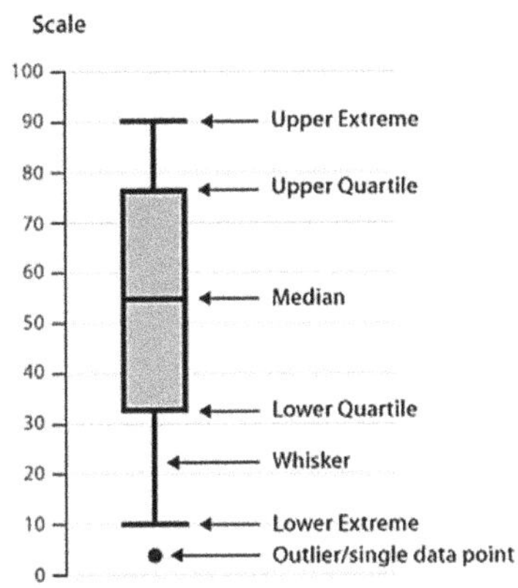

Figure 7. Example of application of box and whisker principle.

In the middle of the diagram in Figure 7, there are two middle quartiles. They are the limits between which half of all measurement values are distributed (twice 25%—two quartiles). In the middle of these quartiles, there is a median–mean value. The remaining values are outside these limits, and the authors did not analyze these values.

The ratio of the total width of the two middle quartiles to the median in estimating the wagon wheel damage magnitudes is calculated as:

$$Q_M = \frac{Q_U - Q_L}{M} \rightarrow min \tag{1}$$

where Q_u—upper quartile; Q_L—lower quartile; M—median (Figure 6).

For simplicity, this indicator name is sometimes abbreviated to indicator Q_M. This methodology estimates the ratio of the wheel-to-rail scattering field to the median Q_M. Such an indicator is important in proving that the size of the scatter field of impact values is less significant at high force values. Therefore, the lower value of the indicator Q_M shows better repeatability of the measured vertical force. The value of Q_M is calculated under different operating conditions of the rolling stock (in this case—different speeds). Comparing the Q_M values in different cases determined the states that are best suited to relate the magnitude of the damage to its vertical impact on the rail.

4. Measurement of Vertical Forces Caused by Wheel Damages

The operation in summer conditions of the standard ATLAS-LG and the special designed IC VEIP measurement systems was compared. The measurement results of the standard system, ATLAS-LG, were compared too. When the wheels of a rolling vehicle act on the rails, each system captures the vertical force's basic (mean) values and the maximal values. Graphical interpretations of test results are provided in Figures 8 and 9.

Since the curves of Figure 8 show the base forces (mean forces around the entire circumference of the wheel), their values are lower than the maximal forces shown in Figure 9. An analysis of the data in Figure 8 shows that the points corresponding to the mean force values are arranged along the hemisphere of the coordinate axes. That means that the measurement results of both systems are very similar, and the systems are equally suitable for research. Figure 8 shows a slightly wider scatter of the maximal load values while the trend remains the same. The field of values is arranged in a hemisphere of coordinates. The scatter of the values needs to be assessed using the appropriate methods presented below. Finally, the measurement results of the considered systems are equivalent, so the type of wayside measurement system does not have a significant impact on the repeatability of the measurement of vertical forces.

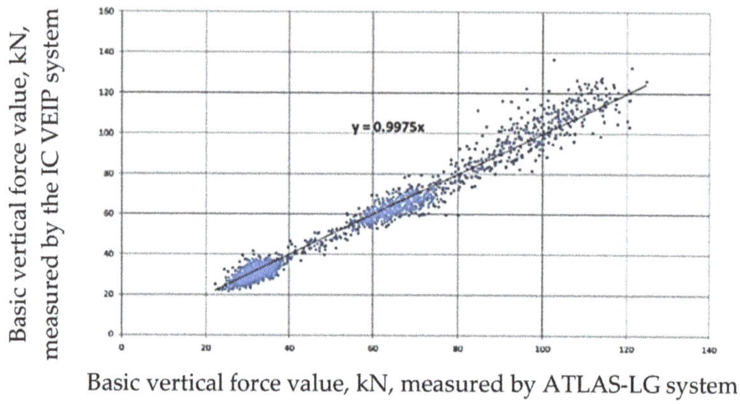

Figure 8. Correlation of measured values of basic vertical forces according to measurement systems.

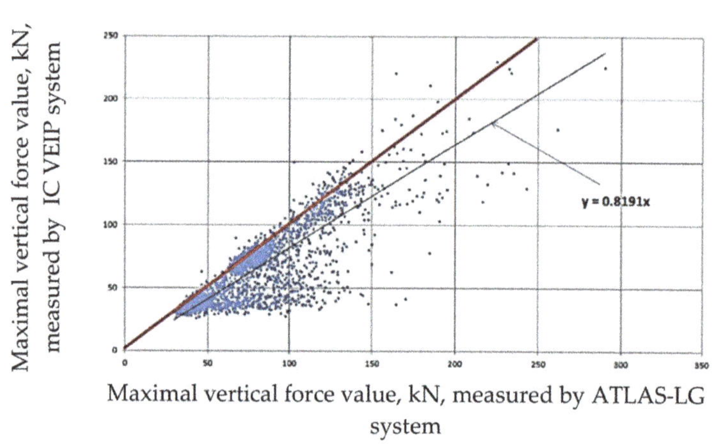

Figure 9. Correlation of measured values of maximal vertical forces according to measurement systems.

5. Distribution of Values of Wheel Vertical Forces

The authors propose to use the ratio of the total width of the two middle quartiles of the scattering field of the vertical force values to the median (indicator Q_M, Formula (1)) to examine the impact of the wheel-rolling surface damage on the measurement repeatability of the vertical forces. Two wheel-rolling surface damage types were selected for this study. For each damage type (shape), the vertical forces caused by the wheel damage were measured in four dimensions. Four data sets were obtained: mean values of vertical forces of unloaded wagon and of loaded; values of maximal vertical forces of unloaded wagon and of loaded wagon.

First, the test was performed with the damaged left wheel of the first wheelset. The wheel has two $20 \times 10 \times 4$ mm and $15 \times 40 \times 3$ mm cracks. Wheel damages provided in Figure 10.

Box and whisker values of the maximal vertical forces of loaded goods wagon are shown Figure 11.

Figure 10. Two cracks in the wagon's first wheelset wheel.

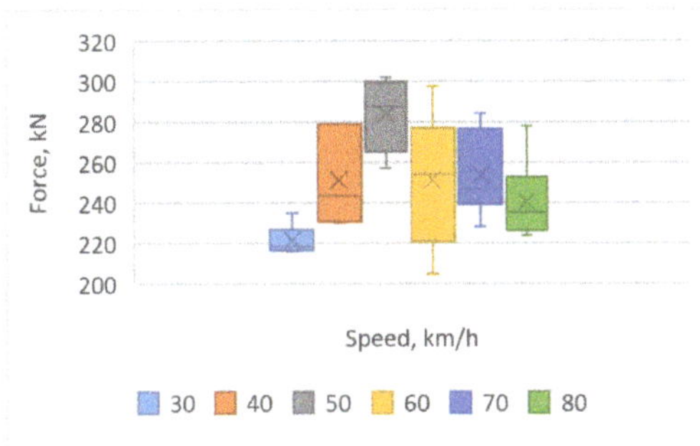

Figure 11. Maximal force values of loaded wagon.

The values of the indicator Q_M, calculated by the Formula (1) of maximal vertical forces of the loaded wagon are provided in Figure 12.

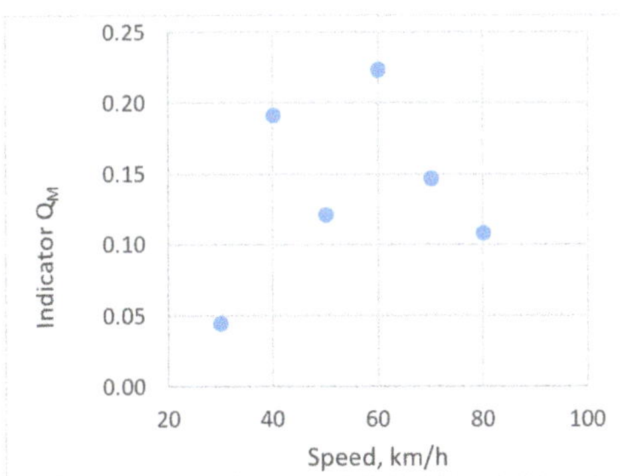

Figure 12. Values of Q_M, considering the maximal forces of loaded wagon.

The data in Figure 12 reveal that the influence of wagon wheel damage on the measurement of repeatability vertical forces is the highest at 30 km/h—lowest Q_M value, and the highest value is at a speed of 60 km/h. When wagon running speed increases, this impact on measurement repeatability decreases. It is inappropriate to perform tests at a speed of fewer than 30 km/h because, in this case, the amplitude of the carriage's oscillations may be lower than when driving at higher speeds, which would distort the adequacy of the results (equipment may not capture all damage). Since the tests are carried out on freight wagons, it is not appropriate to conduct them at speeds higher than 80 km/h. More test results were analyzed to better examine the dependences of wheel damage nature on repeatability. The values of the maximal vertical forces are given in the form of a box and whisker graph in Figure 13 when the goods wagon is unloaded.

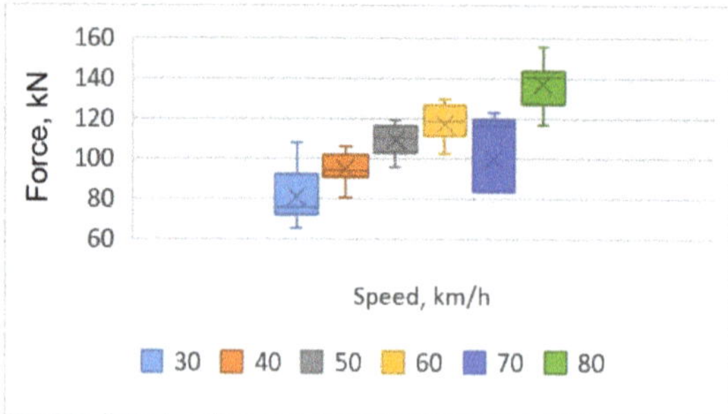

Figure 13. Values of maximal forces of unloaded goods wagon.

The chart of Figure 13 shows that the maximal scatter of the force values is at a speed level of 70 km/h. The values of Q_M, calculated according to Formula (1) for the maximal forces when the goods wagon is unloaded, are given in Figure 14.

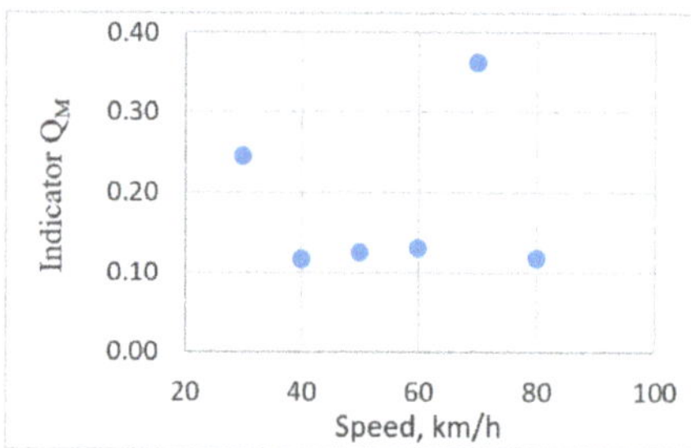

Figure 14. Values of Q_M, considering the maximal forces when unloaded goods wagon.

The data in Figure 14 reveal that the impact of wheel damage on the repeatability of measurement of vertical forces is the best (the lowest values of the indicator Q_M) at a speed of 40–60 km/h. More test results were analyzed to examine these dependencies better. The values of the mean vertical forces of loaded goods wagon are given in Figure 15.

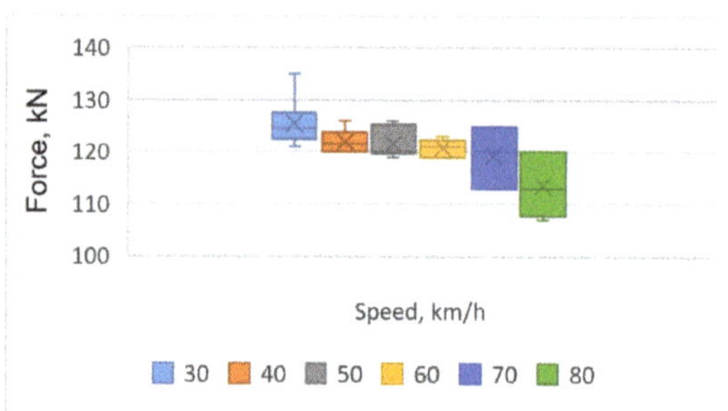

Figure 15. Mean values of vertical force when the wagon is loaded.

The chart of Figure 15 clearly disclose that as the wagon speed increases, the dispersion of the mean vertical forces increases. The values of Q_M with the loaded wagon are shown in Figure 16.

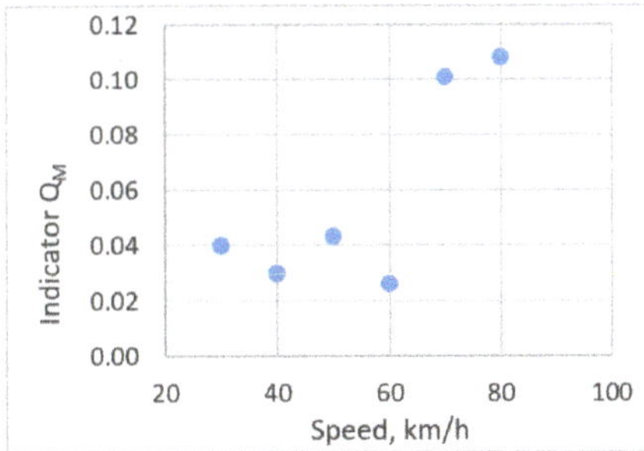

Figure 16. Values of the indicator Q_M, according to the mean vertical forces of loaded wagon.

The data in Figure 16 disclose that the impact of wheel damage on measurement repeatability of vertical forces is best at speeds of 40 and 60 km/h. No clear law is seen here. More test results were analyzed to examine the correlations with measurement repeatability better. The values of the mean forces of unloaded wagon are given in Figure 17.

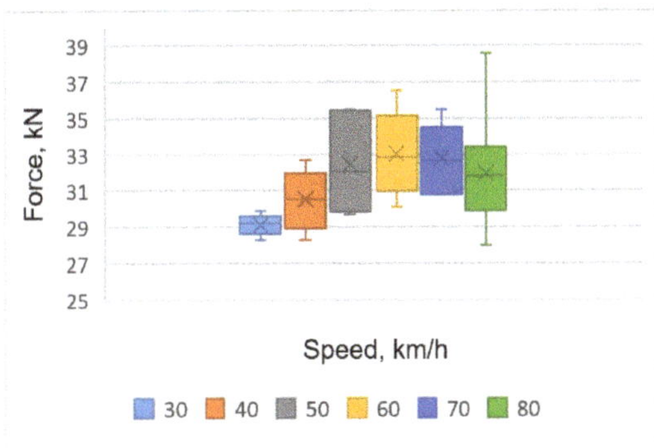

Figure 17. Vertical force mean values of unloaded wagon.

The chart of Figure 17 shows that the force dissipation is lowest at a speed of 30 km/h. The values of Q_M when the wagon is unloaded are given in Figure 18.

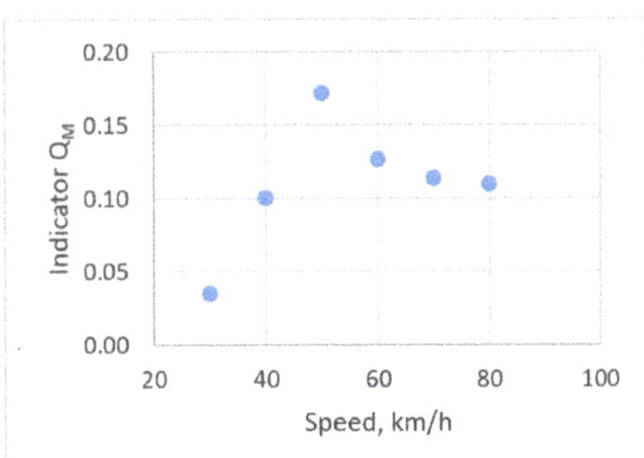

Figure 18. Values of Q_M considering to the mean vertical forces of unloaded goods wagon.

The data of Figure 18 show that the impact of wheel damage nature on the measurement repeatability of the vertical forces is best at a speed of 30 km/h.

For the second iteration of wheel damage testing, the damaged fourth wheel set's (in running direction) left wheel of the first wagon was selected: 87 × 30 × 2.0 mm flat and 30 × 40 × 2 mm cracks. Photos of wheel damage are provided in Figure 19.

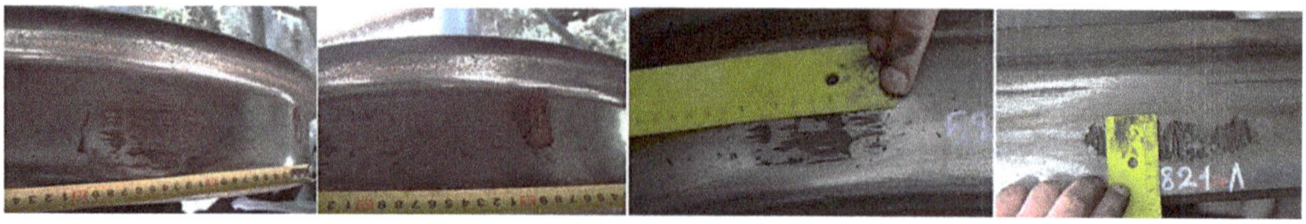

Figure 19. Wheel damage—flat and cracks.

Examples of the distribution of vertical forces caused by wheel damage (flat and cracks) in rail cross-sections are provided in the literature [12].

Vertical force maximal values of loaded wagon are given in Figure 20.

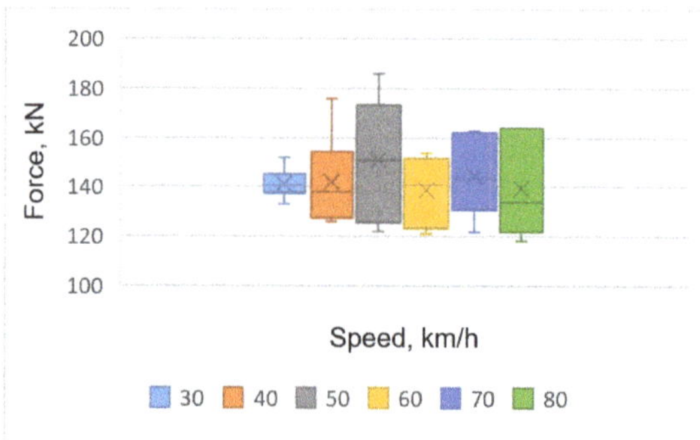

Figure 20. Maximal force values of loaded wagon (the second iteration).

The chart of Figure 20 shows that the scatter of the maximal vertical force values are lowest at the speed 30 km/h. The values of Q_M for the maximal vertical forces of loaded wagon are shown in Figure 20.

The data of Figure 21 reveal that the wagon wheel damage impact on measurement repeatability of vertical forces is best at a speed of 30 km/h. The values of the maximal forces of unloaded wagon are given in Figure 22.

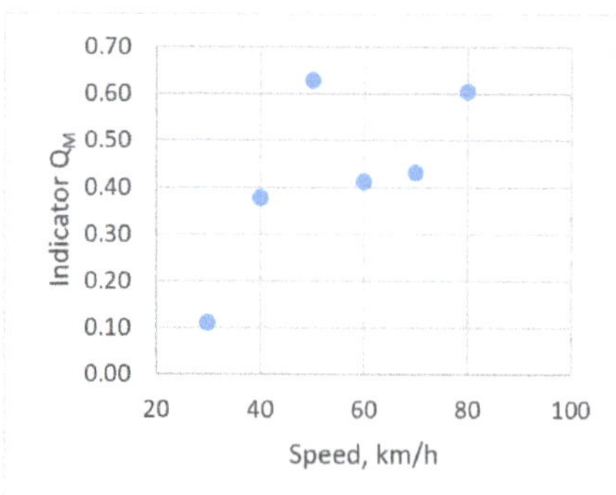

Figure 21. Values Q_M considering the loaded wagon maximal vertical forces.

The data of Figure 23 show that the scattering field of the maximal vertical forces is the lowest at speed of 40 km/h and 60 km/h. The values of Q_M for the maximal vertical forces are given in Figure 22.

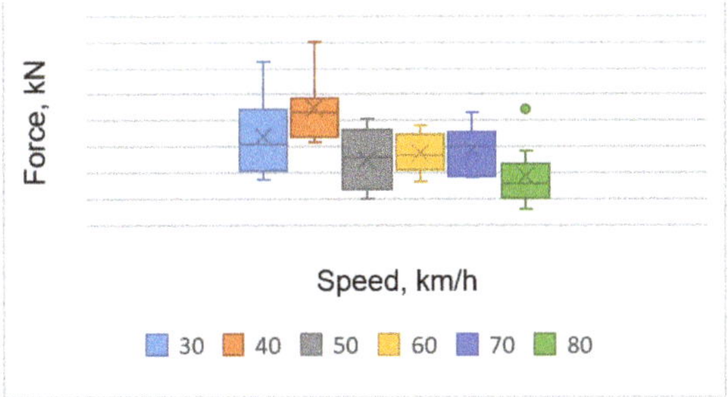

Figure 22. Maximal force values of unloaded goods wagon.

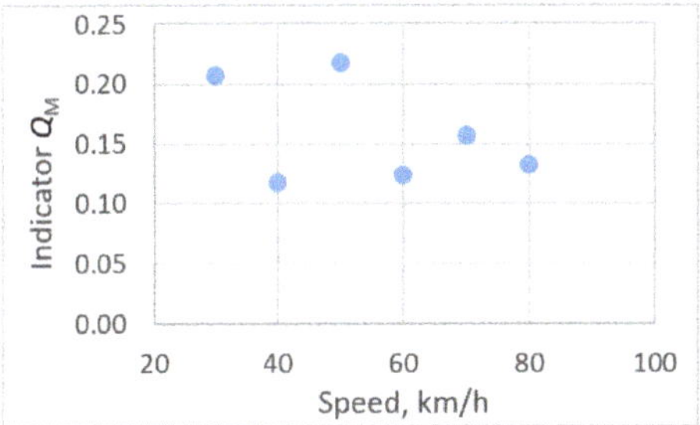

Figure 23. Values of Q_M considering to the maximal vertical forces of unloaded goods wagon.

The data of Figure 23 show that the wheel damage impact on measurement repeatability of vertical forces is best at speed of 40 km/h and 60 km/h. However, no clear pattern is seen here. In the case of the goods wagon being loaded, the vertical force mean values are given in Figure 24.

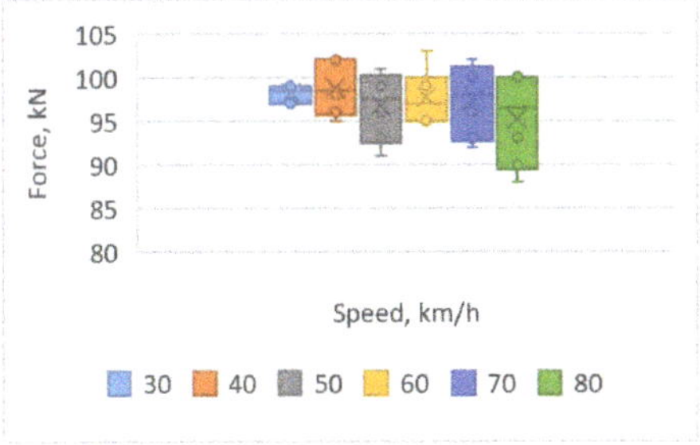

Figure 24. Vertical force mean values of loaded goods wagon.

The chart of Figure 24 shows that the dissipation of the mean forces is lowest when the wagon is at speed 30 km/h. The indicator Q_M values of loaded wagon are presented in Figure 25.

The data of Figure 25 show that the impact of wheel damage on measurement repeatability of vertical forces is best at a speed of 30 km/h. The mean values of the vertical forces of unloaded wagon are given in Figure 26.

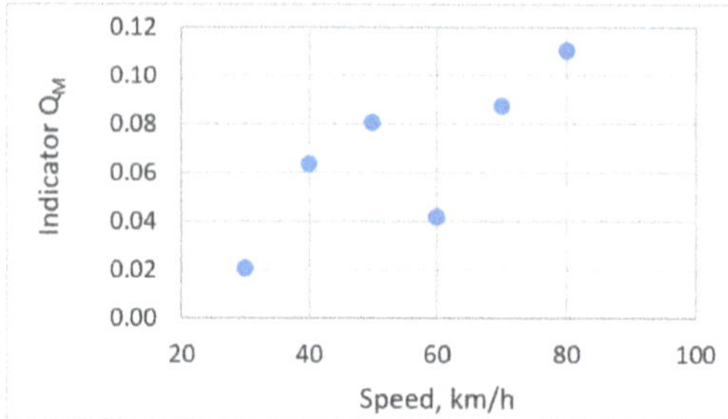

Figure 25. Values of Q_M considering the mean vertical forces of loaded goods wagon.

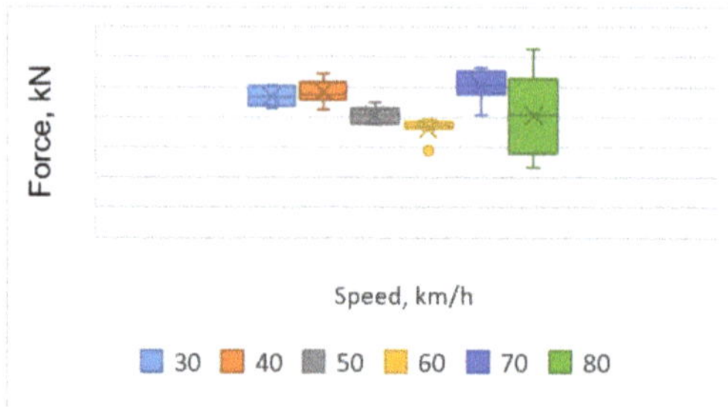

Figure 26. Vertical force mean values of unloaded goods wagon.

The chart of Figure 26 shows that the force dissipation field is the most significant at the speed of 80 km/h. The values of the indicator Q_M unloaded wagon are given in Figure 27.

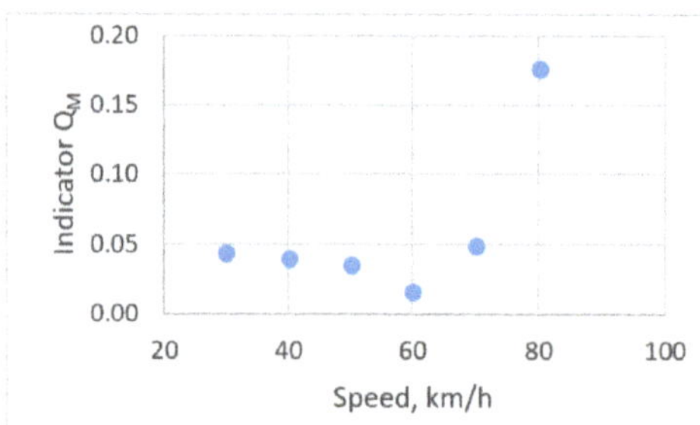

Figure 27. Values of Q_M considering the unloaded wagon mean vertical forces.

The data of Figure 27 show that the impact of wheel damage on measurement repeatability of vertical forces is best at a speed from 30 km/h to 60 km/h.

Summarized Q_M values are compiled based on the performed research. The final gained results are presented in Table 1.

The arithmetic mean of the values is calculated for each row of Table 1 (running speed from 30 km/h to 80 km/h). Normalized values of this mean for appropriate speed are shown in Figure 28.

The indicator Q_M is needed to assess the wheel damage influence on the repeatability of vertical force measurement values (Formula (1)). A lower Q_M value means better measurement repeatability. Summarizing the results, the Q_M values are presented (Figure 28) in normalized form. It can be seen that decrease in running speed from 80 km/h to 50–60 km/h does not improve the correlation between the wheel damage and the vertical forces. To improve this correlation, reducing the speed to 30–40 km/h is needed.

Table 1. Summarized Q_M values.

Speed, km/h	The First Iteration of Force Testing				The Second Iteration of Force Testing			
	Max Forces		Mean Forces		Max Forces		Mean Forces	
	Loaded Wagon	Unloaded Wagon	Loaded Wagon	Unloaded Wagon	Loaded Wagon	Unloaded Wagon	Loaded Wagon	Unloaded Wagon
30	0.045	0.244	0.040	0.034	0.110	0.207	0.021	0.043
40	0.192	0.166	0.030	0.100	0.376	0.117	0.063	0.040
50	0.122	0.124	0.043	0.172	0.629	0.217	0.080	0.035
60	0.224	00.13	0.026	0.126	0.411	0.124	0.041	0.016
70	0.147	0.362	0.101	0.114	0.432	0.156	0.087	0.048
80	0.109	0.118	0.108	0.110	0.606	0.132	0.110	0.176

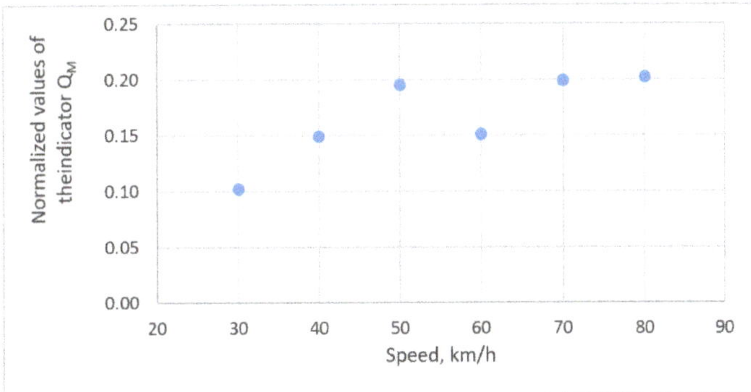

Figure 28. Normalized Q_M values according to the running speed.

6. Conclusions

Rolling stock wheelsets oscillate on the rail track, so the wheel–rail contact points are never on the same circle of the wheel rolling (contacting) surface. As a result of these processes, the wheel damage contacts the rail at different surface points at each moment, and the vertical impact force changes. During rolling-stock operation, it is impossible to determine which of the measured values of the impact force is the most accurate. The scattering field of the values is obtained during the measurements, and a methodology is required to process it. The investigation reveals that the box and whisker method is well-suited for analysing the scattering field of force measurement results.

The ratio of the total width of the two middle quartiles and the median is proven as the suitable indicator for assessing the effect of the car wheel clearance on the repeatability of vertical force measurements.

The authors reveal that of all the various factors that affect the repeatability of the measurement of wheel vertical force, speed is the only factor that has an obvious correlation. For other factors (e.g., type of measuring system, nature of wheel damage), no clear correlation is observed during the study.

The other disclosed aspect is that decreasing the running speed from 80 km/h to 50–60 km/h does not strengthen the correlation between the wheel damage nature and the repeatability of vertical force measurement. Greater reducing levels of train speed were applied to improve the quality of wagon wheel damage nature impact identification. Reduction in the train speed from 80 km/h to 30–40 km/h reduces the value of the indicator Q_M proposed by the authors by 1.5–2 times.

When operating the rolling stock and knowing the conditions under which the values of Q_M are the lowest, the repeatability of measurement of wheel/rail vertical forces is the highest.

In the future, the authors plan to study the repeatability of measurement results of systems operating on other principles. Furthermore, a limitation of this research is the incompatibility of the measuring systems ATLAS-LG and IC VEIP with other diagnostic systems used in railways. This devise incompatibility causes the low reproducibility of the results of the present study.

Author Contributions: Methodology, conceptualization and writing, G.V.; supervision, G.B.; investigation and formal analysis, S.S. All authors have read and agreed to the published version of the manuscript.

Funding: This research received no external funding.

Institutional Review Board Statement: Not applicable.

Informed Consent Statement: Not applicable.

Data Availability Statement: For commercial reasons, more detailed research data is not disclosed.

Conflicts of Interest: The authors declare no conflict of interest.

References

1. Bureika, G.; Vaičiūnas, G.; Shi, D.; Zanuy, A.C. Influence of Track Geometry Condition Monitoring on Railway Infrastructure Maintenance Processing. *Transp. Probl.* **2022**, *17*, 211–220. [CrossRef]
2. Weston, P.F.; Ling, C.S.; Roberts, C.; Goodman, C.J.; Li, P.; Goodall, R.M. Monitoring Vertical Track Irregularity from In-Service Railway Vehicles. *Proc. Inst. Mech. Eng. Part F J. Rail Rapid Transit* **2016**, *221*, 75–88. [CrossRef]
3. Ramalho, A. Wear Modelling in Rail–Wheel Contact. *Wear* **2015**, *330–331*, 524–532. [CrossRef]
4. Meymand, S.Z.; Keylin, A.; Ahmadian, M. A Survey of Wheel–Rail Contact Models for Rail Vehicles. *Veh. Syst. Dyn.* **2016**, *54*, 386–428. [CrossRef]
5. Aalami, M.R.; Anari, A.; Shafighfard, T.; Talatahari, S. A Robust Finite Element Analysis of the Rail-Wheel Rolling Contact. *Adv. Mech. Eng.* **2013**, *5*, 272350. [CrossRef]
6. Thakkar, N.A.; Steel, J.A.; Reuben, R.L. Rail–Wheel Interaction Monitoring Using Acoustic Emission: A Laboratory Study of Normal Rolling Signals with Natural Rail Defects. *Mech. Syst. Signal Process.* **2010**, *24*, 256–266. [CrossRef]
7. Roy, T.; Lai, Q.; Abrahams, R.; Mutton, P.; Paradowska, A.; Soodi, M.; Yan, W. Effect of Deposition Material and Heat Treatment on Wear and Rolling Contact Fatigue of Laser Cladded Rails. *Wear* **2018**, *412–413*, 69–81. [CrossRef]
8. Datsyshyn, O.P.; Panasyuk, V.V.; Glazov, A.Y. Modeling of Fatigue Contact Damages Formation in Rolling Bodies and Assessment of Their Lifetime. *Wear* **2011**, *271*, 186–194. [CrossRef]
9. Seo, J.; Kwon, S.; Lee, D. Effects of Surface Defects on Rolling Contact Fatigue of Rail. *Procedia Eng.* **2011**, *10*, 1274–1278. [CrossRef]
10. Burdzik, R.; Konieczny, Ł.; Deuszkiewicz, P.; Vaskova, I. Application of Time-Frequency Method for Research on Influence of Locomotive Wheel Slip on Vibration. *J. Vibroeng.* **2018**, *20*, 2998–3008. [CrossRef]
11. Dižo, J.; Blatnický, M.; Gerlici, J.; Leitner, B.; Melnik, R.; Semenov, S.; Mikhailov, E.; Kostrzewski, M. Evaluation of Ride Comfort in a Railway Passenger Car Depending on a Change of Suspension Parameters. *Sensors* **2021**, *21*, 8138. [CrossRef] [PubMed]
12. Bureika, G.; Levinzon, M.; Dailydka, S.; Steisunas, S.; Zygiene, R. Evaluation Criteria of Wheel/Rail Interaction Measurement Results by Trackside Control Equipment. *Int. J. Heavy Veh. Syst.* **2019**, *26*, 747–764. [CrossRef]
13. Celiński, I.; Burdzik, R.; Młyńczak, J.; Kłaczyński, M. Research on the Applicability of Vibration Signals for Real-Time Train and Track Condition Monitoring. *Sensors* **2022**, *22*, 2368. [CrossRef] [PubMed]
14. Burdzik, R.; Słowiński, P.; Juzek, M.; Nowak, B.; Rozmus, J. Dependence of Damage to the Running Surface of the Railway Rail on the Vibroacoustic Signal of a Passing Passenger Train. *Vibroeng. Procedia* **2018**, *19*, 226–229. [CrossRef]
15. Wallentin, M.; Bjarnehed, H.L.; Lundén, R. Cracks around Railway Wheel Flats Exposed to Rolling Contact Loads and Residual Stresses. *Wear* **2005**, *258*, 1319–1329. [CrossRef]
16. Popp, K.; Kruse, H.; Kaiser, I. Vehicle-Track Dynamics in the Mid-Frequency Range. *Veh. Syst. Dyn.* **1999**, *31*, 423–464. [CrossRef]
17. Steinfeld, A.; Fong, T.; Kaber, D.; Lewis, M.; Scholtz, J.; Schultz, A.; Goodrich, M. Common Metrics for Human-Robot Interaction. In Proceedings of the HRI 2006: Proceedings of the 2006 ACM Conference on Human-Robot Interaction, Salt Lake City, UT, USA, 2–3 March 2006; pp. 33–40. [CrossRef]
18. Ma, L.; Shi, L.B.; Guo, J.; Liu, Q.Y.; Wang, W.J. On the Wear and Damage Characteristics of Rail Material under Low Temperature Environment Condition. *Wear* **2018**, *394–395*, 149–158. [CrossRef]
19. Fang, Z.; Lou, L.; Tang, K.; Wang, W.; Chen, B.; Wang, Y.; Zheng, Y. A CMOS-Integrated Radar-Assisted Cognitive Sensing Platform for Seamless Human-Robot Interactions. In Proceedings of the IEEE International Symposium on Circuits and Systems, Daegu, Republic of Korea, 22–28 May 2021. [CrossRef]
20. Wang, W.J.; Liu, T.F.; Wang, H.Y.; Liu, Q.Y.; Zhu, M.H.; Jin, X.S. Influence of Friction Modifiers on Improving Adhesion and Surface Damage of Wheel/Rail under Low Adhesion Conditions. *Tribol. Int.* **2014**, *75*, 16–23. [CrossRef]

21. Pieringer, A.; Kropp, W.; Nielsen, J.C.O. The Influence of Contact Modelling on Simulated Wheel/Rail Interaction Due to Wheel Flats. *Wear* **2014**, *314*, 273–281. [CrossRef]
22. Wang, W.J.; Guo, H.M.; Du, X.; Guo, J.; Liu, Q.Y.; Zhu, M.H. Investigation on the Damage Mechanism and Prevention of Heavy-Haul Railway Rail. *Eng. Fail. Anal.* **2013**, *35*, 206–218. [CrossRef]
23. Konowrocki, R.; Chojnacki, A. Analysis of Rail Vehicles' Operational Reliability in the Aspect of Safety against Derailment Based on Various Methods of Determining the Assessment Criterion. *Eksploat. I Niezawodn.* **2020**, *22*, 73–85. [CrossRef]
24. Huang, Y.B.; Shi, L.B.; Zhao, X.J.; Cai, Z.B.; Liu, Q.Y.; Wang, W.J. On the Formation and Damage Mechanism of Rolling Contact Fatigue Surface Cracks of Wheel/Rail under the Dry Condition. *Wear* **2018**, *400–401*, 62–73. [CrossRef]
25. Regularities of Shaping of a Wheel Profile as a Result of Deterioration of the Rolling Surface in Exploitation-Transport Problems-Tom T. 3, z. 4 (2008)-BazTech-Yadda. Available online: https://yadda.icm.edu.pl/baztech/element/bwmeta1.element.baztech-article-BSL7-0032-0017 (accessed on 9 March 2022).
26. Shevtsov, I.Y.; Markine, V.L.; Esveld, C. Design of Railway Wheel Profile Taking into Account Rolling Contact Fatigue and Wear. *Wear* **2008**, *265*, 1273–1282. [CrossRef]
27. Jin, X.; Xiao, X.; Wen, Z.; Guo, J.; Zhu, M. An Investigation into the Effect of Train Curving on Wear and Contact Stresses of Wheel and Rail. *Tribol. Int.* **2009**, *42*, 475–490. [CrossRef]
28. Polach, O. Wheel Profile Design for Target Conicity and Wide Tread Wear Spreading. *Wear* **2011**, *271*, 195–202. [CrossRef]
29. Vaičiūnas, G.; Bureika, G.; Steišūnas, S. Research on Metal Fatigue of Rail Vehicle Wheel Considering the Wear Intensity of Rolling Surface. *Eksploat. I Niezawodn.* **2018**, *20*, 24–29. [CrossRef]
30. Turnia, J.; Sinclair, J.; Perez, J. A Review of Wheel Wear and Rolling Contact Fatigue. *J. Rail Rapid Transit* **2007**, *221*, 271–289. [CrossRef]
31. Bodini, I.; Petrogalli, C.; Faccoli, M.; Lancini, M.; Pasinetti, S.; Sansoni, G.; Docchio, F.; Mazzù, A. Evaluation of Wear in Rolling Contact Tests by Means of 2D Image Analysis. *Wear* **2018**, *400–401*, 156–168. [CrossRef]
32. Tamada, R.; Shiraishi, M. Prediction of Uneven Tire Wear Using Wear Progress Simulation. *Tire Sci. Technol.* **2017**, *45*, 87–100. [CrossRef]
33. Favorskaya, A.; Khokhlov, N. Modeling the Impact of Wheelsets with Flat Spots on a Railway Track. *Procedia Comput. Sci.* **2018**, *126*, 1100–1109. [CrossRef]
34. Chen, Y.; Li, Y.; Niu, G.; Zuo, M. Offline and Online Measurement of the Geometries of Train Wheelsets: A Review. *IEEE Trans. Instrum. Meas.* **2022**, *71*, 3523915. [CrossRef]
35. Wei, X.; Wei, D.; Suo, D.; Jia, L.; Li, Y. Multi-Target Defect Identification for Railway Track Line Based on Image Processing and Improved YOLOv3 Model. *IEEE Access* **2020**, *8*, 61973–61988. [CrossRef]
36. Vaičiūnas, G.; Bureika, G.; Steišūnas, S. Rail Vehicle Axle-Box Bearing Damage Detection Considering the Intensity of Heating Alteration. *Eksploat. I Niezawodn.–Maint. Reliab.* **2020**, *22*, 724–729. [CrossRef]
37. Qu, S.; Wang, J.; Zhang, D.; Li, D.; Wei, L. Failure Analysis on Bogie Frame with Fatigue Cracks Caused by Hunting Instability. *Eng. Fail. Anal.* **2021**, *128*, 105584. [CrossRef]
38. Matsumoto, A.; Sato, Y.; Ohno, H.; Tomeoka, M.; Matsumoto, K.; Ogino, T.; Tanimoto, M.; Oka, Y.; Okano, M. Improvement of Bogie Curving Performance by Using Friction Modifier to Rail/Wheel Interface: Verification by Full-Scale Rolling Stand Test. *Wear* **2005**, *258*, 1201–1208. [CrossRef]
39. Larsen, R.D. Box-and-Whisker Plots. *J. Chem. Educ.* **1985**, *62*, 302–305. [CrossRef]

applied sciences

MDPI

Article

Utilizing Different Machine Learning Techniques to Examine Speeding Violations

Ahmad H. Alomari [1,*], Bara' W. Al-Mistarehi [2], Tasneem K. Alnaasan [2] and Motasem S. Obeidat [3]

1 Department of Civil Engineering, Yarmouk University (YU), P.O. Box 566, Irbid 21163, Jordan
2 Department of Civil Engineering, Jordan University of Science & Technology (JUST), P.O. Box 3030, Irbid 22110, Jordan; bwmistarehi@just.edu.jo (B.W.A.-M.)
3 Department of Computer Science, Yarmouk University (YU), P.O. Box 566, Irbid 21163, Jordan
* Correspondence: alomarish@yu.edu.jo; Tel.: +962-2-7211111 (ext. 4639)

Abstract: This study investigated the potential impacts on speeding violations in the United States, including the top ten states in terms of crashes: California, Florida, Georgia, Illinois, Michigan, North Carolina, Ohio, Pennsylvania, Tennessee, and Texas. Several variables connected to the driver, surroundings, vehicle, road, and weather were investigated. Three different machine learning algorithms—Random Forest (RF), Classification and Regression Tree (CART), and Multi-Layer Perceptron (MLP)—were applied to predict speeding violations. Accuracy, F-measure, Kappa statistic, Root Mean Squared Error (RMSE), Area Under Curve (AUC), and Receiver Operating Characteristic (ROC) were used to evaluate the algorithms' performance. Findings showed that age, accident year, road alignment, weather, accident time, and speed limits are the most significant variables. The algorithms used showed excellent ability in analyzing and predicting speeding violations. The RF was the best method for analyzing and predicting speeding violations. Understanding how these factors affect speeding violations helps decision-makers devise ways to cut down on these violations and make the roads safer.

Keywords: speeding violations; machine learning; Classification and Regression Tree; Random Forest; Multi-Layer Perceptron

Citation: Alomari, A.H.; Al-Mistarehi, B.W.; Alnaasan, T.K.; Obeidat, M.S. Utilizing Different Machine Learning Techniques to Examine Speeding Violations. *Appl. Sci.* **2023**, *13*, 5113. https://doi.org/10.3390/app13085113

Academic Editors: Edgar Sokolovskij and Vidas Žuraulis

Received: 22 March 2023
Revised: 13 April 2023
Accepted: 14 April 2023
Published: 20 April 2023

1. Introduction

Traffic crashes are considered a significant cause of death in the United States (U.S.) [1]. These crashes occur for different reasons, one of them being speeding violations. Speeding can be defined as driving a vehicle at a speed greater than permissible according to the speed limit or prevailing road conditions. Speeding affects the drivers, pedestrians, environment, and vehicles. According to the National Highway Traffic Safety Administration (NHTSA), crashes that result from speeding cost more than $40 billion every year [2]. In 2019, 9478 people were killed by speeding in the U.S. In addition to that, driving a vehicle at high-speed increases air pollution because of increased fuel consumption [3].

There are several negative consequences of speeding. Increasing the speed raises the probability of an accident and its severity. Driving a vehicle at high speed means more time for danger identification and response, or more distance traveled before responding to the threat and stopping the vehicle, which increases the probability of an accident [4,5]. In addition, increasing the speed raises the probability of death because of the high crash energy as the speed increases (the relation between the crash energy and the vehicle's speed is exponential) [6]. Pedestrian death risk redoubled 4.5 times for speed change from 50 to 65 km/h. On the other hand, occupant fatality risk at 65 km/h is 85% for the side-impact accident [7]. In Canada, wrong drivers' behavior leads directly to 65% of crashes and indirectly to 90% of them [8]. The likelihood of an accident and its severity increase with increasing vehicle speed. A 50 km/h or less speed limit should be set in urban areas [9]. In Jordan, it was found that the difference between design speed and speed limit would increase the speed variance and its severity impact when the difference was more than 10 km/h [10,11].

Speeding can be caused by several unusual factors, such as traffic jams, drivers' familiarity with roads, and the drivers having a luxury car, which, in turn, encourages speeding [12]. Several studies worldwide examined speeding violations. In Edmonton, Canada, Gargoum et al. (2016) [13] looked at drivers' compliance with speed limits, considering weather conditions, the shape of the road, and the time of day. Data were analyzed using logistic regression. It was found that traffic volume, the posted speed limit, and the number of lanes positively affected compliance (i.e., a higher speed limit and increasing the number of lanes increased speed compliance). Kanellaidis et al. (1995) [14] investigated the attitudes of Greek drivers toward speed limit violations. Many reasons were assessed to find the most influential ones. The most critical factors were underestimating the risk of speeding, being in a hurry, and overestimating how good you are at driving. In Queensland, Australia, Afghari et al. (2018) [15] studied drivers' speeding behaviors based on traffic characteristics and roadway geometric factors. A panel mixed logit fractional split model was developed for Queensland's speeding data. The results revealed that the number of speeding violations increased as the radius of the horizontal curve increases. On the other hand, speeding violations decreased with increasing speed limits. Huang et al. (2018) [16] studied the effects of different factors on speeding violations by applying a generalized linear model in a comparative study between Shanghai, China and New York City, United States. The speeding rate was utilized to study the violations. Situational and characteristic factors were examined to find their effects on the speeding rate. The results indicated that 16 working hours or less and a driving distance of more than 400 km were the factors that affect speeding violations. In addition to that, it was revealed that speeding behavior had a higher probability in urban areas and in the morning hours (5:00–7:00 AM).

In Asia, Tseng (2013) [17] conducted a study in Taiwan to examine the relationship between speeding violations and the social-economic characteristics of the drivers. A dataset of 8129 drivers was analyzed using logistic regression analysis. It was concluded that the age group (20–29) tends to violate speed. An increasing income level was one of the factors associated with increasing speeding violations. Furthermore, driving for various purposes impacted increasing speeding violations, especially driving for business and leisure. In China, Zhang et al. (2014) [18] studied speeding violations between 2006 and 2010 by examining different risk factors. The independent factors were analyzed using stepwise logistic regression to specify the influencing factors. The analysis illustrated that a high risk of speeding violations was associated with male drivers, the dark, morning time, and accident years. On the other hand, public holidays, days of the week, weather conditions, age, and season did not affect speeding violations. Another study in China conducted by Liang and Xiao (2020) [19] studied speeding behavior using Planning Behavior Theory (PBT). An ANOVA analysis was employed to examine the effect of the demographic and descriptive characteristics on speeding behavior. It was revealed that male drivers, the age group (22–44), and high-income level were the factors that increased the violations. In Kuwait, Al Matawaha et al. (2020) [20] studied the factors that impact drivers' speeding behavior. A *t*-test and an ANOVA analysis were used to analyze data from 536 drivers. Speed Related Score (SRS) values were used to specify the influencing factors. The outcomes showed that male, single, Kuwaiti drivers, and the age group (18–24) were most likely to violate speed limits. In India, Balasubramanian and Sivasankaran (2021) [21] studied speeding violations using logistic regression. The results revealed that male drivers and young ones (less than 25 years old) had a greater speeding violation rate. The environment and road variables revealed that fine weather, daylight, central divider, darkness with no street lighting, light motor vehicles, single-lane roads, and uncontrolled junctions affected speeding behavior. On the other hand, the day of the week and season did not affect speeding violations. In Jordan, Al-Mistarihi et al. (2022) [22] investigated how speeding violations may affect the driver's environment, the vehicle, the road, and the weather. Results showed that speeding violations depended on age, type of vehicle, speed limit, day of the week, season, year of the accident, time of the accident, license category, and condition of lights.

In many areas of engineering branches, learning algorithms, neural networks, and modern adaptive fuzzy systems are used a lot for various tasks and applications [23,24]. Analyzing traffic-related datasets has recently gained interest in machine learning techniques [22–28]. Researchers are interested in these methods because they can deal with vast amounts of data and find correlations between variables that would be hard to find using standard statistical modeling techniques.

Several research studies investigated the effect of different factors on speeding violations, such as age [17–20,22], gender [18–21], weather conditions [13,18,21,22], light conditions [13,16,18,21,22], and area type [16,21,22]. None of the earlier research looked at how the direction of the traffic flow, the condition of the road surface, the grade of the road, or the presence of work zones affected speeding violations. In addition to that, most of these studies used traditional methods of data analysis, such as logistic regression, panel mixed logit, generalized linear model, and analysis of variance (ANOVA). This research aims to study the effects of several variables on speeding violations in the United States using a variety of machine learning techniques, such as Random Forest (RF), Classification and Regression Tree (CART), and Multi-Layer Perceptron (MLP). These factors include age, gender, weather conditions, work zones, traffic way directions, light conditions, road grades, and road surface conditions. Studying all the mentioned factors and comparing them with the previously studied ones contributes to a better understanding of the causes of speeding violations. Understanding the impact of these factors on speeding violations helps decision-makers find suitable solutions that mitigate these violations and improve road safety.

The remainder of this paper is organized as follows: Section 2 introduces the methodology used for modeling speeding violations in the U.S. using machine learning techniques, including CART, RF, and MLP. It also describes the study area and the data used in this work. Section 3 presents the modeling results by comparing the techniques utilized, including CART, RF, and MLP. Finally, Section 4 presents the significant findings of this work.

2. Materials and Methods

The Fatality Analysis Reporting System (FARS) was used to collect information for this study for the years 2015–2019. The National Highway Traffic Safety Administration is responsible for maintaining this database system [29]. FARS is a United States census that provides the National Highway Traffic Safety Administration, Congress, and the American public with annual data on fatalities and injuries in motor vehicle traffic crashes. Specialists collect these data points from police reports, administrative files at the state level, and medical records. In addition, automated error checks and data monitoring ensure that data stay within acceptable ranges [30]. A total of 51,136 records were analyzed using the Waikato Environment for Knowledge Analysis (WEKA) 3.8.4 software [31]. Different variables related to the vehicle, driver, roadway, and environment have been investigated. Table 1 shows these factors and their categories.

As shown in Figure 1, this study selected the top ten states in terms of crashes. These states are California, Florida, Georgia, Illinois, Michigan, North Carolina, Ohio, Pennsylvania, Tennessee, and Texas.

CART, RF, and MLP are different machine learning algorithms used in this work. Classification and Regression Trees (CART) is a machine learning algorithm that explains how the target variable's values can be determined based on other data. It is a decision tree where each fork splits into a predictor variable, and each node ends with a prediction for the target variable. Random Forest (RF) is an ensemble learning-approach for classification, regression, and other tasks. It constructs many decision trees at training time and outputs the class, the mode of the classes (classification), or the mean estimation (regression) of the individual trees. The Multilayer Perceptron (MLP) is a feedforward ANN that generates outputs by passing input data via multiple layers of neurons. Next, what distinguishes the use of these algorithms will be explained in detail.

Table 1. Variables Description for the Studied Data.

Variable	Category	Count	Percent	Variable	Category	Count	Percent
Speeding Violation	Yes No	8175 42,961	16.0 84.0	**Gender**	Female Male	14,275 36,861	27.9 72.1
Traffic Way Direction	One way Two way -	771 50,365 -	1.5 98.5 -	**Road Grade**	Level Ascending Descending	45,181 3214 2741	88.4 6.3 5.4
Holiday	Yes No	1446 49,690	2.80 97.20	**Area Type**	Rural Urban	20,820 30,316	40.7 59.3
Day of The Week	Weekend Weekday	16,539 34,597	32.30 67.70	**Work Zone**	Yes No	1600 49,536	3.1 96.9
Road Alignment	Straight Curve	42,491 8645	83.1 16.9	**Traffic Control**	Yes No	8518 42,618	16.7 83.3
Accident Year	2015 2016 2017 2018 2019	12,716 12,043 7112 10,658 8607	24.9 23.6 13.9 20.8 16.8	**Road Surface Condition**	Dry Wet Snow Mud slush	44,387 6223 366 98 62	86.8 12.2 0.7 0.2 0.1
License Type	Full Intermediate Learner Temporary	50,003 759 333 41	97.8 1.5 0.7 0.1	**Season**	Spring Summer Autumn Winter	12,984 13,090 13,335 11,727	25.4 25.6 26.1 22.9
Accident Time	0:00–6:59 7:00–8:59 9:00–11:59 12:00–16:59 17:00–19:59 20:00–23:59	11,275 3350 5060 12,470 8870 10,111	22.0 6.6 9.9 24.4 17.3 19.8	**Age**	≤25 26–35 36–45 46–55 56–65 ≥66	11,004 10,503 8127 8068 6570 6864	21.5 20.5 15.9 15.8 12.8 13.4
Light Condition	Daylight Dusk Dawn Dark not lighted Dark lighted	25,703 1361 986 13,590 9496	50.3 2.7 1.9 26.6 18.6	**Weather**	Clear Rain Cloudy Fog Snow Other	38,219 3617 8144 661 334 161	74.7 7.1 15.9 1.3 0.7 0.3
Number of Lanes	One Two Three Four Five Six Seven or more - - -	27,828 9333 6836 4511 1611 881 136 - - -	54.4 18.3 13.4 8.8 3.2 1.7 0.3 - - -	**Speed Limit (km/h)**	≤30 30–40 40–50 50–60 60–70 70–80 80–90 90–100 100–110 >110	287 645 3884 5468 3667 9520 14,665 2485 4664 5851	0.6 1.3 7.6 10.7 7.2 18.6 28.7 4.9 9.1 11.4
Vehicle model year	Before 1991 1991–2000 2000–2010 2010–2020	1116 8941 23,878 17,201	2.2 17.5 46.7 33.6				

CART is one of the machine learning techniques, consisting of nodes and branches similar to a tree. It starts with the root (parent) node, which contains all the data; this node starts splitting based on a selected variable according to splitting criteria or impurity function [32]. The impurity function is a splitting criterion used to specify the suitable variable that gives the best subset at the splitting point. Entropy, Gini index, and minimum error are examples of impurity functions aiming to select the split that gives the most significant difference between the parent node's impurity and the impurity of the weighted average of the child nodes [33]. In order to produce a tree of an appropriate size and improve model predictions, a pruning process is conducted after the training phase to remove any unimportant branches [32]. The widely used impurity function is the Gini index, which is calculated using Equation (1), where $p(k \mid t)$ is the dependent variable probability equal to k at node t, and (n) is the classes number. Figure 2 shows the CART structure.

$$\text{Gini index} = 1 - \sum_{k=1}^{n-1} [p(k|t)^2] \quad (1)$$

Figure 1. Top Ten States in Terms of Crashes.

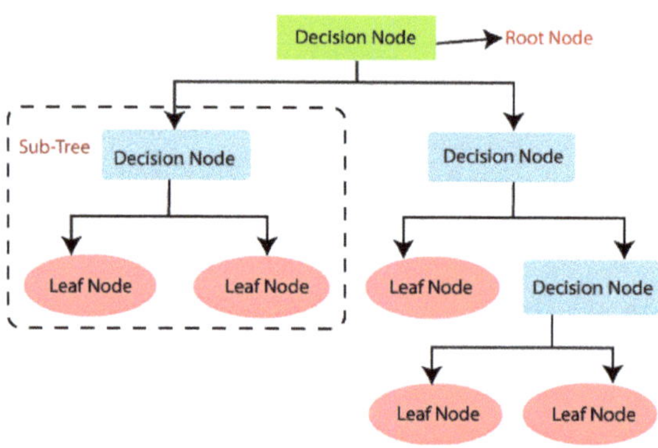

Figure 2. CART structure [34].

RF is a regression and classification machine learning algorithm. It depends on ensemble learning, where many classifiers are combined to solve a complex problem. The RF algorithm contains many decision trees and utilizes a bootstrap aggregation or bagging technique to train the forest and improve its accuracy. The bagging technique randomly resamples the original data to create a training dataset with replacement [35]. Figure 3 shows the RF structure. The final result for RF is the average predictions for regression problems, while classification is the primary vote for individual tree predictions.

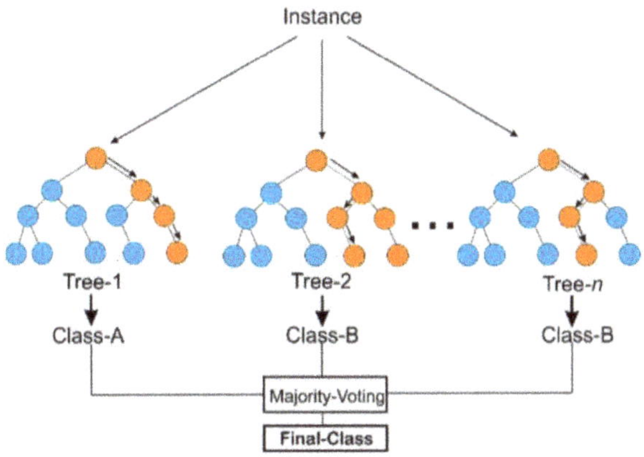

Figure 3. RF Structure [36].

The dataset resulting from the bagging technique is divided into two parts. The first part contains about 2/3 of the dataset and is used for training the individual decision trees. The remaining (1/3 dataset) is called "out-of-bag" and is used for performance evaluation. The decision trees classify out-of-bag elements. The ratio between the misclassified samples and the out-of-bag elements represents the generalization error. The final result for RF is the average of the predictions for regression problems, while the final result for classification problems is the major vote for the predictions.

An MLP is a kind of neural network consisting of different layers (i.e., input, hidden, and output layers), and each layer contains perceptrons or nodes. The first layer is the input layer for the input variables, and the final layer is the output layer for the final result. The hidden layer is located between the input and output layers and contains the activation function. Figure 4 shows the structure of the MLP algorithm.

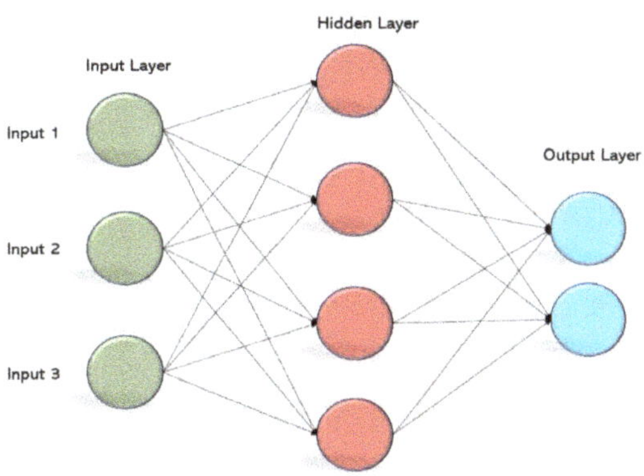

Figure 4. MLP structure [37].

Each neuron accepts the previous layer's output and produces the next layer's output through the activation function. The activation or transfer function finds the node output based on the input from the previous layer [38]. MLP uses the back-propagation technique to improve model performance. It estimates the error for the predicted and correct values; this error is fed back several times to adjust the connections' weights and produce a small error [39].

In this work, the studied dataset is imbalanced; the number of categories for the independent variable is not equal, producing a poor model. To overcome this problem, the Synthetic Minority Oversampling Technique (SMOTE) was used [40]. SMOTE is a technique used to generate new minority class records by interpolating the minority classes. It depends on selecting a random sample from the minority classes and specifying the K nearest neighbors to that sample. One of these K nearest neighbors is selected randomly to create a synthetic sample at a random point between the first selected random sample and the selected K nearest neighbor. In addition, classifiers' performance was examined by using different evaluation metrics. These metrics are Accuracy, Root Mean Squared Error (RMSE), F-measure, Kappa statistic, Receiver Operating Characteristic (ROC) curve, and Area Under the Curve (AUC).

Accuracy is the ratio between the correct classified instances and all instances. In data science, accuracy refers to data free of errors and can be relied upon as a data source [41]. It is the first and most crucial standard/component of the data quality architecture. It is the degree to which a calculated or measured value is close to the actual value [42]. In machine learning, accuracy is a metric used for estimating classification models. It is the fraction of forecasts the model obtained right.

$$\text{Accuracy} = (\text{TP} + \text{TN})/(\text{TP} + \text{TN} + \text{FP} + \text{FN}) \tag{2}$$

Accuracy can also be calculated in terms of positives and negatives, as seen below. Where:

- TP: True-positive classified instances.
- TN: True-negative classified instances.
- FP: False-positive classified instances.
- FN: False-negative classified instances.

TP, FP, TN, and FN form what is called a confusion matrix. It is a matrix used for classifier performance evaluation. It contains the actual and predicted values. Figure 5 shows the layout of the confusion matrix.

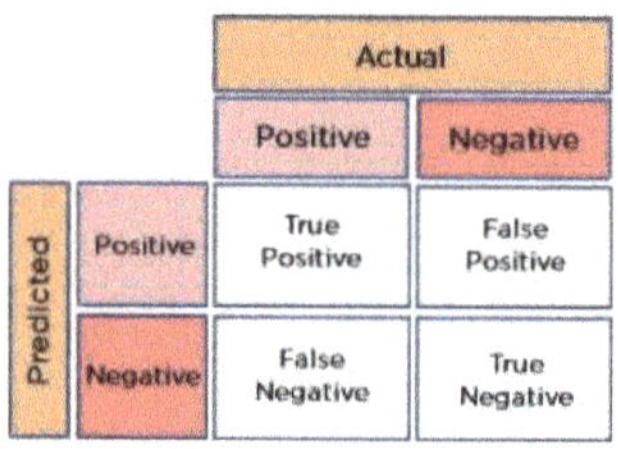

Figure 5. Confusion Matrix [43].

- True positive is the number of times the predicted value is positive, and it is truly positive [43].
- False positive is the number of times the predicted value is positive, but it is truly negative [43].
- True negative is the number of times the predicted value is negative, and it is truly negative [43].
- False negative is the number of times the predicted value is negative, but it is truly positive [43].

RMSE measures the difference between the predicted and observed values.

$$\text{RMSE} = \sqrt{\frac{\sum_{i=1}^{N}(\hat{y}_i - y_i)^2}{N}} \tag{3}$$

where:

- RMSE: Root Mean Squared Error.
- $\hat{y}_i$: ith observation predicted value.
- y_i: ith observation actual value.
- N: Observations number.

The F-measure, also called the F1-score, is used to measure model accuracy on a dataset. It is used to evaluate classification systems that classify examples as 'positive' or 'negative'. The F-score integrates the precision and recall of the model. Equation (4) below for the standard F1-score is the harmonic mean for the precision and recall of the model [44].

$$\mathrm{F-measure} = 2 * \frac{precision * recall}{precision + recall} \tag{4}$$

Precision and recall are two ways to measure how well a classifier works in binary and multiclass classification problems. Precision is calculated by dividing the number of true positives by the sum of true positives and false positives. Recall, in comparison, is the ratio of instances that were correctly classified (True Positives) to the total number of instances that should have been classified as Positive (True Positives + False Negatives) [42]. Where: Precision = TP/(TP + FP), and Recall = TP/(TP + FN).

Kappa statistic is a measure of how close the classified instances are to the true ones.

$$\mathrm{Kappa} = 1 - \frac{1 - p_o}{1 - p_e} \tag{5}$$

where:

- p_o: Observed instances.
- p_e: Expected instances.

ROC curve is a binary classification performance measure. It is a probability curve that plots the True Positive Rate (TPR) against the False Positive Rate (FPR) at various threshold values. The area under the ROC curve indicates a test's overall classification ability. The AUC measures a classifier's ability to distinguish between classes and is used to summarize the Receiver Operating Characteristic (ROC) curve. The greater the AUC, the more accurate the model, discriminating between positive and negative classes.

3. Results

Evaluating machine learning models and algorithms is crucial to any research or project. The quality of a statistical or machine-learning model is measured using evaluation metrics. To examine the classifiers' performance and specify the best one, the evaluation metrics were utilized to compare the classifiers' performance. Table 2 shows the four-evaluation metrics: accuracy, RMSE, Kappa statistic, and F-measure for the three classifiers.

Table 2. Evaluation Metrics.

Classifier	Accuracy%	RMSE	Kappa Statistic	F-Measure
RF	86.42	0.3245	0.728	0.864
CART	81.16	0.3972	0.623	0.811
MLP	73.02	0.43	0.46	0.73

These values for RF, CART, and MLP are 86.42%, 81.16% and 73.02%, respectively. The RF achieved the highest accuracy value, while the lowest value was for the MLP algorithm. Figure 6 shows the accuracy values.

The Kappa statistic measures how close the classified instances are to the true ones. An F-measure is used to measure the model's accuracy. Figure 7 shows the Kappa statistic, F-measure, and RMSE results. The Kappa statistic values for RF, CART, and MLP are 0.728, 0.623, and 0.46, respectively. The F-measure values for RF, CART, and MLP are 0.864, 0.811, and 0.73, respectively. Kappa statistic values for RF and CART are between 0.61 and0.8; substantial agreements, MLP Kappa statistic value is within the range (0.41–0.6); moderate agreement. F-measure values are high and indicate excellent accuracy.

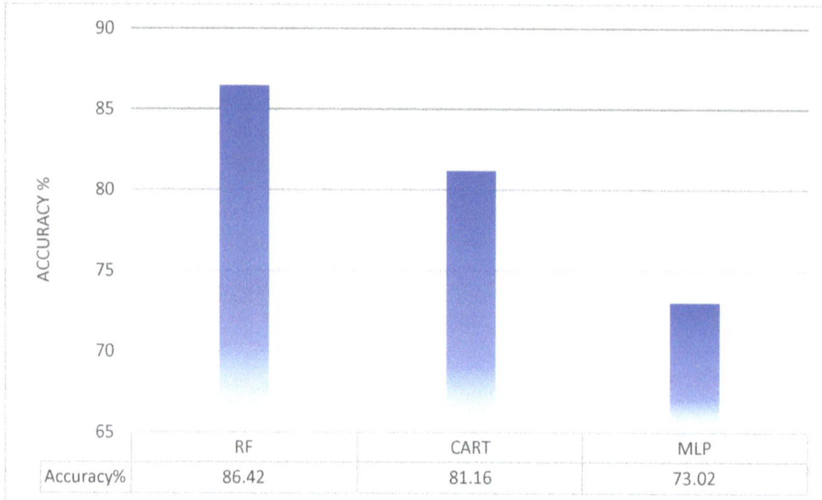

Figure 6. Classifiers' Accuracies.

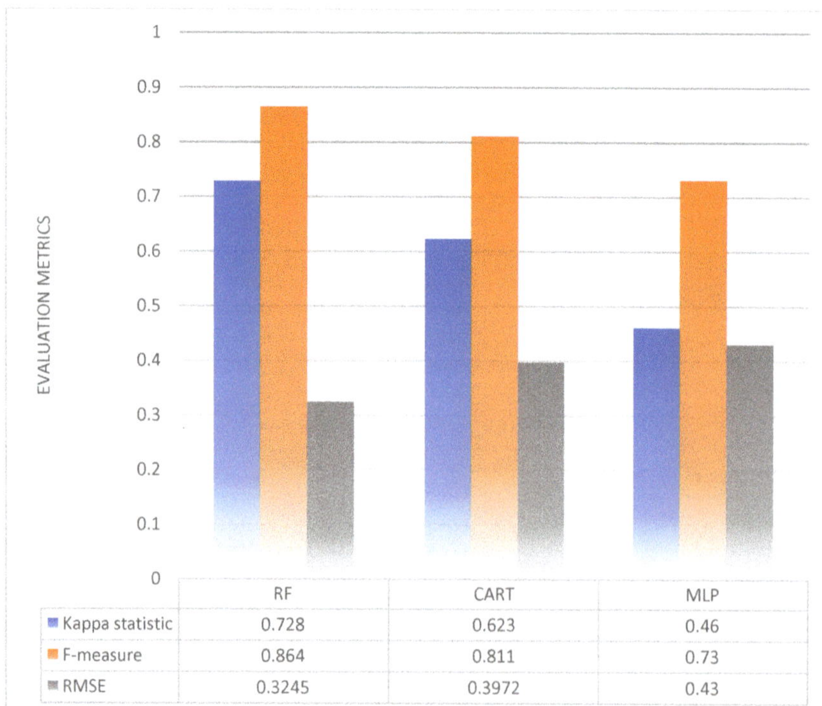

Figure 7. Kappa, F-measure, and RMES Results.

RMSE showed the error in predicting the true values. The RMSE values for RF, CART, and MLP are 0.3245, 0.3972, and 0.43, respectively. RF achieved the lowest value, while MLP achieved the highest value. AUC indicated the ability of the model to distinguish between the classes. ROC showed the relationship between the true-positive rates (*Y*-axis) and false-positive rates (*X*-axis). Table 3 presents AUC and ROC values; the results show that all the AUC values for all classifiers are more than 0.5; the classifiers have a good ability to distinguish classes. RF has the highest values, which indicate excellent performance. Figure 8 shows the ROC curves for the classifiers. The approach of the curve from the top-left corner indicates the good performance of the algorithm. It is obvious that the RF curve is the closest one to the top-left corner (i.e., it gives the best performance).

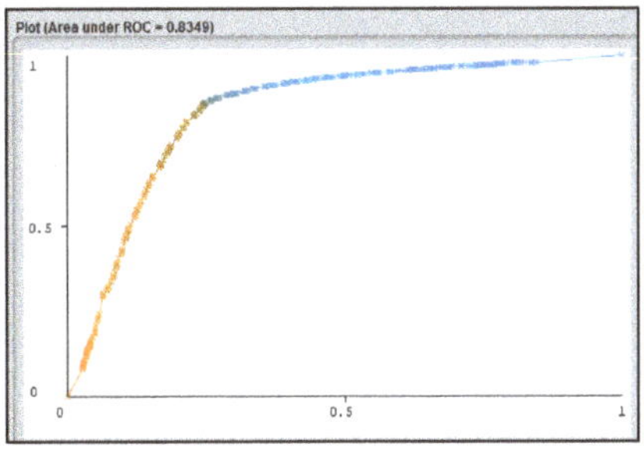

CART (Yes)

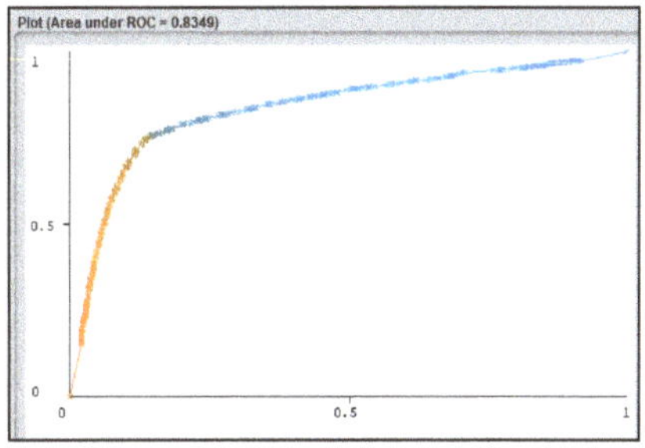

CART (No)

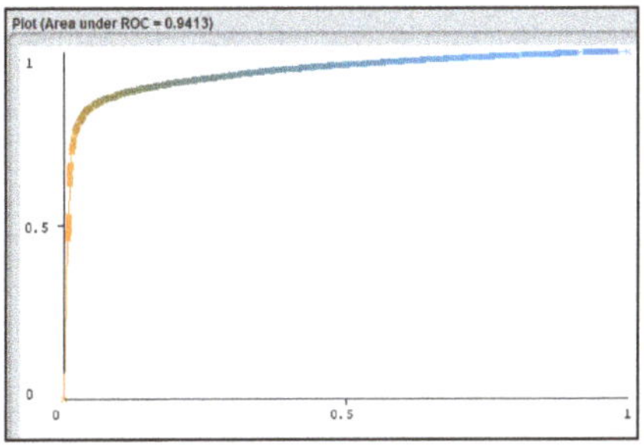

RF (Yes)

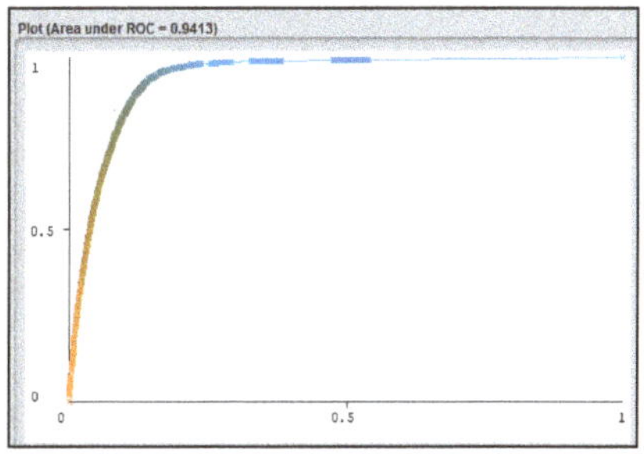

RF (No)

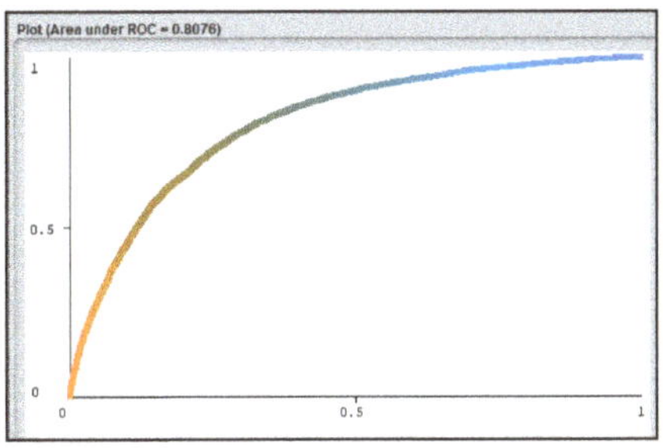

MLP (Yes)

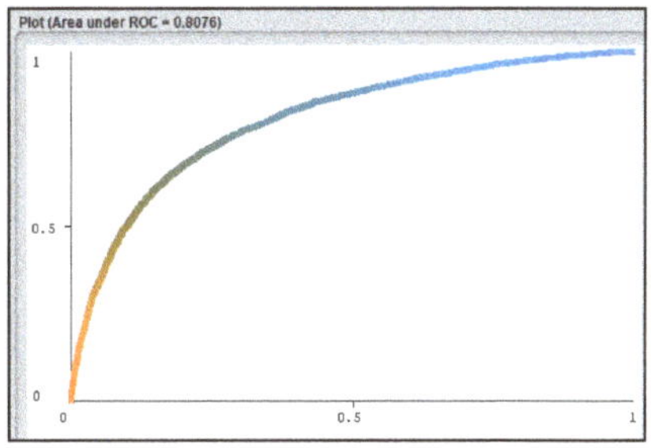

MLP (No)

Figure 8. ROC curves for the two classes: Yes (**left** figures), No (**right** figures).

Table 3. AUC and ROC results.

Classifier	Dependent Variable Classes	AUC Area	ROC Area
CART	Yes	0.834	0.835
	No	0.834	0.835
RF	Yes	0.941	0.941
	No	0.941	0.941
MLP	Yes	0.807	0.808
	No	0.807	0.808

Based on the analysis results, the three classifiers showed good ability for classifying and predicting speeding violations, especially RF. The superior performance of the RF is due to its nature. Its result depends on the majority votes for several decision trees. RF uses the bagging technique, which depends on randomness for data selection. In addition to that, there is no association between the decision trees in RF, which increase the forest accuracy. Furthermore, RF does not need many calculations to build the model and can deal with missing values, unlike the MLP.

Age, road alignment, speed limit, weather, and accident time were the most influential factors for speeding violations in this study. Age groups ($\leq$25) and (26–35) were associated with speeding violations; this result is matched with several previous studies [17,20,21]. As well, young drivers underestimate the speeding risk and overestimate their driving abilities, so they tend to violate speed limits. This conclusion supports several previous studies [17–21,45–47]. However, Zhang et al. (2014) [18] found that age does not affect speeding violations. Another influencing factor in this study was road alignment. The results showed that speeding violations increase at curves. Different factors at curves affect vehicles' speeds, such as the radius of the horizontal curve and the sight distance at vertical curves. Increasing the sight distance and the radius of the horizontal curve encourages driving at high speed [15,48–50]. The speed limit was a significant factor in this study that affected speeding violations. Different age groups violate speed limits at different speeds (30–40, 40–50, 50–60, 60–70, 70–80, and 80–90 km/h). These speed limits may not be appropriate for traffic streams. According to Gargoum et al. (2016) [13], Shawky et al. (2017) [51], and Afghari et al. (2018) [15], speeding violations decrease with increasing speed limits.

In this study, bad weather conditions (i.e., snow, ice, rain) increased speeding violations. In the winter, slippery roads because of ice, snow, or rain at low temperatures increase breaking distance, which increases the probability of speeding. These results matched those of Li et al. (2021) [52] and Ambo et al. (2021) [53]. However, Sutela and Aaltonen (2020) [54] concluded that speeding violations decreased in rainy weather. Nevertheless, another study by Zhang et al. (2014) [18] mentioned that weather conditions did not affect speeding violations. Furthermore, speeding violations were affected by accident time. In this study, night periods experienced high levels of speeding violations, which agrees with Zhang et al. (2014) [18] and Wu and Hsu (2019) [47], who said that a higher probability of speeding violations was associated with midnight and dawn periods. On the other hand, Balasubramanian and Sivasankaran (2021) [21] concluded that speeding violations were associated with daylight more than at night.

Determining the factors that affect speeding violations helps understand the violations more clearly and find the best solutions to mitigate them. In addition, using ML algorithms shows an excellent ability to analyze and predict speeding violations, which encourages using them in further studies and for similar issues.

4. Conclusions

This research aimed to analyze speeding violations in the U.S. using machine learning techniques, including CART, RF, and MLP. A total of 51,136 data records from the FARS were analyzed using WEKA software. Data related to drivers, the environment, vehicles, roads, and accident time were studied. The data were cleaned and balanced using the SMOTE technique. The accuracy, RMSE, F-measure, Kappa statistic, AUC, and ROC evaluation metrics were used to check the performance of the algorithms.

The outcome of this research concluded that young drivers tend to underestimate speeding risks and overestimate their driving abilities. Speeding violations were more common in the age groups (25) and (26–35); the percentages of violators in these two groups were 30.7% and 25.7%, respectively. Bad weather conditions were also associated with a high percentage of speeding violations. Slippery roads due to rain, snow, or ice at low temperatures increase braking distance, which raises speeding violations. The speed limit was a significant factor affecting speeding violations. Different age groups violate speed limits at different speeds (30–40, 40–50, 50–60, 60–70, 70–80, and 80–90 km/h). These speed limits may not be appropriate for traffic streams. Another influencing factor in this study was road alignment. The results showed that speeding violations increase at curves. Different factors at curves affect vehicles' speeds, such as the radius of the horizontal curve and the sight distance at vertical curves. Furthermore, speeding violations were affected by accident time, whereby night periods experienced high speeding violations.

The analysis results indicated that the three classifiers were capable of classifying and predicting speeding violations. The RF was the best method for analyzing and predicting speeding violations. Based on the results of this study, it is recommended to raise the awareness of drivers, especially the young ones, about the dangers of excessive speed, in general, and its dangers under bad weather conditions. In addition, it is urged to apply the law firmly and fairly and increase police enforcement. This work focused on studying speeding violations, in general, using CART, RF, and MLP. For further studies, it is recommended to use different ML algorithms and utilize other programs or programming languages for studying speeding violations. Furthermore, speeding violations can be studied for more specific cases, such as speeding violations for vehicle type (i.e., passenger cars or trucks), area type (i.e., rural, or urban), or any other specific factor. Speeding violations should also be studied with traditional statistical and machine learning methods. Then, compare how well each method can predict speeding violations.

Author Contributions: Conceptualization, A.H.A. and T.K.A. Methodology, A.H.A., B.W.A.-M. and T.K.A. Software, T.K.A. and M.S.O. Validation, A.H.A., T.K.A. and M.S.O. Formal analysis, A.H.A., T.K.A. and M.S.O. Investigation, A.H.A. and T.K.A. Resources, A.H.A. and B.W.A.-M. Data curation, T.K.A. and M.S.O. Writing—original draft preparation, A.H.A. and T.K.A. Writing—review and editing, A.H.A., B.W.A.-M., T.K.A. and M.S.O. Visualization, A.H.A. and T.K.A. Supervision, A.H.A. and B.W.A.-M. Project administration, A.H.A. and B.W.A.-M. All authors have read and agreed to the published version of the manuscript.

Funding: This research received no external funding.

Institutional Review Board Statement: Not applicable.

Informed Consent Statement: Not applicable.

Data Availability Statement: This study obtained data from the Fatality Analysis Reporting System (FARS) for the years 2015–2019. The National Highway Traffic Safety Administration (NHTSA) is responsible for maintaining this database system. Available at: https://www.nhtsa.gov/research-data/fatality-analysis-reporting-system-fars (accessed on 10 January 2023).

Conflicts of Interest: The authors declare no conflict of interest. The roles in the collection, analysis, or interpretation of data; in the writing of the manuscript, and the decision to publish the results are solely by the authors.

References

1. CDC. Centers for Disease Control and Prevention. Road Traffic Injuries and Deaths—A Global Problem. 2020. Available online: https://www.cdc.gov/injury/features/global-road-safety/index.html (accessed on 10 January 2023).
2. Lee, N. Speed-Related Car Accidents Cost the U.S. over $40 Billion Annually. Lowering the Limit May Not Help. Consumer News and Business Channel. Available online: https://www.cnbc.com/2021/12/16/heres-why-speed-limits-arent-working-in-the-us.html (accessed on 10 January 2023).
3. NHTSA. National Highway Traffic Safety Administration. Speeding | NHTSA. Available online: https://www.nhtsa.gov/risky-driving/speeding (accessed on 10 January 2023).
4. Weng, J.; Meng, Q. Effects of environment, vehicle and driver characteristics on risky driving behavior at work zones. *Saf. Sci.* **2012**, *50*, 1034–1042. [CrossRef]
5. Svenson, O.; Eriksson, G.; Slovic, P.; Mertz, C.K.; Fuglestad, T. Effects of main actor, outcome and affect on biased braking speed judgments. *Judgm. Decis. Mak.* **2012**, *7*, 235–243. [CrossRef]
6. IIHS-HLDI Crash Testing and Highway Safety. Speed. 2021. Available online: https://www.iihs.org/topics/speed (accessed on 1 February 2022).
7. WHO, World Health Organization. Road Traffic Injuries. Available online: https://www.who.int/news-room/fact-sheets/detail/road-traffic-injuries#:~:text=Key%20facts,road%20traffic%20crashes%20by%202020 (accessed on 10 January 2023).
8. Sayed, T.; Abdelwahab, W.; Navin, F. Identifying accident-prone locations using fuzzy pattern recognition. *J. Transp. Eng.* **1995**, *121*, 352–358. [CrossRef]

9. WHO, World Health Organization. Urban Speed Limit Range. Available online: https://www.who.int/gho/road_safety/legislation/situation_trends_urban_speed_limit/en/ (accessed on 1 February 2023).
10. Alomari, A.H.; Al-Omari, B.H.; Al-Adwan, M.E.; Sandt, A. Investigating and modeling speed variability on multilane highways. *Adv. Transp. Stud.* **2021**, *54*, 5–16.
11. Alomari, A.H.; Al-Omari, B.H.; Al-Adwan, M.E. Analysis of speed variance on multilane highways in Jordan. In Proceedings of the 1st International Congress on Engineering Technologies, Irbid, Jordan, 16–18 June 2020; CRC Press: Boca Raton, FL, USA, 2021; pp. 206–216. [CrossRef]
12. Build, Price, Option. Top 8 Reasons Why Drivers Speed—Build, Price, Option. 2020. Available online: https://www.buildpriceoption.com/top-8-reasons-why-drivers-speed/ (accessed on 10 January 2023).
13. Gargoum, S.A.; El-Basyouny, K.; Kim, A. Towards setting credible speed limits: Identifying factors that affect driver compliance on urban roads. *Accid. Anal. Prev.* **2016**, *95*, 138–148. [CrossRef]
14. Kanellaidis, G.; Golias, J.; Zarifopoulos, K. A survey of drivers' attitudes toward speed limit violations. *J. Saf. Res.* **1995**, *26*, 31–40. [CrossRef]
15. Afghari, A.P.; Haque, M.M.; Washington, S. Applying fractional split model to examine the effects of roadway geometric and traffic characteristics on speeding behavior. *Traffic Inj. Prev.* **2018**, *19*, 860–866. [CrossRef] [PubMed]
16. Huang, Y.; Sun, D.; Tang, J. Taxi driver speeding: Who, when, where and how? A comparative study between Shanghai and New York City. *Traffic Inj. Prev.* **2018**, *19*, 311–316. [CrossRef]
17. Tseng, C.M. Speeding violations related to a driver's social-economic demographics and the most frequent driving purpose in Taiwan's male population. *Saf. Sci.* **2013**, *57*, 236–242. [CrossRef]
18. Zhang, G.; Yau, K.K.; Gong, X. Traffic violations in Guangdong Province of China: Speeding and drunk driving. *Accid. Anal. Prev.* **2014**, *64*, 30–40. [CrossRef]
19. Liang, Z.; Xiao, Y. Analysis of factors influencing expressway speeding behavior in China. *PLoS ONE* **2020**, *15*, e0238359. [CrossRef]
20. Al Matawaha, J.; Jadaan, K.; Freeman, B. Analysis of speed related behavior of Kuwaiti drivers using the driver behavior questionnaire. *Period. Polytech. Transp. Eng.* **2020**, *48*, 150–158. [CrossRef]
21. Balasubramanian, V.; Sivasankaran, S. Analysis of factors associated with exceeding lawful speed traffic violations in Indian metropolitan city. *J. Transp. Saf. Secur.* **2021**, *13*, 206–222. [CrossRef]
22. Al-Mistarehi, B.W.; Alomari, A.H.; Imam, R.; Alnaasan, T.K. Investigating the Factors Affecting Speeding Violations in Jordan Using Phone Camera, Radar, and Machine Learning. *Front. Built Environ.* **2022**, *8*, 917017. [CrossRef]
23. Mohammadazadeh, A.; Sabzalian, M.H.; Castillo, O.; Sakthivel, R.; El-Sousy, F.F.; Mobayen, S. *Neural Networks and Learning Algorithms in MATLAB*; Springer Nature: Berlin/Heidelberg, Germany, 2022.
24. Mohammadzadeh, A.; Sabzalian, M.H.; Zhang, C.; Castillo, O.; Sakthivel, R.; El-Sousy, F.F. *Modern Adaptive Fuzzy Control Systems (Vol. 421)*; Springer Nature: Berlin/Heidelberg, Germany, 2022.
25. Alomari, A.H.; Khedaywi, T.S.; Jadah, A.A.; Marian, A.R.O. Evaluation of Public Transport among University Commuters in Rural Areas. *Sustainability* **2023**, *15*, 312. [CrossRef]
26. Alomari, A.H.; Khedaywi, T.S.; Marian, A.R.O.; Jadah, A.A. Traffic speed prediction techniques in urban environments. *Heliyon* **2022**, *8*, e11847. [CrossRef]
27. Alomari, A.H.; Abu Lebdeh, E. Smart real-time vehicle detection and tracking system using road surveillance cameras. *J. Transp. Eng. Part A Syst.* **2022**, *148*, 04022076. [CrossRef]
28. Al-Mistarehi, B.W.; Alomari, A.H.; Imam, R.; Mashaqba, M. Using machine learning models to forecast severity level of traffic crashes by r studio and arcgis. *Front. Built Environ.* **2022**, *16648714*, 31. [CrossRef]
29. NHTSA, National Highway Traffic Safety Administration. Fatality Analysis Reporting System (FARS). Available online: https://www.nhtsa.gov/research-data/fatality-analysis-reporting-system-fars (accessed on 10 January 2023).
30. Aljarrah, M.F.; Khasawneh, M.A.; Al-Omari, A.A. Investigating Key Factors Influencing the Severity of Drivers Injuries in Car Crashes Using Supervised Machine Learning Techniques. *J. Eng. Sci. Technol. Rev.* **2019**, *12*, 15–27. [CrossRef]
31. Witten, I.; Frank, E.; Hall, M.; Pal, C. The WEKA Workbench. In *Data Mining: Practical Machine Learning Tools and Techniques*, 4th ed.; Morgan Kaufmann: Burlington, MA, USA, 2016.
32. Loh, W.Y. Classification and regression trees. *Wiley Interdiscip. Rev. Data Min. Knowl. Discov.* **2011**, *1*, 14–23. [CrossRef]
33. Kingsford, C.; Salzberg, S. What are decision trees? *Nat. Biotechnol.* **2008**, *26*, 1011–1013. [CrossRef]
34. Ampadu, H. Decision Trees. 2021. Available online: https://ai-pool.com/a/s/decision-trees (accessed on 25 December 2022).
35. Breiman, L. Random forests. *Mach. Learn.* **2001**, *45*, 5–32. [CrossRef]
36. Nasution, A.K.; Wijaya, S.H.; Gao, P.; Islam, R.M.; Huang, M.; Ono, N.; Kanaya, S.; Altaf-Ul-Amin, M. Prediction of Potential Natural Antibiotics Plants Based on Jamu Formula Using Random Forest Classifier. *Antibiotics* **2022**, *11*, 1199. [CrossRef]
37. Mohanty, A. Multi-Layer Perceptron (MLP) Models on Real World Banking Data. Available online: https://becominghuman.ai/multi-layer-perceptron-mlp-models-on-real-world-banking-data-f6dd3d7e998f (accessed on 10 January 2023).
38. Noriega, L. Multilayer Perceptron Tutorial. School of Computing. Ph.D. Thesis, Staffordshire University, Stoke-on-Trent, UK, 2005.
39. Almeida, L.B. Multilayer perceptrons. In *Handbook of Neural Computation C*; Elsevier: Amsterdam, The Netherlands, 1997. [CrossRef]

40. Chawla, N.; Bowyer, K.; Hall, L.; Kegelmeyer, W. SMOTE. Synthetic Minority Over-sampling Technique. *J. Artif. Intell. Res.* **2002**, *16*, 321–357. [CrossRef]
41. What is Data Accuracy, Why it Matters and How Companies Can Ensure They Have Accurate Data. 2020. Available online: https://dataladder.com/what-is-data-accuracy/ (accessed on 8 April 2023).
42. Towards Data Science. Accuracy, Recall, Precision, F-Score & Specificity, Which to Optimize on? 2019. Available online: https://towardsdatascience.com/accuracy-recall-precision-f-score-specificity-which-to-optimize-on-867d3f11124 (accessed on 8 April 2023).
43. What is a Confusion Matrix in Machine Learning? 2023. Available online: https://www.simplilearn.com/tutorials/machine-learning-tutorial/confusion-matrix-machine-learning#:~:text=A%20confusion%20matrix%20presents%20a,actual%20values%20of%20a%20classifier (accessed on 8 April 2023).
44. DeepAI. What Is the F-Score? 2023. Available online: https://deepai.org/machine-learning-glossary-and-terms/f-score (accessed on 8 April 2023).
45. Zamanov, M.S. Development of a Prediction Model for Speed Limit Violations on Tangent Road Sections. Master's Thesis, Delft University of Technology, Delft, The Netherlands, 2012.
46. Javid, M.A.; Al-Roushdi, A.F.A. Causal factors of driver's speeding behaviour, a case study in Oman: Role of norms, personality, and exposure aspects. *Int. J. Civ. Eng.* **2019**, *17*, 1409–1419. [CrossRef]
47. Wu, Y.W.; Hsu, T.P. Mining characteristics of speeding and red-light running violations using association rules. *J. East. Asia Soc. Transp. Stud.* **2019**, *13*, 2111–2125.
48. Vos, J.; Farah, H.; Hagenzieker, M. How do Dutch drivers perceive horizontal curves on freeway interchanges and which cues influence their speed choice? *IATSS Res.* **2021**, *45*, 258–266. [CrossRef]
49. Poe, C.; Tarris, J.; Mason, J. Operating speed approach to geometric design of low-speed urban streets. *Transp. Res. Circ.* **1998**, *10*, 1–9.
50. Fitzpatrick, K.; Carlson, P.; Brewer, M.; Wooldridge, M. Design Factors That Affect Driver Speed on Suburban Streets. *Transp. Res. Rec.: J. Transp. Res. Board* **2001**, *1751*, 18–25. [CrossRef]
51. Shawky, M.; Sahnoon, I.; Al-Zaidy, A. Predicting speed-related traffic violations on rural highways. In Proceedings of the 2nd World Congress on Civil, Structural, and Environmental Engineering (CSEE'17), Barcelona, Spain, 2–4 April 2017; Volume 117. [CrossRef]
52. Li, Y.; Li, M.; Yuan, J.; Lu, J.; Abdel-Aty, M. Analysis and prediction of intersection traffic violations using automated enforcement system data. *Accid. Anal. Prev.* **2021**, *162*, 106422. [CrossRef]
53. Ambo, T.B.; Ma, J.; Fu, C. Investigating influence factors of traffic violation using multinomial logit method. *Int. J. Inj. Control. Saf. Promot.* **2021**, *28*, 78–85. [CrossRef] [PubMed]
54. Sutela, M.; Aaltonen, M. Effects of temporal characteristics and weather conditions on speeding sanction rates in automatic traffic enforcement. *Police J.* **2021**, *94*, 590–615. [CrossRef]

applied sciences

MDPI

Article

Research Determining the Priority Order of Forces Acting on a Vehicle Transporting Logs

Jonas Matijošius [1], Kristina Čižiūnienė [2], Jūratė Liebuvienė [3] and Edgar Sokolovskij [1,*]

1 Department of Automobile Engineering, Faculty of Transport Engineering, Vilnius Gediminas Technical University, Plytinės Str. 25, 10105 Vilnius, Lithuania; jonas.matijosius@vilniustech.lt
2 Department of Logistics and Transport Management, Vilnius Gediminas Technical University, Plytinės Str. 27, 10105 Vilnius, Lithuania; kristina.ciziuniene@vilniustech.lt
3 Department of Transport Engineering, Klaipeda State University of Applied Sciences, Jaunystės Str. 1, 91274 Klaipeda, Lithuania; j.liebuviene@kvk.lt
* Correspondence: edgar.sokolovskij@vilniustech.lt; Tel.: +370-65-083-509

Abstract: When transporting different types of cargo, the factors affecting the cargo must be assessed to ensure a safe and fast transportation process. However, the specificity of the cargo means that the standard assessment methods do not reflect the existing conditions or are standard at the theoretical and/or recommendation level. Therefore, in the event of a traffic accident, when a vehicle or cargo is damaged or other traffic participants are put at rink, assessment of the situation is difficult and requires an expert assessment. Then, the most optimal means of accomplishing this is to simulate transportation parameters with a specific vehicle and cargo. The transportation of wooden logs is a very specific method of transportation where, in addition to the weight of the load itself, it is necessary to assess how the load affects the dynamics of the vehicle. Therefore, determining the priority of the forces that act on the vehicle carrying wooden logs becomes very important in order to model the transport parameters of this cargo. This article uses an analytical survey, an expert survey and its analysis, and methods to determine correlations. Given the fact that the main forces acting on the load most often manifest during a turn, this study assessed road turns and the respective speed on them as the main parameters. Other parameters, such as road roughness and roundabouts, were not assessed, because they would be an object of long-distance research. The study results revealed that the main forces affecting a vehicle transporting logs manifest in the following priority order: inertial forces, gravity forces, frictional forces and, finally, centrifugal forces.

Keywords: wood log transportation; inertial forces; gravity forces; frictional forces; centrifugal forces; expert survey; correlation; pearson coefficient

Citation: Matijošius, J.; Čižiūnienė, K.; Liebuvienė, J.; Sokolovskij, E. Research Determining the Priority Order of Forces Acting on a Vehicle Transporting Logs. *Appl. Sci.* **2023**, *13*, 6174. https://doi.org/10.3390/app13106174

Academic Editor: Alberto Campagnolo

Received: 6 April 2023
Revised: 8 May 2023
Accepted: 16 May 2023
Published: 18 May 2023

1. Introduction

Freight transportation is an important element of the logistics supply chain [1], especially when taking into account the elements of reliability of supply. There are many aspects that affect the reliability indicators in cargo transportation [2]. This includes the chosen vehicle, the placement of the load itself, its fastening, and its weight. Some aspects are of minor importance in assessing the transportation process itself, e.g., the wear and tear of the fastening elements, which can be minimised by performing the necessary maintenance [3]. Therefore, the arrangement and mass of the cargo, as well as the dynamic indicators of the vehicle itself, are the most important factors [4].

The energy of cargo movement is a threatening force, the impact of which can be avoided only by strictly following certain requirements [5]. The effect of this force on the side structures of the vehicle body on which the loads rest is of particular concern [6]. However, undesirable consequences, which sometimes may even be tragic, can be avoided by using specially adapted vehicle body structures and the reliable fastening of loads [7].

This is especially important when transporting specific cargo [8]. This article examines the reliability of wood log transportation [9].

A rational cargo arrangement scheme is selected in each specific case depending on the type of cargo, container, and vehicle body dimensions [10]. The cargo must be placed in the cargo compartment according to the following requirements [11]:

- The cargo must be placed as evenly as possible throughout the entire floor area of the cargo compartment;
- It must be placed as low as possible. The center of gravity of the cargo must be as close to the line of the longitudinal axis of symmetry of the cargo compartment as possible;
- The cargo must be supported on the front side of the compartment or on the protective structure of the cab, as well as on the sides of the cargo compartment;
- Loads with sharp ends must be placed with the sharp end facing towards the rear.

A load must be secured so that, while the vehicle is moving, it cannot be released by the following forces (Figure 1) [12]:

- Inertial force directed in the direction of movement, the impact of which is equal to 0.8 F_G;
- Lateral and backward inertial force of 0.5 F_G.

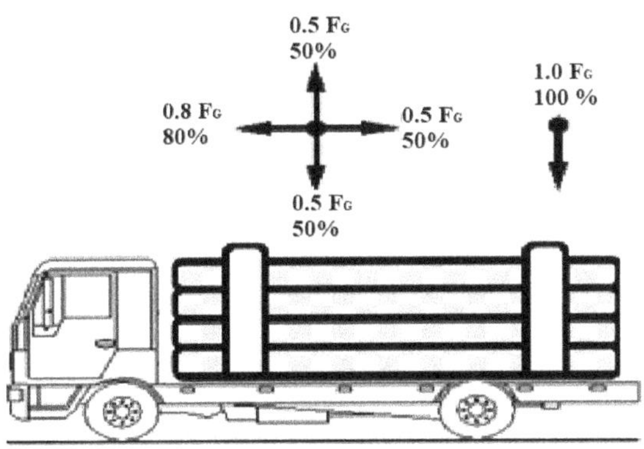

Figure 1. Distribution of inertial forces.

The European Commission states that the top of the logs in the middle should be higher than that of logs on the sides to allow the load to be properly fastened, as shown below [13]. The logs should rest on a conical wedge or a toothed brim [14].

A trailer connector should be installed in the front of the first section of logs, between the driver's cabin and the logs. The strength must meet the requirements of the EN 12642 XL standard [15] and the load should not be higher than the front of the vehicle. Ropes or similar means of fastening should be tied across the top of each load section (stack of logs) to ensure vertical compression of the logs; the number of ropes should be as follows [14,16]:

(a) At least one rope if the cargo section consists of logs with bark and if the length of the log does not exceed 3.3 m;
(b) At least two ropes if the load section is longer than 3.3 m or if the bark has been removed, regardless of the length of the logs.

The overhead lashing ropes must be (transversely) lashed between the front and thee rear pillars of each cargo section, ensuring as much symmetry as possible [17]. More ropes should be used in vehicles that do not have a side of sufficient strength or automatic tensioning means, i.e., using two ropes if logs are not longer than 3 m, three ropes if logs are up to 5 m long, and four ropes if logs are longer than 5 m [18].

Wood is a living product that constantly moves during transportation [19]. During transportation, wood is exposed to external forces, so its fastening equipment must be chosen accordingly. General principles should apply to load distribution, ensuring that the load is stable at the front wall. The front wall should be installed according to the EN 12642 XL class standard, and the load should not protrude the front wall.

Before transporting wood, a loaded stack of wood must be fixed in the vehicle using special belts of at least 50 mm wide [20]. The breaking force indicated on the belt label must be at least 5000 kg. All belts should withstand a minimum tensile force of 1600 daN. A self-tensioner should be used, which makes it easier to fix the wood to the base, ensuring stronger support.

The use of chain or mesh fastening with turnbuckles is recommended, and all belts should be checked and kept taut throughout transport [21]. Each outer piece of wood should be supported by at least two or three perpendicular posts. The strength of the pillars should be sufficient to prevent the vehicle from exceeding the width restrictions.

Logs that are shorter than the distance between the two supporting posts should be loaded at the centre of the load [19]. When there are two pairs of supports supporting wood, their ends should protrude at least 300 mm from the supports. The distance between the wood and the road surface should be at least 0.6 m [22].

When transporting wood in a vehicle with temporarily installed supports, the height of the loaded wood stack cannot exceed 1.6 m from the bottom of the supports [23]. Cushions must be equipped with latches to prevent them from sliding toward the edges [24].

The overhead bracing that creates vertical pressure on the wood should be tightened in each load section at the following points [25]:

(a) Barked wood: at least one strap per load section up to a maximum length of 3.3 m, with two belts if the cargo section is longer than 3.3 m.
(b) Barkless wood: at least two lashing straps per load section.

Wood must be transported within the maximum permitted vehicle dimensions, the maximum axle loads (s), and the maximum allowed gross weight, approved by Order No. 3–66 of the Minister of Transport of 18 February 2002 "On the approval of the maximum permitted dimensions of vehicles, the permitted axle loads (axle), the allowed gross mass" (2002, No. 23–870), and the weight that a vehicle (combination) can carry [26].

To assess the reliability of transport carrying this type of cargo, special instructions are followed, without performing a reliability assessment. The scientific literature does not distinguish any specialized approach to reliability in log transport, which leads to the conclusion that reliability research in log transport is not widespread.

The fact that the most common interaction between cargo fastening and acting forces is assessed at the theoretical level only, and the entire arrangement of the technological process of cargo fastening and transportation is more of a recommendation than a justification for practical application, creates an unprecedented need for this kind of actual, mathematical, calculation. Thus, the aim of this article is to restore the practical situation and provide reliable calculations that can assess the forces acting in a mathematical expression and determine the priority order of the forces acting on the vehicle transporting logs. The obtained results are unique and emphasize the innovative and scientific nature of the obtained results. These calculations were made by taking factors of speed and road turns into account, since this is when most cases of cargo- and/or vehicl—overturning occur, causing damage to the vehicle and cargo, and when the load may also damage the road and its infrastructure elements and put other traffic participants in danger. Such road elements as roundabouts and islands are also important, but they were not considered in the calculations, since, in this case, the speed and the turning angle are significantly lower than in the above-mentioned cases.

2. Materials and Methods

When assessing reliability, the main criteria that have a significant impact on the transportation of logs were distinguished in this article. These include vehicle dynamics' parameters, such as centrifugal force, critical slip, roll, and vehicle speed [27].

The impact of the centrifugal force can be controlled. The centrifugal force is calculated according to the following formula [28]:

$$F = \frac{mv^2}{r} \quad (1)$$

where F is the centrifugal force;

m—mass;
v—speed;
r—radius.

The centrifugal force of the vehicle occurs during [27]:

- A sudden braking;
- Suddenly starting to drive;
- When driving in a corner.

The centrifugal force increases with increasing driving speed and lower road turn radii. With a reduction in the speed at a corner, the centrifugal force also decreases [29].

When turning, the centrifugal force pushes a vehicle away from the centre of the turning curve. The impact on the vehicle depends on which part of the vehicle is affected the most by this force. If it mainly affects the upper part of the vehicle, this lateral movement of the vehicle is difficult to control, and if it affects the lower part, the movement of the chassis and body will be minimal and the vehicle will be more stable [30].

The load's kinetic (movement) force is opposite to the frictional force, which depends on the load's composition, shape, dimensions, and contact with the platform. The drag force is calculated using the coefficient of friction and the weight of the load. The magnitude of the friction force coefficient is usually between 0.01 (load on wet floor) and 0.5 (dry wood on dry wood floor). The difference between weight (mass) and friction forces forms the minimum amount of fastening force that ensures a stable holding of the load [31].

The force of gravity is calculated according to Newton's second law. Table 1 shows the sliding friction of various materials [32], which are used to calculate the load-securing according to a simplified formula. This table was compiled by the author, based on/guided by the "Requirements for load arrangement and/or its attachment in category N and O vehicles" [26].

Table 1. Slip coefficient of the vehicle floor.

Floor of the Cargo Compartment of the Vehicle	Cargo		
	Dry	**Wet**	**Oiled**
Wood/wood	0.20–0.50	0.20–0.25	0.05–0.15
Metal/wood	0.20–0.50	0.20–0.25	0.02–0.10
Metal/metal	0.10–0.25	0.10–0.20	0.01–0.10
Concrete/wood	0.30–0.60	0.30–0.50	0.10–0.20

The critical slip of the vehicle was calculated according to the formula [27,28]:

$$v_{sl} = \sqrt{\frac{(\cos\beta \cdot \phi_s - \sin\beta) \cdot g \cdot R}{\cos\beta + \sin\beta \cdot \phi_s}} m/s \quad (2)$$

where R—the turning radius of the road,
g—free fall, g = 9.81 m/s^2;
φs—coefficient of adhesion to the road surface.

Critical tumble of the vehicle [27,28]:

$$v_T = \sqrt{\frac{(\cos\beta \cdot 0.5 \cdot B - \sin\beta \cdot h)g \cdot R}{\cos\beta \cdot h + \sin\beta \cdot 0.5 \cdot B}} m/s \quad (3)$$

where *R*, the turning radius of the road,

h—height of the centre of gravity (empty/loaded), mm;
B—width of the intercostal space, m;
g—acceleration of free fall, $g = 9.81$ m/s^2;
φ_s—coefficient of adhesion to the road surface.

The centrifugal force is the force that pushes a rotating body away from the centre of rotation. It depends on the object's mass, rotation speed, and distance from the centre [33]. The centrifugal force increases with increasing driving speed and fewer radii of road bends. With a reduction in the speed at a corner, the centrifugal force will decrease. The movement forces are the key thing to take into account when choosing the appropriate fastening equipment [34].

A Volvo FH truck (Gothenburg Sweden) was used in the research. The truck is equipped with an "I-Shift" gearbox (using the "I-SEE" system), an independent front suspension, and a dynamic steering system.

Alucar OY superstructure was chosen for wood transportation. Alucar Oy is a part of Extendo Group, a Swedish company that specialises in the field of timber transport and load fastening, and the design of superstructures and supports for logging trucks. The manufacturers of these superstructures aim to obtain the maximum payload, thus ensuring high quality. The parts of these superstructures are manufactured in such a way that they can be assembled quickly and easily, and require as little welding as possible [35].

Screw connections allow for spare parts to be replaced very quickly. The supports of this superstructure are made of steel, the carrying capacity of which is 7 tons. The supports are cast in conical struts with a length of 2850 mm. The uprights are installed using a steel wedge and attached to the superstructure supports using screws.

The protective wall of the structure is made of double-walled anodised aluminium profiles that are connected to each other using strong profile connections [36]. The structure withstands collisions with logs and their scratching very well [37], as there are no remaining bends. The superstructure is resistant to corrosion, as the supports and struts are coated in a thick zinc primer and painted with powder paint. This superstructure is perfectly adapted to extremely complicated and difficult road conditions.

3. Results

Analytical calculations were performed using roundabout radii. The radius of the roundabout varies from 5.3 to 17 m. A fully loaded vehicle carrying a load of wood (its force action vectors are presented in Figure 2), traveling at a speed of 20 km/h through a 5.3 m roundabout, is affected by the centrifugal force presented in Table 2.

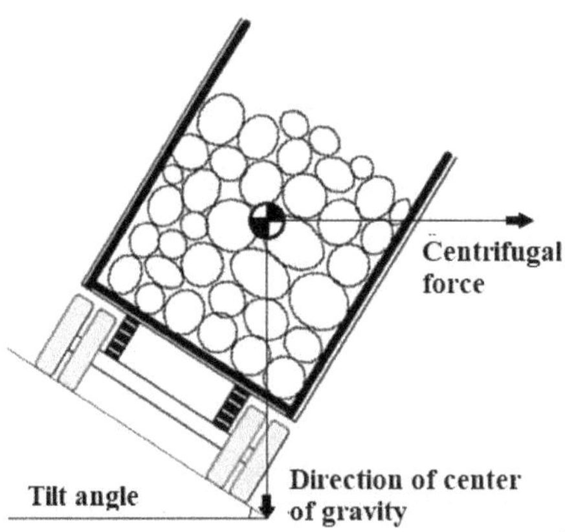

Figure 2. Direction of the force of the centre of gravity under the influence of the centrifugal force.

Table 2. Comparative calculation of forces.

Type of Force	Calculated Value
Centrifugal force	$F = \frac{40{,}000 \cdot 30.8}{5.3} = 232{,}452.8\ N$
Gravity force	$F = 40{,}000 \cdot 9.81 = 392{,}400\ N$

The calculations found that a vehicle driving through a roundabout with a radius of 5.3 m is affected by a centrifugal force of 232,452.8 N.

The force exerted on a vehicle was calculated. When the centrifugal force exceeds the force of gravity, the vehicle rolls over. This in illustrated in the graph of the dependence of centrifugal force on speed, shown below (Figure 3).

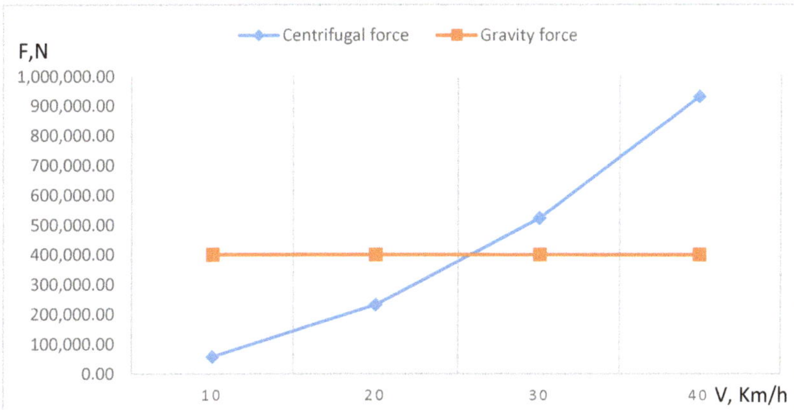

Figure 3. Dependence of centrifugal force on speed.

Figure 3 shows that when a vehicle travels at the speed of up to 25 km/h, it is stable, which means that the effect of the centrifugal force does not exceed the gravity force. However, when a vehicle's speed is greater than 25 km/h, the centrifugal force exceeds the force of gravity, resulting in the truck overturning.

The critical slope of a truck was calculated using the asphalt friction coefficient, the turning radius, and the transverse slope of the road. The asphalt friction coefficient is the most important criterion, and dry asphalt of $\mu = 0.6$, a turning radius of 5.3 m, and a cross-slope of the road of $\beta = 2$ [28] were used in the calculation.

$$v_{sl} = \sqrt{\frac{(0.999 \cdot 0.6 - 0.034) \cdot 9.81 \cdot 5.3}{0.999 + 0.034 \cdot 0.6}} = 5.37\ \text{m/s} \quad (4)$$

A critical truck skid does not occur until the speed reaches 19 km/h [38]. However, when the truck speed exceeds 19 km/h, the truck will start to slide at this asphalt friction coefficient during the turn.

The critical fall of a truck is calculated considering the width of the vehicle track and the height of the load's centre of gravity. The width of the rut of a wooden structure is 2.55 m. The selected height of the centre of gravity of the load is 2.05 m from the ground when the vehicle is loaded, while the selected center of gravity of the unloaded vehicle is 1.2 m [28].

Loaded vehicle:

$$v_T = \sqrt{\frac{(0.999 \cdot 0.5 \cdot 2.55 - 0.034 \cdot 2.05) \cdot 9.81 \cdot 5.3}{0.999 \cdot 2.05 + 0.034 \cdot 0.5 \cdot 2.55}} = 5.83\ \text{m/s} \quad (5)$$

Empty vehicle:

$$v_T = \sqrt{\frac{(0.999 \cdot 0.5 \cdot 2.55 - 0.034 \cdot 1.2) \cdot 9.81 \cdot 5.3}{0.999 \cdot 1.2 + 0.034 \cdot 0.5 \cdot 2.55}} = 7.17 \text{ m/s} \tag{6}$$

The critical rollover speed of a loaded and empty vehicle at a 5.3 m roundabout is estimated in Figure 4. A loaded vehicle will roll over at a speed of more than 21 km/h, and an empty vehicle will roll over at a speed of more than 26 km/h.

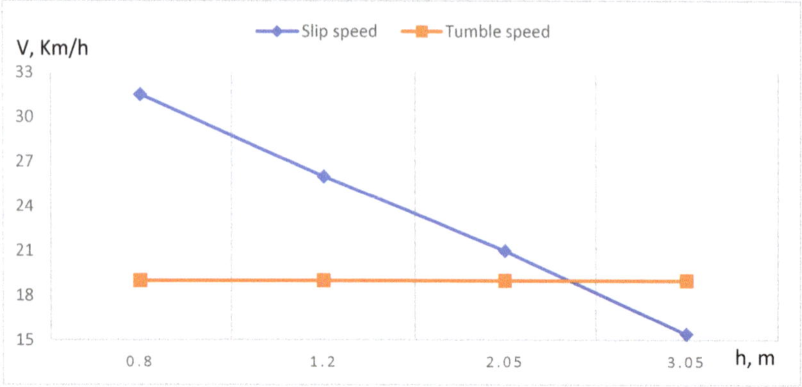

Figure 4. Dependence of the height of the centre of gravity on the speed.

Figure 4 illustrates the dependence of the speed and height of the center of gravity. The critical turning point is where the turning speed and the sliding speed curves meet. In this case, a vehicle will roll over when the height of the centre of gravity of the vehicle is 2.4 m from the ground and the speed of the vehicle is 19 km/h.

4. Discussion

Observations, calculations, and tests were conducted using an inclined plane test bench in the research, analysing the effect of centrifugal force on vehicle dynamics [39]. How the centrifugal force works has been determined, as well as the speed and the turning angle at which a truck can overturn, which are significantly affected by the load's centre of gravity.

Figure 5 shows how the vehicle must be loaded correctly so as not to exceed the axle loads. The most important thing is to properly load the vehicle without exceeding its total permissible weight, properly arranging the load so as not to exceed the axle load, and securing it as required by the European load-securing rules. Properly secured transported loads are less affected by external forces.

The centre of any stack of forest material must not protrude above the supports. The middle part of the stack should be higher than the side, forming a "ridge" in the load and allowing the load to be properly fastened, as shown in the Figure below.

The most dangerous forces affecting heavy transport are as follows:

- Centrifugal force;
- Friction force;
- Gravity force;
- Inertial force.

Figure 5. Correctly and incorrectly loaded round wood.

During log transportation, forces affect both the load and the vehicle. These are usually seen as separate elements, and their priority and the complexity of their interactions are disregarded [40]. Twelve experts participated in the expert evaluation, assessing the impact of the forces (Table 3).

Table 3. Results of expert evaluation rankings.

Formula	Centrifugal Force	Gravity Force	Friction Force	Inertial Force
$\sum_{i=1}^{n} R_{ij} = R_{ij}$	36	28	32	24
$\overline{R}_j = \frac{\sum_{i=1}^{n} R_{ij}}{n}$	3	2.333	2.667	2.000
$\sum_{i=1}^{n} R_{ij} - \frac{1}{2}n(m+1)$	6	−2	2	−6
$\left[\sum_{i=1}^{n} R_{ij} - \frac{1}{2}n(m+1)\right]^2$	36	4	4	36

Balance is the most important factor when making an informed decision. Assessing the compatibility of expert opinions is an essential part of quality research. Therefore, to consistently assess the opinions of two or more experts, Kendall's correlation coefficient (W) is used, ranking experts' assessments (Table 4). The consistency of experts' opinions is defined when W approaches 1. Expert opinions are considered consistent when W1 and inconsistent when W approaches 0 [41].

Table 4. Results of calculations of the concordance coefficient.

Formula	Results
$W = \frac{12S}{n^2(m^3-m)}$	0.1111
$\chi 2 = \frac{12S}{nm(m+1)}$	4.00
$W_{min} = \frac{\chi^2_{v,a}}{n(m+1)}$	0.0162

The obtained results show that W > W_{min}. This suggests that the experts' opinions are consistent, and further calculations can be made, which can evaluate the order of the acting forces (Table 5).

Table 5. Determining the priority order of the forces acting during log transportation.

Indicator Value	Centrifugal Force	Gravity Force	Friction Force	Inertial Force
$\overline{q} = \frac{\overline{R}_j}{\sum_{j=1}^{m} R_j}$	0.3000	0.2333	0.2667	0.2000
$d_j = 1 - \overline{q}_j = 1 - \frac{\overline{R}_j}{\sum_{j=1}^{m} R_j}$	0.7000	0.7667	0.7333	0.8000
$Q_j = \frac{d_j}{\sum_{j=1}^{m} d_j} = \frac{d_j}{m-1}$	0.2333	0.2556	0.2444	0.2667
$\overline{Q}_j = \frac{\sum_{i=1}^{n} B_{ij}}{\sum_{i=1}^{n} \sum_{j=1}^{m} B_{ij}}$	0.2000	0.2667	0.2333	0.3000
Arrangement of factors by importance	4	2	3	1

The obtained results show that the importance of the forces acting during transportation is distributed in the following sequence:

1. Inertial force;
2. Gravity force;
3. Frictional force;
4. Centrifugal force.

It is important to note that acting forces make a vehicle uncontrollable. Gravity, centrifugal force and frictional force are laws of nature that affect vehicle loads. In road traffic, loads are always affected by physical forces.

The results revealed that the forces of inertia and gravity have the greatest impact. However, when evaluating the interaction and its strength among all forces, it is also important to find their dependence (Figure 6).

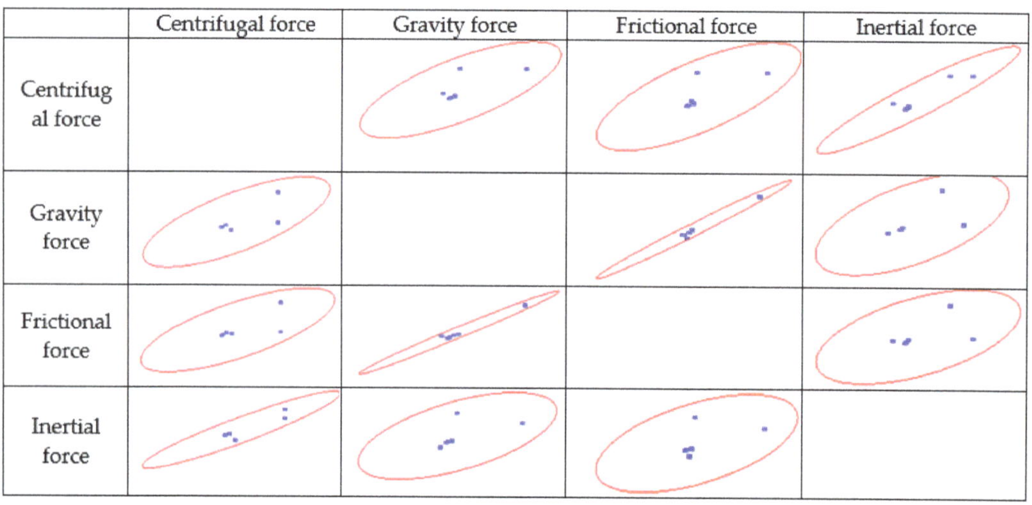

Figure 6. Correlational dependences of forces acting on a vehicle transporting logs.

The analysis of correlations reveals that there is a strong to moderate correlation between all the studied forces (judged by the values of the Pearson's correlation coefficient) (Table 6). A close to very strong correlation between centrifugal and inertial forces (0.93924), and gravity and frictional forces (0.99016), can be distinguished. A strong correlation is also observed between centrifugal and frictional forces (0.72439), and centrifugal and gravity forces (0.7315), while inertial and frictional forces (0.54059) and inertial and gravity forces (0.59001) demonstrate a moderate correlation.

Table 6. Evaluation of correlation dependences of the forces acting on a vehicle transporting logs based on the Pearson correlation coefficient.

	Centrifugal Force	Gravity Force	Frictional Force	Inertial Force
Centrifugal Force	1	0.7315	0.72439	0.93924
Gravity Force	0.7315	1	0.99016	0.59001
Frictional Force	0.72439	0.99016	1	0.54059
Inertial Force	0.93924	0.59001	0.54059	1

However, it should be noted that when examining the correlations between the forces, very strong and strong correlations created by the centrifugal force were observed.

It should be noted that the centrifugal force depends on:

- The height of the centre of gravity, because the lower the center of gravity, the lower the effect of centrifugal force on the vehicle;
- Vehicle speed;
- The turning angle, because the steeper the turning angle, the stronger the effect of centrifugal force.

The stability of the vehicle depends on its speed. The faster the speed, the greater the braking distance, impact force and the turning radius. This means that when making a sharp turn at a speed of 30 km/h, the probability of a vehicle sliding and overturning is minimal, but if the vehicle is traveling at 60 km/h, the probability of risk increases by four times [28] (Figure 7).

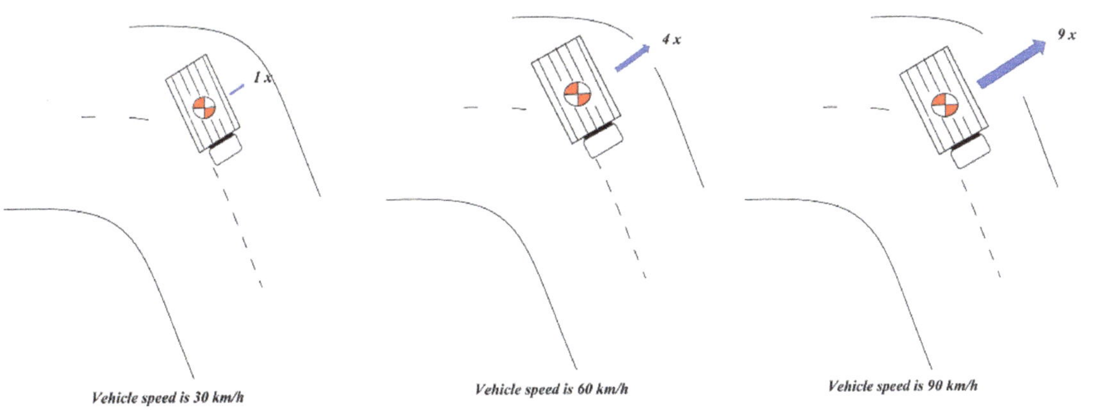

Figure 7. The strength of the centrifugal force at a turn.

When entering a turn at a speed of 90 km/h, the vehicle is exposed to a nine times stronger centrifugal force, which means that the probability of the vehicle overturning is high. The higher the speed, the greater the chance that the driver will lose control and the vehicle will lose control.

The direction of the friction force is opposite to the direction of motion of the body.

The magnitude of the friction force depends on:

- Types of materials that come into contact with each other;
- Treatment of touching surfaces (irregularities);
- The vehicle body weight (mass).

Considering the above information, it is important to evaluate the interdependencies of the constituent criteria of friction and centrifugal forces (Figures 8 and 9) and the correlations (Tables 7 and 8).

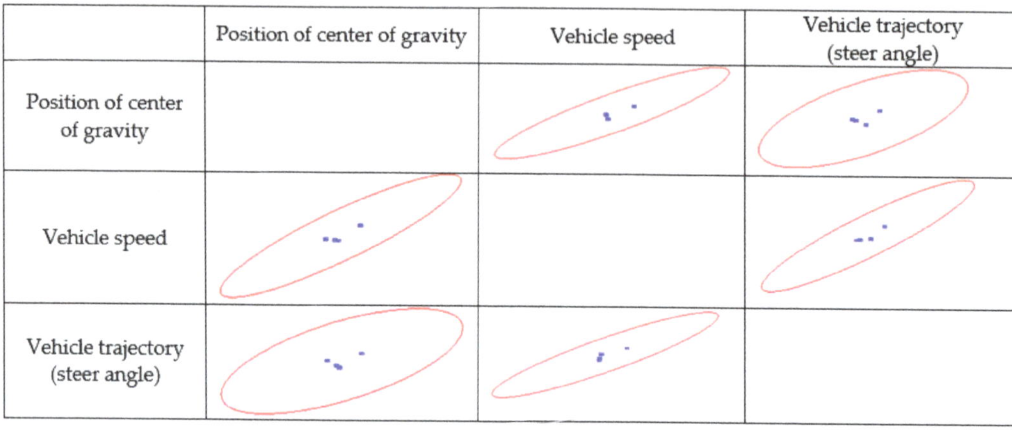

Figure 8. Correlational dependences of components of centrifugal forces.

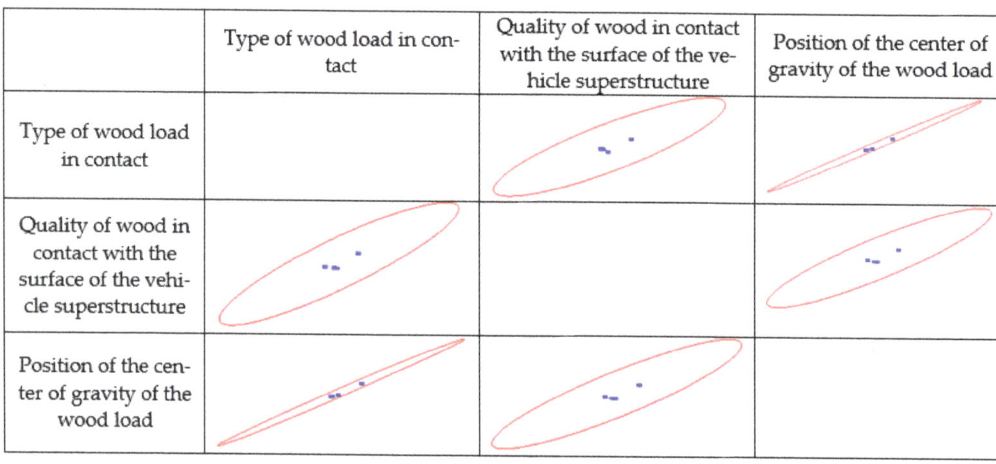

Figure 9. Correlational dependences of components of friction forces.

Table 7. Correlations of components of centrifugal forces.

	Position of the Center of Gravity	Vehicle Speed	Vehicle Trajectory (Steer Angle)
Position of the center of gravity	1	0.89924	0.64308
Vehicle speed	0.89924	1	0.91329
Vehicle trajectory (steer angle)	0.64308	0.91329	1

Table 8. Correlations between components of friction forces.

	Type of Wood Load in Contact	Quality of Wood in Contact with the Surface of the Vehicle Superstructure	Position of the Center of Gravity of the Wood Load
Type of wood load in contact	1	0.86487	0.99665
Quality of wood in contact with the surface of the vehicle superstructure	0.86487	1	0.90302
Position of the center of gravity of the wood load	0.99665	0.90302	1

In summary, according to Pearson correlation coefficient values, a strong and moderate correlation was observed between all the investigated centrifugal force factors. A close to very strong correlation between the vehicle speed and the vehicle trajectory (steer angle) (0.91329), and the vehicle speed and the position of the center of gravity (0.89924), can be distinguished. A strong correlation was also observed between the vehicle trajectory (steer angle) and the position of the center of gravity (0.64308).

An analysis of the correlations (according to the Pearson correlation coefficient values) revealed that a close to very strong and strong correlation between all the investigated factors of friction force. There is a close to very strong correlation between the type of wood load in contact and the position of the center of gravity of the wood load (0.99665), and the quality of wood in contact with the surface of the vehicle superstructure and the position of the center of gravity of the wood load (0.90302), and a strong correlation between the type of wood load in contact and the quality of wood in contact with the surface of the vehicle superstructure (0.86487).

In the transportation for logs, the impact of forces is very important and could be further assessed by modelling this type of transportation. During this study, these criteria were determined according to the scheme presented in Figure 10.

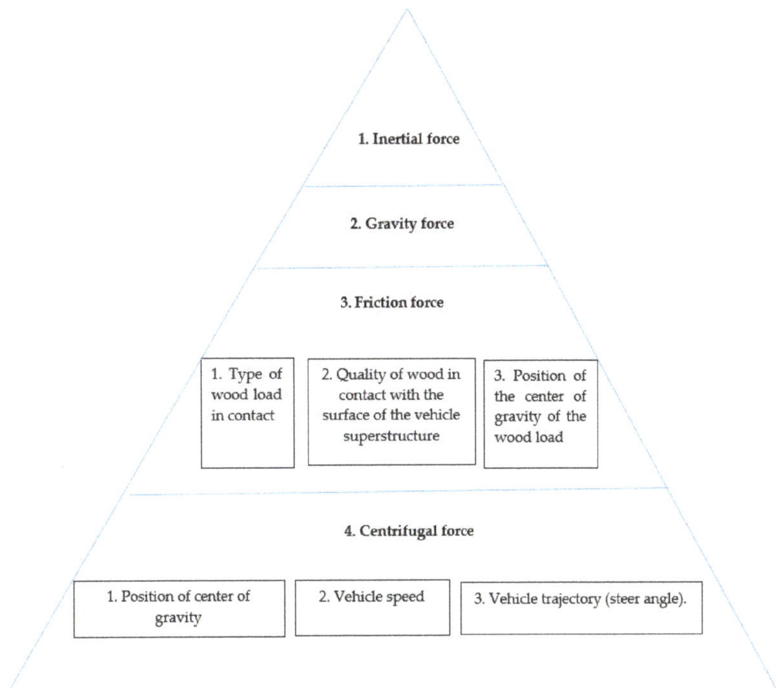

Figure 10. Scheme of priority assessment of forces acting on vehicles transporting logs.

Considering the specificity of the load, the priority of manifesting forces may differ; therefore, when selecting multiple criteria for the importance of loads, this scheme comes in handy when making a reasonable evaluation of the criteria.

5. Conclusions

During the transportation of cargo, the vehicle and the cargo itself are constantly affected by various physical forces, including gravity, centrifugal, frictional and inertial forces. These forces can make a vehicle uncontrollable, manifesting when a vehicle moves on an uneven trajectory, and suddenly stops or accelerates.

During the experimental evaluation, the forces of inertia and gravity were found have the greatest impact during the transportation of logs.

Having determined the interaction and its strength among all forces, a dependence was established, showing a strong to moderate correlation between all the investigated forces. There is a very strong correlation between centrifugal and inertial forces (0.93924) and gravity and frictional forces (0.99016); a strong correlation was also observed between centrifugal and frictional forces (0.72439) and centrifugal and gravity forces (0.7315); a moderate correlation was observed between inertial and frictional forces (0.54059) and inertial and gravity forces (0.59001).

Having analysed the factors on which the trinity and the centrifugal force depend, the interdependence and correlations of the factors were assessed, which showed a strong and medium correlation between all factors of the centrifugal force and a close-to-strong and strong correlation between all the factors of the trinity force.

Future research directions could involve the assessment of different forces by modelling various transportation situations, as well as the assessment of other types of cargo. Road elements such as roundabouts and islands were not considered because their speed and turning angle are significantly lower than in the analysed case. Inherent characteristics of the vehicles chassis and suspension could also be researched as complex road conditions, and they could also be the direction of long-range calculations and practical simulations. Another direction of scientific research could be associated with the evaluation of these parameters through the use of methodologies such as machine learning and fuzzy systems, such as artificial intelligence [42,43], which could improve such calculations.

Author Contributions: Conceptualization, J.M., K.Č., J.L. and E.S.; methodology, J.M., K.Č. and J.L.; software, J.M., K.Č. and J.L.; validation, J.M., K.Č., J.L. and E.S.; formal analysis, J.M., K.Č. and J.L.; investigation, J.M., K.Č. and J.L.; resources, J.M., K.Č., J.L. and E.S.; data curation, J.M., K.Č., J.L. and E.S.; writing—original draft preparation, J.M., K.Č. and J.L.; writing—review and editing, J.M. and K.Č.; visualization, J.M., K.Č. and J.L.; supervision, K.Č. and J.L.; project administration, J.M. and K.Č.; funding acquisition, E.S. All authors have read and agreed to the published version of the manuscript.

Funding: This research received no external funding.

Institutional Review Board Statement: Not applicable.

Informed Consent Statement: Not applicable.

Data Availability Statement: The data could be provided after request to Authors.

Conflicts of Interest: The authors declare no conflict of interest.

References

1. Barrachina, D.G.-L.; Boldizsar, A.; Zoldy, M.; Torok, A. Can Neural Network Solve Everything? In Case Study of Contradiction IN Logistic Processes with Neural Network Optimisation. In Proceedings of the 2019 Modern Safety Technologies in Transportation (MOSATT), Kosice, Slovakia, 28–29 November 2019; pp. 21–24.
2. Petraska, A.; Jarasuniene, A.; Ciziuniene, K. Routing Methodology for Heavy-Weight and Oversized Loads Carried by Rail Transport. *Procedia Eng.* **2017**, *178*, 589–596. [CrossRef]
3. Blatnický, M.; Dižo, J.; Sága, M.; Gerlici, J.; Kuba, E. Design of a Mechanical Part of an Automated Platform for Oblique Manipulation. *Appl. Sci.* **2020**, *10*, 8467. [CrossRef]
4. Kilikevičlenė, K.; Skeivalas, J.; Kilikevičlus, A.; PečEliūNas, R.; Bureika, G. The Analysis of Bus Air Spring Condition Influence upon the Vibration Signals at Bus Frame. *Eksploat. I Niezawodn. Maint. Reliab.* **2015**, *17*, 463–469. [CrossRef]
5. Matijošius, J.; Vasiliauskas, A.V.; Vasilienė-Vasiliauskienė, V.; Krasodomskis, Ž. The Assessment of Importance of the Factors That Predetermine the Quality of a Service of Transportation by Road Vehicles. *Procedia Eng.* **2016**, *134*, 422–429. [CrossRef]
6. Selech, J.; Andrzejczak, K. An Aggregate Criterion for Selecting a Distribution for Times to Failure of Components of Rail Vehicles. *EiN* **2019**, *22*, 102–111. [CrossRef]
7. Jarašūnienė, A.; Čižiūnienė, K. Ensuring Sustainable Freight Carriage through Interoperability between Maritime and Rail Transport. *Sustainability* **2021**, *13*, 12766. [CrossRef]
8. Oliveira, H.D.B.C.L.d.; Campilho, R.D.S.G.; Silva, F.J.G. Design of a Modular Solution for an Autonomous Vehicle for Cargo Transport and Handling. *Procedia Manuf.* **2019**, *38*, 991–999. [CrossRef]

9. Szymczyk, K.; Kadłubek, M. Challenges in General Cargo Distribution Strategy in Urban Logistics—Comparative Analysis of the Biggest Logistics Operators in EU. *Transp. Res. Procedia* **2019**, *39*, 525–533. [CrossRef]
10. Kogler, C.; Rauch, P. Lead Time and Quality Driven Transport Strategies for the Wood Supply Chain. *Res. Transp. Bus. Manag.* **2023**, *47*, 100946. [CrossRef]
11. Feng, X.; Song, R.; Yin, W.; Yin, X.; Zhang, R. Multimodal Transportation Network with Cargo Containerization Technology: Advantages and Challenges. *Transp. Policy* **2023**, *132*, 128–143. [CrossRef]
12. Bai, Y.; He, P.; Zhao, Y.; Ma, S.; Mi, H.; Wei, Z.; Li, Z. Real-Time Online Detection of Trucks Loading via Genetic Neural Network. *Autom. Constr.* **2020**, *120*, 103354. [CrossRef]
13. *White Paper on Transport: Roadmap to a Single European Transport Area: Towards a Competitive and Resource-Efficient Transport System*; European Commission, Ed.; Publications Office of the European Union: Luxembourg, 2011.
14. Xiang, X.; Chen, J.; Yuan, X.; Wan, B.; Zhuang, Y.; Zhang, Y.; Tu, G. Cross-Flow Transition Model Predictions of Hypersonic Transition Research Vehicle. *Aerosp. Sci. Technol.* **2022**, *122*, 107327. [CrossRef]
15. *EN 12642*; Securing of Cargo on Road Vehicles-Body Structure of Commercial Vehicles-Minimum Requirements. iTeh Standards: Toronto, ON, Canada, 2006.
16. Bordón, M.R.; Montagna, J.M.; Corsano, G. Solution Approaches for Solving the Log Transportation Problem. *Appl. Math. Model.* **2021**, *98*, 611–627. [CrossRef]
17. Lee, S.; Kim, B.; Huh, M.; Cho, S.; Park, S.; Lee, D. Mining Transportation Logs for Understanding the After-Assembly Block Manufacturing Process in the Shipbuilding Industry. *Expert Syst. Appl.* **2013**, *40*, 83–95. [CrossRef]
18. Nørstebø, V.S.; Johansen, U. Optimal Transportation of Logs and Location of Quay Facilities in Coastal Regions of Norway. *For. Policy Econ.* **2013**, *26*, 71–81. [CrossRef]
19. Bordón, M.R.; Montagna, J.M.; Corsano, G. An Exact Mathematical Formulation for the Optimal Log Transportation. *For. Policy Econ.* **2018**, *95*, 115–122. [CrossRef]
20. Warguła, Ł.; Kukla, M.; Krawiec, P.; Wieczorek, B. Impact of Number of Operators and Distance to Branch Piles on Woodchipper Operation. *Forests* **2020**, *11*, 598. [CrossRef]
21. Zöldy, M.; Szalay, Z.; Tihanyi, V. Challenges in Homologation Process of Vehicles with Artificial Intelligence. *Transport* **2020**, *35*, 435–446. [CrossRef]
22. Sfeir, T.d.A.; Pécora, J.E.; Ruiz, A.; LeBel, L. Integrating Natural Wood Drying and Seasonal Trucks' Workload Restrictions into Forestry Transportation Planning. *Omega* **2021**, *98*, 102135. [CrossRef]
23. Sibdari, S.Y.; Sepasi, A.H. A Simulation-Optimization Approach for Solving the Forestry Logistics Problem. *IFAC-PapersOnLine* **2022**, *55*, 3178–3183. [CrossRef]
24. Amrouss, A.; El Hachemi, N.; Gendreau, M.; Gendron, B. Real-Time Management of Transportation Disruptions in Forestry. *Comput. Oper. Res.* **2017**, *83*, 95–105. [CrossRef]
25. Jarašūnienė, A.; Sinkevičius, G.; Čižiūnienė, K.; Čereška, A. Adaptation of the Management Model of Internationalization Processes in the Development of Railway Transport Activities. *Sustainability* **2020**, *12*, 6248. [CrossRef]
26. 3-468-(E) Dėl Lietuvos Respublikos Susisiekimo Ministro 2002 m. Vasario 18 d. Įsakymo Nr. 3-66 "Dėl Maksimalių Leidžiamų Transporto Priemonių Matmenų, Leidžiamų Ašies (Ašių) Apkrovų, Leidžiamos Bendrosios Masės Patvirtinimo" Pakeitimo. Available online: https://e-seimas.lrs.lt/portal/legalAct/lt/TAD/3069c2a07b3811e49386e711974443ff (accessed on 6 April 2023).
27. Gillespie, T.D. *Fundamentals of Vehicle Dynamics*; Society of Automotive Engineers: Warrendale, PA, USA, 1992; ISBN 978-1-56091-199-9.
28. Minaker, B.P. *Fundamentals of Vehicle Dynamics and Modelling: B a Textbook for Engineers with Illustrations and Examples*; Automotive series; Wiley: Hoboken, NJ, USA, 2019; ISBN 978-1-118-98007-1.
29. Žuraulis, V.; Sokolovskij, E. Vehicle Velocity Relation to Slipping Trajectory Change: An Option for Traffic Accident Reconstruction. *PROMET* **2018**, *30*, 395–406. [CrossRef]
30. Surblys, V.; Žuraulis, V.; Sokolovskij, E. Estimation of Road Roughness from Data of On-Vehicle Mounted Sensors. *EiN* **2017**, *19*, 369–374. [CrossRef]
31. Deng, L.; Wang, W.; Cai, C.S. Effect of Pavement Maintenance Cycle on the Fatigue Reliability of Simply-Supported Steel I-Girder Bridges under Dynamic Vehicle Loading. *Eng. Struct.* **2017**, *133*, 124–132. [CrossRef]
32. Čygas, D.; Žilionienė, D.; Vitkienė, J. *Automobilių Kelių Projektavimas*; Vilnius Gediminas Technical University: Vilnius, Lithuania, 2009; ISBN 978-9955-28-491-8.
33. Warguła, Ł.; Kukla, M.; Wieczorek, B. The Impact of Wheelchairs Driving Support Systems on the Rolling Resistance Coefficient. *IOP Conf. Ser. Mater. Sci. Eng.* **2020**, *776*, 012076. [CrossRef]
34. Buczaj, A.; Krzysiak, Z.; Pecyna, A.; Caban, J.; Brumercik, F. Safety during Chemical Transport of Dangerous Goods. *Przemysl Chemiczny* **2019**, *98*, 1276–1280. [CrossRef]
35. Nieoczym, A.; Caban, J.; Dudziak, A.; Stoma, M. Autonomous Vans—The Planning Process of Transport Tasks. *Open Eng.* **2020**, *10*, 18–25. [CrossRef]
36. Rebaïne, F.; Bouazara, M.; Rahem, A.; St-Georges, L. Static and Vibration Analysis of an Aluminium and Steel Bus Frame. *WJM* **2018**, *8*, 112–135. [CrossRef]
37. Pravilonis, T.; Sokolovskij, E.; Kilikevičius, A.; Matijošius, J.; Kilikevičienė, K. The Usage of Alternative Materials to Optimize Bus Frame Structure. *Symmetry* **2020**, *12*, 1010. [CrossRef]

38. Kilikevičius, A.; Kilikevičienė, K.; Fursenko, A.; Matijošius, J. The Analysis of Vibration Signals of Critical Points of the Bus Body Frame. *Period. Polytech. Transp. Eng.* **2020**, *48*, 296–304. [CrossRef]
39. Kilikevičius, A.; Kasparaitis, A. Dynamic Research of Multi-Body Mechanical Systems of Angle Measurement. *Int. J. Precis. Eng. Manuf.* **2017**, *18*, 1065–1073. [CrossRef]
40. Maskeliūnaitė, L.; Sivilevičius, H. Identifying the Importance of Criteria for Passenger Choice of Sustainable Travel by Train Using ARTIW and IHAMCI Methods. *Appl. Sci.* **2021**, *11*, 11503. [CrossRef]
41. Sivilevičius, H.; Maskeliūnaitė, L. The Model Assessing the Impact of Price and Provided Services on the Quality of the Trip by Train: MCDM Approach. *E+M* **2019**, *22*, 51–67. [CrossRef]
42. Sabzalian, M.H.; Alattas, K.A.; Aredes, M.; Alanazi, A.K.; Abo-Dief, H.M.; Mohammadzadeh, A.; Mobayen, S.; França, B.W.; Fekih, A. A New Immersion and Invariance Control and Stable Deep Learning Fuzzy Approach for Power/Voltage Control Problem. *IEEE Access* **2022**, *10*, 68–81. [CrossRef]
43. Yan, S.-R.; Tian, M.; Alattas, K.A.; Mohamadzadeh, A.; Sabzalian, M.H.; Mosavi, A.H. An Experimental Machine Learning Approach for Mid-Term Energy Demand Forecasting. *IEEE Access* **2022**, *10*, 118926–118940. [CrossRef]

applied sciences

MDPI

Article

An Approach to Guide the Search for Potentially Hazardous Scenarios for Autonomous Vehicle Safety Validation

Tchoya Florence Koné [1], Eric Bonjour [2,*], Eric Levrat [3], Frédérique Mayer [2] and Stéphane Géronimi [4]

1 Stellantis, Université Lorraine (FR), 54000 Nancy, France
2 Research Team on Innovative Processes (ERPI Laboratory), Université de Lorraine, 8 rue Bastien Lepage, 54000 Nancy, France
3 Research Centre for Automatic Control of Nancy (CRAN Laboratory UMR CNRS 7039), Université de Lorraine, 54506 Vandœuvre-lès-Nancy, France
4 Stellantis, 78140 Vélizy-Villacoublay, France
* Correspondence: eric.bonjour@univ-lorraine.fr

Abstract: Safety validation of Autonomous Vehicles (AV) requires simulation. Automotive manufacturers need to generate scenarios used during this simulation-based validation process. Several approaches have been proposed to master scenario generation. However, none have proposed a method to measure the potential hazardousness of the scenarios with regard to the performance limitations of AV. In other words, there is no method offering a metric to guide the search for potentially critical scenarios within the infinite space of scenarios. However, designers have knowledge of the functional limitations of AV components depending on the situations encountered. The more sensitive the AV is to a situation, the more safety experts consider it to be critical. In this paper, we present a new method to help estimate the sensitivity of AV to logical situations and events before their use for the generation of concrete scenarios submitted to simulators. We propose a characterization of the inputs used for sensitivity analysis (definition of the context of the automation function, generation of functional and logical situations with their associated events). We then propose an approach to set up a distribution function that will make it possible to select situations and events according to their importance in terms of sensitivity. We illustrate this approach by implementing it on the Traffic Jam Chauffeur (TJC) function. Finally, we compare the obtained sensitivity rank with expert judgment to demonstrate its relevance. This approach has been shown to be a promising method to guide the search for potentially hazardous scenarios that are relevant to the simulation-based safety validation process for AV.

Keywords: road safety; autonomous vehicles; safety validation; sensitivity metric; safety of the intended functionality; logical situation; scenario generation; simulation-based validation

Citation: Koné, T.F.; Bonjour, E.; Levrat, E.; Mayer, F.; Géronimi, S. An Approach to Guide the Search for Potentially Hazardous Scenarios for Autonomous Vehicle Safety Validation. *Appl. Sci.* **2023**, *13*, 6717. https://doi.org/10.3390/app13116717

Academic Editors: Edgar Sokolovskij and Vidas Žuraulis

Received: 3 April 2023
Revised: 27 May 2023
Accepted: 28 May 2023
Published: 31 May 2023

1. Introduction

The standard ISO 26262 [1] is the international reference used in the automotive field for functional safety. This standard deals with the electrical and electronic malfunctions that vehicles may face, and which can lead to safety issues. However, in the context of Autonomous Vehicles (AV), ISO 26262 is limited. Indeed, it does not address safety violations which may be caused by performance limitations in the absence of failure. It has also been observed that, with AV, the driver may be out of the decision loop in contrast to human-controlled vehicles. Consequently, we cannot rely on him/her to deal with the deviations resulting from the behavior of other road users or unusual road conditions. However, automotive manufacturers need to ensure that AVs are completely safe before their deployment on roads.

For this reason, another standard known as ISO 21448 [2] was recently proposed. Its objectives are (i) to consider the AV's functional performance limitations, which may occur in the absence of failures, and (ii) to cover the reasonably foreseeable misuse of the vehicle.

This new standard ISO 21448 is to be used complementarily with the ISO 26262 standard. It is commonly called SOTIF (Safety Of the Intended Functionality), which means: "absence of unreasonable risk due to hazards resulting from functional insufficiencies of the intended functionality or from reasonably foreseeable misuse by persons" (ISO 21448, 2022).

The technologies used to develop an AV are specific: sensors, localization systems, communication systems and intelligent control systems (potentially including Artificial Intelligence based algorithms). These technologies are subject to limitations which are mainly due to the environment features in which the vehicle will operate, e.g., rain, fog, snow, degradation of pavements, puddles, curvature of roads, lack of road markings [3,4]. This environment is very complex and its description involves a large number of parameters which must be identified. Therefore, the activities detailed in the standard are intended to define an efficient process for identifying and dealing with scenarios that will be critical for the AV during its operational phase. According to the standard, critical scenarios relate to two areas: the first refers to "known unsafe scenarios" (area 2) and the second to "unknown unsafe scenarios" (area 3). To explore these two areas, simulation-based validation approaches need to be used to complement driving validation [5]. This is all the more true given that Kalra and Paddock [6] have demonstrated that the validation of AV by driving would require covering billions of kilometers. The use of simulation has many advantages, including testing controllability and path planning strategies, simulating different ranges of operational parameters, ensuring reproducibility and efficiency of the tests [7,8], testing specific collision avoidance strategies [9], etc. Previous authors have worked on the validation of automotive models and propose an assessment process that is not only focused on safety validation, but also on the quantification of modeling errors and uncertainties of the simulation compared to real driving [10].

However, the process of implementing simulation-based safety validation must allow sufficient coverage of the operational environment of the AV and meet the "mile of driving" requirement. Thus, there is a need for an approach which can identify the set of test scenarios and contribute to an efficient simulation-based validation process [11]. Such an approach would make it possible to select the scenarios that the AV has to face, in particular those that are considered to be challenging, to achieve the validation objectives more quickly. The main issue is how to identify these scenarios. On the road, frequent and often unchallenging scenarios (considered as nominal) will be encountered but rare and challenging scenarios may not be encountered. The simulation must therefore be oriented to test the AV with critical scenarios that are particularly interesting in order to demonstrate the safety of the vehicle.

Some approaches have been suggested in the industrial domain to help identify these test scenarios [12]: feedback from experience, feedback from users, accidentology databases, information extracted from driving, consortiums and projects such as MOOVE or PEGASUS.

However, despite the efforts made by the car manufacturers, it is difficult to predict all the real-life situations that AV will face, due to many variations in environmental conditions (traffic conditions, weather conditions, infrastructure, behaviors of other road users). As a result, other recommendations to control the scenario generation process have been proposed. The first is to deploy the AV by level of automation. In this case, the vehicle is limited to a number of tactical maneuvers and can carry out its mission in a dedicated and controlled environment called ODD (Operational Design Domain). Another proposal is to structure scenarios into three levels of abstraction in order to gradually identify them: functional, logical and concrete [13]. The functional scenario describes all entities and their relationships in a human-understandable linguistic form. The second level, the logical scenario, uses the functional scenario to describe the state space using parameter ranges. Finally, the concrete scenario allows concrete values to be assigned to the previously defined parameters. This is obtained by choosing a value for each parameter from the range of values defined in the corresponding logical scenario. Structuring scenarios in three levels of abstraction is interesting because it allows a scenario-based approach for the development of AV, while offering a hierarchical approach to structuring the generation process within the infinite space of scenarios.

These approaches seem interesting and helpful in the scenario identification process. However, to the best of our knowledge, there is no approach which proposes a metric to identify potentially critical scenarios in advance, in order to guide scenario generation before the simulation. In the literature, several performance indicators (criteria and metrics) have been proposed for the validation of AV [14–16]. Most of these indicators are used during and after simulation or driving to evaluate the safety level of AV. This means that the proposed indicators can only be used when the scenarios are run, before the measurement is made. In other words, these indicators cannot be used to guide scenario generation, as they only qualify the situation after the simulation.

Therefore, in this paper, we propose an approach to analyze and estimate the sensitivity of AV to logical situations and events, in order to guide the search for scenarios that are relevant to the simulation-based safety validation process.

The rest of the paper is organized as follows. Section 2 is a more detailed review of the state of the art. Firstly, it reviews the performance indicators for scenario generation and simulation-based validation of the safety level of AV. Secondly, it reminds the concepts related to scenario generation strategies in the context of AV. Section 3 provides a quick overview of the proposed approach. Then, as previously mentioned, a structuring of scenarios into functional, logical and concrete levels is used, as proposed by Menzel et al. [13]. Since the sensitivity metric that we want to implement requires knowledge of the operational situations at stake, the ranges of values of the parameters, at the logical level of abstraction, are specified. In Section 4 we include a step to achieve the generation of situations and potentially associated events at the functional and logical levels. Even if we have this structural framework (functional, logical, concrete) for scenario generation, its operationalization is not immediate. Unresolved questions include: which elements are relevant for functional level characterization, how can value ranges of the logical scenarios be defined, or finally, how can these ranges be combined to obtain logical situations? We also identify and characterize the usage context of an automation function in Section 4, as this is necessary for the characterization of functional and logical scenarios. Once logical situations have been defined, Section 5 proposes a sensitivity analysis of AV to logical situations and events in order to prioritize the exploration of those scenarios that pose risks for the AV. The proposed sensitivity analysis is then described in Section 6. Section 7 summarizes the results, compares our results with those of an expert and addresses some discussion points. In Section 8, we conclude this paper and propose further work that can be addressed in the future. Abbreviation part sums up the acronyms and variables used in this paper.

2. Related Work

Different metrics have been proposed in the literature to measure the performance of tests performed for the safety validation of AV. The most widely mentioned metric is test coverage, which is commonly encountered in software engineering. This is defined as follows: "degree, expressed as a percentage, to which specified test coverage items have been exercised by a test case or test cases" (ISO/IEC/IEEE 29119-1 [17]), Indeed, it is necessary to justify that the scenarios used during simulation for the validation of the AV cover its entire operational environment. Such a justification can be based on the fact that driving in real conditions contains "dead periods" which are not useful for the validation. On the other hand, the use of scenarios allows focus only on what is relevant from a validation point of view. However, the large number of parameters influencing the behavior of the AV creates a combinatorial explosion that makes it impossible to test all possible scenarios. For this reason, Amersbach and Winner [15] proposed to determine another indicator: the number of scenarios to be tested. This provides an equivalence to the distance indicator proposed by Kalra and Paddock (2014). Indeed, in order to cover a target distance for the validation of the AV (depending on the level of automation), the corresponding number of scenarios required for testing is determined. Zhou and Re [16] prefer to use directly the coverage indicator of the number of critical scenarios by using databases in which thousands of scenarios are recorded for reference. The reference database used by the authors is the Second Strategic Highway Research Program Naturalistic Driving Study (SHRP 2 NDS). This database includes 775 critical highway events. The idea is to check what percentage of the critical scenarios in the baseline were covered by the scenarios generated by Zhou and Re.

In recent years, advances have also been made regarding the effectiveness of Advanced Driver Assistance System (ADAS) functions in the case of possible road crashes. Authors have provided risk metrics for scenarios where a crash may be inevitable. For instance, Vangi et al. [18] proposed criteria to intervene on braking and steering, based on the injury risk to vehicle occupants. According to Alvarez et al. [19], to estimate the safety benefits of ADAS and active safety systems, "the most frequently used metrics are the number (or reduction in the number expressed as percentage) of avoided accidents and the number of avoided injuries". Gulino et al. [20] defined a Crash Momentum Index (CMI) and demonstrated its effectiveness in assessing the performance of ADAS in critical road scenarios. Smit et al. [21] pointed out the interest of developing appropriate testing methods to assess the safety capabilities of ADAS (avoiding accidents or mitigating the injury severity). They proposed an impact model which aims to predict collision parameters (here delta-v). These are correlated with the risk of becoming injured and based on the re-simulation of real accidents. Such a model is useful to assess the effects of ADAS in terms of injury severity in accidents. However, even if the sensitivity of the model is evaluated for some input parameters, it is not sufficient for Automated Driving Systems (ADS). In fact, the scenario space to be covered for the safety validation of ADS is too large. In addition, there is little feedback on accidents related to these ADS and their behavior, as they have not yet been introduced on the road.

Furthermore, since safety validation is required to ensure that the AV can safely react in all the scenarios it will face, another important consideration is how to measure and assess the hazardousness of a scenario during validation. ISO 26262 assesses the risk associated with a scenario through a combination of severity, exposure and controllability: "combination of the probability of occurrence of harm (§1.56) and the severity (§1.120) of that harm" (ISO 26262-1, 2011). Feng et al. [22] also proposed a definition of critical scenario: "The criticality of a scenario measures the importance in evaluating a performance metric". Here, the performance is related to the following indicators "Safety, Functionality, Mobility, Rider's comfort". Another definition is given by Hallerbach et al. [23], in these words: "Critical scenarios are defined as scenarios that need to be tested, regardless, whether the requirements are functional or non-functional".

However, the criticality referred to by these authors can only be effectively measured once the simulation has been carried out and the reaction of the AV observed. Furthermore, the number of scenarios that need to be tested is very high and the identification of all critical scenarios can be slow, and yet we have not found in the literature a metric to guide the generation of scenarios before simulation, by identifying the potentially critical scenarios beforehand. In this paper, we thus propose a sensitivity metric to fill this gap. The notion of "sensitivity" mentioned here is quite different from the usual meaning, and consists in analyzing how the uncertainty in a system's output (e.g., a failure probability, an injury severity probability) can be attributed to the uncertainty in its input variables [24,25].

Four key concepts appear in the context of scenario generation for the safety validation of AVs: scenario, scene, situation and event. Definitions have been proposed for each of these by several authors [26–29]. The desire to establish a consensus has given rise to reflections, notably in the framework of ISO 21448. The definitions proposed by Ulbrich et al. were considered as references in the preliminary version of this standard. However, Koné et al. [30] found that some terms, such as "scenery" and "self-representations". were still under discussion. They also noted that the definition proposed by Ulbrich et al. for the concept of scene, used to define a scenario, remains unclear: "A scene describes a snapshot of the environment including the scenery and dynamic elements, as well as all actors' and observers' self-representations, and the relationships among those entities. Only a scene representation in a simulated world can be all-encompassing (objective scene, ground truth). In the real world it is incomplete, incorrect, uncertain, and from one or several observers' points of view (subjective scene)".

This definition highlights that a scene can contain a great deal of information and its implementation can be very subjective and vary from person to person. Furthermore, the term "*snapshot*" makes the elements of the environment lose their stable character (no loss of movement or duration of lifespan).

As an alternative, Koné et al. [30] took a closer look at the concept of situation. They found that the term "operational situation" is often considered in scenarios for describing accidents [31] or in methods for determining "hazardous events" [32].

Let us remember that we seek knowledge of the state of the AV and the elements of the situation [33]. We need to know which conditions must be true to solicit a reaction from the AV [28]. As a result, Koné et al. point out that focusing the scenario description on the situation makes it possible to highlight the influencing factors and interactions between the elements of the situation and the AV. This is particularly relevant because the situation also indicates a restrictiveness related to the relevant elements of the scene and is therefore a more interesting level of abstraction than the scene.

In this work, the concepts that will be of interest are scenario, operational situation and event. The definitions proposed by Koné et al. serve as a reference to define these concepts [30]:

Operational situation: set of states of the AV and other entities in an operational environment present for a period of time, defined by the stability of these states.

A scenario: description, in a given time interval, of the temporal sequence of operational situations. The transition from one (initial) situation to another (intermediate) is caused by one or more events.

Event: modification of the state of one or more environmental entities (AV and others) which induces, can induce or must induce a new action from the AV (new behavior). Action is the reaction of the AV following an event (definition proposed from the perspective of the AV).

In summary, this section pointed out the limitations of existing approaches to generate test scenarios and defined key concepts that will be used in the following sections.

The rest of this document develops the steps necessary to set up the AV sensitivity analysis to logical situations and events. The first step is to define the usage context of an automation function and the process for obtaining functional and logical situations, with their associated events.

3. Overview of the Proposed Method

In this section, we present the general structure of the method, which results in the analysis of the sensitivity of AV to situations and events. It is structured according to the following steps:

Define the usage context of the automation function to be tested (Section 4.1)
Generate situations and events at the functional level (Section 4.2)
Generate situations at the logical level (Section 4.3)
Analyze the sensitivity of the AV to logical situations and define the sensitivity distribution function (Section 5.1)
Analyze the sensitivity to events and define the sensitivity distribution function (Section 5.2)

These different steps are detailed in the following sections.

4. Context Definition of an Automation Function, Generation of Situations and Events at the Functional and Logical Abstraction Levels

In this section, we present the steps for generating scenarios at the functional and logical levels. The first step consists in defining the scope of the generation by choosing the automation function concerned and defining its usage context.

4.1. *Define the Usage Context of the Automation Function to Be Tested*

Once the automation function and its automation level have been identified, we need to describe the operational environment in which it will be used. This operational environment is called ODD (Operational Design Domain). It is described in such a way as to emphasize the concept of operational situation. This provides knowledge of the state of the AV (i.e., the expected properties, the level of automation and the corresponding speed range), and of the other entities (i.e., the physical infrastructure, the weather conditions and the other road users). This activity is carried out in support with the functional designer, as he/she has knowledge of the specification of the function and the mission profile related to customer needs.

The expected properties of the AV, such as its functional performances, are among the elements needed to define the usage context. These functional performances must be identified in order to deepen the knowledge of the function and to know what to test during the validation.

Likewise, we also need to specify the operational modes in which the AV may find itself and the maneuvers it may perform during its use. This information is useful to accurately qualify the behavior of the vehicle at the occurrence of any event. Then, these modes and maneuvers must be characterized by specifying their validity criteria. This information helps to qualify the events whose occurrence generates changes or transitions between operational modes and maneuvers.

4.1.1. Example: Choice of the Automation Function

The contributions of this paper will be illustrated on the "Traffic Jam Chauffeur (TJC)" function. The TJC is a level 3 automation function, designed to support autonomous driving on divided lanes (motorway or expressway), in congested conditions, where the speed is limited to a maximum speed Vmax (in the range of 60–70 km/h). The Ego vehicle (the vehicle equipped with the automation function, corresponding here to the AV) follows the vehicle in front of it, keeps to its lane, manages acceleration and braking within the authorized speed limit, and keeps a safe distance. The driver can perform other activities but must remain alert and be able to regain control of the vehicle within 10 s when prompted by the system. The system should allow the driver to activate the function and to intervene at any time, and should give them enough time to do so.

4.1.2. Example of TJC Context Description

Table 1 gives the entities of the ODD, i.e., the type of infrastructure in which the function is to be used, the other users and the weather conditions are defined. Functional constraints, corresponding to the TJC function, are defined in the column "Ego Vehicle". They concern the speed limit for the function (in our example, the speed is limited to 60 km/h), the need for a driver and examples of unauthorized operational maneuvers.

As regards operational modes, several are valid for TJC, including Activation, Give Back and Minimum Risk Maneuver (MRM) modes. The system is in Activation mode when the automation function is activated. The system switches to Give Back mode when the function can no longer guarantee its nominal operation, either because the availability conditions are no longer met or a failure is detected. The MRM mode corresponds to a reduct, ion of risk in the case where the driver does not respond to a request to take over. This may be a vehicle deceleration phase. Moreover, "Car Following" is the main maneuver selected for the TJC function. The validity conditions of AV are as follows: the function must be in Activation mode; a vehicle must be in front of the Ego (this will be named lead vehicle or OtherVehicle.1); the inter-distance between Ego and lead vehicle is greater or equal to the minimum safe distance. However, several functional performances are expected from the function in terms of situation perception, trajectory control and reaction to the situation. For perception, the system should, for example, have the following functionalities: identify the appropriate type of infrastructure; identify the specified weather conditions; identify other relevant objects. Otherwise, certain performance limitations may be observed, including incomplete/limited perception of the situation (e.g., the vehicle does not see all elements of the situation) or misperception of the elements of the situation (e.g., a motorbike is seen as a car).

Table 1. TJC's ODD description.

Physical Infrastructure	Other Road Users	Weather Conditions	Ego Vehicle
• Type of infrastructure: divided lanes (highway, expressway) • Roadway type: separate carriageway • No intersection, no roundabout, no pedestrian crossing, no stop lights, no giving way, no priority entry, no U-turn, no working zone, no tunnel	• No pedestrian, no obstacle, no wrong way driver	• Type of precipitation: not specified • Type of brightness: not specified • Temperature: not specified	• Equipped with TJC • Speed range: $\leq$60 km/h • Presence of Driver = Yes • No lane changing • Vehicle dynamic: no strong deceleration, no strong acceleration, no strong action on the steering wheel • No trailer and no towing vehicle are attached
Other domain constraints			
• On type of lane: Ego must not be on shoulder lane or merging lane or ending lane • No obstacle • Traffic Jam condition: 1 vehicle in front of Ego, speed range $\leq$ 60 km/h, speed difference (between road users) $\leq$ 10 km/h (meaning that traffic density = congested)			

All the information gathered during the definition of the usage context will allow the generation of scenarios according to the three abstraction levels: functional, logical and concrete.

4.2. Generate Situations and Events at the Functional Level

Operational maneuvers allow the description of vehicle behaviors and can be identified so as to be mutually exclusive. Since the aim is to test the behavior of the AV, the operational maneuver seems to be a relevant starting point to address the scenario generation problem. Thus, like most work in the context of the AV, we propose to characterize functional scenarios by the corresponding operational maneuvers. For example, one can have a functional scenario for "car following" or "lane changing". In order to give more details for this characterization, it is necessary to define the functional situations and events which can be found in a functional scenario.

4.2.1. Generate the Functional Situations

Although we defined the situation as describing the state of the AV and of the other entities of its operational environment, when generating the functional situations, we have to remain consistent with the definition of the functional level given by Menzel et al. At this stage of the identification, the detailed information on the parameters and their values is not to be filled in, the aim being simply to know the configurations of situations that will have to be considered when generating the test scenarios. Therefore, the identification of the functional situations proposed here is mainly based on the information collected during the definition of the automation function, its usage context or the operational maneuvers.

Figure 1 shows the process for obtaining the functional situations. These are obtained by combining the information provided by the ODD (type of traffic density, type of infrastructure), and the operational modes and maneuvers.

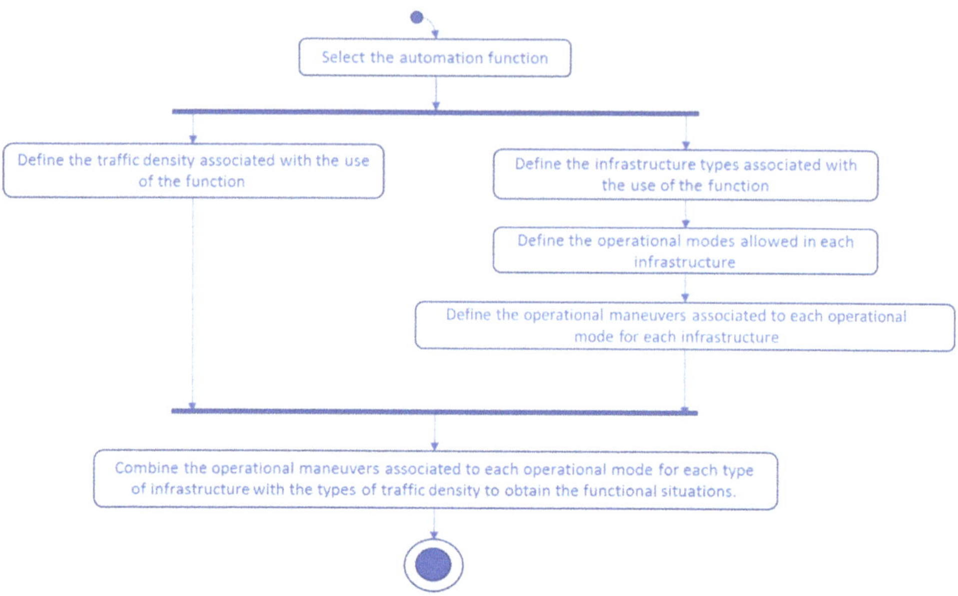

Figure 1. The process used to obtain the functional situations.

Examples for the TJC function: Traffic density can be fluid, congested or accordion-like. It is related to the number of users present in the traffic (i.e., around the AV). If Number_of_users < 3, then traffic is fluid, otherwise traffic is congested or accordion-like.

For the infrastructure, several types can be encountered: divided lanes (motorway, expressway), toll area, work zone, road tunnel, etc.

By considering the TJC function and the description of its usage context, we can deduce the functional situations which need to be considered for the generation. Table 2 illustrates the six functional situations obtained.

In the reminder of this example, we focus on a specific functional situation: Ego in a "Car following" situation in the "Activation" mode on a motorway and in congested traffic.

4.2.2. Types of Events That Can Occur within a Functional Situation

An event is defined as a change in the state of one or more entities in the environment (AV and other entities) that induces, may induce or must induce a new action by the AV (i.e., new behavior). In other words, it means that a parameter acquires a new value in a range different from the range of its initial value. Consequently, the change of the value of a parameter modelling the operational environment of the AV corresponds to the generation of an event. Some events are critical and may increase the risk associated with the scenario, and others are not. We have identified all the factors that can cause limitations in the performance of the AV. They can be organized into events occurring within the ODD and those occurring outside the ODD. The first category includes events related to performance limitations of the automation function, deviations in the behaviors of other road users and transitions in the operational maneuvers of road users. The second category mainly includes events related to the constraints of the use of the function and the transitions between the operational modes of the AV.

Table 2. Example of the obtained functional situations in TJC case.

Type of Traffic Density Associated with the Use of the Function: Congested			List of Functional Situations
The types of infrastructure associated with the function: Motorway, Expressway	The operational modes in the "Motorway" infrastructure: Activation, GiveBack, MRM	The operational maneuver associated with the "Activation" mode for the Motorway infrastructure: Car following	• Ego in "Car following" situation in the "Activation" mode on a Motorway and in congested traffic.
		The operational maneuver associated with the "GiveBack" mode for the Motorway infrastructure: Car following	• Ego in "Car following" situation in "GiveBack" mode on a Motorway and in congested traffic
		The operational maneuver associated with the "MRM" mode for the Motorway infrastructure: Car following	• Ego in a "Car following" situation in the "MRM" mode on a Motorway and in congested traffic.
	The operational modes in the Expressway infrastructure: Activation, GiveBack, MRM	The operational maneuver associated with the "Activation" mode for the Expressway infrastructure: Car following	• Ego in "Car following" situation in "Activation" mode on an Expressway and in congested traffic.
		The operational maneuver associated with the "GiveBack" mode for the Expressway infrastructure: Car following	• Ego in the "Car following" situation in "GiveBack" mode on an Expressway and in congested traffic.
		The operational maneuver associated with the "MRM" mode for the Expressway infrastructure: Car following	• Ego in the "Car following" situation in the "MRM" mode on an Expressway and in congested traffic.

Furthermore, it is important to note that the concept of event considered here covers the notion of "triggering condition" defined in the Standard 21448 SOTIF as follows: "Triggering conditions: specific conditions of a scenario (3.23) that serve as an initiator for a subsequent system reaction, possibly leading to a hazardous behavior". This standard also states that an identification of triggering conditions can be supported by a detailed description of the environment model.

Once the functional situations and event types have been identified, the next section defines the various parameters and identifies their value ranges so they can be used in the rest of the generation process.

4.3. Generate Situations at the Logical Level

The aim is to identify the value ranges of the parameters of all the entities (Infrastructure, Other users, Drivers/Passengers, Atmospheric conditions, AV) observed in the functional situation, or which may be subject to an event. Depending on the way in which the variation spaces are divided into ranges, there will be a greater or reduced number of logical situations and events to be tested at the end of the combinations.

We make the following modelling hypothesis: expert knowledge allows us to partition parameter variation spaces and to define a minimum number of ranges for each parameter, depending on its impact on the AV. The AV will be more or less sensitive to each range.

4.3.1. Define Parameters and Value Ranges

Different parameters (variables) have to be considered for the generation of the scenarios and, more specifically, for the description of the scenarios. For examples, we have qualitative variables (nominal, ordinal, binary) such as "Type of users (pedestrian, cyclist, vehicle)" or "Level of brightness (Low, Medium, High, Very high), and quantitative variables (discrete or continuous) such as "Number of lanes [(1, 2, 3 and more)" or "Vehicle_speed (0–30 km/h).

We define the sensitivity of AV to a range of value as follows: the degree to which the range of value can lure the automation function.

As a situation is defined by a combination of different parameters, the more the AV is sensitive to the range of value of each parameter, the more it is sensitive to this situation.

As these parameters are numerous and most of them have continuous value ranges, we thought it appropriate to distinguish value ranges which make a significant contribution in estimating the achievement of the AV safety performance.

This distinction is made with the help of a safety expert. However, partitioning the value ranges into classes (i.e., exclusive but complementary value ranges) is intended to ensure that each possible situation that the vehicle may encounter and to which it may be sensitive is properly defined. For each parameter, the safety expert is asked to define different classes of variation. For each class, he/she examines the impact of that class on each AV technology. If two close classes have a similar impact, they may be combined into a single class. The number of ranges for each parameter must be determined according to the different sensitivities of the AV technologies, but the safety expert must reduce this number as much as possible.

Consider the example of the "rainfall level" parameter. Its space variation is from 1 to 100 mm/h. We propose to divide it into two classes of variation: Light = 1–7 mm/h and Heavy = 8–100 mm/h. As some sensor technologies are sensitive to heavy rainfall, such a partition makes it possible to highlight the range of heavy rainfall. When combined with other parameters, this range further increases the possibility of logical scenarios that have a high impact on criticality.

4.3.2. Generate the Set of Logical Situations Associated with Each Functional Situation

Logical situations are obtained by combining the value ranges of the parameters describing each functional situation. A systematic combination method is used, i.e., each of the subspaces of variation of each parameter appears in each combination obtained. In addition, special care must be taken when combining value ranges. Constraints between the value ranges may be defined in order to ensure that inconsistencies in the obtained situations are avoided.

Suppose the following parameters are used for the description of the scenarios: Number_lanes, Curvature, Speed_Ego (km/h), Rainfall level and Speed_OtherVehicle.1 (km/h). The value ranges for each parameter are shown in Table 3.

Table 3. Example of parameters and value ranges for obtaining logical situations.

Parameters	Value Ranges	Identification
Number_lanes	2 3 or more	A1_1 A1_2
Curvature	Low High	A2_1 A2_2
Speed_Ego	Low (0, 60 km/h)	A3_1
Rainfall level	Light Heavy	A4_1 A4_2
Speed_OtherVehicle.1 (km/h)	Low (0, 60 km/h)	A5_1

By combining the ranges of these five parameters in a systematic way (2 ranges × 2 ranges × 1 range × 2 ranges × 1 range), we obtain the eight logical situations described in Table 4.

Table 4. Examples of logical situations obtained by combining the variation ranges.

Logical situation 1 (Y1): A1_1, A2_1, A3_1, A4_1, A5_1 = 2 lanes * Curvature.Low * Speed_Ego.Low * Rain.Light * Speed_OtherVehicle.1.Low
Logical situation 2 (Y2): A1_1, A2_2, A3_1, A4_1, A5_1 = 2 lanes * Curvature.High * Speed_Ego.Low * Rain.Light * Speed_OtherVehicle.1.Low
Logical situation 3 (Y3): A1_1, A2_1, A3_1, A4_2, A5_1 = 2 lanes * Curvature.Low * Speed_Ego.Low * Rain.Heavy * Speed_OtherVehicle.1.Low
Logical situation 4 (Y4): A1_1, A2_2, A3_1, A4_2, A5_1 = 2 lanes * Curvature.High * Speed_Ego.Low * Rain.Heavy * Speed_OtherVehicle.1.Low
Logical situation 5 (Y5): A1_2, A2_1, A3_1, A4_1, A5_1 = (3 and more) lanes * Curvature.Low * Speed_Ego.Low * Rain.Light * Speed_OtherVehicle.1.Low
Logical situation 6 (Y6): A1_2, A2_2, A3_1, A4_1, A5_1 = (3 or more) lanes * Curvature.High * Speed_Ego.Low * Rain.Light * Speed_OtherVehicle.1.Low
Logical situation 7 (Y7): A1_2, A2_1, A3_1, A4_2, A5_1 = (3 and more) lanes * Curvature.Low * Speed_Ego.Low * Rain.Heavy * Speed_OtherVehicle.1.Low
Logical situation 8 (Y8): A1_2, A2_2, A3_1, A4_2, A5_1 = (3 or more) lanes * Curvature.High * Speed_Ego.Low * Rain.Heavy * Speed_OtherVehicle.1.Low

The number of logical situations obtained depends on the number of parameters considered and the ranges of values associated with each of them. A combinatorial explosion can be observed as soon as the parameters are numerous and the space of variation of each one is broken down into two or three ranges. This means that the decision to split or not a space of variation into ranges of values, and the identification of the constraints inherited from the ODD and the function specification, must be done carefully, and in collaboration with experts. The number of ranges for each parameter is determined according to the difference of sensitivity of the AV technologies. This procedure aims to limit the number of combinations and then the number of logical situations generated.

Obtaining logical situations is the phase which precedes the generation of concrete scenarios. The high number of concrete scenarios that can be obtained led us to propose a sensitivity analysis of the AV to logical situations and events in order to prioritize the exploration of those that are at risk for the AV.

5. Sensitivity Analysis of AV to Situations and Events for the Generation of Concrete Scenarios

In order to identify the potentially critical scenarios necessary for the simulation-based validation of the AV, we will proceed to an analysis of its sensitivity to logical situations and events. The aim of this analysis is to focus on the logical situations and events most likely to induce a risk for the AV, and to accelerate the validation process.

5.1. Analyze the Sensitivity of AV to Logical Situations and Define the Sensitivity Distribution Function

In this part, we start by analyzing the sensitivity of AV to logical situations. Then, we will establish a distribution function on the sensitivities of the logical situations.

5.1.1. Definition of the Sensitivity of AV to Logical Situations

The calculation of the a priori sensitivity of the AV to logical situations is obtained in two steps. First, we consider the a priori sensitivity of the AV to each value range characterizing the generation parameters. This is done with the functional architecture of the AV composed as follows: perception, decision and actuation. Five sensor technologies ensure perception functionalities and are grouped as follows: radar (long-range and short-range), ultrasonic, lidar, stereoscopic (3D vision) and monocular cameras. The proposed process for determining sensitivity is not tied to specific components. It offers the possibility of being adaptable to the specificities of the components that will be used on a given vehicle. Thus, for each considered range, with the help of the expert, we answer the following question: is the sensor technology under consideration (or decision or actuation) sensitive in this range?

The expected answer from the expert is either yes or no. Then, the scale described in Table 5 makes it possible to determine the level of sensitivity of the AV to the range according to the impacted AV component(s). Next, all the responses obtained will be used to determine the a priori sensitivity of the AV to each logical situation. Indeed, the logical situation is a combination of ranges of values. We propose to calculate the a priori sensitivity of the AV to a combination of ranges (and therefore to a logical situation) by aggregating the sensitivities associated with the ranges.

Table 5. Proposal for a scale to measure a priori sensitivity of AV to a range of values.

Description	Corresponding Sensitivity Value
The range of values misleads the considered sensor technology	0.2
The range of values misleads the five sensor technologies	
The range of values misleads the decision made by the AV (in the absence of any lure to the sensors)	1
The range of values misleads the actuation	

We assume that each component (perception, decision, actuation) makes a maximum contribution of 1 to the overall sensitivity of a logical situation. Since the AV has five main types of sensor technologies, each sensor technology, therefore, is considered to make a contribution of 0.2 towards the sensitivity to the logical situation if it is impacted by at least one range of variation. The decision made by the AV (in the absence of any lure of the sensors) or the actuation makes, respectively, a contribution of 1 in the calculation of sensitivity to the logical situation, when it is misled by at least one of the ranges of values composing the logical situation. Table 6 synthetizes the contribution of each component in obtaining the sensitivity of a logical situation.

Table 6. Contribution of the sensitivity of an AV component in obtaining the sensitivity to a logical situation.

Description	Corresponding Sensitivity Value
The sensor technology under consideration is sensitive to at least one of the value ranges making up the logical situation	0.2
The decision made by the AV (in the absence of any lure to the sensors) is sensitive to at least one of the value ranges making up the logical situation	1
The actuation is sensitive to at least one of the value ranges making up the logical situation	

Furthermore, when generating scenarios, the randomness of their occurrences must be considered. We, therefore, propose the implementation of a sensitivity distribution function.

5.1.2. Sensitivity Distribution Function on Logical Situations

The sensitivity distribution function $q(Y_j)$ is obtained by normalizing the AV sensitivities to the logical situations. This distribution function makes it possible to draw, one after the other, the logical situations which will be used to generate the concrete scenarios for subsequent simulations. As we can see, with such a distribution, the logical situations with high sensitivity values will also have high probability values and will have higher occurrences during the sampling. Thus, the sensitivity distribution function is based on an importance function that allows us to prioritize the sampling of logical situations Y_j for which the a priori sensitivity value of the AV is high.

The probability distribution of $q(Y_j)$ is defined as follows:

$$\begin{cases} q(Y_j) \in [0,1] \\ \sum_{Y_j} q\,(Y_j) = 1 \end{cases}$$

Let Sj be the sensitivity value of the AV to the logical situation Y_j and let $W = \sum_{Yj} S_j$ be the normalisation factor. Thus, the sensitivity or importance distribution function can be defined as follows:

$$q(Y_j) = \frac{S_j}{W}$$

We can check that $\sum_{Y_j} q\,(Y_j) = \sum_{Y_j} \frac{S_j}{W} = 1$.

We have defined how to lead a sensitivity analysis on the logical situations. However, in order to generate scenarios, it is also necessary to create events. In this respect, as for the logical situation, a sensitivity analysis of the events will be introduced in order to draw events according to their potential impact on the safety of the AV.

5.2. Analyze the Sensitivity to Events and Define the Sensitivity Distribution Function

An event is the modification of the state of one or more entities in the environment (the AV and others) which induces, can induce or must induce a new action from the AV. At the logical level, situations are obtained by combining ranges of variation. Thus, the event is formulated at the level of the change of value range for a parameter.

The entry point for identifying events is the initial logical situation. The event can be a change in the value of a parameter remaining in the same range of values with a certain level of sensitivity (the AV remains a priori in the same logical situation). It can also be a change in the value of a parameter from the initial range of values (of the initial logical situation) to another range (of another logical situation).

5.2.1. Sensitivity Analysis and Gradient

We have identified three possible levels of impact of the event:

- Functional impact: once generated, this event leads to the transition from one functional situation to another.
- Logical impact: once generated, this event leads to the transition from one logical situation to another, within the same functional situation, fixed at the beginning.
- Concrete impact: once generated, this event leads to the transition from one concrete situation to another, within the same logical situation set at the start.

Figure 2 illustrates the three impact levels. Consider the logical situation LS1 belonging to the functional situation FS1. The functional impact is the modification of a range of values of LS1 that switches to a logical situation of another functional situation, for instance, FS3. With the logical impact, the modification of a range of values of LS1 causes a switch to another logical situation (for instance, LS3 also belonging to FS1). Finally, at the level of the concrete impact, no range of LS1 is modified (the values that are modified by the event remain in the same value ranges).

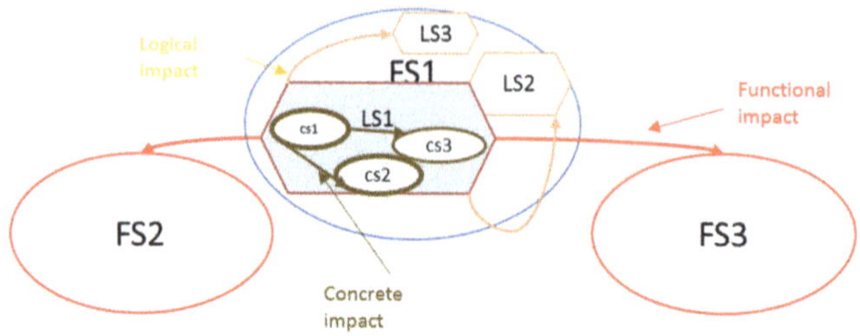

Figure 2. The three impact levels of an event.

A range of LS1 values is considered relevant as an event if it has a non-zero sensitivity level. The drawing of a concrete event value within this range causes the AV to move from one concrete situation to another within LS1 (for instance, from CS1 to CS2). Thus, by assumption, we consider that the other events (which concern the other ranges with a null sensitivity value) are not relevant because the AV is not sensitive to these events.

These three levels of impact of the event suggest the existence of a variation in the associated sensitivity level. When the event concerns a modification of the value of a parameter in the same range of values (impact at the concrete level), its sensitivity is obtained with the a priori sensitivity value of the AV to the range, as defined in the scale in Table 2. It will be noted that Δ(Ax_y), with Ax_y as the value range.

When the event is a modification with a change of the value range (logical or functional impact), it will be noted that Δ(Ax_y, Ax_z). Its sensitivity is then calculated using a concept we call "sensitivity gradient". The sensitivity gradient is defined as the absolute value of the difference in sensitivity between two logical situations. It is thus associated with the event that causes the transition from the initial logical situation to an intermediate situation.

The sensitivity gradient of the event impacting the logical level is obtained by the absolute value of the difference in sensitivity between the initial and the intermediate logical situation (within the same functional situation). Only the level 1 neighborhood of the logical situations is considered, i.e., starting from an initial logical situation (for instance, LS1), the intermediate logical situations to be considered are those which differ by only one range of values. This means that we only consider here the generation of a "single event".

When generated, the event impacting the functional level takes the AV out of the FS under analysis. If this is the case, it means that the new range of values is either outside the validity conditions of the operational maneuver characterizing the FS or outside the ODD. To obtain the sensitivity gradient of this event which impacts the functional level, we are interested in the new range obtained as a result of the modification. We will first calculate the sensitivity associated with the range, using the scale in Table 2, and then add +1 to this value. Indeed, given that such an event induces a change to another FS, an additional difficulty is added to the AV decision component. To consider this, we decide to add +1 to the sensitivity obtained for the change in the range of values.

5.2.2. Sensitivity Distribution Function on Events

To obtain the sensitivity distribution function, the same principles as those used in the case of logic apply. The sensitivity distribution function is obtained by normalizing the sensitivities or sensitivity gradients of events. As already noted, the considered logical situation makes it possible to identify the events that can be associated with it. Thus, for each logical situation Y_j, there is a list E_j of events e_{ji}. The sensitivity value of each event e_{ji} belonging to this list will be normalized to obtain a sensitivity distribution function $q(e_{ji})$.

The probability distribution of $q(e_{ji})$ is defined as follows:

$$\begin{cases} q(e_{ji}) \in [0,1] \\ \sum_{Y_j} q\left(e_{ji}\right) = 1 \end{cases}$$

Let $S(e_{ji})$ be the sensitivity value of an event e_{ji} and let $W_{E_j} = \sum_{e_{ji}} S(e_{ji})$ be the normalization factor for the list E_j of events e_{ji}. Thus, the sensitivity distribution function can be defined as follows:

$$q(e_{ji}) = \frac{S(e_{ji})}{W_{Ej}} \ \forall\, e_{ji} \in E_j$$

We can check that: $\sum_{e_{ji}} q\,(e_{ji}) = \sum_{e_{ji}} \frac{S(e_{ji})}{W_{Ej}} = 1$.

The proposals made in this section on the sensitivity of the AV to logical situations and events aim to provide a layer of risk analysis when declining the scenarios according to the three levels of abstraction: functional, logical and concrete. Such a consideration helps to guide the generation of scenarios by prioritizing the situations to which the AV has a high sensitivity value. In Section 6, we illustrate this principle of sensitivity analysis of the AV to logical situations and events.

6. Results: Application of the Sensitivity Analysis on the "Traffic Jam Chauffeur"

This example consists first in studying the sensitivity of the AV to the logical situations and then in using the result to establish a distribution function of the sensitivity over the logical situations. The types of events faced by the AV, the usage constraints imposed by its ODD and the ranges of variations characterizing the logical situations are defined. Thanks to this information, an identification of the events according to the three types of impact is made for each logical situation. Finally, a sensitivity study is carried out on the events of each logical situation in order to have a sensitivity distribution function per list of events associated with a logical situation.

6.1. Analyze the Sensitivity of the AV to Logical Situations and Define the Sensitivity Distribution Function

Let us consider the eight logical situations obtained in Table 4.

6.1.1. Sensitivity Analysis of AV to the Logical Situations

The sensitivity of the AV to these logical situations is presented in the Appendix A. When the sensor technology (or decision or actuation components) is sensitive to the Ax_x value range, this is indicated by the corresponding value using the scale in Table 5. Otherwise, it means that there is no sensitivity. Then, Table 6 provides the sensitivity value of a logical situation by aggregating the range sensitivity values.

None of the AV components is sensitive to one of the value ranges that characterize the logical situation Y_1 (Table A1).

For Y_2 (Table A2), the AV design experts judge that the A2_2 range (Curvature.High) has an impact on two sensors (contribution of 0.4 to the sensitivity).

In the case of Y_3 (Table A3), we can note that the A4_2 range (Rain.Heavy) can mislead three sensor technologies (contribution of 0.6 to the sensitivity) and actuation (contribution of 1 to the sensitivity) and this leads to a sensitivity of 1.6.

In case of the logical situation Y_4 (Table A4), the same sensitivity is obtained for Y_4. A4_2 range makes the same contribution as for Y_3 and since the range A2_2 may mislead the same sensors than A4_2, this leads to the same sensitivity of 1.6.

In the case of Y_5 (Table A5), we have a sensitivity of 1.4. This is due to the range A1_2 (number of lanes = 3 and more) which can lure both two sensor technologies (contribution of 0.4 to the sensitivity) and the decision (contribution of 1 to the sensitivity calculation). A sensitivity of 1.4 is also obtained for Y_6 (Table A6) since, complementarily to A1_2, the range A2_2 has an impact on the same two sensors.

Finally, Y_7 (Table A7) and Y_8 (Table A8) have the highest sensitivity value. They include both the A1_2 (number of lanes = three and more) and A4_2 (Rain.Heavy) ranges, while Y_8 additionally includes A2_2. The respective effects of each range on the AV components are cumulative, resulting in a sensitivity of 2.6.

In addition, the randomness of the occurrence of the scenarios in relation to the presence of uncertainty in the operational environment of the AV has to be considered when generating the scenarios. Thus, the sensitivity of the AV to a logical situation will be used to set up a sensitivity distribution function. This will allow a random selection of logical situations while prioritizing the occurrence of those to which the AV is the most sensitive.

6.1.2. Sensitivity Distribution Function on Logical Situations

The distribution of the sensitivity of the AV to logical situations, as presented in Section 5.1, is obtained as follows:

$q(Y_j) = \frac{S_j}{W}$ and $\sum_{Y_j} q\,(Y_j) = \sum_{Y_j} \frac{S_j}{W} = 1$, with $W = \sum_{Y_j} S_j$ the normalization factor, Y_j the logical situation j and S_j its sensitivity.

Considering the eight logical situations and their sensitivity, we obtain

$$W = \sum_{j=1}^{8} S_j = 0 + 0.4 + 1.6 + 1.6 + 1.4 + 1.4 + 2.6 + 2.6 = 11.6$$

Thus, for each of the above logical situations, the sensitivity distribution is:

- Logical situation Y_1: $q(Y_1) = \frac{0}{11.6} = 0$
- Logical situation Y_2: $q(Y_2) = \frac{0.4}{11.6} = 3.45 \times 10^{-2}$
- Logical situations Y_5 and Y_6: $q(Y_5) = q(Y_6) = \frac{1.4}{11.6} = 12.07 \times 10^{-2}$
- Logical situations Y_3 and Y_4: $q(Y_3) = q(Y_4) = \frac{1.6}{11.6} = 13.79 \times 10^{-2}$
- Logical situations Y_7 and Y_8: $q(Y_7) = q(Y_8) = \frac{2.6}{11.6} = 22.41 \times 10^{-2}$

Obviously, the value obtained for $q(Y_7)$ and $q(Y_8)$ is greater because the sensitivity of the AV to Y_7 or Y_8 is stronger. Note also that the total sum of probability for the eight situations is equal to 1 by definition.

Therefore, we have five classes, which in terms of priority can be considered as follows:

- Priority 1: $Y_8 = Y_7$
- Priority 2: $Y_4 = Y_3$
- Priority 3: $Y_6 = Y_5$ (rather close to priority 2)
- Priority 4: Y_2
- Priority 5: Y_1

6.2. Sensitivity Analysis of AV to Events and Distribution Function

The events associated with each logical situation are of three types: those impacting the functional level, those impacting the logical level and those impacting the concrete level.

6.2.1. Sensitivity Analysis of AV to Events

For each logical situation, a list of events is associated with it. In order to identify this list, we will rely on the types of events that the AV may face in TJC conditions, the limits of the associated ODD and the value ranges of the logical situations. To illustrate this, let us consider Table 7, which shows the sensitivity of the logical situation Y_8 as defined in the Appendix A.

The events for this situation are as follows:

(1) Events impacting the concrete level concern the ranges to which at least one of the components of the AV is sensitive. Ranges A1_2, A2_2 and A4_2 are concerned. Their sensitivities are defined in Table 8. We note e_{ji} an event i of the logical situation j and $S(e_{ji})$ its sensitivity. Δ(Ax_y) is an event concerning a change of value within the same range of values Ax_y.

(2) Events impacting the logical level consist of a change in the range of values of the logical situation considered and concern what we have called the level 1 neighborhood. By analyzing the eight logical situations, those that differ from Y_8 by a single range of values are: Y_4, Y_6 and Y_7 (they represent the level 1 neighborhood of Y_8). Since

such an event triggers a transition from one logical situation to another, the gradient (i.e., the absolute value of the difference in sensitivity between the initial and final situation) is used to obtain the sensitivity. Table 9 gives the events and their sensitivity in the case of logical situation Y8. Δ(Ax_y, Ax_z) is a range change event.

(3) The events impacting the functional level trigger a transition of the AV out of the "Car following" situation. Considering the conditions of validity of the "Car following" situation and the constraints of the ODD mentioned in paragraph 4.2, the modifications of the value ranges (and thus the events) that make it leave the "Car following" situation are among others:

$e_{8.7}$ = Δ(A1_1, A1_0) (switching from two lanes to one lane: end of Ego lane)
$e_{8.8}$ = Δ(A1_2, A1_0) (switching from 3+ lanes to one lane: end of Ego lane)
$e_{8.9}$ = Δ(A5_1, A5_2) (switching from Speed_OtherVehicle.1.Low to Speed_Other Vehicle.1.High)

Table 7. Sensitivity of the logical situation Y_8.

	Techno 1 (Radar)	Techno 2 (Lidar)	Techno 3 (Ultrasonic)	Techno 4 (Stereo Camera)	Techno 5 (Mono Camera)	Decision	Actuation
Y_8							
A1_2				0.2	0.2	1	
A2_2				0.2	0.2		
A3_1							
A4_2		0.2		0.2	0.2		1
A5_1							
Sensitivity of Y_8/component		0.2		0.2	0.2	1	1
Sensitivity of Y_8				2.6			

Table 8. Sensitivity of events impacting the concrete level for Y_8.

$e_{8.1}$ = Δ(A1_2)	$e_{8.2}$ = Δ(A2_2)	$e_{8.3}$ = Δ(A4_2)
S($e_{8\cdot1}$) = 1.4	S($e_{8\cdot2}$) = 0.4	S($e_{8\cdot3}$) = 1.6

Table 9. Sensitivities of events impacting the logical level for Y8.

		Y_4	Y_6	Y_7
Y_8	Event impacting the logical level	$e_{8.4}$ = Δ(A1_2, A1_1)	$e_{8.5}$ = Δ(A4_2, A4_1)	$e_{8.6}$ = Δ(A2_2, A2_1)
	Sensitivity of the event	S($e_{8.4}$) = ∣2.6 − 1.6∣ = 1	S($e_{8.5}$) = ∣2.6 − 1.4∣ = 1.2	S($e_{8.6}$) = ∣2.6 − 2.6∣ = 0

The range A1_0 does not affect any technology (sensitivity contribution = 0) but its occurrence implies a change of lane for the Ego (addition of a factor of 1 to the sensitivity) and thus causes a change of functional situation; therefore, a sensitivity of 1 is obtained. The sensitivities of these events are shown in Table 10.

Table 10. Sensitivities of events impacting the functional level for Y8.

		Lane Changing Left	GiveBack
Functional Situation of "Car Following"	Event impacting functional situation	$e_{8.7} = \Delta(A1\_1, A1\_0)$ $e_{8.8} = \Delta(A1\_2, A1\_0)$	$e_{8.9} = \Delta(A5\_1, A5\_2)$
	Sensitivity of the event	$S(e_{8.7}) = 1$ $S(e_{8.8}) = 1$	$S(e_{8.9}) = 1$

About the range A5_2, its occurrence leads to a violation of one of the TJC conditions, which is that the speed must be less than a maximum value of 60 km/h (Table 1). This range does not have an impact on any technology but induces a new operational mode (i.e., "GiveBack"). Therefore, the difficulty of this change implies a sensitivity of 1.

The list E_8 of events associated with the logical situation Y_8 groups together the events identified in Tables 8–10 and contains a total of nine events ranging from $e_{8.1}$ to $e_{8.9}$.

As with the logical situations, a distribution function is introduced to generate the events according to their sensitivity.

6.2.2. Sensitivity Distribution Function on Events

To ensure that events with zero sensitivity may be also picked (e.g., this concerns $e_{8.5}$), a minimum non-zero sensitivity value must be assigned to them. We have chosen this value to be the lowest of all the sensitivity values in the list. Thus, we will assign a sensitivity of 0.1 to the event concerned.

The sensitivity distribution function is defined as follows:

$q(eji) = \frac{S(e_{ji})}{W_{Ej}}$ and $\sum_{eji} q\left(e_{ji}\right) = \sum_{eji} \frac{S(e_{ji})}{W_{Ej}} = 1$, where $W_{Ej} = \sum_{e_{ji}} S\left(e_{ji}\right)$ is the normalization factor for the list E_j of events e_{ji} and $S(e_{ji})$ the sensitivity of the event e_{ji}.

Considering the list E_8 of the nine events associated with the logical situation Y_8 and their sensitivity values, we find: $W_{E8} = \sum_{i=1}^{9} S(e_{8i}) = 8.7$.

The sensitivity distribution of the events is given in Table 11. Furthermore, since the sum of all probabilities is 1, we verify that $\sum_{i=1}^{9} q(e_{8i}) = 1$.

Table 11. Distribution of sensitivities of the events in the E_8 list.

Events	$e_{8\cdot1} = \Delta(A1\_2)$	$e_{8\cdot2} = \Delta(A2\_2)$	$e_{8\cdot3} = \Delta(A4\_2)$	$e_{8\cdot4} = \Delta(A1\_2, A1\_1)$	$e_{8\cdot5} = \Delta(A4\_2, A4\_1)$	$e_{8\cdot6} = \Delta(A2\_2, A2\_1)$	$e_{8\cdot7} = \Delta(A1\_1, A1\_0)$	$e_{8\cdot8} = \Delta(A1\_2, A1\_0)$	$e_{8\cdot9} = \Delta(A5\_1, A5\_2)$
Sensitivity values	$S(e_{8.1}) = 1.4$	$S(e_{8.2}) = 0.4$	$S(e_{8.3}) = 1.6$	$S(e_{8.4}) = 1$	$S(e_{8.5}) = 1.2$	$S(e_{8.6}) = 0.1$	$S(e_{8.7}) = 1$	$S(e_{8.8}) = 1$	$E(e_{8.9}) = 1$
Associated probabilities	$q(e_{8.1}) = 0.161$	$q(e_{8.2}) = 0.046$	$q(e_{8.3}) = 0.184$	$q(e_{8.3}) = 0.115$	$q(e_{8.4}) = 0.138$	$q(e_{8.5}) = 0.011$	$q(e_{8.6}) = 0.115$	$q(e_{8.7}) = 0.115$	$q(e_{8.9}) = 0.115$

6.3. *Comparing Our Results with Those of an Expert*

In order to evaluate the relevance of the proposed sensitivity assessment method, we decided to compare our sensitivity results (Table 12) obtained in Section 6.1 with the opinion of an expert.

To do that, we asked an expert to classify the same logical situations (all described in Table 4) from the most critical to the least critical.

The classification is performed using the pairwise comparison with the analogy of the well-known multi-criteria decision-making approach called Analytic Hierarchy Process (AHP). This approach aims to compare criteria and their importance. In our work, the aim is to compare situations according to their criticality. The question is: which criterion (resp. situation) is more important (resp. critical), and how much more on a scale of 1–9? The scale is as follows: 1—Equal Importance, 3—Moderate importance, 5—Strong importance, 7—Very strong importance, 9—Extreme importance.

Table 12. Summary of the sensitivity results obtained with the eight logical situations described in Section 6.1.

Sensitivity Distribution	Priority/Rank	Logical Situations
0	5	Y_1
3.45×10^{-2}	4	Y_2
12.07×10^{-2}	3	Y_6, Y_5
13.79×10^{-2}	2	Y_4, Y_3
22.41×10^{-2}	1	Y_8, Y_7

The first principle intuitively used by the expert to compare two situations was that "one situation is more critical than another if it has a higher number of ranges impacting the automation function". We obtained the results in Figure 3. The logical situations are classed from S1 to S8.

Priorities

These are the resulting weights for the criteria based on your pairwise comparisons:

	Cat	Priority	Rank
1	S1	2.8%	8
2	S2	5.9%	5
3	S3	5.9%	5
4	S4	15.2%	2
5	S5	5.9%	5
6	S6	15.2%	2
7	S7	15.2%	2
8	S8	34.1%	1

Number of comparisons = 28
Consistency Ratio CR = 1.6%

Decision Matrix

The resulting weights are based on the principal eigenvector of the decision matrix:

	1	2	3	4	5	6	7	8
1	1	0.33	0.33	0.20	0.33	0.20	0.20	0.14
2	3.00	1	1.00	0.33	1.00	0.33	0.33	0.20
3	3.00	1.00	1	0.33	1.00	0.33	0.33	0.20
4	5.00	3.00	3.00	1	3.00	1.00	1.00	0.33
5	3.00	1.00	1.00	0.33	1	0.33	0.33	0.20
6	5.00	3.00	3.00	1.00	3.00	1	1.00	0.33
7	5.00	3.00	3.00	1.00	3.00	1.00	1	0.33
8	7.00	5.00	5.00	3.00	5.00	3.00	3.00	1

Principal eigen value = 8.155
Eigenvector solution: 4 iterations, delta = 3.5E-8

Figure 3. The expert's classification of the logical situations using the first principle.

As we can observe, these results give a four-level classification (rank 1 for S_8, rank 2 for S_7, S_6 and S_4, rank 5 for S_5, S_3 and S_2, rank 8 for S_1) which is totally different from our classification results as mentioned in Table 12. Even if the consistency ratio of the expert is quite good as we obtain 1.6% ($0.016 < 0.1$), the problem with such a logic is that the differences between the impact degrees of the ranges are not taken into account: two different ranges affecting the same technology or component are considered as having the same risk level.

Thus, we proposed that another classification be made, but considering the fact that the ranges of values have their proper degree of impact on the sensor technologies or on decision and actuation components. Therefore, the new principle used by the expert is that "A situation is more critical than another if the number of sensor technologies, decision and actuation components that are sensitive to value ranges is more important". Figure 4 shows the resulting classification.

Priorities

These are the resulting weights for the criteria based on your pairwise comparisons:

	Cat	Priority	Rank
1	S1	2.6%	8
2	S2	3.1%	7
3	S3	12.2%	3
4	S4	12.2%	3
5	S5	7.6%	5
6	S6	7.6%	5
7	S7	26.9%	2
8	S8	27.6%	1

Number of comparisons = 28
Consistency Ratio CR = 3.2%

Decision Matrix

The resulting weights are based on the principal eigenvector of the decision matrix:

	1	2	3	4	5	6	7	8
1	1	0.50	0.20	0.20	0.20	0.20	0.20	0.14
2	2.00	1	0.20	0.20	0.33	0.33	0.14	0.14
3	5.00	5.00	1	1.00	2.00	2.00	0.33	0.33
4	5.00	5.00	1.00	1	2.00	2.00	0.33	0.33
5	5.00	3.00	0.50	0.50	1	1.00	0.25	0.25
6	5.00	3.00	0.50	0.50	1.00	1	0.25	0.25
7	5.00	7.00	3.00	3.00	4.00	4.00	1	1.00
8	7.00	7.00	3.00	3.00	4.00	4.00	1.00	1

Principal eigen value = 8.310
Eigenvector solution: 5 iterations, delta = 7.5E-8

Figure 4. The expert's classification of the logical situations using the second principle.

With this new logic, we get a six-level classification (rank 1 for S_8, rank 2 for S_7, rank 3 for S_3 and S_4, rank 5 for S_6 and S_5, rank 7 for S_2, rank 8 for S_1) and the consistency ratio is also quite good ($0.032 < 0.1$).

When comparing with our results (see Table 12), which gave a five-level classification, we can deduce that there is a similarity. In fact, the main difference is that, in our results, situations Y_7 and Y_8 (called S_7 and S_8 by the expert) are considered as belonging to the same class, whereas in the expert's results they have been split into two classes. However, the expert could have gathered them together, as their priority values are very close (26.9% and 27.6% in Figure 4). The other situations belong to the same classes and ranks in both classifications. The experts consider such a classification of sensitivity values for these logical scenarios as relevant and recognize that their former intuitive logic is less rational than the one used subsequently.

7. Discussion

Originality. The two main contributions of this paper are a method for analyzing the sensitivity of AV to logical situations and associated events, and, more specifically, the setting up of a sensitivity distribution function for both logical situations and events. We also define and apply this method on the TJC example to characterize the context of usage of an automation function, as well as the functional and logical abstraction levels. The reasoning is based on the three levels of abstractions (functional, logical and concrete scenarios) proposed by Menzel et al. [13]. Despite its relevance, this structuring cannot obtain all the scenarios corresponding to each level of abstraction. Therefore, we bring an approach that identifies which elements to combine in order to specify each level. Moreover, the sensitivity analysis we introduce offers the possibility of putting forward the ranges of values that can have a high criticality impact on the behavior of the AV.

Practical application. With our approach, the characterization of the logical scenarios is done so as to identify potentially critical scenarios before their submission to the simulator. In contrast to the classical safety metrics, the sensitivity metric proposed here brings an additional tool to the scenario generation strategy. The application to the TJC case study shows that the proposed sensitivity metric is relevant. On the one hand, our metric offers a discriminant classification of logical situations according to the sensitivity of the AV and, on the other hand, the comparison with an expert's classification presents similar results. When facing situations where the number of parameters increases, the expert is no longer able to compare situations 2 by 2 to rank the sensitivity of the logical situations and, consequently, a metric like the proposed sensitivity metric can be used.

Our focus in this paper is on validation and sensitivity analysis for each automation function, here the chosen example is the TJC. The features are the operational modes and maneuvers that define the functional situations (see Figure 1 and Table 2) and the limits of sensor technologies, actuators and decision modules. Our approach does not require a detailed description of the characteristics of the components, the operation and the limits of the AV.

We are aware that the proposed method has some limits that require further development. Each step of the proposed method is based on modelling assumptions and may introduce errors.

The first hypothesis is related to the decomposition of the VA operation by automation function (Section 4.1). This approach is relevant to automation level 3 but may require simulation tests with scenarios integrating all the automation functions to validate levels 4–5.

The second hypothesis is related to an epistemic uncertainty concerning the model of situations and events at the functional level (Section 4.2). The relevant entities and parameters are identified by means of expert knowledge, accident scenario databases, etc. However, relevant entities and parameters describing the environment may be forgotten.

The third hypothesis concerns the division of the value ranges for each parameter with expert knowledge (Section 4.3). These value ranges are influenced by the experts' subjectivity and may vary according to the specific technology that equips the studied VA (e.g., the value range of a "low rain level" may vary depending on the vehicle manufacturer). The process of obtaining logical situations may be subject to the resolution of a combinatorial problem due to the combination among the value ranges of parameters. However, we have chosen to opt for a systematic combination of ranges. The motivation behind this choice is explained by the fact that splitting the ranges according to their sensitivity, carried out with the support of the AV designers and safety experts, limits the number of ranges to the those that are strictly necessary. In addition, although the splitting is done with the support of experts, the relevance of the division into ranges of values for each parameter must be verified, especially as it guides the whole strategy of sensitivity analysis and then the search for critical scenarios.

The fourth hypothesis is related to the choice of the scale to determine the level of sensitivity of each AV technology to logical situations and events (Sections 5.1 and 5.2). When calculating the sensitivity gradient for an event impacting the functional level, we chose to add a value of 1 to consider the difficulty in decision that such an event induced (as it triggers a transition to another functional situation). The relevance of the value "1" may be subject to discussion and to a sensitivity analysis of the proposed metric.

The final limitation is that we only consider the neighborhood of level 1 during the identification of events. This choice is justified by the desire to limit the number of combinations by considering only the unitary events that lead from one logical situation to another.

8. Conclusions and Future Work

In this paper, we propose an approach for measuring the sensitivity of AV to logical situations and events it may face, as well as a characterization of functional and logical scenarios. The aim is to guide the generation of scenarios which are tested for the simulation-based validation of AV, by selecting those which are the most likely to induce a harmful event or accident. Thus, the sensitivity distribution functions make it possible to randomly select situations and events according to their importance in terms of sensitivity for the AV. Thus, during the sampling, the more sensitive logical situations and events will be, the more they will tend to be generated often and then simulated.

Our work contributes to the demonstration of AV safety through scenario generation and the use of simulation. This contribution is a first step to effectively support the deployment of a safe AV.

However, additional work needs to be carried out. First, a key perspective is to verify that highlighting the sensitivity of the scenarios reduces the number of scenarios to be tested while meeting the expected performance criteria. In addition, a heuristic for exploring and generating concrete scenarios should be set up and implemented in order to evaluate the use of the sensitivity analyses we have just proposed. Then, a sensitivity analysis of the proposed sensitivity metric should be carried out either by introducing biases on the sensitivity values estimated by the expert or by modifying the amplitude of the value ranges defined for the parameters. A fuzzy approach for these ranges could be considered and compared to the proposed method that is based on fixed ranges.

The notion of "neighborhood" introduced can be extended to different levels (neighborhood of level X, with $X > 1$). Furthermore, the coupling of the parameters (e.g., constraints between their ranges, types of infrastructure and presence of intersections) should be analyzed and considered in order to further reduce the combinatorial explosion of the generation.

Author Contributions: T.F.K.: Conceptualization; data curation; formal analysis; investigation; methodology; validation; writing—original draft preparation. E.B.: Conceptualization; funding acquisition; methodology; project administration; supervision; visualization; writing—original draft preparation; writing—review & editing. E.L.: Conceptualization; funding acquisition; methodology; supervision; visualization; writing—review & editing. F.M.: Conceptualization; supervision; visualization; review & editing. S.G.: Conceptualization; funding acquisition; methodology; project administration; supervision; validation; review & editing. All authors have read and agreed to the published version of the manuscript.

Funding: This research was funded by ANRT (the French National Association of Research and Technology (Convention CIFRE N° 2017/1246) as well as STELLANTIS (ex GROUPE PSA)).

Institutional Review Board Statement: Not applicable.

Informed Consent Statement: Not applicable.

Data Availability Statement: No new data were created or analyzed in this study. Data sharing is not applicable to this article.

Acknowledgments: This work has been carried out under the financial support of ANRT (the French National Association of Research and Technology (Convention CIFRE N° 2017/1246) as well as STELLANTIS (ex GROUPE PSA).

Conflicts of Interest: The authors declare no conflict of interest. The authors have no competing interests to declare that are relevant to the content of this article. The authors fully respect the "Ethical Responsibilities of Authors" included in the Authors' guidelines of the Journal.

Abbreviations

ADAS	Advanced Driver-Assistance System
ADS	Automated Driving System
AHP	Analytic Hierarchy Process
AV	Autonomous Vehicles
CS	Concrete Situation
FS	Functional Situation
LS	Logical Situation
MRM	Minimum Risk Maneuver
ODD	Operational Design Domain
SOTIF	Safety Of The Intended Functionality
TJC	Traffic Jam Chauffeur

Ax_y	specific value range y of parameter x
Yj	specific logical situation j
E_j	list of events in logical situation j
e_{ji}	event N° i in logical situation j
Sj	sensitivity value of the AV to the logical situation Y_j
$q(Y_j)$	sensitivity distribution function on logical situation Y_j
$S(e_{ji})$	sensitivity value of an event e_{ji}
$q(e_{ji})$	sensitivity distribution function on event e_{ji}
W_{E_j}	normalization factor for the list E_j
$\Delta(Ax\_y)$	modification of the value of parameter x in the same range y of values
$\Delta(Ax\_y, Ax\_z)$	modification of the value of parameter x with a change of the value range

Appendix A

Table A1. Sensitivity of the logical situation Y_1.

	Techno 1 (Radar)	Techno 2 (Lidar)	Techno 3 (Ultrasonic)	Techno 4 (Stereo Camera)	Techno 5 (Mono Camera)	Decision	Actuation
Y1							
A1_1							
A2_1							
A3_1							
A4_1							
A5_1							
Sensitivity of Y_1/component							
Sensitivity of Y_1				0			

Table A2. Sensitivity of the logical situation Y_2.

	Techno 1 (Radar)	Techno 2 (Lidar)	Techno 3 (Ultrasonic)	Techno 4 (Stereo Camera)	Techno 5 (Mono Camera)	Decision	Actuation
Y_2							
A1_1							
A2_2				0.2	0.2		
A3_1							
A4_1							
A5_1							
Sensitivity of Y_2/component				0.2	0.2		
Sensitivity of Y_2				0.4			

Table A3. Sensitivity of the logical situation Y_3.

	Techno 1 (Radar)	Techno 2 (Lidar)	Techno 3 (Ultrasonic)	Techno 4 (Stereo Camera)	Techno 5 (Mono Camera)	Decision	Actuation
Y_3							
A1_1							
A2_1							
A3_1							
A4_2		0.2		0.2	0.2		1
A5_1							
Sensitivity of Y_3/component		0.2		0.2	0.2		1
Sensitivity of Y_3				1.6			

Table A4. Sensitivity of the logical situation Y_4.

	Techno 1 (Radar)	Techno 2 (Lidar)	Techno 3 (Ultrasonic)	Techno 4 (Stereo Camera)	Techno 5 (Mono Camera)	Decision	Actuation
Y_4							
A1_1							
A2_2				0.2	0.2		
A3_1							
A4_2		0.2		0.2	0.2		1
A5_1							
Sensitivity of Y_4/component		0.2		0.2	0.2		1
Sensitivity of Y_4				1.6			

Table A5. Sensitivity of the logical situation Y_5.

	Techno 1 (Radar)	Techno 2 (Lidar)	Techno 3 (Ultrasonic)	Techno 4 (Stereo Camera)	Techno 5 (Mono Camera)	Decision	Actuation
Y_5							
A1_2				0.2	0.2	1	
A2_1							
A3_1							
A4_1							
A5_1							
Sensitivity of Y_5/component				0.2	0.2	1	
Sensitivity of Y_5				1.4			

Table A6. Sensitivity of the logical situation Y_6.

	Techno 1 (Radar)	Techno 2 (Lidar)	Techno 3 (Ultrasonic)	Techno 4 (Stereo Camera)	Techno 5 (Mono Camera)	Decision	Actuation
Y_6							
A1_2				0.2	0.2	1	
A2_2				0.2	0.2		
A3_1							
A4_1							
A5_1							
Sensitivity of Y_6/component				0.2	0.2	1	
Sensitivity of Y_6				1.4			

Table A7. Sensitivity of the logical situation Y_7.

	Techno 1 (Radar)	Techno 2 (Lidar)	Techno 3 (Ultrasonic)	Techno 4 (Stereo Camera)	Techno 5 (Mono Camera)	Decision	Actuation
Y_7							
A1_2				0.2	0.2	1	
A2_1							
A3_1							
A4_2		0.2		0.2	0.2		1
A5_1							
Sensitivity of Y_7/component		0.2		0.2	0.2	1	1
Sensitivity of Y_7				2.6			

Table A8. Sensitivity of the logical situation Y_8.

	Techno 1 (Radar)	Techno 2 (Lidar)	Techno 3 (Ultrasonic)	Techno 4 (Stereo Camera)	Techno 5 (Mono Camera)	Decision	Actuation
Y_8							
A1_2				0.2	0.2	1	
A2_2				0.2	0.2		
A3_1							
A4_2		0.2		0.2	0.2		1
A5_1							
Sensitivity of Y_8/component		0.2		0.2	0.2	1	1
Sensitivity of Y_8				2.6			

References

1. *ISO 26262*; Road Vehicles—Functional Safety. ISO: Geneva, Switzerland, 2011.
2. *ISO 21448*; Road Vehicles—Safety of the Intended Functionality. ISO: Geneva, Switzerland, 2022.
3. Ponn, T.; Muller, F.; Diermeyer, F. Systematic analysis of the sensor coverage of automated vehicles using phenomenological sensor models. In *IEEE Intelligent Vehicles Symposium, Proceedings*; IEEE: Piscataway, NJ, USA, 2019; pp. 1000–1006. [CrossRef]
4. Ignatious, H.A.; El-Sayed, H.; Khan, M.A. Sensor Technology for Autonomous Vehicles. *Encycl. Sens. Biosens.* **2023**, *4*, 35–51. [CrossRef]

5. Li, C.; Sifakis, J.; Wang, Q.; Yan, R.; Zhang, J. *Simulation-Based Validation for Autonomous Driving Systems*; Association for Computing Machinery: New York, NY, USA, 2023; Volume 1. [CrossRef]
6. Kalra, N.; Paddock, S.M. *Driving to Safety*; RAND Corporation: Santa Monica, CA, USA, 2014.
7. Sun, D.; Elefteriadou, L. A Driver Behavior-Based Lane-Changing Model for Urban Arterial Streets. *Transp. Sci.* **2014**, *48*, 184–205. [CrossRef]
8. Thorn, E.; Kimmel, S.; Chaka, M. A Framework for Automated Driving System Testable Cases and Scenarios. Report No. Dot Hs 812 623. 2018. p. 180. Available online: https://www.nhtsa.gov/sites/nhtsa.dot.gov/files/documents/13882-automateddrivingsystems_092618_v1a_tag.pdf (accessed on 27 May 2023).
9. Razzaq, S.; Dar, A.R.; Shah, M.A.; Khattak, H.A.; Ahmed, E.; El-Sherbeeny, A.M.; Lee, S.M.; Alkhaledi, K.; Rauf, H.T. Multi-Factor Rear-End Collision Avoidance in Connected Autonomous Vehicles. *Appl. Sci.* **2022**, *12*, 1049. [CrossRef]
10. Riedmaier, S.; Schneider, D.; Watzenig, D.; Diermeyer, F.; Schick, B. Model validation and scenario selection for virtual-based homologation of automated vehicles. *Appl. Sci.* **2021**, *11*, 35. [CrossRef]
11. Riedmaier, S.; Ponn, T.; Ludwig, D.; Schick, B.; Diermeyer, F. Survey on Scenario-Based Safety Assessment of Automated Vehicles. *IEEE Access* **2020**, *8*, 87456–87477. [CrossRef]
12. Koné, T.F.; Bonjour, E.; Levrat, E.; Mayer, F.; Géronimi, S. Safety Demonstration of Autonomous Vehicles: A Review and Future Research Questions. In *Complex Systems Design & Management*; Springer International Publishing: Cham, Switzerland, 2020; pp. 176–188. [CrossRef]
13. Menzel, T.; Bagschik, G.; Maurer, M. Scenarios for Development, Test and Validation of Automated Vehicles. In Proceedings of the 2018 IEEE Intelligent Vehicles Symposium (IV), Changshu, China, 26–30 June 2018. [CrossRef]
14. Jesenski, S.; Stellet, J.E.; Branz, W.; Zöllner, J.M. Simulation-Based Methods for Validation of Automated Driving: A Model-Based Analysis and an Overview about Methods for Implementation. In Proceedings of the 2019 IEEE Intelligent Transportation Systems Conference, ITSC 2019, Auckland, New Zealand, 27–30 October 2019; pp. 1914–1921. [CrossRef]
15. Amersbach, C.; Winner, H. Defining Required and Feasible Test Coverage for Scenario-Based Validation of Highly Automated Vehicles. In Proceedings of the 2019 IEEE Intelligent Transportation Systems Conference (ITSC), Auckland, New Zealand, 27–30 October 2019; pp. 425–430. [CrossRef]
16. Zhou, J.; Re, L. Reduced Complexity Safety Testing for ADAS & ADF. *IFAC-Pap.* **2017**, *50*, 5985–5990. [CrossRef]
17. *ISO/IEC/IEEE 29119-1*; Software and Systems Engineering—Software Testing—Part 1: Concepts and Definitions. ISO: Geneva, Switzerland, 2013.
18. Vangi, D.; Virga, A.; Gulino, M.-S. Adaptive intervention logic for automated driving systems based on injury risk minimization. *Proc. Inst. Mech. Eng. Part D J. Automob. Eng.* **2020**, *234*, 2975–2987. [CrossRef]
19. Alvarez, S.; Page, Y.; Sander, U.; Fahrenkrog, F.; Helmer, T.; Jung, O.; Hermitte, T.; Al, E. Prospective effectiveness assessment of adas and active safety systems via virtual simulation: A review of the current practices. In Proceedings of the 25th International Technical Conference on the Enhanced Safety of Vehicles (ESV), Detroit, MI, USA, 5–8 June 2017.
20. Gulino, M.S.; Fiorentino, A.; Vangi, D. Prospective and retrospective performance assessment of Advanced Driver Assistance Systems in imminent collision scenarios: The CMI-Vr approach. *Eur. Transp. Res. Rev.* **2022**, *14*, 3. [CrossRef]
21. Smit, S.; Tomasch, E.; Kolk, H.; Plank, M.A.; Gugler, J.; Glaser, H. Evaluation of a momentum based impact model in frontal car collisions for the prospective assessment of ADAS. *Eur. Transp. Res. Rev.* **2019**, *11*, 2. [CrossRef]
22. Feng, S.; Feng, Y.; Sun, H.; Bao, S.; Misra, A.; Zhang, Y.; Liu, H.X. Testing Scenario Library Generation for Connected and Automated Vehicles, Part I: Methodology. *IEEE Trans. Intell. Transp. Syst.* **2021**, *22*, 1573–1582. [CrossRef]
23. Hallerbach, S.; Xia, Y.; Eberle, U.; Koester, F. Simulation-based Identification of Critical Scenarios for Cooperative and Automated Vehicles. In *SAE International Journal of Connected and Automated Vehicles*; SAE Technical Paper 2018-01-1066; SAE International: Warrendale, PA, USA, 2018. [CrossRef]
24. Zhang, F.; Xu, X.; Cheng, L.; Tan, S.; Wang, W.; Wu, M. Mechanism reliability and sensitivity analysis method using truncated and correlated normal variables. *Saf. Sci.* **2020**, *125*, 104615. [CrossRef]
25. García-Herrero, S.; Gutiérrez, J.M.; Herrera, S.; Azimian, A.; Mariscal, M.A. Sensitivity analysis of driver's behavior and psychophysical conditions. *Saf. Sci.* **2020**, *125*, 104586. [CrossRef]
26. Xiong, Z. *Creating a Computing Environment in a Driving Vehicles*; The University of Leeds Institute for Transport Studies & School of Computing: Leeds, UK, 2013.
27. Geyer, S.; Baltzer, M.; Franz, B.; Hakuli, S.; Kauer, M.; Kienle, M.; Meier, S.; Weißgerber, T.; Bengler, K.; Bruder, R.; et al. Concept and development of a unified ontology for generating test and use-case catalogues for assisted and automated vehicle guidance. *IET Intell. Transp. Syst.* **2014**, *8*, 183–189. [CrossRef]
28. Ulbrich, S.; Menzel, T.; Reschka, A.; Schuldt, F.; Maurer, M. Defining and Substantiating the Terms Scene, Situation, and Scenario for Automated Driving. In Proceedings of the IEEE Conference on Intelligent Transportation Systems, Gran Canaria, Spain, 15–18 September 2015. [CrossRef]
29. Bagschik, G.; Menzel, T.; Maurer, M. Ontology based Scene Creation for the Development of Automated Vehicles. In Proceedings of the 2018 IEEE Intelligent Vehicles Symposium (IV), Changshu, China, 26–30 June 2018. [CrossRef]

30. Koné, T.F.; Levrat, E.; Bonjour, E.; Mayer, F.; Géronimi, S. Safety assessment of scenarios for the simulation-based validation process of AV with regards to its functional insufficiencies. In Proceedings of the 30th European Safety and Reliability Conference, ESREL 2020 and 15th Probabilistic Safety Assessment and Management Conference, PSAM, Venice, Italy, 1–5 November 2020; pp. 100–107, ISBN 9789811485930.
31. Jang, H.A.; Kwon, H.M.; Lee, M.K. A Study on Situation Analysis for ASIL Determination. *J. Ind. Intell. Inf.* **2015**, *3*, 152–157. [CrossRef]
32. Mauborgne, P.; Deniaud, S.; Levrat, E.; Bonjour, E.; Micaëlli, J.P.; Loise, D. Operational and System Hazard Analysis in a Safe Systems Requirement Engineering Process-Application to automotive industry. *Saf. Sci.* **2016**, *87*, 256–268. [CrossRef]
33. Rocklage, E. Teaching self-driving cars to dream: A deeply integrated, innovative approach for solving the autonomous vehicle validation problem. In Proceedings of the IEEE Conference on Intelligent Transportation Systems, Proceedings (ITSC), Yokohama, Japan, 16–19 October 2017; pp. 1–7. [CrossRef]

applied sciences

MDPI

Article

Predictive Vehicle Safety—Validation Strategy of a Perception-Based Crash Severity Prediction Function

Roman Putter [1,2,*], Andre Neubohn [1], Andre Leschke [1] and Roland Lachmayer [2]

[1] Volkswagen AG, Vehicle Safety Department, Berliner Ring 2, 38440 Wolfsburg, Germany; andre.neubohn@volkswagen.de (A.N.); andre.leschke@volkswagen.de (A.L.)
[2] Institute of Product Development, Gottfried Wilhelm Leibniz Universität Hannover, An der Universität 1, 30823 Garbsen, Germany; lachmayer@ipeg.uni-hannover.de
* Correspondence: roman.putter@volkswagen.de

Abstract: Traffic accident avoidance and mitigation are the main targets of accident research and vehicle safety development worldwide. Despite improving advanced driver assistance systems (ADAS) and active safety systems, it will not be possible to avoid all vehicle accidents in the near future. Innovative Pre-Crash systems (PCS) should contribute to the accident mitigation of unavoidable accidents. However, there are no standardized testing methods for Pre-Crash systems. In particular, irreversible Pre-Crash systems lead to great challenges in the verification and validation (V&V) process. The reliable and precise real-time crash severity prediction (CSP) is, however, the basic prerequisite for irreversible PCS activation. This study proposes a novel validation and safety assessment strategy for a perception-based crash severity prediction function. In doing so, the intended functionality, safety and validation requirements of PCS are worked out in the context of ISO 26262 and ISO/PAS 21448 standards. In order to reduce the testing effort, a real-data-driven scenario-based testing approach is applied. Therefore, the authors present a novel unsupervised machine learning methodology for the creation of concrete and logical test scenario catalogs based on K-Means++ and k-NN algorithms. The developed methodology is used on the GIDAS database to extract 35 representative clusters of car to car collision scenarios, which are utilized for virtual testing. The limitations of the presented method are disclosed afterwards to help future research to set the right focus.

Keywords: Pre-Crash systems; crash severity prediction; validation and verification; functional safety; accident scenario clustering

Citation: Putter, R.; Neubohn, A.; Leschke, A.; Lachmayer, R. Predictive Vehicle Safety—Validation Strategy of a Perception-Based Crash Severity Prediction Function. *Appl. Sci.* **2023**, *13*, 6750. https://doi.org/10.3390/app13116750

Academic Editors: Edgar Sokolovskij and Vidas Žuraulis

Received: 12 April 2023
Revised: 25 May 2023
Accepted: 26 May 2023
Published: 1 June 2023

1. Introduction

Based on a World Health Organization (WHO) report, traffic accidents cause approximately 1.35 million road deaths every year and rank eighth among the causes of death worldwide. In the ages between 5 and 29 years, traffic accidents are the most common cause of death [1]. To address this problem, WHO has released a global plan for a decade of action for road safety. The United Nations General Assembly Resolution 74/299 recognizes the importance of global road safety and aims to improve it [2]. In addition to legal initiatives, the innovative vehicle safety systems contribute to increasing road safety. Knowledge of the exact accident events enables targeted development and the testing of driver assistance and vehicle safety systems. Passive safety systems, such as passenger airbags and seat belts, are designed to mitigate the consequences of unavoidable accidents for vehicle occupants. Active safety systems, such as autonomous emergency braking (AEB), primarily pursue the goal of accident avoidance. Even if active safety systems cannot prevent every accident, they can contribute to accident mitigation, for example by reducing the collision velocity. However, since not every detected collision can be prevented, further possibilities are being sought to optimize occupant protection. Integrated safety systems are intended to combine the benefits of active and passive vehicle safety and minimize the occupant injury risk.

The current trend in vehicle safety development is moving in the direction of integrated Pre-Crash systems (PCS). PCS are activated before the crash, in the so-called Pre-Crash phase, and require a highly precise and reliable real time prediction of the potential crash [3]. Figure 1 shows the five phases of traffic accidents from regular driving to Post-Crash phase. Each of these phases is addressed by different driver assistance, active, passive or integrated safety systems. Moreover, relevant variables within the five phases, which are used for accident analysis and further test scenario development, are illustrated. A conventional In-Crash restraint activation strategy is based on a measurable technical crash severity, such as a crash pulse or a pressure increase for side impact collisions, during the In-Crash phase of the accident. Acceleration sensors and pressure sensors in the vehicle are conventionally used for this purpose. Pre-Crash safety systems make the decision to trigger based on detected Pre-Crash and real-time predicted In-Crash conditions. For this purpose, data from the environmental sensors must be accessed. The prediction could be performed either individually for each Pre-Crash system or by a central function for crash prediction (CP) and crash severity prediction (CSP). In this paper, the central crash severity prediction function is considered as a prerequisite for PCS activation.

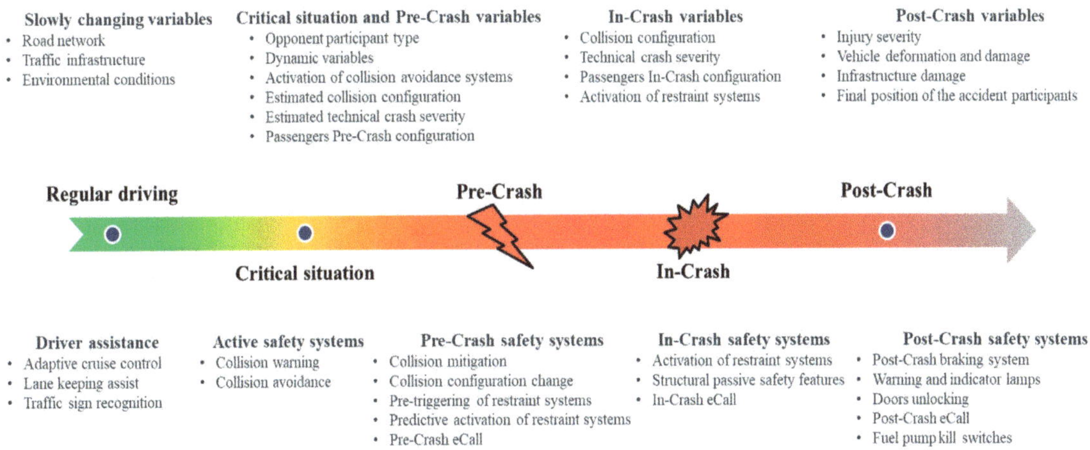

Figure 1. Schematic representation of 5 phases of the traffic accident and addressed driver assistance and safety systems.

In order to release a safety relevant Pre-Crash system, international safety standards for road vehicle systems development and validation must be met. The PCS must be proven at least as safe as conventional safety systems during the validation process. The decisive factor here is the performance of the crash severity prediction function in different situations. For validation of the CSP function, a data-driven scenario-based testing approach is applied. In general, synthetic scenarios can be generated for all conceivable traffic situations within the defined parameter space. However, the phenomenon of parameter space explosion in the scenario generation for highly automated driving systems was identified in the PEGASUS project [4]. It is not achievable to test all possible scenarios within the parameter space ranges [4]. As a consequence, representative and relevant scenarios for the Pre-Crash system under test should be identified to reduce the test effort. Since no standardized testing scope for the Pre-Crash systems based on crash severity prediction performance exist, the authors propose a novel validation strategy.

2. Materials and Methods

2.1. Pre-Crash Systems in Compliance with International Safety Standards

The concept of perception-based Pre-Crash safety systems goes back at least two decades. A few examples of potential Pre-Crash systems are the seat belt pre-tensioner [5], the airbag preset [5], unavoidable collision mitigation through crash constellation optimization [6] and Pre-Crash seat adjustment in highly automated vehicles [7]. Pre-Crash sensing based on a radar sensor is described by Moritz [3]. The necessity to reproduce real world collisions to test a Pre-Crash prediction performance on representative scenarios was recognized by [5]. Due to safety risks and high costs, crash prediction cannot be tested on a sufficiently high number of real crash tests. Reference [5] presents a scenario-based Pre-Crash system validation approach in a fully virtual environment. The predicted parameters in this study are the collision unavoidability, relative collision velocity, impact zone and impact angle, but not the technical crash severity. A real time technical crash severity estimation based on predicted crash constellation was presented by [8]. The approach combines machine learning and a 2D mass–spring–damper model to predict the distribution of expected crash pulses after the crash. Mages, Seyffert and Class show the benefit of a reduced occupant forward displacement for reversible Pre-Crash seat belt pre-pretensioning (PPT) in comparison with a conventional belt system [9]. The effects of automated emergency braking (AEB) in combination with PPT forces of 0, 300 and 600 N, in high severity frontal crashes, were examined by [10]. The rib fracture risk has been found to reduce in proportion to the PPT force increase. Nevertheless, even the non-invasive use of the crash severity prediction function can lead to an increase in road safety. A method of preventive identification of accident black spots using the CSP function and four approaches to road safety increase is presented by Putter [11].

A holistic top-down approach in which occupant protection is paramount was presented by [12]. This was to be achieved through perception-based crash prediction and the optimization of restraint strategies. To test the effectiveness of innovative restraint strategies, relevant accident test cases were extracted from the GIDAS (German In-Depth Accident Study) database. Subsequently, finite element method (FEM) vehicle to vehicle collision simulations and vehicle occupant simulations, including triggering the restraint systems, are carried out with conventional and optimized restraint strategies, which require a predictive estimation of crash severity. By analyzing the injury values, the safety potential of optimized restraint strategies is demonstrated [12]. The question of a reliable validation concept for a crash severity prediction function still remains open.

Safety relevant electrical and electronic (E/E) road vehicle systems are developed to meet international functional safety requirements which are defined by the ISO 26262 standard [13]. The standard considers systematic and random faults on system, hardware and software levels, to determine whether the system meets functional and technical safety requirements. As a part of this process, hazard analysis and risk assessment (HARA) is performed to identify potential hazards and risks in the operating conditions. Based on HARA, the system is classified with an automotive safety integrity level (ASIL). ASIL classification helps to specify the safety requirements of the system and identify the tolerable risk as well as the accepted probability of system failure [13]. Thus, ASIL can be understood as a risk classification, which is a function of the three factors: Severity of possible injuries (S), probability of Exposure (E) and Controllability (C). The lowest ASIL safety relevant classification is A and the highest is D. Additionally, there is a QM level which means that the system has a low safety relevance and the risk should be considered during quality management process.

In addition to ISO 26262, ISO/PAS 21448 focuses on the safety of the intended functionality (SOTIF) of road vehicles and addresses functional insufficiencies of the system on the vehicle level as well as foreseeable misuse [14]. SOTIF standards are not intended to replace the established standards, but to enhance them, especially for advanced driver-assistance systems (ADAS) and autonomous driving (AD) development. It considers system boundaries, in particular sensor, controller and actuator boundaries. For this purpose, the intended functionality of the system is defined and the SOTIF HARA, which can identify additional hazards and risks to ISO 26262, is performed. SOTIF also focuses on scenario-based testing and classifies the relevant use cases into four categories: known safe scenarios, known unsafe scenarios, unknown unsafe scenarios and unknown safe scenarios.

While known safe scenarios should be maximized, known unsafe scenarios should be minimized through functional modification and unknown unsafe scenarios should be identified through test data acquisition [14]. Since Pre-Crash systems are perception-based and safety-relevant systems, development according only to ISO 26262 is not sufficient, and the international standard ISO/PAS 21448 should be taken into account [12].

A Schematic representation of the multi-level crash prediction function and a generic Pre-Crash system is shown in Figure 2. The crash prediction function estimates the probability of the collision, the time to collision (TTC) and the collision configuration for both host and opponent vehicles based on sensors' perceptions and real-time data processing. The collision configuration is described by In-Crash variables at the time t_0, the time of the first impact. In-Crash variables, such as impact zone or impact points, impact angle and collision velocity for both accident participants, define the collision configuration. Crash severity prediction function is designed to predict the technical crash severity, which can be defined by various crash severity parameters, such as crash velocity, crash pulse, delta-v, occupant load criterion (OLC), energy equivalent speed (EES), intrusion zone and depth and others. The reliable and accurate prediction of relevant crash severity parameters enables Pre-Crash activation of irreversible restraint systems. However, activation of QM systems is possible based only on crash prediction without the associated crash severity. Once the targeted PCS is classified as ASIL A+ (ASIL of at least A or higher), system activation is inconceivable without an additional crash severity estimation. Collision constellation optimization based on crash prediction should contribute to the lowest injury severity within the possible collisions. A machine learning approach to estimate the trajectory of the lowest crash severity for unavoidable collisions is shown in [6]. However, the injury severity of the occupants is unknown even for the host vehicle, as there are no real-time models for the predictive estimation of injury severity.

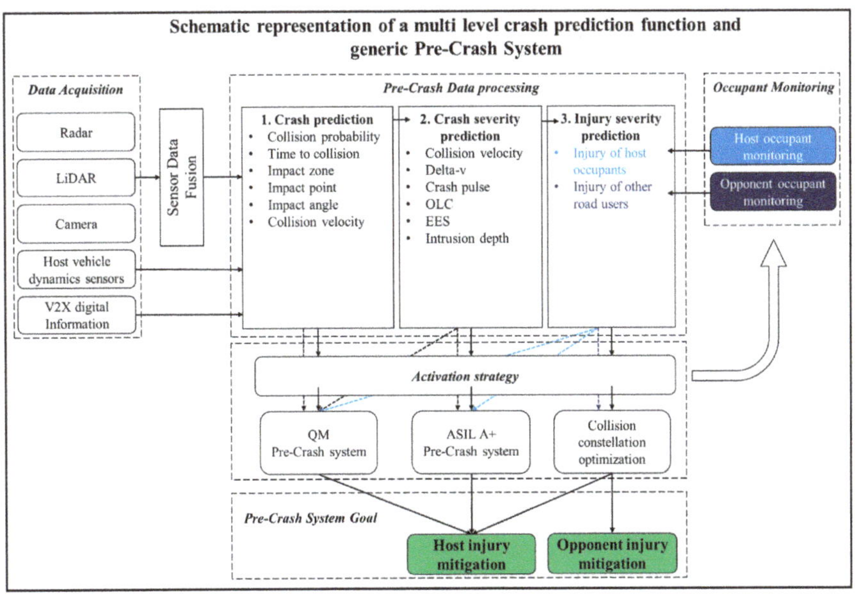

Figure 2. Schematic representation of the multi-level crash prediction function and a generic Pre-Crash system. The illustration is modified and extended according to [5].

In addition, a conflict of interest arises between the safety of host and opponent. The protection of other road users cannot be neglected and is directly addressed by the SOTIF standard. Predicting injury severity distributions for host and opponent occupants for all possible collision constellations in real time is necessary for targeted collision constellation optimization. This is not possible according to the current state of the art and is therefore not part of the testing scope, but could be considered in the future. Additionally, ethical decision-making challenges in automated vehicle crashes were addressed by [15] and remain an unsolved problem to this day.

2.2. Machine Learning Methods for Scenario Extraction

Supervised and unsupervised machine learning methods are particularly relevant in the context of accident data analysis. Commonly used supervised learning algorithms on accident data are classification and regression decision trees (CART), random forest, k-nearest neighbors (k-NN) and support vector machines. Supervised machine learning algorithms applied on accident data frequently have an aim to build a model in order to predict accident-related output features based on given inputs; for example, the prediction of the injury severity of the passengers based on the technical crash severity. Jeong et al. present an approach to the classification of motor vehicle crash injury severity [16]. Assi et al. propose a supervised machine learning model to predict crash injury severity based on 15 crash-related parameters [17]. However, supervised learning algorithms require labeled training data to build a mathematical model.

Unsupervised learning, on the other hand, aims to learn patterns from unlabeled data. In contrast to supervised learning, there is no separation between the training and testing data sets; all data is used for model training. One of the main applications of unsupervised learning on traffic accident data is the extraction of unobserved patterns from accident events. A commonly used unsupervised machine learning technique is cluster analysis or clustering. Cluster analysis is an exploratory data analysis technique, which aims to separate a dataset into groups as different as possible, while the data points within the groups should be as similar as possible. The fundamental concepts of clustering and clustering results validation techniques are presented by Halkidi, Batistakis and Vazirgiannis [18]. The main groups of clustering methods are partitioning clustering, hierarchical clustering, density-based clustering, model-based clustering and spectral methods. The most well-known clustering method is the K-means algorithm, which performs iterative steps to assign data points to clusters based on the previously defined number of clusters k.

Since no labels are used, the evaluation of unsupervised learning results appears to be a challenging problem. The evaluation is not a universal procedure and depends on the used data in the first place. Different evaluation metrics can be applied. The three main approaches to investigate cluster validity are based on external, internal and relative criteria. For external criteria, pre-specified benchmark data are needed, for example externally provided labeling. The internal criteria clustering validity approach involves the vectors of the data set themselves. It is based on characteristics such as cohesion, separation, distortion and likelihood. The common internal metrics are silhouette coefficient or the Davies–Bouldin index. The third, relative criteria approach evaluates a clustering structure by comparing it with other clustering results produced using the same algorithm but with different parameterization, for example a different number of clusters.

In order to understand the distortion and silhouette score algorithms used in chapter 3 to evaluate the clustering results, the mathematical foundations behind them are explained. First of all, the terms cluster cohesion and separation are introduced. Cluster cohesion is an intra-cluster similarity score, which measures how closely related the data points in the cluster are. It is measured by the within-cluster sum of squares (WCSS).

$$\mathrm{WCSS} = \Sigma_k \, \Sigma_i \, (x_i - c_k)^2 \tag{1}$$

with:

- Σ_k: the sum over all clusters k
- Σ_i: the sum over all data points i within cluster k
- x_i: a data point within cluster k

- c_k: the cluster center of cluster k

Cluster separation is an inter-cluster dissimilarity score, which measures how well separated the cluster are. It is measured by the between-cluster sum of squares (BSS).

$$\mathrm{BSS} = \Sigma_k \, n_k \, (c_k - c)^2 \tag{2}$$

with:

- Σ_k: the sum over all clusters k
- n_k: the number of data points in cluster k
- c_k: the cluster center of cluster k
- c: the overall centroid or mean of all data points

The distortion score is the sum of squared distances from each data point to its cluster center. A decreasing distortion score indicates better cluster allocation and helps to identify the optimal cluster number.

$$\text{Distortion Score} = \Sigma_k \, \Sigma_i \, ||x_i - c_k||^2 \tag{3}$$

The silhouette coefficient is an overall measure of how near a data point is to its own cluster center in comparison with other cluster centers [19]. The range of silhouette coefficient goes from its minimal value −1.00 to its maximum value 1.00. A large silhouette coefficient, near to 1.00, means that the clustering has a very strong structure, while 0.00 shows no structure at all. Negative silhouette coefficient indicates when an object has been assigned to the wrong cluster. The silhouette score calculates the mean of all silhouette coefficients within the clustered data.

$$\text{Silhouette Coefficient } (i) = \frac{(b(i) - a(i))}{\max\{a(i), b(i)\}} \tag{4}$$

$$\text{Silhouette Score} = \frac{\sum \text{Silhouette Coefficient}}{n_i} \tag{5}$$

with:

- i : the data points within the cluster
- a(i): the average intra-cluster distance
- b(i) : the smallest inter-cluster distance
- n_i: the number of all data points in the dataset

3. Crash Prediction Testing Methodology

International safety standards explain what steps should be fulfilled, but not how. The proposed validation strategy aims to define concrete validation steps for a perception-based crash severity prediction function. A systems engineering approach is pursued; the CSP function is considered as a safety system in the context of the entire vehicle, thus partner functions are considered in the validation process. In particular, the focus of the proposed method is on mining representative and relevant test scenarios for the specified crash severity prediction function under test. In doing so, the area of action of the function is defined and unsupervised machine learning algorithms are applied on the GIDAS database to extract test scenarios. Figure 3 shows the defined validation strategy steps, which are explained in detail in Sections 3.1–3.8.

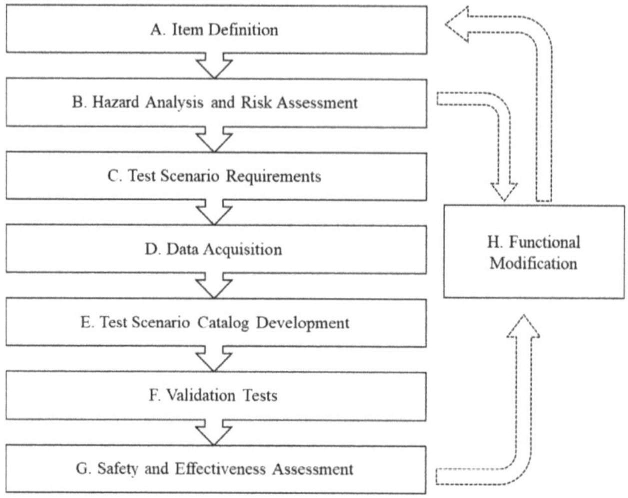

Figure 3. Crash severity prediction function testing strategy.

3.1. *Item Definition*

3.1.1. Definition of the Intended Functionality

The steps of item definition are defined by ISO 26262 and ISO/PAS 21448. They include, among others, the goals and description of the intended functionality. As well as the description, they describe dependencies and interactions with other functions and systems. Menzel, Bagschik and Maurer present three scenario abstraction levels for different process steps defined by the ISO 26262 [20]. Functional scenarios on a semantic level are used for item definition and HARA. Thus, a knowledge-driven approach for the definition of the item and functional scenarios by experts should be utilized.

3.1.2. Definition "Area of Action" or Operational Design Domain (ODD)

The selection of the database for the creation of the test cases catalog is based on the limits within which the tested system is designed to operate. For driving automation systems level one to five, the SAE J3016 defines the operational design domain as "Operating conditions under which a given driving automation system, or feature thereof, is specifically designed to function, including, but not limited to, environmental, geographical, and time-of-day restrictions, and/or the requisite presence or absence of certain traffic or roadway characteristics." [21]. The ODD parameters among others are defined by:

- Road and traffic conditions
- Weather and lighting conditions
- Static (e.g., shape) and dynamic (e.g., velocity) conditions of the host vehicle and other road participants
- Host occupant configuration

For integral safety systems the limits within which the tested system are designed to operate are defined by the area of action. The area of action contains similar parameters to the ODD, with the additional focus on the Pre-Crash and In-Crash variables. Particularly additionally relevant are the following:

- Concrete collision configuration at t_0
- Safety systems configuration
- Passengers Pre-Crash and In-Crash configuration

3.2. Hazard Analysis and Risk Assessment

Hazard analysis and risk assessment is performed based on ISO 26262 and ISO/PAS 21448 guidance. The development is carried out according to ISO 26262-6, whereby the classification of the ASIL level can only take place in the context of the targeted Pre-Crash systems. Here, a distinction is made between QM and ASIL A+ classified target systems. Systems such as predictive eCall or reversible belt pre-pretensioner with a low force level can be classified as QM Pre-Crash systems. However, reversible seat belt with a high force level, which replaces the conventional pyrotechnic seat belt, is classified as ASIL B [22].

Thus, it is not sufficient to consider only whether the Pre-Crash system is reversible or irreversible; the relevant question is if there is a risk of harm for all traffic participants and what risk factors of the systems can lead to it. The target Pre-Crash system hazard and risk should be assessed according to the five proposed risk characteristics:

- Reversibility
- Distraction
- Fallback level
- Risk of injury by the PCS for host vehicle occupants
- Risk of injury by the PCS for opponents

Depending on the ASIL classification of the targeted Pre-Crash system, different requirements arise for the accepted probability of system failure and thus the robustness and accuracy of the crash severity prediction function. Examples of addressed PCS and a classification of risk characteristics are shown in Table 1.

Table 1. Examples of Pre-Crash systems and assignment of the risk characteristics.

PCS	Irreversible	Distracting	Fallback Level	Risk of Injury Host	Risk of Injury Opponent	ASIL Class
Predictive eCall	No	May	Yes	No	No	QM
Reversible seat belt (low-force)	No	May	Yes	No	No	QM
Reversible seat belt (high-force)	No	Yes	May	Yes	May	ASIL B
Pre-Crash airbag activation	Yes	Yes	No (for FP) Yes (for FN)	Yes	May	ASIL D
Collision constellation optimization	Yes	Yes	No	Yes	Yes	ASIL D

While predictive eCall function is classified as QM, the Pre-Crash airbag activation receives ASIL D classification. Fallback level defines a secondary system, which is engaged when the primary system fails. In the case of Pre-Crash airbag activation there is no fallback level if the system is activated as a result of false positive prediction, since the airbag is already deployed. In addition, false positive Pre-Crash activation of airbag can lead to a provoked collision, which leads to injury risk not only for host, but also other traffic participants. For false negative activation, thus no Pre-Crash airbag activation, the conventional In-Crash activation of airbag as a fallback level is still possible.

Thus, a crash prediction function developed and validated according to ASIL B meets the safety requirements for the reversible seat belt deployment with a high force level, but not for the Pre-Crash activation of the airbags or collision constellation optimization. Based on the HARA results, a functional modification of the system can be made before further development and validation steps are performed.

3.3. Test Scenario Requirements

A data-driven approach to scenario generation is chosen. Logical scenarios are utilized to describe the parameter ranges of the state values used for scenario representation and concrete test scenarios for the verification and validation (V&V) tests [20]. In order to describe driving scenarios, a five layer model is proposed by Bagschik et al. [23]. The scenario is described by five levels: road (L1), traffic infrastructure (L2), temporary manipulation of L1 and L2 (L3), Objects (L4) and Environment (L5). Scholtes et al. have proposed a six-layer model by adding digital information as a sixth layer and placing static objects on L2 [24]. To create concrete test scenarios with sufficient data depth for the perception-based crash prediction function, information from level one to five is required. Considering the increasing use of V2X communication in modern vehicles, L6 information is not currently part of this study, but should be used in future testing.

Test scenarios for the CSP function must represent real-world driving situations and are divided into three scenario groups:

- Collision
- Critical situation
- Regular driving

For collision scenarios, all collisions in the area of action of the function should be considered. A critical situation extends from potential conflicts to near accident scenarios. Various criticality metrics are discussed by Hruschka, Töpfer and Zug [25]. The most common metrics with a focus on collision avoidance are time to collision (TTC), time to distance (TTD), time to brake (TTB) and time to steer (TTS). More advanced metrics, such as crash severity and collision configuration distribution, assess the critical situation risk based on the potential accident severity. To validate the Pre-Crash system, various critical situations from a low to high criticality level are required. Regular driving is the test baseline, it represents the different states in the area of action, such as different environmental, road, traffic infrastructure, traffic flow, lighting and other conditions.

Considering the addressed Pre-Crash systems, concrete scenarios are classified by must fire, may fire and no fire scenarios. Exclusively for PCS classified as QM, the condition may fire is established. Considering the crash prediction function, must fire and may fire collisions, as well as may fire nearby collisions, are required to test TP (true positive) prediction. No fire collisions, nearby collisions and regular driving are required to test TN (true negative) crash prediction. Figure 4 shows different test scenario classifications and true prediction classes according to the system under test and their activation requirements. Any prediction assignments other than those shown in Figure 4 lead to false positive (FP) or false negative (FN) results.

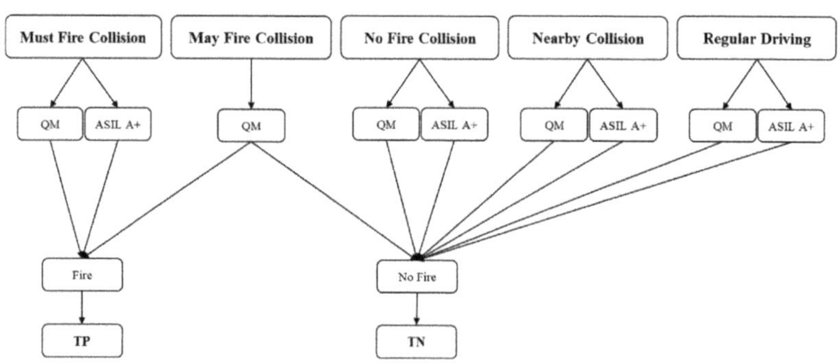

Figure 4. Test Scenario classification for Pre-Crash systems and must fire, may fire and no fire signals.

This results in different requirements for the information that is contained within the different scenario groups. All groups require dynamic and static information about the road network, road users and environment to build the test scenario, as defined for L1 to

L5 [23]. In addition, for collision scenarios, In-Crash information is required. For the testing of ASIL A+ Pre-Crash systems, it is necessary to define the ground truth of the technical crash severity, such as the crash pulse or delta-v values. For QM systems, the ground truth of at least a collision configuration is also sufficient.

3.4. Data Acquisition

Representative test scenarios are required to test Pre-Crash systems. One of the biggest challenges in scenario mining for ADAS/AD and safety systems is the collection of high quality and in-depth real-world data on a large scale. Test scenario data sources on macroscopic, mesoscopic and microscopic levels are presented by Mai et al. [26]. A modified and expanded illustration with additional types of data sources and concrete dataset examples is shown in Figure 5.

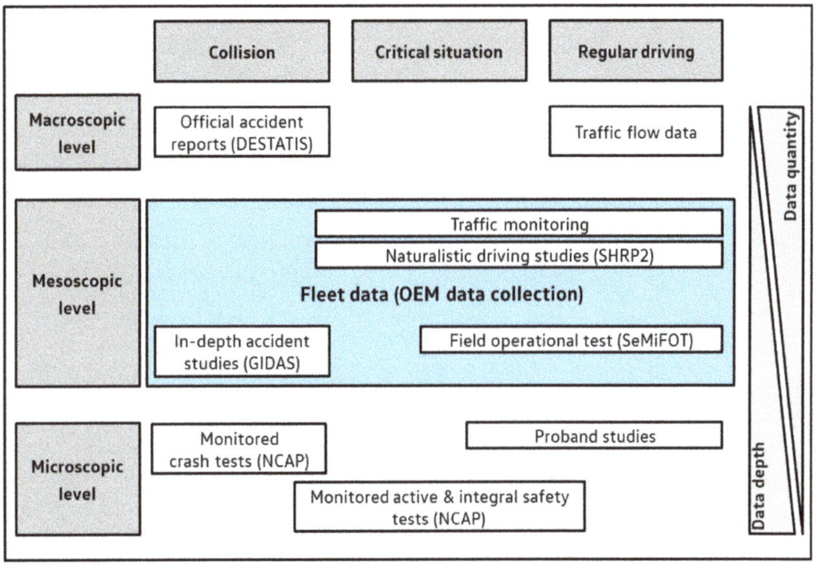

Figure 5. Representation of driving scenario sources and examples of concrete datasets. The illustration is modified and expanded from [26].

A common limitation is the inverse proportionality between the amount of data and the data depth. While the official accident statistics cover all traffic accidents recorded by the police, the depth of data is usually not sufficient to define concrete test scenarios. More important for test scenario extraction is not the amount, but the combination of the representativeness of the data and the data depth. These properties are fulfilled by in-depth accident data bases. The GIDAS database contains detailed information on German traffic accidents with injured participants. The accidents are recorded at the scene of accident and reconstructed in order to obtain information about all five phases of the accident, as shown in Figure 1. The GIDAS PCM (Pre-Crash Matrix) database contains a big sample of reconstructed GIDAS Pre-Crash scenarios in a specific PCM format. It describes the Pre-Crash phase over about 5 s before the collision and contains the information on participants, their dynamics and the environment [27]. Currently, the GIDAS PCM database contains 11,074 collision scenarios, 40% of which are car to car collisions [28]. The PCM scenarios are used for the simulative testing of driver assistance and active safety systems in a virtual environment [29]. However, GIDAS does not include accidents without injured persons, nor near accident scenarios. Naturalistic driving studies and field operational tests collect real world data during regular driving. Nevertheless, critical situations and even a small number of collisions also occur in large traffic data collection campaigns. For example, the

SHRP2 study contains over 36,000 regular driving, near collision and collision events [30]. However, the collisions represent only a small part of the SHRP2 dataset. Modern vehicles are equipped with advanced vehicle dynamics, and interior and environmental sensors, which enables the collection of real-world driving data on a mesoscopic level. The amount and depth of the data is flexible and depends on the vehicle equipment and the definition of the data collection campaign. Thus, the highest possible amount of data corresponds to the entire vehicle fleet, which would vastly exceed conventional data sources. However, the protection of personal data must be ensured.

3.5. Test Scenario Catalog Development

3.5.1. Machine Learning Scenario Extraction

One possibility for the extraction of the relevant test scenarios from the acquired accident database is the explorative machine learning approach. Since our goal is to investigate the occurrence of collision configurations in real-world accident data, no external labels should be defined for the data. Thus, an unsupervised machine learning approach without data labeling was chosen. Furthermore, the property of clustering to organize similar cases into one group makes the extraction of representative test scenarios possible. The clustering aim and input data selection are derived from the test scenario requirements previously defined in step C. The proposed scenario clustering methodology is shown schematically in Figure 6.

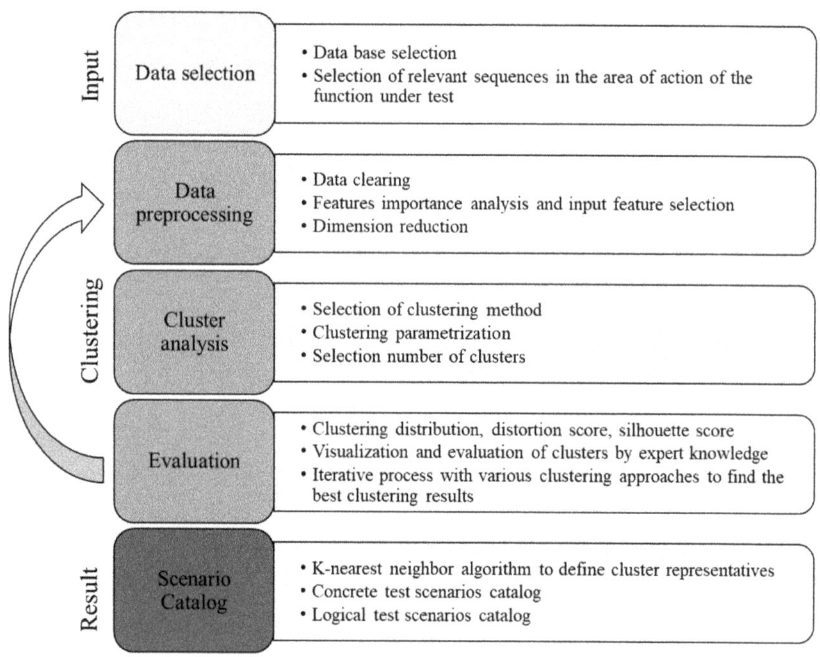

Figure 6. Novel unsupervised machine learning methodology for creating concrete and logical test scenario catalogs.

Depending on the database and the selection of the input features, both accident scenarios and critical situations can be grouped by cluster analysis. However, the selected input features directly determine the clustering results. To test a Pre-Crash function, both the Pre-Crash and In-Crash information in the test scenario are required. The scale level of the different features should be considered. However, the characteristics of the ratio scale features are well suited for cluster analyses since they show rankings and interpretable distances and can be transformed in the preprocessing. Thus, variables of both participants, such as collision velocity, impact angle, floating angle, impact point along the contour of the vehicles, vehicle mass and other In-Crash variables, can be used directly for collision configuration clustering.

Additionally, the k-nearest neighbors (k-NN) algorithm is utilized to determine the representative cases within the clusters. Concrete representative cases from each cluster are selected to create a concrete test scenario catalog. For logical test scenarios, the parameter ranges of features used for scenario representation should be limited by a standard deviation of these features within the clusters. Through k-NN, the selection of more than one cluster representative, as a cluster center, is possible. In this way, test case catalogs with both concrete and logical scenarios can be created.

Proof of concept clustering on the GIDAS dataset of vehicle to vehicle crashes, collected in the years 2000 to 2019, was performed to demonstrate the challenges in the evaluation of clustering results. The goal of this clustering was to identify representative collision configurations and to find the optimal number of clusters, to reduce the test effort of the system without losing the relevant information. The dataset was filtered by the area of action of the function under tests. A sample of 20,239 cases on the passenger level was used for clustering. A K-means++ clustering algorithm, which were introduced by Arthur and Vassilvitskii [31], was utilized. The difference to the conventional K-means algorithm is the initialization method of the cluster centers, which helps to increase the cluster quality and reduces the risk of falling into local data minima. The feature input set consisted of ten overall collision configuration features, such as the impact points, collision velocity, collision angle and technical crash severity dv for both the host and opponent vehicles. Additionally, the input features have been standardized.

In unsupervised learning, no labels are used, which makes the evaluation of the clustering results more complicated. The external clustering validity criteria cannot be used here because no labeling of the generated clusters is available. Relative and internal criteria, such as distortion and silhouette scores, are applied to assess the cluster validity. Figure 7 shows the distortion and silhouette scores for the clustering results, from 1 to 100 clusters. While the distortion score is continuously reduced with an increasing number of clusters, the silhouette score reaches the peak at four clusters and drops rapidly with an increasing number of clusters.

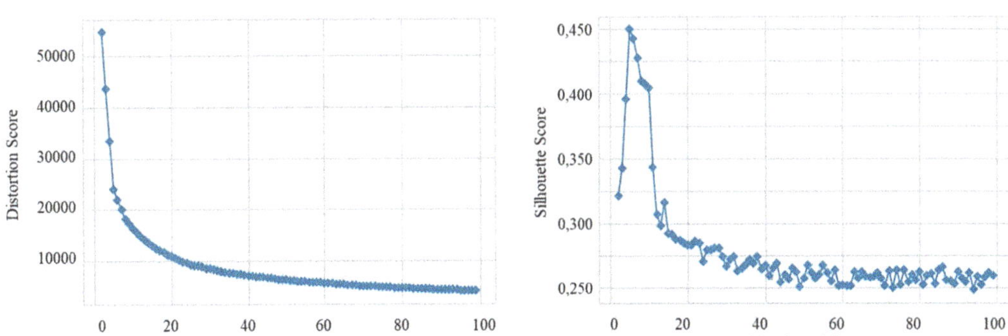

Figure 7. Distortion score (**left**) and silhouette score (**right**) on GIDAS vehicle to vehicle collisions clustering.

According to the elbow method on the distortion score plot, the optimal number of clusters is about 15 clusters. This would also be a very rudimentary representation of the diversity of all vehicle to vehicle collision configurations. From the silhouette score plot, four clusters show the best clustering structure. Obviously, only four clusters cannot represent all relevant vehicle to vehicle collisions. Similar results have been obtained with various clustering algorithms on the same dataset. It is clear that common distance and similarity-based clustering evaluation metrics, such as distortion or silhouette scores, are not sufficient to validate the traffic scenario clustering results in detail. However, it can help to show a tendency. Nevertheless, the evaluation of the clusters by expert knowledge is required.

The current GIDAS clustering results already allow the reduction of thousands of collision configurations to less than 50 representative test cases. For this study, a catalog of 35 clusters representing car to car collision configurations was created. The clustered collision configurations were evaluated by experts based on the system definition and concrete testing requirements. In doing so, the collision velocity, impact points, collision angle, delta-v values and injury severity of the cluster representatives have been examined. To better understand the results, the collision configurations of all the accident cases within the clusters were visualized.

A few examples of the clusters are presented. Figure 8 shows, on the left, the cluster C1 with the highest number of cases, which represents low speed rear-end collisions. The blue box is the host vehicle, and the red boxes are all the opponent vehicles within the cluster. By determining the cluster center, a concrete collision configuration for this cluster is derived. The median velocity speed in this cluster is 15 $\frac{\text{km}}{\text{h}}$, the delta-v is 9 $\frac{\text{km}}{\text{h}}$ and the percentage of serious injured passengers in the host vehicle is 2%. On the right of Figure 8 is the cluster C2, which represents the most severe accidents on the database. The median velocity speed in this cluster is 132 $\frac{\text{km}}{\text{h}}$, the delta-v value is 61 $\frac{\text{km}}{\text{h}}$ and the percentage of serious injured passengers is 52%.

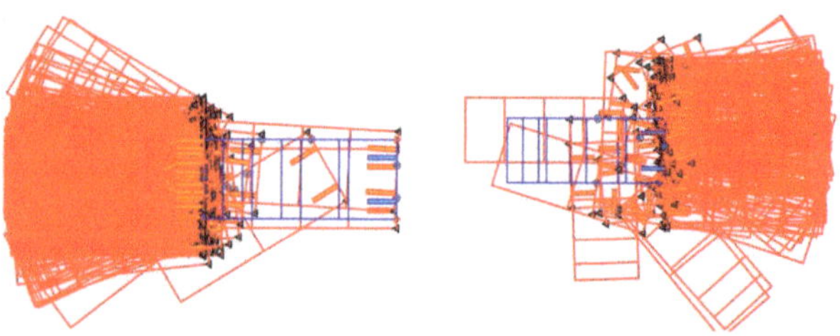

Figure 8. Visualization of clusters C1 (**left**) and C2 (**right**).

However, the implemented clustering algorithm also manages to form unique clusters for some critical accident constellations. This is shown in Figure 9 by cluster C3. This is a cluster with accidents at a very high collision velocity; however, the percentage of serious injuries is in the low–medium range. The median velocity speed in this cluster is 130 $\frac{\text{km}}{\text{h}}$, the delta-v value is 15 $\frac{\text{km}}{\text{h}}$ and the percentage of serious injured passengers is 12%. This can be explained by sliding effects during collision. This cluster is a proof that, by the precise selection of the correct clustering parameterization and the relevant input features, such special cases can be identified as test cases.

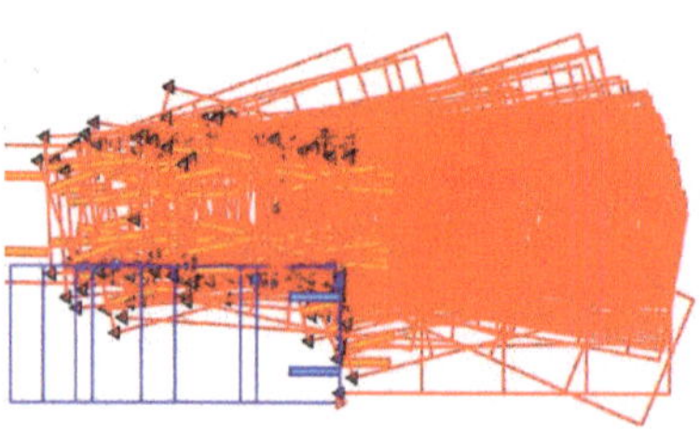

Figure 9. Visualization of cluster C3.

Nevertheless, clustering can still blur some rare and unique scenarios. That is why we used a k-NN algorithm that helps to create the logical test scenario catalog. The use of logical test scenarios in the defined clusters parameter space prevents the disappearance of the corner cases as test cases.

Regular driving scenarios do not necessarily require machine learning analysis. For the baseline, a sufficiently large number of driving hours and the diversity of different driving conditions should be represented. Therefore, a representative assessment of a regular driving dataset is required.

3.5.2. Scenario Relevance Assessment

After the scenario representatives have been extracted, the relevance of each scenario in the testing context was evaluated. The relevance of a test scenario is directly related to the tested system and depends on several factors of influence. Test cases that are legally required for the release of the function receive the highest relevance rating. This is followed by test cases that are rated by car safety performance assessment programs, such as EuroNCAP. In addition to these standardized test scenarios, the relevance of identified collisions, critical and regular driving test sequences in the field data, respectively real world traffic, should be evaluated. For this purpose, a relevance metric is required. For collision scenarios, a field relevance of the cluster can be represented through the combination of the injury severity and the probability of exposure, which is defined as a risk function. Additionally, the technical crash severity (delta-v) of the cluster is considered in the third relevance criteria. The relevance assessment is necessary for potential functional modification decisions.

3.6. Validation Tests

Validation tests are performed in virtual simulations based on the developed test scenario catalog, and contain the crash severity prediction algorithm as well as the virtual vehicles and the environment. For this purpose, the software IPG CarMaker and the clustered GIDAS collision configuration scenarios were utilized. Concrete collision scenarios were extracted as cluster representatives and the case number was searched in the GIDAS PCM database. Since not every GIDAS case contains a GIDAS PCM case, the k-NN algorithm was used here again and the closest cases to the cluster center were considered as representatives of those that can be found in the GIDAS PCM database. This also covered the diversity of the different Pre-Crash phases, since the clustering represents the collision configuration, but the Pre-Crash phase may be different. The GIDAS PCM scenarios were converted into the IPG CarMaker scenario formats. In addition to real accident scenarios from the GIDAS PCM database, generic accident scenarios were created for the identified representative collision configurations.

The tests pursued the goal of verifying the predicated outputs with regard to a defined ground truth. Each predicted variable was compared against the nominal ground truth value from the database, at each prediction timestep in the Pre-Crash phase. This is shown exemplary in Table 2, where VREL is a GIDAS variable for the collision velocity.

Table 2. Simulative assessment of predicted outputs.

Prediction Time	Simulation Nominal Velocity	Simulation Predicted Velocity	Prediction Error
t_1	$VREL_{nominal} = 70\ \frac{km}{h}$	$VREL(t_1) = 50\ \frac{km}{h}$	$\Delta VREL\ (t_1) = 20\ \frac{km}{h}$
t_2	$VREL_{nominal} = 70\ \frac{km}{h}$	$VREL(t_2) = 57\ \frac{km}{h}$	$\Delta VREL\ (t_2) = 13\ \frac{km}{h}$
...	...	...	...
t_{coll}	$VREL_{nominal} = 70\ \frac{km}{h}$	$VREL(t_{\text{coll}}) = 67\ \frac{km}{h}$	$\Delta VREL\ (t_{coll}) = 3\ \frac{km}{h}$

Thus, each predicted variable can be verified as decoupled from others. The time before t_0, known as TTC, for a sufficiently good prediction performance should be determined. However, a purely quantitative assessment of the prediction performance is not sufficient without a connection to the targeted Pre-Crash system. Crash severity prediction contains a predicted distribution of potential collision configurations and their severity values. Methods to generate fire, may fire and no fire signals from these distributions must be developed. As a result, the requirements for prediction performance must be determined by the Pre-Crash systems. Moreover, simulations composed of the complete system under test, including the individual components, such as vehicle sensors, controllers, data transmission and processing, crash severity prediction and Pre-Crash actuators, are required for reliable validation.

3.7. Safety and Effectiveness Assessment

The results of the validation tests were analyzed as a part of the safety and effectiveness assessment argumentation. Figure 10 shows the confusion matrix for PCS activation. True positive predicted cases define the area of efficiency of the system under test, which is usually smaller than the originally defined area of action. The degree of efficiency (DoE) of the system under test is defined by [32] as the quotient of the area of efficiency and the area of action. In addition to the simulative determination of the DoE, a retrospective determination on the basis of the accident data collection is possible. However, DoE shows the efficiency of the system but does not provide proof of safety. The acceptable probabilities for FP and FN cases are defined based on the ASIL classification of the concrete system and must be proven during the testing. For example, when considering FN classification, the main consideration is whether a fallback level is present or not. However, there is a trade-off between the advantage of TP and the disadvantage of the FN and FP cases. Thus, a positive risk balance proof is necessary. TN cases have neither a negative nor a positive effect, while the FP and FN cases lead to potential hazard. For the positive risk balance, the sum of the injury reduction in all the TP cases has to be higher than the sum of injuries potentially caused by all the FP and FN cases.

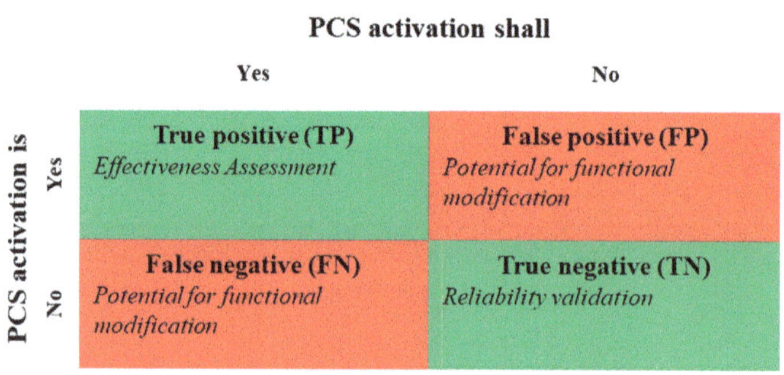

Figure 10. Pre-Crash system triggering confusion matrix.

3.8. Functional Modification

Considering the effectiveness and safety assessment of the function under test and the relevance assessment of the test scenarios, functional modification requirements can be defined to improve the function and increase the safety potential. Functional modifications can lead to a change in the intended functionality or the field of action of the function. Therefore, the presented testing method should be carried out again, after each functional modification, starting with step A.

4. Discussion of Limitations

This chapter discusses the limitations of the methodological steps described above and provides an outlook for future research.

4.1. Crash Severity Validation

A major challenge to the validation of the predicted crash severity is the mining of the ground truth crash severity data. In-depth accident data studies reconstruct traffic accidents, which are an approximation of reality. Additionally, the GIDAS database contains reconstructed scalar values of delta-v, not crash pulses as time series. In order to obtain the crash pulses measured In-Crash, the event data recorder (EDR) signals must be collected.

This has been performed only on a small number of GIDAS cases [33]. Another possibility in estimating the crash pulses for identified representative collision configurations are FEM simulations of car to car collisions; however, these are very cost-intensive.

4.2. Prospective Test Scenarios

Accident databases provide only a retrospective view on traffic accidents. Leledakis et al. present a method for predicting collision configurations in vehicles with crash avoidance systems and identifying the changes in new collision configurations [34]. A lot of prospective studies utilize retrospective accident scenarios to create a new prospective test scenario catalog by using virtual simulations. As addressed by SOTIF, even after the data acquisition and test sequences extraction, there is still a possibility of missing relevant scenarios, especially unknown unsafe scenarios. Therefore, continuous traffic data collection and analysis is still necessary, besides prospective simulations studies.

4.3. Global Traffic Data

International traffic accident differences are caused by different types of roads, infrastructure, traffic rules, road users and their regional behavior. Thus, accident scenarios that are representative for one country cannot be considered as representative for another country without validation. Additionally, the data collection and coding format is not identical for different databases, which makes the analysis and scenario extraction complicated. Therefore, the initiative for the global harmonization of accident data (iGLAD) collects and harmonizes accident data from 12 different countries [35]. PCM created from the iGLAD dataset can be utilized for the simulations of international test scenarios [36]. However, the question of representativeness remains unanswered. The same challenges apply to critical driving and regular driving scenarios.

4.4. Pre-Crash Phase Clustering

While collision configurations clustering has the potential to determine representative retrospective collision configurations, as shown in this study, and even prospective collision configurations, as shown by Leledakis et al. [34], the diversity of different Pre-Crash maneuvers that can lead to the same collision is still unobserved. Information of the Pre-Crash phase should be additionally considered in the clustering algorithm. As shown in Figure 1, Pre-Crash variables consist mainly of the time series of the participants or categorically scaled slowly changing variables. For the clustering of categorical variables, one-hot encoding can be applied to convert the categorical data to numerical form. The encoded categorical variable is removed and new binary variables are added to the dataset. However, this can lead to a strong increase in the number of input features, which increases the risk of the curse of dimensionality.

The first three limitations mentioned above can be completely solved by targeted in-depth fleet data collection. Thus, fleet data has the greatest potential for further representative traffic data collection and is considered as essential for the V&V process of ADAS/AD and novel safety systems.

5. Conclusions

In this paper, the authors have elaborated on the safety potentials and risks of innovative crash severity prediction and Pre-Crash systems. A novel validation strategy for the crash severity prediction function was proposed in the context of ISO 26262 and ISO/PAS 21448 standards. The relevance of a reliable and precise real-time crash severity prediction function was demonstrated using the potential Pre-Crash use cases. Thus, the authors identified fundamental differences in the validation requirements of QM- and ASIL A+-related CSP functions. Test scenario classification criteria for collisions, critical situations and regular driving, combined with must fire, may fire and no fire signals, were presented. In addressing a scenario-based testing approach, unsupervised machine learning was chosen as an exploratory data analysis technique to extract representative and relevant test scenarios from real-world data. A novel, unsupervised machine learning methodology for the creation of concrete and logical test scenario catalogs were developed with K-Means++ and k-NN algorithms. The methodology was implemented on the GIDAS database to create 35 clusters of representative collision configurations in the area of action of the CSP function under test. The presented methodology for machine learning scenario extraction is transferable to other databases and can be carried out for different systems under test. Furthermore, not only the fulfillment of the ASIL-classified probability of failure requirements, but additionally the evidence of a positive risk balance were defined as essential conditions for the proof of safety of Pre-Crash systems. However, the limitations of the presented method have been disclosed.

In the future, the research should focus on the identified limitations. The difficulties in mining representative traffic data with corresponding data depth, such as crash severity information, should be solved as soon as possible. That being the case, fleet data appears as the most promising source for the representative traffic data collection. Moreover, the combination of In-Crash and Pre-Crash variables should be considered in further scenario extraction. One question that still remains unanswered is what injury reduction benefit the Pre-Crash systems can provide through a specific prediction performance. For this reason, methods to generate fire signals from crash prediction distributions must be developed in the future. The complete chain from crash severity prediction to the activation of Pre-Crash systems and passenger injury analysis should be taken into account for a valid effectiveness and safety assessment of the crash severity prediction function.

Author Contributions: Conceptualization, R.P., A.N. and R.L.; methodology, R.P.; investigation, R.P.; resources, A.L.; data curation, R.P.; writing—original draft preparation, R.P.; writing—review and editing, A.N., A.L. and R.L.; visualization, R.P.; supervision, A.N., A.L. and R.L. All authors have read and agreed to the published version of the manuscript.

Funding: This research project is supported by Volkswagen AG. The results, opinions and conclusions expressed in this publication are those of the authors and do not necessarily represent the views of Volkswagen AG.

Institutional Review Board Statement: Not applicable.

Informed Consent Statement: Not applicable.

Data Availability Statement: Roman Putter, Andre Neubohn and Andre Leschke possess a valid data use license for the GIDAS and GIDAS PCM datasets.

Conflicts of Interest: The authors declare no conflict of interest.

Abbreviations

AD	Autonomous Driving
ADAS	Advanced Driver Assistance Systems
AEB	Autonomous Emergency Braking
ASIL	Automotive Safety Integrity Level
ASIL A+	ASIL of at least A or higher
BSS	Between-Cluster Sum of Squares
CP	Crash Prediction
CSP	Crash Severity Prediction
DoE	Degree of Efficiency
E/E	Electrical and Electronic
EDR	Event Data Recorder
EES	Energy Equivalent Speed
EuroNCAP	European New Car Assessment Programme
FEM	Finite Element Method
FN	False Negative
FP	False Positive
GIDAS	German In-Depth Accident Study

HARA	Hazard Analysis and Risk Assessment
iGLAD	Initiative for the Harmonization of Global in-Depth Traffic Accident Data
k-NN	K-Nearest-Neighbor
ODD	Operational Design Domain
OLC	Occupant Load Criterion
PCM	Pre-Crash Matrix
PCS	Pre-Crash Systems
PPT	Pre-Pretensioning
QM	Quality Management
SOTIF	Safety of the Intended Functionality
TN	True Negative
TP	True Positive
TTB	Time to Brake
TTC	Time to Collision
TTD	Time to Distance
TTS	Time to Steer
V&V	Verification and Validation
WCSS	Within-Cluster Sum of Squares
WHO	World Health Organization

References

1. World Health Organization. *Global Status Report on Road Safety 2018*; World Health Organization: Geneva, Switzerland, 2019; Available online: https://www.who.int/publications/i/item/9789241565684 (accessed on 11 November 2022).
2. United Nations General Assembly. Resolution A/RES/74/299 Improving Global Road Safety. 2020. Available online: https://frsc.gov.ng/wp-content/uploads/2021/02/UNGA-Resolution-2020.pdf (accessed on 24 May 2023).
3. Moritz, R. *Precrash Sensing Its Functional Evolution Based on a Platform Radar Sensor*; Society Automotive Engineers Int.: Detroit, MI, USA, 2000.
4. Amersbach, C.; Winner, H. Functional decomposition—A contribution to overcome the parameter space explosion during validation of highly automated driving. *Traffic Inj. Prev.* **2019**, *20*, S52–S57. [CrossRef] [PubMed]
5. Gietelink, O.; Verburg, D.; Labibes, K.; Oostendorp, A. Pre-Crash System Validation with PRESCAN and VEHIL. In *IEEE Intelligent Vehicles Symposium*; IEEE: New York, NY, USA, 2004; pp. 913–918.
6. Müller, M.; Botsch, M.; Böhmländer, D.; Utschick, W. Machine Learning Based Prediction of Crash Severity Distributions for Mitigation Strategies. *J. Adv. Inf. Technol.* **2018**, *9*, 15–24. [CrossRef]
7. Moon, J.; Bae, I.; Kim, S. A pre-crash safety system for an occupant sitting on a backward facing seat for fully automated vehicles in frontal crashes. In Proceedings of the 2017 IEEE International Conference on Vehicular Electronics and Safety (ICVES), Vienna, Austria, 27–28 June 2017; pp. 168–171.
8. Müller, M.; Long, X.; Botsch, M.; Böhmländer, D.; Utschick, W. Real-Time Crash Severity Estimation with Machine Learning and 2D Mass-Spring-Damper Model. In Proceedings of the 2018 21st International Conference on Intelligent Transportation Systems (ITSC), Maui, HI, USA, 4–7 November 2018; pp. 2036–2043.
9. Mages, M.; Seyffert, M.; Class, U. Analysis of the Pre-Crash Benefit of Reversible Belt Pre-Pretensioning in Different Accident Scenarios. In Proceedings of the the 22nd ESV Conference, Washington, DC, USA, 13–16 June 2011.
10. Ekant, M.; Mroz, K.; Pipkorn, B.; Lubbe, N. Effects of Automated Emergency Braking and Seatbelt Pre-Pretensioning on Occupant Injury Risks in High-Severity Frontal Crashes. *Front. Future Transp.* **2022**, *3*, 883951.
11. Putter, R. Poster: Preventive Identification of Accident Black Spots on the Basis of Crash Severity Estimation. In Proceedings of the 2023 IEEE Vehicular Networking Conference (VNC), Istanbul, Turkey, 26–28 April 2023; pp. 155–156.
12. Grotz, B.; Straßburger, P.; Huf, A.; Roig, L. Prädiktive Sicherheit-Wahrnehmungsbasierte Aktivierung von Pre-Crash-Systemen. *ATZ-Automob. Z.* **2021**, *123*, 18–25. [CrossRef]
13. ISO 26262-1:2018(en) Road Vehicles—Functional Safety. Available online: https://www.iso.org/standard/68383.html/ (accessed on 3 January 2023).
14. ISO/PAS 21448:2019 Road Vehicles—Safety of the Intended Functionality. Available online: https://www.iso.org/obp/ui#iso:std:iso:pas:21448:ed-1:v1:en/ (accessed on 3 January 2023).
15. Goodall, N. Ethical decision making during automated vehicle crashes. *Transp. Res. Rec.* **2014**, *2424*, 58–65. [CrossRef]
16. Jeong, H.; Jang, Y.; Bowman, P.; Masoud, N. Classification of motor vehicle crash injury severity: A hybrid approach for imbalanced data. *Accid. Anal. Prev.* **2018**, *120*, 250–261. [CrossRef] [PubMed]
17. Assi, K.; Rahman, S.M.; Mansoor, U.; Ratrout, N. Predicting Crash Injury Severity with Machine Learning Algorithm Synergized with Clustering Technique: A Promising Protocol. *Int. J. Environ. Res. Public Health* **2020**, *17*, 5497. [CrossRef] [PubMed]
18. Halkidi, M.; Batistakis, Y.; Vazirgiannis, M. On Clustering Validation Techniques. *J. Intell. Inf. Syst.* **2001**, *17*, 107–145. [CrossRef]
19. Rousseeuw, P. Silhouettes: A Graphical Aid to the Interpretation and Validation of Cluster Analysis. *J. Comput. Appl. Math.* **1987**, *20*, 53–65. [CrossRef]

20. Menzel, T.; Bagschik, G.; Maurer, M. Scenarios for development, test and validation of automated vehicles. In Proceedings of the 2018 IEEE Intelligent Vehicles Symposium (IV), Suzhou, China, 26–30 June 2018; pp. 1821–1827.
21. Society of Automotive Engineers. *Taxonomy and Definitions for Terms Related to Driving Automation Systems for on-Road Motor Vehicles*; Society of Automotive Engineers: Warrendale, PA, USA, 2021.
22. Infineon: Reversible Seatbelt Pretensioner. Available online: https://www.infineon.com/cms/en/applications/automotive/chassis-safety-and-adas/reversible-seatbelt-pretensioner/ (accessed on 20 December 2022).
23. Bagschik, G.; Menzel, T.; Maurer, M. Ontology based Scene Creation for the Development of Automated Vehicles. In Proceedings of the 2018 IEEE Intelligent Vehicles Symposium (IV), Suzhou, China, 26–30 June 2018; pp. 1813–1820.
24. Scholtes, M.; Westhofen, L.; Turner, L.R.; Lotto, K.; Schuldes, M.; Weber, H.; Wagener, N.; Neurohr, C.; Bollmann, M.; Körtke, F.; et al. 6-layer model for a structured description and categorization of urban traffic and environment. *arXiv* **2020**, arXiv:2012.06319. [CrossRef]
25. Hruschka, C.M.; Töpfer, D.; Zug, S. Risk assessment for integral safety in automated driving. In Proceedings of the 2019 2nd International Conference on Intelligent Autonomous Systems (ICoIAS), Singapore, 28 February–2 March 2019; pp. 102–109.
26. Mai, M.; Baeumler, M.; Lehmann, M.; Siebke, C.; Blenz, K.; Prokop, G.; Bönninger, J.; Höpping, K. Die Dresdner Methode—Ein Baukasten zur ganzheitlichen Bewertung aktiver Sicherheits- und automatisierter Fahrfunktionen. *VDI Ber.* **2022**, *13*, 419–434.
27. Schubert, A.; Erbsmehl, C.; Hannawald, L. Standardized Pre-Crash-Scenarios in Digital Format on the Basis of the VUFO Simulation. 2013. Available online: https://bast.opus.hbz-nrw.de/opus45-bast/frontdoor/index/index/docId/630 (accessed on 31 March 2023).
28. GIDAS-PCM. Available online: https://www.vufo.de/gidas-pcm/ (accessed on 3 January 2023).
29. Stark, L.; Düring, M.; Schoenawa, S.; Maschke, J.E.; Do, C.M. 'Quantifying vision zero: Crash avoidance in rural and motorway accident scenarios by combination of ACC, AEB, and LKS projected to German accident occurrence. *Traffic Inj. Prev.* **2019**, *20*, S126–S132. [CrossRef] [PubMed]
30. SHRP 2 Naturalistic Driving Study. Available online: https://insight.shrp2nds.us/ (accessed on 3 January 2023).
31. Arthur, D.; Vassilvitskii, S. K-Means++: The Advantages of Careful Seeding. In Proceedings of the eighteenth annual ACM-SIAM symposium on Discrete Algorithms; Society for Industrial and Applied Mathematics: Philadelphia, PA, USA, 2007; Volume 8, pp. 1027–1035.
32. Winkle, T.; der Unfallforschung; Maurer, E.A.I.M.; Gerdes, C.; Lenz, B.; Winner, H. (Eds.) *Autonomes Fahren—Technische, Rechtliche und Gesellschaftliche Aspekte*; Springer: Berlin/Heidelberg, Germany, 2015; pp. 351–376.
33. Nutzung Der Erkenntnisse Aus Dem Unfallforschungsprojekt GIDAS Für Unfallgutachter Und Versicherer. Available online: https://www.gidas.org/pdf/Liers_Nutzung_der_Erkenntnisse_aus_GIDAS_print.pdf (accessed on 3 January 2023).
34. Leledakis, A.; Lindman, M.; Östh, J.; Wågström, L.; Davidsson, J.; Jakobsson, L. A method for predicting crash configurations using counterfactual simulations and real-world data. *Accid. Anal. Prev.* **2021**, *150*, 105932. [CrossRef] [PubMed]
35. Initiative for the Global Harmonisation of Accident Data. Available online: http://www.iglad.net/ (accessed on 10 December 2022).
36. Spitzhüttl, F.; Liers, H.; Petzold, M. Creation of pre-crash simulations in global traffic accident scenarios based on the iglad database. In Proceedings of the FAST-Zero'15: 3rd International Symposium on Future Active Safety Technology toward Zero Traffic Accidents, Gothenburg, Sweden, 9–11 September 2015.

applied sciences

MDPI

Article

Road Pavement, Road Pollution, and Sustainability under Climate Change Increased Temperature

Ana Estela Barbosa [1,*], Simona Fontul [2], Ana Cristina Freire [2] and Ana Rita Simões [3]

1 Hydraulics and Environment Department, National Laboratory for Civil Engineering, Av. do Brasil, 101, 1700-066 Lisbon, Portugal
2 Transportation Department, National Laboratory for Civil Engineering, Av. do Brasil, 101, 1700-066 Lisbon, Portugal; simona@lnec.pt (S.F.); acfreire@lnec.pt (A.C.F.)
3 Global Business Services Record to Report Real Estate, Siemens SA, Rua Irmãos Siemens 1, 2720-093 Amadora, Portugal; arbatistasimoes@gmail.com
* Correspondence: aestela@lnec.pt; Tel.: +351-218-443-454

Abstract: This paper presents a multidisciplinary approach to understand the impacts of temperature increase on the retention of particulate pollutants, such as heavy metals and microplastics, by the road pavement material. A soil with a particle size distribution similar to road dust was selected. A Wheel Tracking device was used to assess the permanent deformation behavior of prismatic specimens and the retention of the dust, at controlled temperatures of 40 °C and 60 °C, likely to occur on pavement in the future. The soil representing road pollutants was placed at the top of the slabs prior to the Wheel Tracking Test (WTT), based on the European Standard EN 12697-22:2020. After the WTT, two common methods were used, in order to evaluate the soil retention (pollution accumulation) on road pavement. The results confirm that the viscoelastic behavior of bituminous mixtures under increased temperatures can contribute to particle retention at the pavement. Future studies are needed to understand the phenomena, the retention characteristics by different bituminous mixtures, and the efficiency of pollutants capture. The work opens the opportunity to develop innovative road pavement bituminous mixtures that can reduce the discharge of road particulate pollutants, and have increased resilience and sustainability in extreme weather conditions.

Keywords: climate change; bituminous mixtures; heat effects; new methodologies; road pavement materials; road runoff pollution; sustainable development

Citation: Barbosa, A.E.; Fontul, S.; Freire, A.C.; Simões, A.R. Road Pavement, Road Pollution, and Sustainability under Climate Change Increased Temperature. *Appl. Sci.* **2023**, *13*, 6949. https://doi.org/10.3390/app13126949

Academic Editors: Edgar Sokolovskij and Vidas Žuraulis

Received: 18 April 2023
Revised: 22 May 2023
Accepted: 26 May 2023
Published: 8 June 2023

1. Introduction

1.1. Climate Change Impacts on Road Pavement and Road Runoff

Climate changes are foreseen as scenarios with higher temperatures and extreme rainfall (droughts and floods), as well as sea level rise. Such meteorological conditions will modify the behavior of road pavement materials, the water hydrological cycle conditions, and, consequently, road drainage processes and road runoff discharge. It is predicted that the climate will continue to change during the next century or even longer, depending on human activities alongside the lifecycle of Green House Gases (GHG) and the atmospheric systems that promote climate changes [1].

Several research studies evaluate direct and indirect impacts of extreme rainfall in road pavement, road runoff quantity, and quality and in water resources. The task is not easy due to the uncertainties associated with future scenarios (e.g., [2–9]). Changes in precipitation will change the pattern of diffuse pollution sources, including road runoff discharge, in the coming decades.

Road operators and administration bodies are interested in studies that evaluate the impacts of climate change on road pavement structures. They feel the urge for measures aimed at ensuring both road safety and accessibility in future changed climates. The World Road Association [10], based on questionnaire responses from countries included in the five main Köppen–Geiger Climate Groups, reports impacts that are raising concern. Among them are rainfall variation (both increase or reduction) that can alter moisture balances and influence pavement deterioration. Table 1 contains a summary of reported effects and impacts of climate changes, including temperature increase, on road pavement.

Table 1. Relevant effects of climate change extremes on road pavements.

Extreme Occurrences	Relevant Effects on Road Pavements
Temperature top range are increased	– Flexible pavements experience a higher occurrence of rutting – Bituminous pavements display signs of bleeding – Bituminous layers undergo accelerated aging – Higher demand for maintenance – Higher risk of occurrence of bushfires
Decreased temperatures and augmented freeze–thaw occurrences	– The bearing capacity decreases – Bituminous pavements may suffer from cracking, raveling, and the formation of potholes – The demand for pavement maintenance, including deicing, rises
Higher frequency of storms and higher rainfall intensity	– Water levels tend to rise – Road drainage systems may become inadequate, resulting in flooding – The bearing capacity is diminished – Cracking and permanent deformation may occur – Higher risk of slope failure – Risk of road closure and loss of efficiency in the transportation service
Increase in droughts	– Surface roughness increases – Subgrade moisture tend to decrease – Issues related to salinity arise

According to Table 1, rainfall changes can modify moisture balances and induce pavement deterioration (cf. Figure 1). In addition, temperature can affect the aging of bitumen and, consequently, the pavement behavior. In Australia, the latter was estimated to represent an increase in embrittlement of the surface chip seals of more than 90% of the rural sealed roads. Austroroads [11] considered that the life of bituminous surface treatments will be affected by ambient temperature; the increase in temperature will accelerate the rate of deterioration of seal binders and require earlier surface dressings/reseals, which will lead to higher maintenance costs. Mokoena et al. [12] show concern for the global warming and the need to appropriately decide on bituminous road materials for a reliable performance of asphalt roads in South Africa. Fletcher et al. [13] used observations, output from global climate models, and a statistical model to evaluate the extreme maximum pavement temperature in 17 major Canadian cities, concluding that for most of the cases there will be a projected increased temperature.

The most recent report from the Intergovernmental Panel on Climate Change [14] corroborates the concern, by estimating that the warming up for the period 2081–2100, compared to the years 1850–1900, may vary from 1.4 °C up to 4.4 °C, in the very high GHG emissions scenario.

Li et al. [15] used the MAGICC/SCENGEN tool to simulate pavement performance deterioration over time, for sites in the USA with various climate change levels and pavement structures. Although they established that the modelling tool "was robust and effective", they acknowledged the importance of gathering experimental results. The HDM-III road deterioration model was applied by Chai et al. [16] to model the effects of climate change on flexible pavement deterioration in Queensland, Australia. The study concluded that the maintenance cost is expected to increase by nearly 30% under future climate change conditions. A recent study by Mulholland and Feyen [17] estimated the increase in operation and maintenance costs (O&M) of railway and road infrastructures across the 27 Member States of the EU and the United Kingdom, based on climate projections under RCP4.5 and RCP8.5. The authors concluded that under a 4 °C global warming level, increased levels of extreme heat would cause annual transport O&M costs to rise by €4.8 billion, corresponding to an overall 6.9% rise compared to current values. Some studies dedicated to the impacts of climate change regarding pavement maintenance are mainly focusing on pavement materials and maintenance costs, and the conclusions are based on modeling tools [18,19].

Figure 1. Pavement deterioration, with alligator cracking, after very heavy rainfalls during the winter (Portugal).

On the other hand, when evaluating road runoff, specialists are concerned with the transport and discharge of increased water volumes and increased flow rates during runoff events caused by extreme precipitation (e.g., [20–24]). While precipitation is clearly a critical issue of concern, temperature, expected to change as well, is not addressed much; although, it has already been correlated with the presence and concentration of pollutants (e.g., [25]).

Bieliatynskyi et al. [26] note that asphalt concrete road pavement is sensitive to ambient temperature variations which, together with mechanical impact from vehicle load leads to several damages and material deformations. The thermodynamic behavior of road pavement, as well as the reliability and durability of different mixtures, are of great importance and a concern for research and practice [27,28].

Flexible pavements, under current climate conditions, can reach temperatures between 40 °C and 50 °C in Southern Europe [29]. Therefore, under future climate conditions higher pavement temperatures are expected. It has been pointed out (e.g., [30]) that road pavement softening at elevated ambient temperature may be likely to aggregate particles accumulated at the surface, therefore, acting as a sink of pollutants and contributing to pollution retention and environmental protection.

1.2. The Road Pavement as a Sink and Source of Road Runoff Pollutants

Markiewicz et al. [31] listed sources of organic pollutants in road runoff, namely tire wear, brake lining, integrated vehicle components, car care products, fuels, oils and lubricants, road construction materials, concrete, and road paint. The main sources of emitted Polycyclic Aromatic Hydrocarbons (PAH) were vehicle exhaust gases, followed by tire wear, motor lubricant oils, road surface wear, and brake linings. Contributions from surrounding land use, such as agriculture, industrial, or urban areas, are also acknowledged (e.g., [32]). Solids or dust found in road pavement are originated from distinct sources and materials: pavement wear; construction sites or rehabilitation works; atmospheric fallout; anthropogenic wastes; and vehicles components deterioration—with tire wear being a relevant source [21,33,34].

Road dust is considered to be a very important source of particles in the atmosphere, and its characteristics are studied worldwide. The presence of heavy metals in road dust are related to both human health and ecological risks, due to their persistency in the environment, and they are, for these reasons, the focus of several studies [35].

Road runoff washes off and transports the pollutants accumulated at the road surface. The composition and the concentration of pollutants in stormwater depend on the rainfall pattern, road pavement material and construction, road slope, vehicle characteristics, wheel configuration, ambient conditions, and environmental aspects, among other variables (e.g., [36,37]). Therefore, the presence of particles at the road surface and of associated pollutants must be evaluated, since they are transferred into the environment and biota [34,38–42]. Total suspended solids (TSS) and turbidity are proposed as indicators of highway stormwater quality, for being positively correlated with the presence of pollutants, such as the heavy metals Cu, Pb, Zn, Cr, and Ni [41,43]. Baum et al. [43] proposed that the particular fraction of the TSS < 63 μm could be a surrogate parameter for metal pollution presence in urban stormwater.

Ahmad et al. [38] studied the road runoff dust and particle pollution from road dust in a huge city in Pakistan. The authors concluded that metals such as Cu, Cd, and Mn are attached to road dust and particles, and are likely to pose risks for human health and ecosystems, due to their toxicity and persistency. Cu, Pb, and Zn are commonly found in highway runoff [21,34,36,41,43–45] and are mostly responsible for the overall toxicity of road sediment samples [34].

Pollutants accumulated at the road pavement surface are transported to the surrounding soil and water bodies by precipitation (e.g., [46]); the assessment of impacts depends on a huge number of variables. It has been demonstrated that the discharge of even small concentrations of road runoff pollutants can lead to impacts in confined water bodies (e.g., [47]). It is not just the total concentration that must be evaluated; the fraction of pollutants present in dissolved or particulate form is relevant for impact evaluation and the design of treatment systems [32,48,49]. Several countries have already characterized national road runoff constitution and know which are the most relevant pollutants. It was demonstrated that national, regional, and local site-specific variables control such processes. Crabtree et al. [25], for the purpose of characterizing highway runoff, defined four climate regions in the UK, on the basis of annual average rainfall and annual average winter temperature. The study showed a trend, according to climate, for PAH and dissolved copper (Cu) concentrations. Roads located in different climatic regions, even within a small country as is the case in Portugal, showed different patterns for road runoff quantity and quality (e.g., [44]).

Several strategies are used to deal with road runoff pollution, such as the control at source (reduction of pollutants' emissions) or the construction of treatment systems that receive the runoff from the road and discharge it to the environment, with a reduced concentration of pollutants. The removal of solids and particulate pollutants are among the operations targeted in these treatment systems. Yun et al. [50] indicate that around 80% of highway particles are <0.05 mm. Kayhanian et al. [51] report the difference between particles found at the road pavement surface and the ones that are likely to be washed out by rainfall events (smaller fractions). At least 55% of particles found at a highway and a parking lot surfaces were <0.6 mm. Rommel et al. (2021) [49] studied filter media removal of particles from road runoff and found that, according to several authors, they are predominantly present in the fine fraction (<63 μm).

Drygiannaki et al. [45] studied the content of heavy metals (Cd, As, Cu, Cd, Hg, Ni, Pb, and Zn) in stormwater sediments, and assessed the correlation of metals to finer particle fractions. For instance, Cd was strongly associated with particles sized > 63 μm and Cu to lower than 0.45 μm sizes. Kontchou et al. [34] and Baum et al. [43] agree that the adsorption of metals to road particles < 63 μm is most relevant, and that road sediment must be studied in order to prevent impacts in receiving waters and toxicity in the biota. Mahjoub et al. [52] studied tire and bitumen particles in several size ranges, from 37 up to 5000 μm. The research showed that most of the tire and bitumen particles were from the size range from 37 to 300 μm, and the tire particles were predominantly present in smaller sizes. Lange et al. [53] present similar results for rubber, bitumen, and other microplastic particles in highway runoff—the higher concentrations detected were associated with particles size fractions of 100 μm to 300 μm.

The adaptation of roads to climate change should be used to explore the performance of pavement materials and mix design innovation, as well as to combine it with pollution control strategies developing integrated solutions and responding to the challenges of sustainability. Research focused on the evaluation of the effects of temperature on bituminous mixture viscoelastic properties and its likeliness to affect particle retention at the road surface can contribute to new approaches for adaptation to climate change and for pavement sustainability.

2. Objectives

This study objective is to evaluate a novel hypothesis, namely if the impact of temperature increase on the viscoelastic behavior of bituminous mixtures can contribute to dust and particle bond pollutant retention within the pavement material. The work also aims at testing and proposing new experimental approaches designed to evaluate the formulated premise.

The experimental procedure established was based on a combined knowledge of pavement material characteristics and behavior, road drainage, and road particulate pollution characteristics. The laboratory methodology includes the use of apparatus designed to evaluate the permanent deformation behavior of bituminous mixtures, on the simulation of particulate pollution accumulation, and of particle retention by a bituminous mixture under increased temperature.

3. Materials and Methods

3.1. Soil Selection

A soil with geometrical and physical characteristics comparable to road pollution particles was chosen for the purposes of the test. Figure 2 shows the soil size fraction characterization. Most of the soil particles (89%) are <0.074 mm diameter. This description is within the ranges of particle dust present in road pavements found in the literature [49,51], as well as particulate pollutant fractions [34,41,43,45]. The characteristics of particles found at each road are dependent on site-specific conditions, including surrounding geology, soil type, and land use, as well as on the road pavement materials and vehicle characteristics. This was one of the reasons why sediments from a specific road were not used, as other authors had (e.g., [54]), ensuring that the results would not be biased by specific physical characteristics of particles from a given road location.

3.2. Bituminous Pavement Specimens Preparation and Testing Equipment

Four bituminous mixture slabs of 305 mm × 305 mm × 50 mm thickness were selected for the study. Both sides of each slab were used. The tests were performed at LNEC's Transportation Department. The slabs (specimens) were produced in the laboratory with bituminous mixtures used in flexible pavements in Portugal, with a 35/50 nominal penetration bitumen. This material is typical for pavement-wearing courses nowadays, all over Southern Europe. These specimens were washed with tap water, identified, weighed, and dried in a proper laboratory dry room with a constant temperature of 30 °C. The soil was also kept in the same dry room, to prevent changes in density and water retention.

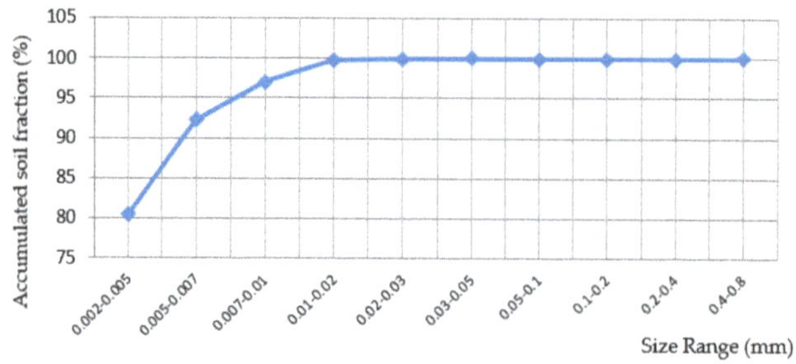

Figure 2. Characterization of the soil mass used in the tests in accordance with the accumulated fraction for different size ranges.

The testing apparatus used was a Wheel Tracking Machine (Figure 3a). This equipment is used for the evaluation of permanent deformation behavior of bituminous prismatic specimens (two at each time) under controlled temperature, producing results of the wheel-tracking slope, in mm per 1×10^3 load cycles, and comprises the application of controlled compression. The European Standard EN 12697-22:2020 [55] procedures were followed for this test. The compression is made by a wheel with 200 mm diameter and 50 mm width of solid rubber, on each specimen, simulating loaded vehicle wheel pressure on road pavement surfaces.

3.3. *Procedures to Evaluate Particles Retention at High Temperature*

The Wheel Tracking Test (WTT) required that the specimens were placed in the Wheel Tracking Machine 6 h before the test start, as specified by the standard. The temperature targeted for the test was programmed. The first 4 h are required for the testing equipment to reach the set temperature, and 2 additional hours are needed for all parts of the slabs to reach the target temperature and to stabilize. This latter stage is called temperature conditioning. After this conditioning, the wheel is rolling and compressing the slab during a period of 6 h, which corresponds to 1×10^3 load cycles, or until a rut depth of 20 mm is reached. During testing, the temperature is constant with variations within ±3 °C and the slab surface shape is measured and recorded by laser, at least at three cross-sections.

After test completion and cooling down to room temperature, the slabs were removed from the equipment.

For the purposes of studying road pollutants and the control of road runoff pollution discharges, a particle-based approach is frequently used to represent the physical processes of pollutant detachment, deposition, build up, and wash-off [22,41]. Furthermore, the simulation of rainfall and wash-off of pollutants from road pavements, in laboratory and field conditions, as well and road dust collection (sweeping, use of vacuum cleaners, wet vacuuming, etc.), are procedures used to study road dust, as well as pollution accumulation and removal from pavements (e.g., [36,40,43].)

Two procedures were designed to assess the retention of particles by the specimens, using the WTT device already described. One was based on spreading an amount of soil all over the slab and, after the WTT, removal of the excess of soil (not retained by the specimen) using water. The second procedure, based on dry removal of the excess of soil, focused only on the strip of the slab that would be compressed during the WTT. The two procedures are described in detail below, and are illustrated in Figure 4.

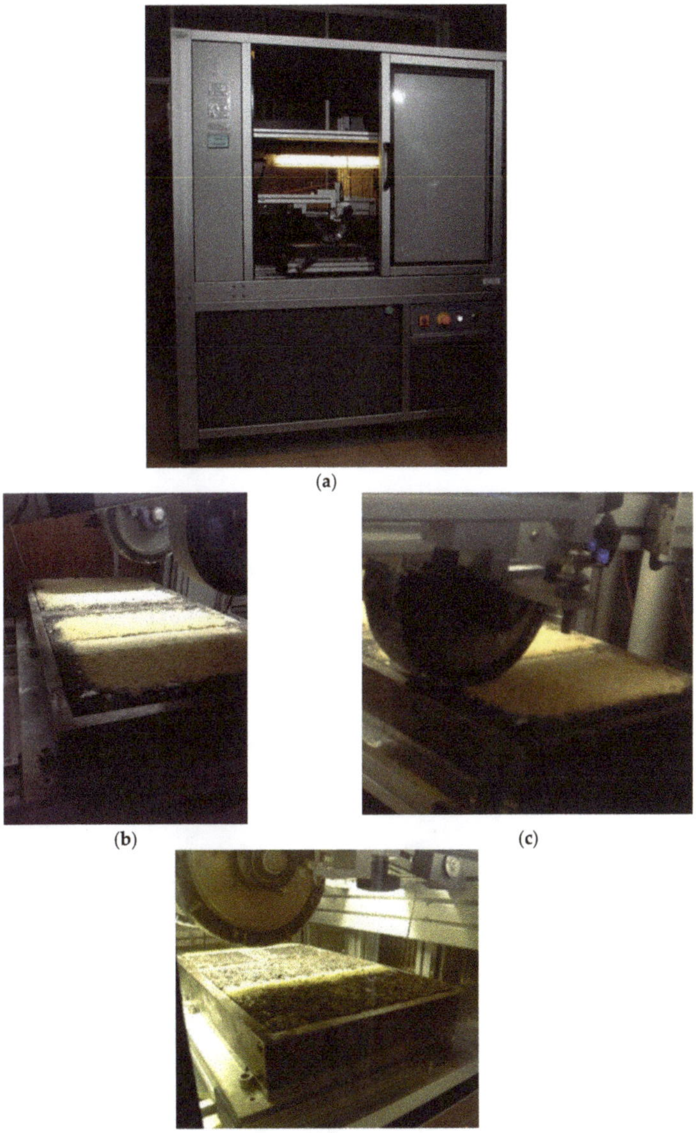

Figure 3. Wheel Tracking equipment (**a**). Test with 30 g of soil over the specimen (**b**,**c**). Test with 5 g of soil over a 5 cm strip (**d**).

i. **Wet removal of excess soil**

A total of 30 g of soil was weighed and manually spread, as homogenously as possible, on the entire surface of the specimen that would undergo the WTT, cf. Figure 3b,c. After the test conclusion and when room temperature was reached, each slab was placed inside an inox container of 35 × 50 cm, supported by a metallic basis with a slope of 2.5%, similar to common pavement cross-section slope. Then, the surface was washed out with 500 mL of deionized water, during 5 min, trying to remove as much of the surface soil as possible. The specimen was taken to the dry room for some days, until it reached constant weight. During the different stages, each specimen was weighted and several pictures were taken.

The inox container was taken to an oven at constant 100 °C until the soil was dry, and at that time the dry soil was weighted.

ii. **Dry removal of excess soil**

A total of 5 g of soil was weighed and manually spread only over the 5 cm strip of the specimen that would be directly under the rubber wheel during the WTT, cf. Figure 3d. After the test conclusion and when room temperature was reached, the slab undertook a process of soil removal: (i) Vacuum aspiration (900 W power device); (ii) Brushing; (iii) Vacuum aspiration, repeated twice. The vacuum equipment was weighed before and after this particle removal process, allowing to estimate the removal of soil from the specimen.

In order to have an indicator of the sensitivity to temperature, tests were performed at 40 °C and 60 °C. The latter aimed to simulate the wearing course behavior under the climate change scenarios. During the different testing stages, the slab specimens were carefully observed, and pictures were taken (Figure 4).

Figure 4. Illustration of steps of the two procedures for the removal of soil after the WTT. *Procedure (i) Wet removal of excess soil:* (**a**) Slab with 30 g of soil all over the surface after Wheel Tracking Test at 60 °C. (**b**,**c**) Slab being washed with deionized water. (**d**) Slab with soil particles retained after being washed and dried. *Procedure (ii) Dry removal of excess soil:* (**e**) Slab with 5 g of soil over the area of the wheel path, after WTT at 60 °C. (**f**) Slab with soil particles retained after being brushed and vacuum cleaned.

4. Results

The WTT test, performed according to the dedicated standard, considers the assessment of maximum rut depth, creep slope, and number of passes. However, taking into consideration the main purpose of the work, the target results focused on assessing the impacts of temperature increase on the pavement material structure and on the retention of particulate pollutants. Tables 2 and 3 present the results for the soil retention by the slabs, under the 40 °C and 60 °C WTT. Figure 5 compares the soil particle retention by the slabs, at 40 °C and 60 °C.

Table 2. Soil retention by the slabs after the wet removal procedures (use of 30 g soil).

Slab	T (°C)	Soil Mass (g)				% Soil Kept	Particles Retained (g/cm^2)
		Placed at the Surface	Collected	Kept within Slab	Average Soil Kept		
BBR1	40	30	26	4	4	13%	0.0043
BBR2	60	30	18	12	11.5	38%	0.012
FOSSA2	60	30	19	11			

Table 3. Soil retention by the slabs after the dry removal procedures (use of 5 g soil).

Slab	T (°C)	Soil Mass (g)				% Soil Kept	Particles Retained (g/cm^2)
		Placed at the Surface	Vacuumed	Kept within Slab	Average Soil Kept		
BBR2	40	5	3.4	1.6	1.6	32%	0.010
FOSSA2	40	5	3.4	1.6			
BBR2	60	5	3.2	1.8	2	40%	0.013
FOSSA2	60	5	2.8	2.2			

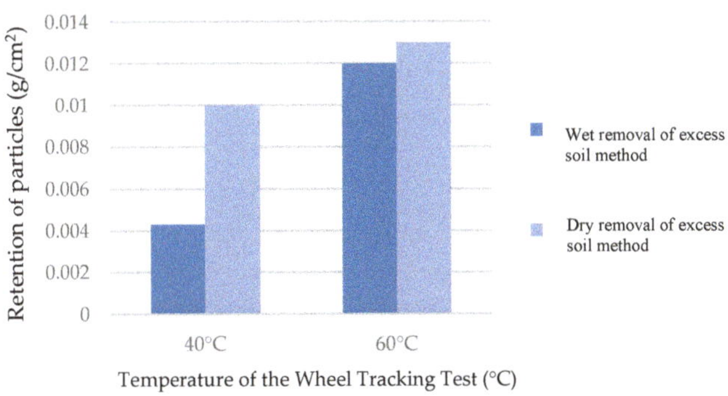

Figure 5. Comparison of soil particles retained within the slab after the two procedures (wet and dry removal).

The results for the wet removal procedure, presented in Table 2, include one test at 40 °C that was used as an indicator. The results show a soil retention by the slab of 13% at 40 °C and of 38% at 60 °C. It is understood that there is a different behavior at each temperature, corroborated by the Table 3 figures. The quantification was not an objective of the study, which is consistent with the limited number of samples, with the same asphalt mixture.

For the dry removal of exceeding soil (Table 3), the quantification of the amount of soil that remained embedded in the slabs should be merely indicative, due to the fact that some soil was lost during the brushing. The vacuum cleaner (itself with a weight of 1.848 kg) contained, respectively, 3.4 (at 40 °C) and 3.2 g or 2.8 g (at 60 °C) of soil at the end of the procedure. Therefore, from the total 5 g of soil, 1.6 g or 2 g remained embedded in the slabs' surfaces—which is visible in Figure 4f, for the 60 °C test.

Table 4 reports the weight variations that allowed the assessment of soil retention by the two slab specimens (BBR2 and FOSSA2) tested at 60 °C, according to the wet removal of exceeding soil procedure. FOSSA2 lost parts of bituminous material during the test procedure. Therefore, the measurement of 8 g of soil retained by this specimen after the WTT must be underestimated. The values of the washed soil are comparable (18 g and 19 g). Visual observation confirmed the presence of soil embedded in the surface of the specimen material—Figure 4d.

Table 4. Weight measurements during the different phases of the wet test at 60 °C.

Slab	Weight (kg)	Weight after 30 g of Soil (kg)	Weight after Test at 60 °C (kg)	Weight after Washing the Soil (kg)	Slab Weight after Dried (kg)	Weight of Soil in the Slab (kg)	Weight of Soil in the Inox Basin (kg)
BBR2	11.267	11.284	11.290	11.286	11.272	0.012	0.018
FOSSA2	10.683	10.97	10.697 *	10.695	10.675	0.08 *	0.019

* Loss of specimen material was observed during this step of the procedure.

The overall results (cf. Figure 5) show that there was retention of soil particles within the bitumen, and that the level of retention is higher at 60 °C compared to 40 °C.

5. Discussion

The presented approach showed itself to be feasible and a promising starting point to develop a methodology to be adopted in the future. The results confirmed that the viscoelastic behavior of bituminous mixtures due to bitumen softening under increased climate temperatures contributes to particles retention at the road surface and, consequently, may support road pollutant imprisonment. To what extent the phenomena occur and which is the strength of the retention, under different ambient temperatures and by distinct mixtures, must be assessed in future studies. It is understood that the results presented cannot be evaluated from a qualitative approach, given the limited samples used, but the confirmation of the hypothesis was possible, and the results are in line with outcomes from road runoff monitoring studies under distinct temperature climate regions. It is acknowledged that some dust particles may have adhered to the tires, during the WTT test. This process is similar to what takes place in real conditions, where not just vehicle tires but also the movement of vehicles passing and wind are processes that resuspend and may remove dust accumulated on the road pavement.

The test did not use road dust, as performed, e.g., by Bauer et al. [54], to avoid the possibility of biased results in the face of particles specific for a road location. Future studies should work with road particles from different locations, with distinct physicochemical characteristics and pollution content, to further understand particle behavior. Road dust can be easily obtained by collecting sediments at road runoff treatment facilities, or by vacuum cleaning sections of selected roads. Nanoparticles from road runoff sediment have distinct content of heavy metals—such as zinc, copper, lead, and cadmium—and microplastics, consequently having distinctive characteristics and causing differentiated impacts in the environment. Thus, the future research should also comprise the assessment of the pollutant content of the dust to be used.

The dry removal procedure used is more straight forward, and likely to allow the evaluation of total particles strongly embedded within the bitumen material. On the other hand, the wet removal method has the potential to be, in the future, fine-tuned to test the degradation of the pavement under cycles of heat/compression and rainfall, likely to take place under climate change scenarios, and can also contribute to understand the removal efficiency of rainfall events, in relation to its intensity and duration characteristics.

The proposed approach uses a standard procedure based on WTT with the application of controlled test conditions which allows a proper behavior comparison. The WTT standard test provides other results, such as the maximum rut depth, creep slope, and number of passes. Since the objective of the study was to validate the causal relation between heated pavement bituminous materials and dust retention at the road surface, the recommendation that was adopted concerned the use of controlled temperature and a rolling wheel. In fact, the procedure presented is an adaptation, considering the purposes of the experiment.

Future studies should consider adjustments and improvements to the procedures, to understand other variables not yet addressed. For instance, additional laboratory tests to characterize the bituminous mixtures should be considered, in order to evaluate asphalt performance under different climate change scenarios and with the incorporation of innovative materials that may enhance the retention of road pollutants [56].

Mokoena et al. [12] pointed out that pavement temperature relies on more than air temperature—it is determined based on various climatic and weather factors as well as on pavement material properties and the overall pavement structure. Future studies should be performed for distinct types of bitumen used in flexible pavements, as the retention potential can vary with the bitumen type. Baensch-Baltruschat et al. [33] recommend future studies to investigate environmental concentrations and the degradation of polymer compounds under environmental conditions. Among the mitigation measures to be implemented, according to the principle of precaution, there is the "successive optimization of road surfaces", in line with the approach of the work presented herein.

6. Conclusions

This study was designed based on multidisciplinary knowledge in the scope of road pavement and road pollution. The motivation was the current necessity for innovative approaches to deal with economic, social, and environmental sustainability.

The most recent report from the Intergovernmental Panel on Climate Change [14] supports the relevance of being prepared for temperature increases. Although the assessment of temperature impacts on road pavement behavior and resilience are the topic of theoretical, experimental, and modeling studies [27,28], there is not much research linking these objectives with road pollution control. In fact, the use of pavement materials to control non-point pollution from roads is not common. There is the example from Huang et al. [56], who studied the possibility of using a permeable reactive road asphalt pavement designed to capture volatile organic compounds produced by vehicles and, therefore, ensure some control of road pollution.

It is relevant to establish combined laboratory procedures to evaluate both pavement behavior at high temperatures, such as its rutting resistance or fatigue behavior, as well as the pollution retention effect. These new methodologies should aim to enable optimization of the final bituminous mixture design. The development of new types of bitumen, capable of ensuring adequate performance and with the capacity to incorporate particulate pollutants and present an improved behavior, would provide an answer to sustainability under climate change, pollution control, and the reduction of toxicity discharged by roads. Given the increased tendency to use cold or temperate bituminous mixtures in order to reduce the pollution during fabrication, transport, and laying of the asphalt, it is also important to study these types of materials and their aptitude to absorb/retain contamination.

This work opened opportunities to combine the reduction of pollutants discharged into the environment with new bituminous mixtures for improved road pavement resilience and sustainability. The future target is to expand and consolidate the work started, through laboratory and field research, facilitating the establishment of normalized tests and adaptation guidelines for road pavement materials, considering the accumulation of pollutants at the road surface, and consequently contributing to minimize economic, social, and environmental impacts.

Author Contributions: Conceptualization, A.E.B. and S.F.; methodology, A.E.B. and A.C.F.; validation, A.E.B., S.F. and A.C.F.; formal analysis, A.E.B. and A.R.S.; investigation, A.E.B. and A.R.S.; resources, A.C.F. and S.F.; writing—original draft preparation, A.E.B.; writing—review and editing,

A.E.B., S.F., A.C.F. and A.R.S.; visualization, A.R.S. and A.E.B.; supervision, A.E.B. All authors have read and agreed to the published version of the manuscript.

Funding: This research received no external funding.

Institutional Review Board Statement: Not applicable.

Informed Consent Statement: Not applicable.

Data Availability Statement: Not applicable.

Conflicts of Interest: The authors declare no conflict of interest.

References

1. Pavement Interactive. Available online: https://pavementinteractive.org/climate-change-impacts-on-pavements-and-resilience/ (accessed on 9 April 2023).
2. Zhang, C.; Tan, Y.; Gao, Y.; Fu, Y.; Li, J.; Li, S.; Zhou, X. Resilience assessment of asphalt pavement rutting under climate change. *Transp. Res. Part D Transp. Environ.* **2022**, *109*, 103395. [CrossRef]
3. Dunn, S.M.; Brown, I.; Sample, J.; Post, H. Relationships between climate, water resources, land use and diffuse pollution and the significance of uncertainty in climate change. *J. Hydrol.* **2012**, *434–435*, 19–35. [CrossRef]
4. Ibeje, A.O. Vulnerability of Road Pavements to Climate Change in Abia and Imo States of Nigeria. *Niger. J. Technol.* **2021**, *40*, 6–12. [CrossRef]
5. Hettiarachchi, S.; Wask, C.; Sharma, A. Can antecedent moisture conditions modulate the increase in flood risk due to climate change in urban catchments? *J. Hydrol.* **2019**, *571*, 11–20. [CrossRef]
6. Pan, H.; Jin, Y.; Zhu, X. Comparison of Projections of Precipitation over Yangtze River Basin of China by Different Climate Models. *Water* **2022**, *14*, 1888. [CrossRef]
7. Comité Technique. *Adaptation Methodologies and Strategies to Increase the Resilience of Roads to Climate Change—Case Study Approach*; Report 2019R25EN; World Road Association (PIARC): Paris, France, 2019; ISBN 978-2-84060-558-4.
8. Comité Technique. *Measures for Improving Resilience of Pavements—A PIARC Collection of Case Studies*; Report 2022R25EN; World Road Association (PIARC): Paris, France, 2022; ISBN 978-2-84060-713-7.
9. Qiao, Y.; Santos, J.; Stoner, A.; Flinstch, G. Climate change impacts on asphalt road pavement construction and maintenance. *J. Ind. Ecol.* **2020**, *24*, 342–355. [CrossRef]
10. Comité Technique D.2 Chaussées Routières. *Dealing with the Effects of Climate Change on Road Pavements*; Report 2012R06EN; World Road Association (PIARC): Paris, France, 2012; ISBN 2-84060-247-4.
11. Austroads. *Impact of Climate Change on Road Infrastructure*; Austroads: Sidney, Australia, 2004; AP–R243/04.
12. Mokoena, R.; Mturi, G.; Maritz, J.; Mateyisi, M.; Klein, P. African Case Studies: Developing Pavement Temperature Maps for Performance-Graded Asphalt Bitumen Selection. *Sustainability* **2022**, *14*, 1048. [CrossRef]
13. Fletcher, C.; Matthews, L.; Andrey, J.; Saunders, A. Projected Changes in Mid-Twenty-First-Century Extreme Maximum Pavement Temperature in Canada. *J. Appl. Meteorol. Climatol.* **2016**, *55*, 961–974. [CrossRef]
14. Intergovernmental Panel on Climate Change. Climate Change 2023, The IPCC Sixth Assessment Report, (IPCC). 2023. 85p. Available online: https://www.ipcc.ch/report/ar6/syr/downloads/report/IPCC_AR6_SYR_LongerReport.pdf (accessed on 17 May 2023).
15. Li, Q.; Mills, L.; McNeil, S. *The Implications of Climate Change on Pavement Performance and Design*; Department of Transportation University Transportation Centers Program, Delaware University: Newark, DE, USA, 2011.
16. Chai, G.; Stadenb, R.; Guana, H.; Kellya, G.; Chowdhury, S. *The Impacts of Climate Change on Pavement Maintenance in Queensland, Australia*; Transport Research Arena: Paris, France, 2014.
17. Mulholland, E.; Feyen, L. Increased risk of extreme heat to European roads and railways with global warming. *Clim. Risk Manag.* **2021**, *34*, 100365. [CrossRef]
18. ITF. *Adapting Transport to Climate Change and Extreme Weather: Implications for Infrastructure Owners and Network Managers*; ITF Research Reports; OECD Publishing: Paris, France, 2016. [CrossRef]
19. López-Montero, T.; Miró, R. Ageing and temperature effect on the fatigue performance of bituminous mixtures. *Mater. Construcción* **2017**, *67*, 327. [CrossRef]
20. de Macedo, M.B.; Júnior, M.N.; Jochelavicius, V.; de Oliveira, T.R.P.; Mendiondo, E.M. Modular Design of Bioretention Systems for Sustainable Stormwater Management under Drivers of Urbanization and Climate Change. *Sustainability* **2022**, *14*, 6799. [CrossRef]
21. Hvitved-Jacobsen, T.; Vollertsen, J.; Nielsen, A.H. *Urban and Highway Stormwater Pollution: Concepts and Engineering*, 1st ed.; CRC Press: Boca Raton, FL, USA, 2010; pp. 1–367. [CrossRef]
22. Kourtis, I.M.; Tsihrintzis, V.A. Adaptation of urban drainage networks to climate change: A review. *Sci. Total Environ.* **2021**, *771*, 145431. [CrossRef]
23. Kumar, S.; Agarwal, A.; Ganapathy, A.; Villuri, V.G.K.; Pasupuleti, S.; Kumar, D.; Kaushal, D.R.; Gosain, A.K.; Sivakumar, B. Impact of climate change on stormwater drainage in urban areas. *Stoch. Environ. Res. Risk Assess.* **2022**, *36*, 77–96. [CrossRef]

24. Kalantari, Z.; Briel, A.; Lyonb, S.W.; Olofssona, B.; Folkeson, L. On the utilization of hydrological modelling for road drainage design under climate and land use change. *Sci. Total Environ.* **2014**, *475*, 97–103. [CrossRef]
25. Crabtree, B.; Dempsey, P.; Moy, F.; Brown, C.; Song, M. *Improved Determination of Pollutants in Highway Runoff—Phase 2. Final Report*; Highways Agency Contract Reference 3/376, Report No. UC7697; WRc Swindon: Swindon, UK, 2008.
26. Bieliatynskyi, A.; He, Y.; Pershakov, V.; Akmaldinova, O.; Krayushkina, K. Pollution of the roadside environment by dust particles from road surface repairs. *Environ. Sci. Pollut. Res.* **2023**, *30*, 35663–35669. [CrossRef]
27. Khroustalev, B.M.; Liu, T.; Aliakseyeu, Y.H.; Li, Z.; Akeliev, V.D.; Minchenya, V.T. Thermodynamic Aspects of Pavement Engineering. *Sci. Tech.* **2022**, *21*, 28–35. [CrossRef]
28. Malitha, C.; Rajapaksha, M.; Athukorallage, B.; Senadheera, S.; James, D. Temporal and spatial temperature predictions for flexible pavement layers using numerical thermal analysis and verified with large datasets. *Case Stud. Constr. Mater.* **2023**, *18*, e02008. [CrossRef]
29. Freire, A.C.; Antunes, M.L.; Picado-Santos, L. Characterization of bituminous mixtures for numerical modelling of permanent deformations. *Road Mater. Pavement Des.* **2006**, *7*, 87–102. [CrossRef]
30. Barbosa, A.E.; Fernandes, J.N.; Simões, A.R.; Fontul, S.; Freire, A.C. New approaches to the assessment of the impacts of roads in climate change scenarios. In *8th National Congress on Impacts Evaluation (CNAI'19), Coimbra*; APAI (Associação Portuguesa de Avaliação de Impactos): Lisboa, Portugal, 2019. (In Portuguese)
31. Markiewicz, A.; Björklund, K.; Eriksson, E.; Kalmykova, Y.; Strömvall, A.; Siopi, A. Emissions of organic pollutants from traffic and roads: Priority pollutants selection and substance flow analysis. *Sci. Total Environ.* **2017**, *580*, 1162–1174. [CrossRef]
32. Huber, M.; Helmreich, B. Stormwater Management: Calculation of Traffic Area Runoff Loads and Traffic Related Emissions. *Water* **2016**, *8*, 294. [CrossRef]
33. Baensch-Baltruschat, B.; Kocher, B.; Stock, F.; Reifferscheid, G. Tyre and road wear particles (TRWP)—A review of generation, properties, emissions, human health risk, ecotoxicity, and fate in the environment. *Sci. Total Environ.* **2020**, *733*, 137823. [CrossRef]
34. Kontchou, J.A.; Baetz, N.; Grabner, D.; Nachev, M.; Tuerk, J.; Sures, B. Pollutant load and ecotoxicological effects of sediment from stormwater retention basins to receiving surface water on Lumbriculus variegatus. *Sci. Total Environ.* **2023**, *859*, 160185. [CrossRef]
35. Vlasov, D.; Ramírez, O.; Luhar, A. Road Dust in Urban and Industrial Environments: Sources, Pollutants, Impacts, and Management. *Atmosphere* **2022**, *13*, 607. [CrossRef]
36. Kayhanian, M.; McKenzie, E.R.; Leatherbarrow, J.E.; Young, T.M. Characteristics of road sediment fractionated particles captured from paved surfaces, surface run-off and detention basins. *Sci. Total Environ.* **2012**, *439*, 172–186. [CrossRef]
37. Huber, M.; Welker, A.; Helmreich, B. Critical review of heavy metal pollution of traffic area runoff: Occurrence, influencing factors, and partitioning. *Sci. Total Environ.* **2016**, *541*, 895–919. [CrossRef]
38. Ahmad, H.R.; Mehmood, K.; Sardar, M.F.; Maqsood, M.A.; Rehman, M.Z.U.; Zhu, C.; Li, H. Integrated risk assessment of potentially toxic elements and particle pollution in urban road dust of megacity of Pakistan. *Hum. Ecol. Risk Assess. Int. J.* **2020**, *26*, 1810–1831. [CrossRef]
39. Ermolin, M.S.; Fedotov, P.S.; Ivaneev, A.I.; Karandashev, V.K.; Fedyunina, N.N.; Burmistrov, A.A. A contribution of naoscale particles of road-deposited sediments to the pollution of urban runoff by heavy metals. *Chemosphere* **2018**, *210*, 65–75. [CrossRef]
40. Jiang, J.; Liang, Q.; Xia, X.; Hou, J. A coupled hydrodynamic and particle-tracking model for full-process simulation of nonpoint source pollutants. *Environ. Model. Softw.* **2021**, *136*, 104951. [CrossRef]
41. Wang, Q.; Zhang, Q.; Dzakpasu, M.; Lian, B.; Wu, Y.; Wang, X.C. Development of an indicator for characterizing particle size distribution and quality of stormwater runoff. *Environ. Sci. Pollut. Res.* **2021**, *25*, 7991–8001. [CrossRef]
42. Zhao, H.; Chen, X.; Hao, S.; Jiang, Y.; Zhao, J.; Zou, C.; Xie, W. Is the wash-off process of road-deposited sediment source limited or transport limited? *Sci. Total Environ.* **2016**, *563–564*, 62–70. [CrossRef]
43. Baum, P.; Kuch, B.; Dittmer, U. Adsorption of Metals to Particles in Urban Stormwater Runoff—Does Size Really Matter? *Water* **2021**, *13*, 309. [CrossRef]
44. Barbosa, A.E.; Fernandes, J.N. Comparison of the pollutant potential of two Portuguese highways located in different climatic regions. In *Urban Environment: Alliance for Global Sustainability Bookseries*; Rauch, S., Morrison, G.S., Eds.; Springer: Dordrecht, The Netherlands, 2012; Volume 19, pp. 263–273. [CrossRef]
45. Drygiannaki, I.; Rao, B.; Dawson, J.A.; Rakowska, M.; Reible, D.D.; Hayman, N.T.; Rosen, G.H.; Colvin, M.A.; Chadwick, B.D.; Pitt, R.; et al. Assessing sediment recontamination from metals in stormwater. *Sci. Total Environ.* **2020**, *737*, 139726. [CrossRef]
46. Baensch-Baltruschat, B.; Koche, B.; Kochleus, C.; Stock, F.; Reifferscheid, G. Tyre and road wear particles—A calculation of generation, transport and release to water and soil with special regard to German roads. *Sci. Total Environ.* **2021**, *752*, 141939. [CrossRef]
47. Vieira, R.; Fernandes, J.N.; Barbosa, A.E. Evaluation of the impacts of road runoff in a Mediterranean reservoir in Portugal. *Environ. Monit. Assess.* **2013**, *185*, 7659–7673. [CrossRef]
48. Kayhanian, M.; Fruchtman, B.; Gulliver, J.; Montanaro, C.; Ranieri, E.; Wuertz, S. Review of highway runoff characteristics: Comparative analysis and universal implications. *Water Res.* **2012**, *46*, 6609–6624. [CrossRef]
49. Rommel, S.H.; Stinshoff, P.; Helmreich, B. Sequential extraction of heavy metals from sorptive filter media and sediments trapped in stormwater quality improvement devices for road runoff. *Sci. Total Environ.* **2021**, *782*, 146875. [CrossRef]

50. Yun, Y.; Park, H.; Kim, L.; Ko, S. Size distributions and settling velocities of suspended particles from road and highway. *KSCE J. Civ. Eng.* **2010**, *14*, 481–488. [CrossRef]
51. Kayhanian, M.; Vichare, A.; Green, P.; Alaimo, C.; Hwang, H.; Signore, J.; Troxler, M.; Jones, D.; Harvey, J. Water Quality Evaluation of Leachate Produced from Pavement Specimens under Controlled Laboratory Conditions. *Road Mater. Pavement Des.* **2010**, *11*, 9–28. [CrossRef]
52. Mahjoub, A.; Hashemi, H.H.; Petroody, S.S.A. The role of baseflow and stormwater in transport of tire and bitumen particles in Tehran city: A dense urban environment. *J. Contam. Hydrol.* **2023**, *256*, 104180. [CrossRef]
53. Lange, K.; Magnusson, K.; Viklander, M.; Blecken, G.T. Removal of rubber, bitumen and other microplastic particles from stormwater by a gross pollutant trap—Bioretention treatment train. *Water Res.* **2021**, *202*, 117457. [CrossRef]
54. Baur, S.; Reemtsma, T.; Stärk, H.-J.; Wagner, S. Surfactant assisted extraction of incidental nanoparticles from road runoff sediment and their characterization by single particle-ICP-MS. *Chemosphere* **2020**, *246*, 125865. [CrossRef]
55. *EN 12697-22:2020*; Bituminous Mixtures—Test Methods for Hot Mix Asphalt—Part 22: Wheel tracking. Instituto Português da Qualidade: Caparica, Portugal, 2020.
56. Huang, S.; Liang, C. A conceptual study on the formulation of a permeable reactive pavement with activated carbon additives for controlling the fate of non-point source environmental organic contaminants. *Chemosphere* **2018**, *193*, 436–446. [CrossRef]

applied sciences

MDPI

Article

Periodical Vehicle Inspections with Smart Technology

Peter Tapak [1,*], Michal Kocur [1], Matej Rabek [1] and Juraj Matej [2]

1 Faculty of Electrical Engineering and Information Technology, Slovak University of Technology in Bratislava, Ilkovicova 3, 812 19 Bratislava, Slovakia; michal.kocur@stuba.sk (M.K.)
2 TESTEK, a.s., Authorized Technical Service for Technical Inspections of Vehicles, P.O. Box 84, Placheho 14, 841 02 Bratislava, Slovakia; juraj.matej@testek.sk
* Correspondence: peter.tapak@stuba.sk

Abstract: This paper presents the major outcomes of the utilization of smart phone applications in the periodical technical inspection process. Road safety is one of the main topics in transportation. Technical defects of vehicles can play an important role in fatal accidents; therefore, periodical vehicle inspection is a common practice amongst many countries across all continents. The authors observed that advancements in smart technologies presented an opportunity to enhance the efficiency and effectiveness of vehicle inspections. They specifically focused on the Slovak Republic and described the transition of technical inspections towards smart technologies in this country. The goal was to utilize new technologies and improvements without increasing costs dramatically. The paper discusses the outcomes of the first two years of using this app at every technical inspection station in the Slovak Republic. The smart phones were used not only to help with process management but the novelty is that they also collected data from the sensors and used their own sensors for the vehicle dynamics measurements. One of the important factors is the successful implementation of low cost devices at such a large scale, country wise.

Keywords: vehicle inspection; smart phone application; on-board diagnostics; vehicle safety; brake testing

Citation: Tapak, P.; Kocur, M.; Rabek, M.; Matej, J. Periodical Vehicle Inspections with Smart Technology. *Appl. Sci.* **2023**, *13*, 7241. https://doi.org/10.3390/app13127241

Academic Editor: Edgar Sokolovskij

Received: 28 April 2023
Revised: 14 June 2023
Accepted: 15 June 2023
Published: 17 June 2023

1. Introduction

According to the United Nations Economic Commission for Europe (UNECE) online workshop [1], vehicle technical defects can relate to 8 to 15% of death accidents in high income countries. In middle income countries the number can be even higher, at 15 to 25%.

Vehicle inspection service stations offer periodical vehicle safety inspections (PTI) to prevent traffic accidents resulting from vehicle malfunction. PTI's significance has been highlighted in [2]. During PTI, vehicles are assessed for compliance with national safety regulations and, in several countries, emission inspections are conducted in conjunction with safety inspections to ensure compliance with emission regulations. The significance of road-worthiness inspections extends beyond verifying the vehicle's proper functioning to include environmental considerations. In the European Union, two forms of evaluation are performed: on-the-spot roadside inspections and periodic checks that necessitate vehicle owners taking their cars to an inspection service station

It is difficult to quantify exactly how much road safety in Europe has improved as a result of periodic technical inspections. PTIs are just one of many factors that can influence road safety and it is difficult to isolate the specific impact of PTIs on road safety trends. Taneerananon et al. in [3] conducted a study on traffic accidents in Thailand and the correlation between vehicle defects and these incidents. The study findings indicate that implementing adequate vehicle inspections could potentially serve as an efficient approach to decrease road accidents. On the other hand, Hoagland and Woolley proposed an opposing view in the paper [4], stating that advancements in modern vehicle technology and reliability have reduced the necessity for effective vehicle inspections. This conclusion yielded from examining the impact of removing vehicle inspections in New Jersey and observing no notable rise in accidents caused by vehicle defects.

That being said, there is evidence that PTIs can help to improve road safety by identifying and correcting problems with a vehicle that could potentially cause accidents. For example, a PTI might identify worn brakes or faulty lighting, which could be repaired before they become a safety hazard. Previous reports have shown a wide variation in the proportion of road crashes caused by vehicle defects. In developed countries, this percentage has been estimated to range from 3% to 19% [5,6], while in developing countries, it was reported to be as high as 27% [3]. The primary basis for the implementation of PTI programs in many countries is the relationship between technical defects in vehicles and road crashes. By detecting such defects, PTIs can help ensure roadworthiness and prevent crashes [6–9]. Studies [10,11] provide further evidence of the positive influence of Periodic Technical Inspections (PTI) on road safety. It explores the relationship between PTI for vehicles and traffic accidents caused by technical defects in Slovakia. The aim was to evaluate the effectiveness of PTI in enhancing road safety by analyzing statistical data on accidents resulting from vehicle technical defects and PTI records.

The study revealed two significant findings. Firstly, as the PTI validity period approaches its end, the probability of accidents caused by technical defects increases. Secondly, an increase in the number of vehicles assessed at PTI stations as temporarily roadworthy or not roadworthy corresponds to a decrease in accidents caused by vehicle technical defects. These findings demonstrate that PTI has a measurable impact on reducing accidents caused by vehicle technical defects, thus contributing positively to road safety and justifying its implementation.

Instead of questioning the PTI systems, it is recommended that national authorities implement more stringent measures to enhance PTI performance, particularly at the EU level. This may include the mandatory accreditation of PTI under ISO 17020, which has been proven to enhance the quality of PTI activities. Additionally, attention should be given to inspections of PTI. The example of the Slovak Republic is cited, where the introduction of camera monitoring effectively controls the work of inspectors. As a result, the rate of temporary roadworthiness and non-worthiness assessments significantly increased, leading to a reduction in accidents caused by vehicle technical defects. PTIs can also help to reduce emissions from vehicles, which can have positive impacts on air quality and public health, especially in urban areas [12]. According to several studies [13–17], measuring PN (particle number) during PTI (periodic technical inspection) can be a significant factor in reducing particle emissions and improving air quality. In general, according to [18,19], road safety in Europe has improved significantly over the past decades due to a combination of factors including improved vehicle safety technologies, changes in traffic laws and enforcement, and improvements in road infrastructure. It is likely that PTIs have contributed to this improvement in road safety to some degree, although it is difficult to quantify the specific impact of PTIs on road safety trends. According to reports by the National Highway Traffic Safety Administration in the USA, the risk of road crashes associated with driving a vehicle manufactured before 2000 is 71% higher than for vehicles manufactured in 2010 and later [20].

The European Union, previously known as the European Community or European Economic Community, has conducted periodical technical vehicle inspections since the late 1970s through directive 77/143/EEC ([21]). As of April 2014, all member states of the European Union are required to conduct periodic safety and emission inspections for most types of motor vehicles according to the EU directive 2014/45 ([22]). Although the specific process for PTI varies among EU countries, inspections generally include the following checks:

Technical inspections, as per EU directive [23], are concerned with the vehicle's condition and maintenance, focusing on the following aspects:

- Axles, brakes, wheels, tyres, and wheel alignment;

- Frame/body, including vehicle identification (such as chassis number and licence plates);
- Exhaust system, with attention to environmental factors such as noise and exhaust emissions;
- Steering;
- Lighting and electrical systems, as well as electrical equipment;
- Windows, mirrors, and visibility;
- Accessories;
- Pedals, seats, and seat belts;
- Electronic safety systems.

The tools, devices, and processes used by the inspectors during the inspection process can differ between countries.

1.1. OBD Reading

OBD (On-Board Diagnostics) is an integrated self-diagnostics system incorporated within vehicles. The OBD-II socket, a 16-pin connector, is typically located near the driver's wheel, under the dashboard. The inception of OBD can be traced back to California, where the CARB (California Air Resources Board) implemented strict regulations and raised awareness regarding emissions control [24]. CARB mandated the use of OBD in all cars manufactured after 1991 to monitor and regulate the emission of harmful gases [25]. OBD-II plays a crucial role in the automotive industry by enabling the logging of various essential parameters of vehicles. It encompasses the analysis and diagnostics of small vehicle systems as well as facilitates comprehensive fleet management and route optimization [25].

OBD-II (On-Board Diagnostics version 2) is a protocol used for identifying, diagnosing, and reporting the health and condition of a vehicle's engine to the user. It operates on the CAN (Controller Area Network) protocol, allowing the vehicle to connect with scanning tools and external hardware known as OBD scanners. These scanners can be connected to PCs or laptops to extract information from the vehicle's onboard computer, providing parameters such as driving speeds, engine speeds, coolant levels, coolant temperature, emission control, engine idle time, and other vital engine information [24].

CAN is a Vehicle Bus Standard designed to facilitate communication between various modules within the vehicle. It enables interaction and coordination among microcontrollers and components, eliminating the need for a dedicated host computer. OBD scanners retrieve CAN data and convert it into a format that is easily understood through presentation and application layers. Look-up tables are used to decipher the received data. OBD functions as the higher layer, handling presentation and application aspects, while the CAN Bus serves as the communication layer, dealing with data transmission and physical layer issues [24].

A wide range of OBD readers can be found in the market, spanning from as low as EUR 10 to several thousand euros in price. The most affordable devices typically utilize the ELM 327 chip [26] or its clones and offer limited functionality, providing only basic diagnostic capabilities. These support only SAE J1850 PWM (41.6 kbit/s), SAE J1850 VPW (10.4 kbit/s), ISO 9141-2 (5 baud init, 10.4 kbit/s), ISO 14230-4 KWP (5 baud init, 10.4 kbit/s), ISO 14230-4 KWP (fast init, 10.4 kbit/s), ISO 15765-4 CAN (11 bit ID, 500 kbit/s), ISO 15765-4 CAN (29 bit ID, 500 kbit/s), ISO 15765-4 CAN (11 bit ID, 250 kbit/s), ISO 15765-4 CAN (29 bit ID, 250 kbit/s), SAE J1939 (250 kbit/s), and SAE J1939 (500 kbit/s) and that makes these unsuitable for heavy duty vehicles. One of the advantages is that there are many mobile/PC apps supporting these low end OBD readers.

Advanced, high-end OBD devices such as [27–29], catering to professionals, are available at a considerable cost, often ranging in the several thousands of euros. These devices are designed to support a wide range of OBD II protocols, ensuring compatibility with most vehicles. Additionally, they often offer proprietary OBD services specifically tailored for advanced diagnostics in repair shops. Alongside the hardware, these devices typically include the necessary software products required for comprehensive vehicle analysis and troubleshooting.

In addition to the low-end and high-end options, there are also mid-range OBD readers available on the market. These devices offer additional functionalities and incorporate sophisticated IoT capabilities. For instance, products such as [30] provide real-time fleet monitoring capabilities, enabling users to track and manage a fleet of vehicles effectively. This trend highlights the direction in which OBD technology is evolving, moving towards connected OBD solutions that leverage connectivity and advanced features for enhanced vehicle monitoring and management [31].

1.2. PTIs in EU Countries

As per the [32] report, Luxembourg holds the highest record of passenger cars per thousand residents. In Luxembourg, all registered motor vehicles, trailers, and articulated lorries are required to undergo regular technical inspections. The prescribed intervals for these inspections for automobiles are as follows: the first periodic inspection occurs four years after the vehicle has been put into service and the second periodic inspection takes place six years after the vehicle has been put into service. Subsequently, inspections are required annually. Technical inspections relate to the condition and maintenance of the vehicle and, in particular, to: vehicle identification (chassis number, licence plates), environmental nuisances (noise, exhaust emissions), the braking mechanisms, lights and reflectors, the electrical equipment, axles, wheels, tires and suspension, the chassis and its accessories, seat belts, steering, and visibility [33].

Italy has the second largest number of passenger cars per thousand inhabitants [32]. All vehicles must undergo a regular inspection at official test centers. This inspection is to ensure the car is legal and roadworthy with safe emissions. According to Article 80 of the Italian Highway Code (Codice della Strada), vehicles must be submitted to a compulsory test for roadworthiness within four years after their first registration and then every alternate year thereafter. This includes passenger cars and transport vehicles, as well as mopeds and motorcycles from 50cc upwards. The tests focus on: brakes, tires, lights, steering, suspension, wheel alignment, transmission, windscreen and wipers, chassis, seat belts, and horn [34].

Finland has the third largest number of passenger cars per thousand inhabitants [32]. The vehicles are inspected within three years after their first registration, then within two years after and then every alternate year thereafter. This includes all passenger cars, vans, trucks, ATVs and for trailers with maximum structural weight of more than 750 kg. The tests focus on: brakes, suspension, emission, and a diesel test [35].

Nevertheless, the largest number of passenger cars in use in the EU is in Germany (more than 48 million according to [36]). All registered vehicles must be periodically tested for safety on the road and adherence to traffic safety regulations. The tests focus on: brakes, wheels, tyres, frame/body, exhaust system, steering, lighting/electrical systems, windows/mirrors, accessories, pedals, seats, seat belts, and electronic safety systems. Passenger cars have to be inspected within three years of their first registration and then every two years thereafter [37].

There are many countries using some kind of mobile/PC application in the PTI process. However, as far as we know, only in Germany, Croatia, Spain, Slovakia, and Belgium are safety-relevant OBD DTCs readings being performed during PTI checks. In the largest market, Germany, a special device named an HU adapter is used to obtain these data [38].

Starting from January 1st, 2019, Croatia has implemented OBD (On-Board Diagnostics) testing methods. Vehicles belonging to emission classes Euro 3 and newer are required to have an OBD connection. The OBD scanning collects various data including the MIL status, readiness-code status, number of Diagnostic Trouble Codes (DTCs), coolant temperature, and engine speed. In addition to the OBD test, vehicles of all Euro classes are also subjected to the classic tailpipe test. Self-developed OBD bluetooth devices have been utilized in the process; for more details see, e.g., [39]. In Spain, an OBD reader [40] has been used at a number of PTI stations. In general, the OBD readers mentioned above are designed to offer a solution, which includes both the hardware device and the accompanying software. However, these closed solutions make it hard to replace the vendor of the reader devices or the software provider. In Belgium, the modified version of the presented app has been utilized from the end of May 2023 with the same kind of OBD readers as in Slovakia.

1.3. PTI in Slovak Republic

Back in 1935, the Slovak Republic, which was then part of Czechoslovakia, made an initial attempt to introduce mandatory periodic technical inspections (PTI). Nowadays, almost all vehicles in Slovakia are required to undergo inspection. This article's main emphasis is on vehicle category M, passenger cars with four wheels, which must undergo the inspection within four years after their first registration and subsequently every two years thereafter. However, vehicles that are used for medical emergency services or mining rescue services, if they are equipped with priority driving rights—ambulance vehicles, vehicles used for the maintenance of gas facilities, and vehicles used for taxi services—have to undergo PTI every year. PTI intervals are comparable among the member states of the European Union and Slovakia is no exception. In Slovakia, the emission check is conducted as a distinct inspection procedure. The age of a vehicle is a crucial determinant of potential vehicle defects as the survey [41] made in Germany which is the EU country with more than 48 millions of registered vehicles, which is the most amongst EU countries. Table 1 shows a notable increase in the significant defect rate as vehicles age. In Slovakia, the average age of the registered vehicle is 14.3 years, which means significant defects appear with more than a 30% rate.

Table 1. PTI results 2020 by DEKRA in Germany.

Vehicle Age [Years]	Approx. Number of Cars	No Defects [%]	Minor Defects [%]	Significant Defects [%]
<3	902,500	92.3	2.94	4.72
3 to 5	975,000	87.42	4.83	7.67
5 to 7	823,000	79.82	7.71	12.31
7 to 9	763,500	71.15	10.89	17.59
9 to 12	1,210,000	59.65	15.05	24.79
>12	2,000,000	42.99	20.88	35.00

To combat fraudulent activities at PTI service stations, the Slovak Republic has implemented several changes to its PTI policies and supervisory measures. According to Table 2, in 2019, over 50% of severe regulation breaches were associated with the service brake, while almost 30% were linked to the parking brake. In Table 3, the data from 2020 indicate that service brake-related violations still account for over 50% of severe regulation breaches, but instances related to parking brakes have decreased to less than 15%.

In Table 4, the data from 2021 show a similar percentage of service brake-related violations of over 50% of severe regulation violations and, again, similarly to 2020, the violations related to parking brake percentage are below 15%.

This paper did not aim to investigate the reasons behind these fraudulent practices or whether the regulations are excessively stringent. However, it is crucial to note that malfunctioning service brakes can result in highly hazardous situations on the road. Consequently, measures to mitigate fraudulent behavior at PTI stations have been implemented in Slovakia. Since 2019, it has been mandatory for all vehicle inspection stations in the Slovak Republic to install surveillance cameras as a preventive measure. The heightened monitoring has resulted in a higher failure rate during PTI inspections, which increased from 6.2% to 13.1% in March 2019, more than doubling the failure rate; the trend in failure rate from 2013 to 2021 is shown in Figure 1. A comprehensive summary of the quality management and assessment of PTI stations in Slovakia and other selected EU countries is presented in [42].

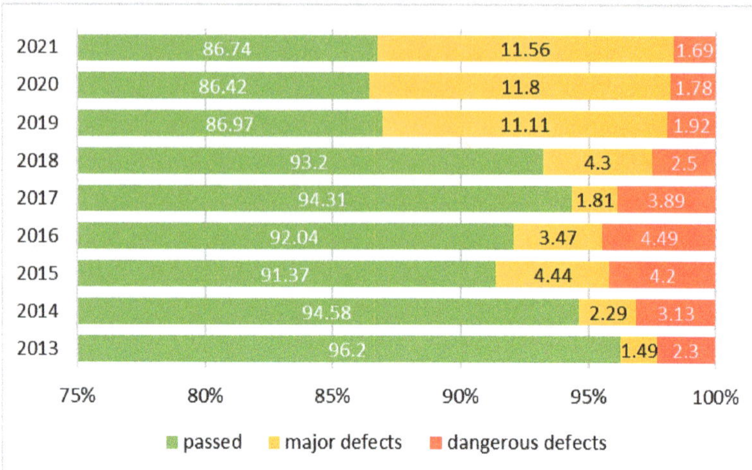

Figure 1. PTI failure rate in Slovakia from 2013 to 2021.

The implementation of a mobile application has been enhancing the supervision of technical vehicle inspection (PTI) in Slovakia since 1 January 2020. A mobile app provided a digital platform that allows inspectors to perform specific tasks on vehicles and record data using their mobile devices.

The app has been utilized to perform the following tasks (Figure 2):

- Taking and uploading a photograph of the odometer to the national PTI server;
- Taking and uploading a photograph of the Vehicle Identification Number (VIN) to the national PTI server;
- Reading diagnostic trouble codes from the car's onboard diagnostics port (OBD) and uploading them to the national PTI server;
- Measuring the vehicle's acceleration during a brake test, calculating the results, and uploading both the measurement and the results to the national PTI server. Since 1 January 2021, the smartphone has become the only device by which the driving brake test can be performed.

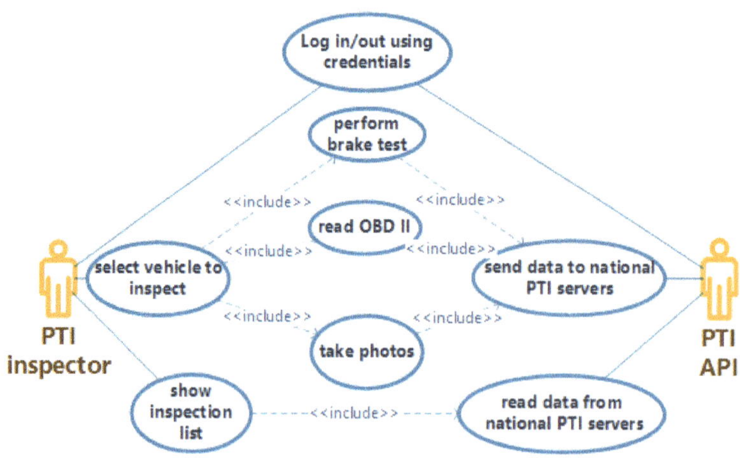

Figure 2. Use case diagram.

Table 2. Serious cases of fraud and intentional manipulation of PTI in 2019 in Slovak Republic (service brake-related violations of regulations are highlighted in gray).

PTI Fraud	%
Service brake test manipulation (measurement on different axle of the same vehicle)	42.78
Service brake test manipulation (measurement on different vehicle)	2.28
Skipped obligatory brake drive test	5.82
Parking brake test manipulation (using of service brake)	14.94
Parking brake test manipulation (measurement on different axle of the same vehicle)	11.90
Parking brake test manipulation (measurement on different vehicle)	2.53
Parking brake test manipulation (simulating wheel lock)	0.51
Skipped headlight alignment test	2.28
Skipped inspection of vehicle bottom	1.52
Wrong conclusion on roadworthiness	15.44

Table 3. Serious cases of fraud and intentional manipulation of PTI in 2020 in Slovak Republic (service brake-related violations of regulations are highlighted in gray).

PTI Fraud	%
Service brake test manipulation (omitting low part of the measurement range)	34.32
Service brake test manipulation (omitting high part of the measurement range)	1.32
Service brake test manipulation (brake power test omitting)	4.95
Service brake test manipulation (brake power test misinterpretation)	0.99
Service brake test manipulation (insufficient brake efficiency)	4.29
Service brake test manipulation (brake symmetry test misinterpretation)	7.92
Service brake test manipulation (measurement on different axle of the same vehicle)	13.86
Service brake test manipulation (brake judder ignoring)	3.30
Parking brake test manipulation (using of service brake)	12.87
Parking brake test manipulation (insufficient braking)	1.65
Parking brake test manipulation (measurement on different axle of the same vehicle)	1.65
Parking brake test manipulation (simulating wheel lock)	0.66
Skipped headlight alignment test	3.52
Ignoring lights defect	15.49
Skipped inspection of vehicle bottom	3.52
Wrong conclusion on roadworthiness	0.7
Wrong or missing photograph	2.11

Table 4. Serious cases of fraud and intentional manipulation of PTI in 2021 in Slovak Republic (service brake-related violations of regulations are highlighted in gray).

PTI Fraud	%
Service brake test manipulation (omitting low part of the measurement range)	38.73
Service brake test manipulation (omitting high part of the measurement range)	5.63
Service brake test manipulation (brake power test omitting)	3.52
Service brake test manipulation (brake power test misinterpretation)	0.7
Service brake test manipulation (insufficient brake efficiency)	2.82
Service brake test manipulation (brake symmetry test misinterpretation)	1.41
Service brake test manipulation (simulating wheel lock)	0.66
Service brake test manipulation (using of parking brake)	3.52
Service brake test manipulation (measurement on different axle of the same vehicle)	3.52
Service brake test manipulation (using of reverse gear)	1.32
Parking brake test manipulation (using of service brake)	9.15
Parking brake test manipulation (insufficient braking)	4.93
Parking brake test manipulation (measurement on different axle of the same vehicle)	2.53
Skipped headlight alignment test	1.35
Ignoring lights defect	9.57

2. Materials and Methods

The mSTK Android mobile application, available at [43], was developed in the Java programming language in partnership with TESTEK a.s., which was established in 2004 due to changes in the legislation of the Slovak Republic. The new law increased the independence criteria required for the central organization responsible for technical vehicle inspections within the national system. Consequently, some specialists in the field founded TESTEK, which won the tender issued by the Ministry of Transport in 2005, and began operating as a "technical service for technical vehicle inspections" in the December of that year. TESTEK is accredited as a Type A Inspection Body in accordance with the ISO/IEC 17020:2012 standard and is a member of the International Motor Vehicle Inspection Committee (CITA). In 2014, TESTEK transformed from a limited liability company to a joint-stock company and a new sister company, TESTEK servis, was established to complement and expand TESTEK's operations.

The utilization of the smart app in the PTI process helps prevent violations of rules and regulations in several ways. Firstly, an internet connection is required to perform any task using the app and the inspectors have to enter their credentials to access it. Secondly, the VIN photograph ensures that the correct vehicle is being tested, while the odometer photograph prevents the manipulation of odometer values in the national information system. The app also sends the location of the taken photographs to prevent tests from being performed outside of the PTI service station area, as the photograph cannot be taken without enabling location services.

Moreover, the app performs double checks on the VIN by reading it from the OBD and reads diagnostic trouble codes via OBD, which prevents the inspectors from ignoring these possible major defects. During the driving brake test, the location of the device is stored and uploaded to the national PTI server to verify that it was conducted at the declared location under the declared conditions.

2.1. App Design

The inspector can perform the actions described in Figure 2 in the application. First of all, the inspector must log into the application Figure 3. Once successfully logged in, the inspector will see the main activity screen of the app, as displayed in Figure 4, which begins with the inspector's name displayed at the top. Beneath the name is the so-called waiting list, which includes a list of assigned inspections. The assigned inspections are identified by the car's registration number, make, and model. When a specific inspection is selected by the inspector, the inspection details are displayed as shown in Figure 4. At this time, the inspector may perform the following actions:

- Take photos of the VIN, odometer and vehicle;
- Read data from the OBD II diagnostic port of the vehicle;
- Perform a brake test of the vehicle.

Data from all three actions are continuously uploaded to the national PTI server. Therefore, the developed application needs a continuous internet connection to function properly. The network architecture of the application can be seen in Figure 5. The application operates in this architecture as a client and communicates with the server through a REST API, from which it retrieves the necessary data about inspections and sends measurements and photos. The communication between the server is encrypted and possible only after authentication and authorization of the user account on the client side (Figure 3). There are two ways to connect a mobile application to the server: using a local Wi-Fi network or mobile data. The second option is advantageous, especially for inspections of agricultural vehicles that are performed outdoors or at the owner's location.

During the process of taking photos and testing brakes, information about the mobile phone's location is also obtained. The choice of navigation system (GPS, GALILEO, etc.) depends on the capabilities of the phone used, but precise location is required. Therefore, localization based on the GSM network is considered insufficient.

When reading data from the OBD II port, the mobile application obtains data from the ELM 327 module through Bluetooth technology.

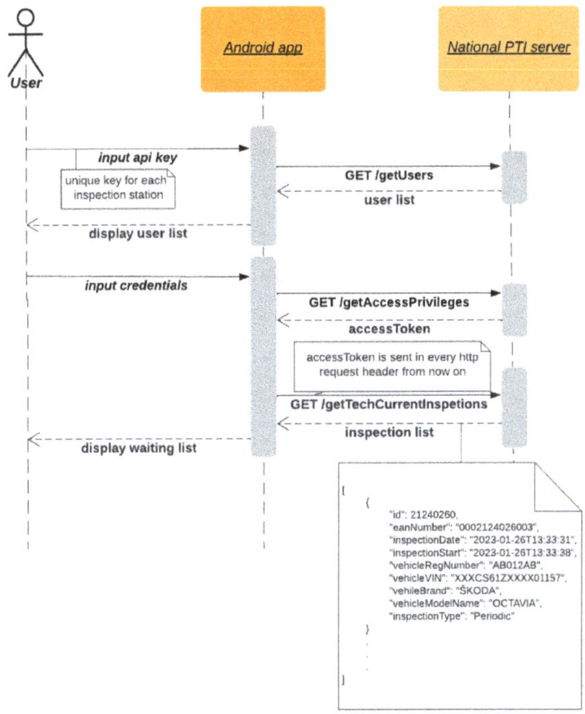

Figure 3. Sequence diagram of sign-in procedure.

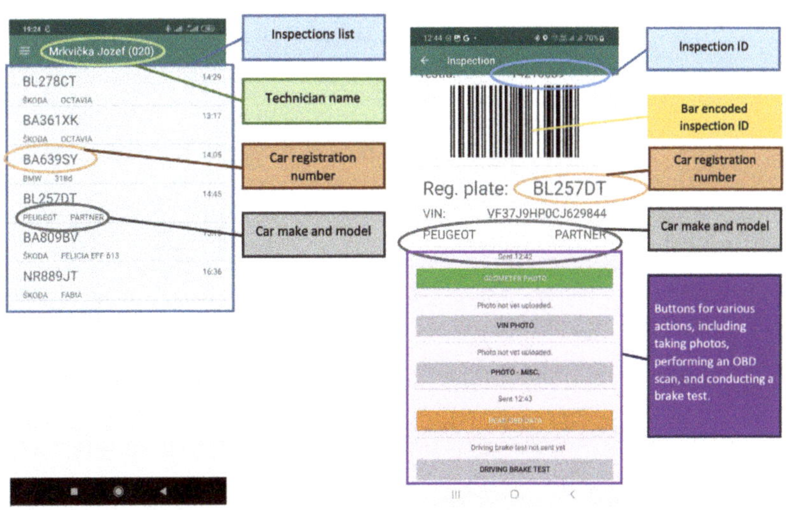

Figure 4. Waiting list (**left**), vehicle detail and inspection progress (**right**).

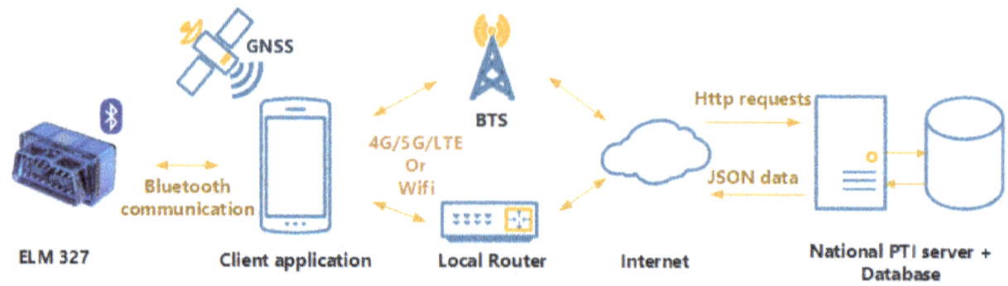

Figure 5. Application network architecture.

2.1.1. Photographs

Using the mobile device, the inspector can utilize any available camera to take pictures. Taking photos is possible only after selecting a specific vehicle for inspection. The inspector can take and send photos in three different categories. However, the process of capturing and sending a photo is the same. The sequence diagram of this action can be seen in Figure 6. After selecting the photo category, the built-in camera of the mobile phone is activated, allowing the inspector to capture a photo.

Once captured, the photo is resized to 640 by 480 pixels and compressed into a JPEG format. To ensure the file size remains below 100 kilo bytes, a compression level is applied. However, to prevent any potential cheating, the files cannot be utilized as photographs. If the snapshot is confirmed, the photo is sent to the server.

The dashboard gauges photograph, depicted in Figure 7, displays the malfunction indicator lamp activated. When encountering such a defect, it should be documented using the miscellaneous photo button to upload a photograph along with a brief description of the issue to the server. It is important to note that this type of defect typically appears on the diagnostic trouble codes list, and the presence of the MIL can be detected by reading the OBD as well.

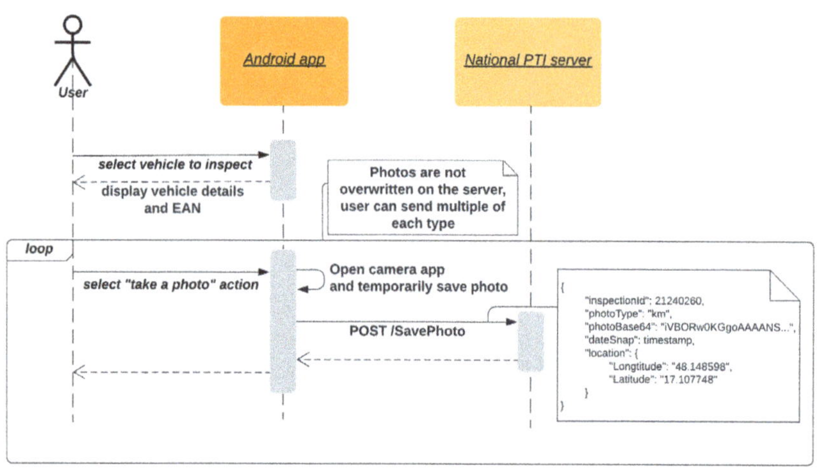

Figure 6. Sequence diagram of the photo capturing process.

Figure 7. Photograph of gauges with malfunction indicator lamp on.

2.1.2. OBD

Although it is possible to gather more data from a vehicle using proprietary hardware and software, the solution being presented aims to be independent of vehicle make or manufacturer by utilizing the OBD II standards outlined in [44]. Special OBD readers, similar to the one described in [45], can be designed and developed as well, but it would result in additional expenses not only for the development and design stages but also for the maintenance of the devices in future. Our solution does not require any additional expenses on equipment or software licenses for PTI service stations since it employs a low-cost and widely available wireless OBD interface, the ELM327 (refer to [26]). The dongles utilized in the study were sourced from the lower end of the price range, approximately EUR 10 per unit. This device can also be used efficiently for research purposes, as demonstrated in [46]. The primary objective was to retrieve diagnostic trouble codes (DTCs) available via the open OBD service. Numerous smartphone applications are available that can read Diagnostic Trouble Codes (DTCs) using an ELM327 On-Board Diagnostic (OBD) reader. However, this particular application incorporates further PTI responsibilities and procedures, with an emphasis on preventing fraud by obtaining the following data:

- ELM327 voltage: This is not the voltage requested from vehicles ECUs using OBD requests, but it is the voltage measured by the ELM 327 reader on the OBD port pins;
- VIN: The vehicle identification number is obtained by the OBD request. After successful parsing, it is compared with the VIN in the information system;
- DTCs: A request is also made for the list of permanent diagnostic trouble codes.

There are additional communication features between the phone and ELM327, such as resetting to defaults, detecting the vehicle OBD protocol, and providing a list of supported services. However, these features are not extensively discussed in this paper. The sequence diagram shown in Figure 8 illustrates the process of gathering OBD data sending it to the national PTI server. After selecting the vehicle designated for inspection, the inspector can press the "read OBD" button. Communication is activated after selecting the correct ELM 327 module among Bluetooth devices. During OBD communication, requests are sent to the vehicle sequentially through the Bluetooth module, and incoming responses are processed accordingly. Every response from the vehicle is logged and transmitted to a central server for additional scrutiny in situations where the inspection has failed, the responses are incorrect, or there are any indications of possible fraudulent activity. Since 2020, these data have also been utilized in debugging and research endeavors, as it marks the initial year that such devices were integrated on a broad scale. Figure 9 illustrates the OBD reading screen screenshot. The results from parsed OBD responses are displayed in the upper section, starting with the VIN if the car does provide it via standardized OBD request according to [44]. Below the VIN number, one can see the status of self tests also referred to as readiness and the status of a malfunction indicator lamp also known as MIL. Then, the permanent DTCs are displayed below. The DTCs are displayed as hexadecimal figures and are subsequently parsed on the server.

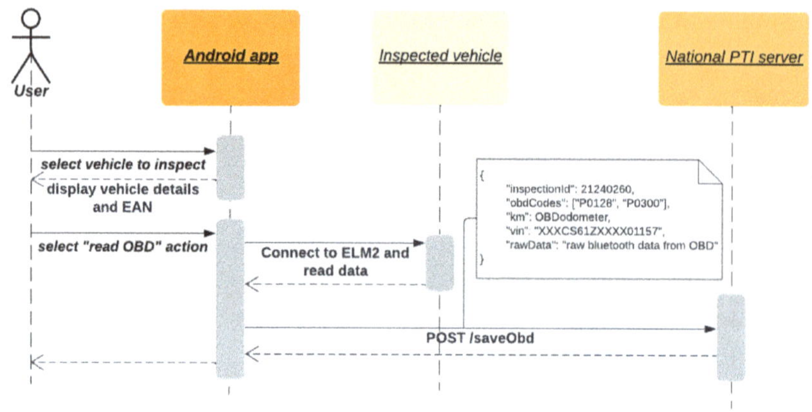

Figure 8. Sequence diagram for gathering and sending OBD data.

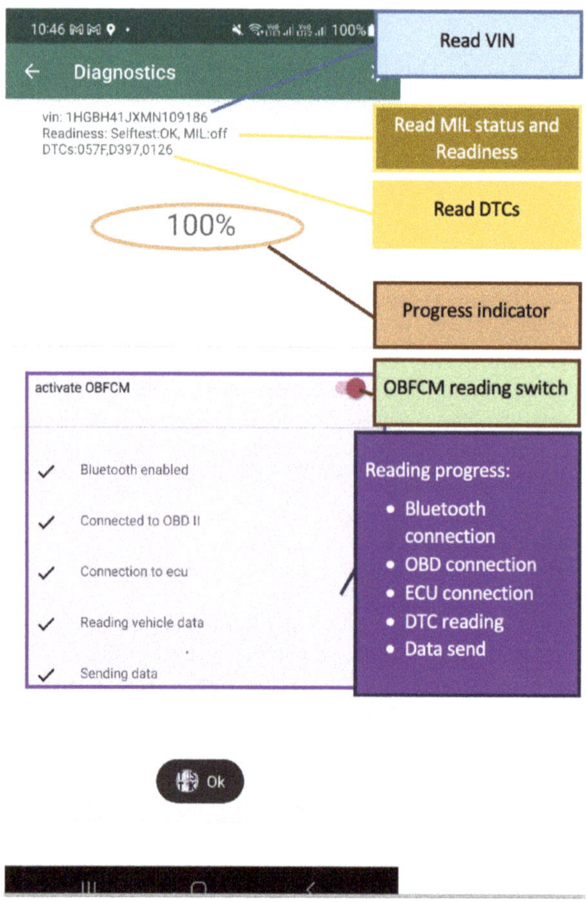

Figure 9. OBD reading activity.

The remaining portion of the screen displays the overall progress percentage, accompanied by check marks that indicated accomplished tasks, providing the inspector with more comprehensive feedback on the ongoing inspection. In case certain readings fail and must be repeated, delays may occur. Occasionally, some vehicles may not be compatible with the ELM327 reader, in which case the inspector must at least attempt to connect via Bluetooth, record the voltage, and submit requests to the vehicle multiple times. All communication is recorded and forwarded to the server and, if the reading is unsuccessful, the OBD reading button in the application changes to orange, as displayed in Figure 4 on the right-hand side.

The rest of the screen shows overall progress in percentage, along with check marks indicating successfully completed tasks to provide the inspector with better feedback on the ongoing inspection. Delays may occur if some readings fail and must be repeated a few times. It is possible that some cars cannot be read by the ELM327 reader, in which case the inspector has to try to connect to the ELM327 via Bluetooth, read the voltage, and send requests to the car several times. All communication is saved and submitted to the server and, if the reading is unsuccessful, the OBD read button in the app is indicated in orange, as shown in Figure 4 on the right.

2.1.3. Drive Brake Testing

Brake failure-related crashes were discussed in [47], presenting an increasing rate of brake failure-related crashes with the age of the vehicle, which corresponds to conclusions in [41] presented in Table 1. Since the average age of a vehicle registered in Slovakia is still over 14 years, the drive brake testing feature has become a crucial aspect of this application.

While roller bench devices are commonly used to test vehicle brakes, not all roller benches are suitable for certain vehicles or situations. For instance, roller benches may not be suitable for 4-wheel drive or agricultural vehicles. According to a survey in [48], dangerous defects related to brakes account for 2–3% of dangerous defects found in the agricultural vehicles category. In Slovakia, agricultural vehicle inspections can be performed on-site to avoid transporting the vehicles to PTI service stations. In such cases, the drive brake test serves as an alternative to the roller brake test, as stipulated by Slovak Republic legislation based on the United Nations Economic Commission for Europe Agreement [49]. Prior to 2021, decelerometers, such as the MAHA VZM300 [50], were used in these tests, which left ample room for fraudulent behavior since there was no way to verify the test results and there was no guarantee that the test had been conducted at all.

According to the regulations, the function of brakes must be determined by either measuring the stopping distance relative to the initial vehicle speed and/or by measuring the mean fully developed deceleration during the test; for further information on the calculation and test procedure one can refer to an ISO standard [51].

There are multiple methods to measure the stopping distance using a mobile phone. One approach involves using GNSS sensors; however, these sensors have a limited update rate of no more than 1 Hz, as demonstrated in [52]. A more preferable alternative is to utilize the phone's inertial measurement unit (IMU), which typically generates data every 4–20 milliseconds, depending on the device's manufacturer and operating system version. To obtain the braking distance, the acceleration sensor data must be double-integrated, which can result in accumulating measurement errors. This positioning approach, referred to as dead reckoning, is unsuitable to be the sole method for localizing a vehicle. The capabilities of IMU positioning are summarized in [53], emphasizing its applicability for indoor use as well, while the effects of the cumulative error are discussed more closely in [54]. Nevertheless, in this scenario, the driving brake test, the measurement takes only a few seconds, which limits the impact of the cumulative error on the results. Therefore, in this mobile application, the measurement is based on IMU data through dead reckoning.

The results provided by the phone measurements are consistent and offer a level of precision that is almost identical to that of the formerly used devices.

Calibrating and testing each device before using it to measure the brakes is necessary for ensuring the accuracy and reliability of the safety system. Without such measures, it would not be feasible to use any device for this critical task. It is necessary to test the accelerometers at steady states every two years to ensure accuracy. Fortunately, the cost and frequency of calibration are comparable to those of previously used devices, so there is no significant increase in cost associated with this process.

The process of performing a driving brake test is described in the sequence diagram shown in Figure 10. After selecting the vehicle designated for inspection, the inspector can press the "driving brake test" button. At this moment, the mobile phone should be securely mounted in a stand. Subsequently, the inspector can initiate the test during which they follow the instructions displayed on the mobile device. After completing the drive, the measured data are evaluated directly on the mobile device and the results are displayed on the device's screen. After confirming the test results, the test evaluation, along with the measured data, are sent to the national PTI server.

The user interface of the application is designed to be simple and user-friendly as it is meant to be used while driving.

The screenshot from drive brake test activity is in Figure 11. It shall be used in the following way:

- The phone should be securely mounted on the vehicle, such as on the windscreen, using a holder that does not cause excessive vibrations. Shorter arms are recommended for this purpose;
- The app requires a GNSS position to be available before the Start button can be pressed and the measurement can begin. Otherwise, the button is grayed out;
- Once the first GNSS position is obtained, the measurement can be started by pressing the Start button. This initiates a self-adjustment measurement to determine the phone's orientation, while the vehicle and phone have to be standstill, the engine can be idling. The yellow light is on during this process;
- If the self-adjustment measurement is successful, the green light turns on. The driver should then accelerate the vehicle to the desired speed and then apply the brake;
- After stopping the vehicle, the Stop button should be pressed to complete the measurement. The results are then evaluated and, if consistent, are displayed. The data can then be submitted.

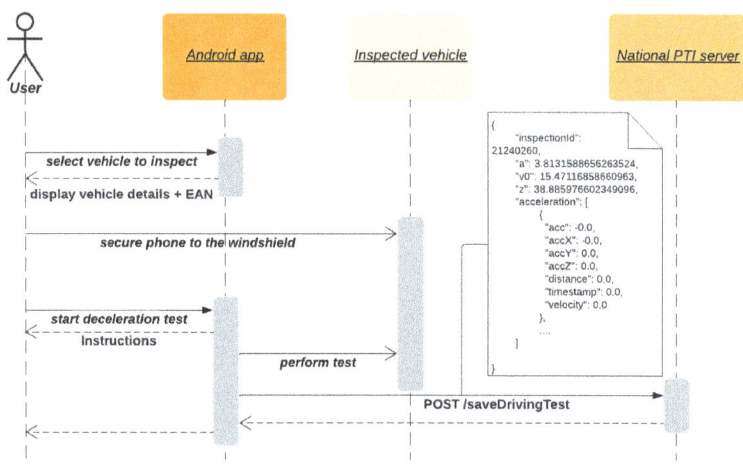

Figure 10. Drive brake testing sequence diagram.

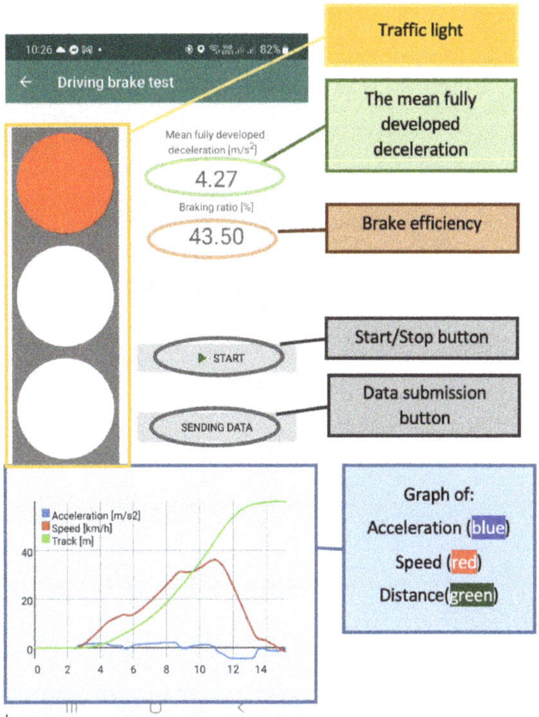

Figure 11. Brake testing activity.

3. Results

This paper presents cumulative data from OBD readings and driving brake tests.

3.1. OBD Readings

The OBD reading via the presented app became mandatory in 2020. By the end of 2020, over 10,000 diagnostic trouble codes (DTCs) had been detected from approximately 1.4 million inspected cars, with this number surpassing 20,000 in 2021 from more than 1.5 million inspections in Slovakia. The majority of these codes were related to emission inspections, which are conducted as a distinct service. Table 5 provides a yearly overview of the DTC readings, with safety-related trouble codes displayed on separate lines. The final line summarizes all DTCs, including both safety and emission-related issues.

3.2. Drive Brake Testing

There have been more than 7000 drive brake tests performed in the Slovak Republic until the end of 2020 during the field testing of the app. The app was rolled out in October 2020. None of the vehicles failed the mobile app braking test in 2020. However, it was voluntary to use the smartphone app for brake testing until 31st of December 2020. It became mandatory from the 1st of January 2021. Therefore, only the data from 2021 brake tests are presented in this paper. Figure 12 shows the number of performed tests by a number of tests per vehicle category. It shows that more than 70% of the vehicles measured by the mobile app are from the M category (for more information on vehicle categories see, e.g., [55]), which corresponds to the vehicles having at least four wheels and used for the carriage of passengers (e.g., standard car with two, three, or four doors). Another significant category with more than 10% is N category corresponding to power-driven vehicles having at least four wheels and used for the carriage of goods. Agricultural and forestry tractors, category T, have been tested in more than 15% of cases as well. Table 6 shows results from the year 2021. It shows a large number of tests performed by the mobile app. It also shows 195 cases with major defects and 36 cases of dangerous defects detected. The failure rate in Table 6 is approximately 0.2%, which is below the long term statistics of brake defects detected at PTI stations. However, this issue should be evaluated after a longer period of use, because the driving test cannot be as sensitive as roller brake devices. In Table 6, the driving brake test failure rate of 12.5% is the highest and it occurs in vehicle category T4a, special purpose wheeled tractors. However, only one major defect has been detected in this category. Other vehicle categories with a failure rate higher than 3% include heavy vehicles such as category N2, a vehicle intended primarily for the transport of goods that features a total mass that exceeds 3.5 tons but does not exceed 12 tons; N3, a vehicle with mass over 12 tons; and O4, a towed vehicle with total mass exceeding 10 tons. The highest dangerous defects ratios have vehicle classes N2, N3, and T1.

Table 5. The DTCs found during the years 2020 and 2021. (Part: refers to the component or area of the vehicle that is affected).

DTC	2020	2021	Part
B0075	0	1	Seat belts
C0020	0	1	ABS
C0040	0	1	EBS
C0048	0	1	EBS
P0504	20	9	EBS
P0555	1	9	EBS
P0572	0	1	EBS
P2299	6	1	EBS
U0121	14	31	ABS
U0128	0	3	EBS
U0151	0	2	SRS
U0241	0	1	Head lamp
U0415	11	14	ABS
U0418	4	2	EBS
P0552	1	0	EPS
P0553	1	0	EPS
P0557	1	0	EBS
C0041	1	0	EBS
P0556	1	0	EBS
P0573	2	0	EBS
Total safety related	63	77	
...	...	...	
Total	10,183	21,174	

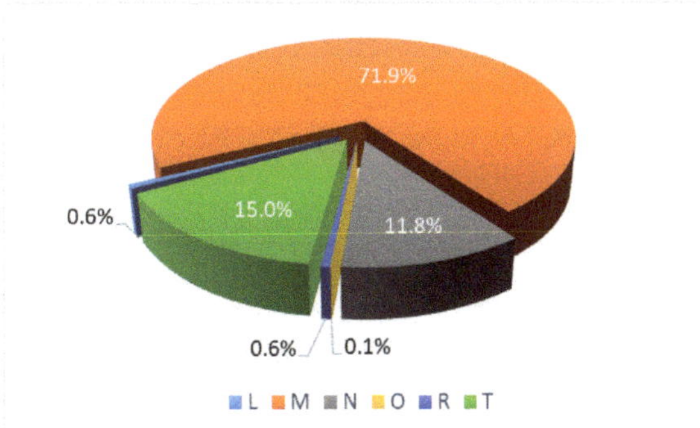

Figure 12. Driving brake tests performed by mobile app in 2021 in the Slovak Republic by vehicle categories.

Table 6. Summary of brake drive test performed by mobile app in 2021 in the Slovak Republic.

Vehicle Class	Number of Tests	Number of Defects	Number of Defects	%	%
		Major	Dangerous	All	Dangerous
M1	78,343	29	4	0.04	0.01
M1G	19,329	6	1	0.04	0.01
N1	8070	19	3	0.27	0.04
N1G	5772	1	0	0.02	0.00
N2	1215	46	10	4.61	0.82
N2G	331	9	0	2.72	0.00
N3	470	19	2	4.47	0.43
N3G	151	6	0	3.97	0.00
O4	17	1	0	5.88	0.00
PS	101	1	0	0.99	0.00
R3a	691	1	0	0.14	0.00
R3b	5	0	0	0.00	0.00
T1	915	8	6	1.53	0.66
T1a	18,181	37	9	0.25	0.05
T1b	158	3	1	2.53	0.63
T2	67	3	0	4.48	0.00
T2a	749	1	0	0.13	0.00
T4a	8	1	0	12.50	0.00
T5	61	4	0	6.56	0.00
Sum	135,913	195	36	0.17	0.03

4. Discussion

The novelty of the approach is in using the ELM327 device for OBD scans at such a large scale successfully. The results in Table 5 show that a large number of malfunctions, 10,183, were read from cars' OBDs at PTI stations in 2020 in Slovakia. Most of these cases were emission related and only 63 were classified as major defects at technical inspection. Nevertheless, it did detect the defects and the cars had to be repaired and inspected again within the 60-day period. By incorporating OBD readings into the PTI process through a mobile app, Slovakia has become the third country in the EU to do so and is in compliance with EU legislation. In 2021, there were similar results to those of 2020 regarding OBD readings, with more than 21,000 DTCs detected, out of which 77 were related to safety.

The number of brake driving tests performed by the presented app show the importance of designing this tool. The fact it became mandatory in 2021 will create a lot of supervision options for preventing fraud in brake assessment at PTI stations. The drive brake testing feature introduced an affordable, very easy way to test vehicle brakes and contributes significantly to road safety, particularly for older cars where major defects are more likely to occur. The ability to detect brake-related issues through OBD readings also helps to prevent brake failure-related accidents, which are known to increase with the age of the vehicle. An innovative aspect of the proposed method is also the utilization of a smartphone's IMU for measuring vehicle deceleration, which detected more than 200 defects in vehicle brake systems, more than 20% of which were on heavy duty vehicles. This technology, combined with regular calibration of the phone's accelerometers, provides a cost-effective solution for accurate drive brake testing and contributes to overall road safety. The app has already been expanded and new features, such as On-Board Fuel Consumption Meter (OBFCM) readings, have been field tested. The OBFCM reading functionality will be the next step in the app's development, contributing to fuel consumption and emission monitoring in the EU passenger cars.

The technology is very affordable, making it suitable for countries where the cost of technical equipment is a barrier to implementing PTIs.

In conclusion, this paper highlighted the significance of utilizing smartphone applications in the PTI process. Recognizing the crucial role of road safety and the impact of vehicle defects on fatal accidents, the authors explored the potential of smart technologies to enhance the efficiency and effectiveness of vehicle inspections. By leveraging the capabilities of smartphones, including data collection from the vehicles and utilizing their own sensors for vehicle dynamics measurements, the authors successfully implemented low-cost devices at a large scale across the country. The outcomes of the first two years of utilizing this application at every technical inspection station in the Slovak Republic were discussed, showcasing the positive advancements made in transitioning to smart technologies while managing costs effectively.

Author Contributions: Conceptualization, P.T.; Data curation, P.T. and J.M.; Formal analysis, P.T., M.K., M.R. and J.M.; Methodology, P.T., M.K., M.R. and J.M.; Software, P.T., M.K. and M.R.; Validation, P.T., M.K., M.R. and J.M.; Writing—original draft, P.T.; Writing—review & editing, P.T., M.K., M.R. and J.M. All authors have read and agreed to the published version of the manuscript.

Funding: This research was funded by the Slovak Grant Agency, grant KEGA 039STU-4-2021 and by Slovak Research and Development Agency, grant No. APVV-21-0125.

Institutional Review Board Statement: Not applicable.

Informed Consent Statement: Not applicable.

Data Availability Statement: The data presented in this study are available on request from the corresponding author. The data are not publicly available due to possible violation of personal data privacy.

Conflicts of Interest: The authors declare no conflict of interest.

Abbreviations

The following abbreviations are used in this manuscript:

OBD	On board diagnostics
ABS	Anti lock braking system
EBS	Electronic braking system
EPS	Electronic power steering
DTC	Diagnostic trouble code
PTI	Periodical technical inspection
OBFCM	On-Board fuel consumption meter

References

1. Mr. Walter Nissler, UNECE, The 1997 Agreement. 2020. Available online: https://www.unece.org/fileadmin/DAM/trans/doc/2020/SafeFITS/SIII_UNECE_WN.pdf (accessed on 15 July 2022).
2. Martín-delosReyes, L.M.; Lardelli-Claret, P.; García-Cuerva, L.; Rivera-Izquierdo, M.; Jiménez-Mejías, E.; Martínez-Ruiz, V. Effect of Periodic Vehicle Inspection on Road Crashes and Injuries: A Systematic Review. *Int. J. Environ. Res. Public Health* **2021**, *18*, 6476. [CrossRef] [PubMed]
3. Taneerananon, P.; Chanwannnakul, T.; Suanpaga, V.; Khompratya, T.; Kronprasert, N.; Tanaboriboon, Y. An evaluation of the effectiveness of private vehicle inspection process in Thailand. *J. East. Asia Soc. Transp. Stud.* **2005**, *6*, 3482–3496.
4. Hoagland, A.; Woolley, T. It's no accident: Evaluating the effectiveness of vehicle safety inspections. *Contemp. Econ. Policy* **2018**, *36*, 607–628. [CrossRef]
5. Rechnitzer, G.; Haworth, N.; Kowadlo, N. The Effect of Vehicle Roadworthiness on Crash Incidence and Severity (No. 164). 1997. Available online: https://www.monash.edu/muarc/archive/our-publications/reports/muarc164 (accessed on 4 May 2023).
6. Cuerden, R.; Edwards, M.; Pittman, M. *Effects of Vehicle Defects in Road Accidents*; TRL: London, UK, 2011.
7. Jarosinski, W. Periodic technical inspections of vehicles and road traffic safety with the number of road accidents involving fatalities; [System okresowych badań technicznych pojazdow a bezpieczenstwo ruchu drogowego i liczba zdarzen drogowych z udzialem ofiar smiertelnych]. *Eksploatacja i Niezawodnosc* **2014**, *16*, 105–111.
8. Cairns, S.; Rahman, S.; Anable, J.; Chatterton, T.; Wilson, R. *Vehicle Inspections—From Safety Device to Climate Change Tool*; TRL: Workingham, UK, 2014.
9. Petit, L. El factor humano en el sistema tránsito y seguridad vial y el modelo interaccional comportamental de tránsito. *Psiencia Revista Latinoamericana de Ciencia Psicológica* **2014**, *6*, 48–54.
10. Hudec, J.; Sarkan, B. Effect of Periodic Technical Inspections of Vehicles on Traffic Accidents in the Slovak Republic. *Commun.-Sci. Lett. Univ. Zilina* **2022**, *24*, A142–A159. [CrossRef]
11. Hudec —Olexa, Effect of Periodic Technical Inspections of Vehicles on Traffic Accidents in the Slovak Republic—Part I. 2022. Available online: https://www.testek.sk/PDF/Effect_of_PTI_on_Traffic_Accidents_in_the_Slovak_Republic_CITA_1.pdf (accessed on 18 May 2023).
12. Gerhard Muller, CITA. The Importance of Periodic Vehicle Emission Tests for Air Quality. 2018. Available online: https://www.euractiv.com/section/transport/opinion/the-importance-of-periodic-vehicle-emission-tests-for-air-quality/ (accessed on 22 May 2023).
13. Boveroux, F.; Cassiers, S.; Buekenhoudt, P.; Chavatte, L.; De Meyer, P.; Jeanmart, H.; Verhelst, S.; Contino, F. *Feasibility Study of a New Test Procedure to Identify High Emitters of Particulate Matter during Periodic Technical Inspection*; Technical Report, SAE Technical Paper; SAE International: Warrendale, PA, USA, 2019.
14. Burtscher, H.; Lutz, T.; Mayer, A. A new periodic technical inspection for particle emissions of vehicles. *Emiss. Control Sci. Technol.* **2019**, *5*, 279–287. [CrossRef]
15. Bainschab, M.; Schriefl, M.A.; Bergmann, A. Particle number measurements within periodic technical inspections: A first quantitative assessment of the influence of size distributions and the fleet emission reduction. *Atmos. Environ. X* **2020**, *8*, 100095. [CrossRef]
16. Kadijk, G.; Spreen, J.; Van der Mark, P. *Investigation into a Periodic Technical Inpsection Test Method to Check for Presence and Proper Functioning of Diesel Particulate Filters in Light-Duty Diesel Vehicles*; TNO: Delft, The Netherlands, 2016.
17. Kadijk, G.; Elstgeest, M.; Ligterink, N.; Van der Mark, P. *Investigation into a Periodic Technical Inspection (PTI) Test Method to Check for Presence and Proper Functioning of Diesel Particulate Filters in Light-Duty Diesel Vehicles. Part 2*; TNO: Delft, The Netherlands, 2017.
18. European Commission. Mobility & Transport-Road Safety, Annual Statistical Report 2022. 2023. Available online: https://road-safety.transport.ec.europa.eu/statistics-and-analysis/data-and-analysis/annual-statistical-report_en (accessed on 14 May 2023).
19. European Commission. Mobility & Transport-EU Road Safety Policy. 2023. Available online: https://road-safety.transport.ec.europa.eu/eu-road-safety-policy/what-we-do_en (accessed on 14 May 2023).
20. National Highway Traffic Safety Administration. *How Vehicle Age and Model Year Relate to Driver Injury Severity in Fatal Crashes*; National Highway Traffic Safety Administration: Washington, DC, USA, 2013; pp. 1–7.
21. 77/143/EEC. 1976. Available online: http://data.europa.eu/eli/dir/1977/143/oj (accessed on 1 January 2020).
22. Directive 2014/45/EU. 2014. Available online: http://data.europa.eu/eli/dir/2014/45/oj (accessed on 18 November 2019).
23. Directive 2014/45/EU of the European Parliament and of the Council. 2014. Available online: https://eur-lex.europa.eu/legal-content/EN/TXT/?uri=CELEX:02014L0045-20140429 (accessed on 16 June 2023).
24. Meenakshi; Nandal, R.; Awasthi, N. OBD-II and Big Data: A Powerful Combination to Solve the Issues of Automobile Care. In *Proceedings of the Computational Methods and Data Engineering*; Singh, V., Asari, V.K., Kumar, S., Patel, R.B., Eds.; Springer: Singapore, 2021; pp. 177–189.
25. OBD2 Explained—A Simple Intro. 2023. Available online: https://www.csselectronics.com/pages/obd2-explained-simple-intro (accessed on 16 June 2023).

26. ELM 327. 2017. Available online: https://www.elmelectronics.com/wp-content/uploads/2017/01/ELM327DS.pdf (accessed on 15 July 2021).
27. MaxiSYS Ultra. 2023. Available online: https://autel.com/mk3/3525.jhtml (accessed on 14 June 2023).
28. ADS 525X Diagnostic Scan Tool. 2023. Available online: https://www.boschdiagnostics.com/products/ads-625-diagnostic-scan-tool (accessed on 14 June 2023).
29. X 431 EURO LINK. 2023. Available online: https://launch-europe.eu/x-431-euro-link (accessed on 14 June 2023).
30. OBD Reader. 2023. Available online: https://www.autopi.io/ (accessed on 22 May 2023).
31. Guardiola, C.; Vigild, C.; De Smet, F.; Schusteritz, K. From OBD to connected diagnostics: A game changer at fleet, vehicle and component level. *IFAC-PapersOnLine* **2021**, *54*, 558–563. [CrossRef]
32. Passenger Cars in the EU. 2023. Available online: https://ec.europa.eu/eurostat/statistics-explained/index.php/Passenger_cars_in_the_EU (accessed on 22 May 2023).
33. The Government IT Centre, The Government of the Grand Duchy of Luxembourg. 2023. Available online: https://guichet.public.lu/en/citoyens/transports-mobilite/transports-individuels/vehicule-motorise/controle-technique-obligatoire-vehicule.html (accessed on 22 May 2023).
34. Ministero Delle Infrastrutture e dei Trasporti. Revisione Periodica. 2023. Available online: https://www.ilportaledellautomobilista.it/web/portale-automobilista/veicoli/revisioni (accessed on 22 May 2023).
35. The Finnish Transport and Communications Agency Traficom. Vehicle Inspection. 2023. Available online: https://www.traficom.fi/en/transport/road/vehicle-inspection (accessed on 22 May 2023).
36. ACEA Report. 2023. Available online: https://www.acea.auto/files/ACEA-report-vehicles-in-use-europe-2023.pdf (accessed on 18 May 2023).
37. TÜV Rheinland. Vehicle Inspection. 2023. Available online: https://www.tuv.com/world/en/vehicle-inspection-(private-customers).html (accessed on 22 May 2023).
38. TUV, HU Adapter. 2015. Available online: https://www.tuvsud.com/de-de/branchen/mobilitaet-und-automotive/autohaus-und-werkstatt/hauptuntersuchung/hu-adapter (accessed on 18 May 2023).
39. Resetar, M.; Pejic, G.; Ilincic, P.; Lulic, Z. Primary results of OBD tests collected during PTI of vehicles in Croatia. In Proceedings of the 23rd International Transport and Air Pollution Conference, Thessaloniki, Greece, 15–17 May 2019; p. 1. [CrossRef]
40. OBD Reader, Applus. 2023. Available online: https://wirelessobdbyapplus.com/ (accessed on 22 May 2023).
41. DEKRA. PTI Results 2020. 2022. Available online: https://www.dekra-roadsafety.com/en/pti-reveals-safety-related-deficiencies (accessed on 4 May 2023).
42. Hudec, J.; Šarkan, B.; Czödörová, R.; Caban, J. The Influence of Quality Management System on the Operation of Periodical Technical Inspection Stations. *Appl. Sci.* **2021**, *11*, 4854. [CrossRef]
43. mSTK. 2020. Available online: https://play.google.com/store/apps/details?id=sk.deletech.uitk&authuser=0 (accessed on 1 January 2020).
44. Society of Automobile Engineers, E/E Diagnostic Test Modes J1979_201702. 2017. Available online: https://www.sae.org/standards/content/j1979_201702/ (accessed on 4 May 2023).
45. Yen, M.H.; Tian, S.L.; Lin, Y.T.; Yang, C.W.; Chen, C.C. Combining a Universal OBD-II Module with Deep Learning to Develop an Eco-Driving Analysis System. *Appl. Sci.* **2021**, *11*, 4481. [CrossRef]
46. Hong, S.; Hwang, H.; Kim, D.; Cui, S.; Joe, I. Real Driving Cycle-Based State of Charge Prediction for EV Batteries Using Deep Learning Methods. *Appl. Sci.* **2021**, *11*, 1285. [CrossRef]
47. Haq, M.T.; Ampadu, V.M.K.; Ksaibati, K. An investigation of brake failure related crashes and injury severity on mountainous roadways in Wyoming. *J. Saf. Res.* **2023**, *84*, 7–17. [CrossRef] [PubMed]
48. Tarancón-Andrés, E.; Santamaria-Peña, J.; Arancón-Pérez, D.; Martínez-Cámara, E.; Blanco-Fernández, J. Technical Inspections of Agricultural Machinery and Their Influence on Environmental Impact. *Agronomy* **2022**, *12*, 907. [CrossRef]
49. Agreement Concerning the Adoption of Harmonized Technical United Nations Regulations for Wheeled Vehicles, Equipment and Parts Which Can Be Fitted and/or Be Used on Wheeled Vehicles and the Conditions for Reciprocal Recognition of Approvals Granted on the Basis of these United Nations Regulations. 2019. Available online: http://www.unece.org/trans/main/wp29/wp29wgs/wp29gen/wp29fdocstts.html (accessed on 17 January 2019).
50. Maha VZM 300. 2021. Available online: https://www.maha.de/en/products/brake-testing-technology/decelerometer/vzm-300~p1111 (accessed on 17 July 2021).
51. SO/TR 13487:1997: Braking of Road Vehicles—Considerations on the Definition of Mean Fully Developed Deceleration. 1997. Available online: https://www.iso.org/obp/ui/#iso:std:iso:tr:13487:ed-1:v1:en (accessed on 4 May 2023).
52. Romero-Andrade, R.; Trejo-Soto, M.E.; Vázquez-Ontiveros, J.R.; Hernández-Andrade, D.; Cabanillas-Zavala, J.L. Sampling Rate Impact on Precise Point Positioning with a Low-Cost GNSS Receiver. *Appl. Sci.* **2021**, *11*, 7669. [CrossRef]
53. Geng, J.; Xia, L.; Xia, J.; Li, Q.; Zhu, H.; Cai, Y. Smartphone-Based Pedestrian Dead Reckoning for 3D Indoor Positioning. *Sensors* **2021**, *21*, 8180. [CrossRef] [PubMed]

54. Murakami, Y.; Ohtsuki, T. Localization calibration using illuminance sensor for pedestrian dead reckoning with smartphones. In Proceedings of the 2014 IEEE 25th Annual International Symposium on Personal, Indoor, and Mobile Radio Communication (PIMRC), Washington, DC, USA, 2–5 September 2014; pp. 2034–2039. [CrossRef]
55. Vehicle Categories. 2021. Available online: https://ec.europa.eu/growth/sectors/automotive/vehicle-categories_nn (accessed on 15 July 2021).

applied sciences

MDPI

Article

Research on the Vehicle Steering and Braking Stability Region

Xianbin Wang [1,*], Weifeng Li [1], Fugang Zhang [2], Zexuan Li [1] and Wenlong Bao [1]

1 School of Civil Engineering and Transportation, Northeast Forestry University, Harbin 150040, China; lwf@nefu.edu.cn (W.L.); lizexuan@nefu.edu.cn (Z.L.); meihao18869@nefu.edu.cn (W.B.)
2 Commercial Vehicle Development Institute of China FAW Jiefang Automobile Co., Ltd., Changchun 130062, China; zhangfugang@fawjiefang.com.cn
* Correspondence: xbwang@nefu.edu.cn; Tel.: +86-139-4605-2235

Abstract: Solving the stability region in the plane motion of vehicles has become a hot research topic in vehicle handling stability under extreme conditions, but there is still a lack of research on the stability region under steering and braking conditions. In this paper, a five-degree-of-freedom (5DOF) nonlinear dynamic model of a vehicle with braking torque introduced is established, and the model is transformed into an equivalent system by using the D'Alembert principle. Then, the equilibrium points of the equivalent system are solved by using an improved hybrid algorithm combining the genetic algorithm (GA) and sequential quadratic programming (SQP) method. According to the bifurcation characteristics of the equilibrium points, the boundary of the stability region at the given initial longitudinal velocity is determined, and the three-dimensional stability region is fitted. Finally, the stability region of the equivalent system and the original system are analyzed by the energy dissipation method, and the stability region determined by the equilibrium point bifurcation method is verified by simulation. The results show that as the braking torque increases, the number of equilibrium points increase to three from one, the equilibrium bifurcation method proposed in this paper can effectively solve the stability region of the equivalent system, and the solution results are consistent with the original system stability region. When the limited braking torque is 500 N·m and the initial longitudinal velocity increases from 30 m/s to 50 m/s, the absolute value of the front wheel steering angle at the boundary point changes from less than 0.02 rad to more than 0.02 rad.

Keywords: vehicle nonlinear dynamics; stability region; equilibrium point bifurcation method; steering and braking; energy dissipation method

Citation: Wang, X.; Li, W.; Zhang, F.; Li, Z.; Bao, W. Research on the Vehicle Steering and Braking Stability Region. *Appl. Sci.* **2023**, *13*, 7806. https://doi.org/10.3390/app13137806

Academic Editors: Edgar Sokolovskij and Vidas Žuraulis

Received: 20 April 2023
Revised: 29 June 2023
Accepted: 30 June 2023
Published: 2 July 2023

1. Introduction

With the application of nonlinear dynamics in the field of vehicle dynamics, the determination of stable regions in the plane motion of vehicles has become an important research field [1]. For the vehicle system, the external input (or driver) affects the dynamic stability of the system through acceleration, deceleration, and steering, which puts the vehicle in single or coupled working conditions. Many scholars have studied the plane motion stability of vehicles under various driving conditions through different nonlinear analysis methods.

Steering stability research of vehicles at a constant velocity was originally based on two-degree-of-freedom (2DOF) nonlinear vehicle model analysis. The steering angle bifurcation caused by tire lateral force saturation has been confirmed by many scholars. The band convergence region in the phase plane has been determined as the stable region of the system, and significant results have been achieved in the stability control of lateral motion [2–5]. Sadri et al. used the Lyapunov function method to analyze the lateral stability of vehicle steering [6], and in a follow-up study, Lyapunov exponents were also used to obtain the lateral stability region of the vehicle, which was an extension of the stability region obtained by using the Lyapunov function in the original study [7]. Rossa et al. combined the 2DOF nonlinear vehicle model with a simple driver model, analyzed the

Hopf bifurcation characteristics of vehicles with understeer and oversteer characteristics under steady steering and found different bifurcation forms of the system through the combination of different vehicle models and driver models, and found chaos in the driving simulator as well [8,9]. Steindl et al. found supercritical Hopf bifurcation of oversteer vehicles when studying the 2DOF nonlinear model [10].

Although the research based on the classical 2DOF nonlinear model of vehicles has achieved many research results, the model assumes that the longitudinal velocity remains unchanged and only considers the lateral force of the tires, without considering the impact of the longitudinal force of the tires on the system stability, which is not in line with the actual force situation of the tires in the actual driving of vehicles. Liu et al. introduced the longitudinal velocity into the vehicle model to establish a three-degree-of-freedom (3DOF) model and removed the assumption that the longitudinal velocity was constant [11]. Through a change of phase space trajectory, they pointed out that the essential feature of vehicle steering instability was the chaotic motion of the vehicle dynamics system. Masouleh et al. estimated the region of attraction (ROA) of the lateral dynamics of a 3DOF vehicle model through the sum of squares (SOS) programming algorithm [12]. Meng et al. put forward an energy dissipation method of vehicle plane motion and calculated the stability region of the 3DOF model. By comparison, it was proved that the energy dissipation method was more intuitive and easier to calculate than the Lyapunov exponents method [1]. In this study, the largest Lyapunov exponents and the sum of the Lyapunov exponents take 895,273 s (about 10 days) for numerical calculation. The total dissipation of energy takes 420 s for calculation.

Based on the research on 3DOF model steering conditions, Wang et al. established a five-degree-of-freedom (5DOF) vehicle nonlinear model incorporating driving torque, analyzed the impact of driving modes on vehicle stability under different steering angles [13], and obtained a two-dimensional bifurcation parameter set (driving stability region) under vehicle steering and driving conditions based on solving the equilibrium point of the 5DOF system [14]. Shi et al. used the Lyapunov exponents method and energy dissipation method to quantitatively analyze the stability region of the system under steering and driving conditions and obtained the stability region of state variables (lateral velocity and yaw rate) [15] and the stability region of system control parameters (front wheel angle and driving torque) [16], respectively. The different characteristics of dynamics between the high-degree-of-freedom model and the 2DOF model were revealed and explained. It was also proved that the maximum deviation of the energy dissipation method from the equilibrium point bifurcation method is less than 1.2%.

Braking is a common condition in vehicle driving, and many scholars have also studied its stability through nonlinear analysis methods. Based on considering the wheel slip dynamics, Olson et al. studied the longitudinal braking dynamics of a two-wheeled vehicle model coupled with the vehicle body and studied the variety of wheel slip rate when the braking torque changes through a bifurcation diagram, showing the whole process of how the system transits from stable braking to locking in one or two wheels, and then locking in two wheels [17]. Horiuchi et al. proposed an analysis method based on constrained bifurcation and continuity methods with the D'Alembert principle introduced, established a vehicle dynamics model including longitudinal velocity, lateral velocity, yaw rate and roll, and analyzed the bifurcation phenomenon caused by braking deceleration [18]. Chen et al. studied the 3DOF dynamic model combined with the Burckhardt tire model, studied the change of system equilibrium point under front/rear/four-wheel braking by the phase plane method, and obtained the stable region corresponding to the slip rate of the front and rear wheels when the vehicle turns to brake [19]. Zhu et al. considered engine braking in the 5DOF model and obtained the bifurcation diagram of the equilibrium point with the change of front wheel angle and longitudinal velocity through the phase plane method and static bifurcation theory [20]. Lai et al. constructed an 18-degree-of-freedom unified dynamic model of a vehicle chassis including three subsystems: a steering subsystem, braking subsystem and suspension subsystem, analyzed the influence of front and rear wheel braking modes on vehicle driving stability by the phase plane analysis method, and found that front wheel braking is more stable than rear wheel braking [21]. A new model based on factors of the driver, vehicle and environment was developed to determine the 'safe stopping distance' of vehicles, and tested on real road conditions by using equipment placed on a vehicle [22].

It can be stated that, as a nonlinear dynamic system with multiple degrees of freedom, the steering instability mechanism of a vehicle has been basically recognized as a bifurcation phenomenon in a nonlinear system, and the instability mechanism under driving and steering conditions can also be explained by bifurcation theory. From the above research, it can be found that due to the particularity of braking dynamics, many scholars have focused on the design of controllers and the analysis of braking mechanisms based on simple vehicle models. When the system dimensions increase, it is necessary to simplify the model or improve the nonlinear analysis methods for analysis. However, there is still a lack of achievements in terms of analyzing or calculating the braking stability region. By systematically studying the vehicle steering and braking stability region, the integrated control theory of vehicle handling stability will be further improved, and the theoretical fundamentals of nonlinear vehicle dynamics will be provided for the longitudinal and lateral coupled control of intelligent vehicles under extreme conditions (mostly steering and braking working conditions) [23–25].

The content of this paper is arranged as follows: Section 2 is the establishment of a 5DOF steering and braking stability analysis model, and the correctness and effectiveness of the model are verified by simulation analysis. In Section 3, the D'Alembert principle is introduced, and the original model is transformed into an equivalent equilibrium system to analyze the bifurcation of the equilibrium point. In Section 4, the two-dimensional bifurcation parameters set under steering and braking condition are solved by using the equilibrium bifurcation method, and the three-dimensional stability region is fitted. In Section 5, the stable regions of the original system and the equivalent equilibrium system are calculated by the energy dissipation method, and the results are compared with those of the equilibrium bifurcation method. Section 6 gives the conclusions.

2. Stability Analysis of 5DOF Steering and Braking Vehicle Model

2.1. Vehicle Model

To study the stability of the plane motion of vehicles, this paper does not consider the influence of suspension systems and load transfers in steering and braking conditions, that is, the vertical, pitch, and roll motions of vehicles are ignored.

This paper mainly analyzes and calculates the dynamic stability region under vehicle braking. If a double-track model is used, the degrees of freedom of the system model will reach seven. Through previous research, we have found that as the degrees of freedom of the system model increases, the difficulty of accurately solving the equilibrium point greatly increases. Moreover, the dynamic situation in this study can also be effectively analyzed through a 5DOF model. In addition, for front wheel steering vehicles, although the actual steering angles of the two tires are different due to the differential (usually the inner tire has a larger steering angle), this is not the focus of this study. We simplify the double-track model to the single-track model, which can better study how changes in steering angles affect the dynamic mechanism of vehicle steering and braking.

Based on the classical 3DOF steering model [11], the longitudinal force and braking torque of the tires are introduced. The schematic diagrams of the single-track model considering the longitudinal force of the tire and the wheel with braking torque are shown in Figure 1.

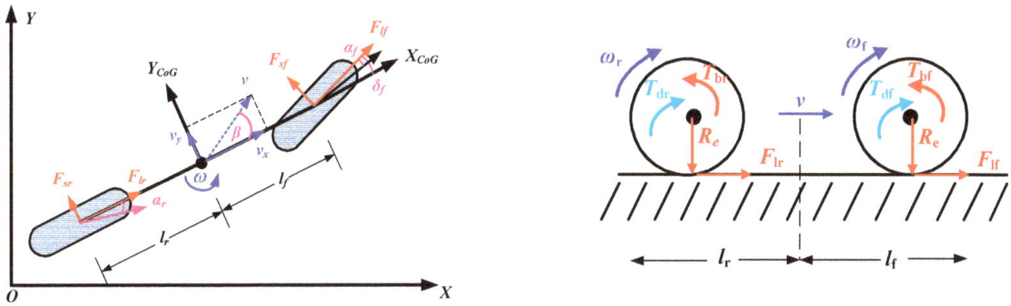

(a) (b)

Figure 1. The 5DOF model: (**a**) Single-track model with longitudinal force added; (**b**) The front and rear wheel braking model.

Where X_{CoG} and Y_{CoG} represent the X-axis and Y-axis of the vehicle coordinate system, and X and Y represent the X-axis and Y-axis of the geodetic coordinate system.

When the vehicle velocity is low, the air resistance is small, and the wheel rolling resistance is the main driving resistance. While the vehicle is in high-velocity conditions, air resistance surpasses wheel rolling resistance and becomes the main resistance [26]. The direction of air resistance is always opposite to the driving direction, so it is necessary to introduce symbolic functions into the model to describe the longitudinal and lateral air resistance of the vehicle body. Finally, the 5DOF dynamic system equation can be given as follows:

$$\begin{cases} \dot{v}_y = -v_x\omega + \frac{F_{lf}\sin\delta_f + F_{sf}\cos\delta_f + F_{sr} - \mathrm{sgn}(v_y)\cdot C_{air\_y}A_{L\_y}\frac{\rho}{2}{v_y}^2}{m} \\ \dot{\omega} = \frac{(F_{lf}\sin\delta_f + F_{sf}\cos\delta_f)l_f - F_{sr}l_r}{I_z} \\ \dot{v}_x = v_y\omega + \frac{F_{lf}\cos\delta_f - F_{sf}\sin\delta_f + F_{lr} - \mathrm{sgn}(v_x)\cdot C_{air\_x}A_{L\_x}\frac{\rho}{2}{v_x}^2}{m} \\ \dot{\omega}_f = \frac{-\mathrm{sgn}(\omega_f)\cdot T_{bf} - R_e\cdot F_{lf}}{J} \\ \dot{\omega}_r = \frac{-\mathrm{sgn}(\omega_r)\cdot T_{br} - R_e\cdot F_{lr}}{J} \end{cases} \tag{1}$$

where m is the mass of the vehicle, I_z is the yaw moment of inertia of the vehicle, v_x is the longitudinal velocity, v_y is the lateral velocity, ω is the yaw rate, ω_f is the angular velocity of the front wheels, ω_r is the angular velocity of the rear wheels, l_f is the distance from the front axle to the mass center, l_r is the distance from the rear axle to the mass center, J is the yaw moment of inertia of the wheels, δ_f is the front wheels' steering angle, C_{air_x} is the longitudinal air resistance coefficient, C_{air_y} is the lateral air resistance coefficient, A_{L_x} is the longitudinal area of the vehicle, A_{L_y} is the lateral area of the vehicle, ρ is the density of air, T_{bf} is the braking torque of the front wheels, T_{br} is the braking torque of the rear wheels, R_e is the rolling radius of the wheels, F_{lf} and F_{lr} are the longitudinal tire forces of the front wheel and the rear wheel, respectively, and F_{sf} and F_{sr} are the lateral tire forces of the front wheel and the rear wheel, respectively.

For the convenience of description, the braking torque T_b is regarded as a positive scalar in this paper. It is mainly applied to the front wheel to prevent the rear wheel from locking and slipping first. The braking torque of the front and rear wheels is distributed according to the following formula:

$$\begin{cases} T_{bf} = \eta T_b \\ T_{br} = (1-\eta)T_b \end{cases} \tag{2}$$

where η is the braking torque distribution coefficient, and η = 0.7.

To improve the stability of braking and make the best use of the road adhesion conditions, a braking mode with both the front and rear wheels locked simultaneously is adopted. On any road surface with an adhesion coefficient μ (μ = 0.3 in this paper), the condition for both the front and rear wheels to lock at the same time is that the sum of the braking forces of the front and rear wheels are equal to the adhesion force, and the braking forces of the front and rear wheels are equal to their respective adhesion forces [27], that is:

$$\begin{cases} F_{\mu f} + F_{\mu r} = \mu F_z \\ F_{\mu f} = \mu F_{zf} \\ F_{\mu r} = \mu F_{zr} \end{cases} \tag{3}$$

where $F_{\mu f}$ is the braking force of the front wheel brake, $F_{\mu r}$ is the braking force of the rear wheel brake, and F_z is the total load of the front and rear wheels of the vehicle. Only plane motion is considered, therefore the front wheel load F_{zf} and the rear wheel load F_{zr} are fixed values, and they are distributed as follows:

$$\begin{cases} F_{zf} = \frac{mgl_f}{l_f+l_r} \\ F_{zr} = \frac{mgl_r}{l_f+l_r} \end{cases} \tag{4}$$

Without loss of generality, the simulation in this paper is carried out on a low adhesion coefficient road. The model parameters of the 5DOF system are selected according to [3,28], as shown in Table 1.

Table 1. Vehicle model parameters.

Component Names and Parameters	Values
vehicle mass m/kg	1500
yaw moment of inertia I_z/kg·m^2	3000
the distance from the front axle to the mass center l_f/m	1.2
the distance from the rear axle to the mass center l_r/m	1.3
yaw moment of inertia of the wheels J/kg·m^2	2.0
longitudinal air resistance coefficient C_{air_x}	0.3
the lateral air resistance coefficient C_{air_y}	0.4
the longitudinal area of the vehicle A_{L_x}/m^2	1.7
the lateral area of the vehicle A_{L_y}/m^2	3.5
the density of air ρ/kg/m^3	1.2258
the rolling radius of the wheels R_e/m	0.224

2.2. Tire Model

To reflect the nonlinear characteristics of tire force, the expression of tire force adopts the classic magic formula [29]:

$$F = \mathrm{D}\sin(\mathrm{C}\arctan(\mathrm{B}x - \mathrm{E}(\mathrm{B}x - \arctan\mathrm{B}x))) \tag{5}$$

where B, C, D, and E are tire parameters, F is the tire steady-state longitudinal force or lateral force, and x is the longitudinal slip rate or sideslip angle.

The calculation of the slip rate is based on the unified formula considering all tire operating conditions in [13].

$$k = \frac{\omega_w R_e - v_{wx}}{|v_{wx}|} \tag{6}$$

The tire velocity in the tire coordinate system is calculated by coordinate transformation.

$$\begin{cases} v_{xf} = v_x \cos\delta_f + (v_y + \omega l_f)\sin\delta_f \\ v_{xr} = v_x \cos\delta_r + (v_y - \omega l_r)\sin\delta_r \end{cases} \tag{7}$$

$$\begin{cases} v_{yf} = -v_x \sin\delta_f + (v_y + \omega l_f)\cos\delta_f \\ v_{yr} = -v_x \sin\delta_r + (v_y - \omega l_r)\cos\delta_r \end{cases} \tag{8}$$

where v_{xf} is the longitudinal velocity of the front wheel in the tire coordinate system, and v_{xr} is the longitudinal velocity of the rear wheel in the tire coordinate system.

The sideslip angle is calculated by using the unified formula which is suitable for the conditions of a large sideslip angle and reverse wheel rotation [1]:

$$\begin{cases} \alpha_f = \arctan\frac{v_{yf}}{v_{xf}} \cdot \mathrm{sgn}\left(v_{xf}\right) \\ \alpha_r = \arctan\frac{v_{yr}}{v_{xr}} \cdot \mathrm{sgn}(v_{xr}) \end{cases} \tag{9}$$

The combined slip of the tire is analyzed and calculated according to the model proposed by Pacejka [30].

$$\begin{cases} F_{lf} = F_{lf0} \cdot G_x \\ F_{lr} = F_{lr0} \cdot G_x \\ G_x = \cos[\arctan\{B_{g,x}(\alpha) \cdot \alpha\}] \\ B_{g,x}(\alpha) = r_{x,1}\cos[\arctan(r_{x,2} \cdot k)] \\ F_{sf} = F_{sf0} \cdot G_y \\ F_{sr} = F_{sr0} \cdot G_y \\ G_y = \cos[\arctan\{B_{g,x}(k) \cdot k\}] \\ B_{g,y}(k) = r_{y,1}\cos[\arctan(r_{y,2} \cdot \alpha)] \end{cases} \tag{10}$$

where F_{lf0}, F_{lr0}, F_{sf0}, F_{sr0} are the longitudinal forces and lateral forces of the front and rear wheels under a steady state, G_x and G_y are the tire force combined slip correction function, and $r_{x,1}$, $r_{x,2}$, $r_{y,1}$, $r_{y,2}$ are the tire force combined slip correction coefficient. The values of the tire model parameters [30] are shown in Table 2.

Table 2. Tire model parameters.

Tire	Longitudinal Tire Parameters				Lateral Tire Parameters				Longitudinal Combined Slip Coefficients		Lateral Combined Slip Coefficients	
	B	C	D	E	B	C	D	E	$r_{x,1}$	$r_{x,2}$	$r_{y,1}$	$r_{y,2}$
Front	11.275	1.56	2574.8	0.4109	11.275	1.56	2574.7	−1.999	35	40	40	35
Rear	18.631	1.56	1749.6	0.4108	18.631	1.56	1749.7	−1.7908				

2.3. Model Validation

In this section, the effectiveness of the model is verified by the phase space trajectory (v_y-ω-v_x) of different single initial conditions, the time series of state variables, and the change of body attitude. See Table 3 for the initial conditions of specific simulations.

Table 3. Initial conditions.

Initial Conditions	v_x (m/s)	v_y (m/s)	ω (rad/s)	δ_f (rad)	T_b (N·m)
1	30	0.1	0.1	0.010	300
2	30	0.1	0.1	0.015	300
3	30	0.1	0.1	0.015	600

Figure 2 shows the simulation results, with a simulation time of 30 s. The blue line, red line, and black line in Figure 2a represent the dynamic behavior of the system under the input of initial conditions 1, 2, and 3, respectively. Under the input of initial condition 1, the phase space trajectory first appears a wide range of semi-circular motion, corresponding to the fluctuation of lateral velocity and yaw rate in Figure 2c for the first 5 s. After that, the phase trajectory inclines downward, which also corresponds to the rapid convergence of lateral velocity and yaw rate in Figure 2c and the linear decrease of longitudinal velocity. It can be seen in Figure 2d from the change of body attitude that the vehicle is in a steady-state steering state and the distance between the two vehicles is decreasing continuously, indicating that the vehicle will eventually be stationary.

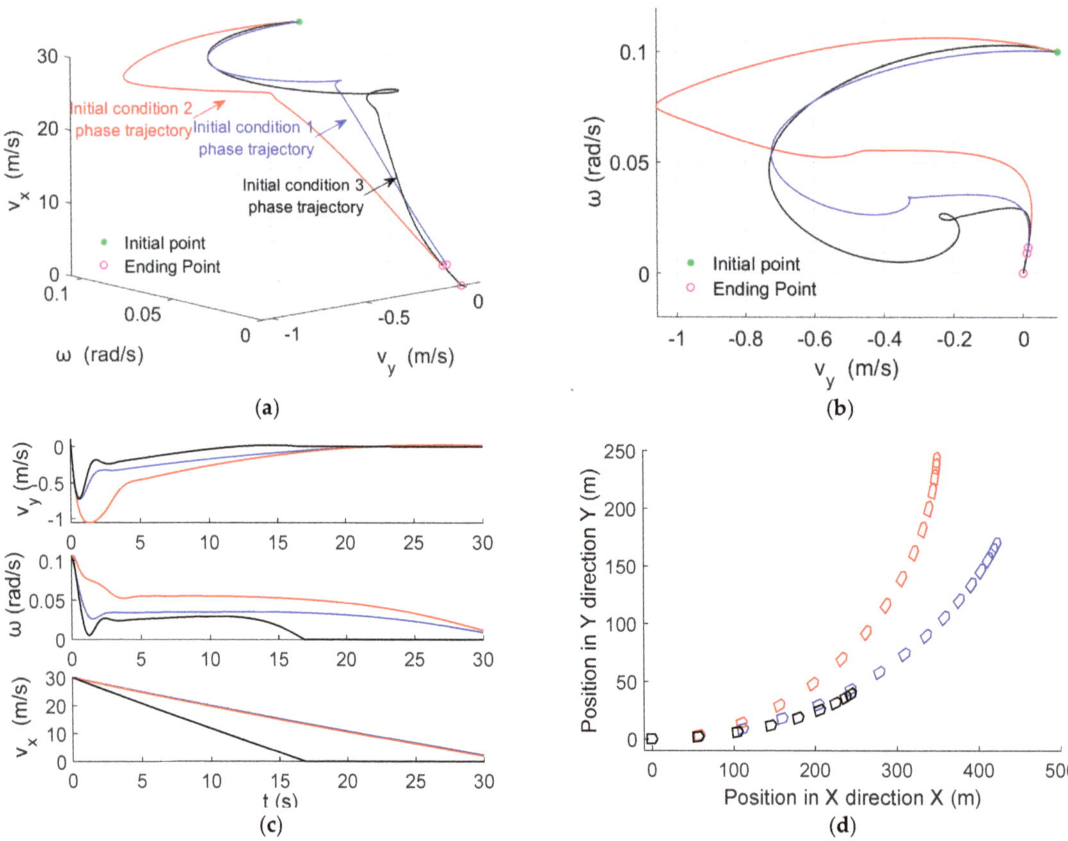

Figure 2. Time-domain responses of single initial conditions and vehicle body attitude: (**a**) Phase space trajectories; (**b**) Top view of a; (**c**) Time-domain responses; (**d**) Vehicle body attitude.

The initial condition 2 increases the front wheel steering angle input compared with the initial condition 1. It can be seen that the motion range of the phase trajectory in the early stage is larger, and the state variables fluctuate more violently in the early stage as well, but it can still be stabilized. We also can find that the turning radius of the vehicle is smaller than that in the initial condition 1.

The initial condition 3 increases the braking torque input compared with the initial condition 2. It can be seen that the motion range of the phase trajectory in the early stage is smaller, and the fluctuation amplitude of state variables is smaller in the early stage too. The longitudinal velocity linearly decreases to 0 when the simulation time is about 16 s. The vehicle can also be observed to be stationary during the body attitude change, and the overall braking distance is significantly reduced compared with the previous two groups of simulations.

From the above analysis, it can be found that the 5DOF model can reflect the motion process of the vehicle under the combined steering and braking conditions, thus providing a basic analysis model for studying the dynamic bifurcation characteristics of the vehicle under the combined steering and braking conditions.

3. Solution of Equivalent Equilibrium Point

3.1. Introducing the D'Alembert Principle into the 5DOF Model

The vehicle dynamics system is a typical nonlinear system, and the common classical analysis method is the static bifurcation theory based on the change of the number of equilibrium points of the nonlinear system. Many scholars have studied vehicle steering stability. Under the premise of constant longitudinal velocity, the dynamic essence of vehicle steering instability can be attributed to the bifurcation of the equilibrium point in nonlinear dynamics [2,3,5]. However, under the steering and braking conditions, because the longitudinal velocity of the vehicle is constantly changing, there is no equilibrium state according to the strict definition. As a result, the mature research method of steering bifurcation cannot be directly applied to the braking process. Therefore, it is necessary to explore new methods to analyze the bifurcation characteristics of braking.

Horiuchi mentioned in [18] that when analyzing the acceleration or braking of a vehicle system, the virtual force in the D'Alembert principle can be introduced so that an accelerated system can be transformed into an equivalent equilibrium system. The conversion method involves applying a virtual external force determined by acceleration to the system. Based on this principle, if a virtual force with the size of $F = ma$ is applied to the vehicle (where the acceleration is acting on the center of mass of the vehicle), the braking process can be regarded as an equilibrium state (called equivalent equilibrium in this paper) and bifurcation analysis can then be carried out.

Figure 3 gives the force analysis diagram of the whole vehicle after adding the virtual force:

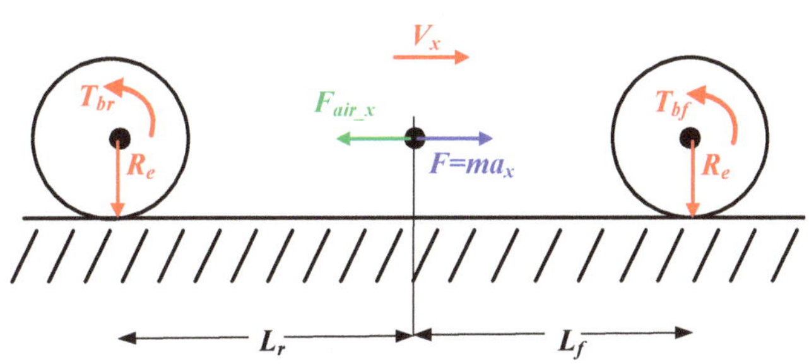

Figure 3. Force analysis diagram with virtual force added.

To maintain the longitudinal force equilibrium of the vehicle, the following can be obtained:

$$ma_x = \frac{T_b}{R_e} + \mathrm{sgn}(v_x) C_{\mathrm{air}\_x} A_{\mathrm{L}\_x} \frac{\rho}{2} v_x^2 \tag{11}$$

The virtual acceleration a_x acting on the center of mass of the vehicle body can be obtained:

$$a_x = \frac{\left(\frac{T_b}{R_e} + \text{sgn}(v_x)C_{\text{air\_}x}A_{\text{L\_}x}\frac{\rho}{2}v_x^2\right)}{m} \tag{12}$$

The equilibrium in the plane motion of the vehicle must satisfy the steady state equation, and it is constructed by making the right side of the dynamic equations equal to zero [31]. The accuracy of the equilibrium point solution is evaluated by the value of the fitness function. That is, the smaller the objective function value, the more accurate the solution. The equation for solving the equivalent equilibrium point of the system is as follows:

$$\begin{cases} \dot{v}_y = -v_x\omega + \frac{F_{lf}\sin\delta_f + F_{sf}\cos\delta_f + F_{sr} - \text{sgn}(v_y)\cdot C_{\text{air\_}y}A_{\text{L\_}y}\frac{\rho}{2}{v_y}^2}{m} = 0 \\ \dot{\omega} = \frac{(F_{lf}\sin\delta_f + F_{sf}\cos\delta_f)l_f - F_{sr}l_r}{I_z} = 0 \\ \dot{v}_x = v_y\omega + \frac{F_{lf}\cos\delta_f - F_{sf}\sin\delta_f + F_{lr} - \text{sgn}(v_x)\cdot C_{\text{air\_}x}A_{\text{L\_}x}\frac{\rho}{2}{v_x}^2 + ma_x}{m} = 0 \\ \dot{\omega}_f = \frac{-\text{sgn}(\omega_f)\cdot T_{bf} - R_e\cdot F_{lf}}{J} = 0 \\ \dot{\omega}_r = \frac{-\text{sgn}(\omega_r)\cdot T_{br} - R_e\cdot F_{lr}}{J} = 0 \end{cases} \tag{13}$$

The equilibrium point solution of a 5DOF system can be regarded as a programming problem constrained by Equation (13):

$$\min f\left(\dot{v}_x, \dot{v}_y, \dot{\omega}, \dot{\omega}_f, \dot{\omega}_r\right) = |\dot{v}_x| + |\dot{v}_y| + |\dot{\omega}| + \left|\dot{\omega}_f\right| + |\dot{\omega}_r| \tag{14}$$

The optimal solution $\left(\dot{v}_x^*, \dot{v}_y^*, \dot{\omega}^*, \dot{\omega}_f^*, \dot{\omega}_r^*\right)$ must satisfy $f\left(\dot{v}_x^*, \dot{v}_y^*, \dot{\omega}^*, \dot{\omega}_f^*, \dot{\omega}_r^*\right) = 0$, and the fitness value of the 5DOF system is as follows:

$$\text{Fitness} = |\dot{v}_x| + |\dot{v}_y| + |\dot{\omega}| + \left|\dot{\omega}_f\right| + |\dot{\omega}_r| \tag{15}$$

To verify the equivalent equilibrium state of the nonlinear system after adding the virtual force, Figure 4 shows the comparison of multiple initial point phase trajectory and single initial point state variables time series of the system without or with the virtual force.

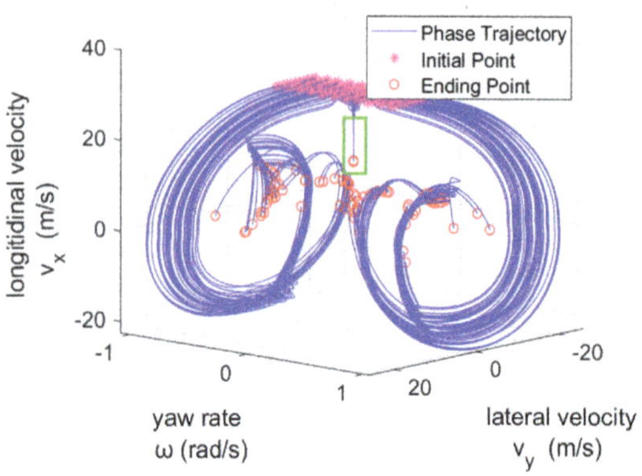

(a)

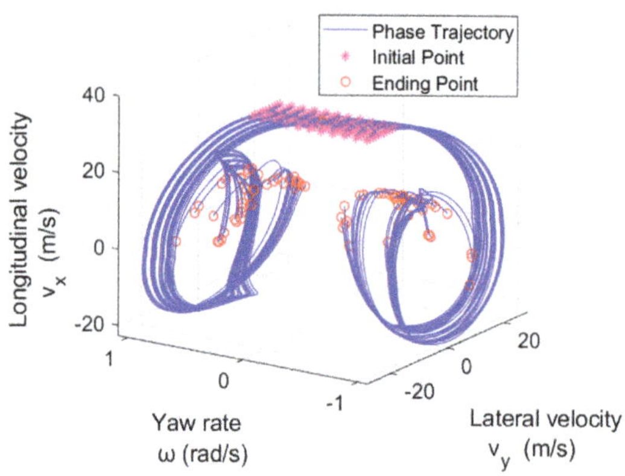

(b)

Figure 4. *Cont.*

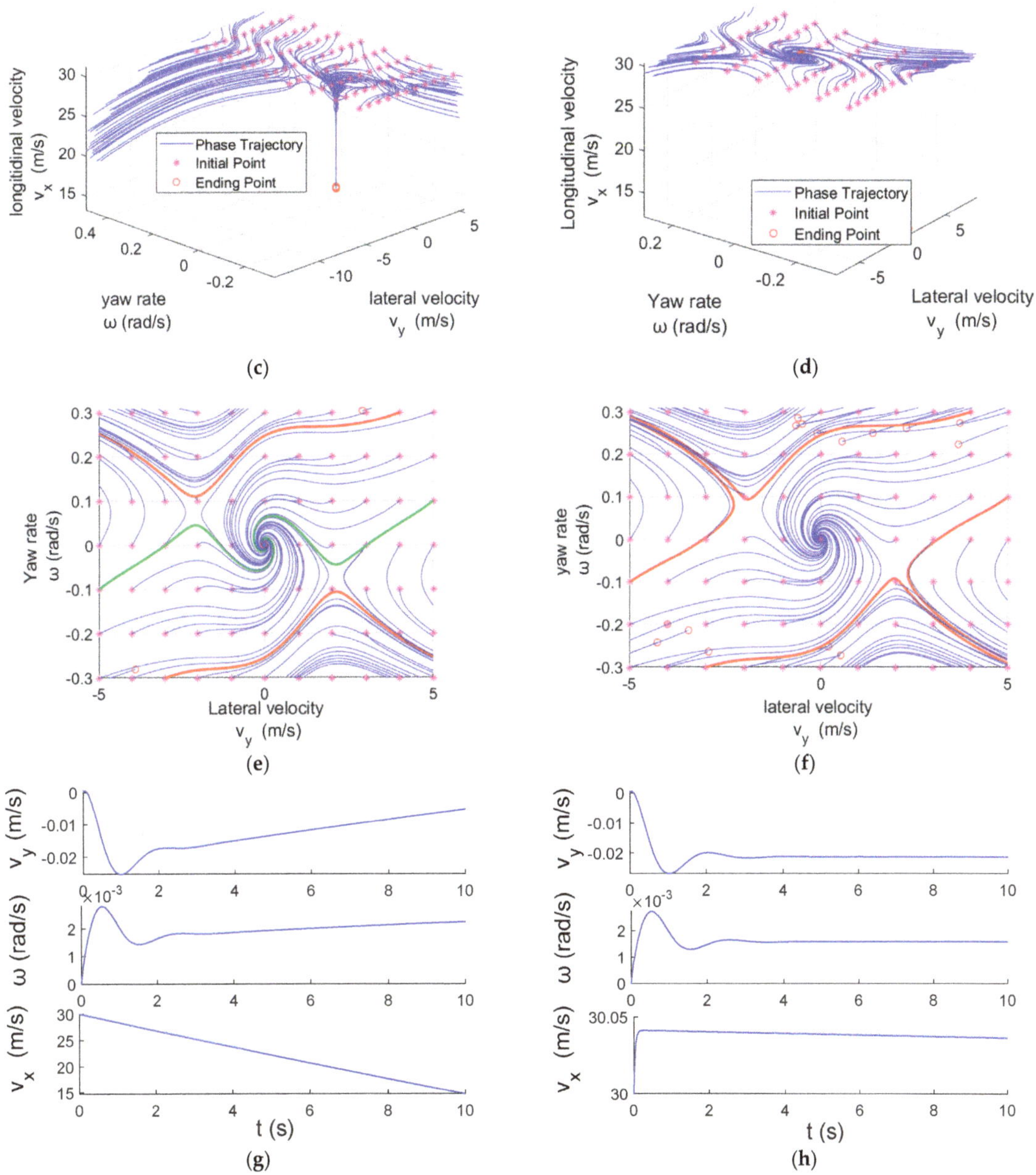

Figure 4. Comparison of the 5DOF model with and without virtual force: (**a**) Phase space without adding virtual force; (**b**) Phase space with virtual force; (**c**) Local view of (**a**); (**d**) Local view of (**b**); (**e**) Phase plane without virtual force; (**f**) Phase plane with virtual force; (**g**) Time domain without virtual force; (**h**) Time domain with virtual force.

The initial conditions of the multi-initial points simulation are: longitudinal velocity $v_x = 30$ m/s, front wheel steering angle $\delta_f = 0.001$ rad, braking torque $T_b = 500$ N·m, lateral velocity v_y ranging from -10 m/s to 10 m/s, with intervals of 1 m/s, and the value range of the yaw rate ω is -0.5 rad/s to 0.5 rad/s, with value intervals of 0.1 rad/s. The single initial point simulation conditions are: longitudinal velocity v_x = 30 m/s, front wheel steering angle δ_f = 0.001 rad, and braking torque T_b = 500 N·m.

Figure 4a,b show the global dynamic characteristics of the system, and it can be seen that the global characteristics of the system have not changed obviously after adding the virtual force. However, by comparing the local diagrams of Figure 4c,d, it can be seen that the downward spiral motion trend of the phase trajectory (phase trajectory in the green rectangle in Figure 4a) disappears and is replaced by the spiral motion in the two-dimensional plane in Figure 4d. This means that the longitudinal velocity of these phase trajectories remains constant.

The attraction domains of Figure 4e,f are almost the same. It is noted that the two green phase trajectories at the edge of the attraction domain in Figure 4f can return to the original point, while they cannot become stable in Figure 4e. It is indicated that the stability of the system is slightly improved.

Figure 4g,h are time series of phase trajectory state variables under stable initial conditions. It can be seen that the longitudinal velocity changes from linear decreasing to basically constant, and the lateral velocity and yaw rate also remain constant during the simulation time. It shows that the system under these initial conditions can achieve the equivalent equilibrium state. Figure 4 confirms the equivalent equilibrium state of the system from the global and local perspectives, so that the static bifurcation analysis of the nonlinear system can be carried out.

3.2. Analysis of Bifurcation Characteristics of Equilibrium Point

In this section, the equilibrium point of the vehicle dynamics equivalent system is solved by an improved hybrid algorithm based on the genetic algorithm (GA) and sequential quadratic programming (SQP) method [32]. The specific solution flow chart is shown in Figure 5.

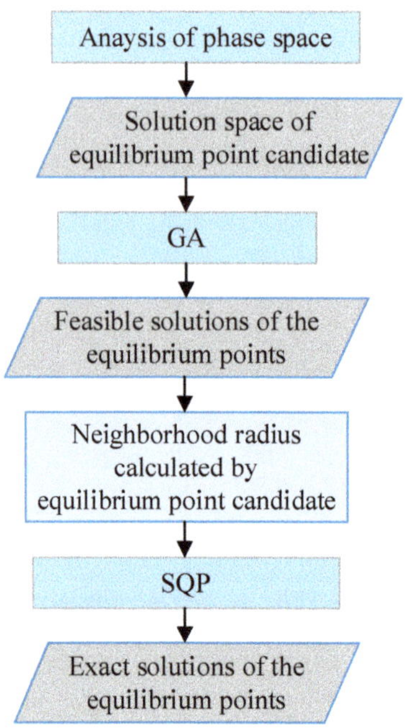

Figure 5. The hybrid solving process.

The solution conditions are: initial longitudinal velocity v_x = 30 m/s, front wheel angle δ_f = 0.015 rad, and the search range of braking torque T_b is [0, 504] N·m, with value intervals of 10.08 N·m (the bifurcation of the equilibrium point is solved to the order of 0.01 N·m). According to the solution process of the equilibrium point in Figure 5, the corresponding virtual acceleration a_x is calculated under each given braking torque T_b, and the equivalent equilibrium state is constructed so as to complete the solution of the equilibrium point.

Figure 6 shows the bifurcation diagram of the equilibrium point with braking torque as the bifurcation parameter. Tables 4–6 show the partial values of the stable equilibrium point and two unstable equilibrium points of the system when the braking torque changes.

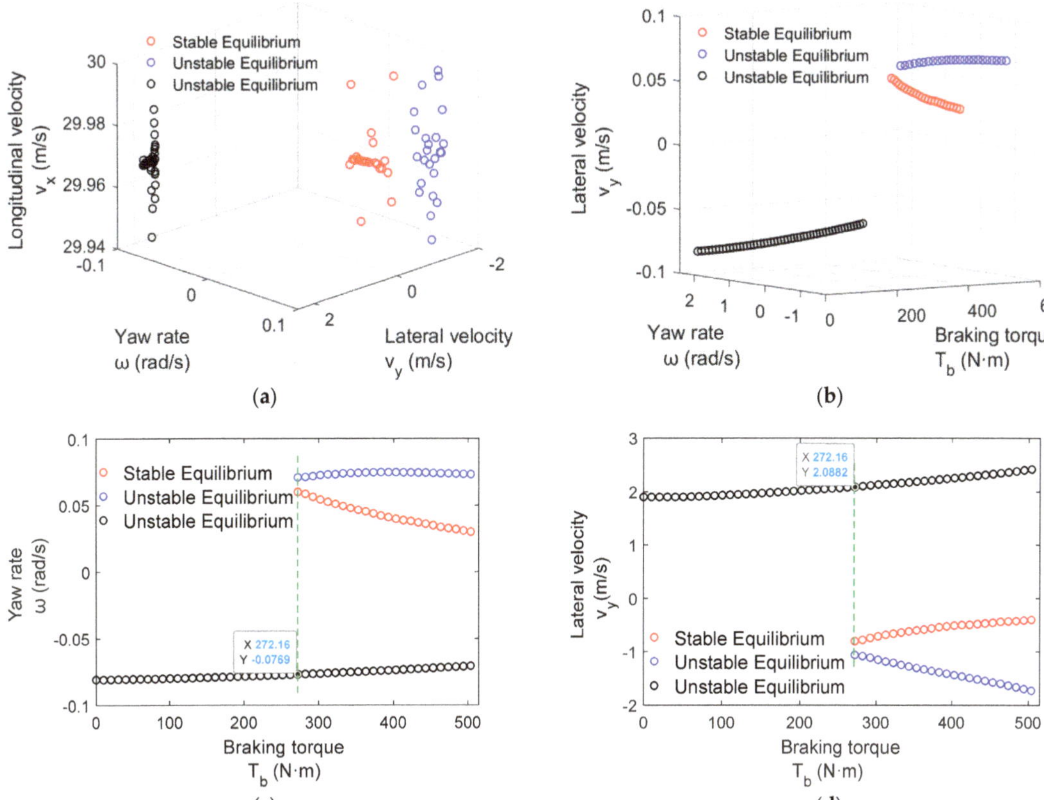

Figure 6. Range of the equilibrium points and bifurcation diagram of equilibrium points with braking torque varying: (**a**) Spatial distribution of equilibrium points; (**b**) Bifurcation space of the equilibrium points; (**c**) Bifurcation diagram of equilibrium points; (**d**) Bifurcation diagram of equilibrium points.

Table 4. Vehicle dynamics partial values for stable equilibrium points.

T_b (N·m)	Vehicle Dynamics Equilibriums					Rate of System Variable Change					Fitness
	v_y (m/s)	ω (rad/s)	v_x (m/s)	ω_f (rad/s)	ω_r (rad/s)	$\dot{v}_y$	$\dot{\omega}$	$\dot{v}_x$	$\dot{\omega}_f$	$\dot{\omega}_r$	
272.16	−0.8008	0.0600	29.9973	132.3378	132.5518	0.0000	0.0023	−0.0621	0.0001	0.0000	0.0646
322.56	−0.6604	0.0509	29.9670	129.5839	132.2900	0.0000	0.0000	−0.0452	0.0000	0.0000	0.0452
372.96	−0.5614	0.0435	29.9687	129.0988	132.1528	0.0000	0.0012	−0.0343	0.0004	0.0000	0.0359
423.36	−0.4939	0.0379	29.9494	128.4523	131.9026	0.0000	0.0000	−0.0270	0.0000	0.0000	0.0271
473.76	−0.4302	0.0325	29.9691	127.9211	131.8235	0.0000	0.0009	−0.0213	0.0000	0.0000	0.0222

Table 5. Vehicle dynamics partial values for one unstable equilibrium point.

T_b (N·m)	Vehicle Dynamics Equilibriums					Rate of System Variable Change					Fitness
	v_y (m/s)	ω (rad/s)	v_x (m/s)	ω_f (rad/s)	ω_r (rad/s)	$\dot{v}_y$	$\dot{\omega}$	$\dot{v}_x$	$\dot{\omega}_f$	$\dot{\omega}_r$	
272.16	−1.0630	0.0709	29.9855	129.7561	132.2881	0.0000	0.0037	−0.0919	0.0000	0.0000	0.0956
322.56	−1.2107	0.0732	29.9939	128.9161	131.9043	0.0000	−0.0014	−0.1057	0.0000	0.0000	0.1071
372.96	−1.3626	0.0746	29.9676	127.8544	131.3307	0.0000	0.0001	−0.1188	0.0000	0.0000	0.1188
423.36	−1.4954	0.0745	29.9688	126.8547	130.8688	0.0000	0.0000	−0.1286	0.0000	0.0000	0.1286
473.76	−1.6361	0.0738	29.9702	125.7211	130.3674	0.0000	0.0000	−0.1378	0.0000	0.0000	0.1378

Table 6. Vehicle dynamics partial values for another unstable equilibrium point.

T_b (N·m)	Vehicle Dynamics Equilibriums					Rate of System Variable Change					Fitness
	v_y (m/s)	ω (rad/s)	v_x (m/s)	ω_f (rad/s)	ω_r (rad/s)	$\dot{v}_y$	$\dot{\omega}$	$\dot{v}_x$	$\dot{\omega}_f$	$\dot{\omega}_r$	
60.48	1.9081	−0.0804	29.9688	132.8727	133.2487	0.0000	0.0000	−0.1343	0.0000	−0.0001	0.1345
120.96	1.9501	−0.0798	29.9851	131.9924	132.7804	0.0000	0.0000	−0.1367	0.0001	0.0000	0.1368
181.44	2.0027	−0.0789	29.9691	131.0290	132.1721	0.0000	0.0000	−0.1392	0.0000	0.0000	0.1392
241.92	2.0624	−0.0777	29.9596	130.1154	131.5962	0.0000	0.0000	−0.1414	0.0000	0.0001	0.1418
302.40	2.1372	−0.0765	29.9692	129.2572	131.0944	0.0000	−0.0030	−0.1452	0.0000	0.0000	0.1483
362.88	2.2020	−0.0749	29.9694	128.3242	130.5514	0.0000	0.0000	−0.1471	0.0000	0.0000	0.1471
423.36	2.2848	−0.0732	29.9698	127.2831	129.9736	0.0000	0.0000	−0.1499	0.0001	0.0000	0.1500
483.84	2.3826	−0.0714	29.9698	126.0794	129.3452	0.0000	0.0000	−0.1532	0.0000	0.0000	0.1532

Figure 6a shows the variation of values of all equilibrium points obtained by the hybrid algorithm. It can be seen that the longitudinal velocity v_x of the equilibrium points is around 30 m/s. Figure 6 shows the bifurcation diagrams of equilibrium points from different perspectives, with red "○" indicating stable equilibrium points, and blue and black "○" indicating unstable equilibrium points.

Through the analysis of the equivalent equilibrium point of the system, it can be concluded that when the value of the braking torque T_b is less than 272.16 N·m, there is only one unstable equilibrium point in the system. As the braking torque increases, the closer the system approaches the stable equilibrium point, the easier it is to stabilize. When the value of braking torque T_b exceeds 272.16 N·m, another unstable equilibrium point and stable equilibrium point of the system appear.

Based on the above analysis, it can be confirmed that the change in the number of equilibrium points of system with the change of braking torque T_b is a typical static bifurcation phenomenon. For this phenomenon, we can visually observe the change in the number of equilibrium points through the phase plane diagram, as shown in Figure 7. The initial conditions for the multi-initial points simulation are: longitudinal velocity v_x = 30 m/s, front wheel steering angle δ_f = 0.015 rad, lateral velocity range of −5 m/s to 5 m/s, with intervals of 1 m/s, and range of yaw rate values ω from −0.5 rad/s to 0.5 rad/s, with intervals of 0.2 rad/s.

When the braking torque is T_b = 262.08 N·m, as shown in Figure 7a, the system only has one unstable saddle point (which can be confirmed as a saddle point based on the phase trajectory manifold around the equilibrium point), and from the phase trajectories in the lower left corner, it can be seen that the phase trajectories in the area are connected. When the braking torque T_b = 272.16 N·m, the phase trajectory in the small graph in the lower left corner separates, resulting in the stable focus point and another unstable saddle point. The change in the number of equilibrium points before and after the increase of braking torque indicates that the system has undergone static bifurcation, which is consistent with the equilibrium point bifurcation diagram in Figure 7.

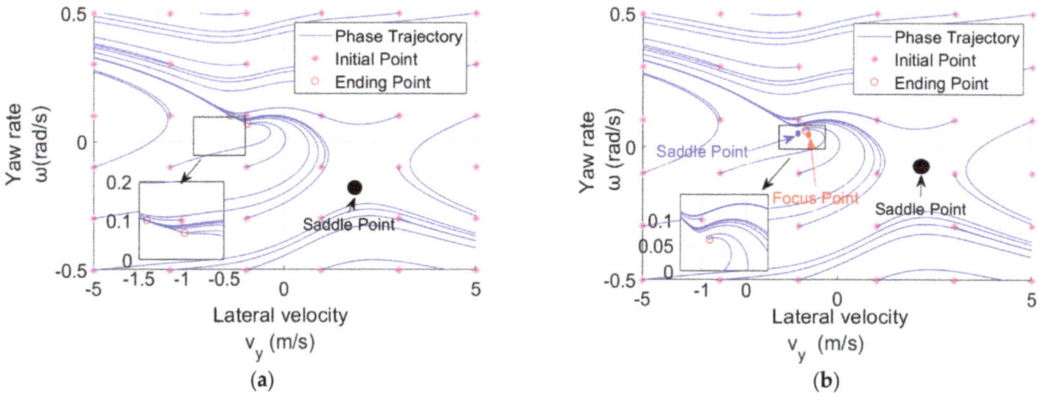

Figure 7. Change of phase plane before and after bifurcation: (**a**) T_b = 262.08 N·m; (**b**) T_b = 272.16 N·m.

4. Solution of Stability Region

4.1. Solution Procedure

Based on the previous achievements of our research team [14] in driving steering stability regions, the idea of constructing a two-dimensional bifurcation parameter set using system input parameters (front wheel angle-braking torque) will be applied again. The specific solution flow is shown in Figure 8, and the specific solving steps are as follows:

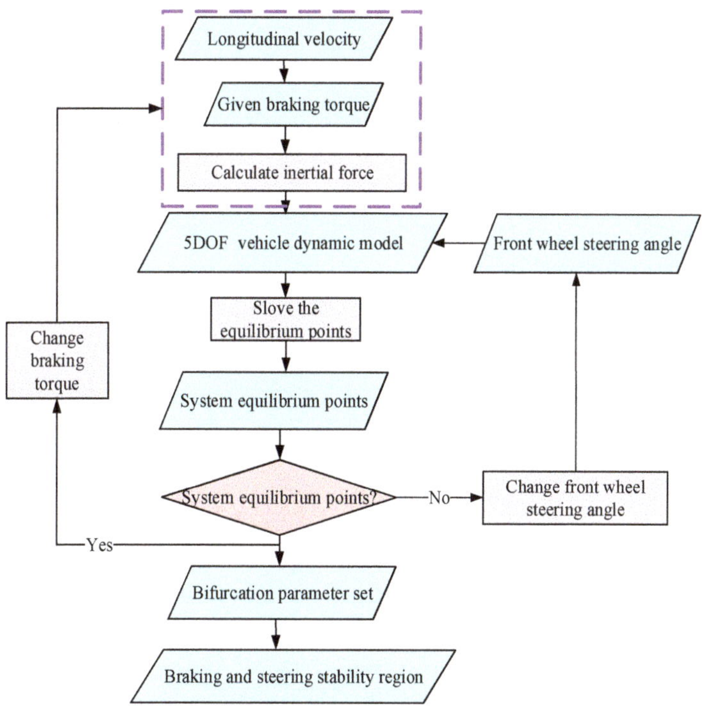

Figure 8. Flow chart for determining the steering and braking stability region.

Step 1. The initial longitudinal velocity range is 10 m/s to 60 m/s, with step sizes of 10 m/s.

Step 2. The initial braking torque is 10.08 N·m and is then continuously increased according to the given step size (the braking torque step size is 10.08 N·m) in each cycle until 604 N·m. After applying the braking torque each time, the corresponding inertia acceleration is calculated through Equation (12).

Step 3. The equilibrium points of the 5DOF equivalent system are solved and it is determined whether it is equilibrium point bifurcation according to the definition of static bifurcation. If so, the front wheel steering angle value of the bifurcation point is recorded and the braking torque is increased to continue solving the system equivalent equilibrium point; if not, searching for the front wheel steering angle value of the bifurcation point when the number of equilibrium points changes continues.

Step 4, The two-dimensional bifurcation parameter set (front wheel steering angle-braking torque) for a given initial longitudinal velocity is obtained.

4.2. Fitting of Three-Dimensional Stable Region

Figure 9 shows the two-dimensional bifurcation parameter set composed of the front wheel steering angle and braking torque at the given velocity, which is the boundary of the steering and braking stability region. In other words, within this boundary the system is stable.

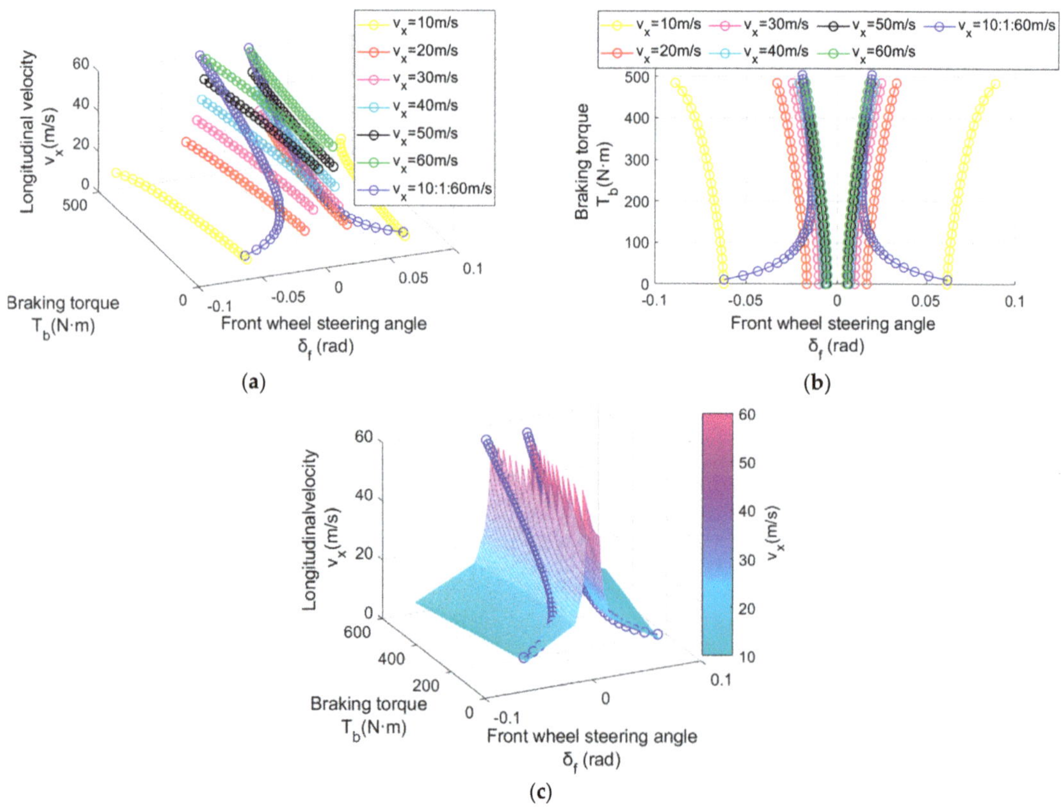

Figure 9. Steering and braking stability region: (**a**) Bifurcation parameter set in δ_f-T_b-v_x space; (**b**) Bifurcation parameter set in the δ_f-T_b plane; (**c**) Steering and braking stability region and surface fitting.

It can be observed in Figure 9a,b that the boundary of the stability region is a hyperbola. With the increase in velocity, the stability region gradually decreases. Especially, when the longitudinal velocity $v_x = 10$ m/s increases to $v_x = 20$ m/s, the stability region is significantly reduced. When the longitudinal velocity $v_x = 50$ m/s increases to $v_x = 60$ m/s, the stability region does not change much and only shrinks slightly. The blue circle line is a set of two-dimensional bifurcation parameters solved at different velocities. It can be seen that it passes through the boundary of the stability region at a given velocity in three-dimensional space, thereby mutually proving the correctness of the solved data.

Figure 9b shows the relationship between the bifurcation parameters at a given initial velocity and different initial velocities: when the initial velocity is the same, as the braking torque increases, the front wheel steering angle at the bifurcation point gradually increases; when the initial velocity continues to increase, as the braking torque increases, the front wheel angle at the bifurcation point gradually decreases, but then gradually increases after decreasing to a certain extent.

When there is more and more data on the two-dimensional bifurcation parameter set, theoretically, the steering and braking stability region should be composed of two symmetrical two-dimensional surfaces and a three-dimensional coordinate system. Figure 9c is a hyperboloid (three-dimensional parameter space) fitted according to six sets of boundary data of the stable region from $v_x = 10$ m/s to $v_x = 60$ m/s, that is, the boundary surface of the stable region. The surface interpolation method adopts the default "linear" method of MATLAB 2022a. The blue circle line in Figure 9 crosses the curved surface obliquely, which also proves the correctness of the fitted hyperboloid.

5. Verification Based on the Energy Dissipation Method

5.1. Analysis of Energy Dissipation Process

During the braking process of vehicles, tire force and air resistance can cause a decrease in system energy. By analyzing the energy dissipation of the steering and braking conditions of the vehicle, the stability region with and without inertia force applied at a given initial longitudinal velocity is calculated. A comparative analysis is also conducted with the stability region obtained from the equilibrium point bifurcation method in Section 4.2.

The energy dissipation process of a vehicle system can be divided into two parts as shown in Figure 10:

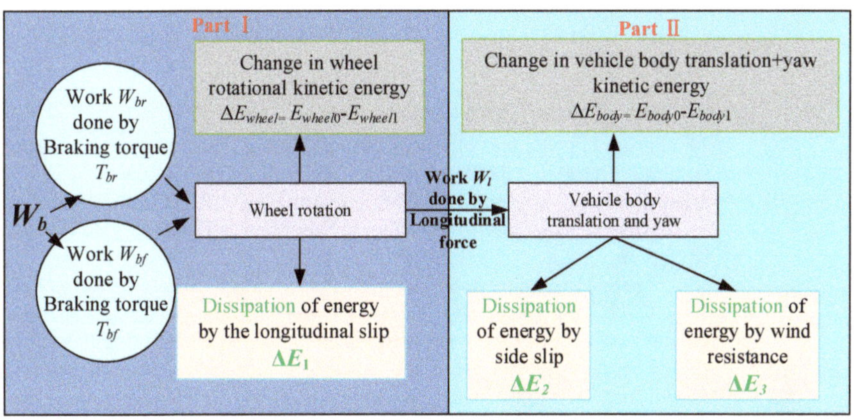

Figure 10. Schematic diagram of the energy dissipation process of the 5DOF vehicle system.

The first part is the dissipation of wheel rotational energy caused by tire longitudinal slip ΔE_1. As soon as the braking torque is applied to the front and rear wheels, the tire and ground experience longitudinal slip. The resulting longitudinal force reacts on the wheel to hinder the vehicle from moving forward. At this time, the effect of the braking torque causes a decrease in the rotational kinetic energy E_{wheel} of the wheel itself.

The second part is the energy dissipation caused by the lateral force generated by the cornering characteristics between the tire and the ground, ΔE_2, and the energy dissipation caused by the air resistance, ΔE_3. The lateral force of the tire and the air resistance act on the vehicle together, causing a change in the sum of the lateral translational kinetic energy and yaw kinetic energy of the body, E_{body}.

Based on the above analysis and energy conversion principle, there is:

$$W_b + \Delta E_1 + \Delta E_2 + \Delta E_3 = \left(E_{\text{wheel}_1} + E_{\text{body}_1}\right) - \left(E_{\text{wheel}_0} + E_{\text{body}_0}\right) \quad (16)$$

In Equation (16), w_b is the work done by the braking torque, E_{wheel0} is the rotational kinetic energy of the vehicle body at the initial simulation time, E_{wheel1} is the rotational kinetic energy of the vehicle body at the end of the simulation time, E_{body0} is the sum of the lateral translational kinetic energy and yaw kinetic energy of the vehicle body at the initial simulation time, and E_{body1} is the sum of the lateral translational kinetic energy and yaw kinetic energy of the vehicle body at the end of the simulation time.

The total dissipated energy ΔE of the 5DOF model can be calculated as:

$$\begin{aligned} \Delta E &= \Delta E_1 + \Delta E_2 + \Delta E_3 \\ &= \left(E_{\text{wheel}_1} + E_{\text{body}_1}\right) - \left(E_{\text{wheel}_0} + E_{\text{body}_0}\right) - W_b \end{aligned} \quad (17)$$

To facilitate the display of calculation results, the calculation of dissipated energy is taken as a positive value, and the calculation expression of dissipated energy is further refined as follows:

$$\begin{aligned} &\Delta E = W_b + \left(E_{\text{wheel}_0} + E_{\text{body}_0}\right) - \left(E_{\text{wheel}_1} + E_{\text{body}_1}\right) = \int_{t0}^{t1} (T_{bf} \cdot \omega_f + T_{br} \cdot \omega_r)dt \\ &+ (\tfrac{1}{2}m{v_{x0}}^2 + \tfrac{1}{2}m{v_{y0}}^2 + \tfrac{1}{2}{\omega_0}^2 {I_z}^2 + \tfrac{1}{2}J{\omega_{f0}}^2 + \tfrac{1}{2}J{\omega_{r0}}^2) - \\ &(\tfrac{1}{2}m{v_{x1}}^2 + \tfrac{1}{2}m{v_{y1}}^2 + \tfrac{1}{2}{\omega_1}^2 {I_z}^2 + \tfrac{1}{2}J{\omega_{f1}}^2 + \tfrac{1}{2}J{\omega_{r1}}^2) \end{aligned} \quad (18)$$

where ω_f and ω_r are the angular velocities of the front and rear wheels, v_{x0}, v_{y0}, ω_0, ω_{f0} and ω_{r0} are longitudinal velocity, lateral velocity, yaw rate, and front and rear wheel angular velocities at the initial simulation time, and v_{x1}, v_{y1}, ω_1, ω_{f1} and ω_{r1} are longitudinal velocity, lateral velocity, yaw rate, and front and rear wheel angular velocity at the end of the simulation.

5.2. Comparison and Verification

Figure 11 provides both the energy dissipation diagram and projection diagram of the system with the inertia force applied. The initial longitudinal velocities are v_x = 30 m/s and v_x = 50 m/s, respectively. The ranges of δ_f are −0.03 rad to 0.03 rad and −0.02 rad to 0.02 rad, with step sizes of 0.0005 rad. The range of T_b is 0 to 504 N·m, with step sizes of 10.08 N·m.

Figure 11a,c show that the energy dissipation results of the system at two initial velocities have positive and negative values, and the energy dissipation increases with the increase of velocity. The occurrence of negative values is due to the increase of energy caused by the positive work of the inertia force. However, it does not affect the hierarchical distribution trend of energy. The blue part in Figure 11 has a lower energy dissipation, resembling a trapezoid, and is the stable region for the steering and braking condition. The purple part has a high energy dissipation and smooth surface distribution trend, which represents the unstable region of the steering and braking condition.

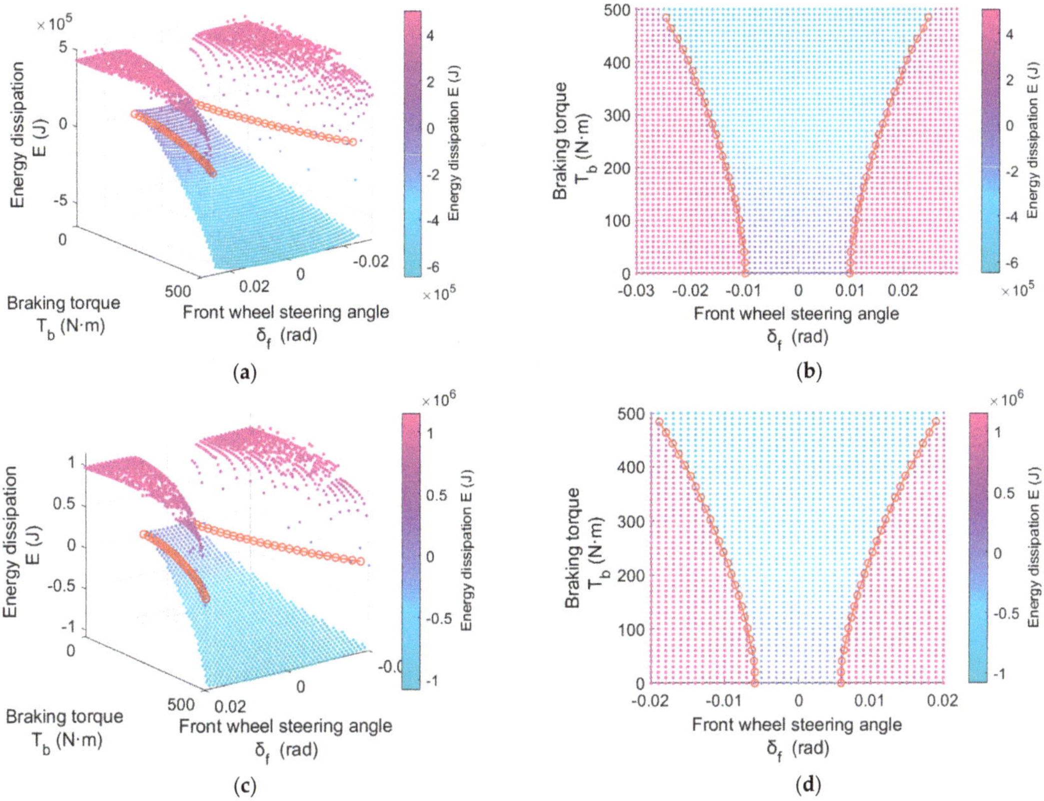

Figure 11. Steering and braking stability region with virtual force: (**a**) v_x = 30 m/s; (**b**) top view of (**a**); (**c**) v_x = 50 m/s; (**d**) top view of (**c**).

Figure 11b,d compare the boundary lines of the stability region obtained by the energy dissipation method (the boundary lines of the purple and blue parts) with the boundary lines obtained by the bifurcation method (the red circle line). It can be seen that both of them are almost identical. Furthermore, taking Figure 11d as an example, we can observe some quantitative phenomena. When limiting the front wheel steering angle or braking torque, we can easily see the range of stability and instability of the system. For example, when the braking torque is limited to 100 N·m, the area where the absolute value of the front wheel angle is greater than 0.006 rad is the instability region, and the area less than 0.006 rad is the stability region. When the front wheel steering angle is limited to 0.01 rad and the braking torque is greater than 231.84 N·m, it is considered a stable region, and, vice versa, it is considered an unstable region.

In order to study the stability region under steering and braking conditions using the classical bifurcation analysis method, the virtual force is applied for converting the original model into an equivalent system. However, due to the difference of dynamic characteristics between the equivalent system and the original system after the addition of the virtual force, the energy dissipation method is used to calculate the steering and braking stability region of the original system.

Figure 12 provides the schematic diagram and projection of the energy dissipation of the system without applying virtual force, which is also compared with the stability region obtained by the equilibrium point bifurcation method.

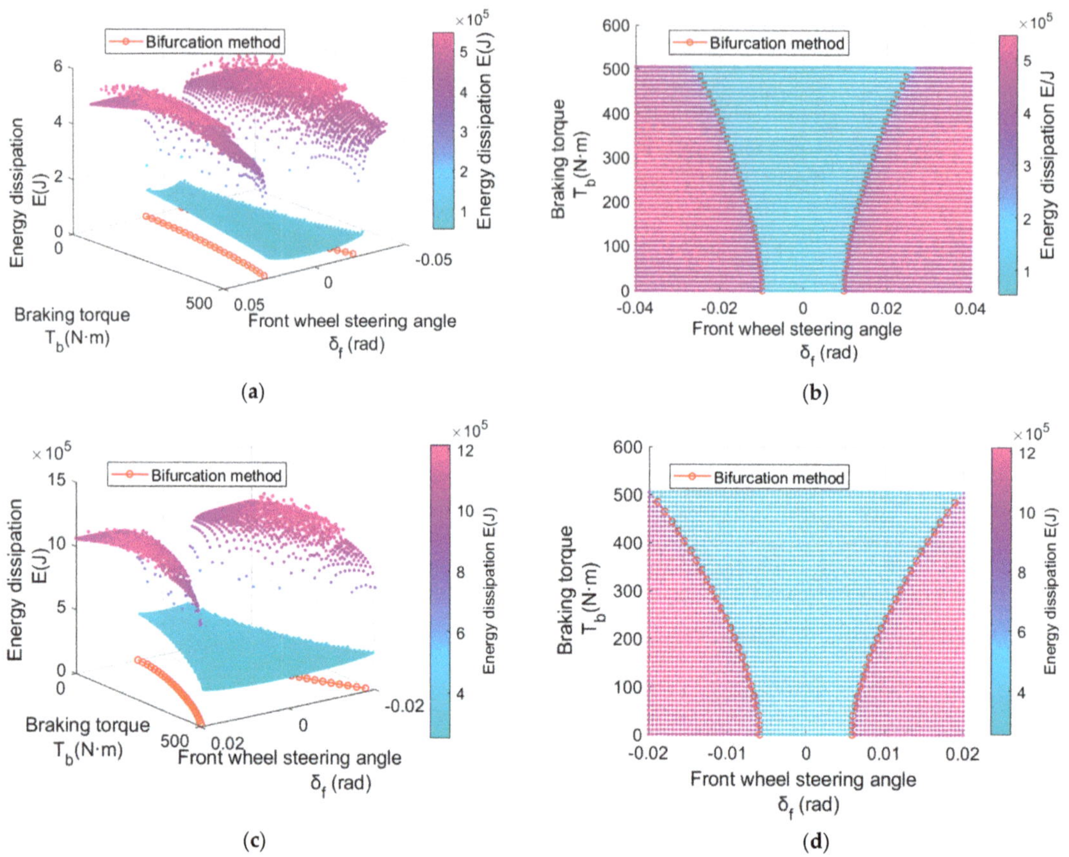

Figure 12. Steering and braking stability region without virtual force: (**a**) v_x = 30 m/s; (**b**) top view of (**a**); (**c**) v_x = 50 m/s; (**d**) top view of (**c**).

From Figure 12, it can be seen that there is no external input of virtual force on the system, and the energy dissipation is all positive, showing a clear layered distribution. The blue part has a lower energy dissipation, resembling a trapezoid, which is the stable region for the steering and braking condition. The purple part has a high energy dissipation and smooth surface distribution trend, which represents the unstable region of the steering and braking condition.

The boundary of the energy dissipation stable region can be seen in Figure 12 (the boundary lines of the purple energy region and the blue energy region are clearer than those of the system with added inertia force). At the same time, they are compared with the boundary obtained by the bifurcation method. It can be seen that when v_x = 30 m/s and v_x = 50 m/s, the results show that the stability regions of the 5DOF system with and without the application of virtual force almost remain the same. From Figure 12b,d, it can be seen that when the braking torque is limited to 500 N·m and the initial longitudinal velocity is 30 m/s, the absolute value of the front wheel steering angle at the boundary point is greater than 0.02 rad. When the initial longitudinal velocity increases to 50 m/s, the absolute value of the front wheel angle at the boundary point is less than 0.02 rad. This means that as the vehicle velocity increases, the stable region shrinks.

Compared with traditional equilibrium point bifurcation based stability region solving methods, the energy dissipation method based stability region verification method has better solving efficiency. The equilibrium point bifurcation method determines the bifurcation point through phase trajectory. It takes approximately 5.12 s to calculate one bifurcation point in MATLAB. Assuming that the solving speed of all bifurcation points is basically the same and taking Figure 12d as an example, there are a total of 50 bifurcation points in Figure 12d and the total solution time is approximately 241.5 s. In addition, it also takes a lot of data processing time. The time required to solve the stability region using the energy dissipation method is approximately 200.668 s. All the simulations were conducted on a hardware device with an Intel i5-10210U CPU.

6. Conclusions

In this paper, the stability region of the steering and braking condition in vehicle planar motion is solved by the equilibrium point bifurcation analysis method. In addition, it is verified using the energy dissipation method. The main work and conclusions are as follows:

(1) A 5DOF vehicle dynamic model has been established that can analyze the stability of vehicles under steering and braking conditions effectively. The application conditions of braking torque, the locking of front and rear wheels, and the limitation of road adhesion are considered in the model. The validation results indicate that this model can describe the dynamic characteristics of vehicle braking and steering conditions accurately. It can serve as the basic model for braking and steering stability analysis.

(2) The original system has been transformed into the equivalent system by using the D'Alembert principle. The equilibrium points of the equivalent system are solved by the hybrid algorithm combining the genetic algorithm and sequential quadratic programming method. The bifurcation diagram of the equilibrium point with the braking torque is calculated. The results indicate that as the braking torque increases, the number of equilibrium points increases to three from one, and the system state changes to stable from unstable. The bifurcation characteristics have been confirmed by the phase portrait.

(3) The process of solving the stability region under steering and braking conditions has been determined, and the two-dimensional bifurcation parameter set (front wheel steering angle and braking torque) has been obtained for different longitudinal velocities. The three-dimensional stability region (longitudinal velocity, front wheel steering angle and braking torque) was fitted and verified.

(4) Stable regions of the system with and without inertia force have been solved using the energy dissipation method, and compared with the stability region obtained by the equivalent static bifurcation method. The results indicate that the equilibrium point bifurcation method proposed in this paper for steering and braking conditions can effectively solve the stability region of the equivalent system. The stability region of the equivalent system is almost the same as that of the original system without applying virtual force. When the limited braking torque is 500 N·m and the initial longitudinal velocity increases from 30 m/s to 50 m/s, the range of the stable region significantly decreases, and the absolute value of the front wheel steering angle at the boundary point changes from less than 0.02 rad to more than 0.02 rad.

Due to the danger of high-speed steering and braking conditions, as well as the extensive experiments required to explore the unstable region of the vehicle system, it has been difficult to achieve actual vehicle experimental verification so far. In the future, we will verify and analyze the results of steering and braking stability regions with a driving simulator.

This paper mainly studies vehicle steering and braking stability in planar motion, without considering the impact of the dynamic characteristics of the pitch and roll. In future research, factors such as pitch and roll will be considered to explore the effects of tire load changes and different braking modes on the system equilibrium point bifurcation and stability region.

Author Contributions: Conceptualization, X.W. and W.L.; methodology, X.W.; software, W.L.; validation, F.Z., Z.L. and W.B.; writing—original draft preparation, W.L.; writing—review and editing, W.L. and X.W. All authors have read and agreed to the published version of the manuscript.

Funding: This research was funded by the Key R&D Plan of Heilongjiang Province, grant number JD22A014 and by National Natural Science Foundation of China, grant number 52175497.

Data Availability Statement: Not applicable.

Conflicts of Interest: The authors declare no conflict of interest.

Nomenclature

T_b	braking torque
ω	actual body yaw rate
ω_e	expected body yaw rate
m	vehicle mass
l_f, l_r	distance from the front and rear wheels to the center of mass
L	sum of the front and rear wheelbases
K	vehicle stability factor
k_f, k_r	cornering stiffness of the front and rear wheels
$\mathrm{B}_f, \mathrm{B}_r$	stiffness factor of front and rear wheels
$\mathrm{C}_f, \mathrm{C}_r$	shape factor of the front and rear wheels
$\mathrm{D}_f, \mathrm{D}_r$	peak factor of front and rear wheels
v_y, v_x	lateral and longitudinal velocity of vehicle
ω_f, ω_r	angular velocity of front and rear wheels
T_{bf}, T_{br}	braking torque of front and rear wheels
F_{lf}, F_{lr}	longitudinal tire force of front and rear wheels
F_{sf}, F_{sr}	lateral tire force of front and rear wheels
R_e	wheel rolling radius
J	moment of inertia of the wheel
I_z	moment of inertia of the vehicle around the Z axis
δ_f, δ_r	steering angle of front and rear wheels
$C_{\mathrm{air\_}x}, C_{\mathrm{air\_}y}$	longitudinal and lateral air drag coefficient
$A_{\mathrm{L\_}x}, A_{\mathrm{L\_}y}$	longitudinal and lateral windward area of the vehicle
F_z	the total load of the front and rear wheels of the vehicle
$F_{\mu f}, F_{\mu r}$	the braking force of the front and rear wheel brake
F_{zf}, F_{zr}	the front and rear wheel load
ρ	air density
μ	adhesion coefficient
η	the braking torque distribution coefficient
g	gravity
x	longitudinal slip rate or sideslip angle
$\mathrm{B, C, D, E}$	stiffness factor, shape factor, peak factor, curvature factor
α	tire sideslip angle
k	longitudinal slip
G_x, G_y	tire force combined slip correction parameters
$F_{lf0}, F_{lr0}, F_{sf0}, F_{sr0}$	longitudinal force and lateral force of front and rear tires in steady state
$r_{x,1}, r_{x,2}, r_{y,1}, r_{y,2}$	tire combined slip correction coefficients
ω_w	wheel rotation angular velocity
v_{wx}	longitudinal velocity at the wheel center in the tire coordinate system
α_f, α_r	sideslip angle of front and rear wheels
v_{xf}, v_{xr}	longitudinal velocity of the front and rear wheels in the tire coordinate system
v_{yf}, v_{yr}	lateral velocity of front and rear wheels in tire coordinate system
K_D	system parameter
v_{x1}, v_{y1}	longitudinal velocity, lateral velocity at the end of the simulation
$\omega_1, \omega_{f1}, \omega_{r1}$	yaw rate, and front and rear wheel angular velocity at the end of the simulation

v_{x0}, v_{y0}	longitudinal velocity, lateral velocity at the start of the simulation
ω_0,ω_{f0}, ω_{r0}	yaw rate, and front and rear wheel angular velocity at the start of the simulation
W_b	the work done by the braking torque
ΔE_1, ΔE_2, ΔE_3	the dissipation of wheel rotational energy caused by tire longitudinal slip, the energy dissipation caused by the lateral force generated by the cornering characteristics between the tire and the ground, the energy dissipation caused by air resistance
E_{wheel1}, E_{wheel0}	the rotational kinetic energy of the vehicle body at the end of the simulation time, the rotational kinetic energy of the vehicle body at the initial simulation time,
E_{body1}, E_{body0}	the sum of the lateral translational kinetic energy and yaw kinetic energy of the vehicle body at the end of the simulation time, the sum of the lateral translational kinetic energy and yaw kinetic energy of the vehicle body at the initial simulation time

References

1. Meng, F.; Shi, S.; Bai, M.; Zhang, B.; Li, Y.; Lin, N. Dissipation of energy analysis approach for vehicle plane motion stability. *Veh. Syst. Dyn.* **2021**, *60*, 4035–4058. [CrossRef]
2. Inagaki, S.; Kushiro, I.; Yamamoto, M. Analysis on vehicle stability in critical cornering using phase-plane method. *Jsae Rev.* **1995**, *2*, 216.
3. Ono, E.; Hosoe, S.; Tuan, H.D.; Doi, S.I. Bifurcation in vehicle dynamics and robust front wheel steering control. *IEEE Trans. Control Syst. Technol.* **1998**, *6*, 412–420. [CrossRef]
4. Liaw, D.-C.; Chiang, H.-H.; Lee, T.-T. Elucidating vehicle lateral dynamics using a bifurcation analysis. *IEEE Trans. Intell. Transp. Syst.* **2007**, *8*, 195–207. [CrossRef]
5. Shen, S.; Wang, J.; Shi, P.; Premier, G. Nonlinear dynamics and stability analysis of vehicle plane motions. *Veh. Syst. Dyn.* **2007**, *45*, 15–35. [CrossRef]
6. Sadri, S.; Wu, C.Q. Lateral stability analysis of on-road vehicles using Lyapunov's direct method. In Proceedings of the Intelligent Vehicles Symposium, Madrid, Spain, 3–7 June 2012.
7. Sadri, S.; Wu, C. Stability analysis of a nonlinear vehicle model in plane motion using the concept of Lyapunov exponents. *Veh. Syst. Dyn.* **2013**, *51*, 906–924. [CrossRef]
8. Della Rossa, F.; Mastinu, G. Straight ahead running of a nonlinear car and driver model—New nonlinear behaviours highlighted. *Veh. Syst. Dyn.* **2018**, *56*, 753–768. [CrossRef]
9. Mastinu, G.; Biggio, D.; Della Rossa, F.; Fainello, M. Straight running stability of automobiles: Experiments with a driving simulator. *Nonlinear Dyn.* **2020**, *99*, 2801–2818. [CrossRef]
10. Steindl, A.; Edelmann, J.; Plöchl, M. Limit cycles at oversteer vehicle. *Nonlinear Dyn.* **2020**, *99*, 313–321. [CrossRef]
11. Liu, L.; Shi, S.; Shen, S.; Chu, J. Vehicle Planar Motion Stability Study for Tyres Working in Extremely Nonlinear Region. *Chin. J. Mech. Eng.* **2010**, *23*, 185–194. [CrossRef]
12. Imani Masouleh, M.; Limebeer, D.J. Region of attraction analysis for nonlinear vehicle lateral dynamics using sum-of-squares programming. *Veh. Syst. Dyn.* **2018**, *56*, 1118–1138. [CrossRef]
13. Wang, X.; Shi, S.; Liu, L.; Jin, L. Analysis of driving mode effect on vehicle stability. *Int. J. Automot. Technol.* **2013**, *14*, 363–373. [CrossRef]
14. Shi, S.; Li, L.; Wang, X.; Liu, H.; Wang, Y. Analysis of the vehicle driving stability region based on the bifurcation of the driving torque and the steering angle. *Proc. Inst. Mech. Eng. Part D-J. Automob. Eng.* **2017**, *231*, 984–998. [CrossRef]
15. Shi, S.; Meng, F.; Bai, M.; Lin, N. The stability analysis using Lyapunov exponents for high-DOF nonlinear vehicle plane motion. *Proc. Inst. Mech. Eng. Part D-J. Automob. Eng.* **2022**, *236*, 1390–1400. [CrossRef]
16. Meng, F.; Shi, S.; Li, Y.; Zhang, B.; Yue, B.; Lin, N. Driving stability region solving method based on dissipation of energy. *Proc. Inst. Mech. Eng. Part D-J. Automob. Eng.* **2022**, *237*, 852–861. [CrossRef]
17. Olson, B.J.; Shaw, S.W.; Stépán, G. Stability and bifurcation of longitudinal vehicle braking. *Nonlinear Dyn.* **2005**, *40*, 339–365. [CrossRef]
18. Horiuchi, S.; Okada, K.; Nohtomi, S. Analysis of accelerating and braking stability using constrained bifurcation and continuation methods. *Veh. Syst. Dyn.* **2008**, *46*, 585–597. [CrossRef]
19. Chen, K.; Pei, X.; Ma, G.; Guo, X. Longitudinal/Lateral Stability Analysis of Vehicle Motion in the Nonlinear Region. *Math. Probl. Eng.* **2016**, *2016*, 3419108. [CrossRef]
20. Zhu, J.; Zhang, S.; Wang, G.; Zhang, W.; Zhang, S. Research on vehicle stability region under critical driving situations with static bifurcation theory. *Proc. Inst. Mech. Eng. Part D-J. Automob. Eng.* **2021**, *235*, 2072–2085. [CrossRef]
21. Lai, F.; Huang, C.; Ye, X. Analysis of Vehicle Driving Stability based on Longitudinal-lateral and Vertical Unified Dynamics Model. *Int. J. Automot. Technol.* **2022**, *23*, 73–87. [CrossRef]

22. Gürbüz, H.; Buyruk, S. Improvement of safe stopping distance and accident risk coefficient based on active driver sight field on real road conditions. *IET Intell. Transp. Syst.* **2019**, *13*, 1843–1850. [CrossRef]
23. Li, Z.; Chen, H.; Liu, H.; Wang, P.; Gong, X. Integrated Longitudinal and Lateral Vehicle Stability Control for Extreme Conditions With Safety Dynamic Requirements Analysis. *IEEE Trans. Intell. Transp. Syst.* **2022**, *23*, 19285–19298. [CrossRef]
24. Wang, Y.; Shi, S.; Gao, S.; Xu, Y.; Wang, P. Active Steering and Driving/Braking Coupled Control Based on Flatness Theory and A Novel Reference Calculation Method. *IEEE Access* **2019**, *7*, 180661–180670. [CrossRef]
25. Wang, J.; Luo, Z.; Wang, Y.; Yang, B.; Assadian, F. Coordination Control of Differential Drive Assist Steering and Vehicle Stability Control for Four-Wheel-Independent-Drive EV. *IEEE Trans. Veh. Technol.* **2018**, *67*, 11453–11467. [CrossRef]
26. Hucho, W.H. *Aerodynamics of Road Vehicles: From Fluid Mechanics to Vehicle Engineering*; Elsevier: Berlin/Heidelberg, Germany, 2013.
27. Yu, Z. *Automobile Theory*; China Machine Press: Beijing, China, 2009.
28. Wang, X.; Shi, S. Vehicle coupled bifurcation analysis of steering angle and driving torque. *Proc. Inst. Mech. Eng. Part D-J. Automob. Eng.* **2021**, *235*, 1864–1875. [CrossRef]
29. Pacejka, H.B.; Bakker, E. The magic formula tyre model. *Veh. Syst. Dyn.* **1992**, *21*, 1–18. [CrossRef]
30. Pacejka, H.B. *Tire and Vehicle Dynamics*; Elsevier: Amsterdam, The Netherlands, 2006.
31. Milani, S.; Marzbani, H.; Jazar, R.N. Vehicle drifting dynamics: Discovery of new equilibria. *Veh. Syst. Dyn.* **2022**, *60*, 1933–1958. [CrossRef]
32. Wang, X.; Shi, S.; Liu, L.; Jin, L. Vehicle Dynamics Equilibriums Solution Search Based on Hybridization of Genetic Algorithm and Quasi-Newton Method. *Chin. J. Mech. Engng.* **2014**, *50*, 120–127. [CrossRef]

applied sciences

MDPI

Article

Proportion-Based Analytical Hierarchy Process for Determining Prominent Reasons Causing Severe Crashes

Md Kamrul Islam [1,*] and Uneb Gazder [2,*]

1 Department of Civil and Environmental Engineering, College of Engineering, King Faisal University, Al-Ahsa 31982, Saudi Arabia
2 Department of Civil Engineering, University of Bahrain, Isa Town P.O. Box 32038, Bahrain
* Correspondence: maislam@kfu.edu.sa (M.K.I.); ugazder@uob.edu.bh (U.G.)

Abstract: Governments and authorities worldwide consider road traffic crashes (RTCs) to be a major concern. These crashes incur losses in terms of productivity, property, and life. For a country to establish its road and action plans, it is crucial to comprehend the reasons for and consequences of traffic collisions. The main objective of this research study was to evaluate and rank the important and supporting factors influencing traffic crashes on the road. To identify the most significant accident causation elements, the proportion-based analytic hierarchy process (PBAHP) was used to order the factors in terms of their relative importance. In this study, the city of Al-Ahsa, located in the eastern province of Saudi Arabia, was used as a case study, since this city is the highest RTC-prone area in the region. PBAHP was used to calculate relative importance/weights for different crash types and reasons in terms of their impact on crash severity. It was found that vehicle-overturned collisions which result in fatal crashes have the most weight, whereas "hit motorcycle" crashes result in serious injury crashes. When vehicles (two or more) collide with one another while they are moving, it appears that the likelihood of a fatality in a collision increases. The highest weights for serious injury crashes came from "driver distraction", "leaving insufficient safe distance", and "speeding", which also generated similar and relatively high weights for fatal crashes. Weights from the PBAHP approach were also used to develop utility functions for predicting the severity of crashes. This approach could assist decision-makers in concentrating on the key elements affecting road traffic crashes and enhancing road safety.

Keywords: road traffic crashes; PBAHP; crash reasons; crash types

Citation: Islam, M.K.; Gazder, U. Proportion-Based Analytical Hierarchy Process for Determining Prominent Reasons Causing Severe Crashes. *Appl. Sci.* **2023**, *13*, 7814. https://doi.org/10.3390/app13137814

Academic Editors: Edgar Sokolovskij and Vidas Žuraulis

Received: 15 May 2023
Revised: 28 June 2023
Accepted: 29 June 2023
Published: 3 July 2023

1. Introduction

Road traffic crashes (RTCs) cause enormous losses of life and property and are a major concern for public safety worldwide. Road crashes have significant and irreversible economic and social impacts on countries, making them a primary concern for transportation managers globally [1]. According to the World Health Organization (WHO), 1.35 million people die because of RTCs each year, and millions more suffer nonfatal injuries, impairments, and other long-term health effects. Very severe crashes can cause major bodily and monetary harm. Between 1975 and 1998, the number of individuals who were either killed or injured in RTCs increased significantly in Malaysia, Colombia, and Botswana. A study on RTCs in Iran has revealed that the number of people injured in these incidents is about ten times higher than the number of fatalities, with approximately 240,000 instances every year. Given such elevated rates, in addition to the substantial volume of passengers commuting, it is essential to conduct a comprehensive investigation into passenger safety. In Saudi Arabia, RTCs are a major public health problem since they result in large human casualties and material losses. With an average of 7000 RTC fatalities per year, the nation has one of the highest rates in the world [2]. A study that analyzed 404 crash reports from various types of RTCs found that the conditions of the road surroundings were a contributing factor in around 14.5% of all crashes. In 2007, Saudi Arabia saw 30 deaths per 100,000 individuals due to RTCs, resulting in 6358 deaths. According to a report by the WHO, RTCs are the main cause of fatality, injury, and disability in the Kingdom of Saudi Arabia, with a projected cost of SAR 652.5 million to treat those who are injured or killed. According to officials in Saudi Arabia, the country experiences one road traffic crash (RTC) every minute, making it one of the most perilous nations in the world for RTCs. With an average of 21 deaths/day, it is ranked as the second most lethal nation in the Gulf Region and the second deadliest country in the Middle East. These crashes have multiple contributing factors. A thorough investigation of several elements, such as road design, driver behavior, vehicle safety features, traffic volume, and weather conditions, is needed to identify the main causes and factors that result in serious crashes. It is crucial to understand the causes of these crashes in order to implement successful methods to reduce RTCs and enhance road safety in Saudi Arabia.

To determine the reasons behind RTCs in Saudi Arabia, several investigations have been carried out. According to a study by Ansari et al. (2014) [3], driving-related variables are the main reasons for traffic crashes in the nation. The study indicated that the most important driver-related variables causing RTCs in Saudi Arabia were speeding, distracted driving, and reckless driving. The survey also found that key causes of RTCs in the nation included aspects of the road, such as inadequate traffic signs and signals and bad road design. Research into the causes of young Saudi drivers' participation in RTCs was undertaken by Hassan et al. (2016) [4]. According to the survey, speeding and being late were to blame for 60% of unsafe driving behaviors and more than 70% of fines in Riyadh, Saudi Arabia. A study by Ramisetty-Mikler and Almakadma (2016) [5] on adolescent drivers in Riyadh, Saudi Arabia, revealed that 40% of drivers engage in "drifting" as a recreational activity with their vehicles, despite being aware of the risks involved. In contrast to less experienced and less educated drivers, Issa's (2016) [6] study indicated that educated and experienced drivers demonstrated a higher likelihood of being engaged in RTCs. Vehicle collisions, multiple-vehicle crashes, and pedestrian collisions are the most common types of crashes in the Al-Ahsa region of Saudi Arabia, according to research by Islam et al. (2022) [7]. Although they occur less frequently, RTCs had a high severity index in the study area, with 73% of severe RTCs involving people between the ages of 15 and 44. The study also discovered that people between the ages of 15 and 44 were more frequently involved in RTCs with a large number of fatalities and injured victims. According to Rahman et al.'s (2022) [8] research, there is a 26% chance of a collision occurring merely as a result of speeding, a rise of 63% from an earlier estimate. Furthermore, the likelihood of a collision rises from 26% to 33% when speeding and brake failure are factored in, more than tripling the initial odds of a collision. Islam et al. (2022) [9] used the random forest approach to evaluate the same research region and discovered that the two most significant factors impacting the severity of injuries in RTCs are the reason for the crash and the kind of collision. They also demonstrated that the research region had substantial spatial dependence, with clustered spatial patterns found within a distance threshold of 500 m, in addition to the findings on the factors determining the severity of injuries in RTCs. The analysis used by the authors, which employed Getis Ord Gi*, the crash severity index, and spatial autocorrelation analysis based on Moran's I, was successful in locating and ranking crash hotspots as well as evaluating the severity of the crashes.

The Saudi Arabian government has implemented various measures to enhance road safety across the country in recent years, including in the region of Al-Ahsa. These measures include increasing the presence of traffic police personnel, improving the road infrastructure, and intensifying awareness programs aimed at promoting safe driving practices. Additionally, the government has established the Saudi Traffic Safety Society, which plays a significant role in fostering road safety through educational campaigns and awareness initiatives. As a result of these concerted efforts, the number of road crashes in Al-Ahsa and Saudi Arabia as a whole has witnessed a decline in recent years [7]. However, it remains crucial for drivers to exercise caution, comply with traffic regulations, and prioritize the safety of themselves and other road users in order to prevent collisions and ensure a safe driving environment.

The study aims to improve our understanding of severe crashes and offer a more useful analytical framework for analyzing crash data by combining these efforts in the following two main points:

1. Providing insight into the types and causes of severe crashes in the Al-Ahsa region:

This study makes a significant addition by offering insightful information about the factors that contribute to serious crashes in the Saudi Arabian province of Al-Ahsa. The study intends to find patterns and trends related to severe crashes using data analysis, taking into account underlying causes such as the type of vehicle, crash types, road conditions, and driver behavior.

2. Employing the newly proposed proportion-based analytic hierarchy process (PBAHP) for crash data analysis:

The development and use of the proportion-based analytic hierarchy process (PBAHP) approach for crash data analysis is another significant contribution of this study. The study suggests a novel strategy that overcomes the methodological limitation associated with the conventional analytic hierarchy process (AHP) method, such as inconsistent pairwise comparisons and reliance on subjective judgments, by allowing the weighting and ranking of items based on their relative occurrence.

The newly developed PBAHP was applied to analyze the factors involved in traffic crashes and then to develop a utility function for predicting the severity of crashes. The implementation of this technique has not been found in the literature; instead, several variants of AHP were used for the data collected through surveys and interviews. In the proposed approach of PBAHP, the proportions of crashes of different types and those because of different reasons were used to calculate the ranking of these factors. Later, these rankings were summarized to calculate the weight/impact of each factor on the severity of crash, resulting in the formation of the utility function.

This study aimed to use these contributions to develop evidence-based road safety actions with the goal of lowering the frequency and severity of serious crashes in Al-Ahsa, which opens the door for improved road safety initiatives and policies in the study area, with the possibility of wider applications in comparable circumstances.

The subsequent sections of this paper are structured as follows. In Section 2, a comprehensive review of recent literature relevant to this research field is presented, where several variants of AHP methods are compared and the proposed PBAHP method is discussed. Section 3 provides detailed information about the data sources and their characteristics, as well as the steps taken for data preparation. The modeling methodology is outlined in Section 4, where the formulation of the proposed PBAHP is presented. Section 5 presents the results and discussion based on the analysis of the crash data. Finally, in Section 6, the key findings of this study are summarized, along with recommendations for authorities.

2. Literature Review

Many investigations have been made into the types and causes of auto crashes. Crash severity analysis, which focuses on identifying the elements that contribute to serious crashes, is one of the most discussed aspects in this area. The ability of conventional crash severity analysis methodologies to offer a thorough grasp of the intricate interactions between the different components that contribute to severe crashes is constrained. Experts employ a variety of techniques to pinpoint the main causes of catastrophic crashes, as well as the sorts of crashes that cause the most severe effects. These approaches incorporate statistical analysis, professional judgment, and machine learning strategies. RTC analysis and prevention using cutting-edge methods have become urgently necessary. Multicriteria decision-making (MCDM) tools have been used in several research works to examine the causes of traffic crashes [10–16].

The analytical hierarchy process (AHP) is a multicriteria decision-making tool that enables decision-makers to specify pertinent safety criteria in a methodical manner. Based on the relative relevance of factors such as crash severity, frequency, road conditions, driving habits, and vehicle attributes in connection to the goals of traffic safety, AHP makes it easier to compare and rank them. For instance, Cheng et al. (2011) [17] used the AHP model in a modeling study to investigate hidden safety issues and determine the root causes of traffic accidents in China. The study's goal was to offer insights into the state of road traffic safety and to suggest preventive measures based on the analysis. Despite the possibility of certain subjective elements being present in the AHP model employed in the study and the possibility that the expert grading may not have been entirely thorough, they ultimately obtained objective materiality levels of road crash causes in China. For the case of Manila, Philippines, Fernandeza et al. (2020) [18] examined how road users' prioritization of the factors that cause traffic crashes is affected by their understanding of traffic signs, and found that a thorough understanding of traffic signs led to a more precise and trustworthy ranking of the causes of traffic crashes. They compared AHP with the forced ranking method in terms of a pairwise comparison of crash factors and found that the AHP performed better in representing the importance of factors with in-depth details. Saifullizan et al. (2022) [19] used the AHP method to rank and weigh car crashes according to the different categories of injury using data from Balai Traffic Batu Pahat Johor, Malaysia, from October 2016 to October 2020. They prioritized wreckage injuries at the top of the hierarchy system, followed by fatal injuries, minor injuries, and serious injuries. They reported that the suitability of the model and technology were key factors in the successful application of AHP, and recommended enhancing the application of AHP, including via parameter determination, reaching consensus, and framework modification, in order to achieve progress.

Based on the conventional AHP method, an Improved Fuzzy-AHP method was suggested by Hu et al. (2009) [20] for evaluating the safety of road traffic. The importance of each assessment index was determined using the scale technique after providing a summary of the important road traffic safety indices. Utilizing a fuzzy consistent matrix, queuing each index according to relevance, and adopting an accident prevention mechanism in accordance with the index, the weight of each index could be confirmed. In order to compare and quantify the driver behavioral attitudes in different traffic cultures, such as in Hungary, Turkey, Pakistan, and China, Farooq et al. (2020) [21] devised the fuzzy-AHP (FAHP) framework built on a three-level hierarchical structure. Based on pairwise comparisons (PCs) of drivers' responses to the driver behavior questionnaire (DBQ), the FAHP process calculated the weight factors and rated the significant driver behavior criteria. According to their study's findings, "violations" were the most important driver behavior standard for level 1 in all nominated regions, with the exception of Hungary. All regions, with the exception of Turkey, observed "aggressive violations" to be the most important driver behavior criterion for level 2. Additionally, level 3 drivers in Hungary and Turkey rated "driving while intoxicated" as the most important driver behavior criterion. Farooq and Moslem (2022) [22] applied the Pythagorean fuzzy analytic hierarchy process (PF-AHP), an expanded method of ordinary fuzzy sets, to evaluate and rank important driver behavior factors built into a hierarchical model using information acquired from drivers' groups observed in Budapest, Hungary. They claimed that the PF-AHP approach was a helpful method to evaluate driver behavior uncertainty while handling complex road safety issues.

Khademi and Choupani (2018) [23] used the analytic network process (ANP) to analyze the inter-organizational complex system's behavior to identify major shortcomings in the lead agency of a country for road safety. They applied their developed ANP model on the Road Safety Commission (RSC) of Iran to identify the synthesized influence of factors on each other and suggested a complementary process to specify institutional improvements to prevent any organizational inefficiencies. Yang et al. (2018) [24] devised an analytic network process (ANP) along with statistical methods to evaluate the overall degree of highway safety for Chinese roadways. They analyzed quantitative and qualitative indices (variables), along with factors that influence safety, such as collisions, intersections, alignments, and other important factors. They reported that ANP can take these elements into account, be used to elicit judgments, and generate ratio scales for use in basic arithmetic operations.

Mirmohammadi et al. (2013) [25] studied the prioritization of various elements that can have an impact on the rate and severity of crash issues. They used 15 different safety indicators with a case study in Iran. They applied three different "multiple criteria decision making" (MCDM) methods, viz., analytical hierarchy process (AHP), technique for order of preference by similarity to ideal solution (TOPSIS), and simple additive weighting method (SAW), and reported that AHP performed better compared to others. Farooq et al. (2021) [26] devised a combination of the best–worst method (BWM) and the analytical hierarchy process (AHP) method to compare and quantify the factors for frequent lane changing behavior in relation to road safety. The driver's replies to a predetermined questionnaire survey were used to prioritize the most important parameters impacting lane changing in order to demonstrate the applicability of the suggested model. They claimed that compared to the traditional AHP, their combined model saw fewer pairwise comparisons (PCs) and more accurate and consistent findings. Table 1 summarizes references for AHP methods reviewed in this paper.

Table 1. Comparative summary of references for AHP methods used in traffic safety analysis.

References	AHP Methods	City/Country	Comments
Cheng et al. (2011) [17]	Traditional AHP	China	Some subjective elements in the AHP model were employed; however, they ultimately obtained objective materiality levels of road crashes causes
Fernandeza et al. (2020) [18]	AHP and forced ranking method	Manila, Philippines	Compared AHP with the forced ranking method and reported that the AHP performed better
Saifullizan et al. (2022) [19]	Traditional AHP	Malaysia	Adequacy of the model and technology were key factors in successful application the AHP
Hu et al. (2009) [20]	Improved fuzzy-AHP	China	Assessment index was determined using the scale technique after a summary of the important road traffic safety indices
Farooq et al. (2020) [21]	Fuzzy-AHP (FAHP)	Hungary, Turkey, Pakistan, and China	The FAHP process calculated the weight factors and rated the significant driver behavior criteria built on a three-level hierarchical structure.
Farooq and Moslem (2022) [22]	Pythagorean fuzzy analytic hierarchy process (PF-AHP)	Budapest, Hungary	Evaluated and ranked important driver behavior factors built into a hierarchical model using information acquired from drivers' groups observed
Khademi and Choupani (2018) [23]	Analytic network process (ANP)	Iran	Identified and synthesized the influence of factors on each other and suggested a complementary process to specify institutional improvements to prevent any organizational inefficiencies
Yang et al. (2018) [24]	Analytic network process (ANP)	China	Analyzed quantitative and qualitative indices (variables) along with factors that influence safety such as collisions, intersections, alignments, and other important factors

Table 1. *Cont.*

References	AHP Methods	City/Country	Comments
Mirmohammadi et al. (2013) [25]	AHP, TOPSIS, and SAW	Iran	Applied AHP, TOPSIS, and SAW methods and reported that AHP performed better compared to others
Farooq et al. (2021) [26]	Best–worst method (BWM) and the analytical hierarchy process (AHP)	Budapest, Hungary	Compared to the traditional AHP, their combined model saw fewer pairwise comparisons (PCs), resulting in more accurate and consistent findings

In Table 2, several variants of the AHP method used in the literature are compared with the proposed PBAHP. Traditional AHP is one of the methods more frequently used in traffic crash analysis to methodically assess the seriousness and significance of factors involved in traffic crashes. It uses pairwise comparisons and determines the relative importance of various variables, including the state of the roads, driving habits, vehicle features, and environmental factors [17–19]. The traditional AHP methods, however, do not consider the actual occurrence or practical impact of each criterion when making a choice, which can produce biased results [27]. FAHP expands AHP by introducing fuzzy logic and helps decision makers deal with the language evaluations and inherent uncertainties that come with traffic crash investigation [20,21]. The FAHP enables a more flexible and thorough assessment of crash variables by utilizing fuzzy numbers and logic to better capture the subjective assessments. Another extension is P-FAHP, which uses Pythagorean Fuzzy subsets, considering the degree of membership and nonmembership of factors [22]. The ANP method adopts a more comprehensive approach by taking into account both the interdependencies and the hierarchy of crash reasons. ANP captures the interconnectedness of crash causes, offering a comprehensive view of crash analysis and facilitating a more in-depth comprehension of system dynamics. Despite the advantages of the AHP's multi-criteria decision-making technique, some limitations are frequently present [28]. In this present study, proportion-based or percentage-based AHP (PBAHP) is proposed, which is a different approach that considers the relative importance of each criterion in the decision-making process and has been offered as a solution to this problem. PBAHP can be promising as a method for identifying the main causes and types of major crashes. More details about this technique are provided in the proceeding sections of this paper.

Table 2. Comparison of variants of AHP with the proposed PBAHP method.

Methods	References	Description	Key Features	Comments
Analytical hierarchy process	Cheng et al. (2011) [17], Fernandeza et al. (2020) [18], Saifullizan et al. (2022) [19]	Use pairwise comparisons to prioritize and rank alternatives	Subjective judgments, consistency checks	May suffer from inconsistency in pairwise comparisons Relies on subjective judgments
Fuzzy analytical hierarchy process	Hu et al. (2009) [20], Farooq et al. (2020) [21]	Extension of AHP that incorporates fuzzy logic to handle uncertainties and vagueness	Fuzzy scale on component weight, fuzzy set theory	Requires expertise in fuzzy logic and fuzzy set theory Complex mathematical calculations

Table 2. *Cont.*

Methods	References	Description	Key Features	Comments
Analytic network process	Khademi and Choupani (2018) [23], Yang et al. (2018) [24]	Extension of AHP that models complex decision structures with interdependencies and feedback loops	Incorporates dependence and feedback relationships, super matrix representation	Requires expert knowledge in structuring and modeling the decision problem More complex mathematical calculations Increased computational complexity
Pythagorean fuzzy analytic hierarchy process (PF-AHP)	Farooq and Moslem (2022) [22]	Extension of AHP that incorporates the concept of Pythagorean fuzzy sets and combines fuzzy logic and the AHP	Pythagorean fuzzy subset	Requires expertise in fuzzy logic and fuzzy set theory Complex mathematical calculations Difficulty in obtaining precise linguistic terms
Proportion-based analytical hierarchy process	Current study	Proposed method that uses proportional comparison for pairwise comparisons	Proportional judgments, addresses ratio bias	Addresses ratio bias in AHP Provides more accurate pairwise comparisons Reduces subjectivity in judgments

Overview of the Analytical Hierarchy Process (AHP) and the Proportion-Based Analytical Hierarchy Process (PBAHP)

AHP is a multicriteria decision-making tool which has been used in a wide range of applications. These applications include social and political sciences, as well as engineering processes [29]. The literature shows that AHP was first proposed by T. Saaty as early as 1972. Since then, the process has adopted many developments, enabling its greater use in different fields. These include the establishment of ranking scales, inconsistency limits and indices, and the development of easy-to-use exclusive software packages for AHP [30].

AHP is a measurement tool which works on pairwise comparisons based on fixed scales. The measuring scale can use discrete ratings of actual measurement of user preferences [31]. The problem is divided into two levels; the first level is the "criteria", and the second level is called the "goal". Criteria refers to the factors that affect the choice or the outcome, while goal is the choice or outcome of the process. There are six steps to formulating AHP methods [32–34]:

First step: problem identification and set up of evaluation standards;
Second step: setting up the AHP hierarchy;
Third step: creating matrices for pairwise comparison of the criteria;
Fourth step: comparison matrices' normalizing;
Fifth step: calculation of the priority vectors;
Sixth step: calculation of consistency ratio (CR).

Following the abovementioned steps for the pairwise comparison of the criteria, their rankings from each comparison are averaged out and normalized to obtain the relative weight of the criteria, as per Equation (1).

$$W = [(\prod_{i=1,n} R_i)^{1/n}]/[\sum_{i=1,n} (\prod_{i=1,n} R_i)^{1/n}] \quad (1)$$

where

R = ranking for each pair of factors;
n = number of pairs of factors.

The pairwise comparison was performed for each goal based on one criterion at a time. The rankings from multiple comparisons of goals were combined, similar to criteria, to obtain their weights. The resulting weight of each goal is the dot product of its weight, calculated for each criterion, and the weight of the specific criterion.

As the weights (as explained above) are calculated using multiple independent comparisons, there is a chance of inconsistency in the rankings. To consider that issue, the consistency ratio (CR) and consistency index (CI) are calculated for the matrix resulting from multiple comparison at each stage. CR is a measure of variation within the data, derived from the eigen vector of the ranking matrix, while CI is a relative ratio for the CR with the standard randomness index given by Saaty (2005) [35]. These values can be determined using Equations (2) and (3).

$$CR = (\lambda - n)/(n - 1) \tag{2}$$

$$CI = CR/RI \tag{3}$$

where

λ = maximum value of eigen vector for the ranking matrix;
n = size of the matrix;
RI = random consistency index, given by [35].

The rankings of criteria and goals are normally taken using the perception or opinion of respondents to a survey [36].

In AHP, a complicated problem requiring decision making is divided into smaller, more manageable components, and each component is assessed in relation to the others. According to their relative importance, the criteria and options are compared pairwise in the conventional AHP.

This research proposes a new proportion-based approach for calculating ratings for each event and crash severity. It is a newly developed variant of AHP that evaluates the components using percentage scores of the requirements that each alternative meets, rather than using ratio scales. PBAHP's capability to consider the comparative relevance/occurrence of each element causing traffic crashes is one of its primary features. PBAHP assesses each factor's significance using percentage scores, enabling a more precise evaluation of the elements that contribute to serious collisions. Furthermore, PBAHP is a straightforward approach that is simple to comprehend and use. It is expected that the current methodology can be adopted in a variety of fields for the application of AHP, wherein the available data are extracted from an experiment or field. To the best of the authors' knowledge, proportion-based ratings have not been used for AHP until now.

3. Study Area, Data Sources, and Data Description

Al-Ahsa is located in the Eastern Province of Saudi Arabia and is bordered by the Persian Gulf to the east. The area extends from 29°20′ N in Kuwait to the southernmost point of the Gulf of Bahrain at 25°10′ N (Figure 1). The area is well known for its agricultural output, historic buildings, and palm trees. It is one of Saudi Arabia's largest regions, and has a vibrant culture and history. The area is also home to several historical attractions, such as the Qara Mountain Caves, which are thought to be more than 7000 years old and are home to prehistoric rock art. UNESCO has recently listed it as a heritage site in Saudi Arabia. The Al-Ahsa region was selected for this study since previous research has shown a high rate of traffic crashes in Al-Ahsa, with 31.9% of all crashes in the Eastern Province occurring in this city between 2009 and 2016. Dammam, Hafr Al-Batin, Qatif, Jubail, Dhahran, and Khobar are the next cities in terms of the number of crashes, with crash rates below 5% in other cities (Figure 2) [37]. These findings indicate that Al-Ahsa is particularly susceptible to traffic crashes in the Eastern Province. Although Al-Ahsa possesses significant potential to thrive as an international tourist attraction because of its recognition as a UNESCO-listed heritage site and a Guinness World Record, its high crash rate may put this at risk, necessitating immediate action. During the period of October 2014 to May 2018, the Road Safety Authority in Dammam provided data on traffic crashes in Al-Ahsa. The data revealed that around 8% of fatal and 30% of serious injury incidents were due to crashes between cars. Fatal crashes refer to those which involve at least one death, while serious injury crashes refer to those crashes in which at least one of the victims is hospitalized, without involving any deaths. The data provided by the Road Safety Authority also included the classification of crashes on the bases mentioned above. Another significant type of crash was vehicle overturning, which contributed to 6.5% of fatal and injury events. Pedestrians were also shown to be vulnerable, as they were involved in 12.5% of injury incidents and 2% of fatal crashes. Although some types of crashes, such as crashes with road guardrails, parked vehicles, motorcyclists, and permanent objects, have low fatality rates, they still pose a risk of harm, as illustrated in Figure 3. Therefore, further analysis and actions are needed to address the issue of traffic crashes in Al-Ahsa.

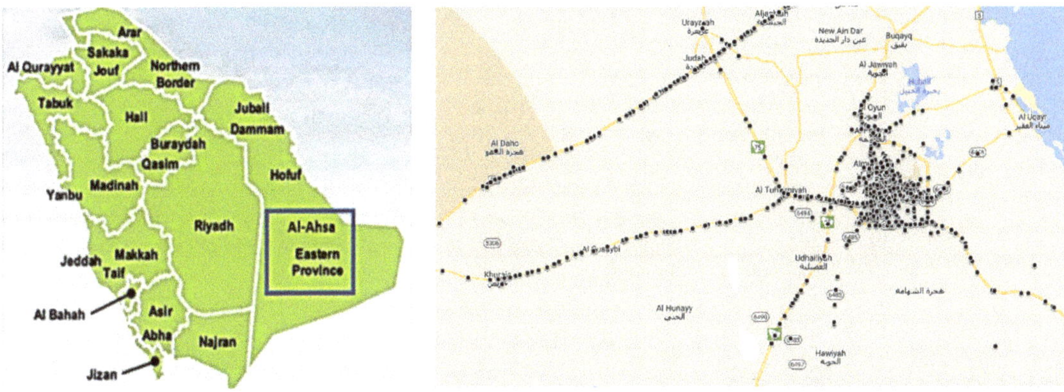

(a) (b)

Figure 1. (**a**) Location of the study area (Al-Ahsa) in Saudi Arabia; (**b**) distribution of traffic crashes in the study area (adapted from [8]).

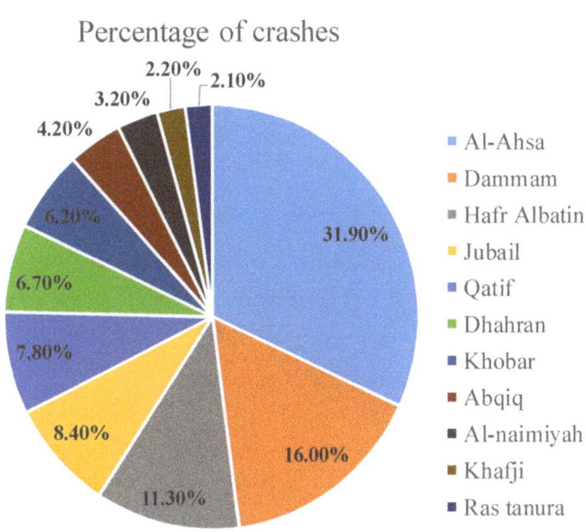

Figure 2. The percentage of road crashes in different cities of the eastern region of Saudi Arabia (adapted from [8]).

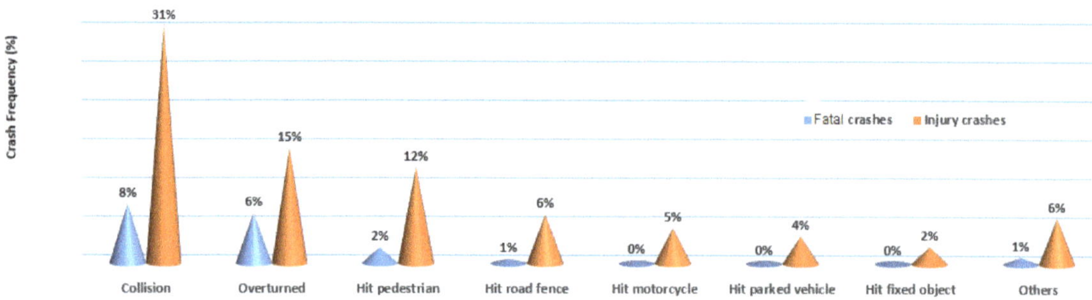

Figure 3. Percentage of crash types in the study area (adapted from [8]).

Data Description

Table 3 shows the number of crashes reported for each crash type, and the reason. Crash type refers to the type of impact which the vehicle had during the crash, while crash reason refers to the cause of the crash leading up to it. The first step was to merge the types and reasons that had insignificant reported crashes into a category of "others". The limit of significance was set as 5%, and therefore all those crash reasons and types with less than 5% crashes were merged in the above category. This limit was set as per the normal practice of statisticians in which the tests for significance were set at the 5% confidence level [38]. The crash types and reasons which had a proportion above 5% were considered individually, and they are highlighted in Table 3, while the rest were merged into the "others" category for each case. They are also shown in Figures 4 and 5.

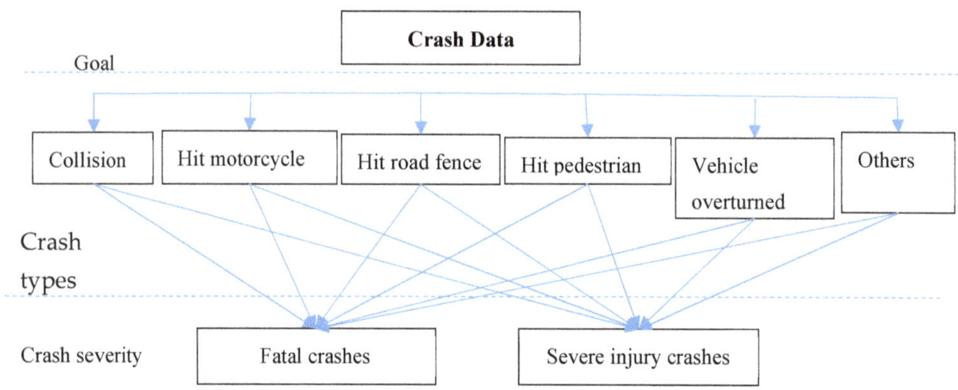

Figure 4. Hierarchy structure for crash types.

Table 3. Distribution of crash types and reasons in the dataset.

Crash Types	Number of Crashes	Frequency of Crashes	Crash Reasons	Number of Crashes	Frequency of Crashes
Collision	1542	38.54%	Sudden lane changing	1918	47.91%
Vehicle overturned	862	21.54%	Speeding	725	18.11%
Hit pedestrian	582	14.55%	Not giving way	615	15.36%
Hit road fence	278	6.95%	Insufficient safe distance	256	6.40%
Hit motorcycle	201	5.02%	Driver distraction	230	5.75%
Hit parked vehicle	154	3.85%	Crossing without pedestrian crossing	39	0.97%
Hit electric post	100	2.50%	Illegal overtaking	22	0.55%
Hit animal	61	1.52%	Red light violation	13	0.32%
Hit bicycle	53	1.32%	Driving opposite to traffic	10	0.25%
Hit roadside barrier	52	1.30%	Not stopping at stop sign	7	0.17%

Table 3. *Cont.*

Crash Types	Number of Crashes	Frequency of Crashes	Crash Reasons	Number of Crashes	Frequency of Crashes
Undefined category	37	0.92%	Drifting	5	0.12%
Hit fixed object	34	0.85%	Falling asleep	2	0.05%
Hit tree	25	0.62%	Getting out of moving vehicle	2	0.05%
Fell off the slope	7	0.17%	Hanging on the outside of vehicle	2	0.05%
Hit plate	4	0.10%	Unsafe road works	2	0.05%
Fell off bridge	4	0.10%	Exhaustion	1	0.02%
Fire on vehicle	3	0.07%	Violating pedestrian sign	1	0.02%
Hit signal	2	0.05%	Downhill	1	0.02%
Hit waste container	0	0.00%	No warning signs	1	0.02%

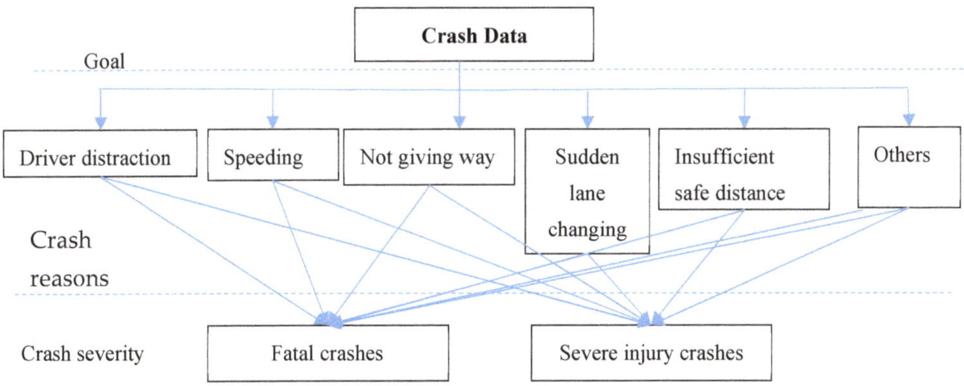

Figure 5. Hierarchy structure for crash reasons.

4. Modeling Methodology

The first step of the AHP is to structure the problem with respect to its criteria and goals. For this research, the criteria were the crash types and reasons, while the goal was the crash severity. The structure can be observed from Figures 4 and 5; the process was applied exclusively for the crash type and then for crash reason. The ranking for each criterion and goal were calculated, as shown by Equations (4) and (5), while the total weight for each crash severity was calculated as per Equation (6).

As stated earlier, this research employed the PBAHP method for the analysis. The difference between this method and traditional AHP is the use of proportions for calculating the ranking of the factors (crash types and reasons, in this case), as shown in Equations (4) and (5). The traditional AHP method uses a survey or interview and asks the respondents to carry out a pairwise comparison [39]. To achieve this, the respondent hypothetically generates the ratio of importance of each factor while answering the comparison questions. The same approach can be applied to quantitative data, such as crash data, by applying a method to evaluate the relative importance of one event over the other, which can be used in place of the ratings given by respondents in the traditional AHP method. The present study used a simple and efficient method to evaluate the relative importance of each crash event (type or reason) using the proportion of its occurrence in the crash data. This method makes use of the available data without the risk of receiving biased responses and does not require complicated or lengthy preprocessing for using the data. In the present research, these rankings were used to determine the relative weights (based upon their occurrence) for the crash severity. Higher rankings/ratios indicated a higher proportion of certain crash types or reasons in comparison to the other, which shows their comparatively higher importance or significance for crash severity. For example, if the ratio is 3 between crash type A and B, then this indicates that type A has 3 times more importance (weightage) than type B, and vice versa, in the available dataset. This is further explained in Section 5.

$$Rij = Pi/Pj \quad (4)$$

where

Rij = Ranking of reason/type 'i' in comparison to "j";
Pi = Proportion of crashes by reason/type "i";
Pj = Proportion of crashes by reason/type "j".

$$Rxyi = Pxi/Pyi \quad (5)$$

where

Rxyi = Ranking of crash severity "x" over "y" based on crash type/reason "i";
Pxi = Proportion of crashes belonging to severity "x" out of those caused by crash type/reason "i";
Pyi = Proportion of crashes belonging to severity "y" out of those caused by crash type/reason "i".

The total weight for each crash severity was the weighted sum of probability of each crash severity based on the contributing factor, as shown in Equation (6). This was used to develop the utility function for the crash severities.

$$TW\ (X) = \sum_{i=1}^{6} W_{xi} W_i \quad (6)$$

where

TW = total weight for crash severity "X";
Wxi = weight of severity "X" with crash type/reason "i" Table 4 or Table 5;
Wi = weight of crash type/reason "i", from Table 5 or Table 6.

Table 4. Ranking and relative weights for crash types.

Crash Types	Collision	Hit Motorcycle	Hit Road Fence	Hit Pedestrian	Vehicle Overturned	Others	Relative Weights
Collision	1.00	7.77	5.54	2.65	1.79	2.88	1.00
Hit motorcycle	0.13	1.00	0.72	0.34	0.23	0.37	0.13
Hit road fence	0.18	1.39	1.00	0.48	0.32	0.52	0.18
Hit pedestrian	0.38	2.94	2.08	1.00	0.68	1.08	0.38
Vehicle overturned	0.56	4.35	3.13	1.47	1.00	1.61	0.56
Others	1.00	7.77	5.54	2.65	1.79	2.88	1.00
CI							0.00

Table 5. Ranking and relative weights for crash reasons.

Crash Reasons	Driver Distraction	Speeding	Not Giving Way	Sudden Turning	Insufficient Safe Distance	Others	Relative Weights
Driver distraction	1.00	0.32	0.37	0.12	0.90	0.89	0.06
Speeding	3.13	1.00	1.18	0.38	2.83	2.80	0.18
Not giving way	2.70	0.85	1.00	0.32	2.40	2.37	0.15
Sudden lane changing	8.33	2.63	3.13	1.00	7.49	7.40	0.48
Insufficient safe distance	1.11	0.35	0.42	0.13	1.00	0.99	0.06
Others	1.12	0.36	0.42	0.14	1.01	1.00	0.06
CI							0.00

Table 6. Ranking and relative weights for the severity of crashes according to crash type.

Crash Severity	Fatal	Serious Injury	Weights
Collision			
Fatal	1	0.25	0.33
Serious injury	4	1	0.67
Hit motorcycle			
Fatal	1	0.09	0.23
Serious injury	11.11	1	0.77
Hit road fence			
Fatal	1	0.1	0.24
Serious injury	10	1	0.76
Hit pedestrian			
Fatal	1	0.16	0.28
Serious injury	6.25	1	0.71
Vehicle Overturned			
Fatal	1	0.43	0.4
Serious injury	2.33	1	0.6
Others			
Fatal	1	0.14	0.27
Serious injury	7.14	1	0.73

5. Results and Discussions

Figure 6 shows the distribution of crashes based on type, while Figure 7 shows the same on the basis of crash reasons. Using Equation (4), rankings were calculated for each crash type and reason, and are presented in Tables 4 and 5. These tables also show the weights of each criterion and the CI for each matrix, calculated as per Equations (2) and (3).

Table 4 shows that "collision" was the highest-contributing type of crash to fatal and serious injury crashes in the study, followed by "vehicle overturned" and "hit pedestrian". Figure 6 also points to the same fact.

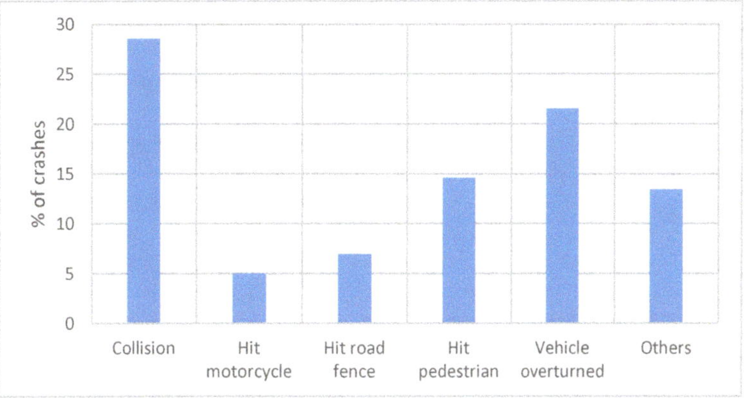

Figure 6. Distribution of crash types.

Inconsistency was "0", as the rankings were derived from the ratios of the crash proportions. This parameter can be an issue when rankings are made by the respondents based on their judgements [40].

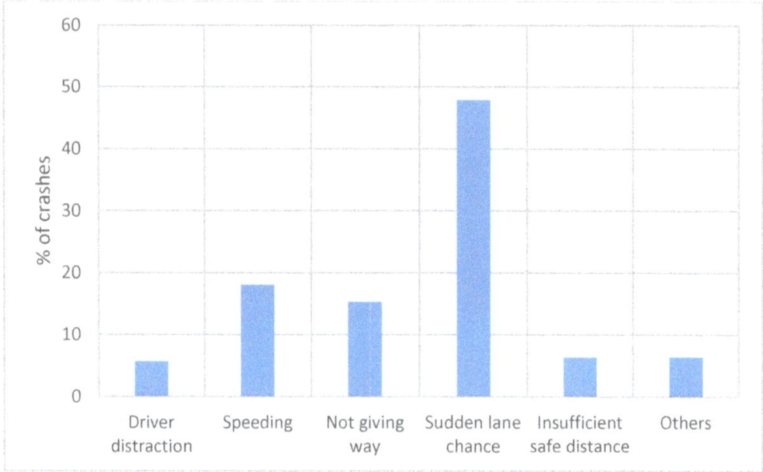

Figure 7. Distribution of crash reasons.

Table 5 shows that "sudden lane changing" was the highest contributing reason for fatal and serious injury crashes, followed by the vehicle overturning. This finding was also corroborated by the fact that the most common type of crash was the "collision", which is bound to happen during the sudden turning of a vehicle. Other major contributing reasons to fatal and serious injury crashes were speeding and not giving way. The trends are also evident in Figure 7.

Inconsistency for these weights was also found to be 0, for the reasons explained above for the crash types.

Figures 8 and 9 show the distribution of crash severities based on crash types and reasons, respectively. The rankings of each severity on the basis of each criterion were calculated as per Equation (5). These rankings are shown in Tables 6 and 7. The weights for each crash severity based on each criterion are also shown in these tables.

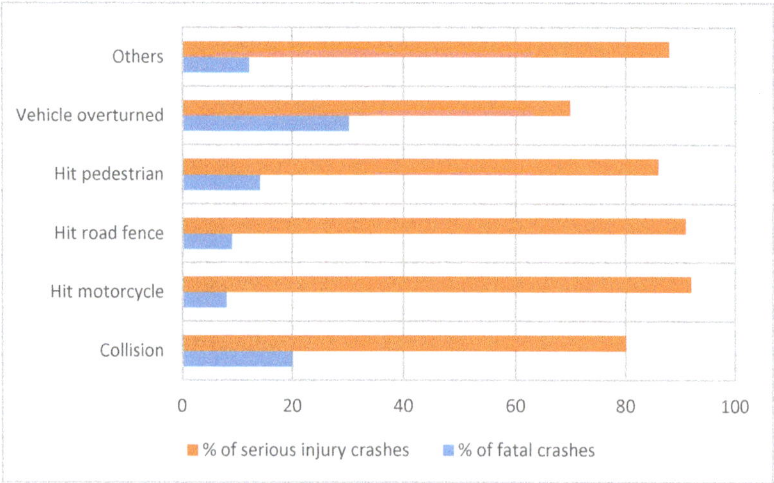

Figure 8. Distribution of fatal and serious injury crashes according to crash types.

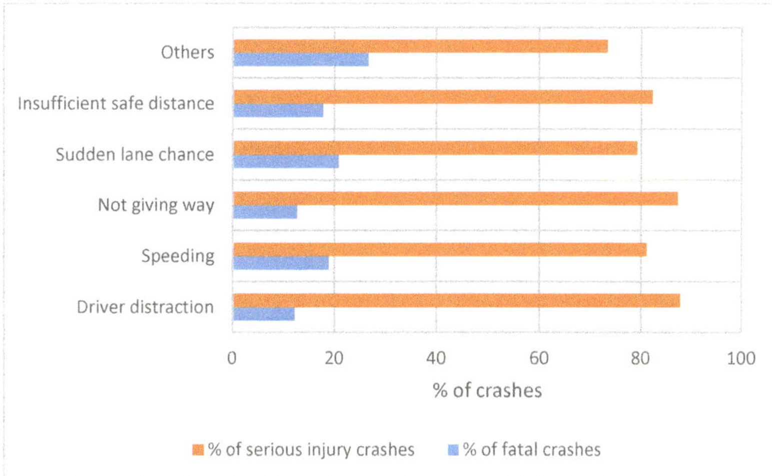

Figure 9. Distribution of fatal and serious injury crashes according to crash reasons.

Table 7. Ranking and relative weights for the severity of crashes according to crash reasons.

Crash Severity	Fatal	Serious Injury	Weights
Driver distraction			
Fatal	1	0.14	0.27
Serious injury	7.14	1	0.73
Speeding			
Fatal	1	0.23	0.32
Serious injury	4.35	1	0.68
Not giving way			
Fatal	1	0.15	0.28
Serious injury	6.67	1	0.72
Sudden lane changing			
Fatal	1	0.26	0.34
Serious injury	3.85	1	0.66
Insufficient safe distance			
Fatal	1	0.22	0.32
Serious injury	4.54	1	0.68
Others			
Fatal	1	0.36	0.38
Serious injury	2.78	1	0.62

The inconsistency check was not required since there were only two outcomes to be compared. The chances of inconsistency are produced through multiple pairwise comparisons. This is the reason that the inconsistency ratios (IRs) for 1 and 2 dimensional matrices are given as "0" [35].

Fatal crashes had the highest weight among "vehicle overturned" crashes, while the highest for serious injury crashes was when "hit motorcycle" occurred. The crash type "collision" also had a relatively high weight for fatal crashes. It should be noted that a higher weight for a certain severity level reduces the weight for the others, as the total of the weights should be "1". It seems that the chances of fatalities in crashes increase when a vehicle is overturned, which could be due to the loss of control by the driver. These observations are also confirmed in Figure 8. The total weights were calculated as per Equation (6) for different crash severities, and they are given in Table 6 based on different crash types and reasons.

Inconsistency calculations were not required, as explained earlier for crash types. The fatal crashes had the highest weight for "sudden lane changing" crashes, while the highest weight for serious injury crash was from "driver distraction", and "leaving insufficient safe distance" and "speeding" also caused similar and relatively high weights for fatal crashes. These trends were corroborated by Figure 9. Hence, it can be said that fatal crashes are more likely to occur when drivers are not following road regulations properly or do not have enough time to respond to events.

It can be observed that the overall weight for each severity type remained the same whether they are taken on the basis of crash types or reasons as shown in Table 8. This was expected, since it would have related to their overall proportion in the available sample. Figures 10 and 11 show the combined weights for the severity of the crash based on each crash type and reason.

Table 8. Total weights for each crash severity type.

Contributing Factor	Crash Severity	
	Fatal	**Serious Injury**
Crash Types	0.322	0.686
Crash Reasons	0.321	0.669

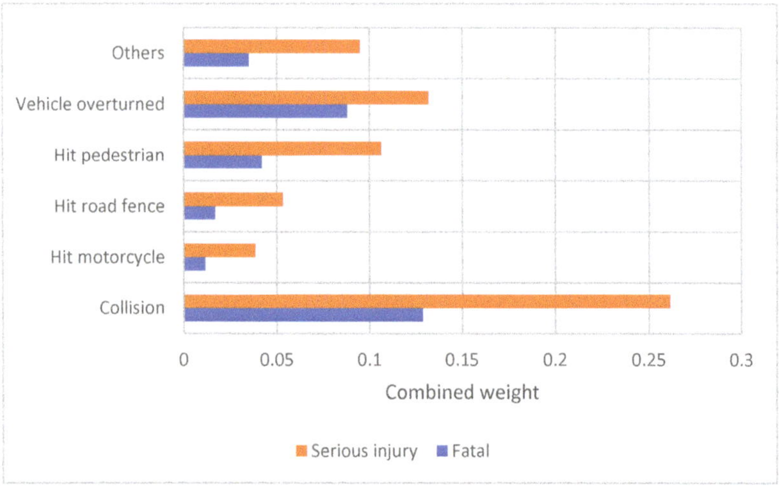

Figure 10. Combined weight for different crash severities based upon crash types.

Figure 10 shows that collisions and vehicles overturning are the two most common types of crashes which cause fatality or serious injury in a crash. These weights were calculated as the cross product of the relative weights of a factor (provided in Table 4 for crash type and Table 5 for crash reason) and the relative weight of factor with the severity level (Table 6 or Table 7). These weights indicate the relative chance of a severity type, among the entire dataset of crashes, with a particular type or reason for the crash. The weight or relative importance of "collision" to cause a serious injury crash was at least 10% higher than any other crash type. Similarly, it also had the highest weight for fatal crashes as well; however, the difference in its weight (approximately 5%), compared to other types, was not so prominent for fatal crashes as it was for serious injury crashes. Motorcycle riders were proven to be vulnerable road users because of their greater involvement in crashes of higher severity [41].

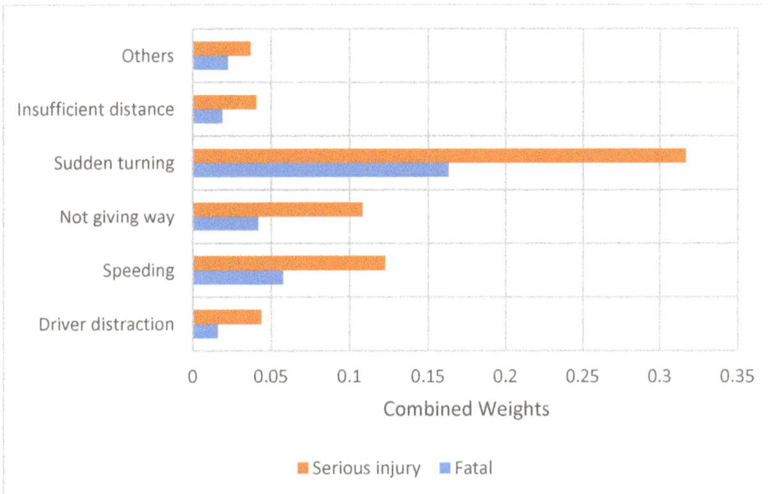

Figure 11. Combined weights for different crash severities based upon crash reasons.

Figure 11 shows that sudden lane changing was the most common reason for serious injury and fatal crashes, followed by speeding. The importance of "sudden turning" was at least 20% higher than any crash reason for serious injury crashes, while it was at least 10% higher than any other crash reason for fatal crashes. The least common reason for serious injury and fatal crashes was driver distraction. This could be because distracted drivers tend to become slower and keep a higher headway between the vehicles, as has been proven by other studies [42]. The combined weights could be used to form a utility function for the prediction of crashes, as proposed by [43]. Hence, the utility function for different crash severities can be listed as Equations (7) and (8). The development of such utility functions would have been unlikely with traditional statistical approaches for the type of data available, because of their statistical constraints.

$$\begin{aligned} U_{Fatal} &= 0.129\,(\text{Collision}) + 0.012\,(\text{Hitmotorcycle}) + 0.017\,(\text{Hitroadfence}) + 0.042\,(\text{Hitpedestrian}) + \\ &0.088\,(\text{Vehicle overturned}) + 0.035\,(\text{Otherstypes}) + 0.016\,(\text{Driverdistraction}) + 0.058\,(\text{Speeding}) + \\ &0.042\,(\text{Notgivingway}) + 0.163\,(\text{Suddenlanechanging}) + 0.019\,(\text{Insufficientsafedistance}) + 0.023\,(\text{Otherstypes}) \end{aligned} \tag{7}$$

$$\begin{aligned} U_{Seriousinjury} &= 0.261\,(\text{Collision}) + 0.038\,(\text{Hitmotorcycle}) + 0.053(\text{Hitroadfence}) + 0.106\,(\text{Hitpedestrian}) + \\ &0.132\,(\text{Vehicleoverturned}) + 0.095\,(\text{Othertypes}) + 0.044\,(\text{Driverdistraction}) + 0.122\,(\text{Speeding}) + \\ &0.108\,(\text{Notgivingway}) + 0.317\,(\text{Suddenlanechanging}) + 0.041\,(\text{Insufficientsafedistance}) + 0.037\,(\text{Otherreasons}) \end{aligned} \tag{8}$$

6. Conclusions

In Saudi Arabia, traffic crashes are a major public health hazard because of their various causes, which include aspects relating to the driver, the road, and the vehicle. Understanding the root causes of traffic crashes in the country is crucial for developing effective strategies to stop these occurrences and improve road safety. Understanding the typical reasons and crash types that lead to severe crashes is a crucial first step in developing effective solutions for severe crash prevention. In this research, the PBAHP technique was used, and its potential in discovering the primary causes and crash types that result in catastrophic crashes was proven. The PBAHP was used to identify and rank the causes of traffic crashes, as well as to assess the efficiency of safety measures.

According to this study, "collision" crashes were the most common cause of fatalities and serious injuries, followed by "vehicle overturned" and "hit pedestrian". This conclusion is further supported by the fact that "collisions", the most frequent sort of crash, are invariably caused by rapid turns of a vehicle. Speeding and failure to yield are two other significant causes of death and serious injury collisions. Using the same process, the combined effects of events on the crash severity (i.e., the interaction of different crash types and reasons) could be analyzed. However, the limitation of this study is that it could not be performed with the present dataset because of the lack of samples for each combination. Thus, future studies should be based on enhanced datasets. As an alternative, the total weights from the PBAHP process were used to develop utility functions for fatal and serious injury crashes. These functions could be used for estimating the likelihood of crashes in different scenarios.

Author Contributions: Conceptualization, U.G. and M.K.I.; methodology and software, U.G. and M.K.I.; validation and formal analysis, U.G. and M.K.I.; resources and data curation, M.K.I. and M.K.I., writing—original draft preparation, review and editing, M.K.I. and U.G.; project administration, M.K.I.; funding acquisition, M.K.I. All authors have read and agreed to the published version of the manuscript.

Funding: This work was financially supported by the Deanship of Scientific Research at the King Faisal University, Saudi Arabia with Grant 2907.

Institutional Review Board Statement: Not applicable.

Informed Consent Statement: Not applicable.

Data Availability Statement: The data that support the findings of this study are available from the corresponding authors, Md Kamrul Islam (maislam@kfu.edu.sa) and Uneb Gazder (ugazder@uob.edu.bh), upon reasonable request.

Conflicts of Interest: The authors declare no conflict of interest.

References

1. Islam, M.K.; Dalhat, M.A.; Al Mamun, A. Road Infrastructure Investment Limits Based on Minimal Accidents Using Artificial Neural Network. *Appl. Sci.* **2022**, *12*, 11949. [CrossRef]
2. World Health Organization. *Global Status Report on Road Safety 2021*; World Health Organization: Geneva, Switzerland, 2021.
3. Ansari, S.; Akhdar, F.; Mandoorah, M.; Moutaery, K. Causes and effects of road traffic accidents in Saudi Arabia. *Public Health* **2000**, *114*, 37–39. [CrossRef]
4. Hassan, H.M. Investigation of the Self-Reported Aberrant Driving Behavior of Young Male Saudi Drivers: A Survey-Based Study. *J. Transp. Saf. Secur.* **2016**, *8*, 113–128. [CrossRef]
5. Ramisetty-Mikler, S.; Almakadma, A. Attitudes and behaviors towards risky driving among adolescents in Saudi Arabia. *Int. J. Pediatr. Adolesc. Med.* **2016**, *3*, 55–63. [CrossRef]
6. Issa, Y. Effect of Driver's Personal Characteristics on Traffic Accidents in Tabuk city in Saudi Arabia. *J. Transp. Lit.* **2016**, *10*, 25–29. [CrossRef]
7. Islam, M.K.; Gazder, U.; Akter, R.; Arifuzzaman, M. Involvement of Road Users from the Productive Age Group in Traffic Crashes in Saudi Arabia: An Investigative Study Using Statistical and Machine Learning Techniques. *Appl. Sci.* **2022**, *12*, 6368. [CrossRef]
8. Rahman, M.M.; Islam, M.K.; Al-Shayeb, A.; Arifuzzaman, M. Towards Sustainable Road Safety in Saudi Arabia: Exploring Traffic Accident Causes Associated with Driving Behavior Using a Bayesian Belief Network. *Sustainability* **2022**, *14*, 6315. [CrossRef]
9. Islam, M.K.; Reza, I.; Gazder, U.; Akter, R.; Arifuzzaman, M.; Rahman, M.M. Predicting Road Crash Severity Using Classifier Models and Crash Hotspots. *Appl. Sci.* **2022**, *12*, 11354. [CrossRef]
10. Furda, A.; Vlacic, L.B. Enabling Safe Autonomous Driving in Real-World City Traffic Using Multiple Criteria Decision Making. *IEEE Intell. Transp. Syst. Mag.* **2011**, *3*, 4–17. [CrossRef]
11. Yan, L.; Li, X. Traffic safety evaluation in the rural-urban continuum based on ANP. In Proceedings of the 2009 Second International Conference on Intelligent Computation Technology and Automation, Changsha, China, 10–11 October 2009; pp. 853–858.
12. Korhonen, P.; Wallenius, J. Behavioral issues in MCDM: Neglected research questions. In *Multicriteria Analysis, Proceedings of the XI International Conference on MCDM, Coimbra, Portugal, 1–6 August 1994*; Springer: Berlin/Heidelberg, Germany, 1997; Volume 5, pp. 412–422.

13. Nanda, S.; Singh, S. Evaluation of factors responsible for road accidents in India by fuzzy AHP. In *Networking Communication and Data Knowledge Engineering, Lecture Notes on Data Engineering and Communications Technologies*; Springer: Singapore, 2018; Volume 3, pp. 179–188.
14. Haghighat, F. Application of a Multi-criteria Approach to Road Safety Evaluation in the Bushehr Province, Iran. *Promet Traffic Transp.* **2012**, *23*, 341–352. [CrossRef]
15. Shi, H. Fuzzy evaluation approach of road traffic safety based on AHP. In Proceedings of the International Conference on Future Bio Medical Information Engineering, Sanya, China, 13–14 December 2009; pp. 394–397.
16. Hermans, E.; Bossche, F.V.D.; Wets, G. Combining road safety information in a performance index. *Accid. Anal. Prev.* **2008**, *40*, 1337–1344. [CrossRef] [PubMed]
17. Cheng, J.; Chen, C. An AHP method for road traffic safety. In Proceedings of the 2011 International Conference on Computational Science and Optimization (CSO), Kunming, China, 19 April 2011; pp. 111–115.
18. Fernandez, J.J.; Paringit, M.C.; Salvador, J.R.; Lucero, P.I.; Galupino, J.G. Driver's Road Accident Factor Prioritization using AHP in Relation to Mastery of Traffic Signs in the City of Manila. *Transp. Res. Procedia* **2020**, *48*, 1316–1324. [CrossRef]
19. Saifullizan, M.B.; Nur, A.F.M.F.; Nazirah, M.A.; Anuar, M.S. Analysis of Severity Level Types and Trends in Road Accident Cases at Johor Inter State Road Using Analythical Hierarchy Process (AHP) And Geographical Information System. *Int. J. Integr. Eng.* **2022**, *14*, 66–73. [CrossRef]
20. Hu, L.; Pei, Y.; Qiu, Z.; Liu, Z. Evaluation of Road Traffic Safety Based Improved Fuzzy-AHP Method. In Proceedings of the ICCTP 2009: Critical Issues in Transportation Systems Planning, Development, and Management, Harbin, China, 5–9 August 2009; pp. 1–7. [CrossRef]
21. Farooq, D.; Moslem, S.; Faisal Tufail, R.; Ghorbanzadeh, O.; Duleba, S.; Maqsoom, A.; Blaschke, T. Analyzing the Importance of Driver Behavior Criteria Related to Road Safety for Different Driving Cultures. *Int. J. Environ. Res. Public Health* **2020**, *17*, 1893. [CrossRef] [PubMed]
22. Farooq, D.; Moslem, S. Estimating Driver Behavior Measures Related to Traffic Safety by Investigating 2-Dimensional Uncertain Linguistic Data—A Pythagorean Fuzzy Analytic Hierarchy Process Approach. *Sustainability* **2022**, *14*, 1881. [CrossRef]
23. Khademi, N.; Choupani, A.-A. Investigating the Road Safety Management Capacity: Toward a Lead Agency Reform. *IATSS Res.* **2018**, *42*, 105–120. [CrossRef]
24. Yang, Y.; Easa, S.M.; Lin, Z.; Zheng, X. Evaluating Highway Traffic Safety: An Integrated Approach. *J. Adv. Transp.* **2018**, *2018*, 4598985. [CrossRef]
25. Mirmohammadi, F.; Khorasani, G.; Tatari, A.; Yadollahi, A.; Taherian, H.; Motamed, H.; Fazelpour, S.; Khorasani, M.; Verki, M.R.M. Investigation of Road Accidents and Casualties Factors with MCDM Methods in Iran. *J. Am. Sci.* **2013**, *9*, 11–20.
26. Farooq, D.; Moslem, S.; Jamal, A.; Butt, F.M.; Almarhabi, Y.; Faisal Tufail, R.; Almoshaogeh, M. Assessment of Significant Factors Affecting Frequent Lane-Changing Related to Road Safety: An Integrated Approach of the AHP–BWM Model. *Int. J. Environ. Res. Public Health* **2021**, *18*, 10628. [CrossRef]
27. Ghorbanzadeh, O.; Feizizadeh, B.; Blaschke, T. An Interval Matrix Method Used to Optimize the Decision Matrix in AHP Technique for Land Subsidence Susceptibility Mapping. *Environ. Earth Sci.* **2018**, *77*, 584. [CrossRef]
28. Munier, N.; Hontoria, E. Shortcomings of the AHP method. In Proceedings of the 1st International Electronic Conference on Mathematics, Online, 1–10 June 2021.
29. Vaidya, O.S.; Kumar, S. Analytic Hierarchy Process: An Overview of Applications. *Eur. J. Oper. Res.* **2006**, *169*, 1–29. [CrossRef]
30. Ishizaka, A.; Labib, A. Review of the Main Developments in the Analytic Hierarchy Process. *Expert Syst. Appl.* **2011**, *38*, 14336–14345. [CrossRef]
31. Saaty, R.W. The Analytic Hierarchy Process—What It Is and How It Is Used. *Math. Modell.* **1987**, *9*, 161–176. [CrossRef]
32. Liu, S.; Zhao, Q.; Wen, M.; Deng, L.; Dong, S.; Wang, C. Assessing the Impact of Hydroelectric Project Construction on the Ecological Integrity of the Nuozhadu Nature Reserve, Southwest China. *Stoch. Environ. Res. Risk Assess.* **2013**, *27*, 1709–1718. [CrossRef]
33. Saaty, T.L. Decision Making with the Analytic Hierarchy Process. *Int. J. Serv. Sci.* **2008**, *1*, 83–98. [CrossRef]
34. Alemdar, K.D.; Kaya, Ö.; Çodur, M.Y. A GIS and Microsimulation-Based MCDA Approach for Evaluation of Pedestrian Crossings. *Accid. Anal. Prev.* **2020**, *148*, 105771. [CrossRef]
35. Saaty, T.L. *Theory and Applications of the Analytic Network Process: Decision Making with Benefits, Opportunities, Costs and Risks*; RWS Publications: Pittsburgh, PA, USA, 2005.
36. Liu, F.H.F.; Hai, H.L. The Voting Analytic Hierarchy Process Method for Selecting Supplier. *Int. J. Prod. Econ.* **2005**, *97*, 308–317. [CrossRef]
37. Jamal, A.; Rahman, M.T.; Al-Ahmadi, H.M.; Mansoor, U. The Dilemma of Road Safety in the Eastern Province of Saudi Arabia: Consequences and Prevention Strategies. *Int. J. Environ. Res. Public Health* **2020**, *17*, 157. [CrossRef]
38. Chakraborti, S.; Li, J. Confidence Interval Estimation of a Normal Percentile. *Am. Stat.* **2007**, *61*, 331–336. [CrossRef]
39. Rahman, F.; Yoshida, S.; Kojima, A.; Kubota, H. Paired comparison method to prioritize traffic calming projects. *J. East. Asia Soc. Transp. Stud.* **2015**, *11*, 2472–2487.
40. Lin, S.W.; Lu, M.T. Characterizing Disagreement and Inconsistency in Experts' Judgments in the Analytic Hierarchy Process. *Manag. Decis.* **2012**, *50*, 185–199. [CrossRef]

41. Yannis, G.; Nikolaou, D.; Laiou, A.; Stürmer, Y.A.; Buttler, I.; Jankowska-Karpa, D. Vulnerable Road Users: Cross-Cultural Perspectives on Performance and Attitudes. *IATSS Res.* **2020**, *44*, 220–229. [CrossRef]
42. Gazder, U.; Assi, K.J. Determining Driver Perceptions about Distractions and Modeling Their Effects on Driving Behavior at Different Age Groups. *J. Traffic Transp. Eng. Engl. Ed.* **2022**, *9*, 33–43. [CrossRef]
43. Accorsi, R.; Zio, E.; Apostolakis, G.E. Developing Utility Functions for Environmental Decision Making. *Prog. Nucl. Energy* **1999**, *34*, 387–411. [CrossRef]

applied sciences

Article

Characteristic Analysis and Decision Model of Lane-Changing Game for Intelligent Connected Vehicles

Dayi Qu, Shouchen Dai *, Aodi Li, Yicheng Chen and Chuanbao Wei

School of Mechanical and Automotive Engineering, Qingdao University of Technology, Qingdao 266520, China; dyqu@263.net (D.Q.)
* Correspondence: dsc2177@163.com

Abstract: To study the lane-change interaction characteristics of intelligent connected vehicles (ICVs) and reduce the risk of vehicle lane-changing decisions, a decision model based on the lane-changing game characteristics of the ICV is proposed in this paper. In the modeling process, the characteristics of vehicle lane-changing interaction behavior are analyzed based on evolutionary game theory and the vehicle game lane-changing payoff functions are quantified. The stability of the game equilibrium points is analyzed by using a dynamic evolution equation, and sensitivity analysis of the main factors affecting vehicle lane changes and the time to the collision of vehicles is conducted. The SUMO software is used to simulate and verify the vehicle game decision model, and the results show that the game decision system converges to different optimal strategy combinations under different traffic conditions, and this model can effectively reduce the decision-making conflict and the collision risk of vehicles.

Keywords: lane change; interaction characteristics; evolutionary game; sensitivity analyses; time to collision

Citation: Qu, D.; Dai, S.; Li, A.; Chen, Y.; Wei, C. Characteristic Analysis and Decision Model of Lane-Changing Game for Intelligent Connected Vehicles. *Appl. Sci.* **2023**, *13*, 8321. https://doi.org/10.3390/app13148321

Academic Editors: Edgar Sokolovskij and Vidas Žuraulis

Received: 25 June 2023
Revised: 9 July 2023
Accepted: 11 July 2023
Published: 18 July 2023

1. Introduction

Vehicle following and lane changing are the two most important research directions in road traffic, though vehicle lane changing has more complex traffic characteristics than vehicle-following behavior, reflecting the competition between vehicles in adjacent lanes for the right of way in time and space [1]. The application of IoT technology has promoted the development of the autonomous driving industry and facilitated the safety and reliability of vehicles [2]. With the application and popularity of IoT devices, information sharing between urban road facilities and vehicles or other facilities is achieved through IoT technology, laying the foundation for the future development of intelligent transportation [3]. Assisted driving or autonomous driving technology has also been developed rapidly as a result, and more and more vehicles are equipped with advanced assisted driving systems to help drivers reduce the driving burden and improve vehicle safety.

Vehicle lane changing is a topical and difficult problem in autonomous driving technology. Vehicle lane changes can usually be divided into free lane changes and mandatory lane changes, and the vehicle in the free lane-change scenario usually pursues a higher driving speed or driving comfort, while the mandatory lane change is lane-change behavior made by the vehicle due to the limitation of road conditions, and the process reflects the best decision made by the driver after perceiving the change of information of the surrounding traffic environment and weighing up the gains and losses. Vehicle decision making is one of the core modules of autonomous vehicles and a process that integrates vehicle travel efficiency and safety [4]. Existing research on vehicle lane changes usually focuses on lane-change model control, trajectory prediction, etc. However, the study of vehicle lane-change interaction characteristics and the study of lane-change game mechanisms have not received sufficient attention. Although many game theory-based vehicle lane-changing models have high decision accuracy and reliability, researchers usually apply the game results directly to the control aspects of the model and lack a deeper analysis of the lane-changing game mechanism.

The objective of this paper is to propose an evolutionary game-based vehicle lane-changing decision model, to study and analyze the characteristics of vehicle lane-change interactions, the convergence direction of decision combinations, and the impact of the evolutionary game lane-change decision on vehicle safety. The evolutionary game is a theoretical approach to analyzing the rationality of decision making through dynamic evolutionary equations, which can consider the limited rationality of decision makers and reduce the assumptions of the model compared with other game models. At present, the research on vehicle lane changes mostly focuses on the highway [5–7] and intersection scenarios [8,9], because these scenarios belong to traffic bottleneck sections, and vehicle lane changing in bottleneck sections is a difficult problem. In this paper, taking the urban intersection road as an example, the forced lane-change interaction characteristics of connected autonomous vehicles are analyzed; the convergence process of vehicle lane-change decisions is dynamically analyzed by an evolutionary game model, and sensitivity analysis of the main factors is conducted to explore the influence of road traffic environment changes on vehicle lane changes; and the vehicle trajectory and output collision time are simulated by SUMO to analyze the collision risk of vehicles; and, finally, further research is summarized and extended.

The organization of this paper is as follows: Section 1 introduces the research on lane changing for autonomous vehicles. Section 2 reviews the relevant studies on the vehicle lane-change game. Section 3 analyzes the game characteristics of vehicle lane changes with the example of an intersection roadway. Section 4 constructs the payoff matrix and dynamic evolution equation for both sides of the game. In Section 5, Simulation experiments on the decision-making algorithm are carried out, sensitivity analysis of the main factors affecting vehicle lane changing is performed, vehicle trajectory and collision risks are compared, and validity threats are explained. The last section provides the conclusions.

2. Literature Review

In previous studies, many researchers have analyzed and optimized the decision-making process of a vehicle lane change in the hope that autonomous vehicles can reduce the driver's driving burden and safety risks and improve traffic efficiency. Traffic safety is the most important concern for road traffic participants, and the study of vehicle lane changes can help reduce the risk of decision conflicts and lane changes for vehicles [10,11]. With the cross development of game theory and the transportation field, researchers solve the decision conflict problem by establishing a game model and framework of vehicle lane changes, which can usually be divided into cooperative and non-cooperative games.

2.1. Non-Cooperative Game Model

The non-cooperative game model is an optimal strategy for maximizing self-interest and weighing up the gains and losses. In the study of vehicle non-cooperative game lane changing, Qu et al. [12] analyzed the dynamic influences of vehicle lane changing by quantifying the decision intention based on collision probability and established a lane-changing model through game theory. Yao et al. [13] analyzed the lane-change game relationship between buses and social vehicles at bus bay stops through an uncooperative mixed strategy game by analyzing the dynamic influencing factors of vehicle lane changes, and the results showed that the game model can effectively predict each other's lane-change strategies. Arbis et al. [14] used a quantitative Nash equilibrium framework to model vehicle lane-change behavior and concluded that lengthening acceleration lanes or lowering the speed limit of ramps can effectively reduce risky conflicts on freeway ramps.

Some researchers have also improved the vehicle game process by taking into account the driver factor to establish a model more consistent with realistic driving scenarios. Dai et al. [15] established a vehicle lane-change game model for the upstream sections of urban intersections by considering driver demand based on a mixed strategy game, but the lane-change model based on the mixed strategy game still has decision conflicts. Hang et al. [16] combined driver style with game theory to construct a payoff matrix for different driving types and validated it with a model predictive control algorithm. Despite the potential of non-cooperative game models in vehicle lane changing, the characteristics of non-cooperative game theory dictate that such models cannot maximize overall benefits and therefore have limitations.

2.2. Cooperative Game Model

Unlike non-cooperative games, cooperative games aim to maximize the benefits of all subjects. Cooperative game theory realizes the benefits of decision making through the allocation process by enabling participants to form binding coalitions or groups.

Sun et al. [17] proposed a two-lane collaborative lane-change model by analyzing the impact of lane-change behavior on the target lane, and the experimental results showed that the method can effectively reduce the oscillation of traffic flow. Zheng et al. [18] proposed a game model for optimizing the decision of lane-change vehicles by considering the influence of surrounding vehicle motion state and vehicle–vehicle communication, which can effectively improve traffic operation efficiency and reduce traffic oscillations. Liu et al. [19] proposed a cooperative LPV/MPC and risk assessment cooperative driving strategy, and the results showed that the algorithm can effectively assist drivers in reducing vehicle risk. Yu et al. [20] established a multi-player dynamic game model based on game theory considering the driving status of surrounding vehicles, which can effectively reflect the driving intention of surrounding vehicles and the impact of different decisions on the vehicle. Pan et al. [21] proposed a distributed structure multi-vehicle cooperative control model that combines game theory and an MPC algorithm to obtain the optimal sequence of multi-lane vehicles and vehicle acceleration.

The cooperative game has certain advantages in terms of overall effectiveness, but it is necessary to consider the willingness of the participants to cooperate, the degree of cooperation of the participants, and the implementability of the cooperative game determined by individual expectations.

2.3. Extension of Game Lane-Change Model

Although the above game models have been widely used in research in the field of transportation, there are still problems that need to be improved. Therefore, several researchers have extended the game-theoretic-based lane-change model, and researchers have improved the adaptability of the game model by changing the lane-change scenario.

Hang et al. [22] proposed a cooperative decision-making method for ICVs to reduce vehicle conflicts at unsignalized intersections by forming a game coalition. Smirnov et al. [23] proposed a game theory-based model of urban intersection lane changes by considering the cooperation level of vehicles in adjacent lanes, which can accurately predict the decision propensity of game participants. Zhang et al. [24] studied lane-change game behavior in a foggy environment by analyzing lane-change intention and considering visual features, which has contributed to the study of lane-change models in complex weather environments. In addition, the optimal strategy that satisfies both individual and overall benefits can be obtained by combining deep learning with game theory [25]. It is also of great importance to study lane-changing behavior using other game branch theories such as repeated games [26].

According to the above literature review, researchers have conducted a large number of studies on vehicle lane changing based on game theory and the game model of lane changes of vehicles has great potential in reducing both decision conflicts and vehicle collisions; these studies focus on making the decision result of vehicle lane changing through game theory and lack research on the game mechanism of lane changing. It is necessary to study the evolution law of vehicle lane-change games, and it is helpful to explore the intrinsic characteristics of vehicle game behavior. We will present the methodology and experiments in the next sections.

3. Characteristics of Vehicle Lane-Changing Game

Vehicle lane changes as a common microscopic driving behavior in road traffic has the characteristic of changing with driving intention, and vehicle lane changes can usually be divided into free lane changes and forced lane changes according to lane-change intention. The urban road is the main scenario of vehicle driving, with the driver generally pursuing a higher driving speed when the vehicle drives through the upstream intersection exit lane; therefore, the decision to change lanes freely is usually made.

When approaching the intersection entrance lanes, conservative drivers tend to change to the target lane early, while drivers who are more confident in their driving skills or have an aggressive driving style tend to choose the lane with a higher speed. Left-turning vehicles choose to enter the left-turning lane early, while straight-traveling vehicles choose to stay in the straight lane, and then all vehicles move into the intersection inlet lanes in turn and wait for the green light to release. This paper assumes that vehicles are not driving aggressively and that left-turning vehicles should move into the left-turn lane early to comply with traffic rules; however, the lane-change conditions are usually restricted by the surrounding vehicles and road traffic, so the vehicle lane-change behavior at this time is usually a mandatory lane-change behavior. Figure 1 represents the lane-change game process for vehicles.

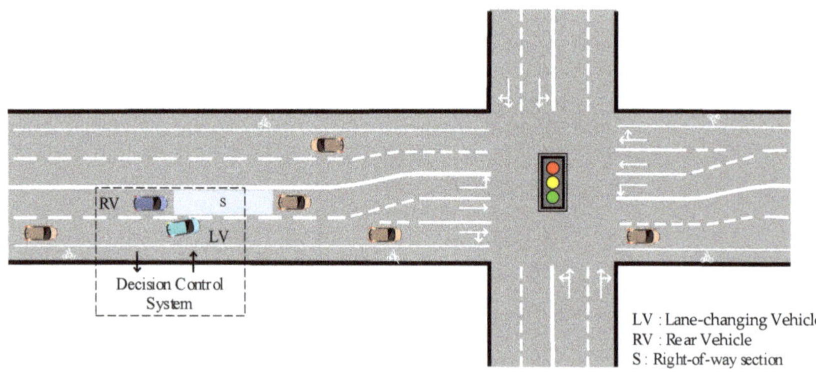

Figure 1. Diagram of vehicle gaming lane changes.

In this paper, we assume that the LV (Lane-changing Vehicle) is the vehicle that needs to turn left and the RV (Rear Vehicle) is the vehicle driving in the left-turn lane. The vehicles in the model are assumed to be ICVs, which are equipped with onboard detectors and sensors, and the vehicles can transmit information with road facilities such as signals. Vehicles obtain decision combinations through the upper decision control system, and the system establishes a gain matrix for each vehicle and determines the optimal driving decision through evolutionary gaming.

The intelligence of the ICVs is reflected in the ability to reduce the driving burden of the driver, accomplish the task of automatic driving, meet the expectations of the driver, and the vehicle can make human-like driving behavior in the face of a certain traffic environment. For example, the vehicle should not be too conservative and increase the driving time in the process of decision-making interaction with surrounding vehicles, and the vehicle should not be too aggressive and reduce the comfort of the driver. To simplify the game behavior and decision-making process, this paper decomposes the driving task of the ICVs, with the driving task of the vehicle in a certain period triggered by changes in the traffic environment, and the vehicle decision-making mechanism established by considering the expectations of passengers.

4. Game Model

4.1. Definition of Right of Way

A vehicle driving in a lane means that it has the right to use a certain length of road in time and space; due to the security requirements of the vehicle in the driving state, the front and rear of the car need a certain length of clearance to prevent the intrusion of other vehicles. Based on the game theory point of view, the purpose of the game between vehicles can be thought of as competing for the next section of right of way, as shown in Figure 1. The priority right-of-way in the area ahead of the target lane is a competing purpose for vehicles. For example, if through the game the LV has priority to enter region S, then the LV achieves driving efficiency and the RV needs to wait for the LV to drive through before entering the region. Since the LV is ahead of the RV in terms of position, the LV not only has priority of the section but also has the priority of subsequent sections. The right-of-way area is considered a dynamically adjusted area to quantify and clarify the vehicle game process, which is represented schematically in Figure 2.

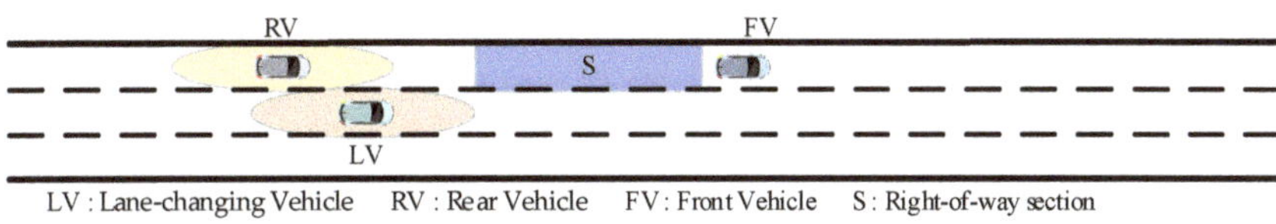

Figure 2. The diagram of the right of way.

The length of the competitive road section is determined by the driving space in front of the vehicle, which can be expressed as:

$$L_{space} = x_{FV} - min\{x_{RV}, x_{LV}\} \tag{1}$$

where L_{space} is the length of the right of way competing for the gaming process; x_{FV} indicates the coordinate position of the vehicle in front of the target lane (this value should be smaller than the effective length of the section); x_{RV} indicates the coordinate position of the vehicle behind the target lane; and x_{LV} indicates the coordinate position of the vehicle changing lanes.

4.2. Game Components

In previous research on the vehicle lane-change game, most researchers defined the efficiency gain of the vehicle as the increase or loss of speed and failed to consider the influence of the characteristics of the scenario. The game process cannot ignore the driving scenario of the vehicle and should construct a gain matrix more in line with the actual situation according to the characteristics of the driving scenario.

4.2.1. Payoffs for Vehicle RV

The efficiency gain of a vehicle should be defined as the increase or loss of time loss through the intersection. When the LV or RV decides to yield, it increases its own passing time and decreases its driving efficiency, and, conversely, when the RV makes the decision not to give way, it decreases its own passing time. The efficiency gains of the RV include travel gains and space gains. When the RV position is ahead of the LV, the RV not only narrows the travel time, but also has the initiative in space relative to the LV, and the same for the LV as well. The driving space gain of the vehicle is expressed as:

$$P_s = \frac{L_{space}}{L_{s,max} - L_{s,min}} \tag{2}$$

where L_{space} indicates the spatial extent obtained with the current state of the vehicle and $L_{s,max}$ and $L_{s,min}$ are the maximum and minimum spatial ranges obtained based on the statistical results of the data.

The efficiency gain of the RV not giving way is expressed as:

$$P_{eff}^{RV} = P_s \cdot \frac{t_{cur}^{RV} - t_{min}^{RV}}{t_{max}^{RV} - t_{min}^{RV}} \tag{3}$$

where t_{cur}^{RV} is the time to cross the intersection from the current position corresponding to the different decisions made by the RV; t_{max}^{RV} is the maximum time for the RV to pass the intersection considering the game with the LV; and t_{min}^{RV} is the time for the vehicle to cross the intersection at a steady speed.

The safety gain of the vehicle is generally measured by the relative distance of the vehicle. Widely used in lane-change research is the minimum safe distance model, which takes into account the dynamic change process of vehicle lane changes when the vehicle meets the requirements of Equation (4), such that the vehicle can be safely changed.

$$S_t = \int_0^t \left(\int_0^\zeta (a_L(\tau) - a_R(\tau)) d\tau d\zeta \right) + \Delta v_0 \cdot t + S_0 \geq S_{min} \tag{4}$$

where a_L and a_R are the accelerations of the LV and RV, respectively; Δv_0 is the initial speed difference of the vehicles; and S_0 is the initial workshop distance. S_t should be greater than the minimum vehicle distance S_{min}.

Since the safety of vehicles interacts with each other, the lane-changing game requires both sides of the game to make corresponding decisions; when the LV changes lanes, it requires the RV to slow down to provide lane-change spacing for the LV. Equation (5) shows the safety gain of the vehicles.

$$P_{safe}^{RV} = \frac{S_{veh} - S_{min}^{veh}}{S_{max}^{veh} - S_{min}^{veh}} \tag{5}$$

where S_{veh} is the longitudinal vehicle distance; S_{max}^{veh} is the maximum gaming distance (when the workshop distance is greater than the maximum gaming distance, the LV can change lanes without gaming); and S_{min}^{veh} is the minimum shop distance for the rear vehicle to take the maximum deceleration to provide the lane-change spacing for the LV.

Considering the prospect of the development of ICVs, we built a game matrix by considering passenger expectations. The ICVs control the vehicle driving through the autonomous decision-making system, and the identity of the driver changes from the driver to the vehicle passenger. The ICVs should meet the passenger expectations under different driving scenarios, under the condition of complying with traffic rules; for example, the vehicle should not be too conservative and increase the travel time, and should not be too aggressive and reduce comfort. In this paper, we consider passengers' expectation requirements for vehicle decisions in the game payoffs by dynamically adjusting the vehicle's weight in terms of efficiency and safety. The game gain of the RV is mainly affected by the LV changing lanes; when the LV changes lanes and drives in front of the RV, it increases the waiting time of the RV through the intersection and decreases the efficiency gain; if the RV does not give way, it will increase the risk of decision conflict and decrease the safety gain. Equation (6) shows the expectation of the RV for efficiency gain.

$$\beta = max\left\{ min\left\{ \frac{t_{tl}^{RV} - t_{min}^{gl}}{t_r^{gl} - t_{min}^{gl}}, 0.7 \right\}, 0.3 \right\} \tag{6}$$

where t_{tl}^{RV} is the travel time of the RV through the intersection; t_r^{gl} is the remaining time of the green light at the intersection; and t_{min}^{gl} is the minimum green light time required for the vehicle to pass through the intersection from its current position. When the travel time is greater than the minimum green light time, the vehicle needs to wait for the next cycle of the green light before it can pass through the intersection.

Using the payoff functions constructed above and the RV's expectation of efficiency gains, the payoff matrix of the RV in different strategy profiles is listed in Table 1.

Table 1. Payoffs of RV in different strategy profiles.

	RV	
LV	**Give Way**	**Do Not Give Way**
Change lanes	$-\beta P_{eff}^{RV} + (1-\beta)P_{safe}^{RV}$	$-\beta P_{eff}^{RV} - (1-\beta)P_{safe}^{RV}$
Do not change lanes	$-\beta P_{eff}^{RV} + (1-\beta)P_{safe}^{RV}$	$\beta P_{eff}^{RV} + (1-\beta)P_{safe}^{RV}$

4.2.2. Payoffs for Vehicle LV

To simplify the model and highlight the focus of the game process, this paper assumes that the task of the game decision is to play between vehicles and complete the lane-change process without considering the maximization of the full trip driving gain of the LV. Therefore, the expected gain of the LV can be regarded as changing lanes to the target lane as soon as possible. The efficiency gain of the LV also includes the trip gain and the space gain, with the efficiency gain expressed as:

$$P_{eff}^{LV} = P_s \cdot \frac{t_{lc}^{LV} - t_{min}^{LV}}{t_{max}^{LV} - t_{min}^{LV}} \tag{7}$$

where t_{lc}^{LV} is the lane-change time corresponding to the different decisions made by the LV; t_{max}^{LV} is the maximum time to consider the decision process of the vehicle game; and t_{min}^{LV} is the shortest lane-change time considering vehicle dynamics when the lane-change process is not constrained.

According to the vehicle safety interactions, the safety gain of the LV can be similarly expressed as:

$$P_{safe}^{LV} = \frac{S_{veh} - S_{min}^{veh}}{S_{max}^{veh} - S_{min}^{veh}} \tag{8}$$

The LV lane change needs to consider the impact of decision moment location: when the LV is closer to the intersection inlet lanes, the expectation of lane changes is a stronger column; when the LV is further away from the intersection, it usually does not show a willingness to change lanes. For the LV and RV game decision, the decision to not change lanes increases the time for the LV to drive into the target lane and reduces the efficiency gain of the LV. Equation (9) shows the LV's expectation of driving efficiency.

$$\alpha = max\left\{ min\left\{ \frac{S_{max}^{LV} - S_{cur}^{LV}}{S_{max}^{LV} - S_{min}^{LV}}, 0.7 \right\}, 0.3 \right\} \tag{9}$$

where S_{cur}^{LV} is the distance from the LV decision moment position to the lane-gradient section; S_{min}^{LV} is the LV minimum desired lane-change distance, which is used to ensure that vehicles can change lanes earlier; and S_{max}^{LV} is the farthest lane-change game position, with vehicles less affected by the intersection bottleneck when the lane-change position exceeds this value.

By creating separate safety and efficiency functions for the vehicle above, the payoff matrix of the LV in different strategy profiles is listed in Table 2.

Table 2. Payoffs of LV in different strategy profiles.

	RV	
LV	**Give Way**	**Do Not Give Way**
Change lanes	$\alpha P_{eff}^{LV} + (1-\alpha) P_{safe}^{LV}$	$-\alpha P_{eff}^{LV} - (1-\alpha) P_{safe}^{LV}$
Do not change lanes	$-\alpha P_{eff}^{LV} + (1-\alpha) P_{safe}^{LV}$	$-\alpha P_{eff}^{LV} + (1-\alpha) P_{safe}^{LV}$

4.3. Game Equilibrium Analysis

The evolutionary game theory originated from biological evolution, which can dynamically analyze the evolutionary process of a group and clarify the evolutionary mechanism of which strategy to choose. With the cross development of game theory and the transportation field, many researchers apply game theory to study a vehicle interaction model [27–29]; compared with other game theories, the evolutionary game takes into account the characteristics of imperfect rationality of game subjects and reflects the evolutionary process of decision making more objectively. The evolutionary game reflects the evolutionary trends of different strategies through dynamic evolution equations, and the expected payoffs of the LV choosing to change lanes and choosing not to change lanes are:

$$E_{lc}^{LV} = (2y-1)\left(\alpha P_{eff}^{LV} + (1-\alpha) P_{safe}^{LV}\right) \tag{10}$$

$$E_{nlc}^{LV} = -\alpha P_{eff}^{LV} + (1-\alpha) P_{safe}^{LV} \tag{11}$$

The overall expected payoff of the LV based on the payoff matrix is:

$$E_{total}^{LV} = xE_{lc}^{LV} + (1-x)E_{nlc}^{LV} = 2\left(\alpha P_{eff}^{LV} + (1-\alpha) P_{safe}^{LV}\right)xy - 2x(1-\alpha)P_{safe}^{LV} - \alpha P_{eff}^{LV} + (1-\alpha) P_{safe}^{LV} \tag{12}$$

where x is the probability that the LV chooses to change lanes and $1-x$ is the probability that the LV chooses not to change lanes.

The evolutionary dynamic equation for the LV selection lane-change decision is:

$$F_{LV} = \frac{\partial x}{\partial t} = x\left(E_{lc}^{LV} - E_{total}^{LV}\right) = 2x(1-x)\left[\alpha y P_{eff}^{LV} + (1-\alpha)(y-1)P_{safe}^{LV}\right] \tag{13}$$

Similarly, the expected payoffs of the RV's choice of giving way and not giving way are:

$$E_{gw}^{RV} = -\beta P_{eff}^{RV} + (1-\beta) P_{safe}^{RV} \tag{14}$$

$$E_{ngw}^{RV} = (1-2x)\left(\beta P_{eff}^{RV} + (1-\beta) P_{safe}^{RV}\right) \tag{15}$$

The expected payoff of the RV with a probability of choosing to give way and with a probability of choosing not to give way is:

$$E_{total}^{RV} = yE_{gw}^{RV} + (1-y)E_{ngw}^{RV} = -2\beta P_{eff}^{RV} y + (2xy - 2x + 1)\left(\beta P_{eff}^{RV} + (1-\beta) P_{safe}^{RV}\right) \tag{16}$$

where y is the probability that the RV chooses to give way and $1-y$ is the probability that the RV chooses not to give way.

The evolutionary dynamic equation for the RV selection yield decision is:

$$F_{RV} = \frac{\partial y}{\partial t} = y\left(E_{gw}^{RV} - E_{total}^{RV}\right) = 2y(1-y)\left[(x-1)\beta P_{eff}^{RV} + (1-\beta)xP_{safe}^{RV}\right] \tag{17}$$

The strategy combinations of the LV and RV are (0, 0), (0, 1), (1, 0), and (1, 1), which are obtained by Equations (13) and (17); the corresponding strategy combinations are (do not change lane, do not give way), (do not change lane, give way), (change lane, do not give way), and (change lane, give way). In addition to the above four strategy combinations, there is another set of solutions for $\left(P_{eff}^{RV}/\left(P_{eff}^{RV}+(1/\beta-1)P_{safe}^{RV}\right), P_{safe}^{LV}/\left((\alpha/1-\alpha)P_{eff}^{LV}+P_{safe}^{LV}\right)\right)$.

The Jacobi matrix is constructed for the dynamic game system and expressed as:

$$\text{Jacobi}=\begin{pmatrix}\frac{\partial F_{LV}}{\partial x} & \frac{\partial F_{LV}}{\partial y}\\ \frac{\partial F_{RV}}{\partial x} & \frac{\partial F_{RV}}{\partial x}\end{pmatrix}=\begin{pmatrix}2(1-2x)\left[\alpha yP_{eff}^{LV}+(y-1)(1-\alpha)P_{safe}^{LV}\right] & 2x(1-x)\left(\alpha P_{eff}^{LV}+(1-\alpha)P_{safe}^{LV}\right)\\ 2y(1-y)\left(\beta P_{eff}^{RV}+(1-\beta)P_{safe}^{RV}\right) & 2(1-2y)\left[(1-\beta)xP_{safe}^{RV}+(x-1)\beta P_{eff}^{RV}\right]\end{pmatrix} \tag{18}$$

According to Equation (18), the determinant and trace of the Jacobi matrix are obtained as:

$$\begin{aligned}|\text{Jacobi}|= &\left[2(1-2x)\left(\alpha yP_{eff}^{LV}+(y-1)(1-\alpha)P_{safe}^{LV}\right)\right]\cdot\left[2(1-2y)\left((1-\beta)xP_{safe}^{RV}+(x-1)\beta P_{eff}^{RV}\right)\right]\\ &-\left[2y(1-y)\left(\beta P_{eff}^{RV}+(1-\beta)P_{safe}^{RV}\right)\right]\cdot\left[2x(1-x)\left(\alpha P_{eff}^{LV}+(1-\alpha)P_{safe}^{LV}\right)\right]\end{aligned} \tag{19}$$

$$\text{tr}(\text{Jacobi})=2(1-2x)\left[\alpha yP_{eff}^{LV}+(y-1)(1-\alpha)P_{safe}^{LV}\right]+2(1-2y)\left[(1-\beta)xP_{safe}^{RV}+(x-1)\beta P_{eff}^{RV}\right] \tag{20}$$

The five solutions obtained from Equations (13) and (17) are substituted into Equations (19) and (20) to obtain the stability of each equilibrium point, as shown in Table 3.

Table 3. Stability of each equilibrium point.

Equalization Point	|Jacobi|	tr(Jacobi)	Stability
$(0,0)$	$4\beta(1-\alpha)P_{eff}^{RV}P_{safe}^{LV}$	$-2\left((1-\alpha)P_{safe}^{LV}+\beta P_{eff}^{RV}\right)$	Stable point
$(0,1)$	$4\alpha\beta P_{eff}^{RV}P_{eff}^{LV}$	$2\left(\alpha P_{eff}^{LV}+\beta P_{eff}^{RV}\right)$	Instability point
$(1,0)$	$4(1-\alpha)(1-\beta)P_{safe}^{LV}P_{safe}^{RV}$	$2(1-\alpha)P_{safe}^{LV}+2(1-\beta)P_{safe}^{RV}$	Instability point
$(1,1)$	$4\alpha(1-\beta)P_{eff}^{LV}P_{safe}^{RV}$	$-2\alpha P_{eff}^{LV}-2(1-\beta)P_{safe}^{RV}$	Stable point
$\left(\frac{P_{eff}^{RV}}{P_{eff}^{RV}+(1/\beta-1)P_{safe}^{RV}},\frac{P_{safe}^{LV}}{\alpha/(1-\alpha)P_{eff}^{LV}+P_{safe}^{LV}}\right)$	$-4\frac{\alpha P_{eff}^{LV}P_{safe}^{LV}}{\alpha/(1-\alpha)P_{eff}^{LV}+P_{safe}^{LV}}\frac{\beta P_{eff}^{RV}P_{safe}^{RV}}{\beta/(1-\beta)P_{eff}^{RV}+P_{safe}^{RV}}$	0	Saddle point

According to the above stability determination, we can achieve the stable equilibrium points (0, 0) and (1, 1), whose corresponding stable strategies are (do not change lane, do not give way) and (change lane, give way), while the unstable points are (0, 1) and (1, 0), whose corresponding strategies are (do not change lane, give way) and (change lane, do not give way), $\left(P_{eff}^{RV}/\left(P_{eff}^{RV}+(1/\beta-1)P_{safe}^{RV}\right), P_{safe}^{LV}/\left((\alpha/1-\alpha)P_{eff}^{LV}+P_{safe}^{LV}\right)\right)$ is the saddle point, and the convergence direction to (0, 0) or (1, 1) is indefinite.

5. Simulation Results and Analysis

In this paper, SUMO (Simulation of Urban Mobility) software is used for simulation validation, simulation control, and decision planning of vehicle game strategies through the Traci interface. SUMO is an open-source and microscopic traffic simulation software for road traffic simulation. It was developed mainly by the staff of the Institute for Transportation Systems at the German Aerospace Center. We are using version sumo-1.17.0, which can be obtained in detail at https://sumo.dlr.de/docs/index.html (accessed on 21 June 2023). These vehicles adopt the CACC following model [30] and use quintic polynomials to plan the horizontal and longitudinal lane-change trajectory, The CACC model is based on vehicle–vehicle communication to obtain information about the surrounding vehicles and has the characteristics of accuracy and smaller time delay. The expression of the CACC model is:

$$e=x_{i-1}-x_i-Tv_i \tag{21}$$

$$v_i(t)=v_i(t-1)+k_1e+k_2\dot{e} \tag{22}$$

where e is the difference between the actual following distance and the desired distance; x_i is the displacement of the vehicle i; v_i is the velocity of the vehicle i; x_{i-1} is the displacement of the vehicle $i-1$; T is the minimum safe headway time distance; and k_1 and k_2 are parameters.

We turn off the default lane-changing model by Traci's command and plan the lane-changing trajectory by using quintic polynomials. Quintic polynomials can obtain the minimum jerk value to satisfy the constraints and meet the comfort of a lane change. The quintic polynomials are used to plan the horizontal and vertical directions, respectively, and track the trajectory points by Traci's command to complete the lane-change process. The lane-change trajectory polynomials are:

$$x(t) = a_0t^5 + a_1t^4 + a_2t^3 + a_3t^2 + a_4t + a_5 \tag{23}$$

$$y(t) = b_0t^5 + b_1t^4 + b_2t^3 + b_3t^2 + b_4t + b_5 \tag{24}$$

where $a_i(i = 0,1,2,3,4,5)$ is the polynomial coefficient of the longitudinal trajectory; $b_i(i = 0,1,2,3,4,5)$ is the polynomial coefficient of the horizontal trajectory; and the trajectory constraints are determined by the constraint of the starting and ending points of the path.

To elucidate the strategy evolution process of the vehicle lane-change game and to perform simulation validation, the road geometry characteristics and vehicle characteristics data of the intersection of Renmin Road and Yongzhou Road in Qingdao City, Shandong Province, were statistically collected and the main parameter values were set in SUMO software, as shown in Table 4. The decision response time indicates the minimum time from the beginning of the vehicle decision to the completion of the decision behavior, and the meaning of the other parameters can be obtained from the variable names.

Table 4. Main parameters of the model.

Variable Name	Variable Meaning	Numerical Value
t_{simu}	Simulation step/s	0.1
t_d	Decision response time/s	3.0
l_{car}	Vehicle length/m	4.0
d_{max}	Maximum brake deceleration/(m·s^{-2})	3.0
a_{max}	Maximum acceleration/(m·s^{-2})	2.3
v_{max}	Road speed limit/(m·s^{-1})	13.89
S_{min}	Minimum longitudinal distance/m	2.5
$gapl_{min}$	Minimum lateral distance/m	1.5
T	Signal cycle length/s	132

5.1. Sensitivity Analysis

The evolutionary convergence process of the decision path is obtained by simulation, and the evolutionary paths are obtained by selecting the initial values of the initial lane-change ratio and the give-way ratio between 0.1 and 0.9, as illustrated in Figure 3. When the initial values of x and y are smaller, the final decision will converge to not lane change and not give way. When the initial values of x and y are larger, the final decision will converge to a combination of the lane-changing and giving-way strategies.

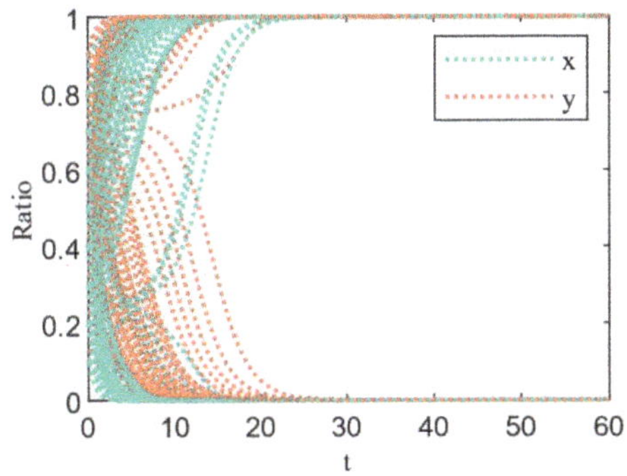

Figure 3. The convergence process of stabilizing strategy evolution.

In constructing the payoff functions of the game vehicles, the effects of lane-change position and the remaining time of green light on the vehicle decision payoff need to be considered, so a sensitivity analysis is conducted to explore the effects of the two main influencing factors mentioned above.

It can be seen in Figure 4 that when the lane-changing position decreases, the evolutionary stabilization strategy evolves from converging on not changing lanes and not giving way to converging on changing lanes and giving way. When the LV is more affected by road traffic and needs to take the lane-changing decision, the increase of the LV's expectation of lane change increases the weight of efficiency gains, and the dynamic evolutionary game results show that the RV should provide a lane-changing gap for the LV based on the goal of revenue maximization; otherwise, it will increase vehicle collision risk and reduce their expected revenue. The convergence rate of the strategy combination increases significantly when the lane-change position increases or decreases, indicating that the vehicles make decisions that match the best strategy for the current traffic environment.

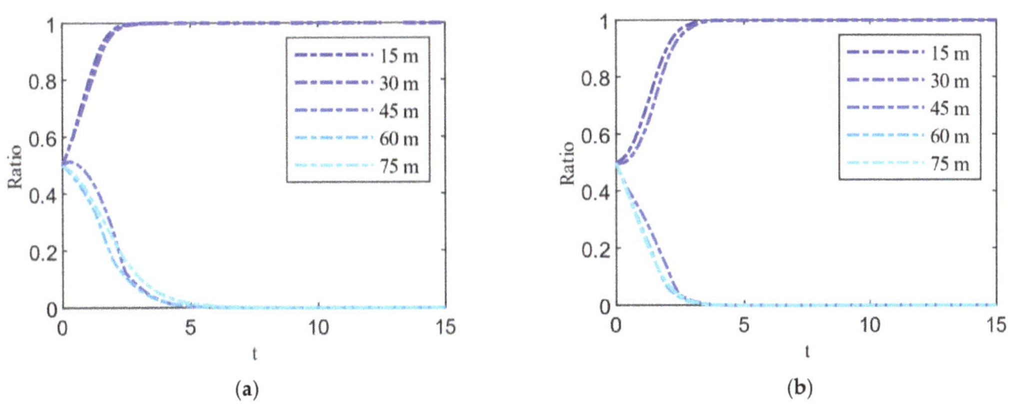

Figure 4. Evolutionary convergence of the proportion of lane-changing strategy under different lane-change locations: (**a**) LV; (**b**) RV.

It can be seen in Figure 5 that when the lane-changing position is certain and the remaining green time increases, the evolutionary game lane-changing stabilization strategy evolves from not changing lanes and not giving way to changing lanes and giving way. When the remaining green time is smaller, the RV expects to pass the intersection as soon as possible so that the expected weight of efficiency increases. When the remaining green light time is greater than 18 s, there is a partial overlap of the convergence curves, but the evolutionary trend remains the same. It can be seen from the dynamic evolutionary game results that the RV makes the decision to not give way, and the LV makes the decision to not change lanes to maximize its benefit, but the convergence rate of the decision of not changing lanes is lower than that of changing lanes. When the remaining time of the green light increases, the convergence speed of the RV and LV strategy combination is faster, indicating that the strategy combination meets the expectations of both.

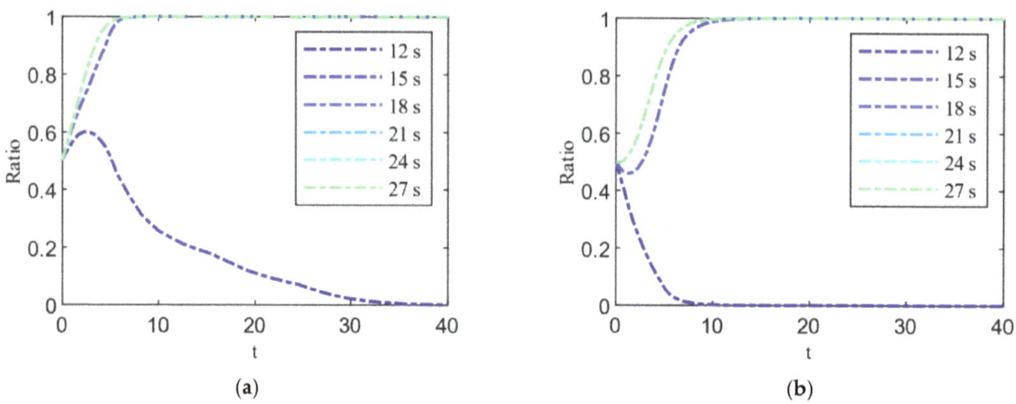

Figure 5. Evolutionary convergence of the proportion of lane-changing strategy under different remaining green time: (**a**) LV; (**b**) RV.

5.2. Analysis of Lane-Change Trajectory

In the simulation scenarios, the center point of the intersection is taken as the origin of the coordinate axis, and the driving direction of the straight vehicle is taken as the positive direction of the x-axis. The lane-changing game scenario is obtained by filtering the vehicle trajectory, and the vehicle trajectory can intuitively reflect the interaction between vehicles and the driving status. In this paper, SUMO's built-in SL2015 lane-change model and the non-cooperative mixed strategy game model are used as comparison models. Among them, the SL2015 lane-change model is the default model for sub-lane scenarios, which can simulate the microscopic lane-change process of vehicles and reflect the interaction state between vehicles. The non-cooperative mixed strategy model is a decision model that maximizes its gain, and previous studies have shown that this model can better predict the decision outcome of the other player [31].

Figure 6 presents the vehicle trajectory curves, the red line in the local zoomed-in figure is the LV trajectory and the black line is the RV trajectory. Figure 6a shows the SL2015 lane-changing model in SUMO, which is a non-game model reflecting the vehicle interaction in the situation where vehicles make decisions as independent individuals, and it can be seen in the figure that the lane-changing process of vehicles will have a large oscillating effect on the traffic flow, making the slope of the curve of the rear vehicle change more. Figure 6b is a non-cooperative mixed strategy game, although the model considers and predicts the other party's decision. The model is aimed at maximizing their interests, so that there is some conflict between the two parties' decisions, as indicated by the amplification curve, with the strategy of the LV lane change forcing the RV to slow down, and the slope of the RV's curve changes drastically. Figure 6c,d are the evolutionary game-based vehicle lane-change models proposed in this paper, with Figure 6c being the lane-changing and giving-way scenario, and Figure 6d being the no-lane-changing and no-giving-way scenario. From the dynamic evolution equations above, it is known that the evolutionary game lane-change model will converge on stable strategy combinations, while non-stable strategy combinations and saddle points are eliminated by the model as non-optimal decisions. The vehicle trajectory further indicates that the vehicle lane-change process of this model has less perturbation to the vehicles behind, and the stable strategy separates the vehicle risk in time and space, which improves the feasibility of vehicle lane changes.

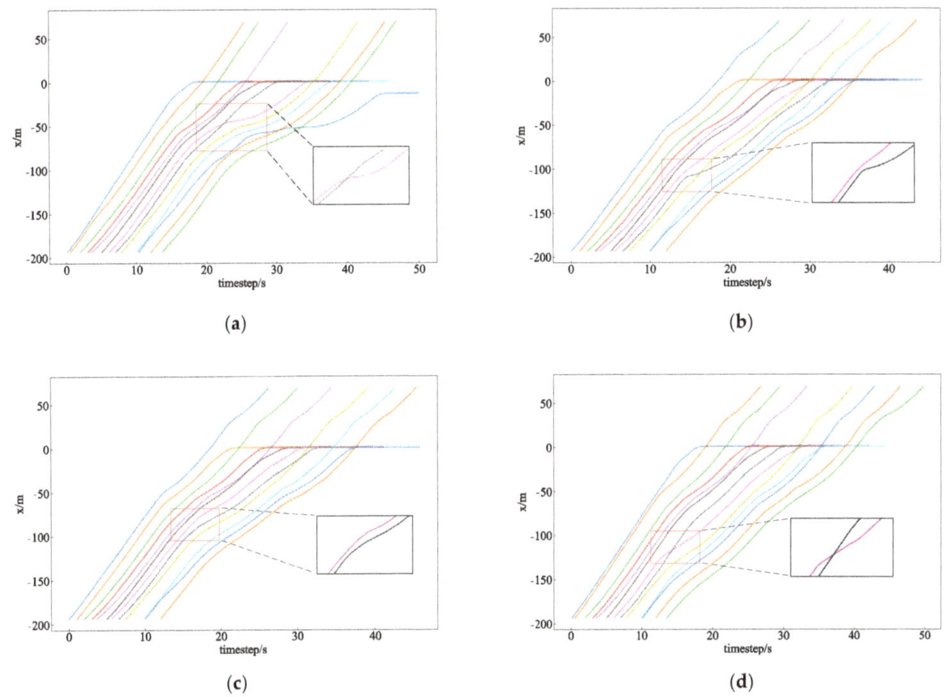

Figure 6. Comparison of lane-change trajectories: (**a**) SL2015; (**b**) The model of mixed strategy game; (**c**,**d**) The model of this paper.

5.3. Analysis of Security

The TTC (Time-To-Collision) [32] indicator is used to portray the collision time between vehicles, reflecting the risk conflict between vehicles; when the rear vehicle speed is greater than the front vehicle, the smaller TTC indicates that the risk collision is more serious.

$$TTC_i = \frac{x_{i-1} - x_i - l_{car}}{v_{i-1} - v_i} \tag{25}$$

where TTC_i is the time of collision between the vehicle and the front vehicle; v is the speed of the vehicle; x is the position of the vehicle; and l_{car} is the length of the vehicle.

To display the trend of the TTC value, we limit the TTC value within [–50, 50]. When the TTC value is larger, the vehicles will not collide; when the TTC value is less than 0, it means that the rear vehicle speed is less than the front vehicle speed, and they will not collide. In Figure 7, the black line represents the SL2015 lane-switching model, the red line represents the decision-conflict situation in the mixed-strategy game model, and the blue line is the evolutionary game model proposed in this paper. It can be seen in Figure 7, in the SL2015 lane-changing model and mixed strategy game of the vehicle lane-changing process, the TTC value quickly decreases to within 15 and tends to 0, indicating that the collision risk between the rear vehicle and the front vehicle is larger, while in the evolutionary game-based lane-changing model, the TTC value is usually greater than 15 or less than 0. When the TTC value is less than 0, it means that the rear vehicle speed is less than the front vehicle speed, and the two sides will not collide when driving at the current speed. In Figure 7, we can see that in the evolutionary game model, the rear car decides to give way to reduce the risk of vehicle collision, the TTC value is reduced to less than 0, and the two sides will not collide.

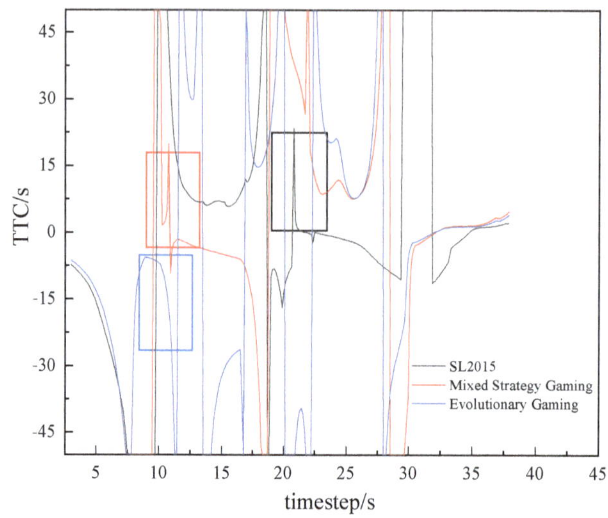

Figure 7. The diagram of Time-To-Collision.

Although the mixed strategy game model has optimal solutions and is usually used to predict each other's decisions, it still has a certain probability of decision conflict, which puts vehicle safety at risk; therefore, the model needs strong constraints to reduce the occurrence of decision conflict. The evolutionary game model, on the other hand, converges on feasible strategies from the perspective of dynamic evolutionary equations and excludes unstable strategies, thus reducing the risk of vehicle conflict.

5.4. Threats to Validity

Validity threats can usually be divided into two parts: internal and external validity. In the proposed model, the possible validity threats include the values of parameters that affect the outcome of vehicle decisions and the results of solving the strategy combinations based on game theory. To study the influence of the change in the values of the main influencing factors on the evolutionary game process of vehicle lane changes, we conducted a sensitivity analysis of the main influencing factors by keeping other parameters consistent and obtained that the change in the values of the remaining time of the green light and the location of the lane-change game affect the convergence direction of the stabilization strategy.

Unlike other game models, the decision results based on the evolutionary game only converge to the optimal strategy combinations, thus reducing the occurrence of uncertainties and unstable strategy combinations, so that the vehicle decision results will converge to stable strategy combinations based on the assumption that the vehicle obeys the decision instructions issued by the upper control system. The modeling thought of the vehicle lane-change model based on game theory is to achieve the optimal strategy by constructing the gain matrix and solving the equation. In constructing the gain matrix, vehicle speed, safety, and comfort are usually considered: vehicle speed gain is expressed in the change of vehicle speed, safety gain is expressed in the safety risk of the vehicle, and comfort gain is expressed in the smoothness of vehicle speed change. In this paper, the gain of the game subject consists of vehicle driving efficiency and safety, comfort by the change of speed comfort is achieved by the trajectory planning part, to build the gain function that considers passengers' expectations of vehicle decision, and to expand the applicability of the model in the face of the traffic bottleneck section and reduce the external validity.

6. Conclusions

In this paper, the vehicle lane-change issue is modeled based on evolutionary game theory. Specifically, in the intersection road scenario, the evolutionary game models of the RV and LV are established by analyzing the game lane-change interaction characteristics of vehicles, and the dynamic evolution equations under different situations are analyzed. The revenue matrix is constructed considering passengers' expectations so that the vehicle decision is neither too aggressive nor too conservative. In addition, the game system converges in different directions when the vehicle lane-change position or the remaining green light time changes, and thus different optimal strategy combinations are obtained.

The results show that the evolutionary game-based vehicle lane-change model can effectively reduce the risk conflict of vehicles. When the decision system converges to lane change and yield, the RV can slow down in advance to provide space for the LV to change lanes; when the decision system converges to no lane change and give way, the LV slows down while the RV maintains speed or accelerates; and when the RV overtakes the LV, the LV changes lanes. Although the evolutionary game is a non-cooperative game model, the results of dynamic evolution show that the system will converge to a stable strategy, which can be regarded as a cooperative decision combination.

Traffic safety is a key issue for future research in the field of transportation, and the risk conflict in the process of vehicle lane changes cannot be ignored; therefore, it is important to analyze and study the characteristics of vehicle decision interactions. In future research, multi-vehicle interaction characteristics of ICVs and a reduction of traffic oscillation are game models worth considering.

Author Contributions: Conceptualization, D.Q. and S.D.; methodology, S.D.; software, A.L.; validation, D.Q., S.D. and Y.C.; investigation, C.W.; writing—original draft preparation, S.D.; writing—review and editing, D.Q. All authors have read and agreed to the published version of the manuscript.

Funding: This research was funded by National Natural Science Foundation of China: 52272311.

Institutional Review Board Statement: Not applicable.

Informed Consent Statement: Not applicable.

Data Availability Statement: Not applicable.

Conflicts of Interest: The authors declare no conflict of interest.

References

1. Liu, T.; Wang, C.; Fu, R.; Ma, Y.; Liu, Z.; Liu, T. Lane-Change Risk When the Subject Vehicle Is Faster Than the Following Vehicle: A Case Study on the Lane-Changing Warning Model Considering Different Driving Styles. *Sustainability* **2022**, *14*, 9938. [CrossRef]
2. Mahmoud, R.; Yousuf, T.; Aloul, F.; Zualkernan, I. Internet of things (IoT) security: Current status, challenges, and prospective measures. In Proceedings of the 2015 10th International Conference for Internet Technology and Secured Transactions (ICITST), London, UK, 14–16 December 2015; pp. 336–341.
3. Bures, M.; Klima, M.; Rechtberger, V.; Ahmed, B.S.; Hindy, H.; Bellekens, X. Review of Specific Features and Challenges in the Current Internet of Things Systems Impacting Their Security and Reliability. In *Trends and Applications in Information Systems and Technologies*; AISC: Chicago, IL, USA, 2021; Volume 1367, pp. 546–556.
4. Sun, K.; Zhao, X.; Gong, S.; Wu, X. A Cooperative Lane Change Control Strategy for Connected and Automated Vehicles by Considering Preceding Vehicle Switching. *Appl. Sci.* **2023**, *13*, 2193. [CrossRef]
5. Van den Beukel, A.P.; Van Driel, C.J.G.; Boelhouwer, A.; Veders, N.; Heffelaar, T. Assessment of Driving Proficiency When Drivers Utilize Assistance Systems—The Case of Adaptive Cruise Control. *Safety* **2021**, *7*, 33. [CrossRef]
6. Ma, S.; Hu, J.; Wang, R.; Qu, S. Study on the Median Opening Length of a Freeway Work Zone Based on a Naturalistic Driving Experiment. *Appl. Sci.* **2023**, *13*, 851. [CrossRef]
7. Kuang, Z.; Zhao, X.; Feng, L. Priority of Emergency Vehicle Dynamic Right-of-Way Control Method in Networked Environment. *Appl. Sci.* **2023**, *13*, 5883. [CrossRef]
8. Wang, C.; Du, Y. Lane-Changing Strategy Based on a Novel Sliding Mode Control Approach for Connected Automated Vehicles. *Appl. Sci.* **2022**, *12*, 11000. [CrossRef]
9. Zhang, L.; Wang, Y.; Zhu, H. Theory and Experiment of Cooperative Control at Multi-Intersections in Intelligent Connected Vehicle Environment: Review and Perspectives. *Sustainability* **2022**, *14*, 1542. [CrossRef]

10. Chen, Q.J.; Xie, Y.X.; Ao, Y.; Li, T.G.; Chen, G.R.; Ren, S.F.; Wang, C. A deep neural network inverse solution to recover pre-crash impact data of car collisions. *Transp. Res. Part C Emerg. Technol.* **2021**, *126*, 103009. [CrossRef]
11. Xie, Y.X.; Wu, C.T.; Li, B.Y.; Hu, X.; Li, S.F. A feed-forwarded neural network-based variational Bayesian learning approach for forensic analysis of traffic accident. *Comput. Methods Appl. Mech. Eng.* **2022**, *397*, 115148. [CrossRef]
12. Qu, D.Y.; Zhang, K.K.; Song, H.; Dai, S.C. Analysis and Modeling of Lane-Changing Game Strategy for Autonomous Driving Vehicles. *IEEE Access* **2022**, *10*, 69531–69542. [CrossRef]
13. Yao, R.H.; Du, X.J.; Zhang, W.S.; Zhang, H. Game Model of Lane Changing for Bus Exiting at Bus Bay Stops. *J. Transp. Eng.* **2022**, *29*, 319–329. [CrossRef]
14. Arbis, D.; Dixit, V.V. Game theoretic model for lane changing: Incorporating conflict risks. *Accid. Anal. Prev.* **2019**, *125*, 158–164. [CrossRef] [PubMed]
15. Dai, S.C.; Qu, D.Y.; Wang, Q.K.; Meng, Y.M.; Yang, Y.F. Characterization and model of Connected and Automated Vehicles lane-changing game on upstream sections of signalized intersections. *Sci. Technol. Eng.* **2023**, *23*, 4408–4416.
16. Hang, P.; Lv, C.; Huang, C.; Xing, Y.; Hu, Z.X.; Cai, J.C. Human-like Lane-Change Decision Making for Automated Driving with a Game Theoretic Approach. In Proceedings of the 2020 4th CAA International Conference on Vehicular Control and Intelligence (CVCI), Hangzhou, China, 18–20 December 2020; pp. 708–713.
17. Sun, K.; Zhao, X.M.; Wu, X. A cooperative lane change model for connected and autonomous vehicles on two lanes highway by considering the traffic efficiency on both lanes. *Transp. Res. Int. Pers.* **2021**, *9*, 100310. [CrossRef]
18. Zheng, Y.; Ding, W.T.; Ran, B.; Qu, X. Coordinated decisions of discretionary lane change between connected and automated vehicles on freeways: A game theory-based lane change strategy. *IET. Intell. Transp. Syst.* **2020**, *14*, 1864–1870. [CrossRef]
19. Liu, R.; Zhao, X.; Zhu, X.C.; Ma, J. A Human-like Shared Driving Strategy in Lane-Changing Scenario Using Cooperative LPV/MPC. *IEEE Trans. Intell. Transp. Syst.* 2023, *in press*. [CrossRef]
20. Yu, Y.W.; Liu, S.K.; Wang, M.X. Multi-Player Dynamic Game-Based Automatic Lane-Changing Decision Model under Mixed Autonomous Vehicle and Human-Driven Vehicle Environment. *Transp. Res. Rec.* **2020**, *2674*, 165–183. [CrossRef]
21. Pan, S.; Wang, Y.; Wang, K. A Game Theory-based Model Predictive Controller for Mandatory Lane Change of Multiple Vehicles. In Proceedings of the 2020 4th CAA International Conference on Vehicular Control and Intelligence (CVCI), Hangzhou, China, 18–20 December 2020; pp. 731–736.
22. Hang, P.; Huang, C.; Hu, Z.X.; Lv, C. Decision Making for Connected Automated Vehicles at Urban Intersections Considering Social and Individual Benefits. *Elec. Eng. Syst. Sci.* **2022**, *23*, 22549–22562. [CrossRef]
23. Smirnov, N.; Liu, Y.; Validi, A.; Morales-Alvarez, W.; Olaverri-Monreal, C. A Game Theory-Based Approach for Modeling Autonomous Vehicle Behavior in Congested, Urban Lane-Changing Scenarios. *Sensors* **2021**, *21*, 1523. [CrossRef]
24. Zhang, X.; Gao, J.; Liao, L.; Wu, G. Expressway Lane Change in Fog Environment by Dynamic Strategic Game. *J. Adv. Transport.* **2022**, *22*, 8612955. [CrossRef]
25. Wang, G.; Hu, J.; Li, Z.; Li, L. Harmonious Lane Changing via Deep Reinforcement Learning. *IEEE Trans. Intell. Transp. Syst.* **2021**, *23*, 4642–4650. [CrossRef]
26. Kang, K.; Rakha, H.A. A Repeated Game Freeway Lane Changing Model. *Sensors* **2020**, *20*, 1554. [CrossRef]
27. Chen, W.; Ren, G.; Cao, Q.; Song, J.; Liu, Y.; Dong, C. A Game-Theory-Based Approach to Modeling Lane-Changing Interactions on Highway On-Ramps: Considering the Bounded Rationality of Drivers. *Mathematics* **2023**, *11*, 402. [CrossRef]
28. Ye, M.; Li, P.; Yang, Z.; Liu, Y. Research on Lane Changing Game and Behavioral Decision Making Based on Driving Styles and Micro-Interaction Behaviors. *Sensors* **2022**, *22*, 6729. [CrossRef] [PubMed]
29. Ji, K.; Orsag, M.; Han, K. Lane-Merging Strategy for a Self-Driving Car in Dense Traffic Using the Stackelberg Game Approach. *Electronics* **2021**, *10*, 894. [CrossRef]
30. Milanés, V.; Shladover, S.E. Modeling cooperative and autonomous adaptive cruise control dynamic responses using experimental data. *Transp. Res. Part C Emerg. Technol.* **2014**, *48*, 285–300. [CrossRef]
31. Sheikh, M.S.; Wang, J.; Regan, A. A game theory-based controller approach for identifying incidents caused by aberrant lane changing behavior. *Phys. A Stat. Mech. Appl.* **2021**, *580*, 126162. [CrossRef]
32. Jiang, Y.S.; Hu, R.; Yao, Z.H.; Wu, P.C.; Luo, X.L. Stability and safety analysis for heterogeneous traffic flow composed of intelligent and connected vehicles. *J. Beijing Jiaotong Univ.* **2020**, *44*, 27–33.

applied sciences

MDPI

Communication

Integral Analysis of a Vehicle and Electric Power Steering Logic for Improving Steering Feel Performance

Seunghoon Woo [1,*], Chanwoo Heo [2], Man-Ok Jeong [3] and Jun-Mo Lee [4]

[1] Department of Automotive Engineering, Kookmin University, Seoul 02707, Republic of Korea
[2] Graduate School of Automotive Engineering, Kookmin University, Seoul 02707, Republic of Korea; dezart159@kookmin.ac.kr
[3] Research & Development Division, Hyundai Motor Company, Hwaseoung-si 18280, Republic of Korea
[4] Hyundai-Kia America Technical Center, Inc., California City, CA 93505, USA
* Correspondence: boltra@kookmin.ac.kr

Abstract: This research aims to investigate steering feel by analyzing a steering system and an electric power steering logic. First, steering feel is defined based on previous research, and methods for evaluating it are discussed. Second, a sensitivity analysis is conducted by modeling our developed vehicle and that of a competitor known for its excellent steering feel via a multi-body simulation. We then propose a straightforward method to determine the parameters associated with steering feel to achieve the desired steering characteristics. Last, by modifying the electric power steering control system, we achieve a steering feel in our vehicle that matches the desired steering characteristics.

Keywords: steering feel; electric power steering; reverse engineering; vehicle dynamics

Citation: Woo, S.; Heo, C.; Jeong, M.-O.; Lee, J.-M. Integral Analysis of a Vehicle and Electric Power Steering Logic for Improving Steering Feel Performance. *Appl. Sci.* **2023**, *13*, 11598. https://doi.org/10.3390/app132011598

Academic Editors: Edgar Sokolovskij and Vidas Žuraulis

Received: 3 September 2023
Revised: 11 October 2023
Accepted: 17 October 2023
Published: 23 October 2023

1. Introduction

When a vehicle is driven in a nominally straight line but encounters large-radius bends at high speeds and low lateral accelerations ($\leq$0.2 g), the handling characteristic perceived by the driver is known as on-center steering feel. According to [1,2], the majority of drivers operate their vehicles within the on-center steering area. Therefore, in general, the steering feel experienced in the on-center region significantly influences the daily driving experience of drivers.

Steering feedback in the on-center driving scenario is influenced by various factors. In [3–5], researchers analyzed the effects of steering mechanism characteristics, nonlinearity of assist torque, friction, and vehicle lateral response. In another study [6], additional analyses were conducted on factors such as the steering ratio, caster angle, steering system stiffness, and tire characteristics. Furthermore, refs. [7,8] focused on analyzing the influence of friction and the angle of the universal joint on the steering mechanism. The aforementioned studies confirmed that steering feedback is influenced by various factors. This implies that even with modifications in the vehicle design based on an analysis of these influencing factors, achieving the desired steering feel may be difficult due to trade-offs with other factors. Studies [9] indicate that design modifications can improve steering feedback. However, changing the design typically incurs a greater expense than software adjustments. Additionally, beyond certain stages of development, implementing design changes can become challenging. Consequently, achieving the desired steering feel requires significant cost and time investments in the design phase.

To expedite the development process and reduce costs, researchers have made various attempts to build a reliable simulation model for analyzing steering characteristics. In [4,8,10], steering systems were implemented and analyzed through mathematical modeling. This approach is relatively simple to implement and suitable for parameter studies. However, it cannot precisely replicate nonlinearities in a real vehicle. In [11,12], nonlinear elements were additionally considered. Moreover, a vehicle model capable of representing the lateral characteristics of the vehicle was implemented, allowing for an interpretation from the perspective of the entire vehicle, rather than one limited to the steering system. However, according to [2], the elasticity of vehicle components influences handling characteristics. Especially in the on-center area, where factors such as the steering column and friction and chassis elasticity have an impact on the steering characteristics, considering these factors in the simulation environment is crucial in establishing reliable simulation models.

In [13], a notable study, the authors implemented a full vehicle model in a multi-body simulation environment and conducted model validation to ensure the reliability of both the vehicle and steering system model. Using the validated model, the electric power steering (EPS) control system was tuned to achieve the same level of steering feel as the target vehicle. However, they did not analyze the target vehicle. This means that they could not suggest which component should be modified in the next stage, as the desired steering feel could not be achieved through EPS controller tuning. Moreover, they did not mention which features of the EPS controller were tuned and how this was performed. Therefore, we extend these previous works, including the one mentioned above, by analyzing the reasons for the differences in steering feel between our vehicle and a competitor's vehicle. Based on this analysis, we suggest a method to design the EPS controller so that its steering feel matches that of the target vehicle. Additionally, by identifying the individual impact of major factors affecting steering feel, we provide various options for modifying the steering feel.

Steering feel refers to the sensation experienced by a driver when performing steering actions. It is subjective and dependent on the driver's preferences. In [12], subjective expressions of steering feel were defined. Specifically, the factors contributing to a trained driver negatively perceiving "inertia feel" remain unclear. Even if engineers identify the factors, how they can be improved and to what extent remain unclear. Therefore, steering feel should be quantified and defined, which will serve as the criteria for setting improvement methods and goals.

In 1984, Kenneth D. Norman proposed experimental evaluation standards [14], known as the steering feeling index for the on-center handling area, by measuring the steering wheel angle, steering torque, and lateral acceleration. The measured values were obtained by inputting steering angles at a frequency of 0.2 Hz, which resulted in a lateral acceleration of approximately 0.2 g. Thereafter, other studies expanded on Norman's proposal by introducing refined indices, as listed in Table 1.

In [15], the authors aimed to quantify vehicle handling by utilizing the time lag between steering angle and yaw rate. However, this approach was not widely adopted in subsequent studies. It was more common to use lateral acceleration to evaluate vehicle response. In [16], the authors interpreted the impact of roll rate on steering feel by considering changes caused by vehicle motion. This consideration was for the interpretation of time lag, and as mentioned in the conclusions of the paper, it did not exhibit a strong correlation. In [17], two novel evaluation methods were introduced. The first was the flick test, where a sharp steering angle is inputted quickly, followed by releasing the steering wheel. This test was performed to verify returnability and assess the torque perceived by the driver during emergency maneuvers. The second method was applied to assess torque/angle in the frequency domain to evaluate vehicle response. In [18], various steering input patterns were presented. While many previous studies derived steering feel indices using sinusoidal steering input, the authors here utilized sine steer but with a more detailed breakdown to effectively examine on-center steering characteristics. Moreover, step steer, which, after sine steer, is another commonly used input pattern, was presented.

Lastly, [19] conducted a correlation analysis to derive indices that exhibit a high correlation with the driver's subjective evaluation from among indices defined by different manufacturers and organizations. This correlation analysis established criteria for assessing whether the steering characteristics of the vehicle under development have been successfully implemented according to the design. Recent studies, [20,21] that utilized driver-in-the-loop simulations still performed evaluations using the indices and maneuvers mentioned in [19] and Table 1. In this study, based on the findings from various research, we define the steering feel indices in Sections 2.1 and 2.2.

Table 1. Steering feel evaluation indices and maneuvers.

Related Previous Work	Index	Maneuver
Kenneth D. Norman, 1984 [14]	SWA *, SWT **, lateral acc.	Sinusoidal angle input 0.2 Hz/approx. 0.2 g
David G. Farrer, 1993 [15]	SWA, SWT, lateral acc., yaw rate, SWA–yaw rate time lag	Sinusoidal angle input 0.2 Hz/approx. 0.1 g
		Slow ramp angle input 2°/s
Akira Higuchi et al., 2001 [16]	SWA, SWT, yaw rate, roll rate SWA–lateral acc. time lag	Sinusoidal angle input 0.2 Hz/approx. 0.25 g
M. Kamel Salaani et al., 2004 [17]	SWA, SWT, lateral acc.	Sinusoidal angle input 0.2 Hz/approx. 0.2 g
		Slow ramp angle input Not specified [°/s]
		Impulse angle input 60° and hand off
	Transfer function (In: SWA/out: SWT)	Sinusoidal angle input ≤1.5~≥2.5 Hz
Andrew Heathershaw et al., 2006 [18]	SWA, SWT, yaw rate gain	Sinusoidal angle input Half sinusoidal angle input Single sinusoidal angle input 0.2~0.33 Hz/approx. 0.2 g
		Slow ramp angle input 10~30°/s
		Step angle input Not specified [deg]/ hold 2~3 s

* SWA (steering wheel angle); ** SWT (steering wheel torque).

This study aimed to evaluate steering feel performance by analyzing our vehicle and a competitor's vehicle, known for its excellent steering feel, using a multi-body simulation model. Constructing a full vehicle model in the simulation environment enables an independent analysis of the various factors that influence steering feel, making it essential for a comprehensive interpretation of steering feel. While previous studies focused on analyzing the individual elements that affect steering feel, we approached the analysis from a component perspective to identify the most efficient direction to modify the steering feel of our vehicle. For example, if achieving the desired steering feel through the least costly method, such as tuning the EPS control system, is not possible, the analysis becomes even more important. It then becomes necessary to consider whether developing a new control system or conducting a suspension design revision with the tuned existing control system would be more effective in achieving the desired steering feel. However, in situations where the influence of the control system alone cannot resolve the issue, the aforementioned choices may be initially unsuitable. Therefore, in this study, we conducted a sensitivity analysis using our vehicle and a target vehicle to determine how to modify the factors and derive the most efficient combination when modifying a single factor does not yield the desired objectives. These benefits outweigh the costs associated with reverse engineering. Based on the aforementioned background, our initial focus was on establishing objective indices for steering feel through subjective–objective correlation studies. Subsequently, we reverse-engineered the competitor vehicles. Lastly, we proposed a design method to overcome the trade-off relationship between reducing initial torque due to friction and increasing stability by damping, enabling us to emulate the desirable steering feel of the reference vehicle.

2. Identifying Steering Feel Factors

2.1. Steering Feel Performance Index

In this study, we utilized the indices commonly selected in previous research defining steering feel. All the indices were measured by the driver or a steering robot while inputting the steering angle. First, two indices could be observed when the steering angle was used as the input, and the corresponding steering torque was measured. As depicted in Figure 1a, the steering torque at a 0° steering angle represents an important aspect related to perceived differences and the change in steering torque with respect to the steering angle. Additionally, the slope of the steering torque at 0° was defined as the stiffness.

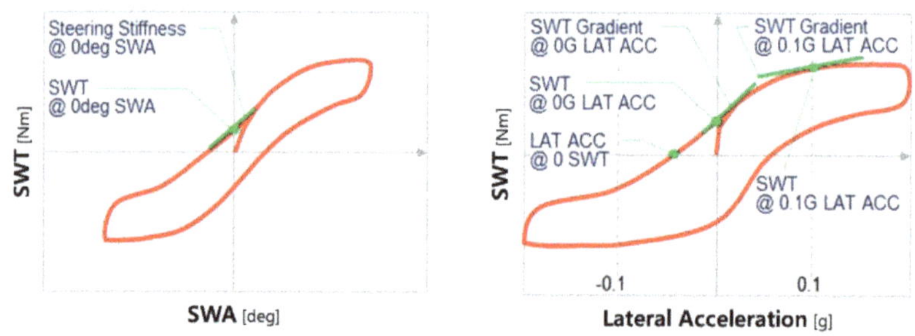

(a) (b)

Figure 1. (**a**) Steering feel performance indices in steering wheel angle and torque graph; (**b**) steering feel performance indices in lateral acceleration and steering wheel torque graph.

Second, by considering the relationship between steering torque and lateral acceleration, multiple indices can be examined, including steering returnability, off-center steering effort, and torque build-up linearity, as shown in Figure 1b. In [15], lateral acceleration and yaw velocity were defined as representing the responsiveness of the vehicle, as illustrated in Figure 2. However, most researchers consistently consider lateral acceleration, but only a few studies take yaw rate into account. According to [19], all the information obtainable from the yaw rate can also be derived from lateral acceleration. Consequently, to verify the reliability of our simulation model and assess any improvements in steering feel, we employed steering angle, steering torque, and lateral acceleration.

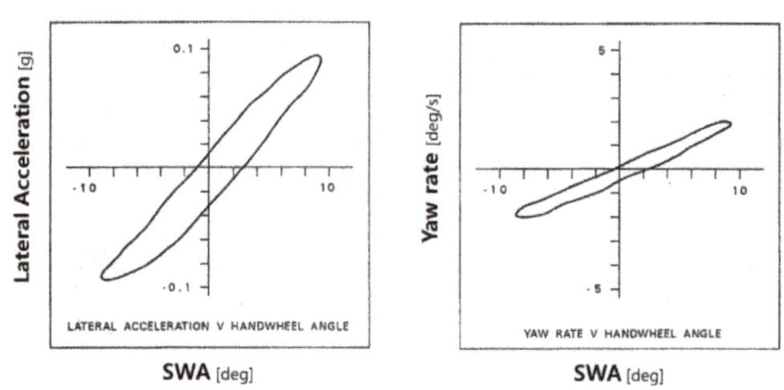

Figure 2. Steering response performance factors (Farrer [15], 1993).

2.2. Test Method of On-Center Handling Performance

The test scenarios to obtain the defined steering performance indices were as follows. One scenario was the weave test, which involves applying a sinusoidal steering angle input of 0.2 Hz at a high vehicle speed (100 km/h) to generate a lateral acceleration of 0.2 g, as shown in Figure 3. Under these conditions, the vehicle reaches a steady state after one cycle, and the quantitative indices can be obtained using the values of the steering angle and steering torque at a specific lateral acceleration. This test scenario is defined in ISO 13674-1 [22].

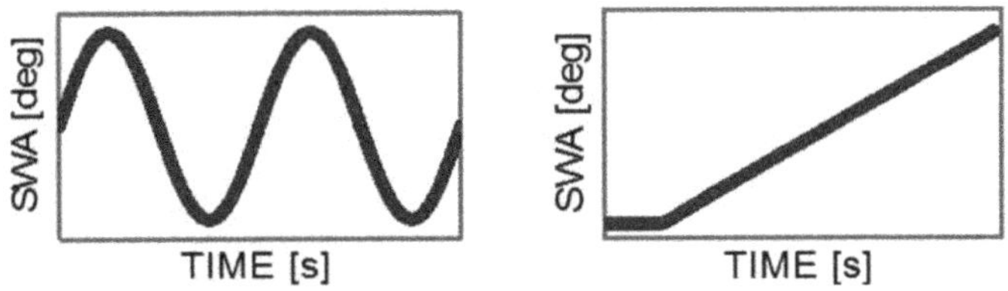

Figure 3. Steering feeling test methods (ISO 13674-1, 2).

The other scenario was the transition test, which involves providing a ramp steering wheel input of 5°/s at a high vehicle speed (100 km/h). Unlike the weave test, this test focuses on capturing the transient response during the early stage of steering and primarily reveals the steady-state response. This test scenario is also defined in ISO 13674-2 [23].

In this study, we classified the handling as on-center and off-center based on the knee point of the transition test result curve. In the on-center section, when the steering angle is inputted, the steering torque response is sharply increased, and when the steering torque is inputted, the steering angle response is slowly increased due to the steering friction. Therefore, the response curves of the on-center section are highly nonlinear as compared to those of the off-center section. The two test scenarios defined by ISO were utilized to verify the abovementioned steering characteristics.

2.3. *Analysis of the Steering Feel Mechanism*

In this section, the free-body diagrams from the steering wheel to the tire are presented as three frames. We utilized them to analyze the factors affecting steering feel and in what direction our vehicle should be designed to achieve the steering feel of the target vehicle [24,25]. Figure 4 shows the components from the steering wheel to the pinion.

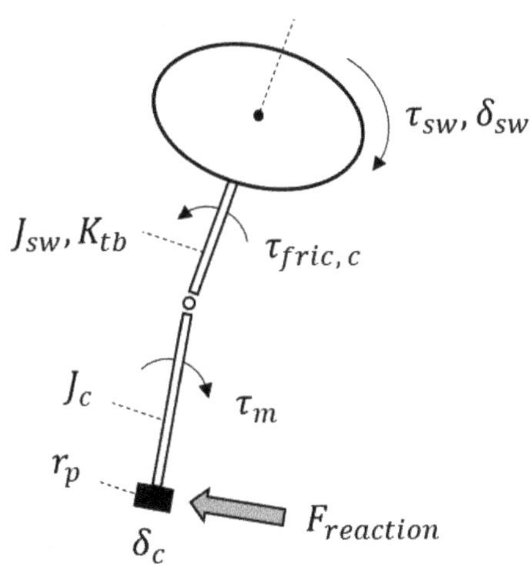

Figure 4. Free-body diagram: from the steering wheel to the pinion.

The equation of the upper steering system is presented in Equation (1), which comprises the driver torque input τ_{sw}, steering wheel inertia J_{sw}, torsion bar stiffness K_{tb}, and steering wheel and column angles δ_{sw}, δ_c. Assume that the column damping is negligible.

The reaction force from the rack to the pinion affects the pinion, as shown in Equation (2). The motor torque τ_m applied at the lower column assists the driver's steering intention in conjunction with the EPS control system. $\tau_{fric,\ c}$, r_p are the friction torque at the column and the radius of the pinion gear, respectively.

$$\tau_{sw} = J_{sw}\ddot{\delta}_{sw} + K_{tb}(\delta_{sw} - \delta_c) \quad (1)$$

$$K_{tb}(\delta_{sw} - \delta_c) + \tau_m - \tau_{fric,c} - J_c\ddot{\delta}_c = F_{reaction} \cdot r_p \quad (2)$$

Second, the relationship between the pinion and tie rod is shown in Figure 5. As shown in Equation (3), the reaction force applied to the rack $F_{reaction}$ results in the movement of the rack x_r against friction and inertia $F_{fric,\ r}$ from the rack and wheel.

$$F_{reaction} = m_r\ddot{x}_r + F_{tie} \cdot \cos\beta + F_{fric,r} \quad (3)$$

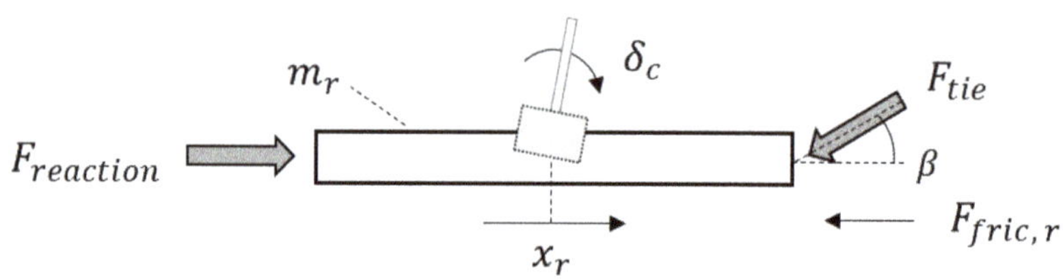

Figure 5. Free-body diagram: from the pinion to the tie rod.

Lastly, the tie rod to the tire is shown in Figure 6. The axis of steering in the tire is mainly determined by the caster angle, which is presented as a red dot in the figure. The point of application of the tie rod and the angle between the wheel rotation axis and tie rod are presented as a blue dot and β, respectively. Equation (4) represents the force from the tie rod F_{tie} to the tire lateral force F_t. J_α, $\tau_{fric,\ s}$ are the inertia of the wheel and the friction at the suspension, respectively. l_e is the effective arm length, which denotes the distance between the steering axis and the tie rod application point. l_t is the trail arm that indicates the distance between the steering axis and the point of action of tire lateral force.

$$F_{tie} \cdot l_e = J_\alpha\ddot{\theta}_t + F_t l_t + \tau_{fric,\ s} \quad (4)$$

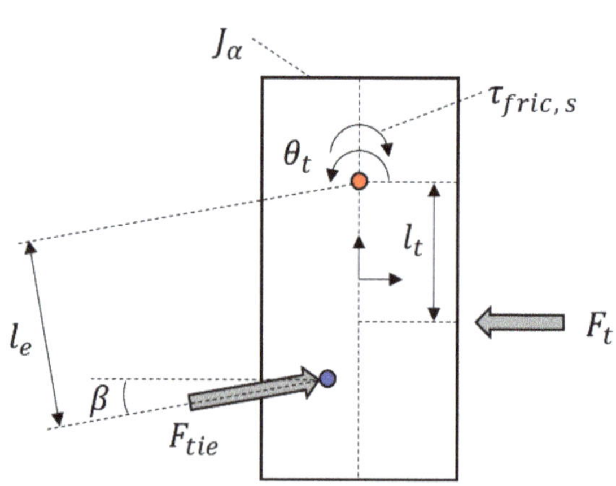

Figure 6. Free-body diagram: from the tie rod to the tire.

Equation (5) can be obtained from Equations (1)–(4). Combining the three aforementioned frames, we can derive an equation for the mechanism of constructing the driver's steering torque.

$$F_t l_t = \frac{l_e}{\cos\beta \cdot r_p}\left(\tau_{sw} - J_{sw}\ddot{\delta}_{sw} + \tau_m - \tau_{fric,\,c} - \left(J_c + r_p^2 m_r\right)\ddot{\delta}_c - \tau_{fric,\,r}\right) - \frac{J_\alpha}{l_e} r_p \ddot{\delta}_c - \tau_{fric,\,s} \tag{5}$$

Assuming that the angular acceleration of the steering wheel and column are the same, Equation (6) can be established.

$$\tau_{sw} = \frac{\cos\beta \cdot r_p}{l_e}\left(F_t l_t + \tau_{fric,\,s}\right) - \left(\tau_m - \tau_{fric,\,c} - \tau_{fric,\,r}\right) + \left(\left(J_{sw} + J_c + r_p^2 m_r\right) + \frac{J_\alpha \cos\beta \cdot r_p^2}{l_e^2}\right)\ddot{\delta}_{sw} \tag{6}$$

As shown in Equation (6), the linearity of the tire lateral force significantly affects the steering feel. Additionally, designing minimal friction at the column, rack, and suspension contributes toward a linear steering feel. This finding aligns with the results of previous studies, which emphasize the influence of friction occurring at various locations in the steering system on the handling characteristics [7]. Particularly, in on-center handling situations with small steering angles, the role of friction cannot be ignored [2]. If friction cannot be minimized, the EPS control system can compensate for the friction elements to improve on-center handling. Furthermore, the angular acceleration term of the steering wheel can be used to derive the characteristics of steering feel during transient steering. A smaller steering gear ratio and caster trail can enhance the clarity of on-center handling. Specifically, an ideal solution for obtaining appropriate steering feedback in the on-center area is utilizing a variable steering gear ratio, as demonstrated in [26].

Based on the above conclusion, we conducted a sensitivity analysis by modifying the chassis and suspension, and we analyzed the effects of variations in key tire parameters. Lastly, we developed a method to improve our EPS logic to achieve a steering feel similar to that of the target vehicle.

3. Simulation Environment for Steering Feel Performance Prediction

3.1. Implementation of Integrated Vehicle Model

We aimed to improve the steering characteristics of our vehicle so that they were similar to those of the target vehicle by utilizing a simulation environment. To achieve this, a reliable virtual vehicle model is essential. As mentioned earlier, the inertia, damping, and stiffness of the steering system, as well as friction and chassis elasticity, among other factors, influence the handling characteristics of the vehicle. Therefore, all these factors should be considered in the simulation environment to obtain a reliable vehicle model as shown in Figure 7.

Accordingly, we modeled our vehicle in a multi-body simulation environment and modeled a separate EPS system to incorporate the steering system characteristics. To enhance the reliability of the steering characteristics, we applied the friction model shown in Figure 8, which includes various components, such as rack gear operation friction and damping, motor gear interlocking friction and damping, the stiffness characteristics of the steering force transmission members, and motor rotation dynamics' characteristics.

The EPS control system has various functions to compensate for factors such as inertia and friction. The basic EPS control system includes a power assist, friction compensation, inertia compensation, and damping [27,28]. These functions help reduce the nonlinear effects, especially in the on-center handling area, where they can have a significant impact on the test results. Furthermore, recent research has focused on developing control systems that are robust to disturbances [29–32]. The steering feel and vehicle handling characteristics can be adjusted depending on the performance of the control system. Therefore, for reliable vehicle handling characteristics, the EPS control system should be integrated into the simulation environment, as depicted in Figure 9.

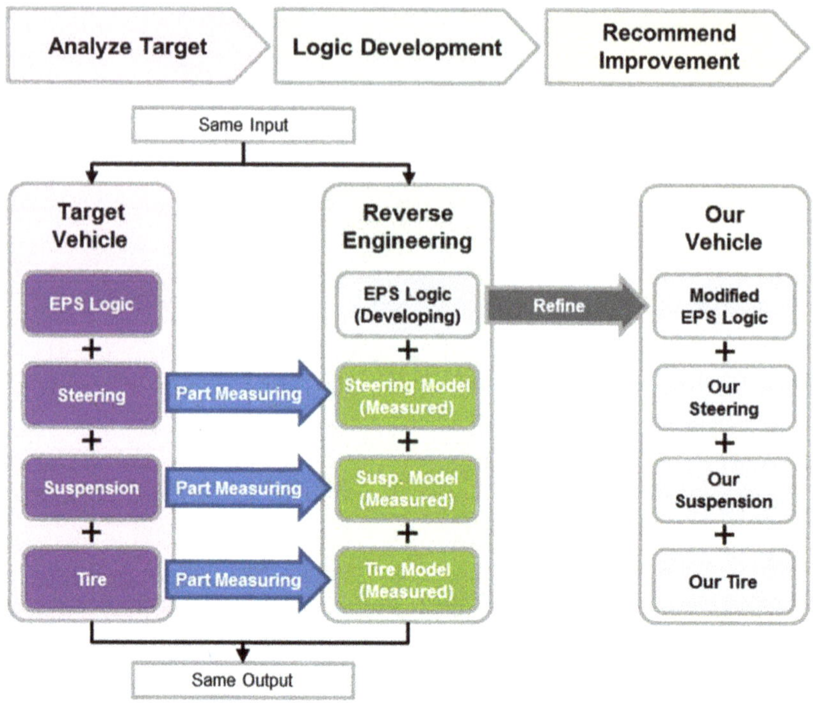

Figure 7. Reverse engineering of control logic through implementing the same input and output.

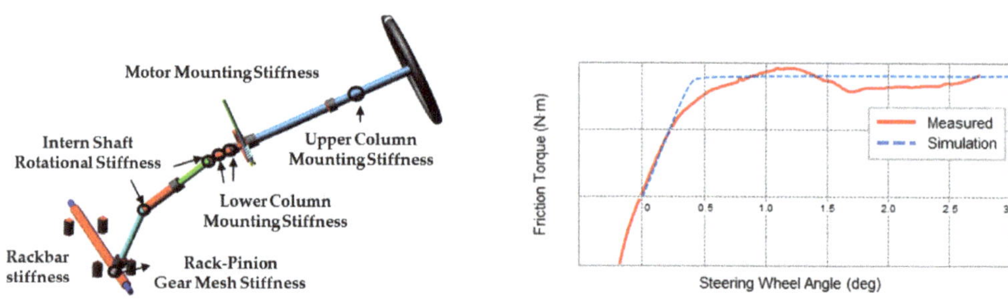

Figure 8. Steering system model and microsection elastic deformation characteristics friction model.

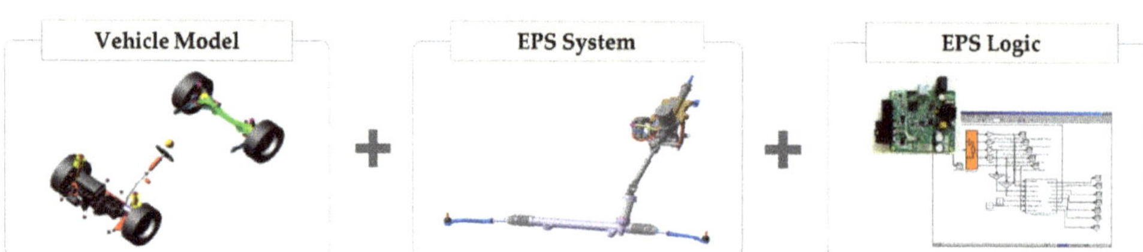

Figure 9. Model for vehicle and control system integration characteristics.

Lastly, to validate the integrated vehicle model, the two aforementioned test methods were performed. As shown in Figures 10 and 11, not only the steering response but also the handling characteristics of our vehicle were accurately captured. Therefore, the integrated vehicle model can also capture the characteristics of the target vehicle. In Section 3.2, we describe the construction process of the target vehicle in the simulation environment.

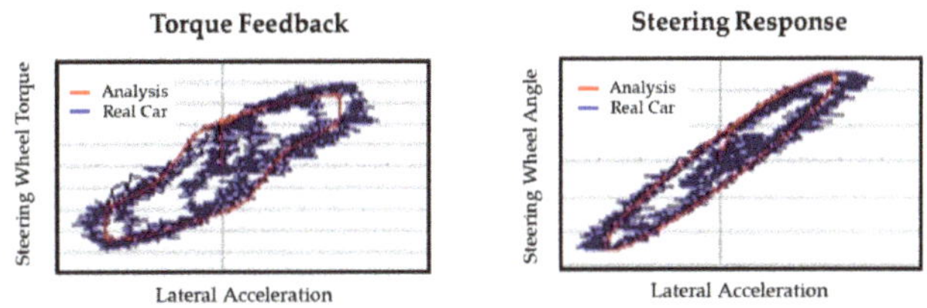

Figure 10. Reliability of steering evaluation at weave test.

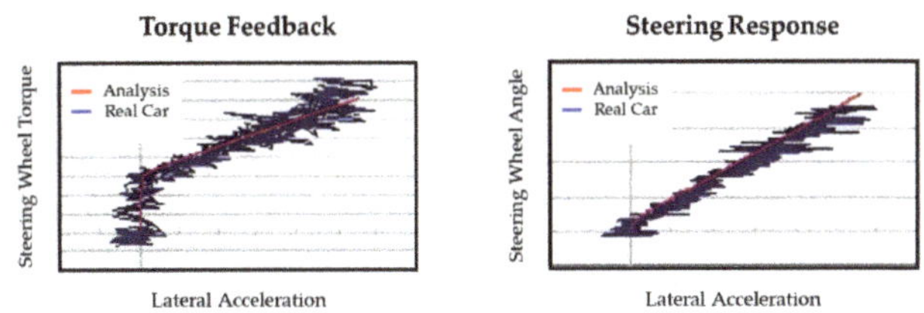

Figure 11. Reliability of steering evaluation at transition test.

3.2. Construction of Target Vehicle Model

In Section 3.1, we described the implementation of our vehicle in a simulation environment and the verification of its reliability. Using a similar approach, we reverse-engineered the parts of the target vehicle and implemented them in the simulation environment.

In modeling the competitor vehicle, we obtained data through testing on a test bench to incorporate the characteristics of the chassis elasticity, friction, and steering system via a similar modeling process as that of our vehicle. To validate the implementation results, the lateral response of the actual competitor vehicle was compared with that of the simulated vehicle, as shown in Figure 12. This validation was performed for the vehicle model while excluding the EPS control system. The steering feel is implemented in Section 3.3 based on the reliable competitor vehicle model obtained.

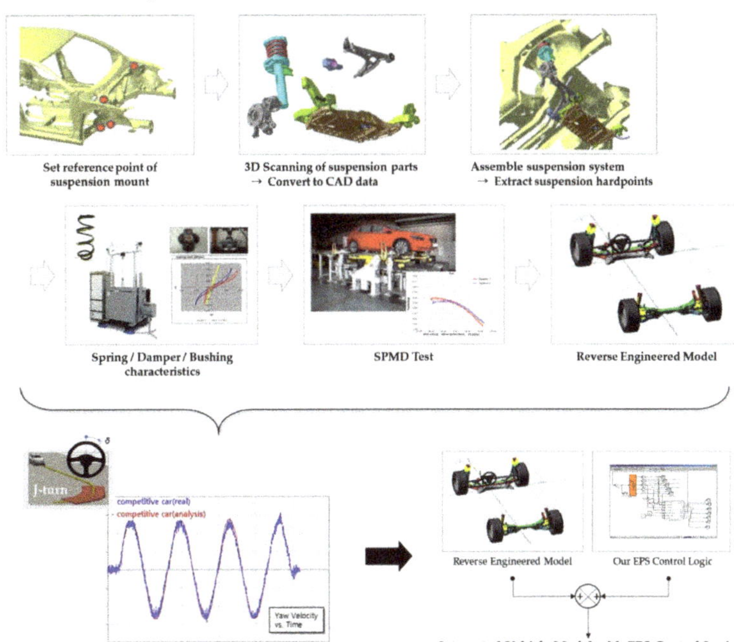

Figure 12. Competitor vehicle modeling by reserve engineering and reliability result.

3.3. Extraction of EPS Control Logic Characteristics of Target Competitor Vehicle

The EPS logic in the target vehicle cannot be directly measured as a separate component, unlike the chassis. Therefore, we performed reverse engineering based on the EPS control system in our vehicle to extract the logic factors. As expected, when applying the EPS control system of our vehicle to the target vehicle, differences in steering feel performance were observed. Therefore, the control system was fine-tuned to achieve the same characteristics as the real car data of the target vehicle. However, only tuning the EPS control system could not achieve the desired outcome. Thus, we modified our EPS control system with regard to the function of damping.

Regarding the transient steering characteristics, the initial steering torque gradient caused by friction is not properly offset, unlike in the measurement data, as shown in Figure 13b. This difference is primarily attributed to friction. To accurately implement the steering characteristics of the competitor vehicle, friction compensation should be enhanced. When significant steering torque is required at the initial steering stage due to friction, one approach is to increase the proportional gain to the high-frequency component generated by applying a high-pass filter to the torque sensor signal, as illustrated by the blue line in Figure 13b. However, too large a high-frequency assist distorts the torque response to a sinusoidal steering input, as shown in Figure 14a. This leads to an overly reduced damping characteristic and diminished friction feel. Additionally, in such cases, the steering wheel becomes more susceptible to disturbances and loses stability, as depicted in Figure 14b. To achieve stability and damping feel, the damping gain should be increased. However, increasing the damping also offsets the enlarged high-frequency assist. Thus, a significant trade-off between these factors exists, making it challenging to set the logic factors to match the steering feel characteristics of the competitor vehicle using high-frequency assist.

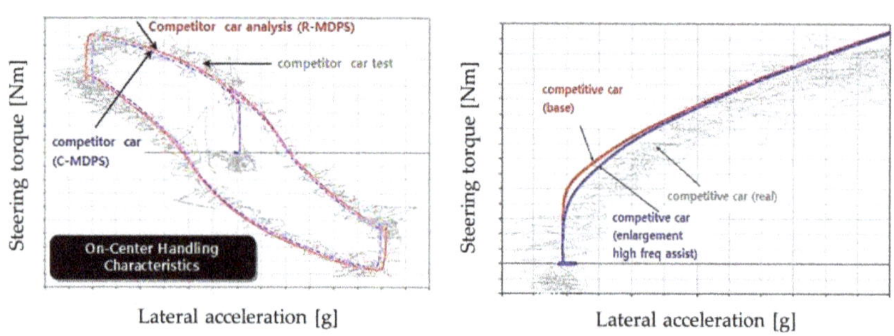

(**a**) (**b**)

Figure 13. (**a**) Comparison of weave test; (**b**) comparison of transition test.

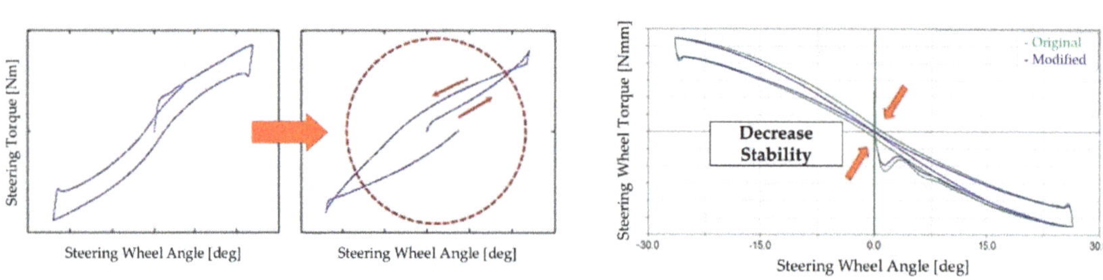

(**a**) (**b**)

Figure 14. (**a**) Steering torque route twist; (**b**) decreasing stability under high-frequency assist.

To solve this problem, we proposed increasing the high-frequency assist with variable damping gain. First, the high-frequency assist gain was increased to improve the offset in the transition test. Then, to fix the torque route twist problem, we modified the damping gain that could be set by steering angular velocity. The detailed parameter settings are as follows:

As shown in Figure 15, the damping parameters were set by dividing the steering angular velocity into three sections. Unlike the conventional spec that defined only one gain regardless of the steering angular velocity, increasing the high-frequency assist gain did not cause the previously defined problem. Before setting up the configurations for each section, the damping gain in Section2 (0.1–0.3 rev/s) was increased to improve the damping feel and route twist that had been degraded by the increase in the high-frequency assist gain. The following section-specific configurations were then implemented.

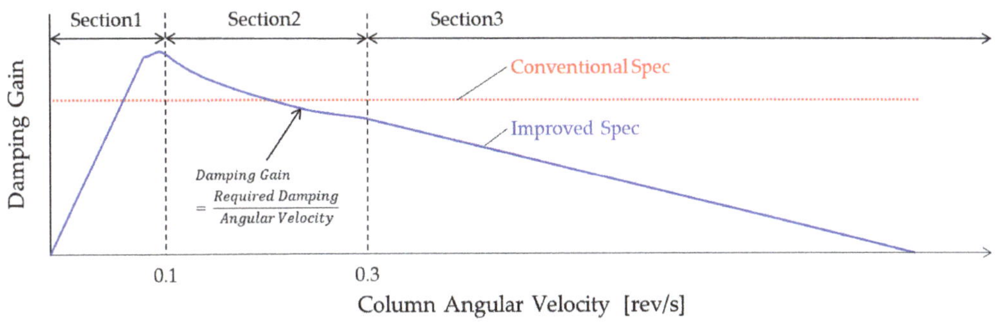

Figure 15. Damping gain setting concept with respect to angular velocity.

- Section1: Set a minimum level of resistance for the initial steering input and small-angle returnability (0–0.1 rev/s).

When setting the damping gain to satisfy damping feel and stability, the level of resistance to driver steering increased, resulting in a delay in the initial response to rapid steering, such as step steer. However, applying the proposed method, the required torque at the initial steering phase was reduced, leading to the tire steering angle increasing, as shown in Figure 16a. In addition, the effect of damping on returnability was observed in hands-off situations, and improved returnability could be confirmed through the proposed methods, as shown in Figure 16b.

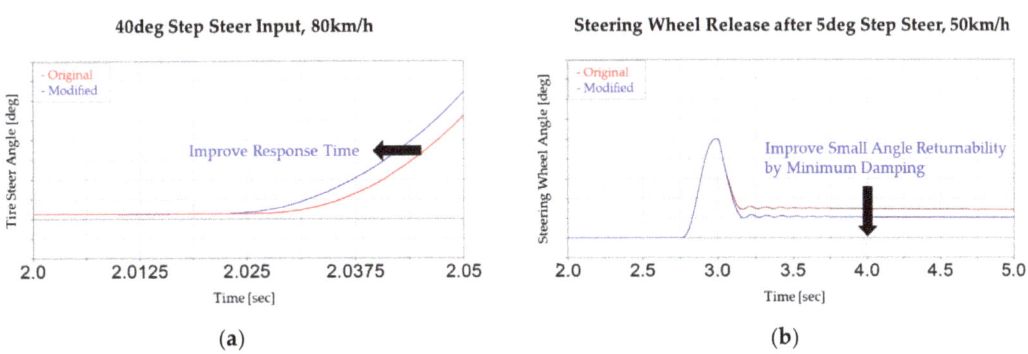

Figure 16. (**a**) Improvement in initial response; (**b**) improvement in steering returnability.

- Section2: Set a consistent damping feel independent of angular velocity in sinusoidal steering (0.1–0.3 rev/s).

The damping gain was set as the required damping divided by the steering angular velocity to achieve a damping feel independent of the input angular velocity. The required damping was set to satisfy the route twist problem and stability, as mentioned earlier, consequently offsetting the effects of the increase in high-frequency assist, as shown in Figure 17.

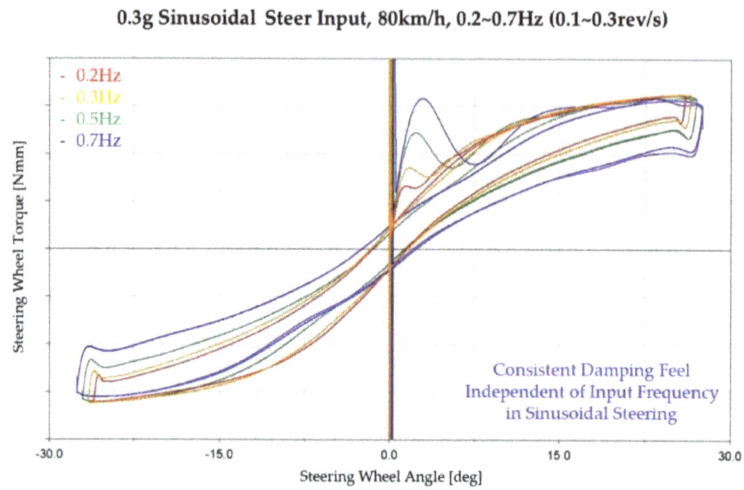

Figure 17. Steering performance at various steering angular velocities.

- Section3: Set the minimum resistance for urgent steering and returnability (~0.3 rev/s).

By reducing the gain in the remaining sections, we enhanced the steering response and returnability, as shown in the blue trajectory in Figure 18. Consequently, we successfully achieved the desired steering performance across various scenarios.

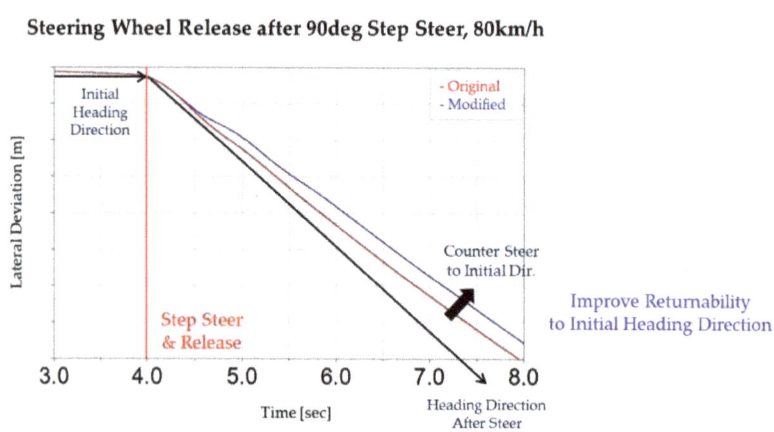

Figure 18. Comparison of steering returnability during high-speed maneuver.

4. Sensitivity Analysis for Improving Steering Feel

4.1. Difference Analysis between the Competitor Vehicle Model and Our Vehicle Model

After obtaining the simulation model for predicting the performance of the competitor vehicle, the factors affecting the steering feeling performance could be analyzed through a sensitivity analysis with respect to the difference between our vehicle and the competitor's. We analyzed the effects of the difference between the two vehicles with respect to the steering system, suspension, and tire (Figure 19).

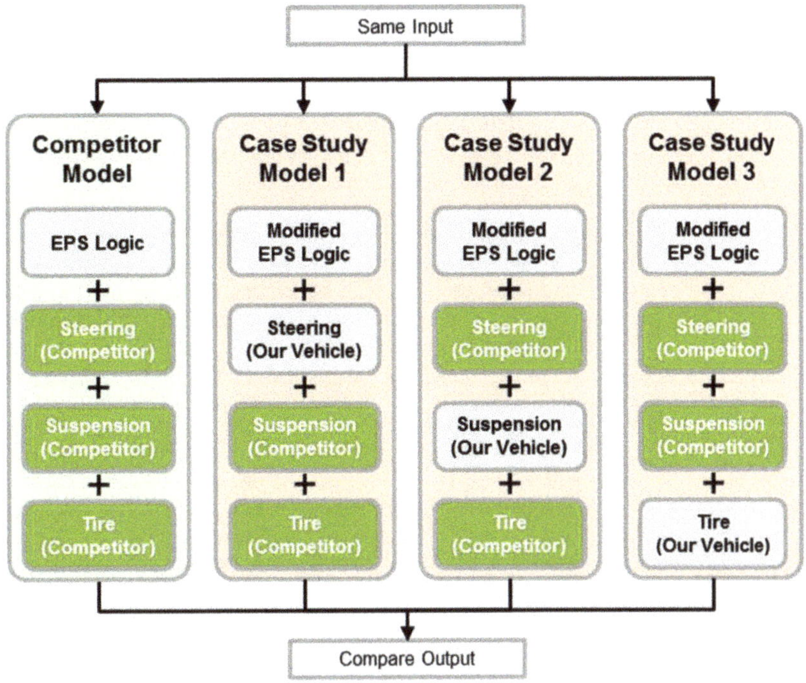

Figure 19. Effect analysis of vehicle characteristic factors.

For reference, the lateral acceleration level appears different in Figure 20, which is the result of the same steering angle input, due to a change in the understeer characteristics.

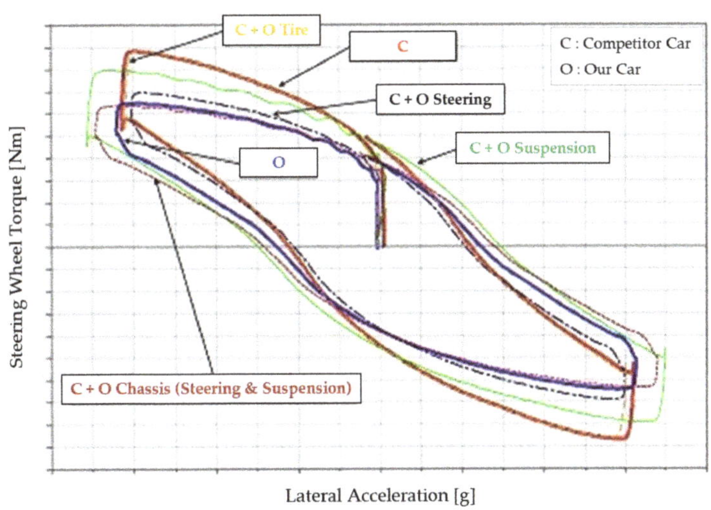

Figure 20. Effect analysis of factors compared between two vehicles.

- Case Study Model 1: Steering

Applying the steering system of our vehicle to the target vehicle (Figure 20, red line vs. black line) revealed an overall decrease in steering torque due to excessive power assist. In addition, a decrease in the gradient was observed in the off-center region, which did not provide proper feedback to the driver.

- Case Study Model 2: Suspension

When applying the suspension (Figure 20, red line vs. green line) and our chassis (including the steering system and suspension) to the target vehicle (Figure 20, red line vs. dotted red line), the steering wheel torque in the off-center area was reduced. This is because the caster trail (l_t in Figure 6) of our development vehicle is smaller, which can be compensated for to a certain extent by reducing the assist torque of the steering system. Moreover, passing through the on-center section, the hysteresis width of the steering wheel torque increased. This is attributable to the friction and damping of the suspension increasing, and compensation to some extent being achieved by changing the damping logic of the steering system.

- Case Study Model 3: Tire

Applying the tire of our vehicle to the target vehicle (Figure 20, red line vs. yellow line) yielded almost the same result. This is because tire cornering stiffness is a factor that determines the lateral acceleration response characteristics by the steering angle input and, thus, has little effect on the steering wheel torque characteristics for lateral acceleration. Therefore, to determine the influence of tire cornering stiffness changes, the lateral acceleration response according to the steering angle input and the resulting steering wheel torque response must be analyzed by accounting for the complexity involved, which will be covered separately in the next section.

4.2. Effect Analysis of the Design Parameters of Tires

The tire characteristics are important in that they can be modified to improve the performance of a vehicle without changing the suspension. Therefore, we conducted a sensitivity analysis on the main parameters of the tires. First, as shown in the lateral acceleration vs. steering angle graph in Figure 21a, an increase in the cornering stiffness of the tire simultaneously results in improved steering sensitivity and maximum lateral acceleration. Furthermore, it decreases the width of the hysteresis curve, improving the initial response characteristics and reducing the lateral acceleration lag time. Second, as the lateral acceleration lag time decreases, the build-up characteristic, as shown in Figure 21b, increases, and the steering wheel torque loss in the on-center section decreases. Lastly, as the pneumatic trail intensified, and the build-up characteristic under the returnability effect, on-center steering wheel torque loss, and lateral acceleration lag increased.

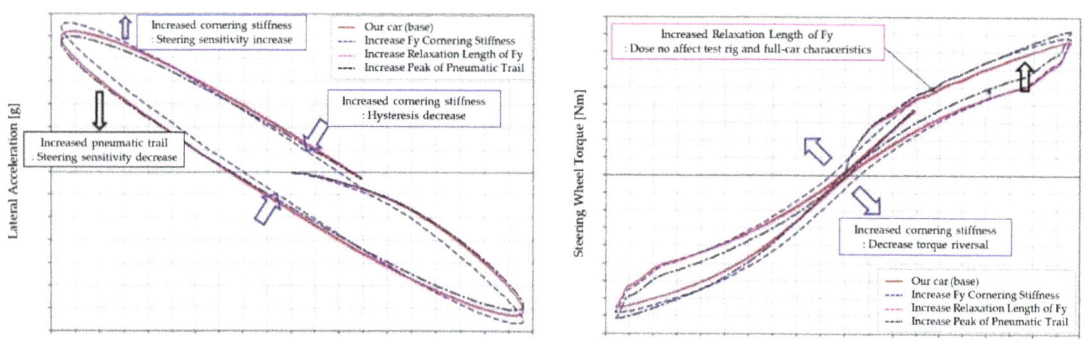

Figure 21. (**a**) Sinusoidal sensitivity according to tire characteristics; (**b**) sinusoidal torque build-up according to tire characteristics.

As the pneumatic trail intensified, the build-up characteristic, as shown in Figure 22, also intensified due to restoration effects, known as the aligning moment. This is the same principle as increasing the caster trail. Tire cornering stiffness determines the lateral acceleration response characteristics under the steering angle input and, therefore, has little effect on the steering wheel torque characteristics under lateral acceleration.

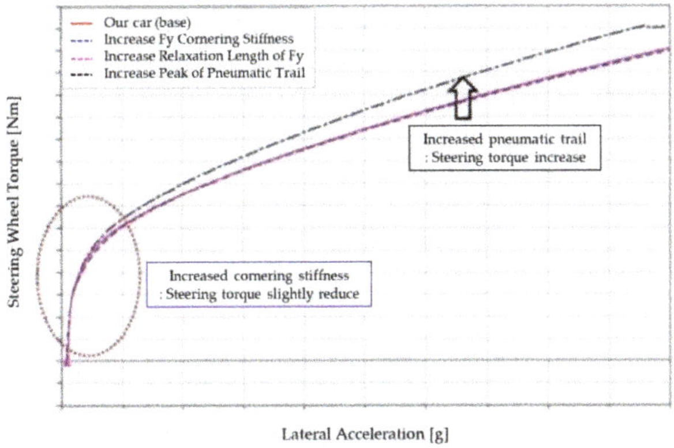

Figure 22. Transition characteristic according to tire characteristics.

4.3. Realization of Steering Feel Performance Improvement by Logic Tuning

If the steering feel performance improves solely through changes in the EPS logic, it should be achieved with minimal costs while minimally affecting other aspects of performance. EPS logic tuning methods to realize the steering feel of the competitor vehicle are the same as the methods in Section 3.3. In this section, based on the previously described methods and the test results of our adjusted vehicle, we demonstrate the outcomes of our proposed approach to improve the steering feel.

Initially, as the transition test result shows (Figure 23, Table 2), we increased the high-frequency gain to decrease the initial torque, which was demonstrated at very low lateral acceleration, by differentiating the torque sensor signal. Our adjusted car exhibited 78% of the initial torque, which was close to that of the competitor car (73%). Additionally, the gradient of the competitor's car was 38% higher than that of our car. Our adjusted car achieved a gradient enhancement of 35%, and it closely mirrored that of the competitor car. In essence, we resolved the lack of a linear increase in the steering torque of our car with increasing lateral acceleration. Meanwhile, we reduced the initial torque so that drivers perceive less friction.

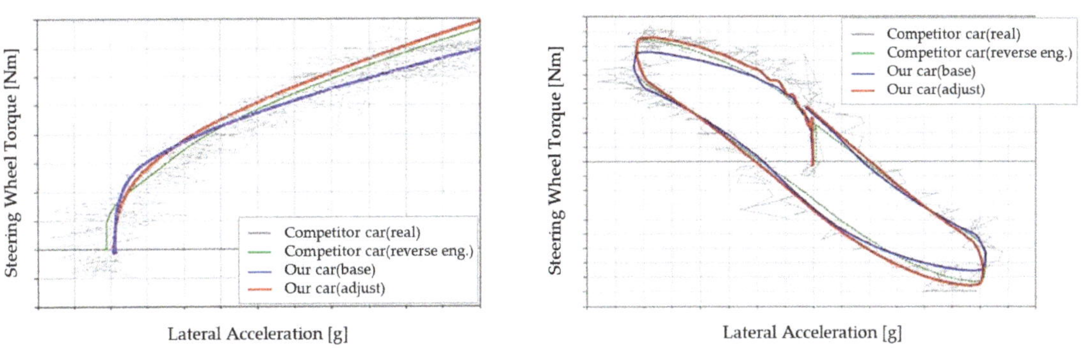

Figure 23. Realizing the sinusoidal characteristic of the competitor vehicle via logic tuning.

Table 2. Steering feel performance improvement results from transition test.

Maneuver (Transition Test)	Initial Torque (Normalized)	Gradient in Linear Range (Normalized)
Our car (base)	1.0	1.0
Competitor car (reverse eng.)	0.73	1.38
Our car (adjusted)	0.78	1.35

Second, as can be seen in the sinusoidal steering graph and Table 3, we reduced the excessive power assist to increase the steering torque at the off-center area. We increased the peak torque of our car by 14%, aligning it closely with the peak torque of the competitor car. Moreover, the torque damping feel, the distance of the steering wheel torque at 0 g, remained unchanged. This means that we managed to maintain stability by applying variable damping gain. As a result, both a decrease in the friction feel in the early stages of steering and an increase in the damping feel during repeated steering, which share a trade-off relationship, were simultaneously achieved.

Table 3. Steering feel performance improvement results from weave test.

Maneuver (Weave Test)	Max Torque (Normalized)	Torque Damping Feel (Normalized)
Our car (base)	1.0	1.0
Competitor car (reverse eng.)	1.12	0.80
Our car (adjusted)	1.14	1.03

5. Conclusions

The handling characteristic perceived by a driver is referred to as the on-center steering feel, which is determined by the nonlinear characteristics of the steering system, the EPS control system, and the lateral force generation characteristics of the tire. To improve this, we analyzed the steering mechanism to identify the key factors affecting steering feel. A target vehicle with good steering feel was then implemented in a multi-body simulation environment. From this, we conducted a sensitivity analysis to determine the important factors influencing steering feeling via systematic approaches. Afterward, we explored ways to improve the steering feel of our vehicle based on changes in the key factors identified in the analysis of the mechanism.

We confirmed that the multi-body simulation environment can capture the steering feeling of the vehicle. Based on this, we constructed a reliable vehicle model by reverse-engineering the steering system, chassis, and tires of the target vehicle without EPS control logic. To achieve the steering feel of the target vehicle, we refined the damping coefficient of our EPS logic by dividing it into three sections. Accordingly, we were able to conduct a sensitivity analysis by applying the components of our vehicle to the target vehicle. Finally, to overcome the differences between our vehicle and the target vehicle, we modified our EPS logic at the lowest possible cost. As a result, we were able to achieve steering characteristics similar to those of the target vehicle.

This study granted us insights into the design of the EPS logic, which are crucial for achieving the desired steering feel. After obtaining these insights, we applied the knowledge to our production vehicle. As a result, our production vehicle exhibited the desired steering feel. Furthermore, as we considered the correlation between subjective and objective assessment, the subjective testers provided positive feedback on the steering feel. The testers' feedback reinforced the validity and reliability of our study.

In future work, we will integrate an actuator, such as a brake system, in a steering system similar to that which was separately modeled. The integrated simulation model will facilitate the analysis of steering feel during interventions from systems other than the steering system. Furthermore, we aim to develop a closed-loop EPS control system to prevent the degradation of steering feel due to interferences from other systems. Through this system, we anticipate a decrease in tuning time, improved disturbance rejection, and an assurance of robust performance.

Author Contributions: Conceptualization, S.W.; methodology, S.W., M.-O.J. and J.-M.L.; software, S.W.; validation, S.W. and J.-M.L.; formal analysis, J.-M.L.; investigation, C.H., M.-O.J. and J.-M.L.; resources, S.W.; data curation, M.-O.J. and J.-M.L.; writing—original draft preparation, C.H. and M.-O.J.; writing—review and editing, S.W. and C.H.; visualization, C.H. and M.-O.J.; supervision, S.W.; project administration, S.W.; funding acquisition, S.W. All authors have read and agreed to the published version of the manuscript.

Funding: This work was supported by the Technology Innovation Program (or Industrial Strategic Technology Development Program—Development of the Core System Technology for Hyper-safe Driving Platform) (20015831, Development of Hyper-safe Driving Platform based on Cooperative Domain Control) funded by the Ministry of Trade, Industry and Energy (MOTIE, Republic of Korea).

Institutional Review Board Statement: Not applicable.

Informed Consent Statement: Not applicable.

Data Availability Statement: Not applicable.

Conflicts of Interest: The authors declare no conflict of interest.

References

1. Harrer, M. Characterisation of Steering Feel. Ph.D. Thesis, University of Bath, Bath, UK, 2007.
2. Ende, K.V.; Kallmeyer, F.; Nippold, C.; Henze, R.; Küçükay, F. Analysis of steering system elasticities and their impact on on-centre handling. *Int. J. Veh. Des.* **2016**, *70*, 211–233. [CrossRef]
3. Peppler, S.A.; Johnson, J.R.; Williams, D.E. Steering system effects on on-center handling and performance. *SAE Trans.* **1999**, *108*, 682–687.
4. Kushiro, I.; Yamomoto, M. Vehicle behavior under the influence of steering dynamics by means of low frequency torque input. *Tech. Rep. SAE Tech. Pap. 2006-01-0557* **2006**. [CrossRef]
5. Pfeffer, P.E.; Harrer, M.; Johnston, D. Interaction of vehicle and steering system regarding on-centre handling. *Veh. Syst. Dyn.* **2008**, *46*, 413–428. [CrossRef]
6. Zheng, H.; Deng, W.; Zhang, S.; Jiang, Y. Studies on the impacts of steering system parameters on steering feel characteristics. In Proceedings of the 2015 IEEE International Conference on Systems, Man, and Cybernetics, Hong Kong, 9–12 October 2015; IEEE: Piscataway, NJ, USA, 2015; pp. 486–491.
7. Li, Y.; Shim, T.; Wang, D.; Offerle, T. Investigation of factors affecting steering feel of column assist electric power steering. In Proceedings of the Dynamic Systems and Control Conference, Minneapolis, MN, USA, 12–14 October 2016; Volume 50701, p. V002T31A005.
8. Gao, J.; Qi, X. Handling performance analysis and optimization of a vehicle with considering the influence of steering shaft parameters. *J. Mech. Sci. Technol.* **2023**, *37*, 569–582. [CrossRef]
9. Choi, J.H.; Nam, K.; Oh, S. Steering feel improvement by mathematical modeling of the Electric Power Steering system. *Mechatronics* **2021**, *78*, 102629. [CrossRef]
10. Lee, D.; Kim, K.S.; Han, M. Model-based analysis of the hysteresis curve of an electric power steering system. In Proceedings of the 2020 20th International Conference on Control, Automation and Systems (ICCAS), Busan, Republic of Korea, 13–16 October 2020; IEEE: Piscataway, NJ, USA, 2020; pp. 490–495.
11. Tao, W.; Gao, B.; Chu, H.; Tian, M.; Chen, H. Studies on steering feeling feedback system based on nonlinear vehicle model. *Tech. Rep. SAE Tech. Pap. 2017-01-1494* **2017**. [CrossRef]
12. Nakano, S.; Hayama, R. Strategy for transfer elemental designing, employing physical characteristic modeling of steering maneuver. *Third Rep. Eval. Method Transf. Charact. Using Steer Wire Syst. Trans. Jpn. Soc. Mech. Eng. Ser. C* **2013**, *79*, 3830–3838. [CrossRef]
13. Camuffo, I.; Caviasso, G.; Pascali, L.; Pesce, M.; Alviano, E. Simulation tools and evaluation criteria for steering wheel feel improvement of an electric power steering system. *Tech. Rep. SAE Tech. Pap. 2002-01-1593* **2002**. [CrossRef]
14. Norman, K.D. Objective evaluation of on-center handling performance. *SAE Trans.* **1984**, *93*, 380–392.
15. Farrer, D.G. An objective measurement technique for the quantification of on-centre handling quality. *Tech. Rep. SAE Tech. Pap. 930827* **1993**. [CrossRef]
16. Higuchi, A.; Sakai, H. Objective evaluation method of on-center handling characteristics. *Tech. Rep. SAE Tech. Pap. 2001-01-0481* **2001**. [CrossRef]
17. Salaani, M.K.; Heydinger, G.J.; Grygier, P.A. Experimental steering feel performance measures. *SAE Trans.* **2004**, *113*, 665–679.
18. Heathershaw, A. Developments in on-centre steering evaluation and testing. *SAE Trans.* **2006**, *115*, 721–728.

19. Zong, C.; Zhang, Z.; Mai, L.; Wang, C.; Wu, Z. Study on objective evaluation index system of on-center handling for passenger car. *Tech. Rep. SAE Tech. Pap. 2013-01-0714* **2013**. [CrossRef]
20. Guan, X.; Zhang, Y.N.; Duan, C.G.; Yong, W.L.; Lu, P.P. Study on decomposition and calculation method of EPS assist characteristic curve. *Proc. Inst. Mech. Eng. Part D J. Automob. Eng.* **2021**, *235*, 2166–2175. [CrossRef]
21. Asperti, M.; Vignati, M.; Sabbioni, E. Driver-in-the-Loop Simulation to Assess Steering Torque Feeling due to Torque Vectoring Control. In Proceedings of the 2022 IEEE Vehicle Power and Propulsion Conference (VPPC), Merced, CA, USA, 1–4 November 2022; IEEE: Piscataway, NJ, USA, 2022; pp. 1–6.
22. *ISO 13674-1*; Road Vehicles—Test Method for the Quantification of On-Centre Handling–Part 1: Weave Test. ISO (International Organization for Standardization): Geneva, Switzerland, 2010.
23. *ISO 13674-2*; Road Vehicles—Test Method for the Quantification of On-Centre Handling—Part 2: Transition Test. ISO (International Organization for Standardization): Geneva, Switzerland, 2016.
24. Düsterloh, D.; Uselmann, A.; Scherhaufer, J.; Bittner, C.; Schramm, D. Objectification of the feedback behavior of the suspension and steering system. In Proceedings of the 9th International Munich Chassis Symposium 2018: Chassis, Tech Plus, Munich, Germany, 12–13 June 2018; Springer: Berlin/Heidelberg, Germany; pp. 505–526.
25. Cianetti, F.; Fabellini, L.; Formica, V.; Ambrogi, F. Development and validation of a simplified automotive steering dynamic model. *Proc. Inst. Mech. Eng. Part D J. Automob. Eng.* **2021**, *235*, 2188–2199. [CrossRef]
26. Heathershaw, A. Optimizing variable ratio steering for improved on-centre sensitivity and cornering control. *SAE Trans.* **2000**, *109*, 1140–1151.
27. Würges, M. New electrical power steering systems. *Encycl. Automot. Eng.* **2013**, 1–17. [CrossRef]
28. Harrer, M.; Pfeffer, P. *Steering Handbook*; Springer: Berlin/Heidelberg, Germany, 2017; Volume 163.
29. Groll, M.V.; Mueller, S.; Meister, T.; Tracht, R. Disturbance compensation with a torque controllable steering system. *Veh. Syst. Dyn.* **2006**, *44*, 327–338. [CrossRef]
30. Mehrabi, N.; Azad, N.L.; McPhee, J. Optimal disturbance rejection control design for electric power steering systems. In Proceedings of the 2011 50th IEEE Conference on Decision and Control and European Control Conference, Orlando, FL, USA, 12–15 December 2011; IEEE: Piscataway, NJ, USA, 2011; pp. 6584–6589.
31. Marouf, A.; Djemai, M.; Sentouh, C.; Pudlo, P. A new control strategy of an electric-power-assisted steering system. *IEEE Trans. Veh. Technol.* **2012**, *61*, 3574–3589. [CrossRef]
32. Na, S.; Li, Z.; Qiu, F.; Zhang, C. Torque control of electric power steering systems based on improved active disturbance rejection control. *Math. Probl. Eng.* **2020**, *2020*, 6509607. [CrossRef]

applied sciences

MDPI

Article

Finite Element Analysis of Tyre Contact Interaction Considering Simplified Pavement with Different Aggregate Sizes

Zhi Li, Weiyong Chen, Yinghui Li and Wenliang Wu *

School of Civil Engineering and Transportation, South China University of Technology, Guangzhou 510640, China; lizhi@scut.edu.cn (Z.L.); 202121010428@mail.scut.edu.cn (W.C.); 202221010106@mail.scut.edu.cn (Y.L.)
* Correspondence: ctwlwu@scut.edu.cn

Abstract: This study considered the effect of pavement aggregate grain size on tyre–pavement contact interaction during the late stages of pavement skid resistance. First, hemispherical shells 7, 9, and 13 mm in diameter were used to simulate coarse pavement aggregates. Subsequently, a three-dimensional finite element tyre–pavement contact model developed using ABAQUS was employed to analyse the contact interaction between each simplified pavement type and the tyre under steady–state rolling and braking conditions. Finally, the concept of occlusal depth was proposed and applied to characterise pavement skid resistance. The results showed that under steady–state rolling conditions, the peak contact stress of the simplified pavement increased with the pavement mean texture depth, whereas the contact area decreased. Under steady–state braking conditions, the effect of the contact interaction between the tyre and simplified pavement aggregates was ranked in order of superiority as aggregate grain sizes of 9, 7, and 13 mm, indicating that aggregate grain size did not exhibit any correlation with tyre–pavement contact interaction. Additionally, the squares of linear correlation coefficients between the pavement cumulative occlusal depth and horizontal braking force reached 0.921, 0.941, 0.889, and 0.894 for vehicle speeds of 30, 60, 90, and 120 km/h, respectively, indicating that they could be used to assess pavement skid resistance.

Keywords: finite element model; aggregate grain size; tyre–pavement contact interaction; occlusal depth

Citation: Li, Z.; Chen, W.; Li, Y.; Wu, W. Finite Element Analysis of Tyre Contact Interaction Considering Simplified Pavement with Different Aggregate Sizes. *Appl. Sci.* **2023**, *13*, 12011. https://doi.org/10.3390/app132112011

Academic Editors: Edgar Sokolovskij and Vidas Žuraulis

Received: 9 October 2023
Revised: 28 October 2023
Accepted: 31 October 2023
Published: 3 November 2023

1. Introduction

The skid resistance of an asphalt pavement is a critical factor affecting road traffic safety [1–3]. Whilst driving, vehicle tyres are in direct contact with the pavement; thus, tyre–pavement contact interaction is a crucial aspect determining asphalt pavement skid resistance. The primary factors affecting tyre–pavement interaction are the tyre characteristics (tyre tread, tyre pressure, driving speed, etc.), pavement characteristics (aggregate properties, aggregate grain size, pavement macrotexture, etc.), and external environment (temperature, water, snow, etc.) [4–6].

Many researchers have studied tyre–pavement contact interactions. Zhang et al. [7,8] used a pressure–sensitive, electrically conductive rubber sensor to measure the contact pressure between the tyre and pavement and successively analysed the corresponding tyre–pavement contact interaction under static and dynamic conditions; however, they ignored the effects of the tyre texture and pavement macrotexture. Chen et al. [9] measured the contact pressure between a tyre and pavements of different macrotextures under static conditions using a pressure film and showed that the pavement macrotexture and tyre load had a greater effect on tyre–pavement contact than the tyre inflation pressure; however, the study lacked a dynamic analysis of tyre–pavement contact.

Indoor testing is a vital research method; however, it can be easily affected by external factors and consumes considerable manpower and material resources. At the same time, tyre–pavement contact analysis involves complex nonlinear mechanical problems. With the rapid development of computers, finite element analysis allows for intuitive and accurate solutions to complex non-linear mechanical problems. Consequently, finite element numerical simulation has become a more effective means of studying tyre–pavement contact interactions. Liu and Al-Qadi [10,11] used a deep learning approach to predict 3D tyre–pavement contact stresses based on finite element generated tyre contact stress datasets and developed ContactNet and ContactGAN models for the fast prediction of 3D tyre pavement stresses. Xie and Yang [12], He et al. [13], and Tang et al. [14] used the ABAQUS finite element-based software package to develop a three-dimensional (3D) tyre–pavement contact model and analyse the impact of driving speed, tyre load, and tyre pressure on tyre–pavement contact interactions under different driving conditions; however, the pavement model applied during their study ignored the pavement macrotexture.

As road researchers, we would prefer to focus on the impacts of factors such as the pavement macrotexture and aggregate properties on tyre–pavement contact interaction. Lin and Wang [15] analysed the effect of fine-aggregate angularity (FAA) on asphalt pavement skid resistance using the British Pendulum Number test. The results showed that the pavement FAA had a considerable effect on skid resistance at the macrotexture level. Xiao et al. [16] analysed the contributions of different texture wavelengths on the aggregate surface to tyre–pavement interaction. The results showed that the correlation between textures with small-scale wavelengths and tyre–pavement interaction decreased with increasing sliding speed; the correlation changed in the opposite direction for textures with large-scale wavelengths. Riahi et al. [17] considered the interaction between the pavement macrotexture and tyre rolling process to propose a model based on the dissipation of energy by the rubber block during the cyclic deformation of the pavement in concave and convex cycles. Yu et al. [18] selected seven parameters to characterise pavement texture features and analysed the impact of those features on pavement skid resistance. Sharma et al. [19] developed a multi-physics tyre–pavement contact model based on the temperature and pavement texture, focusing on the impact of pavement macrotexture on tyre–pavement interaction. Finite element simulation technology can be applied to further analyse tyre–pavement contact interactions; however, in analysing the factors affecting tyre–pavement contact interactions, previous studies considered the tyre factors but ignored the effect of the pavement macrotexture and considered the effect of the aggregate surface properties but paid little attention to the effect of the aggregate grain size after the aggregate texture was abraded.

Consequently, this study focused on the impact of aggregate grain size on tyre–pavement contact interaction. Three groups of simplified pavements comprising single-diameter hemispherical shells were designed. A 3D tyre–pavement contact model developed using ABAQUS 2021 was subsequently employed to analyse the contact interaction between each simplified pavement and a tyre under steady–state rolling and braking conditions. Finally, we summarised the change law of the contact interaction between tyres and aggregates of different grain sizes in simplified pavements and proposed reasonable evaluation indices to guide the design of asphalt pavement skid resistance.

2. Contact Theory

During the normal vehicle driving process, tyre–pavement contact can be divided into three general regions, as shown in Figure 1. Region I denotes before contact (where the tyre and pavement are about to make contact); Region II denotes stable contact (where the tyre and pavement are in full contact); and Region III denotes after contact (where the tyre and pavement experience contact detachment). When the vehicle brakes, the impact force generated in contact Region I and the adhesion friction generated in contact Region III contribute less to the tyre's overall friction, and the vehicle's braking force comes from contact Region II. Consequently, this study focused on the contact behaviour in Region II (full contact state).

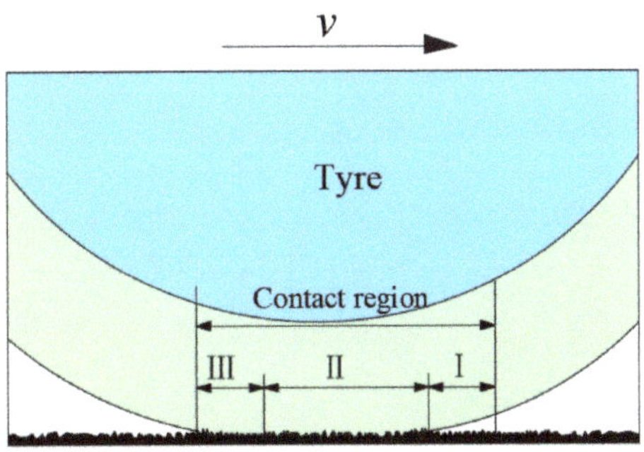

Figure 1. Schematic of the tyre–pavement contact region.

The contact behaviour in Region II primarily comprises tyre–pavement aggregate contact interaction. Because the tyre's internal inflatable structure and tread material stiffness are considerably lower than those of the pavement aggregate, the tyre produces a localised wrapping of the raised aggregate on the pavement during the rolling process. When the vehicle brakes, the speed of the outer edge of the tyre line is less than the vehicle's driving speed; consequently, the tyre and raised aggregate produce an "occlusal braking" effect. This effect can be explained by the hysteresis effect (F_h) of the aggregate as well as the adhesion (F_a) between the tread and pavement materials [20,21], as shown in Figure 2. The hysteresis effect is primarily a result of the resistance provided by the occlusion of the tread with the raised pavement aggregates and the surface friction generated by the relative slip between them. Because the adhesion interaction between materials has a small effect on the overall friction behaviour, its magnitude is negligible [22,23]. Consequently, this study ignored the effect of adhesion and focused on the contact interaction between the tyre and pavement aggregates.

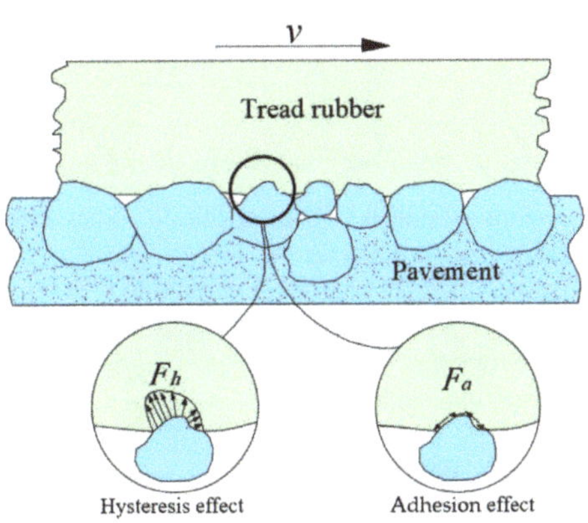

Figure 2. Analysis of tyre–pavement contact interaction.

Additionally, research and practice have shown that during the early stages of road use, different pavement types can provide excellent skid resistance, although their macrotextures vary considerably. With repeated traffic loads, pavement aggregate angles and surface microtextures are repeatedly abraded to present a smoother surface; therefore, the skid resistances of different pavements will exhibit different degrees of attenuation. However, pavement aggregate grain size is a critical factor affecting pavement macrotexture [24–26]. Consequently, attention must be paid to the effect of aggregate grain size on pavement skid resistance. To analyse the contact interaction between a tyre and aggregates of different grain sizes in this study, hemispherical shells were used to simulate abraded coarse aggregates, as shown in Figure 3.

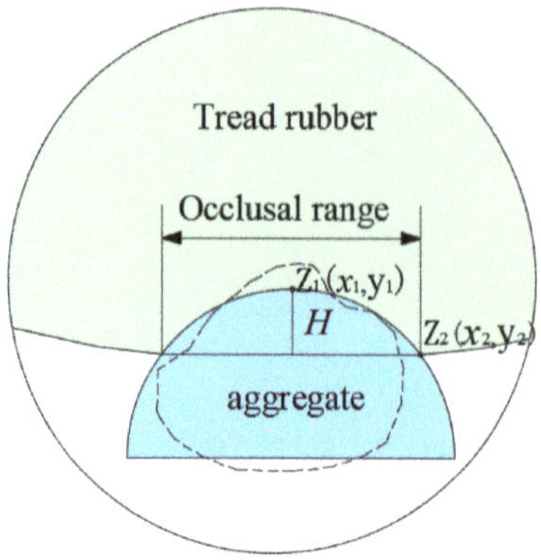

Figure 3. Schematic of occlusal depth using a hemispherical shell.

To model the local package behaviour produced by the process of rolling tyre deformation on pavement raised aggregates, this study proposed the concept of "occlusal depth", which refers to the package depth of a tyre on pavement raised aggregate. Within the occlusal range of the tyre and aggregate, the occlusal depth (H) of a single aggregate is expressed using the difference between the vertex elevation value of the aggregate ($Z_1(x_1, y_1)$) and the critical point elevation of the aggregate ($Z_2(x_2, y_2)$), as shown in Figure 3. It can be calculated as follows:

$$H = Z_1(x_1, y_1) - Z_2(x_2, y_2) \quad (1)$$

where H denotes the occlusal depth of a single aggregate, $Z_1(x_1, y_1)$ denotes the vertex elevation in the occlusal range, and $Z_2(x_2, y_2)$ denotes the critical point elevation in the occlusal range. The mean occlusal depth of the simplified pavement can be calculated as follows:

$$\overline{H} = \frac{\sum_{i=1}^{n} H_i}{n} \quad (2)$$

where $\overline{H}$ denotes the mean occlusal depth of the simplified pavement, $\sum_{i=1}^{n} H_i$ denotes the cumulative occlusal depth of the simplified pavement, and n denotes the number of aggregates in contact with the tyres.

We initially assumed that a greater degree of tyre tread deformation (larger occlusal depth) would result in better contact interaction between the tyre and pavement raised aggregate; this assumption was investigated using the 3D tyre–pavement contact finite element model presented in Section 4.

3. Methods: Finite Element Modelling

3.1. Tyre Model

3.1.1. Tyre Type and Material Parameters

The tyre type selected for this study was a 175/80 R14 radial tyre (where 175 denotes a tyre section width of 175 mm, 80 denotes an aspect ratio of 80%, R denotes a radial tyre type, and 14 denotes a nominal rim diameter of 14 in). A radial tyre is a complex composite comprising rubber and cord skeleton materials. The rubber material includes the tread rubber, crown rubber, and carcass rubber, and the cord skeleton material includes two belt cord layers and one carcass cord layer, as shown in Figure 4.

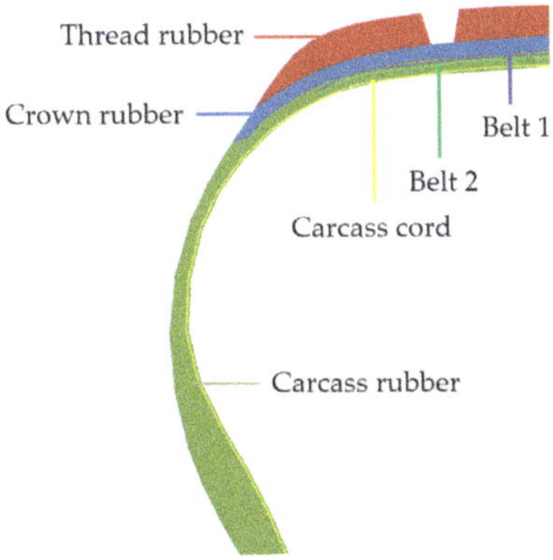

Figure 4. Material composition of the tyre.

The neo-Hookean and linear elasticity intrinsic models were used to numerically simulate the tyre rubber and cord skeleton materials, respectively. The strain energy function is expressed as follows:

$$U = C_{10}(\overline{I_1} - 3) + \frac{1}{D_1}(\overline{J_e} - 1)^2$$

where U denotes strain potential energy, C_{10} denotes the shear characteristics of the material, $\overline{J_e}$ denotes the elastic volume ratio, $\overline{I_1}$ denotes the deformation of the material, and D_1 denotes the compression characteristics of the material.

The tyre industry does not make the exact material properties and structural design of tyres available to the general public. At the same time, the workload and cost of testing for tyre materials are relatively large. Therefore, the mechanical parameters of these materials in Yu et al. [25] are used, as shown in Tables 1 and 2.

Table 1. Rubber material parameters for the 175/80 R14 radial tyre.

Material	Neo-Hookean Model		Density (g/cm^3)
	C_{10} (MPa)	D_1 (MPa^{-1})	
Tread rubber	0.835	0.024	1.12
Crown rubber	0.869	0.04	1.12
Carcass rubber	1.0	0.02	1.15

Table 2. Cord skeleton material parameters for the 175/80 R14 radial tyre.

Material	Cross-Sectional Area (mm^2)	Spacing (mm)	Angle (°)	Young's Modulus (10^5 MPa)	Poisson's Ratio	Density (g/cm^3)
Belt 1	0.212	1.16	70	1.722	0.3	5.9
Belt 2	0.212	1.16	110	1.722	0.3	5.9
Carcass cord	0.421	1.0	0	9.87	0.3	1.5

3.1.2. 2D and 3D Tyre Modelling

This study focused on the contact interaction between the rubber tread material and the different grain size aggregates when the tyre is running in the longitudinal direction; ignoring the transverse tyre tread pattern had little effect on the analysis results. Meanwhile, to save computational time, the tyre modelling ignored the transverse tyre tread pattern and retained only the two longitudinal grooves. As the work of tyre modelling is two-dimensional (2D), the 3D tyre finite element model was obtained by rotating the 2D tyre model. To reduce the modelling workload, only half of the tyre profile was selected for modelling. The specific modelling process employed was as follows:

First, we drew a 2D profile of the tyre based on its size and material composition using AutoCAD and exported the *.iges file format to the HyperMesh 2021 pre-processing software for meshing. Considering that mesh quality has a major influence on the accuracy and convergence of the model results, the rubber material was divided into quadrilateral CGAX4 cells with triangular CGAX3 cells used only at the sharp corners of the material boundaries, and the cord skeleton material was represented by the SFMGAX1 cell type. In ABAQUS 2021, the material mechanical parameters were subsequently set based on the intrinsic tyre model, and a uniform pressure was applied to the inside of the tyre to simulate its inflation pressure. The resulting 2D tyre finite element model is shown in Figure 5.

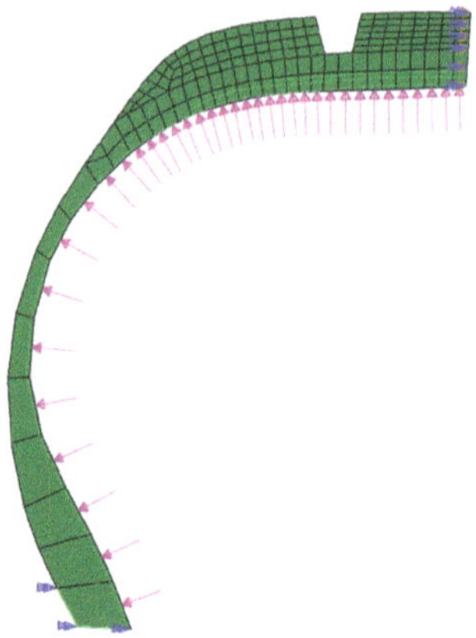

Figure 5. Two-dimensional finite element tyre model.

Second, the 2D finite element model of the tyre was rotated 360° around its rotation axis in ABAQUS 2021 to obtain half of the 3D finite element model, as shown in Figure 6. As the use of a dense tyre mesh can increase model calculation overhead but the use of a sparse tyre mesh can affect the accuracy of the results, mesh encryption was performed only in the contact interval between the tyre and pavement, and the mesh sizes of the tyre and pavement were matched to the greatest extent possible.

Figure 6. Half of the 3D finite element tyre model.

Third, half of the 3D finite element model was mirrored to obtain the completed 3D finite element model of the tyre, as shown in Figure 7. An analytical rigid body was used at the inner edge of the tyre bead to form the rim.

Figure 7. Complete 3D finite element tyre model.

3.2. Simplified Pavement Model

3.2.1. Simplified Pavement Consisting of a Single Grain Size Aggregate

Coarse aggregates larger than 4.75 mm can have a major impact on pavement skid resistance [27,28]; consequently, hemispherical shells with diameters of 7, 9, and 13 mm were used to represent different coarse aggregate grain sizes in three simplified pavements, each comprising a single grain size aggregate. The planar size of this simplified pavement was 200 × 200 mm, and the 7, 9, and 13 mm hemispherical shells were arranged in a matrix such that they numbered 28 × 28, 22 × 22, and 15 × 15, respectively. To ensure the convergence of the model and retain a sufficient pavement height, the heights of the simplified pavements with aggregate grain sizes of 7, 9, and 13 mm were set to 3, 4, and 5 mm, respectively. The 3D models of the simplified pavements were drawn using the CATIA V5R20 software, as shown in Figure 8.

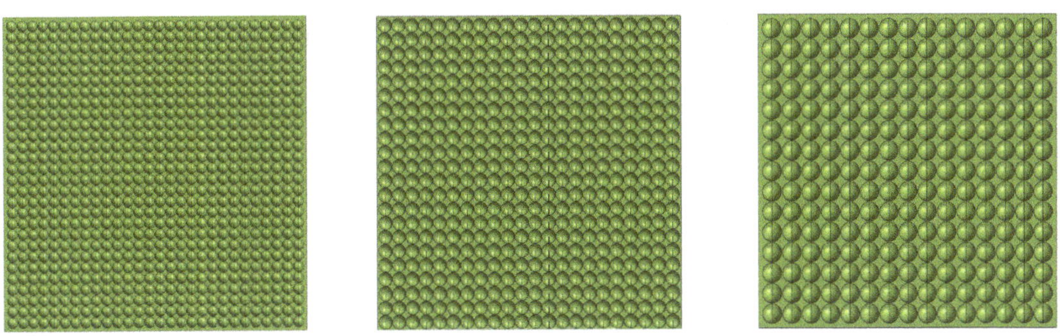

(a) (b) (c)

Figure 8. Three-dimensional model of simplified pavement. (**a**) Aggregate grain size 7 mm. (**b**) Aggregate grain size 9 mm. (**c**) Aggregate grain size 13 mm.

The pavement mean texture depth (MTD) was used to characterise the differences between the macrotextures of the simplified pavements and is expressed as follows:

$$MTD = \frac{V_1 - V_2}{A} \quad (3)$$

where V_1 denotes the cuboid volume within the height of the pavement bulge, V_2 denotes the volume of the pavement bulge, and A denotes the planar area of the pavement.

The pavement *MTD* values corresponding to aggregate grain sizes of 7, 9, and 13 mm were calculated using Equation (3); the results are shown in Table 3.

Table 3. Pavement mean texture depth (*MTD*) with aggregate grain sizes of 7, 9, and 13 mm.

Aggregate Grain Size	V_1 (mm^3)	V_2 (mm^3)	A (mm^2)	*MTD* (mm)
7 mm	120,000	55,417	40,000	1.61
9 mm	160,000	77,040	40,000	2.07
13 mm	200,000	85,412	40,000	2.86

3.2.2. Simplified Pavement Modelling

The primary objective of simplified pavement modelling is to divide the mesh of the pavement model. The simplified pavement model drawn using CATIA V5R20 was output in a saturated file format, then imported into HyperMesh 2021 for meshing. The pavement model was divided into quadrilateral cells to ensure mesh quality, as shown in Figure 9. Finally, the simplified pavement with sufficient meshing was output in the *.inp file format to import into ABAQUS 2021.

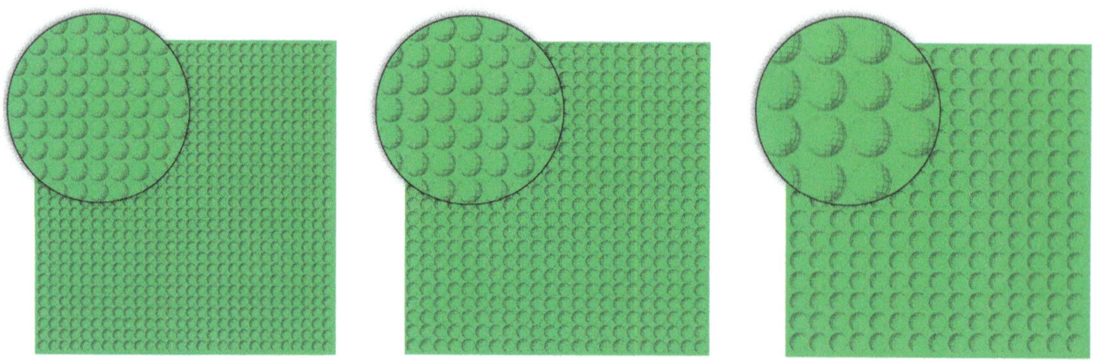

(a) (b) (c)

Figure 9. Three-dimensional mesh model of simplified pavement. (**a**) Aggregate grain size 7 mm. (**b**) Aggregate grain size 9 mm. (**c**) Aggregate grain size 13 mm.

3.3. *Tyre–Pavement Contact Model*

3.3.1. Contact and Boundary Settings

The stiffness of the simplified pavement model used in this study was considerably larger than that of the radial tyre model; consequently, the simplified pavement model could be regarded as a rigid body. The contact constraint of the tyre–pavement system was accordingly set to reflect the finite slip contact between rigid and deformed bodies in ABAQUS 2021. The tyre–pavement contact type was set to face-to-face contact, with the pavement set as the "master surface" and the tyre tread set as the "slave surface". For the contact property setting, the normal property was set as "hard contact", and the tangential property was set as the Coulomb friction model.

In ABAQUS 2021, the pavement was set to be stationary by constraints, and the travel speed of a tyre was simulated by setting the linear and angular velocities at the reference points of the rim. The tyre load was set in the negative direction of the Z-axis at the rim reference point. The 3D contact finite element model of the tyre and simplified pavement is shown in Figure 10.

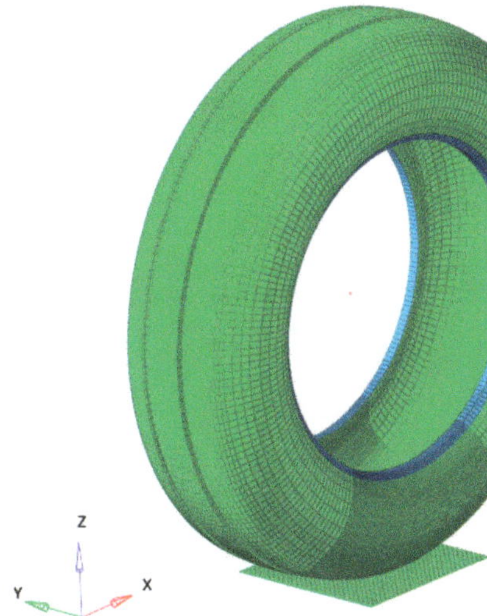

Figure 10. Three-dimensional contact model of the tyre and simplified pavement.

3.3.2. Contact Analysis

This study focused on the steady–state analysis of vehicle drive wheel and simplified pavement. Steady–state transport was used in ABAQUS 2021 to simulate the steady–state dynamic interaction between the tyre and rigid pavement surface. For this, the Eulerian description of rigid-body rotation and Lagrangian description of deformation were used to convert the steady–state moving contact problem into a purely spatially dependent simulation that described the rolling of the tyre as a material flow motion through the mesh.

To conduct a steady–state analysis, the motion state of the tyre can be set to steady–state rolling, braking, or driving. In ABAQUS 2021, the speed at which the tyre travels can be simulated in terms of linear and angular velocities, and the driving state of the tyre can be changed by transforming their combined values [29]; the relationship between the linear and angular velocities is given by

$$\omega = \frac{v}{r} \tag{4}$$

where ω denotes the angular velocity, v denotes the linear velocity, and r denotes the tyre radius of rotation. When the torque of the tyre around the rotation axis is zero, the tyre is in a stable rolling state; when the torque is non-zero, the tyre is accelerating or decelerating.

In the steady–state rolling analysis model, the tyre pressure was 260 kPa, the load was 2600 N, and the driving speed was 60 km/h, corresponding to a linear speed of 16.67 m/s. The angular velocity was initially calculated to be 57 rad/s based on the radius of the tyre. When simulated in ABAQUS 2021, the angular velocity was appropriately adjusted such that the tyre torque was close to zero. Eventually, the linear velocity was determined to be 16.67 m/s and the angular velocity 56.2 rad/s. Thus, when the linear velocity remained constant, tyres with angular velocities less than 56.2 rad/s were decelerating, whereas tyres with angular velocities greater than 56.2 rad/s were accelerating.

In the steady–state braking analysis model, the tyre pressure was 260 kPa, the load was 2600 N, the friction coefficient was 0.3, and the driving initial speeds were 30, 60, 90, and 120 km/h. Using the methods described in the previous paragraph, the linear and angular speeds of the tyres were determined for the different speeds during steady–state braking, and simulation analyses were performed using these values.

4. Results and Discussion

4.1. Steady–State Rolling Analysis

4.1.1. Contact Stress and Contact Area

To analyse the contact stress and contact area of the simplified pavement according to aggregate grain size, the 3D tyre–pavement contact model was used to simulate the steady–state rolling of a vehicle at a driving speed of 60 km/h, as shown in Figures 11 and 12.

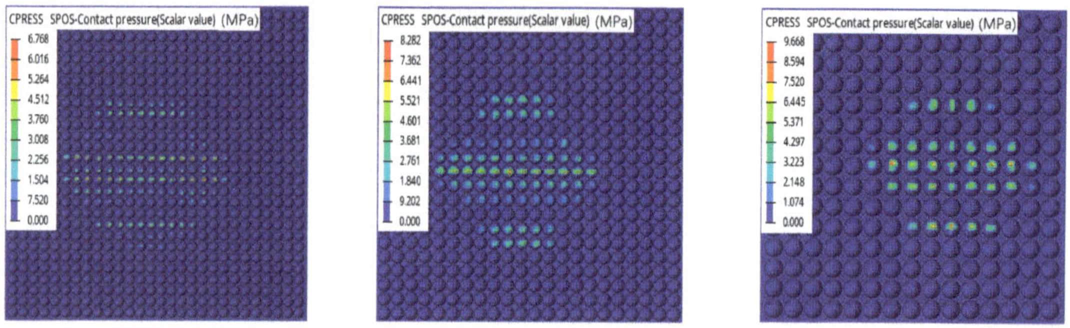

(a) (b) (c)

Figure 11. Contact stresses of simplified pavement during steady–state rolling. (**a**) Aggregate grain size 7 mm. (**b**) Aggregate grain size 9 mm. (**c**) Aggregate grain size 13 mm.

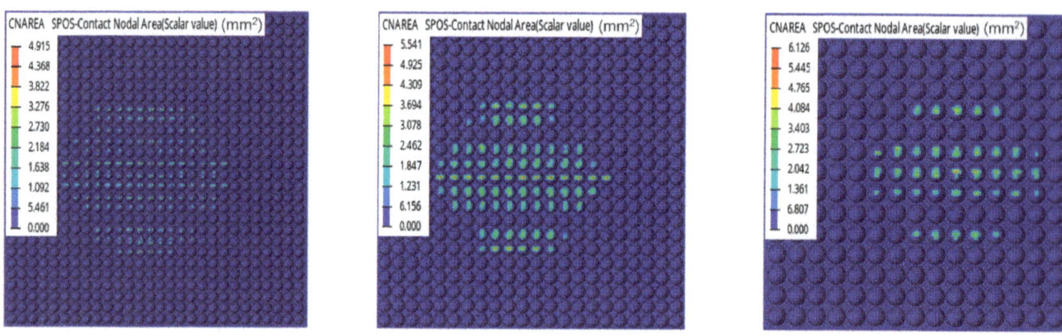

(a) (b) (c)

Figure 12. Contact areas of simplified pavements during steady–state rolling. (**a**) Aggregate grain size 7 mm. (**b**) Aggregate grain size 9 mm. (**c**) Aggregate grain size 13 mm.

As shown in Figure 11a–c, the general extent of contact stresses between the tire and each simplified pavement was similar; however, the different aggregate sizes resulted in different magnitudes of contact stress on each pavement. The peak stresses on the simplified pavements with aggregate grain sizes of 7, 9, and 13 mm were 6.77, 8.28, and 9.67 MPa, respectively. As shown in Table 3, the *MTD* values of the simplified pavements with aggregate grain sizes of 7, 9, and 13 mm were 1.61, 2.07, and 2.86 mm, respectively. These results show that both the *MTD* values and peak contact stresses on the simplified pavements were positively correlated with the aggregate grain size.

As shown in Figure 12a–c, the overall shape of the contact area between the tyre and each simplified pavement was similar; however, the size of the contact area decreased with the change in the macrotexture of the simplified pavement corresponding to the increase in aggregate grain size. To quantitatively simplify the contact area characterisation of the pavement according to aggregate grain size, we outputted the contact area of each simplified pavement in the software post-processing, as shown in Figure 13.

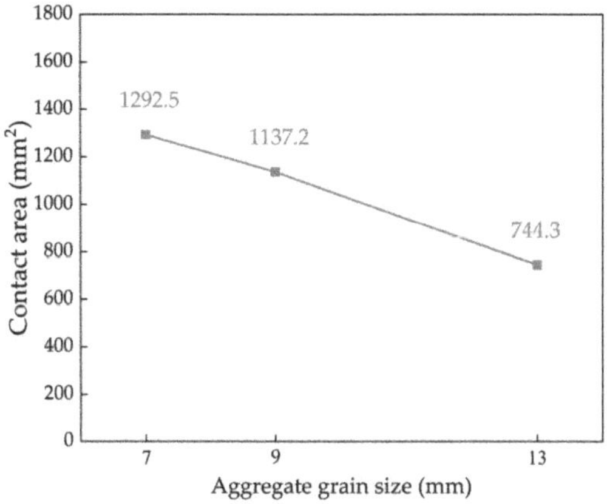

Figure 13. Contact area of each simplified pavement during steady–state rolling.

As shown in Figure 13, the maximum contact areas of simplified pavements with aggregate grain sizes of 7, 9, and 13 mm were 1292.5, 1137.2, and 744.3 mm^2, respectively. These results demonstrate that the contact area of the simplified pavement decreased as its aggregate grain size increased, indicating that pavements with large *MTD* values will have small contact areas.

4.1.2. Occlusal Depth

Based on the modelled contact area between the tyre and pavement, the nodal coordinates of the vertices and contact critical points of the aggregate were output using the post-processing module in ABAQUS 2021. The mean and cumulative occlusal depths of each simplified pavement model were subsequently calculated using Equations (1) and (2) with the results shown in Figures 14 and 15.

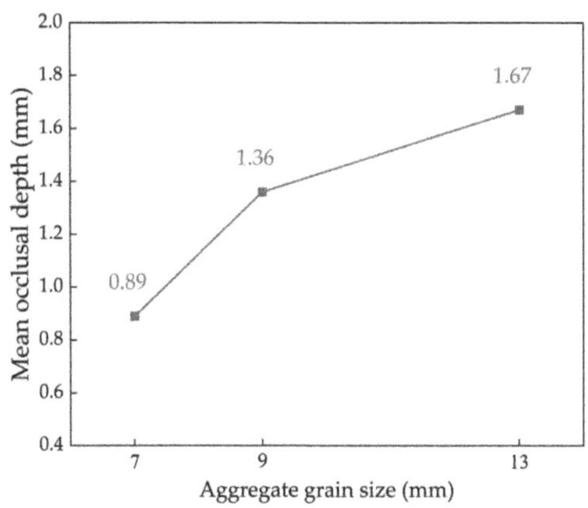

Figure 14. Mean occlusal depth of simplified pavement according to aggregate grain size.

As shown in Figures 14 and 15, the mean occlusal depths of the simplified pavements increased with increasing aggregate grain size, although the magnitude of the cumulative occlusal depth did not exhibit any correlation with aggregate grain size. As shown in Table 3, the *MTD* values of the simplified pavements with aggregate grain sizes of 7, 9, and 13 mm were 1.61, 2.07, and 2.86 mm, respectively. The results indicate that during the late stages of pavement skid resistance, a larger aggregate grain size will increase the *MTD* and mean occlusal depth of the pavement. However, the cumulative occlusal depth of a pavement with a larger aggregate grain size will be smaller because of the fewer points of contact between the pavement and tyre. As the cumulative occlusal depth of the pavement directly reflects the overall deformation of the tyre tread, under tyre rolling conditions, pavement with a smaller aggregate grain size will increase the overall deformation of the tyre tread, thereby producing better contact interaction.

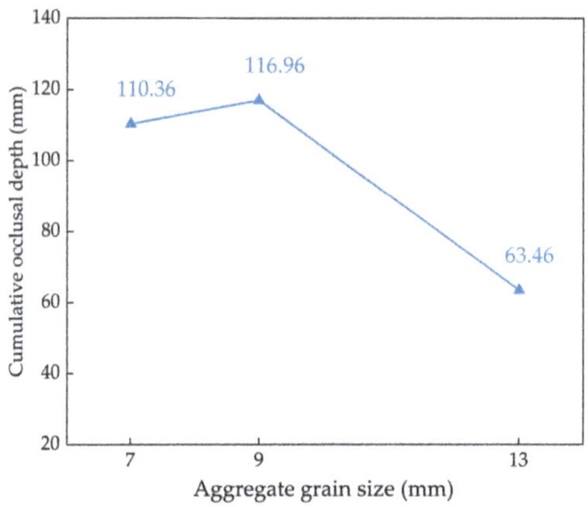

Figure 15. Cumulative occlusal depth of simplified pavement according to aggregate grain size.

4.2. Steady–State Braking Analysis

The effect of the contact interaction between the tyre and simplified pavements with different aggregate grain sizes can be further analysed in terms of the horizontal braking force. In this study, the steady–state braking of a vehicle travelling at initial speeds of 30, 60, 90, and 120 km/h was simulated using the established 3D tyre–pavement contact finite element model. After simulation, the horizontal combined force of the contact force and friction force is output in the post-processing of the software, which is the horizontal braking force, as shown in Figure 16. The cumulative occlusal depth of each simplified pavement for vehicles traveling at different speeds was calculated using Equations (1) and (2), as shown in Figure 17. To verify that the cumulative occlusal depth accurately reflected the contact interaction between the tyre and pavement aggregate, an analysis of the correlation between the cumulative occlusal depth and horizontal braking force of the simplified pavement was conducted, with the results shown in Figure 18.

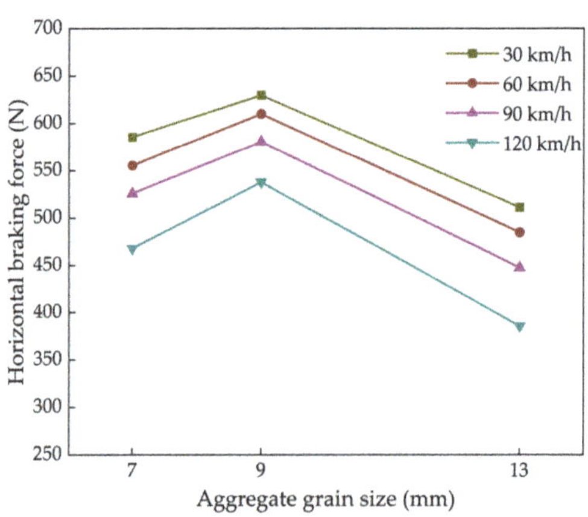

Figure 16. Horizontal braking force of each simplified pavement at different vehicle initial speeds.

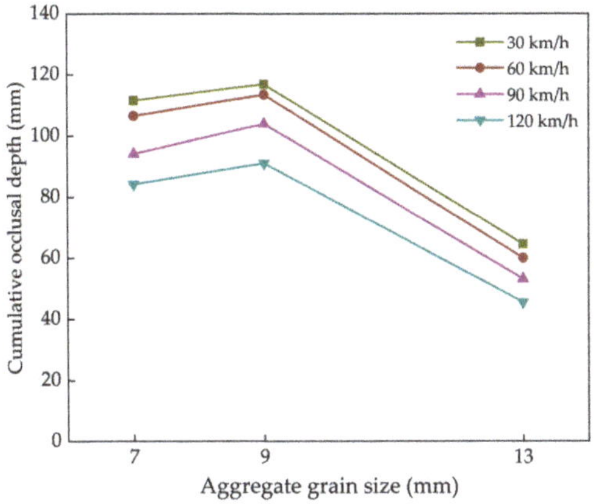

Figure 17. Cumulative occlusal depth of each simplified pavement at different vehicle initial speeds.

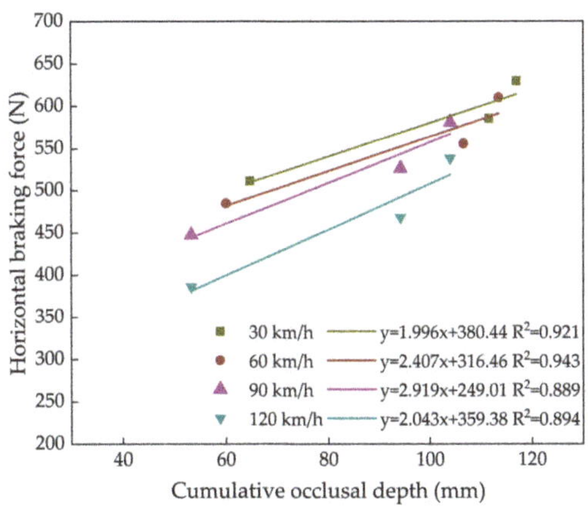

Figure 18. Correlation analysis between horizontal braking force and cumulative occlusal depth.

As shown in Figures 16 and 17, the horizontal braking force and cumulative occlusal depth for each simplified pavement decreased as the vehicle speed increased, with both the horizontal braking force and cumulative occlusal depth for the simplified pavements ranked from largest to smallest as aggregate grain sizes 9, 7, and 13 mm at all four speeds. As shown in Table 3, the *MTD* values of the simplified pavements with aggregate grain sizes of 7, 9, and 13 mm were 1.61, 2.07, and 2.86 mm, respectively. Thus, the effect of the contact interaction between the tyre and simplified pavement aggregates was ranked in order of superiority as aggregate grain sizes of 9, 7, and 13 mm, indicating that the aggregate grain size does not exhibit any correlation with tyre–pavement contact interaction in the later stages of pavement skid resistance. Though the increase in aggregate grain size resulted in a larger *MTD* of the simplified pavement, this does not indicate a better contact interaction effect between the tyre and pavement.

As shown in Figure 18, the squares of linear correlation coefficients between the horizontal braking force and pavement cumulative occlusal depth were 0.921, 0.941, 0.889, and 0.894 for vehicle initial speeds of 30, 60, 90, and 120 km/h, respectively, indicating that using the pavement cumulative occlusal depth to characterise the effect of contact interaction between the tyre and pavement aggregate is accurate and can be used to evaluate asphalt pavement skid resistance.

5. Conclusions

This study considered three simplified pavements comprising aggregates of different grain sizes as the research object to develop a 3D finite element tyre–pavement contact model in ABAQUS 2021. First, the contact stress and contact area distribution characteristics of each simplified pavement and the occlusal depth of the tyre–pavement system were analysed under steady–state rolling conditions. Subsequently, the horizontal braking force and cumulative occlusal depth of each simplified pavement were analysed at different vehicle speeds under steady–state braking conditions, and a correlation analysis between these parameters was conducted. The results of this study can be summarised as follows:

1. Under steady–state rolling conditions, the overall shape of the contact area between the tyre and each simplified pavement was similar; however, the peak contact stresses on the simplified pavement increased with its *MTD*, whereas pavements with larger *MTD* values exhibited smaller contact areas.
2. Under steady–state rolling conditions, the mean occlusal depth of the simplified pavement was positively correlated with the aggregate grain size. However, the cumulative occlusal depth did not exhibit any correlation with the aggregate grain size and ranked from largest to smallest as aggregate grain sizes 9, 7, and 13 mm. These results indicate that a simplified pavement with a smaller aggregate grain size exhibited a larger overall deformation of the tyre tread, implying that the grain size of the pavement aggregate bulge affects pavement skid resistance during the later stages.
3. Under steady–state braking conditions, the effect of the contact interaction between the tyre and simplified pavement aggregates was ranked in order of superiority as aggregate grain sizes of 9, 7, and 13 mm, indicating that during the late stages of pavement skid resistance, the two smaller aggregates provided superior performance. Consequently, to ensure the durability of pavement skid resistance, more consideration should be given to aggregates with grain sizes in the 7–9 mm range during pavement design.
4. The squares of linear correlation coefficients between the horizontal braking force and pavement cumulative occlusal depth were 0.921, 0.941, 0.889, and 0.894 for vehicle speeds of 30, 60, 90, and 120 km/h, respectively, indicating that the cumulative occlusal depth can be confidently used for the assessment of pavement skid resistance.

Note that this study was limited to the use of specific tyre and simplified pavement models. In future research, experimental studies will be conducted on pavements with actual aggregate compositions using different types of tyres. These limitations notwithstanding, the conclusions of this study indicate a clear direction for application and can contribute to the development of skid resistance design methods for asphalt pavements.

Author Contributions: Conceptualization, Z.L. and W.C.; methodology, Z.L.; software, W.C.; validation, W.W., Z.L. and Y.L.; formal analysis, W.C.; investigation, Y.L.; resources, W.W.; data curation, Y.L.; writing—original draft preparation, W.C.; writing—review and editing, Z.L.; visualization, Y.L.; supervision, W.W.; project administration, W.W.; funding acquisition, Z.L. All authors have read and agreed to the published version of the manuscript.

Funding: This study received no external funding.

Data Availability Statement: The data presented in this study are available on request from the corresponding author. The data are not publicly available due to privacy.

Conflicts of Interest: The authors declare no conflict of interest.

References

1. Rasol, M.; Schmidt, F.; Ientile, S.; Adelaide, L.; Nedjar, B.; Kane, M.; Chevalier, C. Progress and Monitoring Opportunities of Skid Resistance in Road Transport: A Critical Review and Road Sensors. *Remote Sens.* **2021**, *13*, 3729. [CrossRef]
2. Fang, T.; Guo, R.; Si, Y. Ensemble Mode and Singular Value Based Road Texture-Skid Study. *Coatings* **2021**, *11*, 947. [CrossRef]
3. Wang, H.; Liu, Y.; Yang, J.; Shi, X.; Xu, X.; Luo, S.; Huang, W. Evaluation of Anti-Skid Performance of Asphalt Mixture Based on Accelerated Loading Test. *Appl. Sci.* **2023**, *13*, 4796. [CrossRef]

4. Sabri, M.; Abda, S. Grip Analysis of Road Surface and Tire Footprint Using FEM. In Proceedings of the 10th International Conference Numerical Analysis in Engineering (NAE), Banda Aceh, Indonesia, 24–25 August 2018; IOP: Bristol, UK, 2018; Volume 308, p. 012018.
5. Huang, W.; Liu, C.; Guo, W.; Wei, Y. A Surface Texture Prediction Model Based on RIOHTrack Asphalt Pavement Testing Data. *Appl. Sci.* **2022**, *12*, 10539. [CrossRef]
6. Dong, Y.; Wang, Z.; Ren, W.; Jiang, T.; Hou, Y.; Zhang, Y. Influence of Morphological Characteristics of Coarse Aggregates on Skid Resistance of Asphalt Pavement. *Materials* **2023**, *16*, 4926. [CrossRef] [PubMed]
7. Zhang, Y.; Allen, A.W.; Yi, J.; Liu, T. Understanding Tire/Road Stick-Slip Interactions with Embedded Rubber Force Sensors. In Proceedings of the IEEE/ASME International Conference on Advanced Intelligent Mechatronics (AIM), Kaohsiung, Taiwan, 11–14 July 2012; IEEE: Washington, DC, USA, 2012; pp. 550–555.
8. Zhang, Y.; Yi, J. Static Tire/Road Stick–Slip Interactions: Analysis and Experiments. *IEEE/ASME Trans. Mechatron.* **2013**, *19*, 1940–1950. [CrossRef]
9. Chen, B.; Zhang, X.; Yu, J.; Wang, Y. Impact of contact stress distribution on skid resistance of asphalt pavements. *Constr. Build. Mater.* **2017**, *133*, 330–339. [CrossRef]
10. Liu, X.; Al-Qadi, I.L. Three-dimensional tire-pavement contact stresses prediction by deep learning approach. *Int. J. Pavement Eng.* **2022**, *23*, 4991–5002. [CrossRef]
11. Liu, X.; Jayme, A.; Al-Qadi, I.L. ContactGAN development-prediction of tire-pavement contact stresses using a generative and transfer learning model. *Int. J. Pavement Eng.* 2022, 1–11, *ahead-of-print*. [CrossRef]
12. Xie, Y.; Yang, Q. Tyre–pavement contact stress distribution considering tyre types. *Road Mater. Pavement Des.* **2019**, *20*, 1899–1911. [CrossRef]
13. He, H.; Li, R.; Yang, Q.; Pei, J.; Guo, F. Analysis of the Tire-Pavement Contact Stress Characteristics during Vehicle Maneuvering. *KSCE J. Civil Eng.* **2021**, *25*, 2451–2463. [CrossRef]
14. Tang, X.; Xie, J.; Xie, H.; Zhang, H. Predictions of three-dimensional contact stresses of a radial truck tire under different driving modes. *Adv. Mech. Eng.* **2022**, *14*, 1992380011. [CrossRef]
15. Lin, C.; Wang, T. Effect of fine aggregate angularity on skid-resistance of asphalt pavement using accelerated pavement testing. *Constr. Build. Mater.* **2018**, *168*, 41–46. [CrossRef]
16. Xiao, S.-Q.; Tan, Y.-Q.; Xing, C.; Zhou, X.-L. Scale demarcation of self-affine surface of coarse aggregate and its relationship with rubber friction. *Road Mater. Pavement Des.* **2021**, *22*, 1842–1859. [CrossRef]
17. Riahi, E.; Do, M.-T.; Kane, M. An energetic approach to model the relationship between tire rolling friction and road surface macrotexture. *Surf. Topogr. Metrol. Prop.* **2022**, *10*, 014001. [CrossRef]
18. Yu, M.; Kong, Y.; Wu, C.; Xu, X.; Li, S.; Chen, H.; Kong, L. The Effect of Pavement Texture on the Performance of Skid Resistance of Asphalt Pavement Based on the Hilbert-Huang Transform. *Arab. J. Sci. Eng.* **2021**, *46*, 11459–11470. [CrossRef]
19. Sharma, A.K.; Bouteldja, M.; Cerezo, V. Multi-physical model for tyre–road contact—The effect of surface texture. *Int. J. Pavement Eng.* **2022**, *23*, 755–772. [CrossRef]
20. Persson, B.N.J. Theory of rubber friction and contact mechanics. *J. Chem. Phys.* **2001**, *115*, 3840–3861. [CrossRef]
21. Lorenz, B.; Persson, B.N.J.; Dieluweit, S.; Tada, T. Rubber friction: Comparison of theory with experiment. *Eur. Phys. J. E* **2011**, *34*, 129. [CrossRef]
22. Heinrich, G.; Klueppel, M. Rubber friction, tread deformation and tire traction. *Wear* **2008**, *265*, 1052–1060. [CrossRef]
23. Lorenz, B.; Oh, Y.R.; Nam, S.K.; Jeon, S.H.; Persson, B.N.J. Rubber friction on road surfaces: Experiment and theory for low sliding speeds. *J. Chem. Phys.* **2015**, *142*, 194701. [CrossRef]
24. Wang, H.; Wang, Z. Evaluation of pavement surface friction subject to various pavement preservation treatments. *Constr. Build. Mater.* **2013**, *48*, 194–202. [CrossRef]
25. Yu, M.; Wu, G.; Kong, L.; Tang, Y. Tire-Pavement Friction Characteristics with Elastic Properties of Asphalt Pavements. *Appl. Sci.* **2017**, *7*, 1123. [CrossRef]
26. Zong, Y.; Xiong, R.; Wang, Z.; Zhang, B.; Tian, Y.; Sheng, Y.; Xie, C.; Wang, H.; Yan, X. Effect of morphology characteristics on the polishing resistance of coarse aggregates on asphalt pavement. *Constr. Build. Mater.* **2022**, *341*, 127755. [CrossRef]
27. Uz, V.E.; Gokalp, I. The effect of aggregate type, size and polishing levels to skid resistance of chip seals. *Mater. Struct.* **2017**, *50*, 126. [CrossRef]
28. Wasilewska, M.; Gierasimiuk, P.; Gardziejczyk, W. Comparison of skid resistance of EAC pavement with different maximum aggregate grain size. *Int. J. Pavement Eng.* **2022**, *23*, 3930–3940. [CrossRef]
29. Yu, L.; Hu, J.; Li, R.; Yang, Q.; Guo, F.; Pei, J. Tire-Pavement Contact Pressure Distribution Analysis Based on ABAQUS Simulation. *Arab. J. Sci. Eng.* **2022**, *47*, 4119–4132. [CrossRef]

applied sciences

MDPI

Review

Retrofitting of Bridge Slabs for Safety Railing Refurbishment in Italy: A State-of-the-Art Review

Giuseppe Santarsiero

School of Engineering, University of Basilicata, Via Dell'ateneo Lucano, 10, 85100 Potenza, Italy; giuseppe.santarsiero@unibas.it

Abstract: The recent accident of Mestre (northern Italy), which caused 21 fatalities due to a bus falling from a bridge, strongly highlighted the safety problem related to the presence of old railings along Italian roads. Bridge railings, also known as guardrails or parapets, serve the crucial function of preventing vehicles from accidentally driving off the edge of a bridge. Performance requirements of safety railings have been recently increased due to laws and technical standards enforced in Italy and Europe. However, many bridges along important roads, such as motorways and highways, are currently equipped with outdated safety railings since they were built before these regulations came into force. Therefore, many people are daily exposed to the risk of heavy accidents due to railing failures as well as vehicles and people eventually present below such structures. This paper aims to outline the technical problems and solutions in bridge refurbishment interventions devoted to increasing traffic safety as, for example, the installation of code-conforming railings, which often require the structural retrofit of bridge elements supporting the railing. To this end, several technical solutions are described and critically compared, and, finally, an economic analysis is reported to highlight the slab retrofit influence on the total intervention cost.

Keywords: bridge; safety railings; guardrail; reinforced concrete; deck slab; retrofit; refurbishment

Citation: Santarsiero, G. Retrofitting of Bridge Slabs for Safety Railing Refurbishment in Italy: A State-of-the-Art Review. *Appl. Sci.* **2023**, *13*, 12051. https://doi.org/10.3390/app132112051

Academic Editors: Edgar Sokolovskij and Vidas Žuraulis

Received: 10 October 2023
Revised: 27 October 2023
Accepted: 3 November 2023
Published: 5 November 2023

1. Introduction

Road bridge railing failures can have serious consequences on the safety of both motorists and pedestrians [1]. In fact, when these railings fail or are inadequate, various safety issues can arise such as vehicle accidents. The primary purpose of bridge railings is to prevent vehicles from veering off the road and falling from the bridge. When railings fail, it increases the risk of vehicles colliding with the bridge structure, potentially causing severe accidents and injuries.

Inadequate bridge railings can lead to fatalities and injuries among occupants of vehicles involved in accidents. The absence of a reliable barrier increases the chances of vehicles plunging off the bridge or striking other vehicles.

When bridges are provided with walkways, effects on pedestrians can occur if the bridge railings on these walkways fail. This puts pedestrians at risk of falling from the bridge, which can result in severe injuries or fatalities.

In some cases, bridge railings may have design flaws that make them less effective in preventing accidents. Poorly designed railings may not provide adequate containment, allowing vehicles to easily breach them during collisions.

Even when design flaws are absent, bridge railings can deteriorate due to factors like weathering, corrosion, and vehicle impacts. Neglected maintenance and poor construction materials can exacerbate this problem. When the railings (or their connection to the bridge) degrade, their ability to withstand vehicle impacts and protect them against accidents diminishes.

In order to cover these gaps, both Italian and European authorities issued new codes pertaining to the evaluation and certification of the adequacy of new safety railings. Several codes on this topic have been issued in Italy since 1992 [2–5], and the European code EN1317 was adopted in Italy and other EU countries [6–8]. This normative framework, which is related to new road infrastructure construction, allows existing infrastructure to adopt old safety railings.

Due to this and the above-mentioned technical issues, poor-quality safety railings on older bridges may be a contributing cause to some of the major accidents observed in the last decade in Italy. For example, the recent accident of Mestre (Italy), which caused 21 fatalities and 15 injuries, is one of the most tragic road accidents that recently occurred in Italy. A bus fell from a bridge (the so-called Vempa overpass) equipped with a very old safety rail. Moreover, in the accident zone, the railing was interrupted, and most probably this facilitated the bus to overcome the parapet and fall. This tragedy strongly highlighted the need of railing retrofitting interventions along Italian roads, especially those with high traffic volume. Neglecting the causes of the vehicle skid, which are under investigation, a simple engineering analysis can easily confirm that the presence of an up-to-date barrier could have avoided or mitigated the tragedy. Another accident happened on 28 July 2013, when another bus fell from the Acqualonga bridge along the A16 motorway in southern Italy. A total of 40 people died and 8 were injured. In this case, the bus lost the brakes, and the bridge barriers failed due to the impact [9].

These two tragedies, similar to each other, caused 62 fatalities, more than the complete collapse of 250 long Morandi bridge span [10] responsible for 43 fatalities. Therefore, accidents due to the collapse of safety railings can have a human life cost similar to great natural disasters, such as earthquakes [11,12], and thus should not be considered minor issues with respect to the global safety of bridges.

Despite this fact, the importance of bridge railings is not fully recognized by the Italian technical code [13]. The recent Italian guidelines on risk classification and management of road bridges [14] are devoted to assigning the "class of attention" (a kind of risk evaluation) to all road bridges through a multilevel evaluation based on documental analysis and accurate inspections. In this code, the focus is the safety of the bridge structural members, while the railings are accounted as accessory elements with lower importance and scarce potential effects on the users' safety.

Several Italian road management bodies have started the requalification of road railings (e.g., Figure 1), even though the extension of the national road network requires huge investments that should be allocated considering precise criteria for prioritization, based on the risk generated by the traffic volume (average daily traffic, representing the exposure) and the adequacy level of installed railings, which represents the vulnerability.

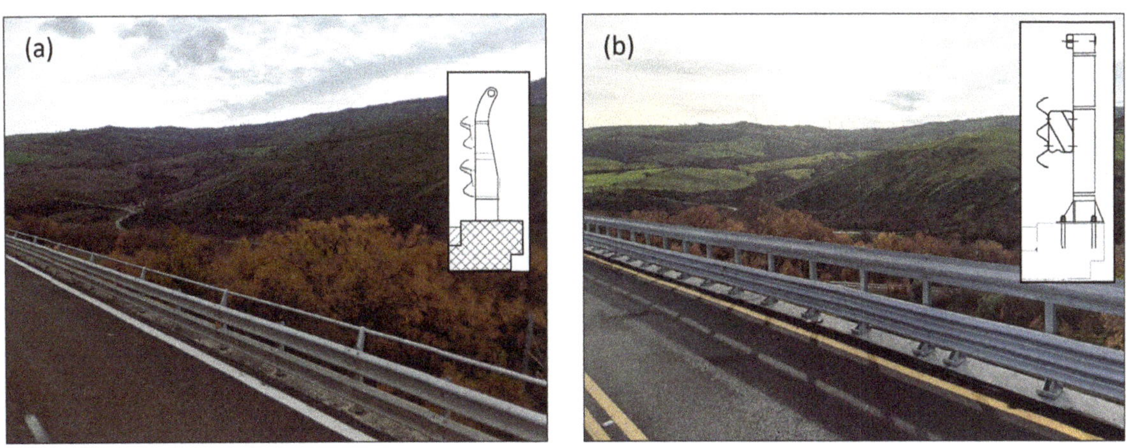

Figure 1. Photo and scheme of pre- (**a**) and post- (**b**) railing refurbishment intervention on a reinforced concrete bridge.

Such investment needs are increased by the inadequacy of supporting elements of bridge barriers. Most available research on the topic has been focused on evaluating the response of safety railings to accidental actions (crashes) [15,16], neglecting the adequacy of supporting elements like the slab.

According to [5], in case a railing is replaced, an H2, H3, or H4 class new railing must be installed depending on the type of road (motorways, extra-urban roads, and urban roads) and traffic volume (also as a percentage of commercial vehicles with mass higher than 3.5 tons). H2, H3, and H4 barriers have increasing containing capacity and are usually installed on bridge edge curbs through chemical anchors. Figure 2 compares old bridge barriers and new ones [17]. These latter (shown only as illustrative models among the wide variety of commercially available devices) are taller than the old ones and are equipped with a 3-wave profile instead of a 2-wave profile. Moreover, connecting elements placed at the top of the barrier are more robust allowing a better collaboration between consecutive posts, thus increasing the containing capacity of the device.

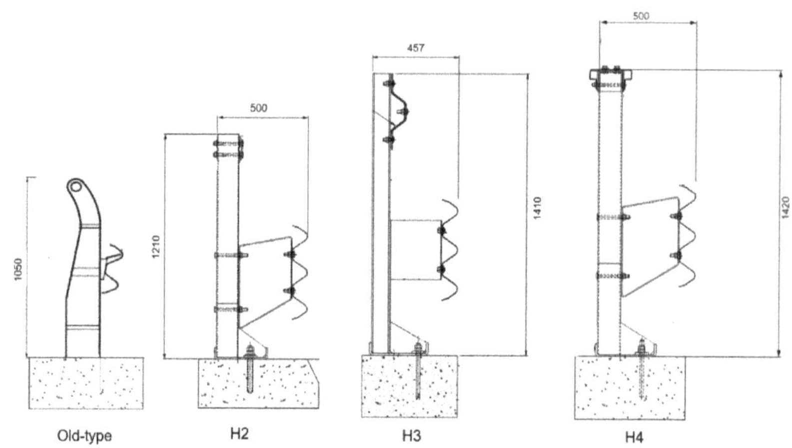

Figure 2. Typical old type and new H2, H3, and H4 bridge railings (dimensions in mm).

In most cases, the bridge slab has a structural capacity insufficient to allow the installation of an up-to-date railing. In fact, the higher the resistance of the barrier, the higher the stresses on the supporting elements. This forces to carry out strengthening interventions on the bridge slab, with a consequent increase in time and cost to carry out works. Furthermore, there can be cases in which the curb is not geometrically suitable for the new railing solution, and works are needed to adapt it. It is worth noting that the bridge edge curb is the part most sensitive to material deterioration possibly impairing the railing effectiveness. Recent studies have shown that this issue can be solved using composite materials not affected by corrosion [18]. However, the structural integrity and durability of bridge edge curbs are not the focus of this study, being devoted to bridge slab structures.

This paper is devoted to illustrating the assessment and design methods as well as the most frequently used strengthening techniques to increase the slab structural capacity making it suitable for the railings' installation in Italy. It is worth noting that possible aerodynamic issues on cable-stayed or suspended bridges due to railing installations are not treated here, and they should be addressed in a specific work. This study deals only with single- and multi-span bridges not subjected to aerodynamic problems, representing by far most Italian bridges.

2. Size of the Problem

According to the WHO (World Health Organization) report released in January 2018, about 1.35 million people are killed annually in road accidents [19,20]. At a European level, data on road safety are collected through CARE [21,22], which is a community database on road accidents (commonly referred to as "crashes") resulting in death or injury, without reporting statistics on damage-only accidents. This allows examining the data related to crashes, disaggregating them, for example, by country, type of vehicle, year, transport mode, etc. Based on CARE, Figure 3 reports the total number of killed people in 2021 across the EU per million inhabitants. As can be seen, the situation in Italy represents a sort of "average" being at the same level as France and Portugal, and significantly better than Eastern Europe countries. However, a high variability is found among European countries. Therefore, Italian authorities dealt with about 2875 killed people in 2021 due to road accidents, which was slightly decreased due to the pandemic restrictions. In fact, in 2022, the number of killed people increased to pre-pandemic levels, with the total being 3159 [23].

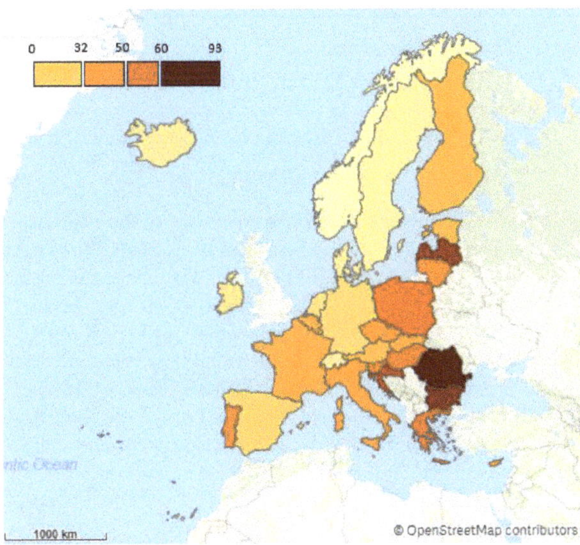

Figure 3. Killed people in road accidents per million inhabitants across EU (including EFTA member states).

To face this problem, the Italian Ministry of Infrastructure and Transportation issued a national plan to strongly reduce the road victims with the objective of halving killed people by 2030 [24]. This plan is consistent with the UN Resolution by the General Assembly 74/299 of 2020 [25] and the EU directive of 2021 [26].

The Italian plan is ambitious and based on the following main steps:

- legislative actions;
- measures to strengthen control and repression;
- interventions to improve the safety of road infrastructures;
- communication and awareness campaigns.

Among the actions on infrastructures, it highlighted the willingness to "Raise awareness among owners or road managers on effects of deficiency maintenance of the infrastructure, on the importance of scheduled maintenance and of the installation of so-called barriers save-motorcyclists". There is no specific reference to the need for a general refurbishment of road railings, even though the main road management bodies are realizing works to modernize safety devices along the nationwide road network. However, the plan estimated the fund allocation needed to deploy its actions, corresponding to EUR 1.4 billion. This amount does not consider the work to be carried out that is much higher.

In fact, Italy has about 167,000 km of roads made of (in decreasing order of importance) motorways; highways (state roads); and regional, provincial, and municipal infrastructures. According to [14], the total number of bridges is estimated around 120,000. Referring to the bridge database reported in [27], the average length of highway bridges is about 220 m. Therefore, a preliminary estimate of the total bridge length in Italy is approaching 26,400 km. To estimate the fund allocation needed to provide these bridges with up-to-date railings, several design documents and price lists issued by road management bodies were examined to find out some indication of an average unit cost [28]. Therefore, after the review of slab-strengthening techniques reported in Section 3, Section 4 shows a rough estimate of intervention costs. However, the number (or the length) of already refurbished bridges is unknown, and, therefore, an evaluation of the national investment to solve this problem cannot be carried out.

3. Slab Assessment and Retrofitting Approaches

3.1. Structural Assessment

When the guardrails on a bridge are replaced with newer and more resistant ones, the forces transferred in case of an accident, especially due to trucks, can be significantly higher. For this reason, the structural assessment of the elements supporting the barriers is mandatory.

When installing a new railing (Figure 4a), checking the needed geometric requirements is necessary. In particular, the dimensions of the edge curb should fit the barrier base plate dimensions. If this check gives a positive result, the effect of the barrier failure in case of a crash must be evaluated. In this case, it can be assumed that the railing posts are subjected to plastic deformation as happens in crash tests performed to approve a safety device. Therefore, the plastic bending moment of the post must be evaluated as a consequence of the truck crash (Figure 4c). Given that they are made of ordinary steel, simple structural engineering methods may be adopted. Based on the post actions, sections 1, 2, and 3 in Figure 5a must be checked. Sections 1 and 2 are subjected to shear, bending moment and axial load, while Section 3 is subjected to shear and overturning moment.

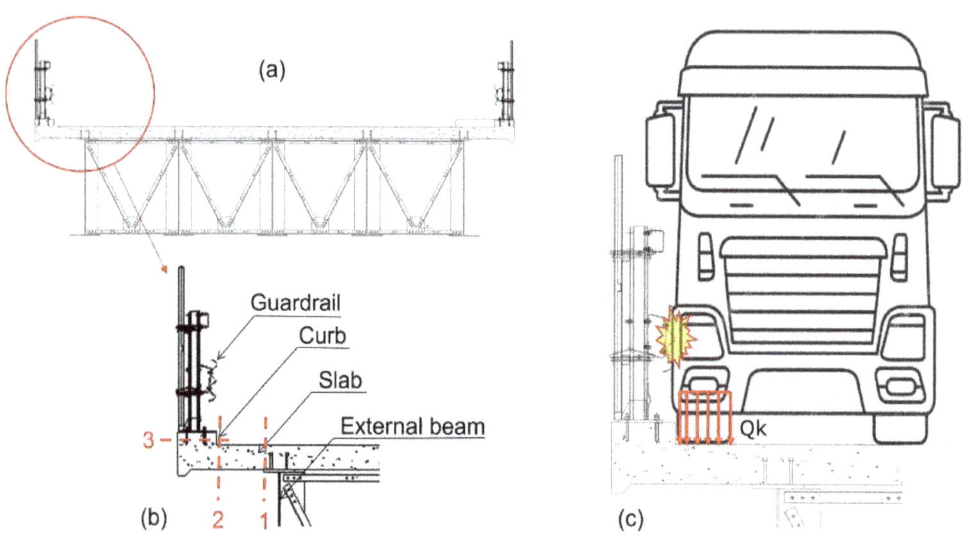

Figure 4. (**a**) Bridge deck section with reinforced concrete slab, (**b**) guardrail supporting elements and critical sections, and (**c**) truck impact position.

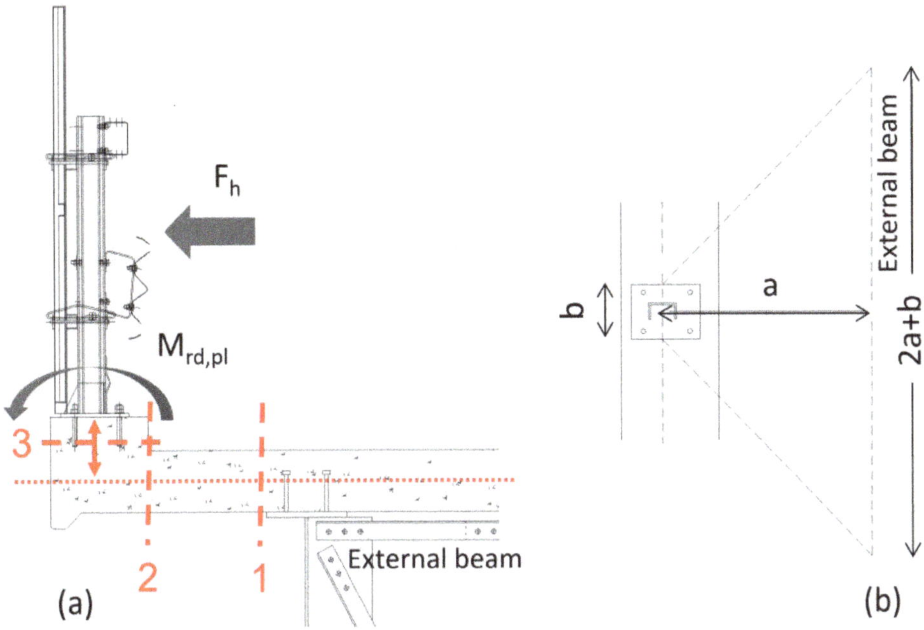

Figure 5. (**a**) Crash actions on the railing and (**b**) plan view of the effective slab sections.

Once the post plastic moment $M_{rd,pl}$ is calculated, the horizontal crash force F_h should be derived taking into account a height of 1.00 m from the road pavement, as reported by the Italian and European codes [29–31]. It is worth noting that F_h represents the tensile axial load acting on slab sections 1 and 2 (Figure 5a). Section 1 has a width equal to $2a + b$, where b is the barrier plate width, and a is the slab cantilever length (Figure 5b). This assumption is based on the commonly adopted 45° stress diffusion, even though more refined methods can be used (e.g., finite-element analyses). Once the crash loads are computed, they must be increased by a factor of 1.5 according to the Italian code and combined with other live and dead loads. Among the live loads, the weight of the truck wheel applied near the curb (Qk in Figure 4c) must be considered (200 kN according to the Italian code [29]). The load combination must be applied to a simple structural scheme represented by the bridge slab cantilever extending from the external beam.

Safety checks in sections 1, 2, and 3 must be performed based on an accurate knowledge of the existing bridge. Information on geometry, material properties, and structural details must be gathered through inspections and in situ and laboratory tests.

Figure 6 shows investigations for the design of a safety railing refurbishment of an RC bridge. Note that the parapet in Figure 6a,b is made of provisional elements installed only for the safety of workers, while the type of guardrail to be installed is similar to that shown in Figure 2 (H2, H3, or H4). In particular, the investigations of Figure 6 are devoted to detecting the type and number of rebars placed in correspondence with the external beam axis, which is the most stressed section of the slab. This must be performed by removing road pavement and concrete cover until reaching the reinforcement. Moreover, further investigation is necessary on the edge curb (Figure 6a). The number of these investigations should be set based on the availability of bridge design documents as well as its total extension. It is worth noting that most bridges built in the period 1950–1970 have no design documents, and, in those cases, the in situ investigations should be more accurate since the structural knowledge must be obtained starting from scratch.

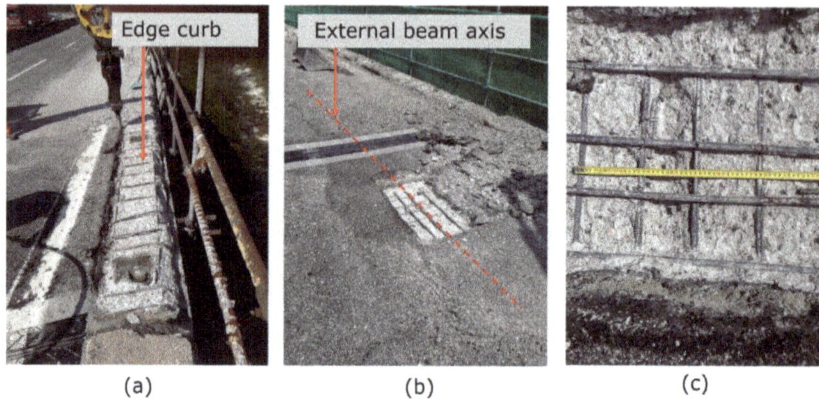

Figure 6. Example of in situ investigations for (**a**) edge curb detailing, (**b**) slab reinforcement across the external beam, and (**c**) detailed survey of reinforcement reported in (**b**).

In the occasion of these investigations, material sampling is also to be performed to determine strength values to be used in safety checks. According to the Italian building code [29,30] and the code for bridge assessment [13], structural investigations can be consistent with three increasing knowledge levels, KL1, KL2, and KL3. Based on the chosen KL, the number of samples to be extracted from the structure is determined. In the case of bridge assessment, the maximum knowledge level should be selected, also based on the indication reported in [13].

Indications on the number of samples and investigations are reported in [29], even though it is mainly related to building structures rather than bridges. Therefore, according to Table 1, in the case of KL3, the building floor area can be converted into a bridge deck area.

Table 1. Minimum number of samples and investigations to be extracted according to KL3.

	Concrete Cores	Rebars
300 m^2 of building floor	2	2
300 m^2 of bridge deck	2	2

Furthermore, it is required to check the reinforcement detailing on 15% of structural elements (e.g., columns and beams). This indication can be applied to bridges converting 15% of structural elements into 15% of the deck surface. Thus, the removal of road pavement and concrete cover to check slab reinforcement details should be made on 15% of the slab's cantilever part.

3.2. *Slab Retrofit Approaches*

Reinforced concrete slabs in bridges are considered secondary elements, and most of the research on retrofitting solution is related to primary elements like girders and piers, e.g., [32,33]. In this case, the focus is the slab strengthening, which can be performed through several techniques based on increasing the amount of reinforcement on the top part of the slab section across the cantilever part [34]. Both the truck wheel load and the crash impact force generate a negative moment that may require additional reinforcement where tensile stresses are present (i.e., at the extrados). Four approaches can be used:

1. Additional steel rebars in correspondence with barrier posts
2. Additional steel rebars along the whole slab length
3. Near surface mounted (NSM) ordinary steel plates connected through chemical anchors
4. NSM bars, laminates, or textile composites

Techniques 1 and 2 differ only in terms of extension of the slab-strengthening intervention and, for this reason, will be described in the same section.

It is worth noting that when performing safety checks under flexure in the presence of new additional reinforcement, the existing concrete properties must be used. In fact, although a new concrete cover made of high-strength cementitious grout is realized on the slab extrados, the intrados (resisting compression stresses) is made of existing concrete. Therefore, a reliable evaluation of its properties through core drilling and laboratory testing is of paramount importance. In the same way, the existing reinforcement should be considered in the safety checks. In order to corroborate the information gathered through investigations, existing sources of information can be used regarding the analyses of databases including historical concrete and steel testing results, which can help in recognizing the concrete and steel grades used at the time of construction [35,36].

3.2.1. Additional Steel Rebars

As can be seen from Figure 7 (showing the section and plan views of the intervention details), the additional reinforcement in terms of rebars is placed across the external beam to provide the necessary development length, obtaining a total width of intervention equal to W. New rebars are installed only in correspondence with the new barrier post axis to limit the invasiveness of works. This requires a precise cut in the slab concrete cover and the hydrodemolition commonly of about 5 cm or until finding the existing rebars. New rebars are anchored in the curb through an inclined branch. Once the main rebars are installed, vertical anchors are realized through vertical holes and epoxy resin. The concrete cuttings are finally filled with high-strength fiber-reinforced grout and a waterproofing layer. Thus, the road pavement is restored to complete the intervention.

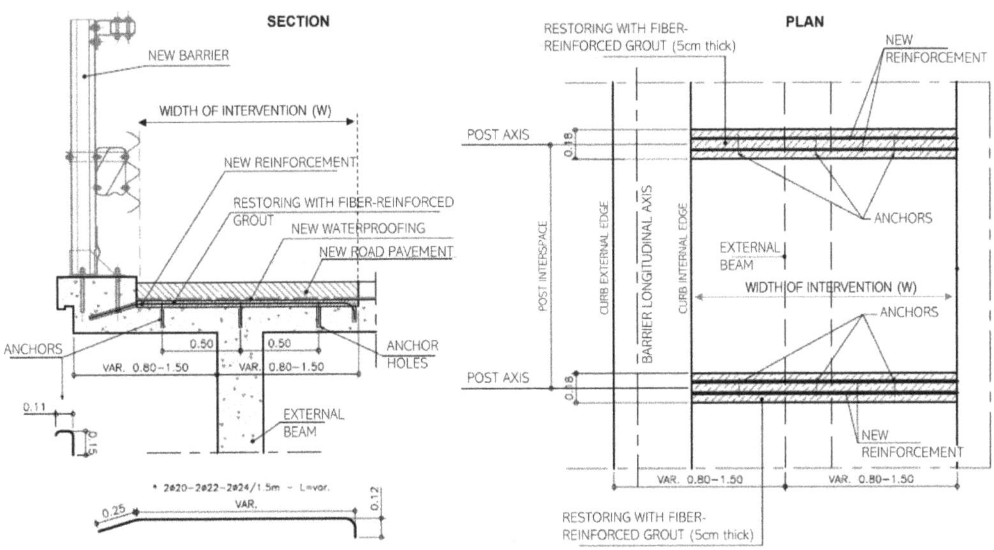

Figure 7. Slab strengthening through additional rebars.

Referring to technique no. 2, the only difference can be found in the fact that additional rebars are distributed throughout the slab length and not only close to the post axis. Therefore, lower-diameter rebars are used, and more cover demolition is necessary. This approach allows neglecting the final position of posts since the slab is uniformly upgraded. However, it is more expensive and impacting on the slab.

3.2.2. NSM Steel Plates

Technique no. 3 is based on the use of steel plates provided with holes that can be coupled with chemical anchors made of threaded rods (Figure 8). This is necessary to make the plates collaborate with the existing concrete. In fact, due to ordinary steel's smoothness, the steel–concrete bond is not effective. Concrete cover cutting has a lower depth (usually 2 instead of 5 cm related to techniques 1 and 2). It requires the complete curb demolition in order to allow the plates to properly extend.

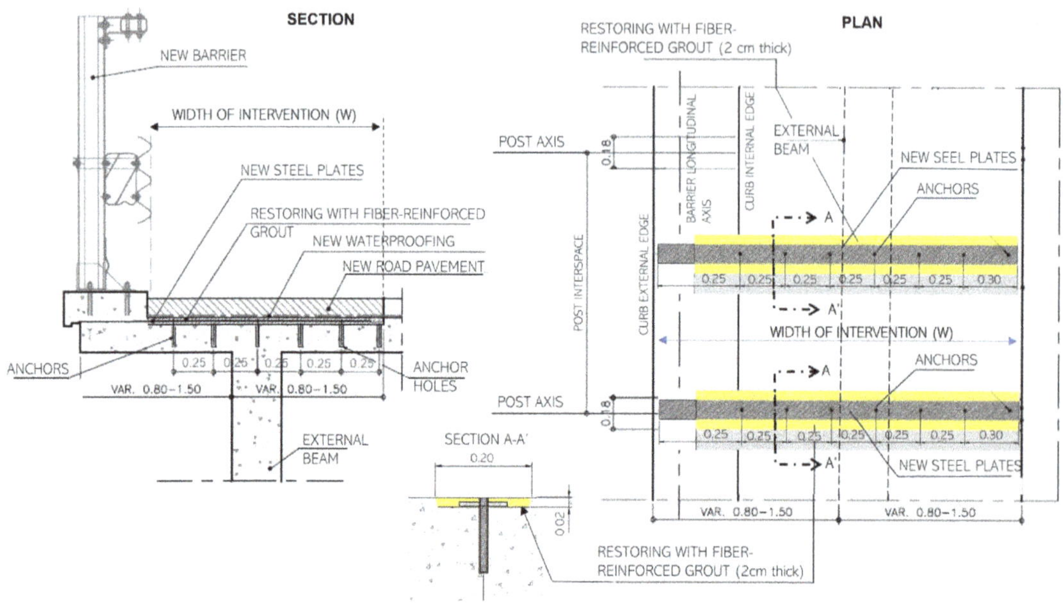

Figure 8. Slab strengthening through NSM steel plates.

Channels created to install the plates are made through hydrodemolition and are following filled with fiber-reinforced cementitious grout. It is worth noting that anchors, in this case, must be installed before the plates, and, therefore, a greater precision is requested during the works. Moreover, ordinary steel used for the plates is provided with a lower yielding stress, forcing to use a larger reinforcement area compared with strengthening options 1 and 2, which adopt deformed bars.

3.2.3. NSM Composites

A typical application of NSM composites is depicted in Figure 9. As can be seen, no anchors are needed due to the epoxy resin connection. A small cut (obtained by milling) in the slab transverse direction is sufficient to install the carbon (or other material) fiber bars after a bed of epoxy resin is created. To strengthen the interface between the curb and slab, in this case, the curb should be demolished and subsequently reconstructed. It is worth noting that concrete members reinforced with composites usually fail by delamination (i.e., concrete failure at the interface with composites) resulting in fragile structural behavior. However, works for this type of intervention are less invasive on the existing slab structure. On the other hand, they are more expensive due to the use of composite bars and a significant amount of epoxy resin, also needing more skilled workmanship.

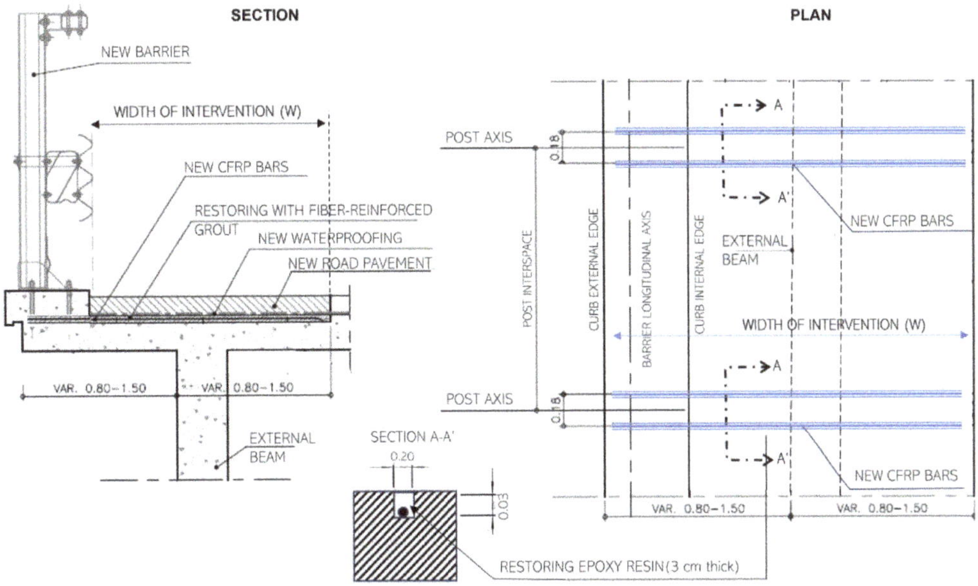

Figure 9. Slab strengthening through NSM composite bars.

4. Discussion

The presented slab-strengthening techniques have been derived from real structural design documents and guide manuals issued by road management bodies [34]. Therefore, they are used in practical applications mainly along motorways and highways in Italy. A critical comparison can be useful to determine their applicability as well as the pros and cons, as reported in Table 2, also referring to durability requirements [37], impact on the existing structure, and costs.

As can be seen, techniques no. 1 and no. 2 are the cheapest since they are based on consolidated techniques of RC elements' jacketing. On the other hand, they are exposed to possible corrosion (like technique no. 3), although the presence of a waterproofing layer on the slab extrados significantly reduces this risk. However, corrosion risk is absent in technique no. 4 due to the absence of metallic materials. Currently, the most used technique is no. 1 due to its versatility. In fact, as the anchors are installed after the additional rebars, small imprecisions in their positioning are admitted and do not compromise the effectiveness of interventions.

Based on the work price lists of the major Italian road management agency (ANAS) [28], the cost of intervention no. 1 per unit bridge length has been computed as a function of the intervention width (W in Figure 6). W depends on both the slab cantilever dimension and the rebar diameter and grade, which determine the development length to be adopted beyond the external beam.

As can be seen from Figure 10, the intervention cost is proportional to the intervention width. Based on inspected design documents of real bridges, the most frequent W value is 2 m corresponding to an intervention cost of EUR 843/m including the cost of new railings and all the works needed for the slab retrofit. These data are strongly susceptible to increase in the near future given the current inflation rates across the EU. Therefore, it should be treated as a temporary estimate.

Table 2. Comparison of slab-strengthening intervention techniques.

#	Description	Advantages	Limitation	Cost
1	Steel rebars on posts	• No curb demolition • Flexible position of anchors	• Possible corrosion effects in the long-term • Need to fix posts' positions before the intervention	Low
2	Distributed steel rebars	• No curb demolition • Flexible position of anchors and rebars	• Possible corrosion effects in the long-term • Hydrodemolition impact on the existing structure	Medium-low
3	NSM steel plates	• Lower impact of hydrodemolition on the existing structure	• Fixed position of vertical rods • Using lower-yielding strength ordinary steel • Curb demolition	Medium
4	NSM composites	• Lowest impact of hydrodemolition on the existing structure • No corrosion • No anchors	• Curb demolition • Possible delamination and fragile structural behavior	High

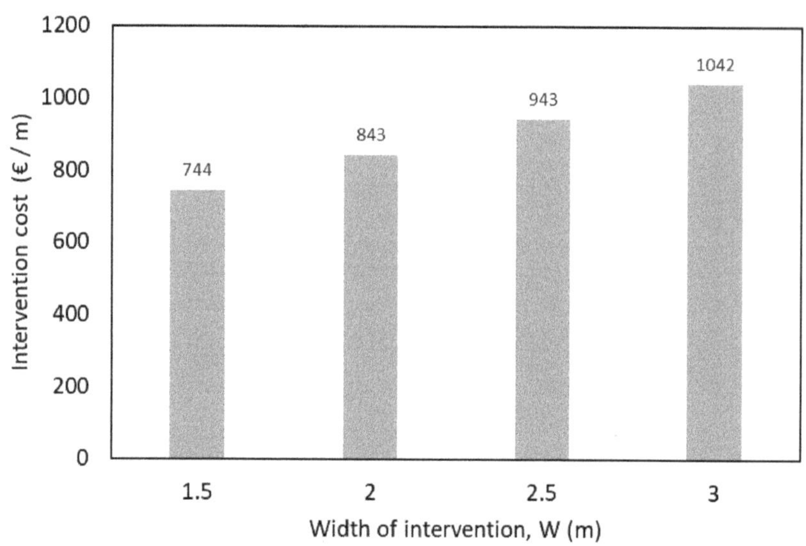

Figure 10. Unit length cost of strengthening solution no. 1.

Looking at Figure 11, the cost of the slab retrofit varies between 50 and 60% of the total intervention cost, being in most cases higher than the new railing cost. This means that low-impact and cheap slab retrofit solutions can greatly reduce the overall cost of interventions of the wide national bridge asset.

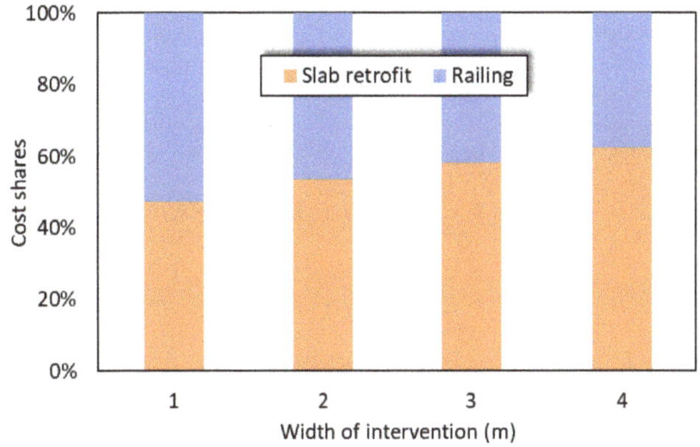

Figure 11. Relative cost of new barrier and slab retrofit for solution no. 1.

5. Conclusions

Although Italy is facing the refurbishment of bridge railings along its road network, some severe accidents still happen, causing huge human life losses. The replacement of old-type railings with new high-performance railing systems, able to avoid vehicles falling from bridges, often requires the reinforced concrete slabs' structural retrofit, which has not been properly investigated in the current literature. This paper wants to provide an overview of the issues related to the presence of weak retaining systems placed along edge bridge curbs, designed according to outdated codes, and working on high-traffic roads. After a description of the problem size at the Italian and European levels and an illustrative comparison of old type and new bridge railings, the general description of the assessment and retrofitting procedures available for the bridge slab retrofitting is reported along with a critical comparison. The more feasible is technique no. 1, where new steel reinforcement is installed on the slab extrados only in correspondence with posts. This technique is based on a concrete jacketing approach widely used in reinforced concrete building strengthening interventions. This technical option does not require curb demolition, resulting in the less expensive of the analyzed approaches. Among the other techniques, the installation of composite materials provides more durable interventions even though more expensive. In summary, the assessment and retrofit procedures here described may be useful to practitioners, to road management agencies for road maintenance activities, and to researchers investigating new solutions able to reduce cost while improving safety.

Funding: This research received no external funding

Institutional Review Board Statement: Not applicable.

Informed Consent Statement: Not applicable.

Data Availability Statement: No new data were created or analyzed in this study. Data sharing is not applicable to this article.

Acknowledgments: The author would like to express his gratitude to the three anonymous reviewers who helped to improve the paper quality.

Conflicts of Interest: The author declares no conflict of interest.

References

1. Roberts, H. Italy's bridge disaster: An inquest into privatization. *Financial Times*. 2018. Available online: https://www.ft.com/content/874b7e4c-ac3f-11e8-94bd-cba20d67390c (accessed on 6 August 2023).
2. *DM 18/02/1992 n. 223*; Regolamento Recante Istituzioni Tecniche per la Progettazione, L'omologazione e L'impiego Delle Barriere Stradali di Sicurezza (G.U. 16/03/1992, n.63). Ministry of Infrastructure: Roma, Italy, 1992.

3. *DM 03/06/1998*; Istruzioni Tecniche Sulla Progettazione, Omologazione e Impiego Delle Barriere Stradali di Sicurezza. Prescrizioni Tecniche per le Prove ai Fini Dell'omologazione (G.U. 29/10/1998, n.253). Ministry of Infrastructure: Roma, Italy, 1998.
4. *Norma UNI 01/05/2003 UNI EN 1317-4*; Barriere di Sicurezza Stradali—Classi di Prestazione, Criteri di Accettazione per la Prova D'urto e Metodi di Prova per i Terminali e Transizioni di Barriere di Sicurezza. UNI: Milan, Italy,, 2003.
5. *DM 21/06/2004 n. 2367*; Aggiornamento del Istruzioni Tecniche per la Progettazione, L'omologazione e L'impiego Delle Barriere Stradali di Sicurezza e le Prescrizioni Tecniche per le Prove Delle Barriere di Sicurezza Stradale (G.U. 06/08/2004, n. 182). Ministry of Infrastructure: Roma, Italy, 2004.
6. *Norma UNI 2010 UNI EN 1317-1*; Barriere di Sicurezza Stradali. Terminologia e Criteri Generali per I Metodi di Prova. UNI: Milan, Italy, 2010.
7. *Norma UNI 2010 UNI EN 1317-2*; Sistemi di Ritenuta Stradali Parte 2: Classi di Prestazione, Criteri di Accettazione Delle Prove D'urto e Metodi di Prova per le Barriere di Sicurezza Inclusi i Parapetti Veicolari. UNI: Milan, Italy, 2010.
8. *Norma UNI 2010 UNI EN 1317-3*; Barriere di Sicurezza Stradali—Classi di Prestazione, Criteri di Accettabilità Basati Sulle Prove di Impatto e Metodi di Prova Degli Attenuatori D'urto. UNI: Milan, Italy, 2010.
9. Incidente stradale del viadotto Acqualonga. In *Wikipedia*, 4 October 2023. Available online: https://it.wikipedia.org/wiki/Incidente_stradale_del_viadotto_Acqualonga (accessed on 27 July 2023).
10. Rymsza, J. Causes of the collapse of the Polcevera Viaduct in Genoa, Italy. *Appl. Sci.* **2021**, *11*, 8098. [CrossRef]
11. Masi, A.; Santarsiero, G.; Chiauzzi, L.; Gallipoli, M.R.; Piscitelli, S.; Vignola, L.; Bellanova, J.; Calamita, G.; Perrone, A.; Lizza, C.; et al. Different damage observed in the villages of Pescara del Tronto and Vezzano after the M6.0 August 24, 2016 central Italy earthquake and site effects analysis. *Ann. Geophys.* **2016**, *59*, fast track 5. [CrossRef]
12. Masi, A.; Santarsiero, G.; Digrisolo, A.; Chiauzzi, L.; Manfredi, V. Procedures and experiences in the post-earthquake usability evaluation of ordinary buildings. *Boll. Di Geofis. Teor. Ed Appl.* **2016**, *57*, 199–200.
13. Ministry of Infrastructure and Transportation - High Council of Public Works (CSLP). *Guidelines on Risk Classification and Management, Safety Assessment and Monitoring of Existing Bridges*; Ministry of Infrastructure: Rome, Italy, 2020.
14. Santarsiero, G.; Masi, A.; Picciano, V.; Digrisolo, A. The Italian Guidelines on Risk Classification and Management of Bridges: Applications and Remarks on Large Scale Risk Assessments. *Infrastructures* **2021**, *6*, 111. [CrossRef]
15. Thanh, L.; Itoh, Y. Performance of curved steel bridge railings subjected to truck collisions. *Eng. Struct.* **2013**, *54*, 34–46. [CrossRef]
16. Zuluaga, C.M.; Albert, A. Preventing falls: Choosing compatible Fall Protection Supplementary Devices (FPSD) for bridge maintenance work using virtual prototyping. *Saf. Sci.* **2018**, *108*, 238–247. [CrossRef]
17. Buildtech, M. Technical Specification of Road Railings. 2023. Available online: https://www.marcegagliabuildtech.it/barriere-sicurezza/ (accessed on 28 August 2023).
18. Ibrahim Sallam, A.A.M.A. BEC—Bridge Edge Curbs—GFRP Rebar Reinforcement. Master's Dissertation, University of Naples "Federico II", Naples, Italy, 2023.
19. WHO (World Health Organization). Global Status Report on Road Safety 2018. Available online: https://www.who.int/publications/i/item/9789241565684 (accessed on 30 August 2023).
20. Wachnicka, J.; Palikowska, K.; Kustra, W.; Kiec, M. Spatial differentiation of road safety in Europe based on NUTS-2 regions. *Accid. Anal. Prev.* **2021**, *150*, 105849. [CrossRef] [PubMed]
21. 93/704/EC; Council Decision of 30 November 1993 on the Creation of a Community Database on Road Accidents. Available online: https://eur-lex.europa.eu/eli/dec/1993/704/oj (accessed on 12 September 2023).
22. DG MOVE—CARE Database on Road Crashes Leading to Death or Injury. Available online: https://dashboard.tech.ec.europa.eu/qs_digit_dashboard_mt/public/extensions/MOVE_CARE_public/MOVE_CARE_public.html (accessed on 10 September 2023).
23. Italian national institute of Statistics. *Incidenti Stradali Anno 2022*. 2023. Available online: https://www.istat.it/it/files/2023/07/REPORT_INCIDENTI_STRADALI_2022_IT.pdf (accessed on 14 September 2023).
24. *Delibera CIPESS 14 aprile 2022, n. 13*; Piano Nazionale Sicurezza Stradale 2030. MIMS. Ministry of Infrastructure and Sustainable Mobility: Rome, Italy, 2022.
25. United Nations General Assembly. Resolution Adopted by the General Assembly on 31 August 2020. 2021. Available online: https://frsc.gov.ng/wp-content/uploads/2021/02/UNGA-Resolution-2020.pdf (accessed on 14 September 2023).
26. EU. REPORT on the EU Road Safety Policy Framework 2021–2030—Recommendations on Next Steps Towards 'Vision Zero'. 2021. Available online: https://www.europarl.europa.eu/doceo/document/A-9-2021-0211_EN.html (accessed on 7 September 2023).
27. Santarsiero, G.; Albanese, P.; Picciano, V.; Ventura, G.; Masi, A. Level 3 Assessment of Highway Girder Deck Bridges according to the Italian Guidelines: Influence of Transverse Load Distribution. *Buildings* **2023**, *13*, 1836. [CrossRef]
28. ANAS, Listino prezzi. (2022). New construction and scheduled maintenance. Available online: https://www.stradeanas.it/it/elenco-prezzi (accessed on 26 September 2023). (In Italian)
29. *n.29 del 4/2/2008*; NTC2018—Ministry of Infrastructure, DM 17 gennaio 2018: Aggiornamento delle Norme tecniche per le costruzioni, Suppl. or. n.30 alla G.U. Ministry of Infrastructure: Roma, Italy, 2018. (In Italian)
30. *CIRCOLARE 21 Gennaio 2019, n. 7 C.S.LL.PP*; Istruzioni per L'applicazione dell'«Aggiornamento Delle "Norme Tecniche per le Costruzioni"» di cui al Decreto Ministeriale 17 Gennaio 2018. Ministero delle Infrastrutture e dei Trasporti: Roma, Italy, 2018.
31. *EN 1992-2*; Eurocode 2: Design of Concrete Structures—Part 2: Concrete Bridges—Design and Detailing Rules. European Committee for Standardization: Brussels, Belgium, 2005.

32. Nagy-Gyorgy, T.; Sas, G.; Daescu, A.C.; Barros, J.A.; Stoian, V. Experimental and numerical assessment of the effectiveness of FRP-based strengthening configurations for dapped-end RC beams. *Eng. Struct.* **2012**, *44*, 291–303. [CrossRef]
33. Santarsiero, G.; Picciano, V.; Masi, A. Structural rehabilitation of half-joints in RC bridges: A state-of-the-art review. *Struct. Infrastruct. Eng.* **2023**, 1–24. [CrossRef]
34. ANAS. Quaderni tecnici per la salvaguardia delle infrastrutture. *Quad. Tec.* **2019**, *17*, 76.
35. Verderame, G.; Manfredi, G.; Frunzio, G. Le proprietà meccaniche dei calcestruzzi impiegati nelle strutture in cemento armato realizzate negli anni '60. In Proceedings of the X Convegno Nazionale "L'Ingegneria Sismica in Italia", Potenza-Matera, Italy, 9–13 September 2001; pp. 9–13. (In Italian)
36. Verderame, G.; Stella, A.; Cosenza, E. Le proprietà meccaniche degli acciai impiegati nelle strutture in ca realizzate negli anni'60. In Proceedings of the X Convegno Nazionale L'Ingegneria Sismica in Italia, Potenza-Matera, Italy, 9–13 September 2001. (In Italian)
37. Angst, U.M. Predicting the time to corrosion initiation in reinforced concrete structures exposed to chlorides. *Cem. Concr. Res.* **2019**, *115*, 559–567. [CrossRef]

applied sciences

MDPI

Article

The Effects of Different Drivers' Steering Inputs on the Response of Heavy Ground Vehicles to Crosswind Disturbances

Tural Tunay [1,*], Lars Drugge [2,3] and Ciarán J. O'Reilly [2,3]

1 Department of Mechanical Engineering, Faculty of Engineering, Adana Alparslan Turkes Science and Technology University, Adana 01250, Turkey
2 Department of Engineering Mechanics, KTH Royal Institute of Technology, Teknikringen 8, 100 44 Stockholm, Sweden; ciaran@kth.se (C.J.O.)
3 The Centre for ECO2 Vehicle Design, KTH Royal Institute of Technology, Teknikringen 8, 100 44 Stockholm, Sweden
* Correspondence: ttunay@atu.edu.tr

Featured Application: The control, stability, or even safety of a vehicle can be influenced by crosswinds. The findings of the current study can be helpful for the reliable design of ground vehicles with less wind sensitivity early in their development processes. The development of lateral disturbance compensation algorithms and autonomous vehicles can also benefit from the results of the study.

Abstract: The general approach in the previous studies was to ignore the driver's steering contribution to a vehicle while investigating the interactions between crosswind and vehicle. Therefore, the goal of this study is to find out how steering inputs by drivers affect a heavy-ground vehicle's dynamic reaction to crosswinds. In the investigation, a two-way interaction between vehicle dynamics and aerodynamic simulations was employed. The steering inputs of drivers were modelled using a driver model taken from the previous literature that is able to reproduce the steering responses of a human driver. The study's findings demonstrated that the steering inputs made by drivers significantly impacted how the vehicle responded to crosswinds. For instance, the greatest lateral displacement of the least skilled driver (Driver 1) was around 1.53 times the greatest lateral displacement of the most skilled driver (Driver 3) at the delay time of $t_{\delta,delay}$ = 0.5 s in the steering input. Additionally, the maximum lateral displacement results of Driver 1 and Driver 3 at $t_{\delta,delay}$ = 1.0 s became 1.39 and 1.56 times greater than their maximum lateral displacement results at $t_{\delta,delay}$ = 0.5 s. Similarly, the total steering inputs of Driver 1 and Driver 3 at $t_{\delta,delay}$ = 1.0 s were 1.4 and 2.2 times greater than their total steering inputs at $t_{\delta,delay}$ = 0.5 s, respectively. In general, the results of a driver who is more skilled than Driver 1 (Driver 2) fall in between the respective results of Driver 1 and Driver 3. On the other hand, each driver's total steering inputs at $t_{\delta,delay}$ = 0.5 s were roughly the same as their total steering inputs at $t_{\delta,delay}$ = 0 s. In all delay scenarios for the start of the driver's steering inputs, the drivers' steering inputs amplified the yaw moment applied to the vehicle. Meanwhile, they diminished the lateral force and roll moment.

Keywords: coupled simulation; crosswind; driver model; driver behaviour; heavy-ground vehicle

Citation: Tunay, T.; Drugge, L.; O'Reilly, C.J. The Effects of Different Drivers' Steering Inputs on the Response of Heavy Ground Vehicles to Crosswind Disturbances. *Appl. Sci.* **2024**, *14*, 270. https://doi.org/10.3390/app14010270

Academic Editors: Edgar Sokolovskij and Vidas Žuraulis

Received: 1 December 2023
Revised: 18 December 2023
Accepted: 18 December 2023
Published: 28 December 2023

1. Introduction

Vehicles moving on roads might be subjected to unsteady dynamic crosswind forces caused by roadside obstacles, turbulence in nature, or interactions between the vehicles' wakes [1]. Unfortunately, the previous studies on improving vehicle aerodynamics to lower the drag coefficients resulted in more crosswind-sensitive vehicles because they tended to move the centre of aerodynamic pressure towards the forward part of the vehicles [2].

The presence of crosswinds significantly influences the operations and susceptibility to roll-over accidents of heavy ground vehicles, given their substantial lateral surface area and the comparatively elevated positioning of the centre of gravity in contrast to passenger cars [3]. Trigell et al. [4] asserted that safety-critical situations in heavy ground vehicle accidents, such as loss of control leading to roll-overs and compromised lateral stability, represent a significant portion of all reported incidents involving heavy vehicles. Juhlin and Eriksson [5] investigated the directional stability of buses subjected to crosswinds. They determined that the key factors include the extent of the yaw moment overshoot upon entering a gust and the inherent characteristics of a bus, such as weight distribution, coupled with the position of the aerodynamic pressure centre. Consistent with the aforementioned studies, the current investigation focused on examining the dynamic characteristics of heavy-ground vehicles when they are subjected to crosswinds.

When a crosswind hits a vehicle, a complex interaction between the crosswind, vehicle, driver, and on-road conditions occurs, in which the characteristics of each can have important effects. For example, the frequency or maximum velocity magnitude of crosswinds, types of vehicles, driver's skills, and floating bridges or winter roads can exert a significant impact on the response of a vehicle to crosswinds. Inherently, a multidisciplinary approach is required to comprehensively understand the effects of each characteristic. Thus, the number of studies aimed at elucidating the knowledge of the effects of crosswinds on vehicles has increased in recent years [6–15]. For instance, Tunay [6] investigated the effects of various crosswind frequencies. Additionally, Tunay et al. [7–9] investigated the accuracy of different numerical approaches (for example, including the roll component to vehicle dynamics simulation, one- or two-way coupling between the aerodynamics and vehicle dynamics simulations, and turbulence models) in the investigation of the effects of crosswinds on vehicles. Furthermore, Tunay et al. [10] studied how a vehicle reacts to crosswinds on winter roads. Brandt et al. [11] studied the characteristics of a passenger car at high speeds when it is subjected to crosswinds. Sekulic et al. [12] conducted a study to investigate the impact of wind loads and the movements of a floating bridge on the lateral stability of a bus in an actual on-road scenario. Moreover, comprehending the sensitivity of heavy ground vehicles to crosswinds has become the subject of many studies in recent years [13–15].

Crosswinds can cause various unfavourable effects on vehicles' driving performance, which range from comfort problems causing the fatigue of drivers to additional safety issues that cause accidents. Additionally, driving manoeuvres like high-speed turning and lane changes become riskier when there are crosswinds [2]. Driver's reactions to crosswind excitations are critical because crosswinds can affect handling, stability, and sometimes a vehicle's safety. When a vehicle encounters crosswinds, the dominant motions perceived by a driver are the yawing motion followed by lateral motion. Theissen [16] stated that, in the critical frequency range of crosswinds from 0.5 to 2 Hz, a driver might magnify the vehicle motion, which overlaps with the eigenfrequencies of typical passenger vehicles. Transient aerodynamic excitations, such as those induced by crosswind gusts, might be either amplified or attenuated depending on the dynamic properties of the vehicle and the driver's behaviour [16]. Wanner et al. [17] stated that drivers act as adaptive controllers and significantly affect the vehicle's stability when there are external excitations on the vehicle, like crosswind gusts. Nevertheless, Drugge and Juhlin [18] stated that, regardless of the driver's efforts, in some cases, due to either the vehicle deviating significantly from its initial path before the driver reacts or the vehicle not responding to the driver's attempts at control, an accident might happen.

Wagner and Wiedemann [19] aimed to simulate an adaptive, virtual driver for analysing the crosswind behaviour of the vehicle by considering the driver as a perceptive sensor. They suggested that comparing the crosswind behaviour between the vehicles' successive models using objective evaluation criteria might not reveal significant differences because of the slight variations between them. However, the drivers might perceive even slight variations as a reduction in comfort. Maruyama and Yamazaki [20] presented that a vehicle's responses to crosswind disturbances can be predicted accurately if the driver model's parameters are appropriately determined. Wanner et al. [17] created a driver model that is sensitive to failures, utilizing experimental data gathered from a driving simulator study. They demonstrated that their driver model accurately replicates realistic human behaviour in response to a vehicle failure leading to an undesired brake torque on one wheel. Similar behaviour occurs when a vehicle runs through a crosswind passage, in which the crosswind forces make the vehicle deviate from its desired path. Consequently, Winkler et al. [21] adjusted the gain parameters of the driver model originally provided by Wanner et al. [17] for bus geometry and incorporated it into their research on crosswind effects.

The primary approach in the earlier studies to examine the sensitivity of ground vehicles to crosswind disturbances was to ignore the driver's inputs and focus only on the interactions between the vehicle and the crosswind. On the other hand, in the current study, the impacts of varying drivers' steering inputs on the reaction of a heavy vehicle to crosswinds were investigated. Two-way coupled aerodynamics and vehicle dynamics simulations were employed. The objective was to comprehend the impacts of drivers' steering performance, e.g., driving skill level and delay time in their steering inputs, to mitigate the adverse effects of crosswinds on the vehicle's dynamics. The next section provides an explanation of the numerical methods and the vehicle model utilized in the current study. Furthermore, the outcomes of vehicle dynamics and aerodynamics are elaborately presented and discussed in the Results and Discussion sections, respectively. Finally, a summary of the key findings and their implications is given in the Conclusion section.

2. Numerical Methods

2.1. Vehicle Dynamics

The vehicle dynamics were simulated utilizing an improved version of the single-track model that included the roll component of motion [3]. Figure 1 shows visualizations of the single-track model, including the roll degree of freedom and an overview of the vehicle motion in the crosswind.

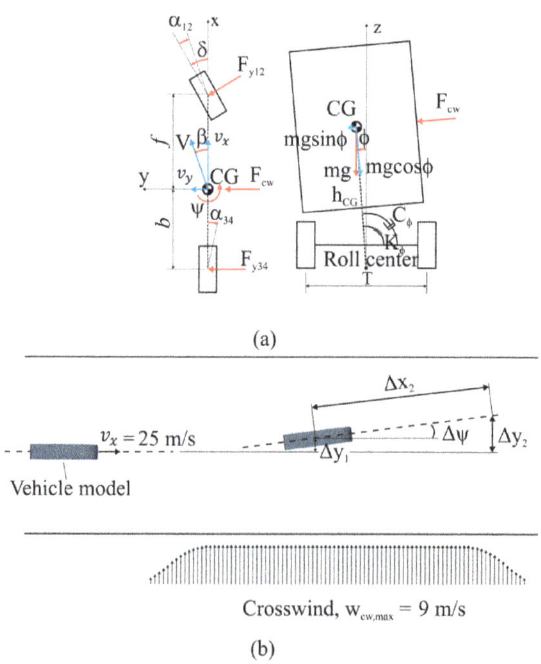

Figure 1. The graphic representations of (**a**) the vehicle dynamics model and (**b**) the simulated vehicle motion in the crosswind.

Equations (1)–(6) provide the equations that govern the dynamics of the vehicle. The vehicle's longitudinal velocity was set at $v_x = 25$ m/s. Utilizing a linear tire model, both front and rear tyre forces, F_{y12} and F_{y34}, were calculated. The symbols, M_{ax}, M_{az}, and F_{ay}, respectively, stand for the aerodynamic roll moment, yaw moment and lateral force that crosswind applies to the vehicle. The qualitative descriptions of F_{ay}, M_{az}, and M_{ax} are given in Figure 2. The other details of the parameters in Equations (1)–(6) are given in Table 1. The effects of suspension moment as a resistance to the body roll motion were represented by the suspension roll stiffness, K_ϕ, and damping, C_ϕ, components in the vehicle model.

$$\dot{x} = Ax + B\delta + C \tag{1}$$

In Equation (1), δ is the steering angle and the other parameters, e.g., A, x, B, and C, are explained in Equations (2)–(6).

$$A = \begin{bmatrix} -\frac{\lambda I_{xeq}}{m v_x I_{xx}} & \frac{\tau I_{xeq}}{m v_x I_{xx}} - v_x & -\frac{hC_\phi}{I_{xx}} & \frac{h(mgh-K_\phi)}{I_{xx}} \\ \frac{\tau}{v_x I_{zz}} & -\frac{\sigma}{v_x I_{zz}} & 0 & 0 \\ -\frac{\lambda h}{v_x I_{xx}} & \frac{\tau h}{v_x I_{xx}} & -\frac{C_\phi}{I_{xx}} & \frac{mgh-K_\phi}{I_{xx}} \\ 0 & 0 & 1 & 0 \end{bmatrix}, \; x = \begin{bmatrix} v_y \\ \dot{\psi} \\ \dot{\phi} \\ \phi \end{bmatrix},$$
$$B = \begin{bmatrix} \frac{C_{y12} I_{xeq}}{m I_{xx}} \\ \frac{C_{y12} f}{I_{zz}} \\ \frac{C_{y12} h}{I_{xx}} \\ 0 \end{bmatrix}, C = \begin{bmatrix} \frac{F_{ay}}{m} \\ \frac{M_{az}}{I_{zz}} \\ \frac{M_{ax}}{I_{xx}} \\ 0 \end{bmatrix} \tag{2}$$

$$\lambda = C_{y12} + C_{y34} \tag{3}$$

$$\tau = C_{y34} b - C_{y12} f \tag{4}$$

$$\sigma = C_{y12} f^2 + C_{y34} b^2 \tag{5}$$

$$I_{xeq} = I_{xx} + mh^2 \tag{6}$$

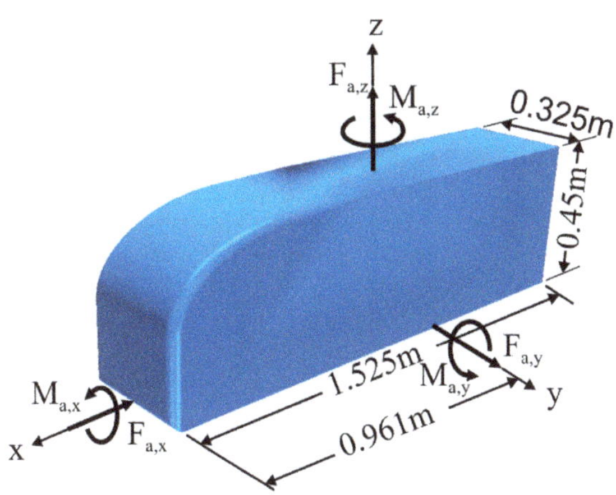

Figure 2. A descriptive explanation of the aerodynamic forces and moments as well as the geometric features of the Ground Transportation System (GTS) [22] vehicle model. $F_{a,x}$, $F_{a,y}$, $F_{a,z}$, and $M_{a,x}$, $M_{a,y}$, $M_{a,z}$ are aerodynamic forces and moments in x-, y- and z-directions, respectively.

Table 1. Features of the vehicle employed in the study.

Vehicle's Parameter	Symbol	Data	Unit
Length	L	12.2	[m]
Height	h	3.6	[m]
Width	w	2.6	[m]
Track width	T	2.25	[m]
Length of the wheelbase	L_{wb}	5.9	[m]
Distance from the front axle to CG	f	3.7	[m]
Distance from the rear axle to CG	b	2.2	[m]
Mass	m	13,650	[kg]
Cornering stiffness for the front tyre	C_{y12}	250	[kN/rad]
Cornering stiffness for the rear tyre	C_{y34}	450	[kN/rad]
Roll damping of the suspension	C_{ϕ}	100	[kN/rad]
Roll stiffness of the suspension	K_{ϕ}	1000	[kNm/rad]
Yaw moment of inertia	I_{zz}	200,000	[kgm^2/rad]
Roll moment of inertia	I_{xx}	30,000	[kgm^2/rad]

2.1.1. Driver Model

The driver model utilized in the current investigation was derived from the research conducted by Winkler et al. [21], in which the crosswind response of a bus was investigated. They modified the constants for proportional gains provided by Wanner et al. [17] to suit the geometry of a bus based on vehicle dynamics data for the bus using the wheel angle, δ, computed via the understeer gradient, K_{us}, the curvature radius, R, and the length of the wheelbase, L_{wb}, with the assumptions of linear tyre behaviour and no-load transfer. Equation (7) provides the driver model, including proportional gain parameters for preview lateral distance, k_l, yaw angle, k_{ψ}, and lateral displacement, k_y.

$$\delta = k_l \Delta y_2 + k_{\psi} \Delta \psi + k_y \Delta y_1 \tag{7}$$

The driver's steering input is denoted by δ in Equation (7). Additionally, Δy_2, $\Delta \psi$, Δy_1 are the preview lateral distance, yaw angle, and change in lateral displacement of the vehicle, respectively. Computations of Δy_1 and $\Delta \psi$ were determined during the solution of vehicle dynamics equations. Additionally, the calculation of the preview lateral distance, Δy_2, was based on the assumption of a one-second preview time, resulting in a length of 25 m, considering a longitudinal velocity of $v_x = 25$, as illustrated in Figure 1. Thus, $\Delta y_2 = \Delta y_1 + v_x \times t \times \sin(\Delta \psi)$ was utilized to calculate the preview lateral distance.

Sets of constants for proportional gains, which were employed to represent three different drivers' attitudes, are given in Table 2. The first set of constants, denoted as Driver 1, were the proportional gains in the driver model utilized in the study of Winkler et al. [21]. The values of the constants in other sets, i.e., Driver 2 and Driver 3, were proportionally increased relative to the constants in their preceding sets. Therefore, the driver became more skilled as the number of drivers increased. In other words, Driver 1 represented the least skilled driver, Driver 2 represented the more skilled driver than Driver 1, and Driver 3 represented the most skilled driver in the study.

Table 2. Sets of proportional control gain parameters indicate three different driver's attitudes.

	k_y	k_{ψ}	k_l
Driver 1	1.2	30	0.9
Driver 2	1.6	40	1.2
Driver 3	2.0	50	1.5

2.2. Aerodynamics

Incompressible, unsteady Navier-Stokes and continuity equations were used for the solution of turbulent flow around the vehicle. Several approaches are available for the solution of the governing equations of turbulent flows. These approaches include directly solving all flow scales, e.g., direct numerical solution (DNS), modelling all or certain scales of flow, e.g., Reynolds-averaged Navier-Stokes (RANS) and Large Eddy Simulation (LES), or approaches using hybrid RANS-LES methods, e.g., detached eddy simulation (DES). The search for less computationally expensive yet sufficiently accurate methods for solving turbulent flows around vehicles remains a critical need [23–25]. For that purpose, hybrid RANS-LES methods, such as DES, were proposed, in which elements of both RANS and LES were combined to provide a balance between computational efficiency and accuracy in simulating turbulent flows. In DES, the simulation dynamically switches between RANS and LES in different regions of the flow, allowing for a more accurate representation of turbulent phenomena. Specifically, RANS is utilized in regions where the flow is predominantly attached and steady, while LES is employed in areas with unsteady and separated flow. However, due to some disadvantages of DES, such as its grid sensitivity in boundary layers and log-layer mismatch, new methods named delayed detached eddy simulation (DDES) and improved delayed detached eddy simulation (IDDES) were proposed [26]. IDDES integrates DDES with an enhanced RANS-LES hybrid model, specifically designed to address wall modelling in LES by adapting to grid resolution [27]. Additionally, Tunay et al. [7] stated that studies employing two-way coupling between aerodynamics and vehicle dynamics in the previous literature commonly utilized scale-resolving turbulence models such as LES and IDDES in aerodynamic simulations, known for their accuracy in turbulent flow solutions compared to RANS equations. Therefore, in the present study, IDDES was used to solve the turbulent flow around the vehicle model.

The Ground Transportation System (GTS), a 1/8-scaled simplified heavy vehicle model, was utilized [22]. Numerous studies [21,22,28–31] confirmed the applicability of GTS in aerodynamic studies of heavy vehicles. Figure 2 provides the features of the GTS, including the coordinate system, along with qualitative descriptions of aerodynamic force and moments.

The computations utilized a deterministic crosswind velocity profile characterized by a cosine inlet and exit profile with a flat top, as depicted in Figure 3. The mathematical formulation of the crosswind's velocity profile is defined in Equation (8).

$$w_{cw}(x) = \begin{cases} 0 \; for \; x < x_0 - \frac{\Delta x_{slope}}{2} \\ w_{cw,max}\frac{1}{2}\left(1 - cos\left[\frac{x-\left(x_0-\frac{\Delta x_{slope}}{2}\right)}{\Delta x_{slope}}\pi\right]\right) \; for \; x_0 - \frac{\Delta x_{slope}}{2} \le x \le x_0 + \frac{\Delta x_{slope}}{2} \\ w_{cw,max} \; for \; x_0 + \frac{\Delta x_{slope}}{2} < x < x_0 + \Delta x_{cw} - \frac{\Delta x_{slope}}{2} \\ w_{cw,max}\frac{1}{2}\left(1 - cos\left[\frac{x-\left(x_0+\Delta x_{cw}+\frac{\Delta x_{slope}}{2}\right)}{\Delta x_{slope}}\pi\right]\right) \; for \; x_0 + \Delta x_{cw} - \frac{\Delta x_{slope}}{2} \le x \le x_0 + \Delta x_{cw} + \frac{\Delta x_{slope}}{2} \\ 0 \; for \; x > x_0 + \Delta x_{cw} + \frac{\Delta x_{slope}}{2} \end{cases} \quad (8)$$

In this instance, the crosswind's maximum velocity magnitude is indicated by $w_{cw,max}$, the length of the transient inlet and exit sections of the crosswind are denoted by Δx_{slope}, equivalent to 1.5 times the GTS's length. The symbol, x_0, signifies the distance between the crosswind's starting location and the origin of the coordinate system, while Δx_{cw} represents the crosswind's length, set at five times the GTS's length. As illustrated in Figure 4a, utilizing a velocity inlet boundary condition on the right-hand side of the flow domain brought the crosswind into computation.

Figure 4 provides the geometric details of the flow domain, the corresponding computational boundary conditions, and the mesh used in the aerodynamics simulation. The pressure outlet boundary condition was applied to the inlet, outlet, and left lateral surfaces, assuming atmospheric pressure on these boundary surfaces. No-slip boundary conditions were applied to the GTS surfaces and the flow domain's ground. In addition, the symmetry boundary condition was applied to the top surface of the flow domain. The velocity of the

GTS was $v_x = 91.6$ m/s. Overset mesh was utilized to let the motion of the GTS in the computational domain. The flow had the Reynolds number of $Re_w = 2 \times 10^6$, calculated using GTS width. Prior to being used in the vehicle dynamics calculations, the force and moment data from the aerodynamics simulation were scaled back.

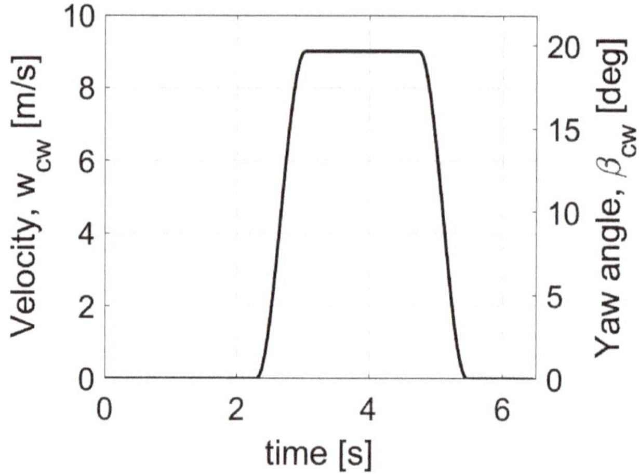

Figure 3. Velocity profile of the crosswind.

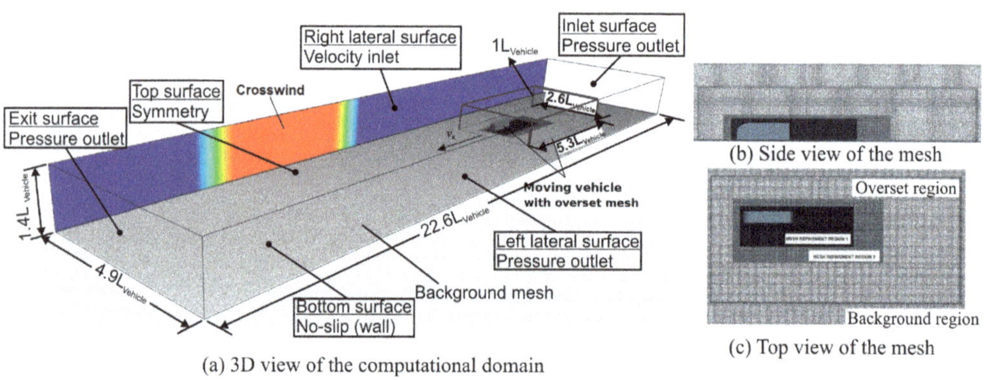

Figure 4. (**a**) The computational flow domain in three dimensions and (**b**,**c**) view of the mesh structure surrounding the GTS and the overset mesh, respectively. Colours on the right lateral surface of the flow domain represent the profile of the velocity inlet; such as blue indicates zero velocity and red indicates the maximum velocity of 9 m/s.

StarCCM+ version 13.02 [32], the commercial computational fluid dynamics (CFD) software, was utilized for mesh generation and the solution of the governing equations. As shown in Figure 4, the mesh structures comprised two regions: an overlap region and a background region. The sizes of the overlap region were 5.3 × 2.6 × 1 vehicle lengths, while the background mesh was 22.6 × 4.9 × 1.4 vehicle lengths. The mesh comprised unstructured hexahedral cells totalling approximately 22 million, with approximately 40% (9 million) specifically utilized in the vicinity of the vehicle model, i.e., the overlap region. Twenty prism layers with a total thickness in the wall-normal direction of 0.0141 m, stretching of 1.25 m, and length of 0.0044 m were applied in order to resolve the boundary layer around the GTS. The dimensionless wall distance, y^+, was maintained at a value less than 15.

In the current investigation, the GTS's drag coefficient, C_D, in the absence of a cross was 0.251. This number was reasonably close to the numerical and experimental values reported in the literature. For instance, at Re_w = 1.6 × 10^6, the C_D values found in the experimental investigations by Croll et al. [28] and Storms et al. [29] were 0.247 and 0.249, respectively. Additionally, the C_D value obtained in the numerical study of Unaune et al. [30] at Re_w = 2 × 10^6 was 0.253. Also, the mesh used in this investigation was the same as the mesh used in Winkler et al.'s [21] work, which also looked into the impact of crosswind on the GTS. As a result, it was decided that the mesh was suitable, and no more refinement was done.

The finite volume approach and segregated solver were utilized in the aerodynamic simulations. Convective discretization of the governing equations was accomplished using a hybrid-bounded central differencing (BCD) scheme. For time discretization, a second-order implicit unsteady scheme was employed. The resulting scalar system of equations was solved with a linear solver called Algebraic Multigrid (AMG). In the simulation, a time step of 6 × 10^{-5} s was employed, and within each time step, the computation involved five inner iterations.

2.3. Coupled Simulation

By exchanging the force and moment results of the aerodynamics simulation with the vehicle dynamics simulation and the velocity results of the vehicle dynamics simulation with the aerodynamics simulation at each computational time step, a two-way coupling between aerodynamics and vehicle dynamics was achieved. These coupled simulations were executed using high-performance computers (HPC) at the PDC Centre for High-Performance Computing, hosted by the KTH Royal Institute of Technology. The computations utilized 24 nodes with 768 cores.

3. Results

This section provides the study results in three distinct parts. Firstly, the impact of three different drivers' steering inputs on the vehicle's response to crosswind disturbances is detailed. Subsequently, the second part delves into the effects of the delay time, $t_{\delta,delay}$, in the drivers' steering inputs on the vehicle's response to crosswind. Lastly, the third part elucidates the effects of various drivers' steering inputs on the characteristics of aerodynamic forces and moments exerted by crosswind on the vehicle.

3.1. The Effects of Drivers' Steering Inputs with Different Driving Skills on the Vehicle's Reaction to the Crosswind

Figure 5 presents the comparisons between the vehicle dynamics results obtained with three drivers, each possessing different driving skills and identified as Driver 1, Driver 2, and Driver 3 in the study. These results are juxtaposed with those obtained without any driver's steering input. As explained previously in Section 2.1.1, Driver 1 represented the least skilled driver, while Driver 3 represented the most skilled driver in the study. Additionally, in this part of the study, after the onset of the crosswind, i.e., $t_{cw,start}$ = 2.67 s, all drivers' steering inputs began with a delay of $t_{\delta,delay}$ = 0.5 s. The crosswind ceased at $t_{cw,end}$ = 5.1 s. The simulation without a driver's steering input was finished at t = 6.45 s, whereas the simulations with a driver's steering input were continued until t = 10 s. Aerodynamics and vehicle dynamics simulations were coupled in two-way until t = 6.45 s for both cases, including and not including the driver's steering inputs.

The greatest variations between the lateral displacement results for all drivers occurred at their first peaks, as shown in Figure 5a, i.e., the maximum lateral displacement of Driver 1 is about 1.53 times larger than that of Driver 3. Moreover, the greatest variations between the other results, e.g., yaw and roll angles, happened at their second and third peaks, as shown in Figure 5b,c. Figure 5 indicated that the patterns of yaw angle and lateral displacement findings exhibited a reversal after the initiation of drivers' steering inputs, in contrast to the results obtained without any steering input from the driver. On the contrary, the drivers' steering inputs caused the roll angle results to have the same trends as the results obtained without any steering input from the driver. Moreover, the introduction of drivers' steering inputs led to an increase in the absolute maximum magnitudes of the roll angle results, in contrast to the results obtained without any steering inputs from the driver. Additionally, as expected, the magnitudes of the oscillations observed in the vehicle dynamics results after the crosswind ceased were the least in the results of Driver 3, who is the most skilled in the simulations.

Figure 6 compares the percentage variations, $|\epsilon|$, between the results obtained with steering inputs from the driver and without any steering input from the driver in the first Δt = 0.5 s period after the steering inputs started. The lateral displacement results given in Figure 6 showed larger differences than the roll and yaw angle results at the beginning, i.e., $|\epsilon| \sim 2.5\%$ at t_δ = 3.35. However, the differences in the lateral displacement results showed less increase than the corresponding differences in the roll and yaw angle results at a later instant, e.g., at t = 3.57 s. On the other hand, the percentage variations between the roll angle results obtained with steering inputs from the driver and without any steering input from the driver were larger than those of lateral displacement and yaw angle at the end of this period of the steering inputs, which was $|\epsilon| \sim 15\%$. Moreover, the discrepancies among the outcomes of drivers with varying skill levels increased significantly after t = 3.5 s.

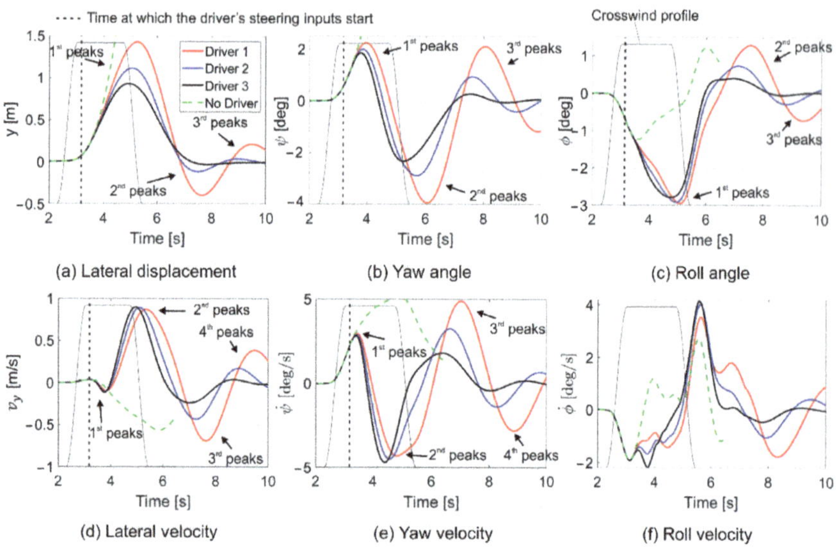

Figure 5. Comparison of the vehicle dynamics results of three drivers with different driving skills and without any steering input from the driver.

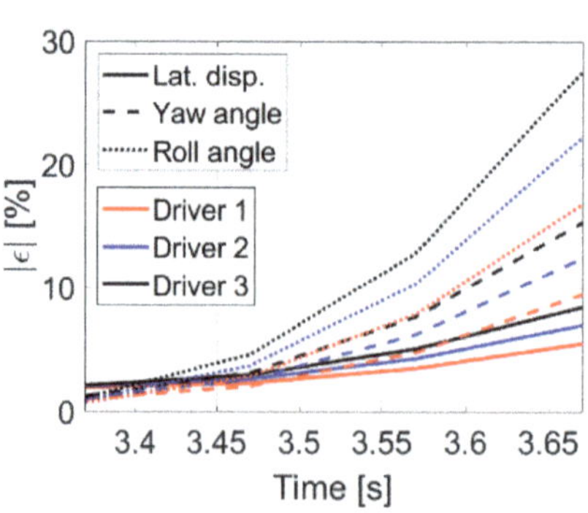

Figure 6. Percentage variations between the results of the simulations with and without drivers' steering inputs in the first 0.5 s after the steering inputs start, i.e., $t_{\delta,start}$ = 3.17.

Figure 7 shows the steering inputs of three different drivers to the vehicle exposed to crosswind. As presented in Figure 7a, the maximum magnitude of the driver's steering inputs at their initial peaks increased with an escalation in their driving skills. However, this situation reversed at subsequent peaks, e.g., at the second and third peaks. For example, Driver 3, i.e., the most skilled driver, had the fewest total steering inputs, whereas Driver 1, i.e., the least skilled driver, had the most total steering inputs; see also Figure 8. Figure 7b presents the breakdown of the steering inputs into their proportional gains. Their results showed that there were time shifts between the contributions of the proportional gains. It is expected that a time shift or phase delay between the gains of proportional control might lead to several undesired effects on the control system, e.g., instability and reduced performance. However, in the present situation, the time shifts between the proportional gains, especially the gains from the lateral displacement and the yaw angle, compensated for each other, which caused the reduced amount of steering to bring the vehicle back on its original track. Additionally, as presented in Figure 7c, the maximum steering wheel rate after the first steering manoeuvres of all drivers was smaller than $\dot{\delta}$ = 240 deg/s. Here, the steering wheel angular velocity was calculated by taking the time derivative of the steering angle at the tyres multiplied by a steering gear ratio of 20.

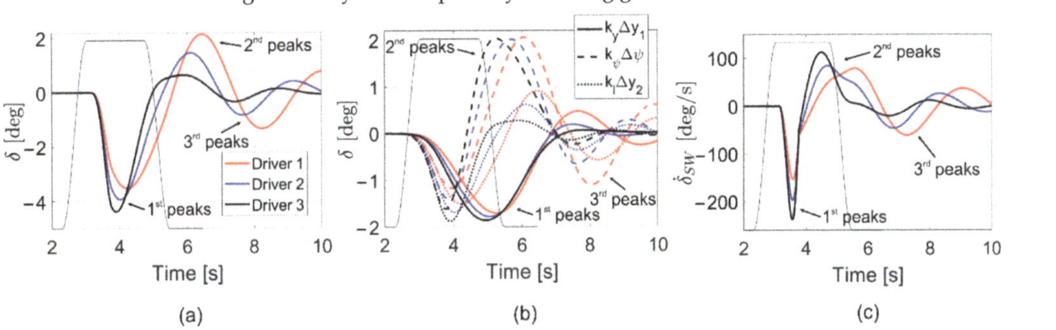

Figure 7. (**a**) The steering inputs of three drivers, δ, who had different driving skills, to the vehicle exposed to crosswind. (**b**) The decomposition of the steering inputs in their proportional gains. (**c**) The steering wheel angular velocity, $\dot{\delta}_{SW}$.

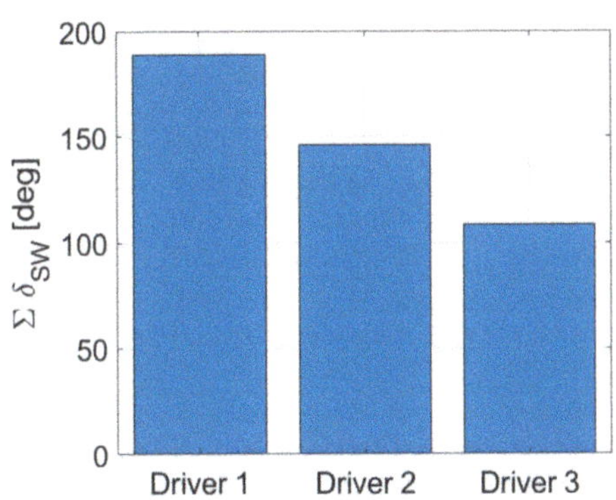

Figure 8. Total steering wheel input, $\sum \delta_{sw}$, of three drivers with different driving skills.

Figure 8 estimates the steering efforts of drivers by integrating the absolute value of the steering wheel angle over the simulation. According to the data shown in Figure 8, the steering effort of Driver 1, who was the least skilled driver in the study, was around 1.7 times the steering effort of Driver 3, who was the most skilled driver in the study.

The temporal evolution of lateral acceleration and yaw velocity in relation to the drivers' steering inputs is depicted in Figure 9. The delayed correlations between the drivers' steering inputs and the lateral acceleration and yaw velocity results are exhibited in Figure 9a. Hence, the cross-correlation results between lateral acceleration, yaw velocity, and drivers' steering inputs, as depicted in Figure 9b, indicated that the vehicle's yaw velocity responded more promptly to the driver's steering inputs compared to the vehicle's lateral acceleration.

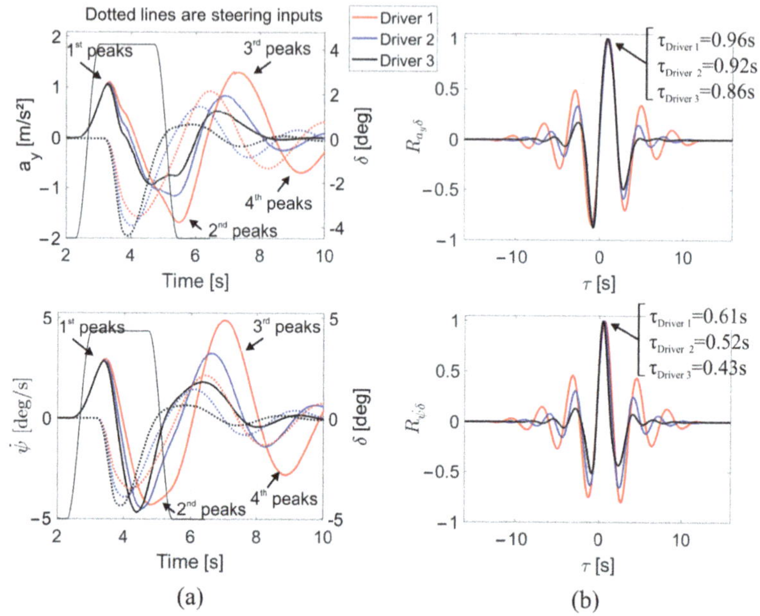

Figure 9. (**a**) Comparisons of lateral accelerations, a_y, and yaw velocity, $\dot{\psi}$, of the vehicle (solid lines) with drivers' steering inputs, δ, (dotted lines). (**b**) Cross-correlations between the lateral acceleration and the steering input by drivers, $R_{a_y\delta}$, and yaw velocity and steering inputs by drivers, $R_{\dot{\psi}\delta}$.

3.2. The Effects of Time Delay in Drivers' Steering Inputs on the Vehicle's Reaction to the Crosswind

In this part of the study, the researchers explored the impacts of delay times in the drivers' steering inputs on the dynamics of vehicles exposed to crosswind. The lateral displacement, y, yaw angle, ψ, and roll angle, ϕ, results of three drivers with different driving skills are presented in Figure 10 at three different delay times, i.e., $t_{\delta,delay}$ = 0 s, 0.5 s, and 1.0 s. An ideal steering response time of $t_{\delta,delay}$ = 0 s was employed to assess the effects of different delay times in the steering inputs.

As the delay time in the drivers' steering inputs increased, the largest deviations in the yaw angles, roll angles, and lateral displacements of the vehicle due to the crosswind increased. For example, at their initial peaks, the maximum values of the lateral displacements for Driver 1 and Driver 3 at the delay time of $t_{\delta,delay}$ = 1.0 s were 1.39 and 1.56 times higher than their respective maximum lateral displacement results at $t_{\delta,delay}$ = 0.5 s. On the other hand, the results obtained at $t_{\delta,delay}$ = 0 s and 0.5 s were close to each other.

The yaw angle experienced the most significant differences from the delays in drivers' steering inputs because it had the largest percentage variations between the results obtained at $t_{\delta,delay}$ = 0 s and $t_{\delta,delay}$ = 1.0 s, as presented in Figure 11. The maximum absolute relative percent difference in the results of the drivers between $t_{\delta,delay}$ = 0 s and 1.0 s was at least more than 29%, whereas the highest relative percentage variances in the results of the drivers between $t_{\delta,delay}$ = 0 s and 0.5 s were less than 30%. Consequently, the impact of delay time on the vehicle's response to crosswind disturbances became crucial as the delay time increased from $t_{\delta,delay}$ = 0.5 s to $t_{\delta,delay}$ = 1.0 s.

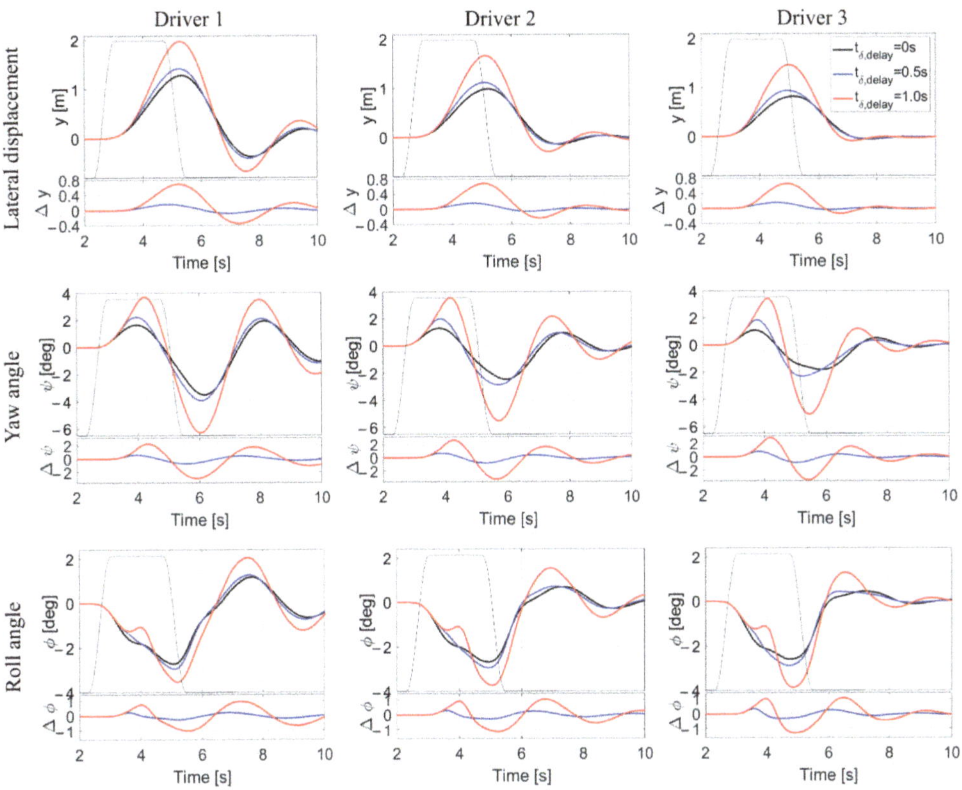

Figure 10. Lateral displacement, y, yaw angle, ψ, and roll angle, ϕ, results of the vehicle obtained in the cases of three delay times, i.e., $t_{\delta,delay}$ = 0 s, 0.5 s, and 1.0 s, in the steering inputs of three drivers with different driving skills. The subfigure in each figure shows the instant differences between the results obtained for $t_{\delta,delay}$ = 0 s, and $t_{\delta,delay}$ = 0.5 s, and 1.0 s.

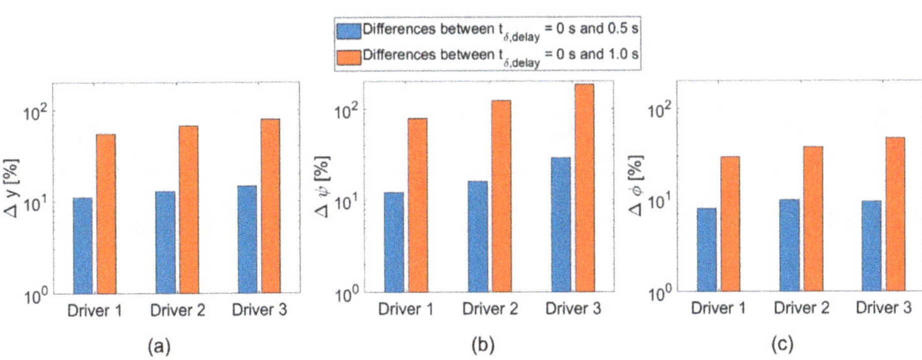

Figure 11. The highest percentage variations in the (**a**) lateral displacement, (**b**) yaw angle, and (**c**) roll angle results of different drivers between delay times of $t_{\delta,delay}$ = 0 s and $t_{\delta,delay}$ = 0.5 s and 1.0 s at their first peaks.

Figure 12 depicts the driver's steering inputs in the instances of three delay times. The magnitudes of the steering inputs of drivers at their initial peaks heightened with both the delay in the steering inputs of drivers and the level of driving skills. For example, the maximum magnitude of the steering input of Driver 3 at the first peak was 1.4 times larger than that of Driver 1 at the delay time of $t_{\delta,delay}$ = 1.0 s. Additionally, in the case of Driver 3, the magnitude of the steering input at the first peak at $t_{\delta,delay}$ = 1.0 s was around 3 times greater than the corresponding one at $t_{\delta,delay}$ = 0 s. Furthermore, the steering efforts needed to bring the vehicle on its original track increased significantly as the delay time in the steering input of the driver increased to $t_{\delta,delay}$ = 1.0 s.

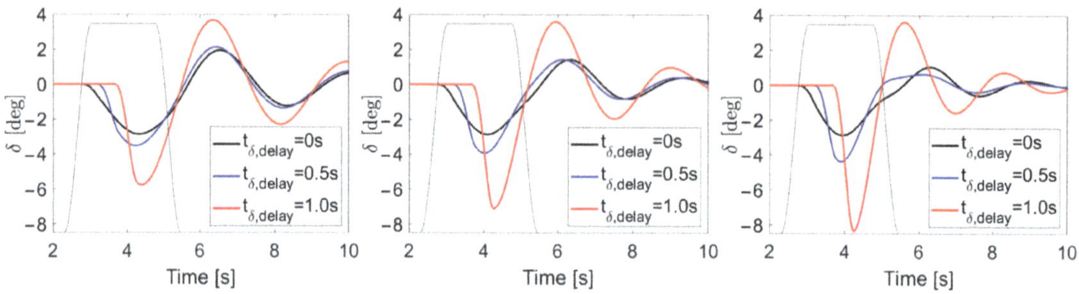

(a) Driver 1 (b) Driver 2 (c) Driver 3

Figure 12. The steering inputs, δ, of three drivers with different skills at three delay times of $t_{\delta,delay}$ = 0 s, 0.5 s, and 1.0 s in the steering inputs.

The total steering wheel inputs of each driver at three distinct delay times are compared in Figure 13. On an individual basis, each driver's total steering inputs were roughly the same at $t_{\delta,delay}$ = 0 s and 0.5 s. Conversely, at $t_{\delta,delay}$ = 1.0 s, the total steering inputs of drivers were at least 1.4 times higher than those at $t_{\delta,delay}$ = 0 s and 0.5 s. Wagner and Wiedemann [16] specified the importance of the promptness of drivers in controlling the different crosswind sensitivity of vehicles. The findings of the current study unequivocally demonstrated that the delay time of $t_{\delta,delay}$ = 0.5 s was important in controlling the vehicle motion after the crosswind hit it. Additionally, there were significant differences between the drivers' steering inputs. For example, the steering wheel inputs of Driver 1 were 1.7 and 1.3 times larger than those of Driver 3 at $t_{\delta,delay}$ = 0 s and 1.0 s, respectively. In conclusion, there were both significant differences between the drivers' steering inputs at certain delay times and significant increases in the drivers' steering efforts as the delay time extended to $t_{\delta,delay}$ = 1.0 s.

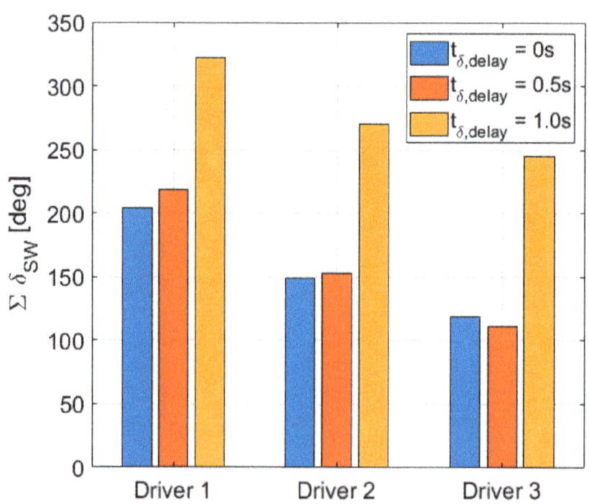

Figure 13. Total steering wheel inputs of drivers with different driving skills, $\sum \delta_{sw}$, at three different delay times, $t_{\delta,delay}$ = 0 s, 0.5 s and 1.0 s.

3.3. The Effects of Drivers' Steering Inputs on the Aerodynamic Force and Moments Acting on the Vehicle Subjected to Crosswind

Previous research studies indicated the significance of the aerodynamic yaw moment, lateral force, and roll moment in influencing the vehicle's response to crosswind [8,21,33]. The findings demonstrated in the current study revealed that the steering inputs made by the drivers had an impact on the aerodynamic forces and moments acting on the vehicle as a result of the crosswind. The characteristics of lateral force, yaw moment, and roll moment acting on the vehicle due to the crosswind, with and without steering inputs from the driver, are depicted in Figure 14. They revealed that both the drivers' driving skills and the delay times in their steering inputs caused differences between the aerodynamic forces and moments exerted on the vehicle due to the crosswind. Thus, the differences in the amount of force and moments arose after the steering started, i.e., $t_{\delta,delay}$ = 0. For all delay times, the drivers' steering inputs increased the yaw moment at various amounts towards the end of the crosswind, whereas they reduced the lateral force and roll moment. Rapid alterations in the vehicle's yaw angle resulting from the drivers' steering inputs led to notable variations in the yaw moment outcomes. For example, the yaw moment results at $t_{\delta,delay}$ = 0 s and 0.5 s reached their maximum values at around t∼3.9 s, while the yaw moment results at $t_{\delta,delay}$ = 1.0 s reached their maximum at a later time of t∼4.5 s.

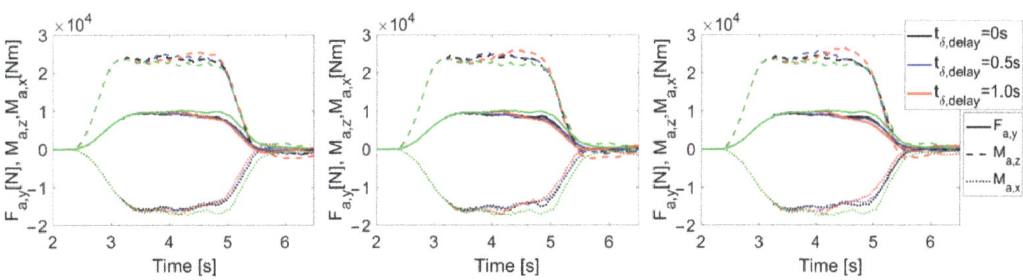

(a) Driver 1 (b) Driver 2 (c) Driver 3

Figure 14. Lateral force, $F_{a,y}$, yaw moment, $M_{a,z}$, and roll moment, $M_{a,x}$, obtained in the case of three drivers with different skills at three delay times in their steering inputs, i.e., $t_{\delta,delay}$ = 0 s, 0.5 s, and 1.0 s. The green lines depict the corresponding forces and moments obtained when there is no driver's steering input.

Furthermore, the other outcomes of aerodynamics simulation, such as the analysis of velocity fields, were not included in the present study due to the specific scope outlined for this research. However, the detailed analysis of aerodynamics simulation of similar crosswind-vehicle interaction studies can be found in Tunay [6] and Tunay et al. [7,9].

4. Discussion

When investigating the crosswind sensitivity of vehicles, previous research studies often used a traditional approach that ignored the driver's contributions and concentrated only on the vehicle-crosswind interactions. Wagner and Wiedemann [19] stated that the response of a less crosswind-sensitive vehicle to crosswind might be different when including the driver's reactions. Thus, in the current study, the responses of a heavy vehicle to a crosswind were investigated by including the steering inputs of three drivers who had different driving skills. Accordingly, the vehicle dynamics results were acquired both with and without the driver's steering inputs for comparative analysis. Comprehensive discussions regarding the vehicle's response to crosswind without any driver's steering input were elaborated in previous studies by Tunay et al. [7–9].

When examining the lateral dynamic properties of vehicles, yaw velocity and lateral acceleration and their relationship with steering-wheel angle were generally specified as important characteristics. For instance, Huemer et al. [33] reported that lateral acceleration, yaw velocity, and the time lag between those parameters are the primary factors influencing the driver's perception of vehicle movements, next to perception limits. In this perspective, the corresponding results are compared in Figure 9, given in the previous section. However, the outcomes presented in Figure 9 demonstrated an intricate interaction among the crosswind, vehicle, and driver's steering inputs, which made it difficult to compare the results of different driver's steering efforts and draw any general conclusions regarding them. For example, in Figure 9a, the second peak of the lateral acceleration of Driver 3 occurred inside the crosswind, whereas the second peak of the lateral acceleration of Driver 1 occurred just after the crosswind ceased. That means the larger second peak of the lateral acceleration of Driver 1 could also be due to the reduction in crosswind, not necessarily only due to the steering input. Thus, it is more appropriate to talk about the vehicle-driver system instead of considering their individual effects in the study of the crosswind sensitivity of vehicles [34].

In line with the aim of investigating the driver-vehicle-crosswind interactions, a driver model that reproduced a realistic human driver's steering response was used in the study. The maximum steering wheel rate after the first steering manoeuvres of all drivers was smaller than $\dot{\delta}$ = 240 deg/s, given in Figure 7c, indicating that the steering responses of drivers in the study corresponded to normal human driving conditions as described by Blundell and Harty [35] and Wang [36]. Based on empirical observations, Blundell and Harty [35] identified three driving regimes described as "normal", "spirited", and "accident avoidance", which were characterised by steering rates of 0 to 400 deg/s, 400 to 700 deg/s, and 700 to 1200 deg/s, respectively. Also, Wang [36] stated that normal driving requires an average steering rate of 500 deg/s.

In general, the study's findings indicated that the vehicle dynamics became more stable after the onset of crosswind with higher levels of driver driving skill. The drivers' steering inputs, aimed at mitigating the unfavourable effects of crosswind on vehicle motion, decreased with increasing driving skills. These outcomes were anticipated, given that the driver model employed in the study was based on proportional gain parameters. However, the variations in the magnitude of the first peaks in the vehicle dynamics results of different drivers were nonlinear, especially in cases where there were time delays, as shown in Figures 10 and 11.

When compared to the outcomes obtained with no steering input, the driver's initial steering manoeuvre against the crosswind decreased the unfavourable rise in the vehicle's lateral displacement and yaw angle while increasing the roll angle. The changing rate of the roll angle was the highest in the first 0.5 s of the steering input, whereas the changing rate of the lateral displacement was the lowest. Additionally, the increase in the magnitude of the drivers' first steering inputs was primarily due to the fast response of the vehicle's yaw motion to the crosswind disturbances. For example, as presented in Figure 7b, the contributions of the proportional gains obtained from both yaw change, $k_{\psi}\Delta y_{\psi}$, and the change in the preview lateral distance, $k_l\Delta y_2$, to the first steering manoeuvre of drivers are prompt and higher than the contribution of the change in the lateral displacement, $k_y\Delta y_1$.

Finally, as the results of the on-road crosswind scenario chosen in the study showed, the vehicle's dynamics, such as lateral displacements, exceeded the acceptable limits set by the road's lateral margins. This could potentially lead to hazardous situations or accidents.

5. Conclusions

The effects of three different drivers' steering inputs on the response of a bus to crosswind were investigated. Two-way coupled aerodynamics and vehicle dynamics simulations were employed. The results showed that the maximum lateral displacement of the least skilled driver (Driver 1) was 1.53 times greater than that of the most skilled driver (Driver 3) at $t_{\delta,delay}$ = 0.5 s. Additionally, while the level of driving skills increased, the amounts of both lateral displacements and the oscillating motions of the vehicle decreased after the crosswind ceased. The vehicle's roll angle gave more significant responses to the first steering manoeuvre of drivers than the yaw angle and the lateral displacement.

The effects of delay time in the driver's steering inputs showed that the delay time of $t_{\delta,delay}$ = 0.5 s was important in alleviating the disturbances caused by the crosswind on the vehicle. For example, the vehicle dynamics results obtained in the cases of $t_{\delta,delay}$ = 0 s and $t_{\delta,delay}$ = 0.5 s were close to each other, whereas the results obtained for $t_{\delta,delay}$ = 1.0 s were significantly larger than those of $t_{\delta,delay}$ = 0 and 0.5 s. Additionally, the drivers' total steering inputs at $t_{\delta,delay}$ = 1.0 s were at least more than 1.4 times their total steering inputs at $t_{\delta,delay}$ = 0.5 s. But each driver's total steering inputs at $t_{\delta,delay}$ = 0.5 s were roughly the same as their total steering inputs at $t_{\delta,delay}$ = 0 s. In conclusion, as the delay time increased, the highest percentage variations between the results of different drivers became smaller, but the influence of the long delay time became critical.

Finally, the drivers' steering inputs increased the yaw moment at various amounts at all delay times towards the end of the crosswind, whereas they reduced the lateral force and roll moment.

Author Contributions: Conceptualization, T.T., L.D. and C.J.O.; methodology, T.T., L.D. and C.J.O.; software, T.T.; validation, T.T., L.D. and C.J.O.; formal analysis, T.T., L.D. and C.J.O.; investigation, T.T., L.D. and C.J.O.; data curation, T.T.; writing—original draft preparation, T.T.; writing—review and editing, T.T., L.D. and C.J.O.; visualisation, T.T.; supervision, L.D. and C.J.O.; funding acquisition, T.T., L.D. and C.J.O. All authors have read and agreed to the published version of the manuscript.

Funding: The authors would like to gratefully acknowledge the co-funding they have received for this work from the Swedish Innovation Agency Vinnova (grant number 2017-03391), the Centre for ECO2 Vehicle Design at KTH (itself funded by Vinnova grant number 2016-05195), and the strategic research area TRENoP. The computations were performed on resources provided by the Swedish National Infrastructure for Computing (SNIC) at the PDC Centre for High-Performance Computing (PDC-HPC).

Institutional Review Board Statement: Not applicable.

Informed Consent Statement: Not applicable.

Data Availability Statement: The data presented in this study are available within the article.

Conflicts of Interest: The authors declare no conflict of interest.

References

1. Sims-Williams, D. Crosswinds and transients: Reality, simulation and effects. *SAE Int. J. Passeng. Cars. Mech. Syst.* **2011**, *4*, 172–183. [CrossRef]
2. Willumeit, H.P.; Müller, K.; Dödlbacher, G.; Matheis, A. Method to Correlate Vehicular Behaviour and Driver's Judgement Under Side Wind Disturbances. *Veh. Syst. Dyn.* **1988**, *17*, 508–524. [CrossRef]
3. Lee, S.; Kasahara, M.; Mori, Y. Roll Damping control of a heavy vehicle under the strong crosswind. In Proceedings of the 7th IFAC Symposium on Advances in Automotive Control, The International Federation of Automatic Control, Tokyo, Japan, 4–7 September 2013. [CrossRef]
4. Trigell, A.S.; Rothhämel, M.; Pauwelussen, J.; Kural, K. Advanced Vehicle Dynamics of Heavy Trucks with the Perspective of Road Safety. *Veh. Syst. Dyn.* **2017**, *55*, 1572–1617. [CrossRef]
5. Juhlin, M.; Eriksson, P.A. Vehicle parameter study on crosswind sensitivity of buses. In Proceedings of the SAE Commercial Vehicle Engineering Congress and Exhibition, Chicago, IL, USA, 26–28 October 2004.
6. Tunay, T. Two-way coupled aerodynamics and vehicle dynamics simulations of a heavy ground vehicle subjected to crosswind of various frequencies. *Proc. Inst. Mech. Eng. D J. Automob. Eng.* **2023**, *237*, 1406–1422. [CrossRef]
7. Tunay, T.; Drugge, L.; O'Reilly, C.J. Assessment of a Two-Equation Eddy-Viscosity Turbulence Model in Crosswind Simulation of a Heavy Ground Vehicle. *SAE Int. J. Commer. Veh.* **2022**, *15*, 81–96. [CrossRef]
8. Tunay, T.; O'Reilly, C.J.; Drugge, L. The significance of roll on the dynamics of ground vehicles subjected to crosswind gusts by two-way coupled simulation of aero- and vehicle dynamics. In *Advances in Dynamics of Vehicles on Roads and Tracks, Lecture Notes in Mechanical Engineering, IAVSD 2019*; Klomp, M., Bruzelius, F., Nielsen, J., Hillemyr, A., Eds.; Springer: Cham, Switzerland, 2020. [CrossRef]
9. Tunay, T.; Drugge, L.; O'Reilly, C.J. On coupling methods used to simulate the dynamic characteristics of heavy ground vehicles subjected to crosswind. *J. Wind Eng. Ind. Aerodyn.* **2020**, *201*, 104194. [CrossRef]
10. Tunay, T.; Drugge, L.; O'Reilly, C.J. Two-way coupled aerodynamics and vehicle dynamics crosswind simulation of a heavy ground vehicle in winter road conditions. In *Advances in Dynamics of Vehicles on Roads and Tracks II, Lecture Notes in Mechanical Engineering, IAVSD 2021*; Orlova, A., Cole, D., Eds.; Springer: Cham, Switzerland, 2021. [CrossRef]

11. Brandt, A.; Jacobson, B.; Sebben, S. High speed driving stability of road vehicles under crosswinds: An aerodynamic and vehicle dynamic parametric sensitivity analysis. *Veh. Syst. Dyn.* **2022**, *60*, 2334–2357. [CrossRef]
12. Sekulic, D.; Vdovin, A.; Jacobson, B.; Sebben, S.; Johannesen, S.M. Effects of wind loads and floating bridge motion on intercity bus lateral stability. *J. Wind Eng. Ind. Aerodyn.* **2021**, *212*, 104589. [CrossRef]
13. Zhang, Q.; Su, C.; Zhou, Y.; Zhang, C.; Ding, J.; Wang, Y. Numerical Investigation on Handling Stability of a Heavy Tractor Semi-Trailer under Crosswind. *Appl. Sci.* **2020**, *10*, 3672. [CrossRef]
14. Du, X.; Wang, G. Analysis of Operating Safety of Tractor-Trailer under Crosswind in Cold Mountainous Areas. *Appl. Sci.* **2022**, *12*, 12755. [CrossRef]
15. Liu, H.; Liu, C.; Hao, L.; Zhang, D. Stability Analysis of Lane-Keeping Assistance System for Trucks under Crosswind Conditions. *Appl. Sci.* **2023**, *13*, 9891. [CrossRef]
16. Theissen, P. Unsteady Vehicle Aerodynamics in Gusty Crosswind. Ph.D. Thesis, Technische Universitat München, München, Germany, 2012.
17. Wanner, D.; Drugge, L.; Edren, J.; Stensson Trigell, A. Modelling and experimental evaluation of driver behaviour during single wheel hub motor failures. In Proceedings of the 3rd International Symposium on Future Active Safety Technology Towards Zero Traffic Accidents (FASTzero'15), Gothenburg, Sweden, 9–11 September 2015.
18. Drugge, L.; Juhlin, M. Aerodynamic loads on buses due to crosswind gusts: Extended analysis. *Veh. Syst. Dyn.* **2010**, *48*, 287–297. [CrossRef]
19. Wagner, A.; Wiedemann, J. Crosswind behavior in the driver's perspective. In Proceedings of the SAE 2002 World Congress, Detroit, MI, USA, 4–7 March 2002.
20. Maruyama, Y.; Yamazaki, F. Driving simulator experiment on the moving stability of an automobile under strong crosswind. *J. Wind Eng. Ind. Aerodyn.* **2006**, *94*, 191–205. [CrossRef]
21. Winkler, N.; Drugge, L.; Stensson Trigell, A.; Efraimsson, G. Coupling aerodynamics to vehicle dynamics in transient crosswinds including a driver model. *Comput. Fluids* **2016**, *138*, 26–34. [CrossRef]
22. Gutierrez, W.T.; Hassan, B.; Croll, R.H.; Rutledge, W.H. Aerodynamics overview of the ground transportation systems (GTS) project for heavy vehicle drag reduction. In Proceedings of the SAE International Congress and Exposition, Detroit, MI, USA, 26–29 February 1996.
23. Aultman, M.; Wang, Z.; Auza-Gutierrez, R.; Duan, L. Evaluation of CFD methodologies for prediction of flows around simplified and complex automotive models. *Comput. Fluids* **2022**, *236*, 105297. [CrossRef]
24. Chode, K.K.; Viswanathan, H.; Chow, K.; Reese, H. Investigating the aerodynamic drag and noise characteristics of a standard squareback vehicle with inclined side-view mirror configurations using a hybrid computational aeroacoustics (CAA) approach. *Phys. Fluids* **2023**, *35*, 075148. [CrossRef]
25. Zhang, J.; Guo, Z.; Han, S.; Krajnovic, S.; Sheridan, J.; Gao, G. An IDDES study of the near-wake flow topology of a simplified heavy vehicle. *Transp. Saf. Environ.* **2022**, *4*, 1–18. [CrossRef]
26. Saini, R.; Karimi, N.; Duan, L.; Sadiki, A.; Mehdizadeh, A. Effects of NearWall Modeling in the Improved-Delayed-Detached-Eddy-Simulation (IDDES) Methodology. *Entropy* **2018**, *20*, 771. [CrossRef]
27. Shur, M.L.; Spalart, P.R.; Strelets, M.K.; Travin, A.K. A hybrid RANS-LES approach with delayed-DES and wall-modelled LES capabilities. *Int. J. Heat Fluid Flow* **2008**, *29*, 1638–1649. [CrossRef]
28. Croll, R.H.; Gutierrez, W.T.; Hassan, B.; Suazo, J.E.; Riggins, A.J. Experimental investigation of the ground transportation systems (gts) project for heavy vehicle drag reduction. In Proceedings of the SAE International Congress and Exposition, Detroit, MI, USA, 26–29 February 1996.
29. Storms, B.L.; Ross, J.C.; Heineck, J.T.; Walker, S.M.; Driver, D.M.; Zilliac, G.G. An Experimental Study of the Ground Transportation System (GTS) Model in the NASA Ames 7-by 10-ft Wind Tunnel. NASA TM-2001-209621, 1 February 2001.
30. Unaune, S.V.; Sovani, S.D.; Kim, S.E. Aerodynamics of a Generic Ground Transportation System: Detached Eddy Simulation. In Proceedings of the 2005 SAE World Congress, Detroit, MI, USA, 11–14 April 2005.
31. Ghuge, H. Detached Eddy Simulations of a simplified Tractor-Trailer Geometry. Master's Thesis, Auburn University, Auburn, AL, USA, 2007.
32. CD-Adapco Inc. *STAR-CCM+ 11.0 User Guide*; CD-Adapco Inc.: Melville, NY, USA, 2016.
33. Huemer, J.; Stickel, T.; Sagan, E.; Schwarz, M.; Wall, W.A. Influence of unsteady aerodynamics on driving dynamics of passenger cars. *Veh. Syst. Dyn.* **2014**, *52*, 1470–1488. [CrossRef]
34. Cioffi, A.; Prakash, A.R.; Sabbioni, E.; Vignati, M.; Cheli, F. Heavy-Vehicle Response to Crosswind: Evaluation of Driver Reactions Using a Dynamic Driving Simulator. *Vehicles* **2023**, *5*, 344–366. [CrossRef]
35. Blundell, M.; Harty, D. *The Multibody Systems Approach to Vehicle Dynamics*, 2nd ed.; Butterworth-Heinemann: Coventry, UK, 2015.
36. Wang, H. Robust Control for Steer-by-Wire Systems in Road Vehicles. Ph.D. Thesis, Swinburne University of Technology, Melbourne, Australia, 2013.

applied sciences

Article

Comparison of the Relative Importance of Factors Affecting the Conveyance of Bulk and Liquid Cargo

Kristina Čižiūnienė [1], Jonas Matijošius [2], Jūratė Liebuvienė [3] and Edgar Sokolovskij [4,*]

[1] Department of Logistics and Transport Management, Vilnius Gediminas Technical University, Plytinės Str. 25, LT-10105 Vilnius, Lithuania; kristina.ciziuniene@vilniustech.lt
[2] Mechanical Science Institute, Vilnius Gediminas Technical University, Plytinės g. 25, LT-10105 Vilnius, Lithuania; jonas.matijosius@vilniustech.lt
[3] Department of Transport Engineering, Klaipeda State University of Applied Sciences, Jaunystės Str. 1, LT-91274 Klaipeda, Lithuania; j.liebuviene@kvk.lt
[4] Department of Automobile Engineering, Faculty of Transport Engineering, Vilnius Gediminas Technical University, Plytinės g. 25, LT-10105 Vilnius, Lithuania
* Correspondence: edgar.sokolovskij@vilniustech.lt

Abstract: The transportation of bulk or liquid cargo is crucial for ensuring the delivery of such cargo to areas lacking well-established infrastructure or favorable natural conditions. These areas may lack sufficient rail or air transport accessibility, or may not have suitable embankments along rivers, lakes or seas with appropriate loading and unloading facilities. Regarding distinctive characteristics, we would like to emphasize the following: (1) Efficient freight accessibility without the need for more infrastructure that is tailored to the existing natural circumstances; (2) An efficient and cost-effective mode of transportation for the movement of small quantities of goods; (3) The cargo philosophy is designed to meet customer expectations by providing a convenient and personalised service that is focused on delivering goods directly to their door. This research employed theoretical methodologies to characterize the primary force exerted by vehicles in the transportation of bulk and liquid commodities (the relationship between centrifugal force and speed and the relationship between the center of gravity and speed). Furthermore, by using expert assessment, the significance indicators pertaining to the individual criteria impacting the transportation of bulk and liquid cargoes were ascertained, along with the identification of their interrelationships (comparison of the different forces in the transportation process of bulk and liquid cargoes, comparison of the criteria of centrifugal force in the transportation process of bulk and liquid cargoes, comparison of the dangerous factors in the transportation process of bulk and liquid cargoes, comparison of the critical level of tank filling in the transportation process of bulk and liquid cargoes, and comparison of the factors that can affect the dynamics of a vehicle in the transportation process of bulk and liquid cargoes).

Keywords: bulk cargo; liquid cargo; transportation by road transport; Pearson correlation coefficient

Citation: Čižiūnienė, K.; Matijošius, J.; Liebuvienė, J.; Sokolovskij, E. Comparison of the Relative Importance of Factors Affecting the Conveyance of Bulk and Liquid Cargo. *Appl. Sci.* **2024**, *14*, 1151. https://doi.org/10.3390/app14031151

Academic Editor: Suchao Xie

Received: 23 December 2023
Revised: 23 January 2024
Accepted: 24 January 2024
Published: 30 January 2024

1. Introduction

The development and efficacy of transport firms heavily relies on their investment strategy, which is significantly influenced by external factors such as the direct and indirect investment environment [1]. The effectiveness of an entity is enhanced through its association with overarching strategies such as marketing [2], market analysis [3], financial planning, and advertising [4]. The cargo transportation sector holds a prominent position within this strategic framework [5]. One important factor is the calculation of the upper limit of acceptable duration for transporting perishable cargoes via road transport, which is done with a specific focus on the influence of external variables on the heat transfer coefficient and temperature differential. Increased vehicle velocities have been observed to significantly decrease the duration of delivery, hence reducing delivery time [6]. On the other hand, the occurrence of an external heat source results in an increase in the transfer of heat, which subsequently affects the timeliness of deliveries [7]. The objective is to develop formal models and methodologies for efficiently managing cargo turnover in rail transport, with a special emphasis on maintaining stability in the presence of unpredictable factors. [8]. Additionally, it suggests the incorporation of risk assessment methodologies to facilitate the implementation of these models and methods [9]. The a multi-agent system (MAS)-based distributed control architecture and hierarchical controller for connected and automated vehicles (CAVs) effectively manage complex traffic scenarios. The upper layer considers potential states and transition rules, while the lower layer includes motion controllers, utilizing the Pareto-optimal method for safety and stability [10].

Bulk cargo refers to a substantial quantity of uniform or multiple small segments possessing similar attributes that cannot be transported individually [11]. The defining attribute of these loads is their composition, which are comprised of desiccated particles and fragments of diverse dimensions. These loads exhibit characteristics such as natural degradation, moisture content, and hardness. Bulk cargo, such as ore, coal, grain, soy, salt, and fertilizer, is commonly carried in substantial numbers without the need for individual packaging. An analysis of data pertaining to tank trucks carrying dangerous liquid products, with the aim of enhancing their inherent safety measures, was presented in [12]. This paper proposes a framework consisting of eight components that have a significant impact on the occurrence of accidents, with a specific emphasis on the inherent safety features of vehicles. The construction of critical safety information and the establishment of a vehicle safety risk evaluation model are accomplished through the utilization of the matter-element extension model. The model is implemented in practical applications.

The delivery of bulk materials via road vehicles has distinct characteristics that necessitate a high level of responsibility and thorough preparation [13]. The movement of bulk freight must not present any potential hazards to fellow road users, the surrounding environment, or the existing infrastructure. The process of loading bulk cargo involves the utilization of a crane or a conveyor system to facilitate the transfer of goods from an elevated position. In order to provide optimal distribution, dry bulk cargo within the cargo compartment should be positioned in a manner that promotes evenness across the entire floor area [14]. Specifically, the center of the cargo should align with the longitudinal line of the axis of symmetry.

Nets employed for the purpose of securing or providing support to specific categories of bulk cargo can be fabricated using woven straps or ropes composed of either natural or synthetic fibers. Band nets are commonly employed as partitions to segregate the cargo area into distinct compartments. Lightweight nets have the potential to serve as suitable coverings for open trucks and containers that do not necessitate a more substantial enclosure. It is imperative to emphasize the importance of ensuring the integrity of the metal components of the nets, as they should be free from any signs of damage or corrosion. Additionally, it is crucial to ascertain that the net remains intact without any cuts, and that all seams are in optimal shape.

Semi-trailer awning is considered to be the most secure and durable method for safeguarding large quantities of cargo. The tarpaulin is secured to the sides of the tipper using fastening mechanisms. This method of preparing the vehicle serves to shield the bulk cargo from the impact of atmospheric conditions, ensuring that the cargo remains unaffected in terms of its physical and mechanical characteristics.

Bulk materials are used in industrial and agricultural production, but their interaction with structures can cause structural wear and dust pollution. A multiphase data fusion simulation approach is proposed for discrete material transfer dispatch in digital twin systems. The Discrete Element Method (DEM) is used to obtain bulk material movement and force data on structures, accurately calculating force data using the Finite Element Method (FEM). This method helps predict structure wear and manage the operating environment [15]. The enhancement of bulk freight transportation efficiency can be achieved by the establishment of technical standards pertaining to the loading of cargo into wagons. Different methods of transporting bulk cargo over railways, such as utilizing bags and big-bags, have been suggested as a means to enhance the capacity of wagon loads. A rationale is needed for selecting a specific style of wagon for the transportation of bulk products via road transport, drawing upon established approaches for efficient freight transportation. Recommendations play a crucial role in the selection of appropriate vehicle types and the optimization of carrying capacity based on transportation conditions, hence enhancing the efficiency of bulk freight transportation [16].

Liquid bulk and gaseous cargoes are types of cargoes that necessitate the utilization of specialized containers for their transportation and storage. The aforementioned items include glass containers, metal containers, and a variety of balloons. The magnitudes of these loads are determined using calculations that consider either their volume or mass. Bulk liquid cargoes refer to substances that possess the capability to be efficiently loaded and unloaded via pipelines, as well as stored and transported within tanks. A range of cargoes, including fuels, water, liquefied gases, and non-liquefied gases, might be associated with this category. For instance, the transfer of liquid items is a domain of cargo transportation that necessitates both specialized transportation and meticulous preparation. To mitigate the risk of tipping, which poses a significant hazard to transportation safety, the tanks are partitioned into several portions. Each portion is designed with a graduated capacity of 2000 L. Initially, one section is occupied, followed by another, and ultimately the remaining sections are filled. In order to promote secure transportation, it is recommended that the sections be filled to a maximum capacity of 80%, as this practice effectively mitigates the risk of spilling.

The analysis of data pertaining to tank trucks carrying hazardous liquids is necessary to enhance safety measures. The research aims to identify eight elements that have a substantial influence on the frequency of accidents, with particular focus on the inherent safety characteristics of vehicles [17]. Researchers have focused on accident risk assessment in hazardous chemicals road transport accidents, analyzing data from China over the past five years. They used a Bayesian network structure and parameter learning to predict the risk probability value of accidents. The model predicts "rear-end-leakage" accidents and identifies the most likely disposal time within 3–9 h. The study demonstrates the model's ability to predict accident scenarios and determine risk probabilities under different parameters [18].

When examining the transportation of bulk and liquid cargoes, transport company managers and logistics specialists often rely on their expertise and employ sophisticated and costly evaluation methods to effectively satisfy the unique requirements associated with such transportation. By employing expert evaluation, it becomes feasible to systematically assess the distinctive characteristics of transporting bulk and liquid cargo, enabling the anticipation of requisite precautionary measures through situational assessment, and facilitating the development of transportation process models.

The liquid or bulk cargoes referred to primarily consist of loads that meet the maximum allowable load of the transportation combination. Numerous carriers and their drivers demonstrate adherence to load securing regulations. However, an examination of accident causation in the previous year has identified a prevalent repeating factor contributing to traffic incidents, namely the technical state of the vehicle's chassis, encompassing aspects such as brake functionality, tire wear, and tread depth.

Simulations employing the Monte Carlo method, as described in [19], are commonly utilized for the purpose of simulating the actual transport process. In this particular scenario, a comparison is made of the selection of components' significance. This comparison facilitates the refinement of subsequent modeling procedures for the transportation of liquid and bulk cargoes under more appropriate rough circumstances, after an assessment of the unique characteristics associated with said cargo transportation. The utilization of these methodologies enables a more comprehensive evaluation of the variables influencing road freight transportation [20]. This approach enables the mitigation of transportation risks, the optimization of vehicle design for transportation processes, and the simultaneous attainment of economic benefits (through the identification of transportation costs) and environmental protection (by evaluating the characteristics of individual cargo transportation and potential traffic incidents that may result in environmental harm). The articles analyze logistical, social [21], and economic [22] difficulties, while the influence of forces is limited to theoretical and recommended levels, with a lack of practical calculations. Nevertheless, when confronted with a traffic collision or the necessity to execute diverse professional tasks, we encounter a dearth of data to facilitate these computations [23]. Moreover, logistics terminals employ specialized software to enhance the efficiency of the multimodal transportation planning process. The optimization method entails the utilization of a sophisticated mathematical framework, which can be enhanced by employing precise data to enhance the efficacy of the transfer process [8,24]. Hence, the examination of the relative significance of various types of cargoes, such as liquid and bulk cargo, holds considerable academic interest in enhancing the aforementioned logistics initiatives.

2. Materials and Methods

The subject matter of this research is limited to examining the job experience of transport logistics professionals. The mathematical tools used are tailored to suit the unique topic. The paper aims to assess the attributes of liquid and bulk cargo transportation via road. This adaptation relies on a dual implementation of mathematical tools. The research methodology consists of two research stages: In the first stage, the forces acting on the transported cargo are calculated, and in the second stage an expert assessment is carried out in order to assess the analysis of these forces at a practical level. Both mathematical calculations and expert assessment are performed on the example of Lithuania.

The purpose of using this model is to examine the forces at play and their primary correlation. In the subsequent phase, expert evaluation is used to ascertain the practical (adaptive) qualities of these forces. Given the utilization of a diverse mathematical apparatus, the authors endeavored to derive the specific capabilities of this mathematical tool. The objective of this is to determine the magnitudes of the forces involved via calculations. This will allow for a more accurate identification of the most suitable method of transporting the cargo, using additional mathematical tools.

While performing force calculations, it is important to consider not only the force equation itself, but also to establish the critical velocities for various levels of cargo capacity. These calculations enable the formulation of the most suitable questions for experts, facilitating the determination of the order of important criteria in subsequent stages via the use of mathematical tools for expert evaluation.

2.1. *Mathematical Calculation Methodology*

Bulk and liquid cargoes have a center of gravity of only 1.8 m from the ground, while others are 2 to 2.5 m high. Considering the fact that the center of gravity of bulk and liquid cargo is not a height, it is not mathematically assessed. Only the centrifugal force, gravity force, and critical slip of the tug are evaluated.

The analytical computations (Equations (1) and (2) [25,26]) were conducted with a circular radius of 17 m. This radius was chosen because, taking into account the methodological instructions for the design of roundabouts, it is stated that the turning radius R of a roundabout with entrances and exits of two traffic lanes is between 12 and 18 m [27]. A vehicle carrying a significant amount of freight, moving at a velocity of 20 km/h, encounters a centrifugal force while traversing a roundabout with a diameter of 17 m (Figure 1):

$$F = \frac{m_v v^2}{r}, \mathrm{N} \tag{1}$$

The force of gravity is computed:

$$F = m_v \cdot g, \mathrm{N} \tag{2}$$

when driving around a bend in the road, under the influence of centrifugal force, the wheels of the vehicle may lose their grip on the road surface and begin to slide in the transverse direction. The critical traction speed in a corner indicates the maximum speed at which a vehicle can pass a corner without losing the grip of the wheels on the road surface and without starting to slide in the transverse deflection. Stability is a key criterion for assessing a truck's appropriateness for employment. When a vehicle's transverse stability is compromised near a curve in the road, it either skids or flips over. In order to maintain the vehicle's stability, the conditions of preventing rolling and sliding are established [26].

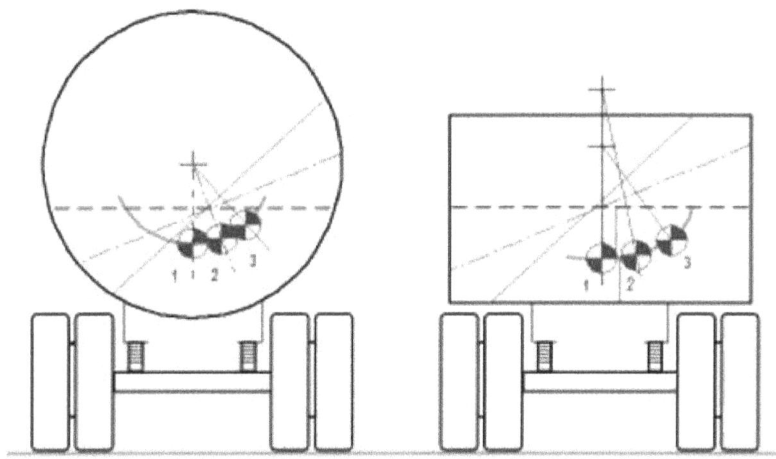

Figure 1. Variation of the center of gravity of bulk and liquid cargoes.

The critical driving speed in the event of a truck skidding is calculated according to the formula [25]:

$$v_{sl} = \sqrt{g \cdot r_m \cdot \mu},\ \text{N} \tag{3}$$

where: r_m—turning radius; g—acceleration of free fall coefficient of adhesion of tires to the road surface.

When driving on a corner, a car may lose stability due to the forces acting upon it, and turn in relation to the longitudinal axis. In this case the following formula can be used to calculate the critical driving speed for the turning condition:

$$v_{ov} \leq \sqrt{g \cdot r_m \cdot \frac{\frac{b}{2} + h_{sc} \cdot tg\beta}{h_{sc} - \frac{b}{2} \cdot tg\beta}},\ \text{N} \tag{4}$$

where: r_m—turning radius, m; g—acceleration of free fall, m/s^2; β—the angle of inclination of the transverse road; h_{sc}—the height of the vehicle's center of mass (Figure 2), m; b—the width of the vehicle's wheelbase, m.

The calculation of the critical slip of the vehicle [25,26] is performed using a specified coefficient of friction of $\mu = 0.6$ [28], specifically for dry asphalt in this scenario [29,30]. The turning radius of the vehicle is measured to be 17 m, while the cross slope of the road is denoted as $\beta = 2$ [31].

Cross road slope for sliding:

$$tg\beta = \mu$$

The height h of the vehicle's center of gravity can be calculated using the formula:

$$h_{sc} = \frac{b}{G_p} \cdot \left(R'_{zd} - R_{zd}\right) \cdot ctg\beta,\ \text{m} \tag{5}$$

where: G_p—front axle load of the vehicle, kN; R_{zd}—support reaction force to the wheel of the front axle, kN; R'_{zd}—the support reaction force on the wheel of the front axle when weighing the leaning vehicle, kN; β—the angle of inclination of the transverse road.

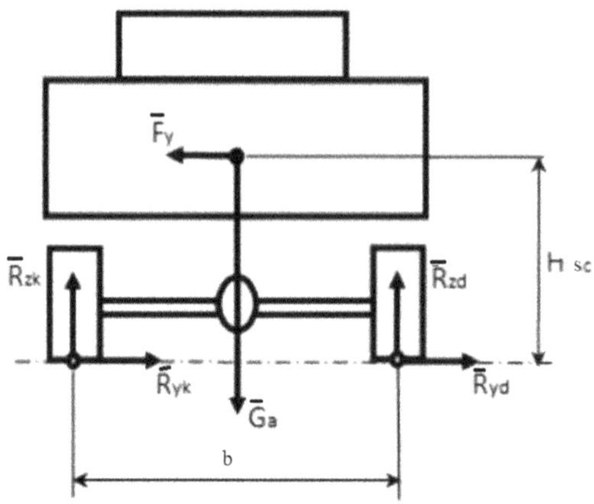

Figure 2. Principle calculation scheme of transverse stability of a truck in a turn.

From the dependencies presented in the formulas, it can be seen that, in order to increase the stability of the vehicle, it is necessary to reduce the height of the vehicle's center of gravity and/or to increase the length of the track [25].

The stability of a truck depends on the height of the center of gravity and the ratio of the rut. Since the rut ratio varies little between different standard trucks, the stability of heavy vehicles depends on the height of the center of gravity. The height of the car's center of gravity is calculated according to the formula [26]:

$$h_{sc} = \frac{h_a \cdot m_a + h_k \cdot m_k}{m_a + m_k}, \text{ m} \tag{6}$$

where: h_a—the height of the center of gravity of the unloaded vehicle, m; h_k—the height of the center of gravity of the transported load above the road surface, m; m_a ir m_k—the mass of the truck and the transported cargo, respectively, in kg.

The formula for calculating the critical speed of a truck for overturning (m/s):

$$v_{ov} = \sqrt{\frac{g \cdot b \cdot r_m}{2h_{sc}}}, \text{ m/s} \tag{7}$$

At a curve in the road, a truck has the potential to not only tip over, but also skid. A truck's wheel may move both parallel and perpendicular to its plane of rotation. From a traffic safety perspective, the lateral sliding of the automobile poses a far greater risk, since it considerably hampers the ability to maneuver away from the truck. The automobile, in turn, is subject to forces resulting from its weight and attempting to displace it off its intended path of travel, due to the centrifugal force. Curiously, the truck's capacity to maintain stability when turning in a road curve is not influenced by the coefficient of adhesion. Instead, it is determined by the height of the center of mass and the ratio of the rut. Given that the rut remains constant for a particular vehicle, the stability of the truck during a rollover will only rely on the elevation of its center of mass. This elevation may fluctuate depending on factors such as the kind of cargo being carried.

When assessing the transportation of bulk and liquid cargoes, certain key elements that impact the safety of the transportation process can be identified. This involves determining the relative importance of the forces influencing the cargo structure, the transport capacity's fill level, which directly affects the transported structure, and other variables influencing the transport structure's dynamics. Taking this into account, an expert assessment was carried out.

2.2. Expert Evaluation Methodology

Considering the fact that the expert assessment was carried out in Lithuania. The selection of experts was carried out based on the following logical sequence:

1. In Lithuania, the total number of companies providing transport services in 2022 was 8570 companies, of which 6200 companies are engaged in freight road transport. After analyzing the volume of cargo transportation for pay and at own expense in thousands of tons, it was determined that in 2022, 100,752 thousand tons of such cargo were transported, of which 36,828.8 thousand tons were transported by domestic transport. This is 36.55 percent. Considering that in the analyzed case, specific cargo is transported, i.e., liquid and bulk cargo, we can claim that such companies makes up about 15 percent of the total.
2. Taking into account that according to the methodological requirements of the study, the evaluation of eight experts is considered as a valid evaluation, the following requirements were imposed on the selected experts:
   (1) the period of the company's activities is 10–15 years,
   (2) carrying out the transportation of specific cargoes—liquid cargoes, bulk cargoes for 8–10 years;
   (3) vehicle park, spec. transport 20–30 percent of the entire vehicle fleet.

Each group of criteria was to be ranked from most important to least important. Each criterion had to be given a score and could not be 0. Criterion m, scored by experts n.

The formulas presented in Table 1 were used for expert evaluation calculations.

Table 1. The formulas description used for the expert survey results calculations [12,13,32,33].

Parameter Name	Formula	Explanation of Meanings
Kendall's coefficient of concordance	$\sum_{i=1}^{n} R_{ij} = R_{ij}$	
Sums of R squares S (dispersion)	$S = \sum_{j=1}^{m} (R_j - \overline{R})^2$	
Total average	$\overline{R} = \frac{\sum_{j=1}^{m} R_{ij}}{m} = \frac{\sum_{i=1}^{n}\sum_{j=1}^{m} R_{ij}}{m}$	
Sum of m indicator rankings of all n experts	$\sum_{i=1}^{n} R_{ij} = \frac{1}{2}n(m+1)$	
Total average	$\overline{R} = \frac{1}{2}n(m+1)$	
Average rank of each criterion R	$\overline{R}_j = \frac{\sum_{i=1}^{n} R_{ij}}{n}$	R—the rank given to the i-th criterion of the j-th expert, n—the number of experts
Coefficient of concordance when there are no associated ranks	$W = \frac{12S}{n^2(m^3-m)}$	
Sum of squares S of deviations of ranks R of each criterion from the average rank	$S = \sum_{j=1}^{m} \left(\sum_{i=1}^{n} R_{ij} = \frac{1}{2}n(m+1) \right)^2$	m—number of criteria ($i = 1, 2, \ldots, m$); n—number of experts ($j = 1, 2, \ldots, n$)
Sum of squares S in an ideal approved case	$S_{\max} = \frac{n^2 m(m^2-1)}{12}$	
Pearson's χ^2 criterion	$\chi 2 = \frac{12S}{nm(m+1)}$	
Lowest value of the coefficient of concordance W_{min}	$W_{min} = \frac{\chi^2_{v,a}}{n(m+1)}$	
Reverse value of the criterion q	$d_j = 1 - \overline{q}_j = 1 - \frac{\overline{R}_j}{\sum_{j=1}^{m} R_j}$	The sum of d estimates is equal to n—1
Criteria importance indicators	$Q_j = \frac{d_j}{\sum_{j=1}^{m} d_j} = \frac{d_j}{m-1}$	
Importance of criteria of the object being evaluated by experts	$Q'_j = \frac{(m+1)-\overline{R_j}}{\sum_{j=1}^{m} R_j}$	m—the number of indicators showing the characteristics of the object, R—the average rank of the jth criterion

3. Results

3.1. Results of Mathematical Calculations

Equation (8) specifies the centrifugal force, which is computed based on Equation (1):

$$F = \frac{35,000 \cdot 30.8}{17} = 63,535, \mathrm{N} \tag{8}$$

Upon conducting calculations, it was ascertained that a motor vehicle traversing a circular intersection with a radius measuring 17 m at a velocity of 20 km per hour experiences the influence of a centrifugal force amounting to 63,535 N.

The gravitational force, as described by Equation (9), is computed based on Equation (2):

$$F = 35,000 \cdot 9.81 = 343,350, \mathrm{N} \tag{9}$$

Figure 3 illustrates that the vehicle maintains stability when its speed does not exceed 45 km/h. At this threshold, the centrifugal force remains within the limits of gravitational force. Nevertheless, when the velocity of the vehicle above 45 km/h, the centrifugal force surpasses the gravitational force, leading to the overturning of the vehicle.

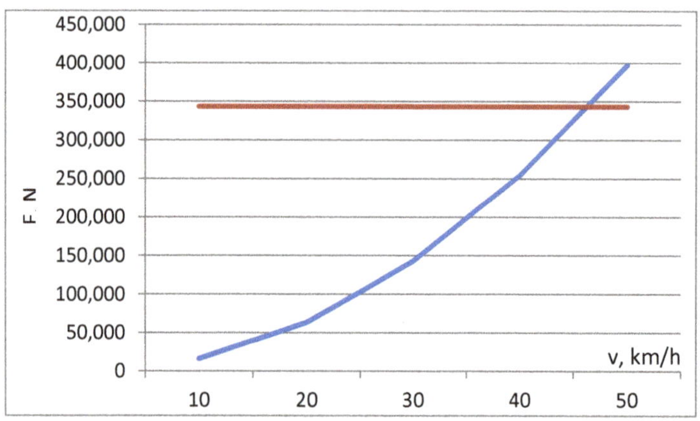

Figure 3. The relationship between centrifugal force and speed (red curve is the gravitational force, blue curve is the centrifugal force).

To analyze the stability of a cargo vehicle, it is crucial to determine the height of the car's center of mass, denoted as hsc. The manufacturer provides the location of the center of mass for both an empty truck and a fully loaded truck in the technical data of the vehicle. The center of mass of an unladen freight truck is consistently displaced towards the front of the vehicle. When an automobile is loaded, the location of its center of mass shifts in relation to the center of mass of an empty car and may move in many directions [29].

The vehicles used for the investigation include a dump truck designed for moving large quantities of bulk freight, as well as a tank truck specifically designed for transporting liquid cargo. The trucks' technical specifications, categorized by type and model, include the following: mass of 20,500 kg and 23,540 kg, respectively, height of the center of mass of 1.624 m and 1.741 m, respectively, and track width of 2.174 m and 2.04 m, respectively. Dynamic stability calculations were conducted to ascertain the threshold speed at which the vehicle would maintain balance and prevent both overturning and sliding. Figure 4 depicts a graphical depiction of the truck's roll and slide in relation to the driving speed and center of gravity. The vertical location of the center of gravity was adjusted within the range of 0.60–3.00 m, with increments of 0.6 m, as seen in Figure 4.

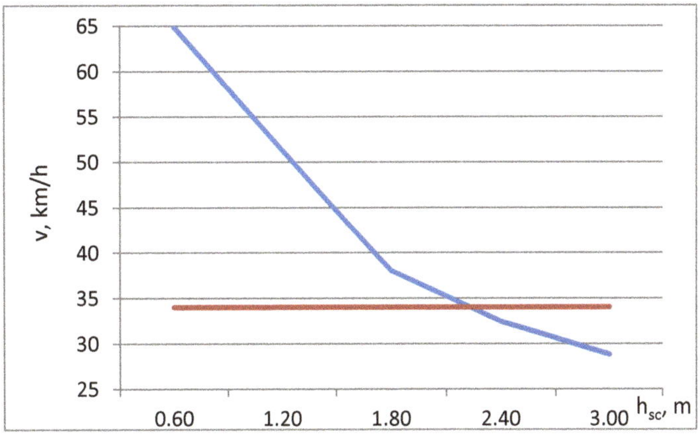

Figure 4. The relationship between centrifugal force and speed (red curve is gravitational force, blue curve is centrifugal force).

The occurrence of crucial skidding in the vehicle is not expected until the velocity exceeds 34 km/h. Nevertheless, once the velocity of the vehicle surpasses 34 km/h, it will initiate a skidding motion when navigating a bend.

The critical roll of the vehicle is calculated by considering the width of the vehicle track and the height of the center of gravity of the load. The width of the semi-trailers for bulk and skid products is 2.55 m. The selected height of the center of gravity of the load is 1.8 m from the ground when the vehicle is loaded, while the center of gravity of the unloaded transport vehicle is selected at a height of 1.2 m.

The essential turning speed of both loaded and empty vehicles was determined for a roundabout with a diameter of 17 m. A vehicle that is carrying a significant amount of weight will experience a tipping motion when its speed exceeds 38 km/h, while a vehicle that is not carrying any additional weight will undergo a tipping motion at speeds exceeding 51 km/h.

Figure 4 provides a visual representation of the relationship between the speed and the height of the center of gravity. The critical juncture occurs when the intersection of the turning speed and sliding speed curves is reached. In this scenario, if the height of the vehicle's center of gravity is 2.4 m above the ground and the vehicle is traveling at a speed of 34 km/h, a rollover event is expected to transpire.

When assessing the transportation of bulk and liquid cargoes, certain key elements that impact the safety of the transportation process are identified. This involves determining the relative importance of the forces influencing the cargo structure, the transport capacities' fill level, which directly affects the transported structure, and other variables influencing the transport structure's dynamics [34,35]. Therefore, it is important to assess not only how these factors work individually [36], but also how they interact with each other and what their importance is. For that purpose, the results of the expert evaluation are presented below.

3.2. Results of Expert Assessment

In the process of transporting bulk and liquid cargoes, several forces are encountered that affect both the cargo and the vehicle. Nevertheless, it is commonly observed that these parts are typically regarded as distinct entities, with little regard given to their relative importance or the intricate nature of their interrelationships.

The survey findings, which compare all of the forces involved in the transportation of bulk and liquid commodities, are shown in Table 2.

Table 2. The expert survey results comparison of the different forces in transportation process of bulk and liquid cargoes.

Formula		Centrifugal Force	Gravity Force	Friction Force	Inertial Force
$\sum_{i=1}^{n} R_{ij} = R_{ij}$	Bulk cargoes	17	26	26	11
	Liquid cargoes	16	24	30	10
$\overline{R}_j = \frac{\sum_{i=1}^{n} R_{ij}}{n}$	Bulk cargoes	2.125	3.250	3.250	1.375
	Liquid cargoes	2	3.000	3.750	1.250
$\sum_{i=1}^{n} R_{ij} - \frac{1}{2}n(m+1)$	Bulk cargoes	−3	6	6	−9
	Liquid cargoes	−4	4	10	−10
$\left[\sum_{i=1}^{n} R_{ij} - \frac{1}{2}n(m+1)\right]^2$	Bulk cargoes	9	36	36	81
	Liquid cargoes	16	16	100	100

For decisions to be taken, the experts' opinions must be balanced. In the case of two or more experts, the consistency of their opinions may be determined by applying a concordance factor. In the calculation of the Kendall's concordance coefficient (W), the experts' assessments are ranked. The experts' opinions are considered to be concordant when W→1, discordant when W→0 (Tables A1, A2, A5, A6 and A9). The next step is to calculate the importance of the criterion using expert survey data (Tables A3, A4, A7, A8 and A10). The obtained results show the priority of the forces acting on bulk and liquid cargoes. The results of the assessment of the forces arranged by the significance of expert opinions from various sources in evaluating the transportation of bulk and liquid cargoes are shown in Table 3.

Table 3. The results of the calculation of the criterion arrangement by importance of expert opinions by the different forces in the evaluation of the transportation of bulk and liquid cargoes.

Sequence of Forces Acting on Bulk Cargo during Transportation		Sequence of Forces Acting on Liquid Cargo during Transportation	
1	Inertial force	1	Inertial force
2	Centrifugal force	2	Centrifugal force
3	Gravity force	3	Gravity force
4	Friction force	4	Friction force

The survey of experts has been conducted to compare the criteria of centrifugal force in the transportation process of bulk and liquid cargoes. The findings of this comparison may be found in Table 4.

Table 4. The expert survey results comparison of the criteria of centrifugal force in transportation process of bulk and liquid cargoes.

Formula		Position of Centre of Gravity	Vehicle Speed	Vehicle Trajectory (Steer Angle)
$\sum_{i=1}^{n} R_{ij} = R_{ij}$	Bulk cargoes	17	14	17
	Liquid cargoes	18	15	15
$\overline{R}_j = \frac{\sum_{i=1}^{n} R_{ij}}{n}$	Bulk cargoes	2.125	1.750	2.125
	Liquid cargoes	2.25	1.875	1.875
$\sum_{i=1}^{n} R_{ij} - \frac{1}{2}n(m+1)$	Bulk cargoes	1	−2	1
	Liquid cargoes	2	−1	−1
$\left[\sum_{i=1}^{n} R_{ij} - \frac{1}{2}n(m+1)\right]^2$	Bulk cargoes	1	4	1
	Liquid cargoes	4	1	1

The results of the assessment of the criterion arranged by the significance of expert judgements about the criteria of centrifugal force in the evaluation of bulk and liquid cargo transportation are shown in Table 5.

Table 5. The results of the calculation of the criterion arrangement by importance of expert opinions by the criteria of centrifugal force in the evaluation of the transportation of bulk and liquid cargoes.

Sequence of Forces Acting on Bulk Cargo during Transportation		Sequence of Forces Acting on Liquid Cargo During Transportation	
1	Vehicle speed	1	Vehicle trajectory (steer angle)
2	Position of centre of gravity	2	Vehicle speed
3	Vehicle trajectory (steer angle)	3	Position of centre of gravity

The outcomes demonstrate that the two most powerful forces are centrifugal and inertial. It is crucial to ascertain the interdependence of all forces, however, in order to assess the strength of the interaction between them.

Determining the most hazardous elements involved in the transportation of liquid and bulk commodities is crucial.

The comparison of dangerous factors in the transportation of bulk and liquid cargoes is shown in Table 6, based on the findings of the expert survey.

Table 6. The expert survey results comparison of the dangerous factors in transportation process of bulk and liquid cargoes.

Formula		Road Surface	Road Turn Curve	Sudden Braking	Vehicle Not Fully Loaded
$\sum_{i=1}^{n} R_{ij} = R_{ij}$	Bulk cargoes	28	16	11	25
	Liquid cargoes	32	19	16	13
$\overline{R}_j = \frac{\sum_{i=1}^{n} R_{ij}}{n}$	Bulk cargoes	3.5	2.000	1.375	3.125
	Liquid cargoes	4	2.375	2.000	1.625
$\sum_{i=1}^{n} R_{ij} - \frac{1}{2}n(m+1)$	Bulk cargoes	8	−4	−9	5
	Liquid cargoes	12	−1	−4	−7
$\left[\sum_{i=1}^{n} R_{ij} - \frac{1}{2}n(m+1)\right]^2$	Bulk cargoes	64	16	81	25
	Liquid cargoes	144	1	16	49

The data collected indicate that while moving bulk and liquid commodities, the riskiest elements have distinct priorities.

The results of the assessment of the criteria arrangement by significance of expert views on the dangerous factors in the evaluation of the transportation of bulk and liquid cargoes are shown in Table 7.

Table 7. The results of the calculation of the criterion arrangement by importance of expert opinions by the dangerous factors in the evaluation of the transportation of bulk and liquid cargoes.

Sequence of Forces Acting on Bulk Cargo during Transportation		Sequence of Forces Acting on Liquid Cargo during Transportation	
1	Sudden braking	1	Vehicle not fully loaded
2	Road turn curve	2	Sudden braking
3	Vehicle not fully loaded	3	Road turn curve
4	Road surface	4	Road surface

The findings indicate that the ranking of the most hazardous elements varies depending on the nature of the cargo [37–39]. When assessing the relationship and its magnitude among several components, it is crucial to determine their interdependence.

Given the consideration that the primary risk element in the transportation of liquid cargo is an inadequately loaded vehicle, it is imperative to ascertain the critical level of tank filling that poses the highest degree of danger.

The comparison of expert survey findings about the crucial level of tank filling during the transportation of bulk and liquid cargoes is shown in Table 8.

Table 8. The expert survey results comparison of the critical level of tank filling in transportation process of bulk and liquid cargoes.

Formula	Fully Filled Cistern	Two-Thirds Full Cistern	One-Third Full Cistern	Empty Cistern
$\sum_{i=1}^{n} R_{ij} = R_{ij}$	23	11	14	32
$\overline{R}_j = \frac{\sum_{i=1}^{n} R_{ij}}{n}$	2.875	1.375	1.750	4.000
$\sum_{i=1}^{n} R_{ij} - \frac{1}{2}n(m+1)$	3	−9	−6	12
$\left[\sum_{i=1}^{n} R_{ij} - \frac{1}{2}n(m+1)\right]^2$	9	81	36	144

The obtained results show that when transporting liquid cargoes, the danger of filling the tank is arranged in the following order:

1. Two-thirds filled cistern;
2. The cistern is one-third full;
3. Fully filled tank;
4. Empty tank.

Also, during the research, it was important to find out the factors that can affect the dynamics of a vehicle with bulk and liquid cargo during road transport.

The survey of experts has yielded data that compare the elements influencing the movement of a vehicle during the transportation of bulk and liquid goods. These results are provided in Table 9.

Table 9. The expert survey results comparison of the factors that can affect the dynamics of a vehicle in transportation process of bulk and liquid cargoes.

Formula		Vehicle Characteristics	Air Temperature and Atmospheric Conditions	Road Condition	Loading of Cargo	Driver Skill and Attention
$\sum_{i=1}^{n} R_{ij} = R_{ij}$	Bulk cargoes	18	30	28	32	12
	Liquid cargoes	22	33	28	25	12
$\overline{R}_j = \frac{\sum_{i=1}^{n} R_{ij}}{n}$	Bulk cargoes	2.25	3.75	3.500	4.000	1.500
	Liquid cargoes	2.75	4.125	3.500	3.125	1.500
$\sum_{i=1}^{n} R_{ij} - \frac{1}{2}n(m+1)$	Bulk cargoes	−6	6	4	8	−12
	Liquid cargoes	−2	9	4	1	−12
$\left[\sum_{i=1}^{n} R_{ij} - \frac{1}{2}n(m+1)\right]^2$	Bulk cargoes	36	36	16	64	144
	Liquid cargoes	4	81	16	1	144

The obtained results show that when transporting bulk and liquid cargoes, the factors influencing the dynamics of the vehicle have different priorities. Only the most important skills and attention of the driver coincide, and the least important are air temperature and atmospheric conditions.

Table 10 presents the results of the computation of the criteria arrangement by significance of expert opinions on the aspects that might influence the dynamics of a vehicle in the assessment of the transportation of bulk and liquid cargoes.

The results show that the driver's skill and attentiveness are probably the most important factors influencing the dynamics of the vehicle. However, when evaluating the interaction and its strength among all forces, it is important to find out their dependence as well.

Table 10. The results of the calculation of the criterion arrangement by importance of expert opinions by the factors that can affect the dynamics of a vehicle in the evaluation of the transportation of bulk and liquid cargoes.

Sequence of Forces Acting on Bulk Cargo during Transportation		Sequence of Forces Acting on Liquid Cargo during Transportation	
1	Driver skill and attention	1	Driver skill and attention
2	Road condition	2	Loading of cargo
3	Loading of cargo	3	Road condition
4	Vehicle characteristics	4	Vehicle characteristics
5	Air temperature and atmospheric conditions	5	Air temperature and atmospheric conditions

4. Discussion

The discussion employed parameter correlation analysis to assess the features of bulk and liquid cargoes and ascertain the interdependencies among their attributes.

In relation to the determination of the significance of the acting forces, a robust association is detected between the acting forces as per Pearson's correlation analysis [40,41]. A more pronounced association was seen in the case of bulk shipments (Figure 5). A significant association was seen exclusively in the context of bulk cargo transit. Upon analyzing the transportation of liquid cargoes, it was seen that the correlation, as measured by Pearson's correlation analysis, had a lesser association. Specifically, a robust correlation was found solely between frictional and inertial forces, while the overall correlation was classified as moderate (Figure 6).

		Centrifugal force	Gravitational force	Frictional force	Inertia force
Centrifugal force	Pearson Corr.	1	0.92236	0.92236	0.98278
	Sig.	--	0.07764	0.07764	0.01722
Gravitational force	Pearson Corr.	0.92236	1	1	0.83509
	Sig.	0.07764	--	--	0.16491
Frictional force	Pearson Corr.	0.92236	1	1	0.83509
	Sig.	0.07764	--	--	0.16491
Inertia force	Pearson Corr.	0.98278	0.83509	0.83509	1
	Sig.	0.01722	0.16491	0.16491	--

2-tailed test of significance is used

Figure 5. Values of Pearson's correlation coefficient when assessing the importance of the forces acting on bulk cargoes transportation.

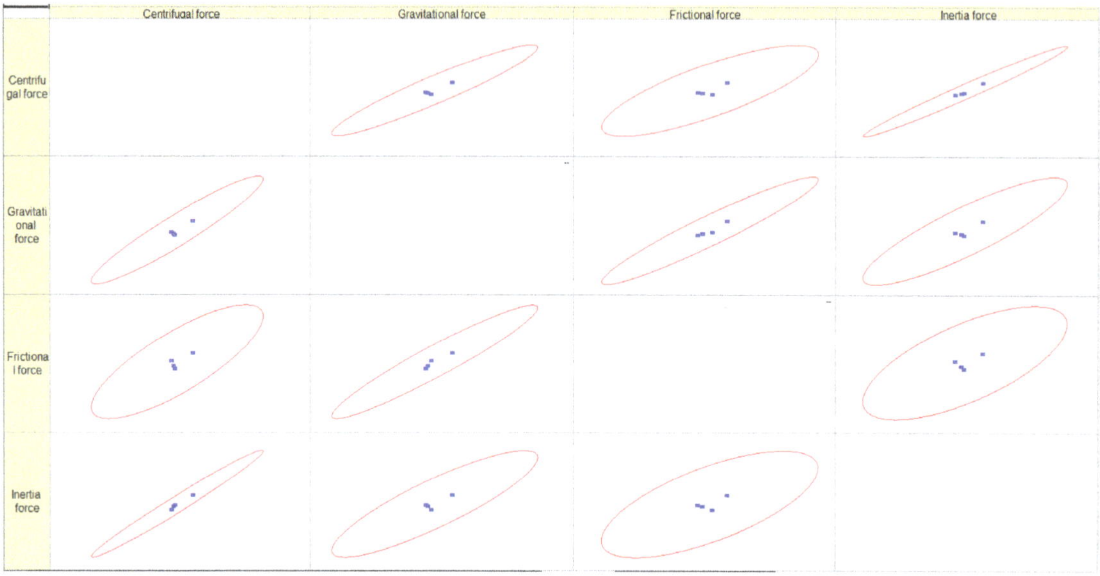

		Centrifugal force	Gravitational force	Frictional force	Inertia force
Centrifugal force	Pearson Corr.	1	0.94536	0.80063	0.9845
	Sig.	--	0.05464	0.19937	0.0155
Gravitational force	Pearson Corr.	0.94536	1	0.95223	0.87351
	Sig.	0.05464	--	0.04777	0.12649
Frictional force	Pearson Corr.	0.80063	0.95223	1	0.68312
	Sig.	0.19937	0.04777	--	0.31688
Inertia force	Pearson Corr.	0.9845	0.87351	0.68312	1
	Sig.	0.0155	0.12649	0.31688	--

2-tailed test of significance is used

Figure 6. Values of Pearson's correlation coefficient when assessing the importance of the forces acting on liquid cargoes transportation.

The presence of centrifugal forces can lead to instability in both bulk and liquid cargo, as a result of abrupt shifts in the center of gravity, the vehicle's high speed, and the trajectory of travel of both the vehicle and its cargo. Hence, it is necessary to assess these inter-factors by calculating their Pearson's correlation coefficient. In both bulk and liquid cargo transit, the coefficient exhibited a consistent similarity and approached approximately 0.95 across all examined instances, indicating a robust association (see Figure 7).

The impact of the road plays a significant role in determining the acting forces involved in transportation. Consequently, it is crucial to emphasize the interconnectedness of the road's influencing components when moving both bulk and liquid commodities. In Figures 8 and 9, it is shown that Pearson's correlation coefficient exhibits highly comparable values. It can be argued that in the majority of instances, there exists a significant or substantial link between the components. One notable distinction between bulk and liquid cargo becomes apparent when analyzing the parameter of "Vehicle not fully loaded." In this regard, the correlation coefficient values are comparatively lower for liquid cargo.

The most intriguing correlational relationships are observed in the context of general variables that impact freight transportation (Figures 10–12). In both instances, the Pearson correlation coefficient had identical values, indicating a robust interparametric association.

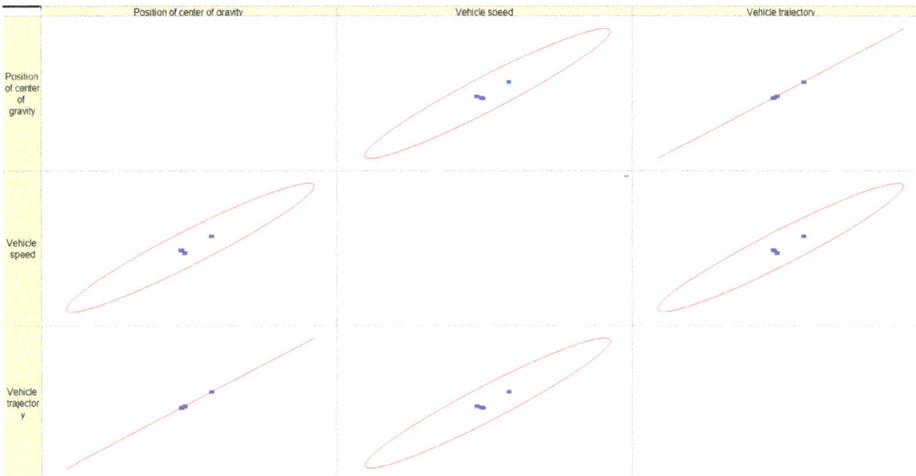

		Position of center of gravity	Vehicle speed	Vehicle trajectory
Position of center of gravity	Pearson Corr.	1	0.95481	1
	Sig.	--	0.04519	--
Vehicle speed	Pearson Corr.	0.95481	1	0.95481
	Sig.	0.04519	--	0.04519
Vehicle trajectory	Pearson Corr.	1	0.95481	1
	Sig.	--	0.04519	--

2-tailed test of significance is used

Figure 7. Values of Pearson's correlation coefficient when assessing the importance of the centrifugal force factors acting on bulk cargoes transportation.

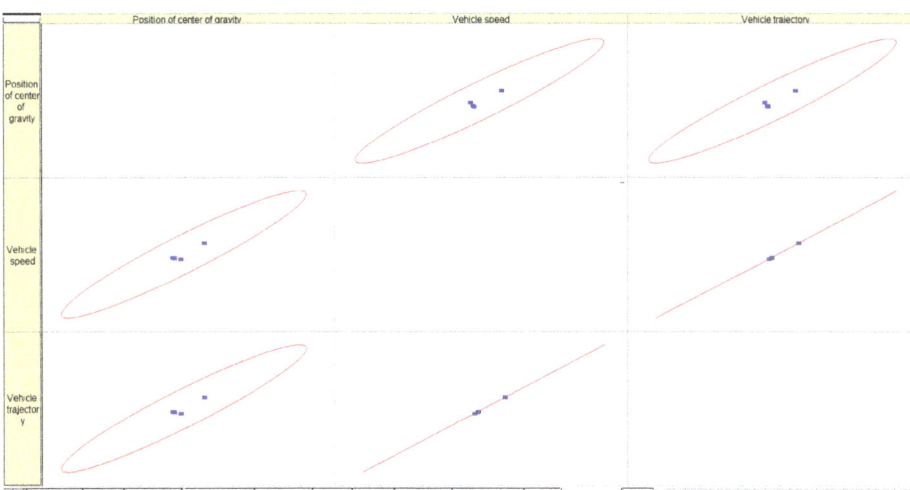

		Position of center of gravity	Vehicle speed	Vehicle trajectory
Position of center of gravity	Pearson Corr.	1	0.94489	0.94489
	Sig.	--	0.05511	0.05511
Vehicle speed	Pearson Corr.	0.94489	1	1
	Sig.	0.05511	--	--
Vehicle trajectory	Pearson Corr.	0.94489	1	1
	Sig.	0.05511	--	--

2-tailed test of significance is used

Figure 8. Values of Pearson's correlation coefficient when assessing the importance of the centrifugal force factors acting on liquid cargoes transportation.

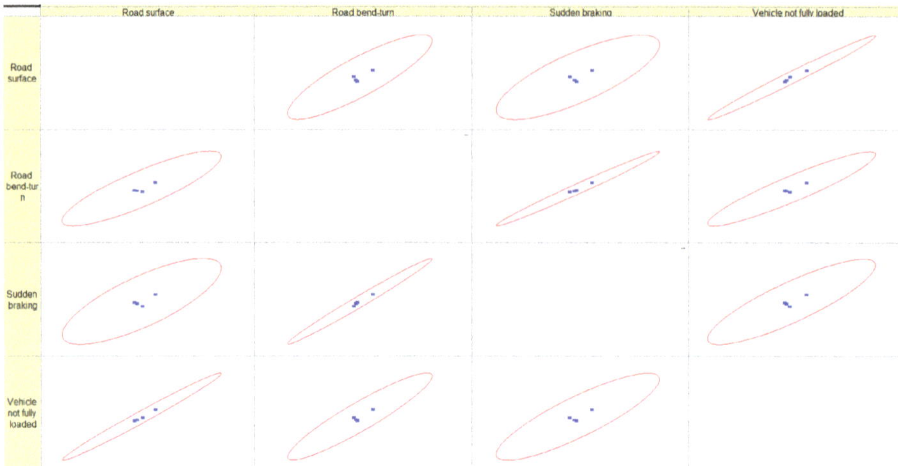

		Road surface	Road bend-turn	Sudden braking	Vehicle not fully loaded
Road surface	Pearson Corr.	1	0.85979	0.77339	0.98806
	Sig.	--	0.14021	0.22661	0.01194
Road bend-turn	Pearson Corr.	0.85979	1	0.98867	0.92819
	Sig.	0.14021	--	0.01133	0.07181
Sudden braking	Pearson Corr.	0.77339	0.98867	1	0.86181
	Sig.	0.22661	0.01133	--	0.13819
Vehicle not fully loaded	Pearson Corr.	0.98806	0.92819	0.86181	1
	Sig.	0.01194	0.07181	0.13819	--

2-tailed test of significance is used

Figure 9. Values of Pearson's correlation coefficient when assessing the importance of the road condition factors acting on bulk cargo transportation.

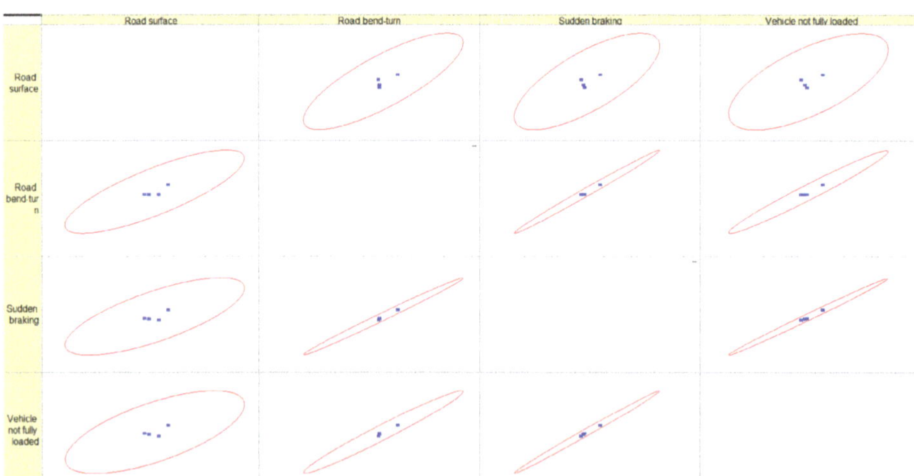

		Road surface	Road bend-turn	Sudden braking	Vehicle not fully loaded
Road surface	Pearson Corr.	1	0.80326	0.73241	0.6644
	Sig.	--	0.19674	0.26759	0.3356
Road bend-turn	Pearson Corr.	0.80326	1	0.99386	0.97884
	Sig.	0.19674	--	0.00614	0.02116
Sudden braking	Pearson Corr.	0.73241	0.99386	1	0.99548
	Sig.	0.26759	0.00614	--	0.00452
Vehicle not fully loaded	Pearson Corr.	0.6644	0.97884	0.99548	1
	Sig.	0.3356	0.02116	0.00452	--

2-tailed test of significance is used

Figure 10. Values of Pearson's correlation coefficient when assessing the importance of the road condition factors acting on liquid cargo transportation.

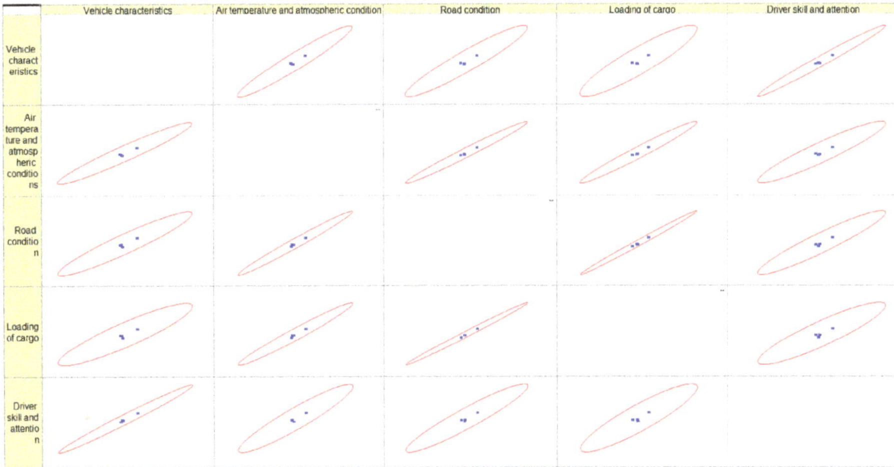

		Vehicle characteristics	Air temperature and atmospheric conditions	Road condition	Loading of cargo	Driver skill and attention
Vehicle characteristics	Pearson Corr.	1	0.96247	0.94368	0.90806	0.98672
	Sig.	–	0.03753	0.05632	0.09194	0.01328
Air temperature and atmospheric conditions	Pearson Corr.	0.96247	1	0.98562	0.97807	0.93537
	Sig.	0.03753	–	0.01438	0.02193	0.06463
Road condition	Pearson Corr.	0.94368	0.98562	1	0.99451	0.94192
	Sig.	0.05632	0.01438	–	0.00549	0.05808
Loading of cargo	Pearson Corr.	0.90806	0.97807	0.99451	1	0.90162
	Sig.	0.09194	0.02193	0.00549	–	0.09838
Driver skill and attention	Pearson Corr.	0.98672	0.93537	0.94192	0.90162	1
	Sig.	0.01328	0.06463	0.05808	0.09838	–

2-tailed test of significance is used

Figure 11. Values of Pearson's correlation coefficient when assessing the importance of the vehicle dynamic factors acting on bulk cargo transportation.

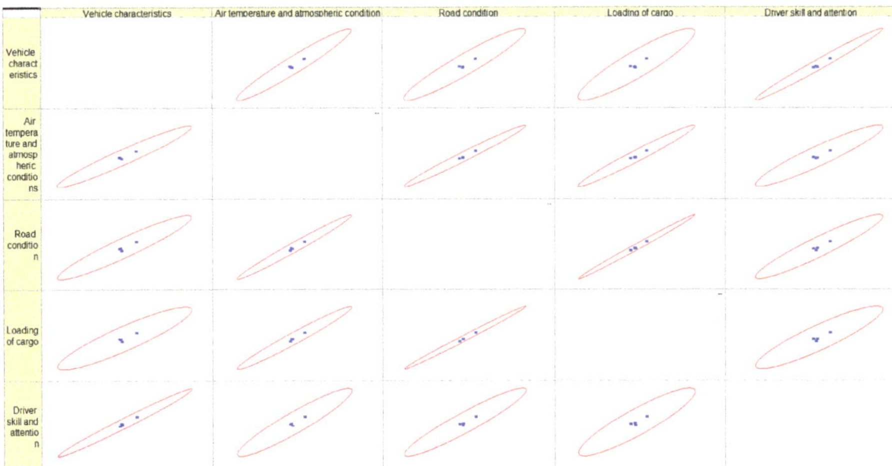

		Vehicle characteristics	Air temperature and atmospheric conditions	Road condition	Loading of cargo	Driver skill and attention
Vehicle characteristics	Pearson Corr.	1	0.96247	0.94368	0.90806	0.98672
	Sig.	–	0.03753	0.05632	0.09194	0.01328
Air temperature and atmospheric conditions	Pearson Corr.	0.96247	1	0.98562	0.97807	0.93537
	Sig.	0.03753	–	0.01438	0.02193	0.06463
Road condition	Pearson Corr.	0.94368	0.98562	1	0.99451	0.94192
	Sig.	0.05632	0.01438	–	0.00549	0.05808
Loading of cargo	Pearson Corr.	0.90806	0.97807	0.99451	1	0.90162
	Sig.	0.09194	0.02193	0.00549	–	0.09838
Driver skill and attention	Pearson Corr.	0.98672	0.93537	0.94192	0.90162	1
	Sig.	0.01328	0.06463	0.05808	0.09838	–

2-tailed test of significance is used

Figure 12. Values of Pearson's correlation coefficient when assessing the importance of the vehicle dynamic factors acting on the transportation of liquid cargoes.

Taking into account the fact that cargo transportation is a complex process, during which the safe transportation of cargo must be ensured, therefore drivers transporting this type of cargo must be trained [7,42]. For example, in Lithuania, the training program for professional drivers has been updated, which will provide an opportunity to acquire a professional qualification that is mandatory throughout the European Union (EU) and to provide cargo and passenger transport services in any EU country [43,44]. The new procedure has been prepared taking into account the proposals submitted by businesses and associations uniting cargo and passenger transport companies, as well as associations of driving training institutions, the main focus of which is safe and economical driving.

In order to obtain the necessary qualifications and obtain a qualification certificate, drivers will have to complete 280 h of training and pass a driver qualification exam (competency assessment). In this training course, at least 20 h will have to be devoted to individual practical driving training, of which up to 8 h will be spent learning to drive in a special area, in specially equipped autodromes or using a high-class simulator.

5. Conclusions

The transportation of bulk and liquid cargoes involves assessing the safety of the process by determining the importance of forces influencing the cargo structure, transport capacities, and other variables. The two most powerful forces are centrifugal and inertial, with interdependence of all forces crucial. Hazardous elements in transportation depend on the nature of the cargo and the critical level of tank filling. Most importantly, the skills and attention of the driver are crucial, while air temperature and atmospheric conditions are less important.

The critical skidding of the truck on a dry asphalt surface ($\mu = 0.6$) when driving a turn with a radius of 17 m will not occur until the speed reaches 34 km/h, but after reaching a speed of 34 km/h and above, the truck will start to slide in the turn.

The essential turning speed of both loaded and empty cars was determined for a roundabout with a diameter of 17 m. A vehicle that is carrying a significant amount of weight will experience a tipping motion when its speed exceeds 38 km/h, while a vehicle that is not carrying any additional weight will undergo a tipping motion at speeds exceeding 51 km/h.

The critical turning point is when the turning speed and sliding speed curves meet. In the analyzed case, when the height of the center of gravity of the vehicle is 2.4 m from the ground and the vehicle speed is 34 km/h, the vehicle overturns.

Pearson's correlation analysis reveals a robust association between acting forces in bulk shipments and bulk cargo transit. The correlation is more pronounced in bulk shipments and is particularly significant in bulk cargo transit. However, the correlation for liquid cargoes is moderate. The presence of centrifugal forces can cause instability in both bulk and liquid cargo due to abrupt shifts in the center of gravity, high speed, and travel trajectory. The Pearson correlation coefficient demonstrates a consistent similarity across all examined instances, indicating a robust interparametric association between the components.

Considering the obtained results, it can be stated that the transportation of specific cargoes, including bulk and liquid cargoes, requires specific skills from the drivers. In order to develop those skills in the training of cargo vehicle drivers, an understanding of the dynamics of the movement of vehicles in a turn, namely about the means of power, their redistribution, the change of the center of gravity when transporting special cargoes (bulk and liquid), and when moving on road bends. This article provides a scientifically sound basis for the development of methodological tools for the training of drivers of this type, which would help them understand what forces and what transport factors affect the movement of vehicles in a turn, how to increase the stability and safety of the movement of vehicles in a turn, etc.

The authors intend to focus on enhancing the acceptability of these criteria and their applicability to multimodal transport, particularly with regards to road transport, also

16. Scherbakov, A.; Lunyakov, M.; Smirnov, V.; Kaigorodova, V.; Verbova, N. Options for Placement and Transport of Bulk Goods by Rail. *Transp. Res. Procedia* **2022**, *63*, 1498–1504. [CrossRef]
17. Liu, H.; Ren, C.; Zhang, G.; Zhou, W.; Zhang, X. Research and Evaluation on Safety Critical Information of Road Tanker for Dangerous Liquid Goods Based on Multi-Source Data. *Procedia Comput. Sci.* **2022**, *208*, 79–87. [CrossRef]
18. Wu, J.; Lu, Y.; Shi, S.; Zhou, R.; Liu, Y. Research on the Prediction Model of Hazardous Chemical Road Transportation Accidents. *J. Loss Prev. Process Ind.* **2023**, *84*, 105103. [CrossRef]
19. Jitmitsumphan, S.; Sripetdee, T.; Chaimueangchuen, T.; Tun, H.M.; Chinkanjanarot, S.; Klomkliang, N.; Srinives, S.; Jonglertjunya, W.; Ling, T.C.; Phadungbut, P. Unveiling the Molecular Origin of Vapor-Liquid Phase Transition of Bulk and Confined Fluids. *Molecules* **2022**, *27*, 2656. [CrossRef]
20. Günay, G. Shipment Size and Vehicle Choice Modeling for Road Freight Transport: A Geographical Perspective. *Transp. Res. Part A Policy Pract.* **2023**, *173*, 103732. [CrossRef]
21. Mahpour, A.; Farzin, I.; Baghestani, A.; Ashouri, S.; Javadi, Z.; Asgari, L. Modeling the Impact of Logistic Performance, Economic Features, and Demographic Factors of Countries on the Seaborne Trade. *Asian J. Shipp. Logist.* **2023**, *39*, 60–66. [CrossRef]
22. Russo, F.; Musolino, G.; Assumma, V. Ro-Ro and Lo-Lo Alternatives between Mediterranean Countries: Factors Affecting the Service Choice. *Case Stud. Transp. Policy* **2023**, *11*, 100960. [CrossRef]
23. Jiménez, P.; Gómez-Fuster, J.M.; Pavón-Mariño, P. HADES: A Multi-Agent Platform to Reduce Congestion Anchoring Based on Temporal Coordination of Vessel Arrivals—Application to the Multi-Client Liquid Bulk Terminal in the Port of Cartagena (Spain). *Appl. Sci.* **2021**, *11*, 3109. [CrossRef]
24. Andrzejczak, K.; Selech, J. Quantile Analysis of the Operating Costs of the Public Transport Fleet. *Transp. Probl.* **2017**, *12*, 103–111. [CrossRef]
25. Gillespie, T.D. *Fundamentals of Vehicle Dynamics*; Society of Automotive Engineers: Warrendale, PA, USA, 1992; ISBN 978-1-56091-199-9.
26. Minaker, B.P. *Fundamentals of Vehicle Dynamics and Modelling: A Textbook for Engineers with Illustrations and Examples*; Automotive series; Wiley: Hoboken, NJ, USA, 2019; ISBN 978-1-118-98007-1.
27. V-106 Dėl Žiedinių Sankryžų Projektavimo Metodinių Nurodymų MN ŽSP 12 Patvirtinimo [Concerning the Approval of Methodological Instructions for the Design of Roundabouts MN ŽSP 12]. Available online: https://e-seimas.lrs.lt/portal/legalAct/lt/TAD/TAIS.424377/asr (accessed on 23 December 2023). (In Lithuanian)
28. Ataei, M.; Khajepour, A.; Jeon, S. A General Rollover Index for Tripped and Un-Tripped Rollovers on Flat and Sloped Roads. *Proc. Inst. Mech. Eng. Part D J. Automob. Eng.* **2019**, *233*, 304–316. [CrossRef]
29. Žuraulis, V.; Sokolovskij, E. Vehicle Velocity Relation to Slipping Trajectory Change: An Option for Traffic Accident Reconstruction. *Promet* **2018**, *30*, 395–406. [CrossRef]
30. Jin, M.; Zhang, M.; Wang, G.; Liang, S.; Wu, C.; He, R. Analysis and Simulation of Wheel-Track High Clearance Chassis of Rape Windrower. *Agriculture* **2022**, *12*, 1150. [CrossRef]
31. Žuraulis, V.; Sokolovskij, E.; Matijošius, J. The Opportunities for Establishing the Critical Speed of the Vehicle on Research in Its Lateral Dynamics. *Sci. Technol.* **2013**, *15*, 312–318.
32. Sivilevičius, H. Application of Expert Evaluation Method to Determine the Importance of Operating Asphalt Mixing Plant Quality Criteria and Rank Correlation. *Balt. J. Road Bridge Eng.* **2011**, *6*, 48–58. [CrossRef]
33. Maskeliūnaitė, L.; Sivilevičius, H. Identifying the Importance of Criteria for Passenger Choice of Sustainable Travel by Train Using ARTIW and IHAMCI Methods. *Appl. Sci.* **2021**, *11*, 11503. [CrossRef]
34. Burdzik, R. Novel Method for Research on Exposure to Nonlinear Vibration Transferred by Suspension of Vehicle. *Int. J. Non-Linear Mech.* **2017**, *91*, 170–180. [CrossRef]
35. Barrachina, D.G.-L.; Boldizsar, A.; Zoldy, M.; Torok, A. Can Neural Network Solve Everything? Case Study of Contradiction in Logistic Processes with Neural Network Optimisation. In Proceedings of the 2019 Modern Safety Technologies in Transportation (MOSATT), Kosice, Slovakia, 28–29 November 2019; pp. 21–24.
36. Stoma, M.; Dudziak, A. Eastern Poland Consumer Awareness of Innovative Active and Intelligent Packaging in the Food Industry: Exploratory Studies. *Sustainability* **2022**, *14*, 13691. [CrossRef]
37. Buczaj, A.; Krzysiak, Z.; Pecyna, A.; Caban, J.; Brumercik, F. Safety during Chemical Transport of Dangerous Goods. *Przem. Chem.* **2019**, *98*, 1276–1280. [CrossRef]
38. Borucka, A.; Kozłowski, E.; Antosz, K.; Parczewski, R. A New Approach to Production Process Capability Assessment for Non-Normal Data. *Appl. Sci.* **2023**, *13*, 6721. [CrossRef]
39. Borucka, A.; Kozłowski, E.; Parczewski, R.; Antosz, K.; Gil, L.; Pieniak, D. Supply Sequence Modelling Using Hidden Markov Models. *Appl. Sci.* **2022**, *13*, 231. [CrossRef]
40. Bowerman, B.L.; O'Connell, R.T. *Applied Statistics: Improving Business Processes*; The Irwin series in statistics; Irwin: Chicago, IL, USA, 1997; ISBN 978-0-07-114907-5.
41. Weisberg, S. *Applied Linear Regression*, 4th ed.; Wiley series in probability and statistics; Wiley: Hoboken, NJ, USA, 2014; ISBN 978-1-118-38608-8.
42. EN 12642:2006; Securing of Cargo on Road Vehicles—Body Structure of Commercial Vehicles—Minimum Requirements. Available online: https://standards.iteh.ai/catalog/standards/cen/c7f52fc5-051f-488c-be40-2a98be6b1023/en-12642-2006 (accessed on 17 May 2023).

43. Zaranka, J.; Matijosius, J.; Radvilaite, U.; Caban, J.; Dudziak, A. Establishing Emergency Sections on Land Roads in Order to Improve the Quality of Transport Services, Creating Comfortable Conditions for International and Local Traffic. *Adv. Sci. Technol. Res. J.* **2023**, *17*, 75–85. [CrossRef]
44. Zaranka, J.; Pečeliūnas, R.; Žuraulis, V. A Road Safety-Based Selection Methodology for Professional Drivers: Behaviour and Accident Rate Analysis. *Int. J. Environ. Res. Public Health* **2021**, *18*, 12487. [CrossRef]

MDPI
St. Alban-Anlage 66
4052 Basel
Switzerland
www.mdpi.com

Applied Sciences Editorial Office
E-mail: applsci@mdpi.com
www.mdpi.com/journal/applsci

www.ingramcontent.com/pod-product-compliance
Lightning Source LLC
Chambersburg PA
CBHW061510220926
51850CB00004B/222